BirdLife Conservation Series No. 15

IMPORTANT BIRD AREAS IN THE CARIBBEAN

Key sites for conservation

Editors
David C. Wege and Verónica Anadón-Irizarry

With special assistance from
Mayra Vincenty, Verónica Mendez, Christian Devenish, Mark Balman,
David Diaz, Richard Elsam, Rob Clay and Ian Davidson

Supported by

and the Olewine Family

Recommended citation
BirdLife International (2008) *Important Bird Areas in the Caribbean: key sites for conservation.* Cambridge, UK: BirdLife International. (BirdLife Conservation Series No. 15).

Wellbrook Court, Girton Road, Cambridge CB3 0NA, United Kingdom
Tel: +44 1223 277318 Fax: +44 1223 277200 Email: birdlife@birdlife.org
Internet: www.birdlife.org

BirdLife International is a UK-registered charity no. 104215

ISBN: 978-0-946888-65-8

British Library-in-Publication Data
A catalogue record for this book is available from the British Library

First published 2008 by BirdLife International

Designed and produced by the NatureBureau Limited, 36 Kingfisher Court, Hambridge Road, Newbury, Berkshire RG14 5SJ, United Kingdom

Printed by Information Press, Oxford, United Kingdom

Available from the Natural History Book Service Ltd, 2–3 Wills Road, Totnes, Devon TQ9 5XN, United Kingdom
Tel: +44 1803 865913 Fax: +44 1803 865280 Email: nhbs@nhbs.co.uk
Internet: www.nhbs.com/services/birdlife.html

IMPORTANT BIRD AREAS IN THE CARIBBEAN
Key sites for conservation

NATIONAL COORDINATORS AND CHAPTER AUTHORS

Important Bird Areas in **Anguilla**
Karim Hodge, Steve Holliday and Farah Mukhida
Department of Environment, Government of Anguilla – Royal Society for the Protection of Birds – Anguilla National Trust

Important Bird Areas in **Antigua and Barbuda**
Joseph Prosper, Victor Joseph, Andrea Otto and Shanee Prosper
Environmental Awareness Group

Important Bird Areas in **Aruba**
Adrian del Nevo
Applied Ecological Solutions Inc.

Important Bird Areas in the **Bahamas**
Predensa Moore and Lynn Gape
Bahamas National Trust

Important Bird Areas in **Barbados**
Wayne Burke

Important Bird Areas in **Bermuda**
Andrew Dobson and Jeremy Madeiros
Bermuda Audubon Society
Department of Conservation Services, Bermuda Government

Important Bird Areas in **Bonaire**
Jeff Wells and Adolphe Debrot
Boreal Songbird Initiative
CARMABI Foundation

Important Bird Areas in the **British Virgin Islands**
Clive Petrovic, Esther Georges and Nancy Woodfield Pascoe
British Virgin Islands National Parks Trust

Important Bird Areas in the **Cayman Islands**
Patricia E. Bradley, Mat Cottam, Gina Ebanks-Petrie and Joni Solomon
Cayman Islands Department of Environment

Important Bird Areas in **Cuba**
Susana Aguilar Mugica
Centro Nacional de Áreas Protegidas
Empresa para la Protección de la Flora y la Fauna
Instituto de Ecología y Sistemática
Museo Nacional de Historia Natural
Facultad de Biología de la Universidad de la Habana
Centro Oriental de Ecosistemas y Biodiversidad

Important Bird Areas in **Curaçao**
Adolphe Debrot and Jeff Wells
CARMABI Foundation
Boreal Songbird Initiative

Important Bird Areas in **Dominica**
Stephen Durand and Bertrand Jno. Baptiste
Forestry, Wildlife and Parks Division

Important Bird Areas in **Dominican Republic**
Laura Perdomo and Yvonne Arias
Grupo Jaragua

Important Bird Areas in **Grenada**
Bonnie L. Rusk
Grenada Dove Conservation Programme

Important Bird Areas in **Guadeloupe**
Anthony Levesque and Alain Mathurin
AMAZONA

Important Bird Areas in **Haiti**
Florence Sergile
Société Audubon Haiti

Important Bird Areas in **Jamaica**
Catherine Levy and Susan Koenig
Windsor Research Centre

Important Bird Areas in **Martinique**
Société d'Etude, de Protection et d'Aménagement de la Nature de Martinique

Important Bird Areas in **Montserrat**
Geoff Hilton, Lloyd Martin, James 'Scriber' Daley and Richard Allcorn
Royal Society for the Protection of Birds
Department of Environment, Montserrat
Flora and Fauna International

Important Bird Areas in **Navassa**
Joseph Schwagerl and Verónica Anadón-Irizarry
US Fish and Wildlife Service
BirdLife International

Important Bird Areas in **Puerto Rico**
Verónica Méndez-Gallardo and José A. Salguero-Faría
Sociedad Ornitológica Puertorriqueña, Inc.

Important Bird Areas in **Saba**
Natalia Collier and Adam Brown
Environmental Protection in the Caribbean

Important Bird Areas in **St Barthélemy**
Anthony Levesque, Alain Mathurin and Franciane Le Quellec
AMAZONA
Réserve Naturelle de St Barthélemy

Important Bird Areas in **St Eustatius**
Natalia Collier and Adam Brown
Environmental Protection in the Caribbean

Important Bird Areas in **St Kitts and Nevis**
Natalia Collier and Adam Brown
Environmental Protection in the Caribbean

Important Bird Areas in **St Maarten**
Natalia Collier and Adam Brown
Environmental Protection in the Caribbean

Important Bird Areas in **St Martin**
Natalia Collier and Adam Brown
Environmental Protection in the Caribbean

Important Bird Areas in **St Lucia**
Donald Anthony and Alwin Dornelly
Forestry Department, Ministry of Agriculture

Important Bird Areas in **St Vincent and the Grenadines**
Lystra Culzac-Wilson
AvianEyes

Important Bird Areas in **Trinidad and Tobago**
Graham White
Trinidad and Tobago Field Naturalists Club

Important Bird Areas in the **Turks and Caicos Islands**
Mike Pienkowski
UK Overseas Territories Conservation Forum
Turks and Caicos National Trust

Important Bird Areas in the **US Virgin Islands**
Jim Corven
Bristol Community College

CONTENTS

ACKNOWLEDGEMENTS

The Important Bird Area program in the Caribbean started in early 2001, and BirdLife International would like to thank the many people and organisations that have contributed to it over the past eight years. The implementation of the program across the region, through support for national coordinators and workshops, was made possible due to a number of sponsors, both nationally and regionally.

The program was launched in Jamaica where field surveys were undertaken to provide new data for the identification of IBAs, and conservation actions undertaken at the highest priority sites (such as Cockpit Country). This work was funded by the **National Fish and Wildlife Foundation**, and the **Environmental Foundation of Jamaica**, and catalysed the regional IBA program. Comprehensive initiatives – including the piloting of IBA Site Support Groups – were undertaken in the Bahamas, Dominican Republic and Puerto Rico with support from the **John D. and Catherine T. MacArthur Foundation** through the project *Biodiversity conservation in the insular Caribbean: catalyzing site action through local, national and regional partnerships*. A long-term project, *Eastern Cuba: saving a unique Caribbean wilderness*, funded by the **British Birdwatching Fair**, enabled the Cuban national IBA program to develop in parallel with an intensive field research program focused on the east of the country (which has resulted in proposals for additional protection for many of the eastern Cuba IBAs). Between 2003 and 2007, the IBA regional program consolidated around the project *Sustainable conservation of globally important Caribbean bird habitats: strengthening a regional network for a shared resource*, funded by the **Global Environment Facility** through the **United Nations Environment Programme (UNEP)**. Support from the **US Fish and Wildlife Service** through the **Neotropical Migratory Bird Conservation Act (NMBCA)** to implement the project *Building a baseline foundation for conserving Important Bird Areas for Neotropical migratory birds in Central America and the Caribbean* helped facilitate the Important Bird Area inventories for Haiti, Trinidad and Tobago and the Lesser Antilles, as did funding from the **Dutch Ministry of Foreign Affairs/Development Cooperation** under their theme-based financing program (TMF). Support from the **UNEP Caribbean Environment Program** and the **SPAW Protocol Regional Activity Centre** has been critical to completing the publication of this book.

We greatly acknowledge the **Olewine Family** whose support since 2000 has been fundamental to the success of the IBA program throughout the Americas and the Caribbean. Recent support from the **Aage V. Jensen Charity Foundation** has helped the program develop essential conservation actions at IBAs. Important Bird Areas in the Caribbean's UK Overseas Territories were published in 2006 (and revised in this publication) through support from the **Royal Society for the Protection of Birds** (RSPB, BirdLife in the UK). The Honorary Presidents of BirdLife's Rare Bird Club – Margaret Atwood and Graeme Gibson – have provided invaluable and impassioned assistance in the development of the Caribbean Program, especially as it relates to Cuba, but also in other countries. BirdLife International would like to take this opportunity to express its appreciation to the above supporters.

This book is the result of an eight-year project of the BirdLife Caribbean program. The data which have been used in its preparation have been collected over a much longer period by a very large number of amateur birdwatchers and professional ornithologists, conservationists and others interested in the biodiversity and natural environment of the Caribbean islands. It is of course impossible to acknowledge all of these people here, but their work provides the foundation of this analysis of Important Bird Areas. There have been contributions to this book from every country and territory in the Caribbean region. National networks of organisations and individuals were mobilised to identify and collect information on IBAs in some places, and in others the national IBA compilers or compilation teams were the principle sources of data. All of these people (many of whom are acknowledged in the national chapters) are to be congratulated on their efforts to identify, map and document 283 sites throughout the Caribbean region. Many of them are already involved in the next, and most important phase of the project, namely to strengthen the protection of these most important sites for bird conservation.

We would like to offer special thanks to all of the national authors (see Pp. iii–iv)—our friends—who have not only survived our persistent requests, but have always responded with a level of professionalism and enthusiasm that has been inspirational. This ground-breaking publication is theirs (although the inevitable mistakes are ours, the editors'). At a personal level the editors would like to thank our respective families for their help, support, patience and understanding, especially during the latter, intense months of work on this project.

Thanks are due for other critical contributors to the project (with apologies for any omission). As always, colleagues at BirdLife Secretariat Offices have provided vital support to the project. In the BirdLife Americas Office in Quito, Ecuador, the Caribbean Program team would like to thank the following: Rob Clay, Ian Davidson, Christian Devenish, David Diaz, Santiago Llore, Amiro Perez-Leroux, Amanda Tapia and Itala Yepez for extensive technical, practical and moral support over the course of the project and especially in the final months of book production. In the BirdLife Secretariat Office in Cambridge, UK, Mark Balman has provided extensive assistance with IBA mapping; Mike Evans, Lincoln Fishpool, Ian May and Martin Sneary have provided invaluable advice and support on technical IBA criteria and database issues; Ben Lascelles on marine IBA issues; and Nigel Collar continues to provide (frustratingly significant) editorial mentorship.

The book would never have been completed without the significant translation and drafting efforts of Mayra Vincenty and Verónica Mendez. Alison Duncan and Bernard Deceuninck provided invaluable assistance with the French Overseas Territories and with translations, complementing some exceptional translation work by Ian Fisher. Anthony Levesque gave constant (and much needed) help and advice for the French islands. Sarah Sanders helped facilitate a comprehensive update and review of the IBAs in the UK Overseas Territories, and Bert Denneman and Kalli de Meyer gave constant support for the Netherlands Antilles chapters. Lisa Sorenson has provided invaluable advice, feedback, contacts and endless support throughout the project. Richard Elsam volunteered a significant amount of time helping compile various pieces of essential information and data. Others who have provided important feedback, contributions and support include: Greg Butcher, John Cecil, Eladio Fernández, Jim Goetz, Jim Kushlan, Craig Lee, Carole McCauley, Matt Morton, Herb Raffaele, Chris Rimmer, Frank Rivera and Jennifer Wheeler. We would also like to acknowledge the Society for the Conservation and Study of Caribbean Birds and Waterbird Conservation for the Americas for their invaluable input and support to the evolution of the IBA program in the Caribbean.

Most of the painted images used to illustrate "characteristic birds" in each IBA are from *A Guide to the Birds of the West Indies* (Princeton University Press), made available through the US Fish and Wildlife Service, Wildlife Without Borders program. BirdLife would like to thank Herb Raffaele for enabling their use here. The painting of the Grenada Dove was adapted by Madeleine Smith from artwork in *A Guide to the Birds of the West Indies*. We would also like to thank Lynx Edicions for allowing us to use *Handbook to the Birds of the World* (HBW) paintings of Little Egret (F. Jutglar: HBW1), Trinidad Piping-guan (F. Jutglar: HWB2), Cayenne Tern (I. Willis: HBW3), Bare-eyed Pigeon (H. Burn: HBW4), Yellow-shouldered Amazon (À Jutglar: HBW4), and White-tailed Sabrewing (H. Burn: HBW5). Tracy Pedersen kindly allowed us to use her paintings of the Cayman and Brac Parrots *Amazona leucocephala*.

Photographers are acknowledged in each chapter, but we would like to thank especially Ricardo Briones, Bert Denneman, Eladio Fernández, Jim Goetz, Gregory Guida, Pete Morris, Allan Sander, Olga Stokes and Lance Woolaver for giving us so much excellent material to choose from.

Finally, we would like to say a special thank you to Peter and Barbara Creed and Justine Pocock at the NatureBureau for the design and layout of this directory, including the production of the maps (from our raw GIS files) and graphics. The production was done under almost impossible time pressure, but they have created a beautiful book for which we are truly grateful.

INTRODUCTION

They paved paradise, put up a parking lot
With a pink hotel, a boutique, and a swinging hot spot—
Don't it always seem to go
That you don't know what you've got 'til it's gone?
They paved paradise, put up a parking lot

Joni Mitchell (1970) *Big Yellow Taxi*

"Pink hotel" on Paradise Island, the Bahamas: this would have been an Important Bird Area up until the late-1990s. (PHOTO: DAVID C. WEGE)

BACKGROUND

The aim of this publication is simple—in a rational, scientifically robust way, it puts the spotlight on a Caribbean network of internationally important biodiversity sites—"Important Bird Areas" (IBAs). In a region that is exceptionally rich in endemic birds, seabirds, waterbirds, and species already at risk of extinction, IBAs are an objective expression of which places in the Caribbean are the most important for these birds and why. By highlighting the significance of IBAs, the goal is to secure their long-term conservation—to ensure that these remnants of paradise are not lost.

The need for site-based conservation action has never been greater. Biodiversity is facing unparalleled threats resulting from human activities. Without focused and rapid conservation interventions, many islands will experience a second wave of human-induced extinction crises in the coming years. The resources available for conservation efforts are scarce and need to be invested in strategic, priority-driven ways to ensure that they make the greatest contribution to preserving global biodiversity.

In the early 1980s, BirdLife International created the Important Bird Areas (IBAs) concept. IBAs are places of the highest global priority for bird and biodiversity conservation. They are irreplaceable "hotspots" and, because of their limited size, always potentially vulnerable. Identified nationally from data gathered locally, but according to internationally standardised scientific criteria, IBAs form a worldwide network of sites for the conservation of nature. Since birds are excellent indicators of overall biodiversity, IBAs are also important for many other species of fauna and flora. Moreover, IBAs provide vital environmental services, such as the provision of freshwater and forest products, and prevention of floods and other environmental disasters, including those likely to result from global climate change. As such their conservation contributes to the broader agenda of environmental management, sustainable development, and poverty eradication.

The IBA concept is now actively and widely used by civil society groups, governments, NGOs (including all members of the BirdLife network), international agencies and scientists all over the world as a priority-driven framework for achieving conservation action. IBAs are widely recognised by local communities, national legislation and policy, international agreements and regulations, and multilateral instruments and organisations.

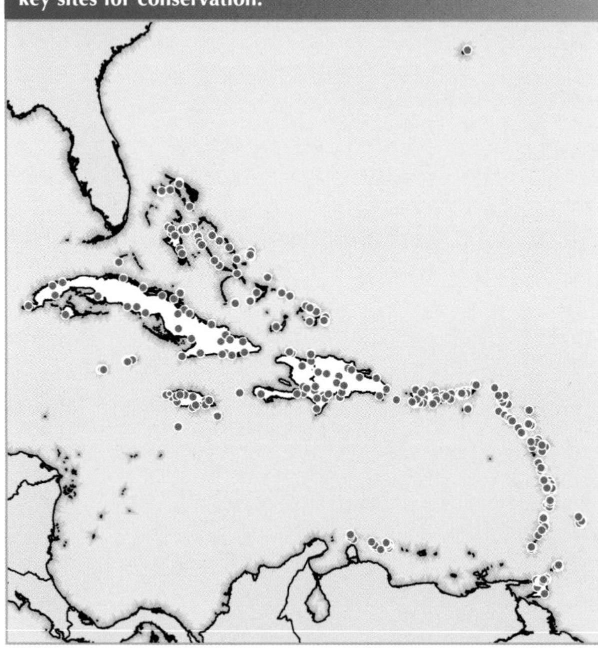
Figure 1. Location of Important Bird Areas in the Caribbean: key sites for conservation.

In the Caribbean, 283 internationally significant IBAs have been identified. This network of key sites faces a large number of diverse threats, exacerbated by the fact that 43% are wholly outside of formal protected areas. There is a critical need for a comprehensive, Caribbean-wide program of coordinated conservation by all sectors of society. This directory aims to facilitate the further development of such a program by presenting information on the region's IBAs, compiled by the BirdLife Caribbean network since 2001, in a standardised and clear format.

THE CARIBBEAN

The Caribbean region as covered by this book is based on a practical, programmatic definition. It includes Bermuda and all islands of the Bahamas, Greater Antilles (Cuba, Hispaniola, Jamaica and Puerto Rico), Virgin Islands, Cayman Islands, Lesser Antilles, Netherlands Antilles, and Trinidad and Tobago. This represents 13 independent nations, and six UK, six Dutch, four French, and three US overseas territories. It does not cover the Central and South American islands on the fringes of the Caribbean Sea. The IBAs of Colombia and Venezuela (including their Caribbean islands) were covered in BirdLife International and Conservation International (2005), and those of the Central American countries will be treated in BirdLife International (in prep.).

The islands of the insular Caribbean (as defined above) are exceptionally important for global biodiversity conservation. They support a wide diversity of ecosystems which are home to populations of endemic plants and vertebrates amounting to at least 2% of the world's total species complement. Species endemism is very high, yet the land area of the insular Caribbean is a mere 233,780 km^2 (c.90% of which is accounted for by Cuba, Hispaniola, Jamaica and Puerto Rico) or about the same size as the state of Arizona, US. Of the 770 bird species occurring in this region, 148 are endemic to it, with 105 confined to single islands.

With only c.10% of the region's original habitat remaining, it seems that most of the major habitat loss has already occurred. However, the human population is c.37.5 million, and many of the region's countries have annual population growth rates in excess of 2.5%, which means that what little habitat remains is at risk from both human activity and natural disasters. Thus the region's biodiversity is at serious risk from species extinctions (currently there are 54 globally threatened birds in the Caribbean, 12 of which are Critically Endangered[1]: see Appendix 1b), even through the destruction of relatively small patches of habitat.

BIRDLIFE IN THE CARIBBEAN

With almost 90% of extinctions having happened on islands, the BirdLife Partnership has made a long-term commitment to biodiversity conservation on the major island groups of the world. The insular Caribbean is one of these, and since 2001 BirdLife has been developing a coherent, integrated program focused on saving threatened species and protecting important sites for biodiversity.

BirdLife International is a global network of non-governmental conservation organisations (NGOs) with a special focus on birds, which is currently present in over 110 countries and territories worldwide. In the Caribbean, BirdLife is formally represented by the Bahamas National Trust (BirdLife in **Bahamas**), Centro Nacional de Áreas Protegidas (BirdLife in **Cuba**), Grupo Jaragua (BirdLife in **Dominican Republic**), Sociedad Ornitológica Puertorriqueña (BirdLife in **Puerto Rico**), Royal Society for the Protection of Birds (BirdLife in the **UK Overseas Territories**), Vogelbescherming Nederland (BirdLife in the **Netherlands Antilles**), Ligue pour la Protection des Oiseaux (BirdLife in the **French Overseas Territories**), and National Audubon Society (BirdLife in the **US Overseas Territories**[2]). The development of the BirdLife Partnership and program is overseen by Regional Committees (comprised solely of BirdLife Partners) in each of the six BirdLife regions. BirdLife Partners from the Caribbean are represented on the Americas Regional Committee. This committee oversees the development and approval of a 4-year program of work to save species, protect sites, conserve habitats and empower people. Progress with the program of work is reviewed by the committee at its annual meetings.

The BirdLife Caribbean Program is also working with a range of organisations in a number of other Caribbean islands (including Barbados, Haiti and Jamaica) and maintains contacts in all countries and territories where there is currently no Partner or Affiliate organisation. At the regional level BirdLife works closely with the Society for the Conservation and Study of Caribbean Birds (SCSCB), which provides support for many bird-focused conservation groups throughout the Caribbean.

BirdLife's approach in the Caribbean is thus based on partnership, and is characterised by an emphasis on strengthening, empowering and coordinating civil society actors at grassroots, national and regional levels. The BirdLife Caribbean Program recently established a small coordinating office in Puerto Rico supported by the Americas Regional Secretariat. The BirdLife Partners, wider programmatic network and Secretariat form a dynamic team for implementing the regional conservation program by working at the local level, whilst also drawing upon the technical and financial resources and support from within the team. This team has met every year (since 2003) to discuss, develop and evaluate shared Caribbean program priorities, policies and actions.

1 Factsheets for globally threatened birds are online at www.birdlife.org/datazone/species

2 National Audubon Society represents BirdLife in US Overseas territories where there is no existing BirdLife Partner

SAVING BIODIVERSITY IN THE CARIBBEAN

THE PROBLEM DEFINED

The arrival of Europeans in the Caribbean during the 1490s marked the beginning of a wave of ecosystem destruction that has left no more than 23,000 km² or 10% of the islands' original vegetation in a pristine state. Forest clearance for sugarcane plantations started in the sixteenth century and escalated through the subsequent centuries, leading to widespread deforestation throughout the region. Sugarcane is still the Caribbean's most important crop.

Prior to the arrival of Europeans, the human population of the Caribbean is estimated to have been 750,000. The regional population grew (after the decimation of its indigenous peoples) to 2.2 million in 1800, but in the following 200 years it grew exponentially to its current level of c.37.5 million. This growing population has put the region's ecosystems under extreme pressure. Forests are being lost to clear-felling and conversion to agriculture (especially cacao, coffee, and tobacco), infrastructure and other land uses, and degraded by timber extraction, livestock grazing, shifting cultivation and over-exploitation of non-timber forest products. Mining for bauxite, sand, and gravel, as well as the production of charcoal from natural vegetation to meet energy needs, also pose threats to the native flora and fauna. Wetlands on all islands are being degraded, drained, and developed. Tourism development continues to put extreme pressure on natural ecosystems (both forests and wetlands alike) on many islands through the alteration of landscapes with golf courses, non-native vegetation, roads, and tourist infrastructure and facilities.

With the arrival of Europeans in the Caribbean came a wave of alien invasive species introductions which now represent one of the greatest threats to the region's biodiversity. Even before the arrival of Europeans, people in the Caribbean were transporting species that they used for food from one island to another. However, the introduction of the mongoose *Herpestes auropunctatus* in 1872 (to control rodents and poisonous snakes) had a devastating effect on native populations of reptiles, amphibians and birds, and led to the extinction of many species. Rats, cats, dogs, goats, donkeys, monkeys, tilapia, and catfish also pose a serious threat to native fauna. The native flora is also being outcompeted by invasive plants (especially some trees and ferns). Furthermore, global climate change threatens to have significant, if as yet uncertain, impacts on the region's ecosystems and biodiversity.

Introduction of the mongoose to the Caribbean has had a devastating effect on the native fauna. (PHOTO: LEO DOUGLAS)

Sugarcane is grown even in the remotest areas at the expense of widespread forest clearance across the region. (PHOTO: DAVID C. WEGE)

Efforts to stem the loss of the Caribbean's biodiversity have centred on the development of national systems of protected areas. While considerable resources have been invested in this area and significant success has been achieved, there are often severe weaknesses in management, and many important areas remain outside the network. Moreover, there has been limited progress in mainstreaming biodiversity into broader socio-political agendas. As a result, on-the-ground conservation action continues to be undermined by incompatible programs and policies of governments and donors, such as infrastructure, housing, industrial, mining and tourism developments.

As natural ecosystems are degraded, natural resources depleted and biodiversity lost, their capacity to support human livelihoods and contribute to the wider socio-economic development of the region (e.g. through ecosystem services including climate regulation, water supply and food production) is being diminished. If the contribution of these ecosystems to human livelihoods is to be maintained, and the full diversity of the region's habitats, communities and species conserved, it is essential that the principle of environmentally sustainable development is adopted on all islands and throughout all sectors of society. A key aspect of such development is avoiding further growth inconsistent with the conservation of the most biologically important sites, and ensuring that conservation action is targeted at the highest priority sites. The Caribbean IBA Program is a contribution to this end.

THE OVERARCHING STRATEGY

The BirdLife Partnership's global strategy is based around four inter-related themes: saving species; protecting sites; conserving habitats; and empowering people. Objectives and activities under these themes need to be fully integrated to achieve BirdLife's aims of:

- Preventing the extinction of any bird species.
- Maintaining and where possible improving the conservation status of all bird species.
- Conserving and where appropriate improving and enlarging sites and habitats important for birds.
- Helping, through birds, to conserve biodiversity and to improve the quality of people's lives.
- Integrating bird conservation into sustaining people's livelihoods.

Each of the themes provides a common thread throughout this strategy and a contribution to achieving the aims. Important Bird Areas are the priorities on which the "protecting sites" objectives and activities are focused. They are a recurring currency that permeates throughout the broader program strategy, and their utility is outlined below.

PROTECTING SITES

■ Sites are an effective unit for conservation action

IBAs are sites which are discrete areas of habitat that can be delineated and, at least potentially, managed for conservation. Effective protection of sites can address over-exploitation (including habitat destruction), one of the major causes of biodiversity loss. Protection of a network of sites can represent a cost-effective approach to conservation, because a relatively small network can support a large proportion of the species, communities and/or habitats within a particular region. In addition, protection of a network of sites is consistent with sustainable development and poverty alleviation agendas, because it allows a significant degree of human use of landscapes. For these reasons, sites are a major focus of conservation investment by government, donors and civil society. In particular, they form the basis of most protected area networks.

■ Birds are a good basis for site networks

The Caribbean IBA Program uses birds as the basis for a region-wide site network. As a group, birds have several features that make them good indicators (relative to most other taxonomic groups):

- They contain high numbers of threatened and restricted-range species.
- They have well understood distributions and habitat requirements.
- They are relatively easy to record and identify in the field.
- They are good indicators of habitat condition and human disturbance.
- They can act as flagships for conservation.

More information is available on the status and distribution of birds in the Caribbean than for any other major taxonomic group. Studies in other regions have shown that birds can be a highly effective means of setting geographical priorities for conservation in the absence of detailed data on other taxa. Consequently, conservation of the IBA network can be expected to make a major contribution to the conservation of taxonomic groups other than birds, and identification of IBAs can contribute to the identification of a network of globally important sites for species conservation, termed "Key Biodiversity Areas" (see Box 1).

Box 1. Key Biodiversity Areas.

Key Biodiversity Areas (KBAs) are sites of global significance for the conservation of biodiversity. They are identified nationally using simple, globally standardised criteria and thresholds. As the building blocks for implementing an ecosystem approach and maintaining effective ecological networks, KBAs are the starting point for landscape-level conservation planning. Governments, inter-governmental organisations, NGOs, the private sector and other stakeholders can use KBAs as a tool to identify and augment national systems of globally important sites for conservation.

Key Biodiversity Areas extend the IBA concept to other taxonomic groups and KBAs are now being identified in many parts of the world, by a range of organisations. Examples include Important Plant Areas (IPAs), Prime Butterfly Areas, Important Mammal Areas, and Important Sites for Freshwater Biodiversity, with prototype criteria developed for freshwater molluscs and fish and for marine systems.

The aim of the KBA approach is to identify, document and protect networks of sites that are critical for the conservation of global biodiversity. Here a "site" means an area (whatever the size) that can be delimited and, potentially, managed for conservation. As with IBAs, KBAs are identified based on populations of threatened or geographically concentrated species. All IBAs, including those in this directory, are KBAs, but some KBAs are not IBAs (i.e. they are significant for the conservation of other taxa, but not birds).

Data on species status and distribution are still very scanty for most taxonomic groups, apart from birds. As such, the IBA network has proved a good approximation to the overall network of KBAs, as it includes the bulk of other target taxa as well as the most significant sites for threatened and restricted-range species. Important Bird Areas are thus an excellent starting point for immediate conservation planning and action—other sites can be added to complete the KBA network as data become available.

The Vulnerable Exuma Island Iguana *Cyclura cychlura figginsi* is present in the Exuma Cays Land and Sea Park (Bahamas)—one of many threatened taxa present in the region's IBAs. (PHOTO: OLGA STOKES)

Box 2. The Site Support Group approach.

Tree nursery in Cockpit Country, Jamaica: used by the Local Forest Management Committee ("Site Support Groups" in Jamaica) for IBA reforestation.

Implementing effective conservation actions at the sites described in this directory is essential to fulfilling the objectives of the Caribbean IBA Program. Building local capacity for implementing these actions is also a vital step in this process. By involving the local community, it hoped that biodiversity conservation can go hand in hand with achieving real improvements in local livelihoods. The Site Support Group approach aims to harness this potential.

Site Support Groups (SSGs, sometimes referred to as Local Conservation Groups) are groups of voluntary individuals who, in partnership with relevant stakeholders, help promote conservation and sustainable development at IBAs and other key biodiversity sites. The volunteers, despite varied backgrounds, ages, and occupation, have similar interests and a good understanding of natural resources and the local context in which they are managed. The SSGs provide a mechanism by which limited resources can by utilised efficiently and equitably. SSG volunteers live in or adjacent to IBAs and are drawn from the local community. They may include the unemployed, students, farmers, teachers, and others; all have a passion for conservation, sustainable development, and responsible citizenship. Membership is generally open to all, and the groups are an excellent means of engaging the local community in IBA conservation. SSGs provide a link between local communities and national institutions, such as NGOs, government agencies and researchers.

The BirdLife Partnership's mission is "to conserve birds, their habitat and global biodiversity, working with people towards sustainability in the use of natural resources". SSGs complement this approach; they are role models in promoting conservation and sustainable development, working at priority IBAs throughout the Caribbean in partnership with other stakeholders.

The aim of this approach is to build a network of local constituencies working to protect the most threatened biodiversity sites in the Caribbean. SSGs play a fundamental role in negotiating, strengthening, and consolidating links and partnerships among local communities, NGOs, local and national governments. They present an entry point for building local capacity for effective biodiversity conservation, management, monitoring and sound decision-making. They help stimulate sustainable development that addresses local people's needs, while conserving their natural resources. They provide a mechanism for developing self-confidence and empowerment, thus allowing people to take control, manage and benefit from their own natural resources and to plan their own livelihoods.

SSGs are already active at some IBAs in the Bahamas, Dominican Republic, Jamaica and Puerto Rico. Projects and conservation actions being implemented include bird monitoring, infrastructure development for site visitation (e.g. boardwalks, observation towers), establishment of native tree nurseries and restoration with native vegetation, awareness raising and education, eradication and management of invasive alien species, fire management, nature guide programs and agro-forest farms.

■ Sites can help to address gaps in protected areas systems

In the region to date, the majority of investment in site-based conservation by national governments and donor agencies has been in the development of protected areas systems. However, protected areas systems are rarely developed systematically. Because IBAs are identified according to objective, scientific criteria, irrespective of current protection status, many of them (43%) lie outside of existing national protected areas systems. Therefore, the IBA network can be used as a tool to review existing national protected areas systems, identify gaps in coverage, and identify candidate sites for expansion or designation of protected areas to address these gaps.

■ Non-formal approaches to site protection are required

In the Caribbean region, it is not feasible to designate every IBA as a formal protected area, owing to factors such as resource limitations, conflicting land ownership, and high opportunity costs in productive landscapes (e.g. lowland forests and coastal zones). Moreover, formal protected area designation may not necessarily be the most effective approach to site-based protection, especially where an IBA supports a large resident human population and/or high levels of human use. Indeed, in some circumstances, formal protected area designation could be counter-productive to conservation objectives, particularly where protected area regulations restrict traditional land- and resource-use practices that are compatible with or contribute to the biological value of a site.

Therefore, there is a need to develop alternative approaches to site-based protection of IBAs, in addition to formal protected areas, such as private protected areas, voluntary agreements with land owners, and community-led management through Site Support Groups (see Box 2). In many cases, these approaches may be more cost-effective and/or engage support from non-traditional sources. Moreover, they may provide greater opportunities for sustainable human use of natural resources and, therefore, make a greater contribution to poverty alleviation among people for whom natural resources form a critical component of their livelihood strategies.

SAVING SPECIES

The BirdLife Caribbean Program has focused primarily on protecting sites as the strategy for saving globally threatened species and a broad range of other biodiversity. Of the Caribbean's 283 IBAs, 62% (176) support significant populations of threatened birds, and many of them are home to two or more such species. For example, the Zapata Swamp IBA in Cuba supports populations of 17 threatened bird species. Protecting the habitat in these IBAs will often be sufficient to ensure the long-term survival of these birds, and this approach is an efficient and cost-effective one for their conservation. However, species-specific action is sometimes needed, especially for birds threatened by a combination of factors, such as hunting, persecution, capture for the pet trade, and invasive alien predators. With this in mind, the BirdLife Caribbean Program has engaged in efforts for a number of the Critically Endangered species, e.g. Cuban Kite *Chondrohierax wilsonii* (field surveys), Grenada Dove *Leptotila wellsi* (working with the government and developers); Puerto Rican Nightjar *Caprimulgus noctitherus* (surveys, monitoring, lobbying); Ivory-billed Woodpecker *Campephilus principalis* (searches in eastern Cuba); and Montserrat Oriole

Box 3. The BirdLife Preventing Extinctions Programme.

The Caribbean has 12 Critically Endangered bird species, 11 of which trigger the A1 criterion in 20 IBAs described in this directory. Most of these sites are either partially protected or unprotected. These are the species most threatened by extinction in the region and 10 are exclusive to the Caribbean, placing a global responsibility on the region to implement immediate actions in order to prevent them from being lost for ever.

Recognising this need to act *now* for globally threatened bird species, especially Critically Endangered species, BirdLife has launched a major new initiative: the Preventing Extinctions Programme. This is spearheading greater conservation action, awareness and funding support for the world's most threatened birds, through appointing Species Guardians (to implement the priority actions) and Species Champions (to provide the resources).

BirdLife Species Guardians are individuals or organisations who take on a responsibility to implement and/or stimulate conservation action for a particular threatened species in a defined geographical area, usually a particular country. They also monitor the status of the species and identify the key actions needed. Species Guardians' activities typically include some of the following:

- Implementing priority actions for the species
- Developing a Species Action Plan, if one does not yet exist
- Facilitating the implementation of priority actions by other individuals or organisations
- Liaising and communicating with other individuals and organisations involved in carrying out research and taking action for the species
- Advocating for appropriate conservation measures to relevant authorities and institutions
- Monitoring the status of the species and the implementation and effect of actions by all parties

SOPI is the BirdLife Species Guardian for the Critically Endangered Puerto Rican Nightjar. (PHOTO: ALCIDES MORALES)

BirdLife Species Champions are a new global community of businesses, institutions and individuals who are stepping forward to provide the funding required to carry out the vital conservation measures BirdLife International has identified to help prevent bird extinctions.

Through the generous support of our first global program sponsor—*The British Birdwatching Fair*—and an ever-growing number of Species Champions and program donors, Critically Endangered species are already benefitting, with new funding flowing directly to their BirdLife Species Guardians.

In fact, Birdfair is the first Species Champion in the Caribbean, supporting the first regional Species Guardian for Puerto Rican Nightjar *Caprimulgus noctitherus*: Sociedad Ornitológica Puertorriqueña, Inc (SOPI), the BirdLife partner in Puerto Rico.

For more information, visit the website: www.birdlife.org/extinction

Icterus oberi (surveys, monitoring, invasives control, action plan development) amongst others.

BirdLife is the world authority on the conservation status of birds and is the Red List Authority for the world's birds on the IUCN Red List, providing IUCN with the categories of extinction risk[1] and extensive associated documentation for all species. In the Caribbean there are 54 globally threatened birds (12 Critically Endangered, 19 Endangered, 23 Vulnerable, and another 25 Near Threatened), factsheets for each species are available online[2]. Recognising that some species, especially those considered Critically Endangered, require individual attention, BirdLife has developed the Preventing Extinctions Program to ensure the right protection is put into place through the appointment of Species Guardians—organisations or individuals best placed to protect the bird—for each threatened species (see Box 3).

CONSERVING HABITATS

■ IBAs can be building blocks of networks

While protecting individual sites can be an effective approach to conservation for many species, at least in the medium term, the long-term conservation of all species requires the protection of interconnected networks of sites. This is particularly important for species with large home ranges, low natural densities and/or migratory behaviour, for which individual sites cannot support long-term viable populations. In addition, inter-connected networks may be less susceptible to the impacts of global climate change, as species are better able to "track" changes in habitat distribution. Therefore, site-level approaches to conservation must be complemented by landscape-level approaches, which maintain or establish habitat connectivity among individual sites.

Landscape-level conservation initiatives typically involve the identification and integration into broader socio-political agendas of interconnected networks of core areas, linked by habitat corridors, protected by buffer zones and, in some cases, further developed by restoration areas. Within the Caribbean, the IBA network represents the most comprehensive assessment of internationally important sites for conservation. Thus, IBAs could be adopted as core areas for such networks, with additional core areas being identified for other taxonomic groups as available data permit (see Box 1).

■ IBAs are a tool to mainstream biodiversity into other policy sectors

Incompatible land-use and development schemes are major threats to biodiversity in the Caribbean. These threats typically arise from insufficient integration of biodiversity conservation objectives into the plans and policies of other sectors, leading to site-based conservation efforts being undermined by incompatible development projects and patterns of land use. Consequently, there is a need to "mainstream" biodiversity into other policy sectors, particularly tourism, housing, energy, agriculture, mining, water, forestry, fisheries and transport. In order to do this, it is essential that accurate, up-to-date information on the conservation importance of sites is made available to decision-makers in government and donor agencies. It is also essential that such information is based on clear, objective and universally accepted criteria. Consequently, IBAs represent a valuable tool for integrating biodiversity into policy and planning.

1 See Appendix 1a+b
2 Factsheets for globally threatened birds are online at www.birdlife.org/datazone/species

IBAs support national commitments under multilateral environmental agreements

Conservation of the Caribbean IBA network will assist national governments and donor agencies to meet their commitments under multilateral environmental agreements. These agreements include the Convention on Biological Diversity (CBD), the Convention on Wetlands of International Importance especially as Waterfowl Habitat (Ramsar Convention), the Convention on Migratory Species (CMS), and the SPAW Protocol of the Cartagena Convention (see Promoting IBAs). Because of the strong similarities between the criteria for identifying important sites for conservation under multilateral environmental agreements and the IBA criteria, many IBAs meet the criteria for designation under these agreements. Consequently, one way in which IBAs can support national commitments under multilateral environmental agreements is by identifying candidate sites for designation.

EMPOWERING PEOPLE

Building local and national institutional capacity

BirdLife aims to empower, mobilise and expand a worldwide constituency of people who care for birds and their natural environment. The BirdLife Partnership is comprised of "indigenous" civil society organisations. As an "NGO family", it believes in supporting and strengthening the capacity of national organisations to implement effective conservation at the local level, and views this as a long-term commitment to ensuring their sustainable development. Capacity development of BirdLife's broad network in the Caribbean is achieved by the BirdLife Secretariat supporting and facilitating the exchange of advice, expertise, training, mentorship and information between network organisations within the Caribbean and globally. In parallel with this, the Caribbean partners (and other network institutions) focus on building grassroots capacity for site level conservation through the establishment and engagement of community-based Site Support Groups (SSGs: see Box 2). SSGs are firmly established at IBAs in several Caribbean countries.

IBAs contribute to socio-economic development

IBAs are not only important for birds and biodiversity but also for socio-economic development at local and national levels. The ecosystem goods and services provided by IBAs often contribute significantly to human livelihoods. For example, coastal IBAs may be a source of marine products for fishing communities, while forest IBAs can be a source of non-timber forest products, such as fuelwood and food, for rural communities. Green tourism (nature tourism, birdwatching etc.) is growing in the sector and many of the IBAs represent the best natural areas in their respective territories, and are thus potential focal areas for sensitively developed green tourism activities. In addition, conservation of the Caribbean IBA network would bring significant benefits to national economies, because many IBAs provide high-value ecosystem services, such as water catchment protection and flood control. Consequently, provided that the socio-economic benefits of IBAs can be equitably shared, and their biological values simultaneously maintained, IBA conservation should be an objective shared by conservationists, local communities and governments alike.

THE CARIBBEAN IBA PROGRAM

Aims and objectives

The Caribbean IBA Program started in 2001 with the aim of identifying, conserving and promoting a global network of internationally important sites for birds and biodiversity. BirdLife Partners and network organisations in the region are working to achieve this aim by focusing on the following long-term objectives:

- To provide a basis for the development of national conservation strategies and protected areas programs;
- To highlight areas which should be safeguarded through wise land-use planning, national policies and regulations, and by the grant-giving and lending programs of international banks and development agencies;
- To provide a focus for the conservation efforts of civil society including national and regional NGO networks;
- To highlight sites which are threatened or inadequately protected so that urgent remedial measures can be taken;
- To guide the implementation of global and regional conservation conventions and migratory bird agreements.

Identifying IBAs

A participatory process

Caribbean IBAs have been identified through a process that has engaged local stakeholders (BirdLife Partners, network organisations, government agencies and local experts) from each island. The methodology followed to identify IBAs is detailed in the next chapter, and this has formed the foundation for IBA identification and documentation at the national level. However, the process followed has varied. In the Bahamas, Cuba, Dominican Republic, Jamaica and Puerto Rico, IBA identification was a fully participatory effort undertaken over a period of years, and often in parallel with targeted field surveys to fill information gaps and site-based conservation actions at some of the important sites. Each of these countries held national workshops for all relevant stakeholders to contribute ideas and information and to learn about the process. These workshops resulted in the creation of lists of "potential" IBAs. Information on the birds (and their populations) present in these areas, and about the areas themselves, was collated by national IBA coordinators, and entered nationally into the online World Bird Database, through which information on each site is publicly available[3]. Subsequent workshops and additional information refined the list of IBAs to those that met the IBA criteria (see Methodology). Once completed, these IBAs were confirmed by the BirdLife Secretariat, thereby ensuring global consistency in the national application of the IBA criteria.

The process in the remaining Caribbean islands was different. National expert authors were contracted to draft IBA chapters to a standardised format, and the data they collated was entered into the World Bird Database by the BirdLife Secretariat. The chapters produced in this way were circulated widely to other experts within NGOs and governments, by the authors themselves and by the BirdLife Caribbean Program team. Thus, while the process was not entirely a participatory one, it was inclusive and the authors are confident that the information presented is the best available at the time of publication.

3 Factsheets for IBAs are online at www.birdlife.org/datazone/sites

Several participatory IBA workshops, such as this one in Cuba, were held in each of five Caribbean countries.

Box 4. Important Bird Areas in the Caribbean.

Caribbean IBAs were first treated, in part, in Sanders (2006) which detailed the IBAs from all of the UK Overseas Territories around the world. This includes six territories in the Caribbean. The inventories for these territories have been significantly updated in their presentation here. The process of identifying IBAs has led a number of other islands to publish their own national language IBA books, building on the information they collated in their national languages in the World Bird Database (available online at www.birdlife.org/datazone/sites). These publications are referenced below and cover Cuba, Dominican Republic, Guadeloupe, Puerto Rico and St Barthélemy. A summary inventory for the whole of the Americas (North, Central and South America and the Caribbean) is due for publication in 2009.

- Aguilar, S. ed. (2008) *Áreas Importantes para la Conservación de las Aves en Cuba.* Havana: Centro Nacional de Áreas Protegidas.
- BirdLife International (in prep.) *Important Bird Areas Americas: a blueprint for conservation.* Quito, Ecuador: BirdLife International. (BirdLife Conservation Series No.16).
- Levesque, A. and Mathurin, A. (2008) *Les Zones Importantes pour la Conservation des Oiseaux en Guadeloupe.* Le Gosier, Guadeloupe: AMAZONA (AMAZONA contribution 17).
- Levesque, A., Mathurin, A. and Le Quellec, F. (2008) *Les Zones Importantes pour la Conservation des Oiseaux à Saint-Barthélemy.* Le Gosier, Guadeloupe: AMAZONA (AMAZONA contribution 18).
- Mendez, V. (2008) *Áreas Importantes para la Conservación de las Aves en Puerto Rico.* San Juan, Puerto Rico: Sociedad Ornitológica Puertorriqueña.
- Perdomo, L., Arias, Y., León, Y. and Wege, D. C. (2008) *Áreas Importantes para la Conservación de las Aves en la República Dominicana.* Santo Domingo: Grupo Jaragua, Inc.
- Sanders, S. M. ed. (2006) *Important Bird Areas in the United Kingdom Overseas Territories.* Sandy, U.K.: Royal Society for the Protection of Birds.

The IBA book

This book, *Important Bird Areas in the Caribbean*, which presents information on the region's 283 IBAs, is an advocacy tool through which the BirdLife Caribbean network aims to achieve further recognition, protection, and monitoring of IBAs throughout the region. A decision was made to publish only in English for a number of reasons: a trilingual book would be impractically large; national IBA publications have been written (in Spanish) for Cuba, Dominican Republic and Puerto Rico, and (in French) for Guadeloupe and St Barthélemy (see Box 4); the primary language of the multilateral environmental agreements and of the major funding agencies is English; and the information presented in this book is available in its original language in the IBA factsheets available through BirdLife's online Data Zone.

The World Bird Database is a critical component of the IBA program. The information in this book will go out-of-date as the conservation landscape evolves: new bird populations will be discovered; revised or new population estimates gathered; new threats or developments will impinge on the status of IBAs; and new protection measures will be put in place. The IBA profiles will be regularly updated online in the database by the national IBA coordinators and the BirdLife Caribbean Program team, and this is available publicly for people to use and review.

■ Conserving IBAs

On-the-ground site action for IBAs lies at the heart of BirdLife's Caribbean Program work. Interventions are underway at a wide range of sites including: Bahoruco-Jaragua-Enriquillo Biosphere Reserve (Dominican Republic); Zapata Swamp, Delta del Cauto, Alejandro Humboldt National Park, and Sierra Maestra (Cuba); Cockpit Country and Mount Diablo (Jamaica); Inagua, Abaco, Central Andros,

IBAs are being conserved through a variety of interventions, often in collaboration with Site Support Groups, including protected area designation and management.

and Harrold and Wilson Ponds national parks (Bahamas); Ashton Lagoon, Union Island (St Vincent and Grenadines); Mount Hartman Estate (Grenada); Sierra Bermeja, Cabo Rojo and Caño Tiburones (Puerto Rico); Centre Hills (Montserrat), and many others. The full range of conservation approaches and interventions is being used, including action planning, micro-enterprise development, reforestation, green-tourism, education awareness provision, invasive species control and eradication, fire management etc. Many of the site projects are being implemented in collaboration with Site Support Groups: independent, mostly volunteer, community-based organisations at high-biodiversity sites working to manage their local natural resources sustainably. This local-level engagement is enabling BirdLife Partners (and other network organisations) to address environmental governance issues, often the root causes of poverty and unsustainable development. By promoting IBAs to governments and NGOs and through multilateral environmental agreements, the BirdLife network aims to catalyse conservation interventions across the full range of the region's IBAs.

■ Promoting IBAs

National governments in the Americas are party to a number of multilateral environmental agreements, established to promote biodiversity conservation and sustainable use of natural resources. In addition, there are a number of other mechanisms that promote international cooperation for the conservation of biodiversity and natural resources. This section briefly describes each agreement and mechanism, and highlights the main ways in which conservation of the Caribbean IBA network would assist national governments and donors to meet their commitments under them.

Table 1. State of ratification of selected multilateral environmental agreements in the Caribbean.

	Convention name (see text)					
Country	Ramsar	SPAW	CBD	WHC	CMS	CITES
Anguilla (to UK)	y	n	n	y	n	n
Antigua and Barbuda	y	n*	y	y	y	y
Aruba (to Netherlands)	y	y	y	y	y	y
Bahamas	y	n	y	n	n	y
Barbados	y	y	y	y	n	y
Bermuda (to UK)	y	n	n	y	y	y
Bonaire (to Netherlands)	y	y	y	y	y	y
British Virgin Islands (to UK)	y	y	y	y	y	y
Cayman Islands (to UK)	y	y	y	y	y	y
Cuba	y	y	y	y	y	y
Curaçao (to Netherlands)	y	y	y	y	y	y
Dominica	n	n	y	y	n	y
Dominican Republic	y	y	y	y	n	y
Grenada	n	n	y	y	n	y
Guadeloupe (to France)	y	y	y	y	y	y
Haiti	n	n	y	y	n	n
Jamaica	y	n*	y	y	n*	y
Martinique (to France)	y	y	y	y	y	y
Montserrat (to UK)	y	n	n	y	y	y
Navassa (to USA)	y	y	n*	y	n	y
Puerto Rico (to USA)	y	y	n*	y	n	y
Saba (to Netherlands)	y	y	y	y	y	y
St Barthélemy (to France)	y	y	y	y	y	y
St Eustatius (to Netherlands)	y	y	y	y	y	y
St Kitts and Nevis	n	n	y	y	n	y
St Lucia	y	y	y	y	n	y
St Maarten (to Netherlands)	y	y	y	y	y	y
St Martin (to France)	y	y	y	y	y	y
St Vincent and the Grenadines	n	y	y	y	n	y
Trinidad and Tobago	y	y	y	y	n	y
Turks and Caicos Islands (to UK)	y	y	n	y	y	n
US Virgin Islands (to USA)	y	y	n*	y	n	y
Total y	27	22	25	31	17	29
Total n/n*	5	10	7	1	15	3

y = signed and ratified; n = not signed; n* = signed but not ratified

Ramsar Convention

The Ramsar Convention, officially known as the Convention on Wetlands of International Importance especially as Waterfowl Habitat, was adopted in 1971 and came into force in 1975. As of August 2008, the convention had 158 parties. The only Caribbean countries not currently parties to the convention are Dominica, Grenada, Haiti, St Kitts and Nevis, and St Vincent and the Grenadines (Table 1). The convention provides a framework for international cooperation for the conservation and wise use of wetlands, and parties have a commitment to promote the wise use of all wetlands in their territory, to designate suitable sites for inclusion on the List of Wetlands of International Importance (Ramsar Sites), and to promote their conservation. As of August 2008, the parties had designated 1,759 Ramsar sites globally, with 313 sites in the Americas, of which 35 are in the Caribbean. Five of these sites have been specially designated for Ramsar criteria matching the A4 criteria in IBA designation. The first Ramsar sites in the Caribbean were designated in 1980 in Bonaire and Aruba. Ramsar sites cover a total area of 13,802 km^2 in the region.

One of the tools available through the convention to promote the conservation of Ramsar sites is the Montreux Record, a register of sites on the List of Wetlands of International Importance where changes in ecological character have occurred, are occurring, or are likely to occur as a result of technological developments, pollution or other human interference. The Montreux Record should be employed to promote IBAs for positive national and international conservation attention. IBA monitoring programs at designated Ramsar sites can help identify sites that should be listed on the Montreux Record, and identify the conservation actions which are required.

The SPAW Protocol

The SPAW Protocol to the Cartagena Convention (officially the Protocol Concerning Specially Protected Areas and Wildlife to the Convention for the Protection and Development of the Marine Environment of the Wider Caribbean Region) aims to protect marine and coastline resources that many Caribbean countries depend on for economically important activities such as tourism and fishing. Among its objectives are to increase the number and effectiveness of protected areas, develop regional capacity for biodiversity conservation, and coordinate activities with other multilateral environmental agreements. The protocol came into force in 2000. Thirteen of a possible 28 countries have ratified the protocol to date, including 22 territories as detailed in this book. Two additional territories are signatories without having ratified (Table 1).

By recognising the importance of protecting rare, fragile ecosystems and habitats as a means to protecting threatened species within them, the protocol was innovative in that its ecosystem approach pre-dated other international environmental agreements, including the CBD (see below). As opposed to global treaties, SPAW was developed by and for governments in the region. In fact, it is the only legal agreement to address biodiversity conservation issues exclusively in the Wider Caribbean (i.e. islands within, and countries bordering the Caribbean Sea). The protocol is particularly strong in providing guidance and operational support to parties.

The SPAW Program of the UNEP Caribbean Environment Program is responsible for coordinating supporting activities, such as promotion of best practices and training for sustainable tourism, coral reef monitoring, technical assistance to establish and strengthen protected areas, implementation of guidelines and recovery plans for species conservation and education, among others. The program also contributes to the regionalisation of global conventions and initiatives such as the CBD, CITES, Ramsar and the International Coral reef Initiative (ICRI). For example, a memorandum of cooperation exists between the Ramsar Convention and SPAW in an attempt to maximise joint efforts. The agreement aims to assist the identification of wetlands as potential Ramsar sites or protected areas under the SPAW protocol. Using Caribbean IBAs meeting appropriate A4 criteria (concerning congregatory species) as a shadow list for both Ramsar and SPAW in this context could also help to avoid duplication of efforts in site identification.

Species conservation is considered through the provision of three annexes to the Protocol. Annex II, containing 41 bird species, is a list of marine and coastal fauna protected by the protocol, for which parties have agreed to ensure total protection and recovery, prohibiting all capture and trade in these species. Annex III, including 12 bird species, is a list of flora and fauna to be maintained at sustainable levels. Thirty-two and seven bird species from annexes II and III, respectively, trigger IBA criteria in the Caribbean. Of these species, 24 have been confirmed in more than 150 IBAs in 23 territories throughout the Caribbean. Therefore, ensuring protection at these IBAs could help parties meet their obligations with regard to the protection and recovery of these species, taken on at ratification.

Convention on Biological Diversity (CBD)

The CBD was adopted in 1992 and came into force in 1993. As of August 2008, the convention had 191 contracting parties. The only country in the Americas not party to the convention is the USA. In the Caribbean, 25 territories have ratified the convention; those not having ratified are either US or UK territories (Table 1). The convention's objectives are the conservation of biological diversity, the sustainable use of its components, and the fair and equitable sharing of the benefits arising out of the utilisation of genetic resources. The convention has a focus on *in situ* conservation, and Article 8(a) commits national governments to establish "a system of protected areas or areas where special measures need to be taken to conserve biological diversity".

The Seventh Conference of the Parties of the CBD, held in Kuala Lumpur in 2004, adopted a Program of Work on Protected Areas (Decision VII/28) with "the objective of the establishment and maintenance by 2010 for terrestrial and by 2012 for marine areas of comprehensive, effectively managed, and ecologically representative national and regional systems of protected areas". This Program of Work is comprised of four elements (implementation, governance and equity, enabling activities, and monitoring), each consisting of several specific goals. The first goal of the first element—"to establish and strengthen national and regional systems of protected areas integrated into a global network as a contribution to globally agreed goals,"—requires the identification of sites of global biodiversity significance in each country to determine which sites are currently not represented in protected area systems, and prioritisation of conservation actions among sites. The Caribbean IBA network provides an objective scientific basis for the review and expansion of protected areas networks in the region. Gap analyses are also necessary for reporting on the "coverage of protected areas" indicator, which was provisionally adopted by the Parties for measuring progress towards the 2010 target of reducing biodiversity loss (Decision VII/30). At the Ninth Conference of the Parties (in Bonn in 2008), a commitment was made to complete protected areas gap analyses by 2009. The IBA network provides a comparative baseline for such analyses, including in marine environments, where BirdLife's work to identify marine IBAs is at the forefront of efforts to identify criteria for the identification of priority marine areas.

At the national level, implementation of the CBD is guided by National Biodiversity Strategies and Action Plans (NBSAPs), which set out priorities for biodiversity conservation. These documents are used, in part, to guide investment from the Global Environment Facility and other funding sources. NBSAPs have been completed for 33 Americas countries, and are under preparation for three more. No information is available regarding the status of an NBSAP for St Kitts and Nevis. NBSAPs present an opportunity for official recognition of national IBA networks within national conservation plans.

National reports provide information on the measures taken for the implementation of the Convention (as identified in the NBSAPS) and the effectiveness of these measures. The Eighth Conference of the Parties (Curitiba, 2006) decided (Decision VIII/14) that "the fourth [due in March 2009] and subsequent national reports should be outcome-oriented and focus on the national status and trends of biodiversity, national actions and outcomes with respect to the achievement of the 2010 target and the goals of the Strategic Plan of the Convention, and progress in implementation of national biodiversity strategies and action plans". Information on the conservation status of IBAs, compiled through national IBA monitoring programs, can make an important contribution to reporting on progress towards achieving the 2010 (and other) CBD targets.

The CBD is increasingly emphasising the links between conservation and poverty reduction as a major consideration in the development of a revised strategic plan and post-2010 target. The IBA approach to site conservation, working with local communities (including forming local conservation groups) and building local and national institutional capacity, provides valuable models for effective site conservation helping to meet local, national and global conservation and development goals.

World Heritage Convention (WHC)

The WHC was adopted in 1972 and came into force in 1975. As of August 2008, the convention had 185 parties globally, including 37 in the Americas region (only Bahamas is not a party) (Table 1). The aim of the convention is to identify and conserve cultural and natural monuments and sites of outstanding universal value. Parties to the WHC have a commitment to nominate suitable sites for recognition by the United Nations Educational, Scientific and Cultural Organisation (UNESCO) as World Heritage Sites. As of August 2008, a total of 878 World Heritage Sites had been designated worldwide, including 160 in the Americas. In the Caribbean, 17 sites have been designated, of which 13 are for cultural values and four for natural values. As in the Caribbean, the majority of World Heritage Sites have been nominated for their cultural values (99 of the Americas sites are exclusively cultural, three are mixed cultural and natural). In order to redress this imbalance, the WHC wishes to see more natural monuments of outstanding value nominated. As a significant number of the Caribbean IBAs have outstanding biological and other natural values, information on IBAs can be used to assist parties to identify candidate sites for nomination as World Heritage Sites.

Convention on Migratory Species of Wild Animals (CMS)

The CMS, also known as the Bonn Convention, was adopted in 1979 and came into force in 1983, with the aim of protecting migratory species that cross international borders. As of August 2008, the CMS had 109 parties, including 17 territories in the Caribbean, as defined in this directory. One further country (Jamaica) is a signatory to the convention but has not ratified yet.

The CMS has two appendices. Migratory species threatened with extinction are listed on Appendix I. CMS Parties strive towards strictly protecting these species, conserving or restoring the places where they live, mitigating obstacles to migration and controlling other factors that might endanger them. Besides establishing obligations for each state joining the convention, CMS promotes concerted action among the range states of many of these species. Information on IBAs can assist parties to meet this commitment by identifying important habitats for migratory bird species in need of conservation.

Migratory species that need or would significantly benefit from international cooperation are listed on Appendix II. The convention encourages the range states to conclude global or regional agreements for these species. Again, information on

IBAs can assist parties to meet this commitment by identifying suitable areas of habitat for each bird species covered by international agreements. The agreements may range from legally binding treaties (called Agreements) to less formal instruments, such as Memoranda of Understanding, and can be adapted to the requirements of particular regions. To date, one Agreement and two MoUs of direct relevance to bird conservation in the Americas have been developed under the auspices of CMS, although these are not relevant to Caribbean species.

Convention on International Trade in Endangered Species of Fauna and Flora (CITES)
CITES was adopted in 1973 and came into force in 1975. As of August 2008, the convention had 173 parties globally, including 37 in the Americas region. In the Caribbean, 29 territories are party to the convention, only Haiti and two UK overseas territories are not party. The aim of CITES is to regulate international trade in wildlife and wildlife products through international cooperation, while recognising national sovereignty over wildlife resources. CITES has three appendices:

- Appendix I, which lists species threatened with extinction and for which trade is prohibited except in exceptional circumstances that cannot be traded commercially;
- Appendix II, which includes species not necessarily threatened with extinction, but in which trade must be controlled in order to avoid utilisation incompatible with their survival. Parties are expected to have trade management regulations in place to ensure this; and,
- Appendix III, which contains species that are protected in at least one country, which has asked other CITES Parties for assistance in controlling the trade.

As of July 2008, CITES lists 152 bird species in Appendix I (plus 11 subspecies and 2 populations), 1,268 species on Appendix II (plus 6 subspecies and 1 population), and 35 species on Appendix III. On the Appendix I there are 17 Caribbean bird species (of which 10 are globally threatened), and on Appendix II there are 114 Caribbean birds (eight of which are globally threatened). Many of these listed species occur in important populations at the IBAs detailed in this publication. The Caribbean IBA network thus provides a focus for the efforts of national governments to implement CITES by identifying sites with significant populations of bird species threatened by the wildlife trade, which may require strengthened enforcement, public awareness raising, and other targeted conservation actions.

Western Hemisphere Shorebird Reserve Network (WHSRN)
WHSRN is a site-focused shorebird conservation strategy launched in 1985. During the last 20 years, over 8.4 million hectares of shorebird habitat have been brought under the auspices of WHSRN. WHSRN works to:

- Build a strong network of key sites used by shorebirds throughout their migratory ranges.
- Develop science and management tools that expand the scope and pace of habitat conservation at each site within the Network.
- Establish local, regional and international recognition for sites, raising new public awareness and generating conservation funding opportunities.
- Serve as an international resource, convener and strategist for issues related to shorebird and habitat conservation.

The network currently has 69 sites in 10 countries, from Alaska in the north to Tierra del Fuego in southern South America, although as yet there are no WHSRN sites in the Caribbean. There are three categories of Sites and one of Landscapes, according to their importance for shorebirds:

- **Sites/Landscapes of Hemispheric Importance:**
  — at least 500,000 shorebirds annually, or
  — at least 30% of the biogeographic population for a species
- **Sites of International Importance:**
  — at least 100,000 shorebirds annually, or
  — at least 10% of the biogeographic population for a species
- **Sites of Regional Importance:**
  — at least 20,000 shorebirds annually, or
  — at least 1% of the biogeographic population for a species

The thresholds for identifying sites of regional importance are similar to those for identifying IBAs for congregatory species. As such, Caribbean sites identified as IBAs for their shorebirds are candidates for designation as WHSRN Sites. The three categories of WHSRN sites also provide a convenient scheme for setting priorities among IBAs identified for shorebirds based on biological importance.

■ Monitoring IBAs

The Caribbean IBA Program aims to conserve a network of internationally important sites for birds and biodiversity. To do this it is vital to know what the general conservation status of IBAs is (using birds as biodiversity indicators) on a regular basis. The BirdLife Partnership has developed an IBA monitoring framework to ensure permanent cycles of monitoring can be put in place and that information from sites is periodically updated. As such, IBA monitoring is one element of a wider framework for monitoring progress towards conserving birds, their habitats and global biodiversity. This includes monitoring of species, sites and habitats.

The IBA monitoring framework provides a standardised way to assign scores for the condition of IBAs ("State"), the threats to IBAs ("Pressure") and conservation actions taken at IBAs ("Response"). This scoring system makes it possible to integrate a wide range of information, which may often be qualitative rather than quantitative. Being standardised, it allows national data to be compiled regionally and globally, and this is proving a powerful tool for international conservation advocacy and fundraising. Ideally, all IBAs in a country or territory should be regularly monitored. However,

Box 5. Caribbean Birds, part of the WorldBirds global program.

Caribbean Birds is part of the WorldBirds global programme—a joint initiative supported by BirdLife, the RSPB and Audubon, linking together Internet systems to collect and report on bird populations and movements in different countries around the world.

Caribbean Birds has been available (www.worldbirds.org/caribbean) since 2006, capturing bird observations recorded throughout the Caribbean. The system aims to collect data from both local observers and visiting tourists, with an emphasis on empowering citizen scientists and community monitoring groups. It also collects data for conservation work and provides a strong focus for birders and birding groups, and the initiative enables users to store and manage their own observations, print or download maps, extract reports for species and locations, and see what is happening to birds across the region.

As the volume of data increases, the system will provide valuable information on bird species at Important Bird Areas (IBAs). More broadly, data collected through all WorldBirds systems will contribute significantly to common bird monitoring, through tracking changes in species represented on birders' daylists. WorldBirds data will become a key source of information from which we can develop indicators for common bird species, which will then help indicate changes in the conditions of the wider environment.

We encourage everyone throughout the Caribbean to participate so that **Caribbean Birds** becomes a huge conservation success, both in strengthening the BirdLife Partnership and provide much needed information on the state of bird populations.

sustainability is also very important, so monitoring must be kept simple, robust and inexpensive. The minimal data required are simple and mainly qualitative. They can usually be collected on site by management authority or project staff, Site Support Group (SSG) members, or other volunteers (e.g. those that contribute to "Caribbean Birds": see Box 5), and thus it is built on the principle that monitoring is participatory. Data is held and owned by the organisations that collect them. National results feed up further to the regional and global levels, coordinated by the BirdLife Secretariat.

Information from the field and from other sources (such as remote sensing) is synthesised by a national BirdLife Partner or network organisation who assign indicator scores and hence overall status scores for each site. More in-depth monitoring can be used to inform the "state" assessment of a site (e.g. through establishing an accurate population estimate for a key bird species). If monitoring is "institutionalised" within these organisations, so that it becomes part of their routine work, then direct costs can be kept low.

KEY REFERENCES

ANDERSON, S. (2002) *Identifying Important Plant Areas*. London, UK: Plantlife International.

BIRDLIFE INTERNATIONAL (2004) A strategy for birds and people: responding to our changing world. Future directions of the BirdLife Partnership 2004–2015. Cambridge, U.K.: BirdLife International.

BIRDLIFE INTERNATIONAL (in prep.) *Important Bird Areas Americas: a blueprint for conservation*. Quito, Ecuador: BirdLife International. (BirdLife Conservation Series No.16).

BIRDLIFE INTERNATIONAL AND CONSERVATION INTERNATIONAL (2005) *Áreas Importantes para la Conservación de las Aves en los Andes Tropicales: sitios prioritarios para la conservación de la biodiversidad.* Quito, Ecuador: BirdLife International. (BirdLife Conservation Series No.14).

BROOKS, T., BALMFORD, A., BURGESS, N., HANSEN, L.A., MOORE, J., RAHBEK, C., WILLIAMS, P., BENNUN, L.A., BYARUHANGA, A., KASOMA, P., NJOROGE, P., POMEROY, D. AND WONDAFRASH, M. (2001) Conservation priorities for birds and biodiversity: do East African Important Bird Areas represent species diversity in other terrestrial vertebrate groups? *Ostrich* 15: 3–12.

DARWALL, W. AND VIÉ, J.C. (2005) *Identifying important sites for conservation of freshwater biodiversity: extending the species-based approach*. Gland, Switzerland and Cambridge, U.K.: IUCN.

EDGAR, G. J., LANGHAMMER, P. F., ALLEN, G., BROOKS, T. M., BRODIE, J., CROSSE, W., DA SILVA, N., FISHPOOL, L. D. C., FOSTER, M. N., KNOX, D. H., MILLER, A. J. K. AND MUGO, R. (in press) Key Biodiversity Areas as globally significant target sites for marine conservation. *Aquatic Conservation*.

EKEN, G., BENNUN, L., BROOKS, T.M., DARWALL, W., FISHPOOL, L.D.C., FOSTER, M., KNOX, D., LANGHAMMER, P., MATIKU, P., RADFORD, E., SALAMAN, P., SECHREST, W., SMITH, M.L., SPECTOR, S. AND TORDOFF, A. (2004) Key biodiversity areas as site conservation targets. *BioScience* 54: 1110–1118.

LANGHAMMER, P. F., BAKARR, M. I., BENNUN, L. A., BROOKS, T. M., CLAY, R. P., DARWALL, W., DE SILVA, N., EDGAR, G. J., EKEN, G., FISHPOOL, L. D. C., DA FONSECA, G. A. B., FOSTER, M. N., KNOX, D. H., MATIKU, P., RADFORD, E. A., RODRIGUES, A. S. L., SALAMAN, P., SECHREST, W. AND TORDOFF, A.W. (2007) *Identification and gap analysis of key biodiversity areas: targets for comprehensive protected area systems*. Gland, Switzerland: IUCN (Best Practice Protected Area Guidelines Series 15).

LINZEY, A.V. (2002) *Important Mammal Areas: A US pilot project*. In: *Society for Conservation Biology. 16th Annual Meeting: Programme and Abstracts*. Canterbury, U.K.: Durrell Institute of Conservation and Ecology.

MITTERMEIER, R. A., MYERS, N. AND MITTERMEIER, C. G. EDS. (1999) *Hotspots: earth's biologically richest and most endangered terrestrial ecoregions*. Mexico City: Sierra Madre.

PAIN, D. J., FISHPOOL, L., BYARUHANGA, A., ARINAITWE, J. AND BALMFORD, A. (2005) Biodiversity representation in Uganda's forest IBAs. *Biological Conservation* 125: 133–138

PLANTLIFE INTERNATIONAL. (2004) *Identifying and protecting the world's most Important Plant Areas: a guide to implementing target 5 of the global strategy for plant conservation*. Salisbury, U.K.: Plantlife International.

TUSHABE, H., KALEMA, J., BYARUHANGA, A., ASASIRA, J., SSEGAWA, P., BALMFORD, A., DAVENPORT, T., FJELDSA, J., FRIIS, I., PAIN, D., POMEROY, D., WILLIAMS, P. AND WILLIAMS, C. (2006) A Nationwide Assessment of the Biodiversity Value of Uganda's Important Bird Areas Network. *Conservation Biology* 20(1): 85–99

VAN SWAAY, C.A.M. AND WARREN, M.S. (2003) *Prime Butterfly Areas in Europe: priority sites for conservation*. Wageningen, Netherlands: National Reference Center for Agriculture, Nature and Fisheries; Ministry of Agriculture, Nature Management and Fisheries.

METHODOLOGY

IBA CRITERIA

The aim of the IBA program is to identify and protect a network of sites critical for the long-term viability of naturally occurring bird populations, across the ranges of bird species for which a site-based approach is appropriate. The selection of IBAs is achieved through the application of standard, internationally recognised criteria, as far as possible based upon accurate, up-to-date knowledge of bird species distributions and populations. The use of standard criteria worldwide means that IBAs are a "common currency", with sites consistent and comparable at national, regional and global levels.

The criteria used to select IBAs in the Caribbean derive from those initially used in the first European IBA inventory (in 1989), which in turn took account of IBA criteria developed earlier for use in the European Community. The 1989 IBA criteria had, however, been developed specifically for application in Europe. When the IBA program was expanded to cover other regions of the world, these criteria had to be adapted, first for the Middle East IBA program and subsequently, following extensive consultation within the BirdLife International Partnership and beyond, further developed and standardised for application worldwide. Using these global criteria, IBAs are selected based on the presence of:

- Species of global conservation concern
- Assemblages of restricted-range bird species
- Assemblages of biome-restricted bird species
- Globally important congregations of birds

These standardised criteria are designed to identify IBAs of global significance, and thus permit meaningful comparison between sites within and between regions of the world. In a number of regions (e.g. Europe, North America, Caribbean), IBAs have been identified at the regional level. The respective regional BirdLife Partnerships decide whether to identify IBAs below the global level, and this was the case in the Caribbean where the strong focus on wetlands and seabirds warranted the identification of regionally (i.e. Caribbean) significant IBAs for waterbirds and seabirds. These regional IBAs are of international conservation significance. Both global and regional IBAs are covered in this publication, and in the individual IBA profiles the distinction is made by reference to populations of species being either globally or regionally significant. Regional criteria are based on the global IBA categories and do not introduce new elements to the criteria framework.

The IBA criteria address the two key issues of concern in site conservation: vulnerability and irreplaceability. The four categories of criteria thus cover globally threatened species (vulnerability), and three classes of geographically concentrated species (irreplaceability), namely restricted-range, biome-

Box 1. Major sources of IBA information.

The Caribbean IBA Program has been built on a wealth of published and unpublished work, some of which is referenced in the national chapters. However, past and on-going research by BirdLife and others, and certain publications have provided the cornerstones on which the IBAs are based.

- For threatened species, *Threatened birds of the Americas: the ICBP/IUCN Red Data Book* (Collar *et al.* 1992), even 16 years after its publication remains the definitive work on threatened species throughout the Americas, with detailed accounts for 327 species including 40 from the Caribbean. *Threatened birds of the world* (BirdLife International 2000) and the BirdLife Data Zone (http://www.birdlife.org/datazone/species/) present annually updated status assessments and details for all of the region's threatened birds.
- The BirdLife Biodiversity Project, summary results of which appeared in *Putting biodiversity on the map* (ICBP 1992) and full details in *Endemic Bird Areas of the world* (Stattersfield *et al.* 1998). This project identified Endemic Bird Areas (EBAs) and Secondary Areas (SAs), and provided lists of the restricted-range bird species which were used to identify IBAs under the restricted-range species category (A2).
- Of key importance for defining biome-restricted species in the Neotropics have been the Parker *et al.* (1996) databases, accompanying the publication *Neotropical birds: ecology and conservation* (Stotz et al. 1996).
- Wetlands International has published estimates of the population sizes of waterbirds in the Americas (Rose and Scott 1997; Wetlands International 2002, 2006), which were used to develop 1% thresholds for waterbirds (see below). The Society for the Conservation and Study of Caribbean Birds' publication *Status and conservation of West Indian seabirds* (Schreiber and Lee 2000) was similarly the basis for establishing 1% thresholds for the region's seabirds. Also, Ducks Unlimited's (unfortunately no longer active) Latin America and Caribbean waterfowl surveys provided invaluable information on waterfowl populations at particular IBAs in the Caribbean.
- The *Birds of the West Indies* (Raffaele *et al.* 1998) guide has been a daily source of reference, and is the foundation for so much of the work that has gone into this publication and is being done to conserve the region's birds. Paintings from this guide—made available through U.S. Fish and Wildlife Service's Wildlife Without Borders Program—have been used to illustrate each IBA profile.

restricted and congregatory species. For each of these categories a list of species is drawn up, together with population thresholds, where appropriate. Species on these lists are referred to here as "key species", and they can trigger IBA criteria for a particular site. Populations of these species form the basis for the identification of particular sites as IBAs. Various information sources have helped establish population estimates, thresholds, key species lists and even the presence of species in particular countries, and have been critical to the development of the IBA program. These invaluable references are outlined in Box 1.

■ Species of global conservation concern (criterion A1)

The site regularly holds significant numbers of a globally threatened or "Near Threatened" species.

The Vulnerable Fernandina's Flicker.
(PHOTO: PETE MORRIS)

Under this criterion, sites are identified for globally threatened species, that is, Critically Endangered (CR), Endangered (EN) and Vulnerable (VU) as defined by IUCN categories and criteria (see Appendix 1a). Globally Near Threatened (NT) species are also included here, in an attempt to prevent them from becoming up-listed to higher threat categories in the future. These species, collectively known as "species of conservation concern" (but generically referred to throughout this publication as "threatened") are listed in Appendix 1b. Fact files, including global population numbers, country distributions and threats, among other information can be viewed in BirdLife's Data Zone[1].

Just the presence of a threatened species is not always enough to trigger IBA criteria at a site. Certain population thresholds must be met, depending on the threat category of the species in question. However, for Critically Endangered and Endangered species, the regular presence of just one individual is enough to justify IBA designation. For Vulnerable and Near Threatened species, the following population thresholds must be met in order for a site to meet global IBA criteria:

- **Vulnerable** All species 10 pairs or 30 individuals
- **Near Threatened** Non-passerines 10 pairs or 30 individuals
  Passerines 30 pairs or 90 individuals

In the absence of accurate population estimates, inference and expert opinion have been used to apply the criteria. Generally, the term "regular" presence of a species in the criterion definition is intended to exclude species which are vagrants or on the limits of their distributional range as well as historical records without recent confirmation. In these cases, although a threatened species may have occurred at a site, its presence is judged to be unimportant for the survival of the population. However, "regular" includes seasonal presence, as is the case for many migratory species in the Caribbean.

This category also allows for the inclusion of sites with the potential to hold species of global conservation concern after habitat restoration or re-introduction programs, e.g. the Río Abajo IBA in Puerto Rico is the focus of a re-introduction program for the Critically Endangered Puerto Rican Amazon *Amazona vittata*.

■ Assemblage of restricted-range species (criterion A2)

The site is known or thought to hold a significant component of the restricted-range species whose breeding distributions define an Endemic Bird Area.

IBAs are designated under this criterion for groups of species within Endemic Bird Areas (EBAs) or Secondary Areas. EBAs are priority regions for conservation where the breeding ranges of two or more restricted-range species coincide. A restricted-range species has been defined as having an original global distribution of 50,000 km^2 or less. Species with current distributions of less than this area due to habitat loss or other pressures were not considered within the EBA analysis. EBAs and Secondary Areas (usually defined for just one restricted-range species) cover a relatively small proportion of the world's land area but support a major part of the global avifauna and other terrestrial biodiversity. Of the 218 global EBAs, six EBAs have been defined for the Caribbean, as well as four Secondary Areas with a total of 135 extant restricted-range species (Figure 1, and species lists in Appendix 2). Restricted-range species account for 81% of all globally threatened birds, and as a group of species (over 25% of the world's birds) they are inherently vulnerable.

Barbuda Warbler, endemic to Barbuda, and one of the 38 species restricted to the Lesser Antilles Endemic Bird Area.
(PHOTO: ANDREW DOBSON)

1 http://www.birdlife.org/datazone/species/

Figure 1. Endemic Bird Areas and Secondary Areas in the Caribbean[2].

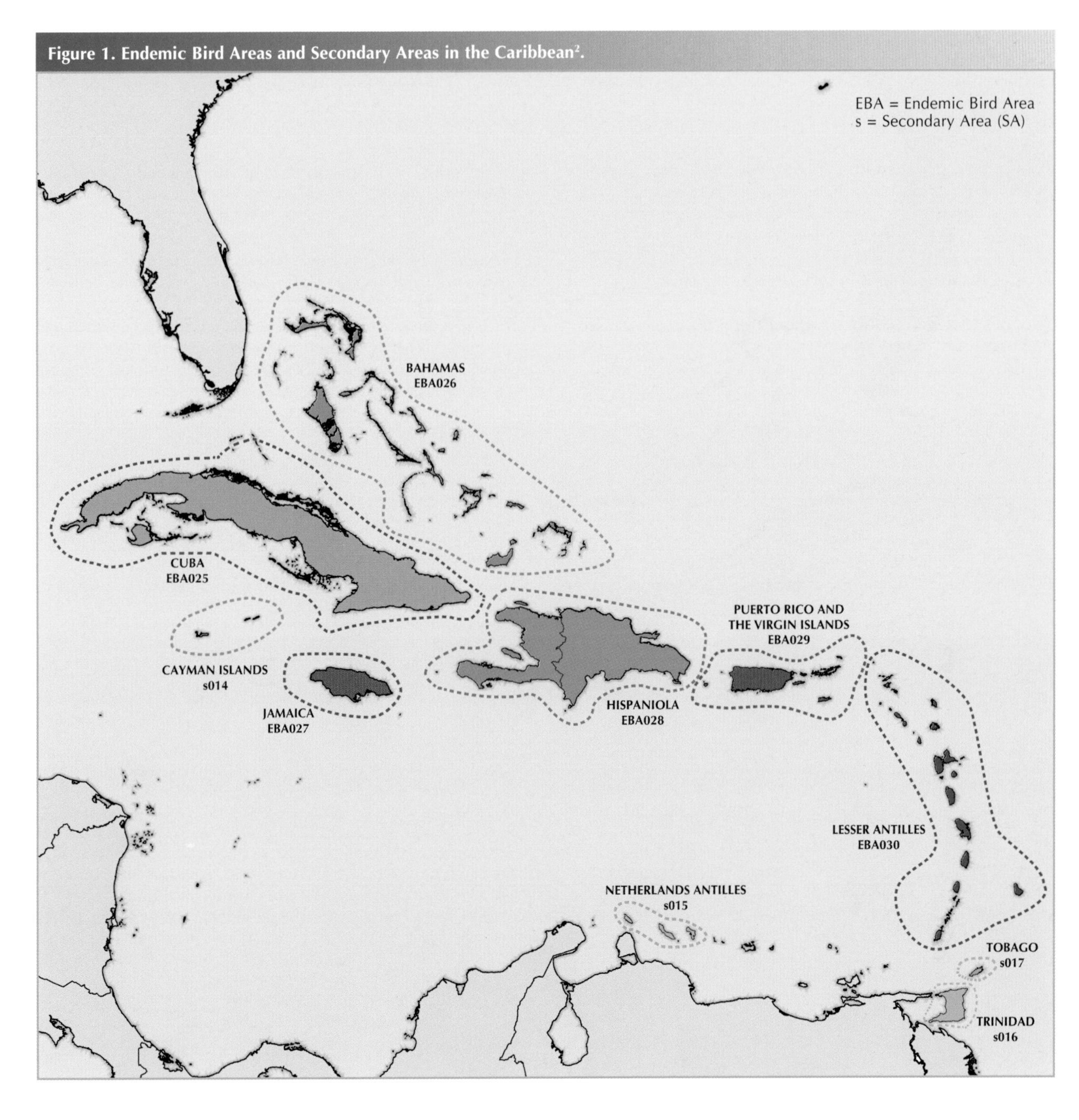

In the application of the A2 criterion, the idea is to select sites which are representative of the EBA. The term "significant component" is normally interpreted to mean that the IBA should hold at least 33% of the species in a particular EBA of those present within the boundaries of the country when an EBA spans two or more political entities. This attempts to avoid site selection for just one or two species. However, in order for all species in an EBA within a country to be represented in the IBA network, it is also necessary to perform a complementarity analysis, and sites may be selected for a small number of species if they are the only sites supporting a particular bird. A geographic balance within the EBA is also sought between different sites meeting this criterion.

■ Assemblage of biome-restricted species (criterion A3)

The site is known or thought to hold a significant component of the group of species whose distributions are largely or wholly confined to one biome.

This criterion is applied in a similar fashion to A2, but with biome-restricted species, that is, groups of species with largely

The Bee Hummingbird is endemic to the Greater Antilles biome, and also to Cuba. However, its range within Cuba is too large to qualify as a "restricted-range" (A2) species. (PHOTO: TIM STEWART)

2 http://www.birdlife.org/datazone/EBAs/

Figure 2. Biomes in the Caribbean, following Stotz *et al.* (1996).

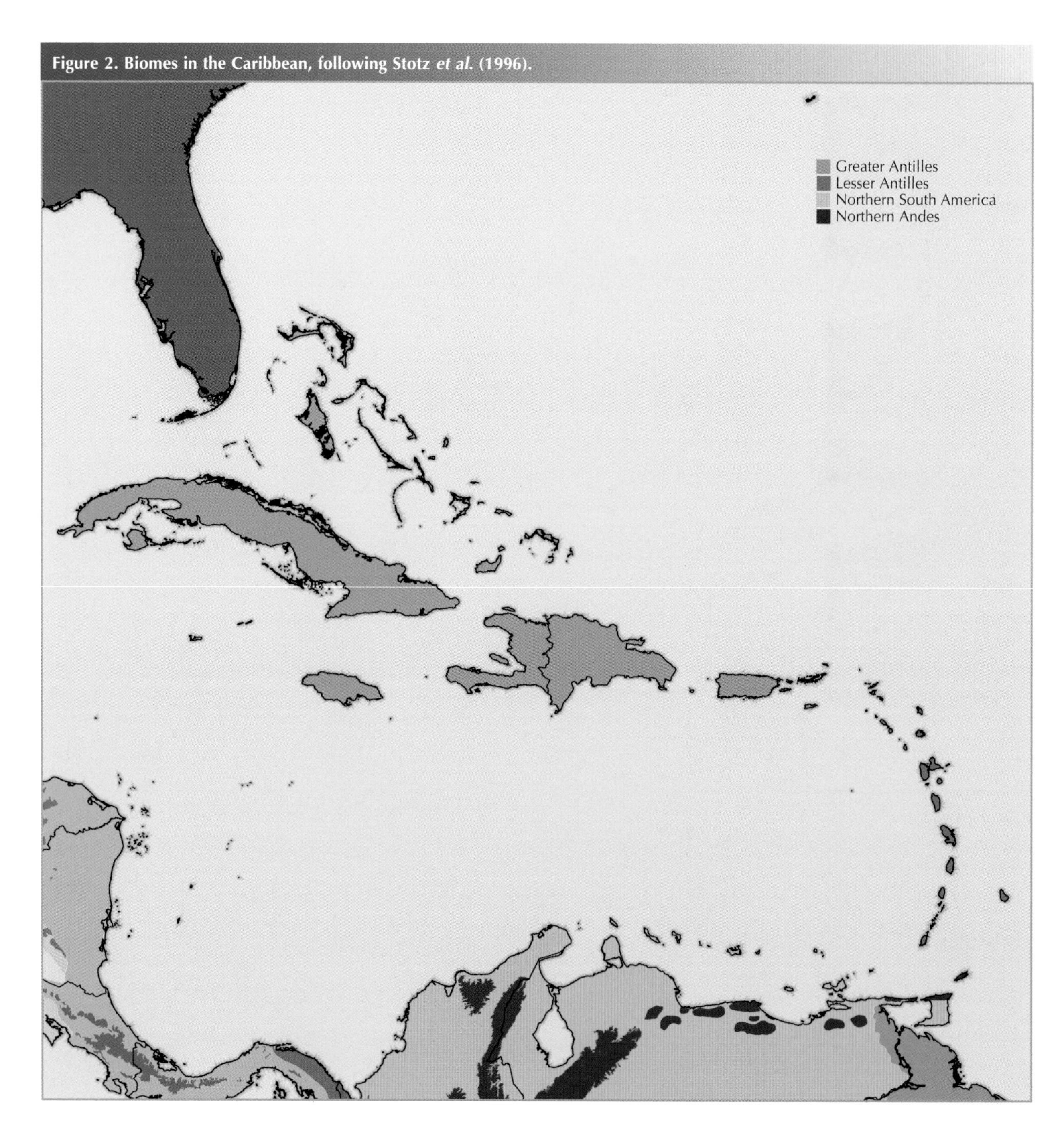

shared distributions whose entire global breeding distributions lie wholly or mostly within a particular biome. A biome is defined as a major regional ecological community, characterised by distinctive life forms and principal plant species. In the case of the Caribbean, definitions of biomes by Stotz *et al.* (1996) have generally been followed (see Figure 2). However, owing to the insular nature of the Caribbean, national lists of (A3) biome-restricted species are almost identical to the national lists of (A2) restricted-range birds. As a general rule, this criterion was not applied in the Caribbean, the exceptions being Cuba, Trinidad and Tobago, Aruba, Bonaire and Curaçao. In Cuba, where most of the island endemics historically had ranges larger than 50,000 km^2, IBAs have been designated for the Greater Antilles biome-restricted species. On islands sharing South American continental biomes off the coast of Venezuela, IBAs have been designated for Northern Andes biome-restricted species (Trinidad and Tobago) and Northern South America biome-restricted species (Aruba, Bonaire and Curaçao, and Trinidad and Tobago). Species lists for each of the Caribbean biomes are given in Appendix 3.

■ Congregations (criterion A4)

The site is known or thought to hold, on a regular basis, a globally significant concentration of one or more congregatory species.

Congregatory species are those that gather in globally significant numbers at a particular site at a particular time in their life cycle for feeding, breeding or resting (e.g. during migration). Such species therefore tend to have specialised ecological requirements due to their dependency on a relatively small proportion of the total territory. Their congregatory behaviour (even for short periods of time) makes these species inherently vulnerable at the population level. This criterion aims to define IBAs (including "marine" IBAs: see Box 1) to help protect against this vulnerability. To trigger this criterion, species congregations must pass the critical thresholds for any of the four sub-criteria outlined in Table 1.

The term waterbird used in criterion A4i is used in the same sense as that used for "waterfowl" in the Ramsar Convention, and is made up of the list published by Wetlands International (2006). This definition uses a whole family approach and thereby includes some species which are not dependent on

Species that congregate, such as these "Cayenne" Terns, are inherently vulnerable at the population level. (PHOTO: ADRIAN DEL NEVO)

Table 1. Summary of global criteria for selection of Important Bird Areas.

Category	Criterion	Notes
Species of global conservation concern (A1)	The site regularly holds significant numbers of a globally threatened species, or other species of global conservation concern.	The site qualifies if it is known, estimated or thought to hold a population of a species categorised as Critically Endangered or Endangered. Population-size thresholds for Vulnerable and Near Threatened species are set regionally, as appropriate, to help in site selection.
Assemblage of restricted-range species (A2)	The site is known or thought to hold a significant component of the restricted-range species whose breeding distributions define an Endemic Bird Area (EBA) or Secondary Area (SA).	The site has to form one of a set selected to ensure that, as far as possible, all restricted-range species of an EBA or SA are present in significant numbers in at least one site and, preferably, more.
Assemblage of biome-restricted species (A3)	The site is known or thought to hold a significant component of the group of species whose distributions are largely or wholly confined to one biome.	The site has to form one of a set selected to ensure that, as far as possible, all species restricted to a biome are adequately represented.
Congregations (A4)	**(i)** The site is known or thought to hold, on a regular basis, ≥1% of a biogeographic population of a congregatory waterbird species. *or*	This applies to waterbird species as defined by Wetlands International (2006). Thresholds are generated in some instances by combining flyway populations within a biogeographic region, but for other species that lack quantitative data, thresholds are set regionally or inter-regionally, as appropriate. In such cases, thresholds will be taken as estimates of 1% of the biogeographic population.
	(ii) The site is known or thought to hold, on a regular basis, ≥1% of the global population of a congregatory seabird or terrestrial species. *or*	This includes those seabird species not covered by Wetlands International (2006).
	(iii) The site is known or thought to hold, on a regular basis, ≥20,000 waterbirds or ≥10,000 pairs of seabirds of one or more species. *or*	For waterbirds, this is the same as Ramsar Convention criteria category 5.
	(iv) The site is known or thought to exceed thresholds set for migratory species at bottleneck sites.	Numerical thresholds are set regionally or inter-regionally, as appropriate.

wetlands, such as marine species in the families Phalacrocoracidae (cormorants) and Laridae (gulls and terns). These few anomalies are thought to be outweighed by the convenience of a whole-taxon approach. Not all waterbirds are by nature congregatory and this criterion only applies to those species which fall within the above definition. The critical threshold for A4i is taken as 1% of the biogeographic population of a species of waterbird which, in the case of the Caribbean, corresponds to the Neotropics. Thus all population estimates for the Neotropics are combined to form the "biogeographic population", from which the 1% critical threshold is calculated.

For criterion A4ii, the critical threshold is 1% of the global population of non-waterbirds, including seabirds of several families, such as Procellariidae (petrels and shearwaters), Phaethontidae (tropicbirds), Sulidae (gannets and boobies) and Stercorariidae (skuas and jaegers). The inconsistency between A4i and A4ii in terms of critical thresholds (namely biogeographic versus global) is justified biologically by the way that many migratory waterbird species are distributed and split into well-defined, discrete flyway populations.

The sub-criteria A4iii and A4iv are applied at the site level only (not by species), with the former for large concentrations of waterbirds or seabirds of one or more species at a particular site, and the latter for sites over which migrants concentrate (often referred to as bottlenecks), such as narrow sea crossings, along mountain ranges or through mountain passes. The list of potential congregatory Caribbean waterbirds and seabirds (and a small number of other migrants) with their total population estimates and threshold values is given in Appendix 4.

IBAs have also been identified based on regionally significant congregations (i.e. those significant at the Caribbean level) of waterbirds and seabirds. This has been done following exactly the same methodology outlined above, but using 1% of the Caribbean (as defined in this publication) population of a species to trigger the regional equivalents to criteria A4i and A4ii. The Caribbean populations and thresholds are also given in Appendix 4.

DEFINING THE BOUNDARIES OF AN IBA

An IBA is defined so that, as far as possible:

- It is different in character or habitat or ornithological importance from the surrounding area.
- It exists as an actual or potential protected area, with or without buffer zones, or is an area which can be managed in some way for nature conservation.
- It is, alone or with other sites, a self-sufficient area which provides all the requirements of the birds, when present, for which it is important.

Where extensive tracts of continuous habitat occur which are important for birds, only the latter two characteristics apply. Practical considerations on how best the site may be conserved are the foremost consideration. Features such as watersheds, ridge-lines and hilltops can be used to delimit site borders in places where there are no obvious discontinuities in habitat (transitions of vegetation or substrate). Boundaries of land ownership are also relevant, while simple, conspicuous boundaries such as roads can be used in the absence of other features.

There is no fixed maximum or minimum size for IBAs—the biologically sensible has to be balanced with what is practical for conservation. Neither is there a clear-cut answer about how to treat cases where a number of small sites lie next to each other. Whether these are best considered as a series of separate IBAs, or as one larger site containing areas of low ornithological significance, depends upon the local situation with regard to conservation and management.

IBA boundaries have been defined by the national IBA coordinators and authors, in consultation with stakeholders nationally, and on occasion with the BirdLife Caribbean Program team. They have been mapped nationally or by BirdLife based on national inputs, with GoogleEarth often providing the best means of efficiently mapping, reviewing and sharing spatial data. The mapped boundaries signify nothing more than an expression of biological importance. They represent no judgements made concerning land ownership or status, and it is clearly recognised that should conservation efforts focus on a particular IBA, more detailed maps will need to be developed as part of a participatory conservation planning process.

In delimiting marine IBAs (see Box 2), seaward extensions to seabird breeding colonies can be applied to allow for areas required for foraging and other maintenance behaviours. In the absence of known foraging ranges, this has been standardised to a distance of 1 km for Caribbean seabird IBAs (based on some multi-genus studies published in McSorley *et al.* 2003). Using a preliminary distance is useful to create awareness of the need to include a minimum foraging area and can be refined as species- or site-specific data become available.

Box 2. The marine IBA concept.

Extending the IBA program to the oceans is a logical and significant development but poses both conceptual and practical challenges. The term "marine IBA" is used as shorthand for those IBAs regarded as marine in nature because of the seabird populations they contain, but this is not intended to imply that they are fundamentally distinct from other IBAs. "Seabirds" in this context include those birds that rely upon the marine environment at some point in their life-cycle, such as petrels and shearwaters, tropicbirds, boobies, skuas and jaegers, but also cormorants, gulls and terns, and some ducks and shorebirds.

Marine IBA selection criteria were reviewed and adapted in Europe, resulting in the recognition of four types of marine IBAs, shown below. Studies continue on how existing criteria and boundary delimitation guidelines need to be adapted to accommodate aspects of seabird life-cycles in order to assess the extent to which they are amenable to site-based conservation. Delimiting IBAs to include a 1-km at-sea extension from a seabird breeding colony was the first step in identifying marine IBAs in the Caribbean. However, this is just one of four types of marine IBA contemplated:

The four types of marine IBA

1. **Seaward extensions of breeding colonies** which are used for feeding, maintenance behaviours and social interactions. Breeding colonies already identified as IBAs in the Caribbean have had their boundaries extended to include a 1-km seaward extension. However, this seaward boundary should, as far as possible, be colony and/or species-specific, based on known or estimated foraging and maintenance ranges.
2. **Non-breeding (coastal) concentrations** include sites, usually in coastal areas, holding feeding and moulting concentrations of seabirds.
3. **Migratory bottlenecks** are sites which, because of their geographic position, seabirds fly over or around in the course of regular migration. These sites are normally determined by topographic features, such as headlands and straits.
4. **Areas for pelagic species** comprise remote marine areas remote from land where pelagic seabirds regularly gather in large numbers to feed or for other purposes. These areas usually coincide with specific oceanographic features, such as shelf-breaks, upwellings and eddies, and their biological productivity is invariably high.

Challenges and opportunities for marine IBA conservation

With an increasing number of activities occurring in the marine environment (e.g. shipping, windfarms, ocean resource exploration, fishing activities) pressure on marine resources is being felt the world over. As a result, an expanding number of BirdLife Partners are working on identifying marine IBAs to feed into maritime planning and management initiatives with the goal of protecting key sites for seabirds. The identification and subsequent protection of marine IBAs will make a vital contribution to global initiatives to gain greater protection and sustainable management of the oceans, including working towards the identification of Marine Protected Areas (MPAs). Among the future activities contemplated in the Caribbean IBA Program with regard to marine IBAs are to:

- Obtain data on foraging and maintenance ranges for seabirds.
- Complete the inventory by identifying sites for all four types of marine IBA.
- Integrate marine IBAs into national and international protection legislation.

KEY REFERENCES

BirdLife International (2007) *The BirdLife checklist of the birds of the world with conservation status and taxonomic sources*. Version 0. Available at www.birdlife.org/datazone/species/downloads/BirdLife_Checklist_Version_0.xls

Collar, N. J., Gonzaga, L. P., Krabbe, N., Madroño Nieto, A., Naranjo, L. G., Parker, T. A. and Wege, D. C. (1992) *Threatened birds of the Americas: the ICBP/IUCN Red Data Book*. Cambridge, U.K.: International Council for Bird Preservation.

Evans, M. I., ed. (1994) *Important Bird Areas in the Middle East*. Cambridge, U.K.: BirdLife International (BirdLife Conservation Series No.2).

Fishpool, L. D. C., Heath, M. F., Waliczky, Z., Wege, D. C. and Crosby, M. J. (1998) Important bird areas—criteria for selecting sites of global conservation significance. In N. J. Adams and R. H. Slotow, eds. Proc. 22 Int. Ornithol. Congr., Durban. *Ostrich* 69: 428.

Fishpool, L. D. C. and Evans, M. I., eds (2001) *Important Bird Areas in Africa and associated islands: priority sites for conservation*. Newbury and Cambridge, U.K.: Pisces Publications and BirdLife International.

Grimmett, R. F. A. and Gammell, A. B. (1989) *Inventory of Important Bird Areas in the European Community*. (Unpublished report prepared for the Directorate-General of the Environment, Consumer Protection and Nuclear Safety of the European Community, study contract B6610-54-88.) Cambridge, U.K.: International Council for Bird Preservation.

Grimmett, R. F. A. and Jones, T. A. (1989) *Important Bird Areas in Europe*. Cambridge, UK: International Council for Bird Preservation (Techn. Publ. 9).

Howgate and Lascelles, B. (2007) *Candidate marine IBAs: global status and progress*. Cambridge, U.K.: BirdLife International. (Unpublished BirdLife report).

ICBP (1992) *Putting biodiversity on the map: priority areas for global conservation*. Cambridge, U.K.: International Council for Bird Preservation.

Langhammer, P. F., Bakarr, M. I., Bennun, L. A., Brooks, T. M., Clay, R. P., Darwall, W., De Silva, N., Edgar, G. J., Eken, G., Fishpool, L. D. C., Fonseca, G. A. B. da, Foster, M. N., Knox, D. H., Matiku, P., Radford, E. A., Rodrigues, A. S. L., Salaman, P., Sechrest, W., and Tordoff, A. W. (2007) *Identification and Gap Analysis of Key Biodiversity Areas: Targets for Comprehensive Protected Area Systems*. Gland, Switzerland: IUCN.

López-Lanús, B. and Blanco, D. E. eds. (2005) *El censo Neotropical de aves acuáticas 2004: una herramienta para la conservación*. Buenos Aires, Argentina: Wetlands International. (Global Series No.17).

McSorley, C. A., Dean, B. J., Webb, A. and Reid, J. B. (2003) *Seabird use of waters adjacent to colonies: implications for seaward extensions to existing breeding seabird colony Special Protection Areas*. Peterborough, U.K.: Joint nature Conservation Committee. (JNCC Report 329).

Osieck, E. R. and Mörzer Bruyns, M. F. (1981) *Important bird areas in the European community*. Cambridge, U.K.: International Council for Bird Preservation.

Osieck, E. R. (2004). *Towards the identification of marine IBAs in the EU: an exploration by the Birds and Habitats Directives Task Force*. Cambridge, U.K.: BirdLife International.

Parker, T. A., Stotz, D. F. and Fitzpatrick, J. W. (1996) *Ecological and distributional databases for Neotropical birds*. Chicago, USA: Chicago University Press.

Raffaele, H. Wiley J., Garrido, O., Keith, A. and Raffaele, J. (1998) *A guide to the birds of the West Indies*. Princeton, New Jersey: Princeton University Press.

Rose, P. M. and Scott, D. A. (1997) *Waterfowl population estimates*. Second edition. Wageningen, Netherlands: Wetlands International (Publ. No.44).

Schreiber, E. A. and Lee, D. S. eds. (2000) *Status and conservation of West Indian seabirds*. Ruston, USA: Society of Caribbean Ornithology (Spec. Publ. 1).

Stattersfield A. J., Crosby, M. J., Long, A. J. and Wege, D. C. (1998) *Endemic Bird Areas of the World: priorities for biodiversity conservation*. Cambridge, UK: BirdLife International (BirdLife Conservation Series No.6).

Stotz, D. F., Fitzpatrick, J. W., Parker, T. A. and Moskovits, D. K. (1996) *Neotropical birds: ecology and conservation*. Chicago, USA: Chicago University Press.

Wetlands International (2002). *Waterbird population estimates—Third Edition*. Wageningen, Países Bajos: Wetlands International. (Global Series No.12).

Wetlands International (2006) *Waterbird population estimates—Fourth Edition*. Wageningen, The Netherlands: Wetlands International.

OVERVIEW OF RESULTS

IBA COVERAGE BY TERRITORY

This book documents a total of 283 Important Bird Areas (IBAs) in the 32 countries and territories of the Caribbean region[1] (see Figure 1). These IBAs cover a total of 47,876 km^2 including their marine extensions—their combined land area is c.38,612 km^2 or 16.5% of the region's land area (233,781 km^2). This is higher than for other regions of the world where IBA analyses have been undertaken, namely: Africa, where IBAs represent 7% of the land area; Europe, 7%; Middle East, 5%; and Asia, 7.6%.

The network of Caribbean IBAs documented here is a comprehensive list of sites of international importance for bird conservation based on the available information relating to the presence and abundance of the key bird species listed in Appendices 1–4. This network will change as the status of sites changes, or as new information becomes available. For example, in Haiti and Jamaica additional sites were identified as "possible" IBAs—areas thought to be important but for which there was not enough quantitative information available to evaluate them against the IBA criteria. Targeted surveys at these, and indeed other sites across the region, may result in new IBAs being recognised within the network, or indeed existing sites being removed.

The number of IBAs identified per territory varies from one in Bermuda, Navassa and Saba, to 39 in the Bahamas, while the total area of the IBA network in each territory ranges from 2 km^2 in Barbados to 23,123 km^2 in Cuba (Table 1). At the regional level there is no correlation between the number and area of IBAs in a territory and the territory's land area.

1 It includes Bermuda and all islands of the Bahamas, Greater Antilles (Cuba, Hispaniola, Jamaica and Puerto Rico), Virgin Islands, Cayman Islands, Lesser Antilles, Netherlands Antilles, and Trinidad and Tobago. This represents 13 independent nations, and six UK, six Dutch, four French, and three US overseas territories. It does not cover the Central and South American islands on the fringes of the Caribbean Sea.

Figure 1. Location of Important Bird Areas in the Caribbean: key sites for conservation.

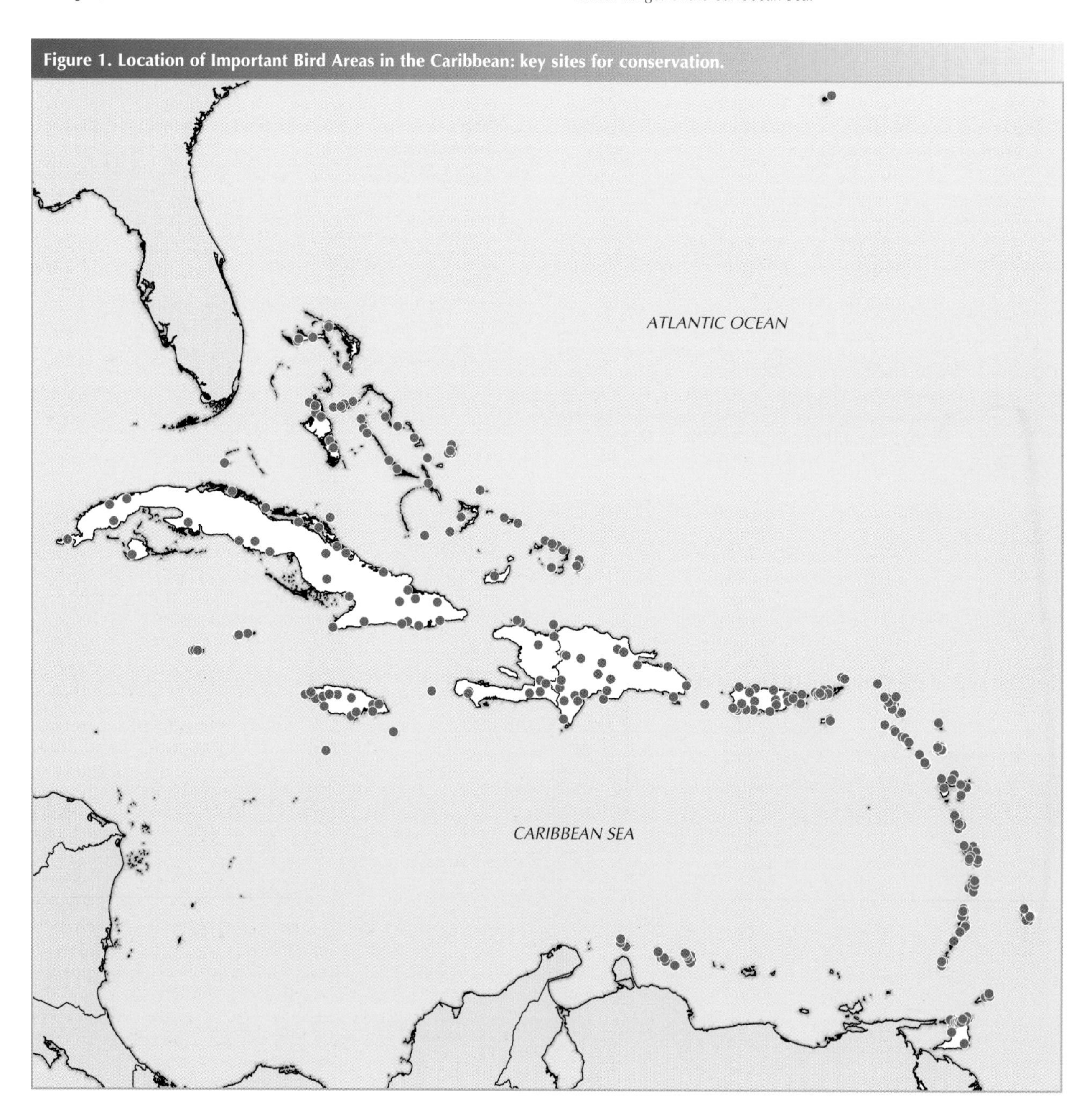

Table 1. The number and total area of IBAs, and the number of IBAs qualifying under each criterion.

						No. of IBAs qualifying under category***				
Country	Number of IBAs	Country land area (km^2)	Total area of IBAs (km^2)**	Total IBA land area (km^2)	Percentage of national land area within IBA network (%)	Globally threatened species (A1)	Restricted-range species assemblages (A2)	Biome-restricted species assemblages (A3)	Congregations (Global) (A4)	Congregations (Regional) (B4)
Anguilla	7	98	52	8	8	0	4	0	4*	6
Antigua and Barbuda	12	441	196	88	20	11	7	0	8	5
Aruba	4	193	6	2	1.2	2	0	2	2	1
Bahamas	39	13,940	4,699	2,370	17	20	14	0	20	28
Barbados	7	431	2	1	0.11	0	6	0	5*	2
Bermuda	1	53	7	1	1	1	0	0	1	0
Bonaire	6	288	238	158	55	4	5	5	2	4
British Virgin Islands	3	153	53	15	10	0	3	0	3	3
Cayman Islands	10	262	61	3	1	10	8	0	3	3
Cuba	28	109,886	23,123	20,878	19	28	17	24	13*	7
Curaçao	5	444	162	107	24	2	2	3	4	2
Dominica	4	754	106	98	13	2	3	0	2	2
Dominican Republic	21	48,730	7,212	6,822	14	20	17	0	5*	3
Grenada	6	344	21	22	6.25	5	6	0	0	0
Guadeloupe	9	1,713	505	325	19	2	4	0	3	6
Haiti	10	27,750	232	278	1	10	9	0	0	0
Jamaica	15	10,829	3,112	2,707	25	13	13	0	6*	5
Martinique	10	1,100	545	429	39	6	6	0	4*	4
Montserrat	3	102	16	16	16	3	3	0	0	0
Navassa	1	5	1,481	5	100	1	0	0	1	1
Puerto Rico	20	8,870	1,971	1,951	22	14	18	0	5*	8
Saba	1	13	21	5	42	0	1	0	1	1
St Barthelemy	3	25	10	0.1	0.4	0	0	0	2	3
St Eustatius	2	21	14	9	41	0	2	0	1	0
St Kitts and Nevis	3	261	65	63	24	0	1	0	1	2
St Lucia	5	616	178	154	25	5	4	0	1*	4
St Maarten	5	33	8	12	36	1	3	0	2	2
St Martin	3	56	8	3	6	0	3	0	1	2
St Vincent and the Grenadines	15	389	178	136	35	7	12	0	3*	4
Trinidad and Tobago	7	5,128	1,061	1,795	35	3	3	4	4*	3
Turks and Caicos Islands	9	500	2,471	75	15	4	6	0	6*	7
US Virgin Islands	9	353	62	77	21.8	2	7	0	2	5
Total	**283**	**233,781**	**47,876**	**38,612**	**16.5%**	**176**	**187**	**38**	**115**	**123**

Notes
*= total includes IBAs with > 20,000 waterbirds
** = total includes land and marine areas
*** = IBAs often qualify under more than one criterion.

Caribbean IBAs range in size from less than 1 ha, to the 530,695-ha Ciénaga de Zapata IBA in Cuba. Over 37% of IBAs are in the 1,000–9,999 ha size class, and 29% are between 100 and 999 ha in area (Figure 2). The median IBA size is 1,672 ha although the mean or average is 16,923 ha, reflecting the disproportionate contribution made by a few very large IBAs to the overall total. Five IBAs in Cuba alone comprise 30% of the total area of the Caribbean IBA network. Just four IBAs are greater than 200,000 ha, and all of them are in Cuba (the Caribbean's largest island). At the other end of the spectrum there are very few really small IBAs: sites less than 9 ha in size represent 2% of the region's IBAs, two of them being in Barbados. However, irrespective of size, these IBAs are all internationally important for the bird populations they support.

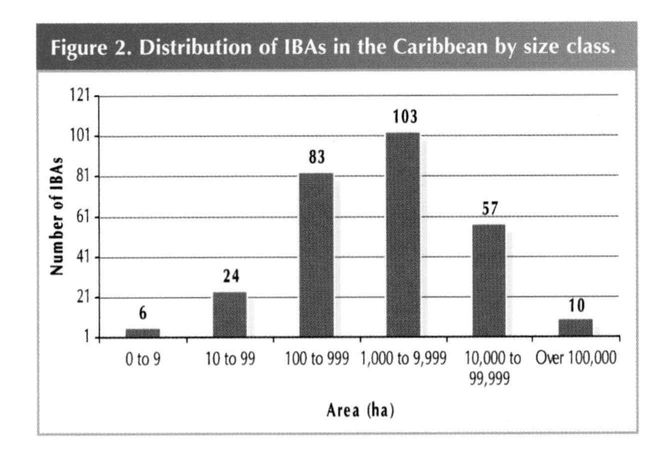

Figure 2. Distribution of IBAs in the Caribbean by size class.

IBA COVERAGE BY CRITERIA AND SPECIES

All 283 of the Caribbean's IBAs are of global importance for bird conservation. Of these, just 11 (3.8%) qualify solely under the "Caribbean" regional congregations criteria (meeting thresholds set against the regional population estimates as detailed in Appendix 4). Over half of the Caribbean's IBAs qualify under the species of global conservation concern criterion (A1)—176 IBAs (62% of the total) support populations of these species. However, the largest number of IBAs—187 (66%)—support restricted-range species (A2), followed by regionally significant congregations (B4) with 123 IBAs (43%) and globally significant congregations (A4) with 115 IBAs (40%) (see Figure 3 and Table 1). The proportion of IBAs qualifying for biome-restricted birds (A3) is just 13%. However, this criterion was not used to identify IBAs except in Cuba, Aruba, Bonaire, Curaçao, and Trinidad and Tobago, and thus the coverage is not regionally comprehensive (see Methodology). A total of 249 "key bird species" have been

Figure 3. The number of IBAs in the Caribbean qualifying under each criterion.

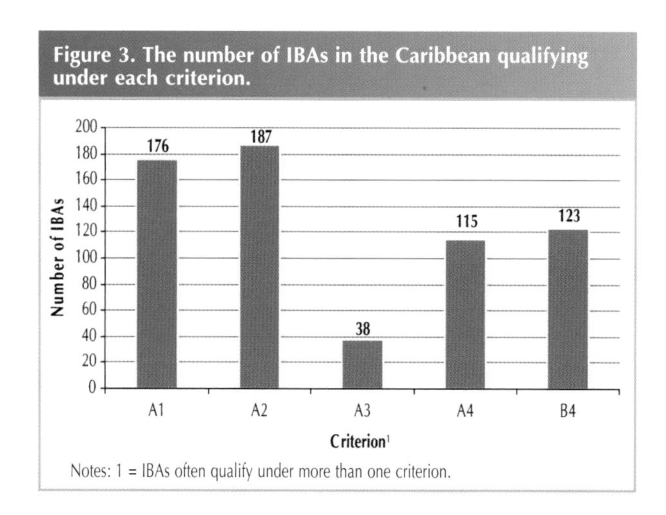

Notes: 1 = IBAs often qualify under more than one criterion.

used to identify the region's IBAs against the four criteria. Details of each of the IBAs and the species within them are presented for each criterion below.

■ Species of global conservation concern (A1)

Excluding vagrants and species marginal to the region, a total of 54 globally threatened species occur in the Caribbean, comprising 12 Critically Endangered, 19 Endangered and 23 Vulnerable species. Of these species, 51 (11 Critical, 18 Endangered and 22 Vulnerable species) are represented within the Caribbean IBA network. The Caribbean is not currently known to support viable populations of the Critically Endangered Bachman's Warbler *Vermivora bachmanii*, the Endangered Red Siskin *Carduelis cucullata* or Vulnerable Cerulean Warbler *Dendroica cerulea* and these were excluded from the IBA analysis. A further four Critically Endangered species are listed at IBAs on the basis of historical records and/or recent unconfirmed reports: Jamaica Petrel *Pterodroma caribbaea*, Jamaican Pauraque *Siphonorhis americana*, Ivory-billed Woodpecker *Campephilus principalis* and Semper's Warbler *Leucopeza semperi* (Table 2). These species may already be extinct, but while realistic hope of their survival persists, they are high priorities for targeted surveys to confirm their existence. Twenty-two (of the 25) Near Threatened species recorded for the Caribbean were included in the IBA analyses and are well represented in the IBA network; the three species with insufficient regional populations to trigger the IBA criteria thresholds are Golden-winged Warbler *Vermivora chrysoptera*, Olive-sided Flycatcher *Contopus cooperi* and Buff-breasted Sandpiper *Tryngites subruficollis*. Ten threatened species (each of which is a national endemic), are thought to occur in just one IBA (Table 2). These IBAs are clearly of critical importance for the long-term survival of these species and must be a priority for targeted conservation actions or surveys as appropriate.

Table 2. Threatened Caribbean island endemics occurring in just a single IBA.

IUCN	Scientific name	Common name	Country	IBA name	IBA code
CR ■	*Pterodroma caribbaea*	Jamaica Petrel	Jamaica	John Crow Mountains	JM014
CR ■	*Chondrohierax wilsonii*	Cuban Kite	Cuba	Alejandro de Humboldt	CU027
CR ■	*Buteo ridgwayi*	Ridgway's Hawk	Dominican Republic	Los Haitises	DO018
CR ■	*Siphonorhis americana*	Jamaican Pauraque	Jamaica	Hellshire Hills	JM011
CR ■	*Leucopeza semperi*	Semper's Warbler	St Lucia	Government Forest Reserve	LC002
EN ■	*Pterodroma cahow*	Bermuda Petrel	Bermuda	Cooper's Island and Castle Islands	BM001
EN ■	*Cyanolimnas cerverai*	Zapata Rail	Cuba	Ciénaga de Zapata	CU006
EN ■	*Corvus minutus*	Cuban Palm Crow	Cuba	Sierra del Chorrillo	CU017
EN ■	*Ferminia cerverai*	Zapata Wren	Cuba	Ciénaga de Zapata	CU006
NT ■	*Dendroica subita*	Barbuda Warbler	Antigua and Barbuda	Codrington Lagoon and the Creek	AG002

The Endangered Giant Kingbird triggers the threatened species (A1) criterion in 11 Cuban IBAs. (PHOTO: PETE MORRIS)

The Bahama Mockingbird is a restricted-range species, primarily confined to the Bahamas and Turks and Caicos EBA. In the Bahamas it is one of seven restricted-range birds, assemblages of which have triggered the A2 criterion for 14 IBAs. (PHOTO: OLGA STOKES)

■ Assemblages of restricted-range species (A2)

The Caribbean IBA network is strongly representative of the region's restricted-range species (A2) with 187 IBAs (66%) identified for these inherently vulnerable birds. Each of these IBAs supports a significant component of the restricted-range species whose breeding ranges define an Endemic Bird Area (EBA) or Secondary Area of endemism (SA). Stattersfield *et al.* (1998) define six EBAs in the Caribbean region, and four SAs (see Methodology and Appendix 2). Only Navassa and Bermuda are not included in any of the Caribbean EBAs or SAs. The Lesser Antilles EBA species are represented in 65 IBAs distributed across 16 countries/territories: this IBA network supports populations of all 38 restricted-range species that occur within the EBA. Coverage of each of the EBAs is good within the IBA network: Puerto Rico and the Virgin Islands EBA species are represented within 28 IBAs; Hispaniola EBA species in 26 IBAs; Cuba EBA in 17 IBAs; Bahamas EBA in 14 IBAs; Jamaica EBA in 13 IBAs; and Bahamas EBA in six IBAs. The Secondary Areas are represented by eight IBAs in the Cayman Islands (SA014); seven IBAs in Netherlands Antilles (SA015); one IBA in Trinidad (SA016) and two IBAs in Tobago (SA017).

■ Assemblages of biome-restricted species (A3)

IBAs were only identified for biome-restricted species in Cuba, Aruba, Bonaire, Curaçao, Trinidad and Tobago. Of the 50 IBAs evaluated, 38 were identified as supporting significant components of the respective biome-restricted birds: Greater Antilles biome (24 IBAs in Cuba); Northern South America (two IBAs in Aruba, five in Bonaire, three in Curaçao, and two in Trinidad and Tobago); and Northern Andes biome (three IBAs in Trinidad and Tobago, one of which also has Northern South America biome-restricted species in). A complete analysis was not undertaken for the Greater Antilles and Lesser Antilles biomes (the two main biomes covering the insular Caribbean) as, owing to the insular nature of the region, national lists of the biome-restricted species are almost identical to the national lists of restricted-range birds, and the resultant IBA would be essentially the same (see Methodology).

The Vulnerable Yellow-shouldered Amazon is one of two Northern South America biome-restricted species found in Bonaire. (PHOTO: SAM WILLIAMS)

■ Congregations (A4/B4)

A total of 115 Caribbean IBAs (40% of the total) qualify under the global congregations criterion (A4) and 123 (43%) under the regional congregations criterion (B4) (Table 1, Figure 3), and are thus important for supporting significant populations of congregatory waterbirds, seabirds or migratory bird species. Of the four criteria in the global congregations category (A4), 96 of the IBAs meet the A4i criterion, supporting, on a regular basis, ≥1% of a biogeographic population of a waterbird species; there are 22 IBAs defined against criterion A4ii (supporting ≥1% of the global population of a seabird species) and A4iii (supporting ≥20,000 waterbirds or ≥10,000 pairs of seabird of one or more species). Only one site meets the A4iv criterion as a migratory bottleneck, namely Siboney-Juticí IBA (CU026) where thousands (6,143 birds) of Osprey *Pandion haliaetus* have been recorded as passage migrants. Of the 123 IBAs that qualify under the regional congregations criteria (B4), 84% of these IBAs meet the B4i criterion (supporting, on a regular basis, ≥1% of the Caribbean population of a waterbird) and 38% meet the B4ii criterion (≥1% of the regional population of a seabird species). Almost all territories in the Caribbean have identified at least one IBA that qualifies under the congregatory species criteria (A4 or B4), the exceptions being Grenada, Haiti and Montserrat. The territories with the most congregatory species IBAs are the Bahamas (with 32) and Cuba (with 14).

Of the 126 congregatory waterbird species in the Caribbean that could potentially trigger the A4i criterion (Appendix 4), IBAs have been identified for 56 species (44% of the total).

Globally significant breeding colonies of Brown Booby have triggered the seabird congregations (A4ii) criterion in 25 IBAs across the Caribbean. (PHOTO: WILF POWELL)

Figure 4. The total number of IBAs in the Caribbean meeting each A4/B4 criterion.

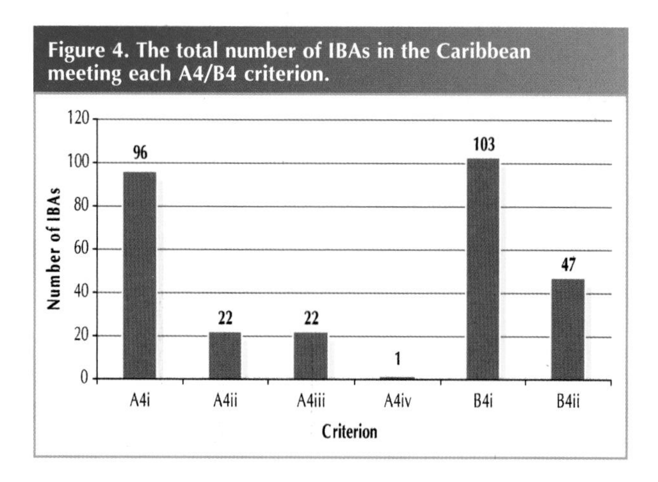

These birds have helped define between one and 51 IBAs per species, with the West Indian Whistling-duck *Dendrocygna arborea* being responsible for triggering this criterion in the largest number of IBAs. Of the 22 seabird species in the Caribbean that could potentially trigger the A4ii criterion (Appendix 4), IBAs have been identified for 10 of them (45% of the total). Brown Booby *Sula leucogaster* is responsible for triggering this criterion in the largest number of IBAs (25). The Caribbean network of IBAs identified for these congregatory species meets most of the life-cycle needs of the species concerned.

■ Marine IBAs

As part of a global marine IBA analysis, all IBAs which may already be considered as marine IBAs on the basis of their seabird breeding colonies have been identified. In the Caribbean, 138 IBAs have been identified as "marine" (Table 3). There are four distinct types of marine IBAs (as detailed in the Methodology, Box 2). The majority of Caribbean marine IBAs represent seaward extensions to breeding colonies, although a small number comprise non-breeding (coastal) concentrations of birds. Further research and analysis will be needed to document all of the region's marine IBA types.

Table 3. Marine IBAs in the Caribbean (see Methodology, Box 2 for description of different marine IBA types).

Country/territory	Marine IBAs	Type 1	Type 2	Type 3	Type 4
Anguilla	7	7	2	0	0
Antigua and Barbuda	9	9	0	0	0
Aruba	2	2	0	0	0
Bahamas	27	27	3	0	0
Barbados	1	1	0	0	0
Bermuda	1	1	0	0	0
Bonaire	4	4	0	0	0
British Virgin Islands	3	3	0	0	0
Cayman Islands	3	3	0	0	0
Cuba	9	9	0	0	0
Curaçao	4	4	0	0	0
Dominica	2	2	0	0	0
Dominican Republic	4	4	0	0	0
Grenada	0	0	0	0	0
Guadeloupe	6	6	0	0	0
Haiti	1	1	0	0	0
Jamaica	6	6	0	0	0
Martinique	4	4	0	0	0
Montserrat	0	0	0	0	0
Navassa	1	1	0	0	0
Puerto Rico	8	8	0	0	0
Saba	1	1	0	0	0
St Barthélemy	3	3	0	0	0
St Eustatius	1	1	0	0	0
St Kitts and Nevis	2	2	0	0	0
St Lucia	4	4	0	0	0
St Maarten	3	3	0	0	0
St Martin	2	2	0	0	0
St Vincent and the Grenadines	4	4	0	0	0
Trinidad and Tobago	4	4	0	0	0
Turks and Caicos Islands	7	7	2	0	0
US Virgin Islands	5	5	1	0	0
Caribbean total	138	138	8	—	—
North America	326	—	—	—	—
Central America	38	—	—	—	—
South America	166	—	—	—	—
Americas total	668	—	—	—	—

DATA PRESENTATION

The national inventories which form the major part of this book have been prepared in a standardised format, consisting of the sections that are explained and illustrated below, and with maps to show the location and indicative boundaries of the IBAs. One or more photographs are used to illustrate the chapters, selected to illustrate sites, conservation issues, activities or birds typical of the country or territory.

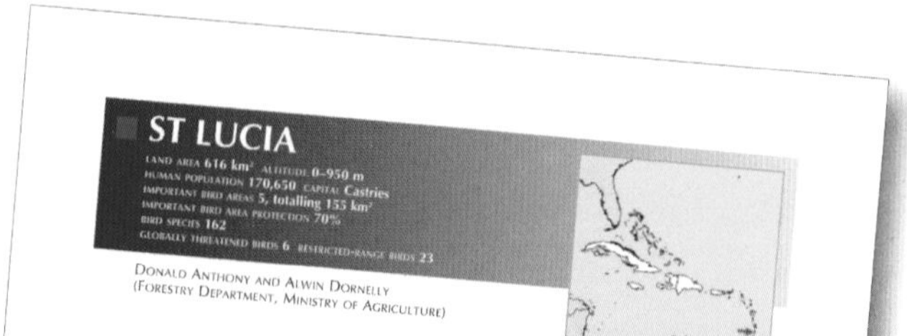

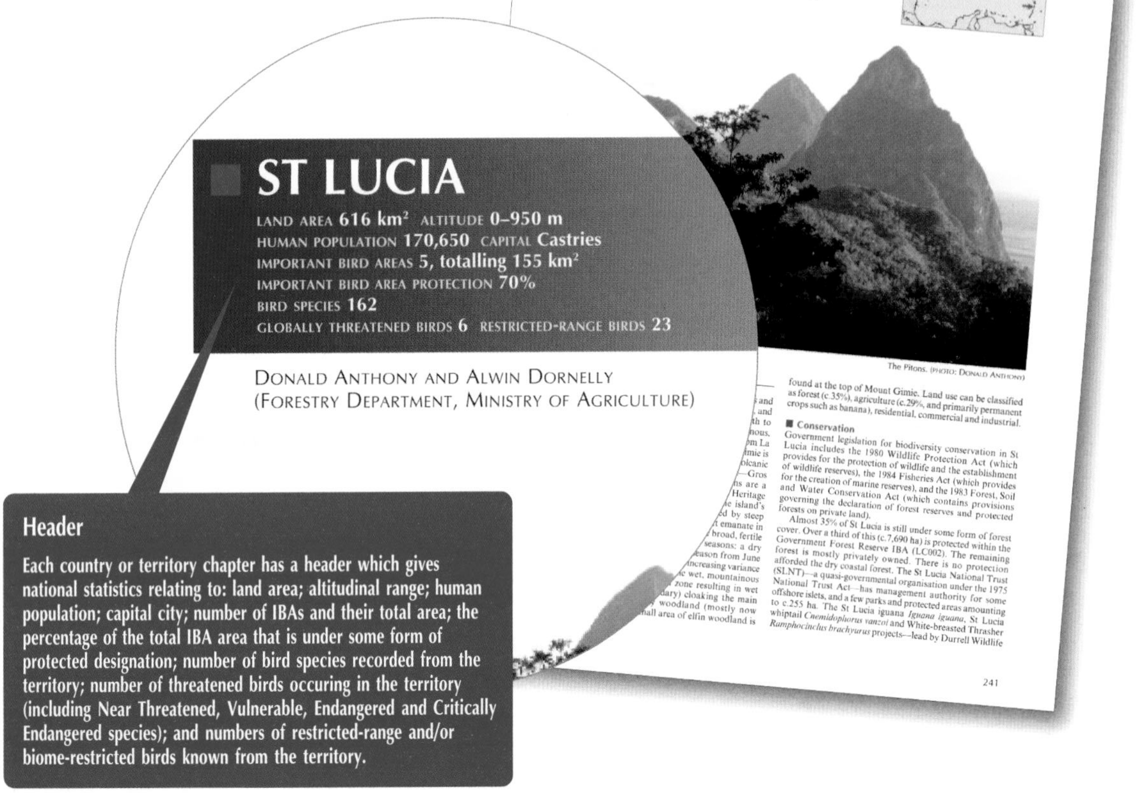

ST LUCIA

LAND AREA 616 km² ALTITUDE 0–950 m
HUMAN POPULATION 170,650 CAPITAL Castries
IMPORTANT BIRD AREAS 5, totalling 155 km²
IMPORTANT BIRD AREA PROTECTION 70%
BIRD SPECIES 162
GLOBALLY THREATENED BIRDS 6 RESTRICTED-RANGE BIRDS 23

DONALD ANTHONY AND ALWIN DORNELLY
(FORESTRY DEPARTMENT, MINISTRY OF AGRICULTURE)

Header

Each country or territory chapter has a header which gives national statistics relating to: land area; altitudinal range; human population; capital city; number of IBAs and their total area; the percentage of the total IBA area that is under some form of protected designation; number of bird species recorded from the territory; number of threatened birds occuring in the territory (including Near Threatened, Vulnerable, Endangered and Critically Endangered species); and numbers of restricted-range and/or biome-restricted birds known from the territory.

INTRODUCTION

This section covers general information to set the context for the rest of the chapter by describing, in summary format, subject areas such as geography, landscape, climate, vegetation, centres of population, history, and major national features or influences on the environment.

■ Conservation

The territorial infrastructure for conservation is described, covering areas such as nature-conservation legislation (particularly laws or programs relevant to site protection and management), government bodies responsible for conservation, the conservation NGO-sector, and the protected area system and its management. This section includes a summary of the main threats to sites, habitats and birds in the territory, and ongoing conservation initiatives.

■ Birds

This section describes the territory's importance for birds, especially in terms of the key bird species that trigger the various IBA categories and criteria. Details are presented on threatened species, restricted-range birds (and biome-restricted species where relevant) and congregatory birds, highlighting species, populations or subspecies for which the territory is particularly notable. Explanations of why particular species have or have not been considered in the IBA analysis are also given along with a brief discussion of bird-specific threats and conservation projects.

IMPORTANT BIRD AREAS

This section presents an overview of the IBA network in the territory, giving information on the total area covered by the IBAs (also given as a percentage of national land area) and the numbers of IBAs that qualify under the various criteria. Details are given as to which threatened species, restricted-range species, biome-restricted assemblages and congregatory species are represented within the territory's IBAs. The coverage afforded to the IBA network by the national protected areas system is evaluated, with proposals made for new protected areas to address any significant gaps, and proposals to address specific threats.

KEY REFERENCES

Citations are listed of the primary publications used in the compilation of the IBA chapter. For countries with larger IBA networks, this reference list is not necessarily exhaustive in terms of the data sources used to compile species lists and populations for each IBA. These references are however maintained by the national IBA coordinators in the World Bird Database.

National IBA tables

Summary data on every IBA in a territory are given in a table that lists all of the key bird species triggering IBA criteria in at least one site. This table does not detail species' presence at IBAs *per se*. It records a species as present (possibly with associated population information) only if that species (alone or in conjunction with other species, such as additional restricted-range birds) triggers the criteria for that site.

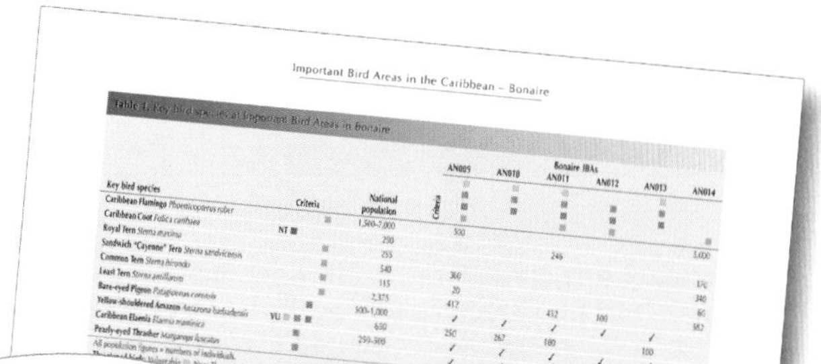

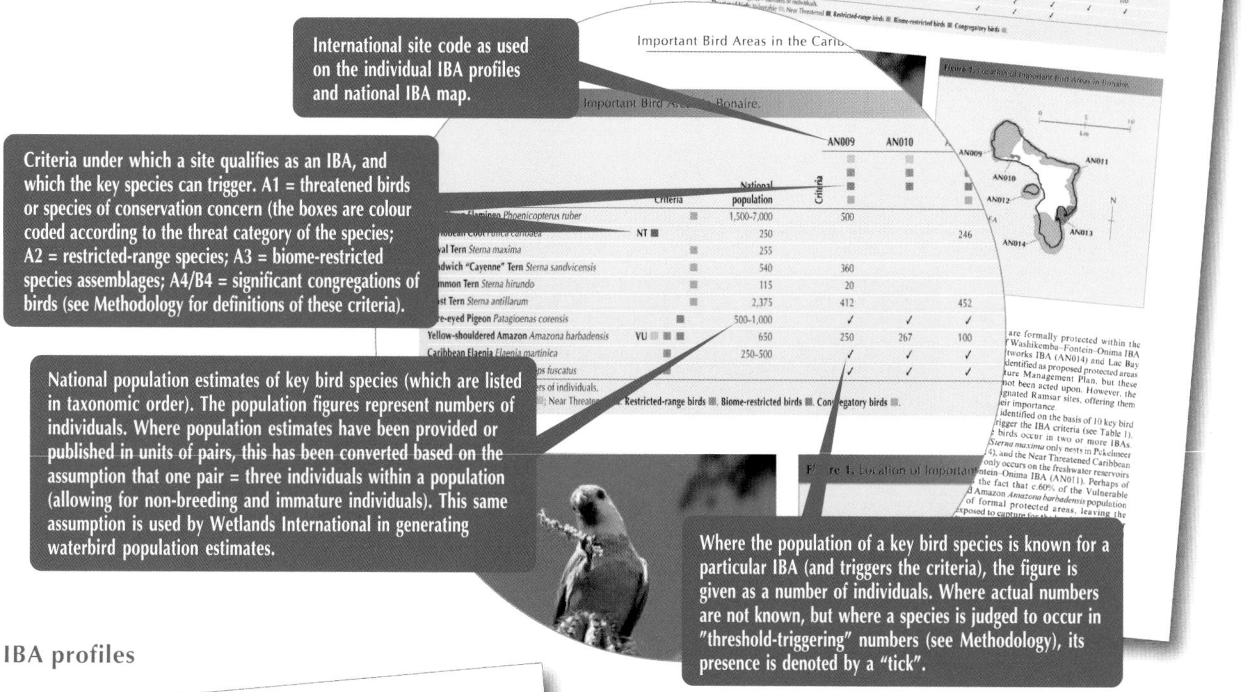

International site code as used on the individual IBA profiles and national IBA map.

Criteria under which a site qualifies as an IBA, and which the key species can trigger. A1 = threatened birds or species of conservation concern (the boxes are colour coded according to the threat category of the species; A2 = restricted-range species; A3 = biome-restricted species assemblages; A4/B4 = significant congregations of birds (see Methodology for definitions of these criteria).

National population estimates of key bird species (which are listed in taxonomic order). The population figures represent numbers of individuals. Where population estimates have been provided or published in units of pairs, this has been converted based on the assumption that one pair = three individuals within a population (allowing for non-breeding and immature individuals). This same assumption is used by Wetlands International in generating waterbird population estimates.

Where the population of a key bird species is known for a particular IBA (and triggers the criteria), the figure is given as a number of individuals. Where actual numbers are not known, but where a species is judged to occur in "threshold-triggering" numbers (see Methodology), its presence is denoted by a "tick".

IBA profiles

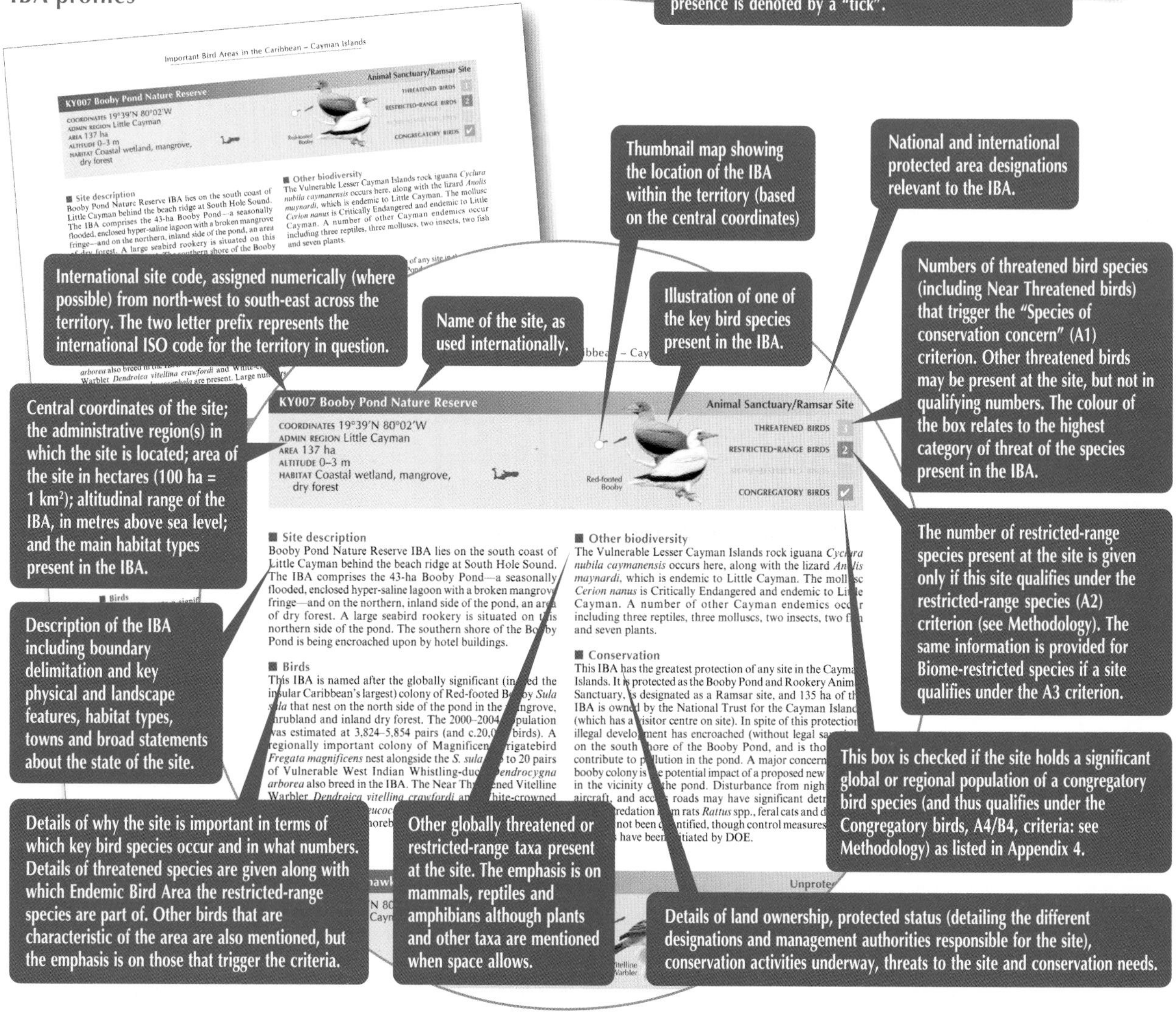

International site code, assigned numerically (where possible) from north-west to south-east across the territory. The two letter prefix represents the international ISO code for the territory in question.

Name of the site, as used internationally.

Thumbnail map showing the location of the IBA within the territory (based on the central coordinates)

Illustration of one of the key bird species present in the IBA.

National and international protected area designations relevant to the IBA.

Numbers of threatened bird species (including Near Threatened birds) that trigger the "Species of conservation concern" (A1) criterion. Other threatened birds may be present at the site, but not in qualifying numbers. The colour of the box relates to the highest category of threat of the species present in the IBA.

Central coordinates of the site; the administrative region(s) in which the site is located; area of the site in hectares (100 ha = 1 km²); altitudinal range of the IBA, in metres above sea level; and the main habitat types present in the IBA.

The number of restricted-range species present at the site is given only if this site qualifies under the restricted-range species (A2) criterion (see Methodology). The same information is provided for Biome-restricted species if a site qualifies under the A3 criterion.

Description of the IBA including boundary delimitation and key physical and landscape features, habitat types, towns and broad statements about the state of the site.

This box is checked if the site holds a significant global or regional population of a congregatory bird species (and thus qualifies under the Congregatory birds, A4/B4, criteria: see Methodology) as listed in Appendix 4.

Details of why the site is important in terms of which key bird species occur and in what numbers. Details of threatened species are given along with which Endemic Bird Area the restricted-range species are part of. Other birds that are characteristic of the area are also mentioned, but the emphasis is on those that trigger the criteria.

Other globally threatened or restricted-range taxa present at the site. The emphasis is on mammals, reptiles and amphibians although plants and other taxa are mentioned when space allows.

Details of land ownership, protected status (detailing the different designations and management authorities responsible for the site), conservation activities underway, threats to the site and conservation needs.

ANGUILLA

LAND AREA **98 km²** ALTITUDE **0–65 m**
HUMAN POPULATION **13,480** CAPITAL **The Valley**
IMPORTANT BIRD AREAS **7, totalling 53 km²**
IMPORTANT BIRD AREA PROTECTION **7.9%**
BIRD SPECIES **139**
THREATENED BIRDS **0** RESTRICTED-RANGE BIRDS **4**

KARIM HODGE (DEPARTMENT OF ENVIRONMENT, GOVERNMENT OF ANGUILLA), STEVE HOLLIDAY (ROYAL SOCIETY FOR THE PROTECTION OF BIRDS) AND FARAH MUKHIDA (ANGUILLA NATIONAL TRUST)

Sombrero, one of Anguilla's larger outlying islands.
(PHOTO: STEVE HOLLIDAY/RSPB)

INTRODUCTION

Anguilla, a UK Overseas Territory, is at the northernmost end of the Lesser Antilles, to the east of the Virgin Islands and just 8 km north of the island of St Martin. It is an archipelago of 22 islands, with the main island (Anguilla) being c.26 km long, 5 km at its widest point, and covering c.91 km². The largest of the outlying islands include Anguillita, Dog Island, Prickly Pear Cays, Scrub and Little Scrub islands, Seal Island, Sombrero and Sandy Island. Anguilla and its many outer islands and cays are low-lying and flat. They are mostly rocky, with limestone, corals and sandstone predominating, and have generally thin, poor soils. Habitats range from coral reefs to coastal cliffs, degraded evergreen woodland with scattered areas of grassland and scrub, to small areas of mangrove. Anguilla and three of the offshore cays have brackish coastal lagoons and a few ponds on the mainland are fed by springs from the water table. Porous rocks capture rainfall as groundwater and these have traditionally provided the island's water resources, although recent concerns over water quality have led to the development of a new desalination plant.

With generally poor soils, Anguilla is largely unsuitable for agriculture (although several pockets of rich soil are cultivated). However, the mainland and a few of the offshore islands have extensive beaches, clear seas and inshore coral reefs, providing a rich basis for Anguilla's primary industries of tourism and fishing. The offshore banking, finance and insurance industries are also important for the territory's economy.

Anguilla has a tropical dry climate. Average annual rainfall is 970 mm but can range from 460 to over 2,050 mm. The wet season extends from June to November and coincides with the Atlantic hurricane season, although most of the island's annual rainfall can fall within a few weeks, causing localised flooding in low-lying areas. Anguilla is periodically hit by hurricanes—such as Luis in 1995 and Lenny in 1999—and these can result in extensive wind damage, torrential rain and flooding.

Although Anguilla is a small territory, it is culturally and ethnically diverse in its own right. This is a result of the migratory habits of people throughout the Caribbean region, from slavery times to the present day. The majority of the native population are those of the African Diaspora, but there is a small population of Irish descendants. The rest is made up of immigrants from the Caribbean region, and some expatriates from North America and Europe. The population is supplemented by more than 150,000 visitors a year, helping to make tourism the most important economic activity. However, this focus on tourism is placing many pressures on the environment, from destroying grassy savannahs and limestone pavements for new development to building golf courses and draining and dredging wetlands for marina developments. As a UK Overseas Territory, the islands are governed by the locally elected Government of Anguilla with a governor representing the UK Government.

■ Conservation

The Anguilla National Trust (ANT) was established by law in 1988 to act as custodian of Anguilla's heritage, preserving and promoting the island's natural environment and its archaeological, historical and cultural resources for present and future generations. It has worked with a growing number of stakeholders and partners on a range of collaborative projects that help further its aims. The Trust has maintained particularly close working relationships with the Government of Anguilla in their efforts to implement national, regional and international environmental policy. However, effective conservation efforts are impeded by the lack of appropriate environmental legislation and land use planning. National parks have been designated through the Anguilla National Trust Act, and five marine parks were legally demarcated in 2007 through the 1991 Marine Parks Act. Environmental legislation has been drafted and is currently being reviewed by government departments and statutory bodies before a final public review and their submission to Executive Council. This draft legislation is quite comprehensive and includes environmental protection, biodiversity and cultural conservation, fisheries management, and environmental health.

However, as tourism expands and development pressures increase, any decisions relating to land use (including the establishment of a protected areas system) will remain a challenge as only c.3% of land is in government ownership. Although the government has increased the resources available to the Environment Department there is a need to invest further in recruitment, training and retention of staff. Establishment of a national conservation fund would greatly assist such investment in the territory's long-term, sustainable future.

Biodiversity on Anguilla is under more serious threat than ever before. A surge in development for housing and tourism-related activities, and an increase in population growth on the island and from immigration have placed severe pressures on an increasingly stressed environment. These pressures will be exacerbated by any impacts of global climate change. The main problems facing biodiversity conservation on Anguilla are habitat loss due to an increase in economic activity, and a lack of public awareness and appreciation about the importance of the environment to the island. Despite the many ecological services Anguilla's dry evergreen forest, wetlands and coral reefs provide, many are being destroyed or compromised for the sake of economic development. Other threats that are less widespread but capable of creating similar environmental problems include invasive alien species, and disturbance to important breeding or feeding areas for a range of species. There are also potential threats from pollution incidents and the growth in domestic and industrial waste from high levels of tourism and its related activities. The ANT has been working to address some of these threats through actions such as environmental education programmes; assisting with the development of marine park management plans; on-going wetland and terrestrial bird monitoring programmes; and a rat eradication study of Dog Island—a joint OTEP project with Royal Society for the Protection of Birds (RSPB), and with assistance provided by the Department of Environment, Government of Anguilla.

■ Birds

To date, 139 bird species have been recorded on Anguilla, of which 38 are recorded as breeding and a further 101 occur as non-breeding Neotropical migrants. Four (of the 38) Lesser Antilles EBA restricted-range birds occur in Anguilla, none of which is endemic to the territory. The four species are Green-throated Carib *Eulampis holosericeus*, Caribbean Elaenia *Elaenia martinica*, Pearly-eyed Thrasher *Margarops fuscatus* and Lesser Antillean Bullfinch *Loxigilla noctis* all of which are widely dispersed across the Anguilla mainland. A number of other Lesser Antilles restricted-range birds have occurred on the island: Antillean Crested Hummingbird *Orthorhyncus cristatus* disappeared from the island in the wake of Hurricane Luis, but a few individuals have recently been recorded; Purple-throated Carib *Eulampis jugularis* is a rare visitor; and Antillean Euphonia *Euphonia musica* was reported in the past but is only known from one recent record.

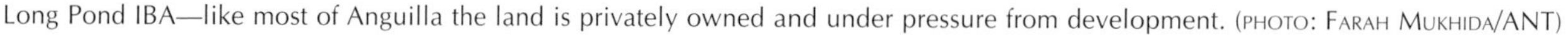

Long Pond IBA—like most of Anguilla the land is privately owned and under pressure from development. (PHOTO: FARAH MUKHIDA/ANT)

The restricted-range Caribbean Elaenia (left) and Pearly-eyed Thrasher (right) occur throughout Anguilla. (PHOTOS: GILLIAN HOLLIDAY)

Four globally threatened birds have been recorded in Anguilla but their status is such that they have not been considered in the IBA analysis. The Vulnerable West Indian Whistling-duck *Dendrocygna arborea* is a rare visitor from other Caribbean populations with recent records in 2003 and 2006. The remaining three species are all Near Threatened: Piping Plover *Charadrius melodus* is a rare migrant with at least three individuals seen in recent years; Caribbean Coot *Fulica caribaea* is a scarce visitor and rare breeding bird with 1–2 birds seen annually; and White-crowned Pigeon *Patagioenas leucocephala* is a rare visitor (one was seen in 2005) following a local extinction in the 1960s through hunting and habitat loss.

Anguilla is important for seabirds. At least 15 species breed, with a further two species—Audubon's Shearwater *Puffinus lherminieri* and Black Noddy *Anous minutus*—reported as former or possible breeding species. Surveys of Anguilla and its outer islands during 1999–2000 found more than 10,000 nesting pairs of gulls and terns, and over 2,000 nesting pairs of boobies *Sula* spp., tropicbirds *Phaethon* spp. and Magnificent Frigatebirds *Fregata magnificens*. Surveys in 2007 documented significantly higher numbers of nesting terns. Although one of the smallest island groups in the West Indies, Anguilla holds up to 10% of the West Indian Masked Booby *Sula dactylactra*

Brown Boobies on Prickly Pear East. (PHOTO: FARAH MUKHIDA/ANT)

Table 1. Key bird species at Important Bird Areas in Anguilla.

			Anguilla IBAs						
			AI001	AI002	AI003	AI004	AI005	AI006	AI007
		Criteria				■	■	■	■
Key bird species	**Criteria**	**National population**	■	■	■	■	■	■	■
Red-billed Tropicbird *Phaethon aethereus*	■	129		45	36				
Magnificent Frigatebird *Fregata magnificens*	■	930		930					
Brown Pelican *Pelecanus occidentalis*	■	87			87				
Masked Booby *Sula dactylatra*	■	210	84	126					
Brown Booby *Sula leucogaster*	■	6,669	1,158	3,801	1,710				
Laughing Gull *Larus atricilla*	■	9,870		1,095	7,617				600
Royal Tern *Sterna maxima*	■	381							345
Roseate Tern *Sterna dougallii*	■	630							210
Least Tern *Sterna antillarum*	■	978			87	165	343	45	60
Bridled Tern *Sterna anaethetus*	■	1,390	810	138	270				
Sooty Tern *Sterna fuscata*	■	340,000		339,000					
Brown Noddy *Anous stolidus*	■	1,815	930	573					
Green-throated Carib *Eulampis holosericeus*	■	75				✓	✓	✓	
Caribbean Elaenia *Elaenia martinica*	■						✓	✓	✓
Pearly-eyed Thrasher *Margarops fuscatus*	■								✓
Lesser Antillean Bullfinch *Loxigilla noctis*	■					✓			

All population figures = numbers of individuals.
Restricted-range birds ■. Congregatory birds ■.

and Bridled Tern *Sterna anaethetus* populations, the region's largest population of Sooty Tern *S. fuscata* and 30% of the Brown Boobies *Sula leucogaster*. Anguilla's seabirds mostly breed on seven small, uninhabited islands. These are all easily accessible from the mainland except for the remote rocky outcrop of Sombrero (IBA AI001). The mainland currently holds small populations of White-tailed *Phaethon lepturus* and Red-billed *P. aethereus* tropicbirds, nesting in holes on low cliffs, and several colonies of Least Terns *Sterna antillarum* breeding around coastal lagoons and ponds. Anguilla's pond network provides important habitats for wintering and passage of Neotropical migratory shorebirds including plovers and sandpipers, and also resident breeding and migratory waterbirds. National population estimates for the waterbirds/ seabirds that meet the IBA criteria are given in Table 1.

IMPORTANT BIRD AREAS

Anguilla's seven IBAs—the territory's international priority sites for bird conservation—cover 53 km² (including marine areas), and about 8% of the islands' land area. The IBAs have been identified on the basis of 17 key bird species (listed in Table 1) that variously trigger the IBA criteria. These 17 species comprise all four restricted-range species (that have viable populations in Anguilla), and 13 congregatory seabirds. The Near Threatened Caribbean Coot *Fulica caribaea*, Piping Plover *Charadrius melodus* and White-crowned Pigeon *Patagioenas leucocephala* have not been considered in the analysis due to the nature of their current status on the islands (see above).

Anguilla's IBAs are focused on four of the larger offshore islands (Sombrero, Dog Island, Prickly Pear and Scrub Island),

Like many of Anguilla's larger offshore islands, Scrub Island IBA supports large numbers of breeding seabirds. (PHOTO: JACKIE CESTERO)

all of which support globally significant seabird colonies. Dog Island IBA (AI002) with 113,000 pairs of Sooty Tern *Sterna fuscata* and many other species of breeding seabird is one of the largest seabird colonies in the insular Caribbean. These seabirds are afforded little legal protection as Anguilla's marine parks only cover only aquatic (not terrestrial) areas, and Dog Island, Scrub, and Prickly Pear East are privately owned. However, for any sort of development to occur on any of the cays, licenses and permission must technically be granted by the Government of Anguilla. The three IBAs on mainland Anguilla are all coastal ponds supporting a range of waterbirds,

Figure 1. Location of Important Bird Areas in Anguilla.

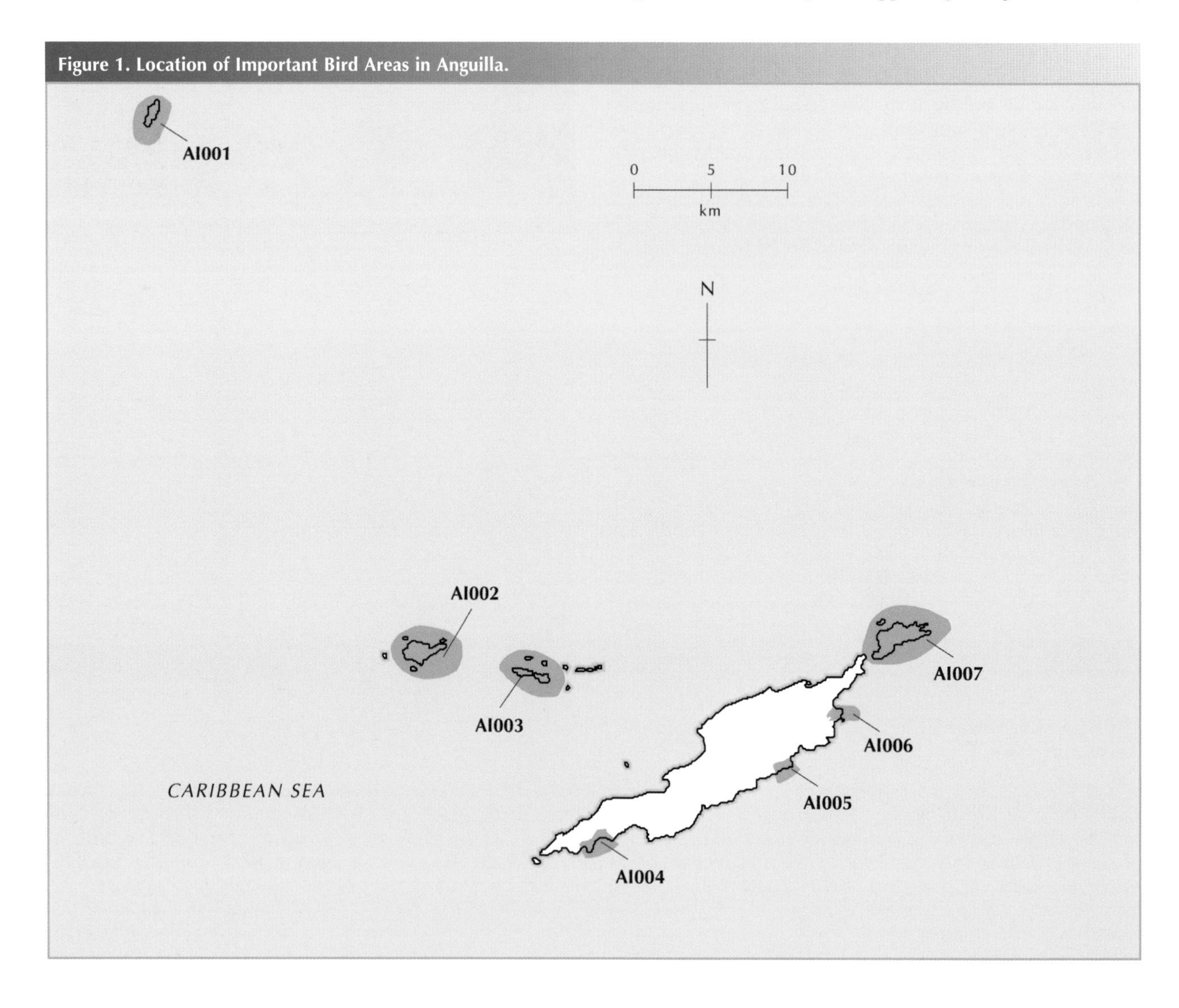

but particularly breeding colonies of Least Tern *S. antillarum*. None of these IBAs are protected, and all face varying degrees of pressure from encroaching, inappropriate, or the threat of development.

State, pressure and response scores at each IBA should be monitored annually to provide an objective status assessment and highlight management interventions that might be required to maintain these internationally important biodiversity sites. This basic site status monitoring would be best informed by regular survey results focused on the key bird species, in particular the seabirds, listed in Table 1. Such surveys should include any additional "potential" IBAs such as Anguillita and Little Scrub Island, both of which support seabird colonies that in time could meet regionally significant population thresholds.

KEY REFERENCES

Bryer, M., Fisher, I., Holliday, S. H. and Hughes, J. (2000) Birds of Anguilla and outer islands, November 1999 to June 2000. Sandy, U.K.: Royal Society for the Protection of Birds (Unpublished sabbatical report).

Collier, N. and Brown, A. C. (2004) Anguilla's offshore islands: seabird census and nest monitoring, May–June 2004. Riviera beach, USA: Environmental Protection in the Caribbean (Unpublished report to Anguilla Min. Env.: EPIC contribution 25).

Hilton, G. M., Bowden, C. G. R., Ratcliffe, N., Lucking, V. and Brindley, E. (2001) Bird conservation priorities in the UK Overseas Territories. Sandy, U.K.: Royal Society for the Protection of Birds (RSPB Research Report 1).

Hodge, K. V. D., Censky, E. J. and Powell, R. (2003) *The reptiles and amphibians of Anguilla, British West Indies*. Anguilla: Anguilla National Trust.

Hodge, K. V. D. and Holliday, S. H. (2006) Anguilla. Pp.9–18 in S. M. Sanders, ed. *Important Bird Areas in the United Kingdom Overseas Territories*. Sandy, U.K.: Royal Society for the Protection of Birds.

Holliday, S. H. and Hodge, K. V. D. (in press) Anguilla. In Bradley P. E. and Norton, R. L. eds. *Breeding seabirds of the Caribbean*. Gainesville, Florida: Univ. Florida Press.

Holliday, S. H., Hodge, K. V. D. and Hughes, D. E. (2007) *A guide to the birds of Anguilla*. Sandy, U.K.: Royal Society for the Protection of Birds.

ICF Consulting (1998) Biological surveys on Sombrero Island. Unpublished report to Beal Aerospace.

ICF Consulting (1999) Supplemental biological surveys on Sombrero and other Anguillian islands. Unpublished report to Beal Aerospace.

Pritchard, D. (1990) The Ramsar Convention in the Caribbean with special emphasis on Anguilla. Sandy, U.K.: Royal Society for the Protection of Birds (Unpublished report).

Procter, D. and Fleming, L. V., eds. (1999) *Biodiversity: the UK Overseas Territories*. Peterborough, U.K.: Joint Nature Conservation Committee.

Wilkinson, C., Pollard, M. and Hilton, G. (in prep.) A survey of breeding seabirds on Anguilla in 2007. (To be submitted to *J. Carib. Orn.* in 2008).

ACKNOWLEDGEMENTS

The authors would like to thank Sir Emile Gumbs (Chief Minister, 1981–94), Judith Dudley (UN Volunteer, 1996–97), Ian Fisher, Melanie Bryer, Julian Hughes, Mike Pollard and Colin Wilkinson (RSPB, BirdLife in the UK), Natalia Collier and Adam Brown (EPIC), and Jackie Cestero (ANT volunteer) and all the volunteers who have provided data for wetland bird counts and other surveys.

AI001 Sombrero

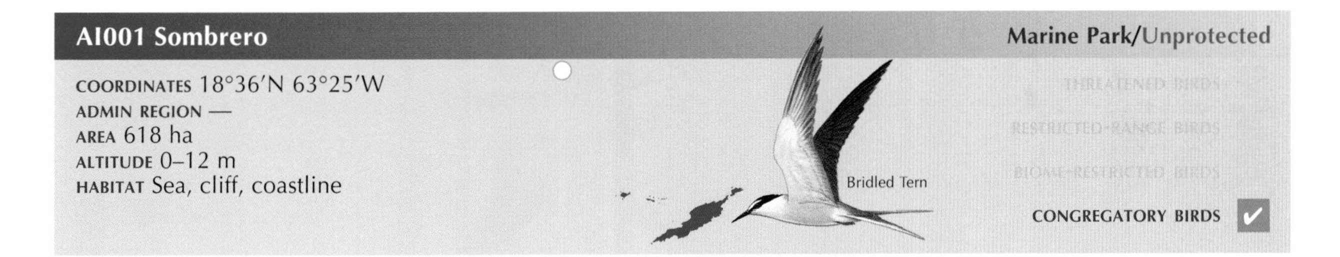

■ Site description

Sombrero IBA is a remote, 38-ha flat-topped rocky island lying 65 km north-west of Anguilla. The island is currently stark and bare following damage by Hurricane Luis in 1995 when large areas of cacti and other plants were destroyed. The vegetation is now in an early stage of recovery. Extensive phosphate deposits were mined in the nineteenth and early twentieth centuries leaving the island's surface pitted with craters up to 10 m deep, and also the remains of industrial buildings. A manned lighthouse with associated buildings was in use until 2002 when it was replaced with an automated light.

■ Birds

The cliffs and rocky areas of this IBA support a significant mixed-species seabird colony. Seven seabird species currently breed, and the numbers of Bridled Tern *Sterna anaethetus* are globally important, while those of Brown Booby *Sula leucogaster* and Brown Noddy *Anous stolidus* are regionally so. This remote rocky outpost has long been important for breeding seabirds with confirmed (historic) reports of 14 species breeding suggesting that with adequate protection and management this island can support even more birds.

■ Other biodiversity

The endemic Sombrero ground lizard *Ameiva corvina* is widespread and easily seen on the island. A recently discovered dwarf gecko *Sphaerodactylus* sp. has been tentatively named Sombrero dwarf gecko.

■ Conservation

Sombrero IBA is crown land. The Government of Anguilla has approved that the island (the terrestrial area and a portion of the marine environment) shall be designated a Marine Protected Area. However, the park boundaries (which need to make provision for the continued use of traditional fishing grounds) have not yet been determined. A few stark buildings from the phosphate industry remain alongside those from Sombrero's long- time use as a lighthouse station. Until recently the island was permanently inhabited by a small number of lighthouse staff who were transported by small boat from mainland Anguilla. With the installation of the unmanned lighthouse, visitation has been limited to the occasional fisherman and biologists engaged in fieldwork. The principal threat to Sombrero's seabirds in recent years was an application in 1999 to build a satellite-launching station on the island, though this has since been withdrawn. Rats *Rattus* spp. and mice *Mus musculus* are not thought to be present on the island, although this possibility should be monitored along with the breeding seabirds.

AI002 Dog Island

Marine Park/Unprotected

COORDINATES 18°16′N 63°15′W
ADMIN REGION —
AREA 1,333 ha
ALTITUDE 0–30 m
HABITAT Sea, shrubland, cliff, coastline, wetlands

THREATENED BIRDS
RESTRICTED-RANGE BIRDS
BIOME-RESTRICTED BIRDS
CONGREGATORY BIRDS ✓

■ Site description

Dog Island IBA is situated 13 km north-west of Anguilla. It is an uninhabited, 202-ha low rocky island with three smaller cays off its west and north coasts. The coastline is characterised by low cliffs interspersed with sandy beaches. Weathered limestone rocks reach sea-level on parts of the west and north-east coast, and two large ponds lie inside beaches at Spring Bay and Stoney Bay. The centre of the island is covered in impenetrable, low, thorny scrub and prickly pear cacti. A herd of c.200 feral goats are a remnant of previous periods of livestock farming.

■ Birds

The cliffs and inland areas of scrub in this IBA support Anguilla's largest seabird colonies with nine species breeding. Surveys in May 2007 documented a globally-significant 113,000 pairs of Sooty Tern *Sterna fuscata* breeding in the IBA. Populations of Brown Booby *Sula leucogaster* and Laughing Gull *Larus atricilla* are also globally important, while those of Red-billed Tropicbird *Phaethon aethereus*, Magnificent Frigatebird *Fregata magnificens*, Masked Booby *Sula dactylatra*, Bridled Tern *Sterna anaethetus* and Brown Noddy *Anous stolidus* are regionally so.

■ Other biodiversity

The Endangered green turtle *Chelonia mydas* nests, and the ground lizard *Ameiva plei* (endemic to Anguilla, St Martin and St Barthélemy), little dwarf gecko *Sphaerodactylus parvus* (an Anguilla Bank endemic) and island dwarf gecko *Sphaerodactylus sputator* (northern Lesser Antilles endemic) all occur on Dog Island.

■ Conservation

Dog Island IBA is privately owned and the surrounding waters are protected within a marine park. At least two development proposals, for tourism and a coastguard station, have been proposed since the early 1990s, but have been rejected by the owners. Disturbance is currently minimal (due ot the lack of habitation) and is restricted to the west end where, however, c.16% of the *Sula leucogaster* breed and where *Phaethon aethereus* nest among boulders almost down to sea-level. It is unknown if the area of scrub in the centre of the island is expanding now that grazing pressure from goats has been reduced. The low scrub affords protection to huge population of nesting *Sterna fuscata*. However, the *Sula* spp. nest in a narrow band on bare cliff-tops where encroachment by cacti may limit the colony. Rats *Rattus* spp. could be eradicated from the island according to a 2007 feasibility study.

AI003 Prickly Pear (East and West)

Marine Park/Unprotected

COORDINATES 18°15′N 63°10′W
ADMIN REGION —
AREA 973 ha
ALTITUDE 0–10 m
HABITAT Coastline, sea, shrubland, rocky areas, wetlands

Brown Booby

THREATENED BIRDS
RESTRICTED-RANGE BIRDS
BIOME-RESTRICTED BIRDS
CONGREGATORY BIRDS ✓

■ Site description

Prickly Pear (East and West) IBA is 8 km north of mainland Anguilla. Prickly Pear East is a low, rocky island with sandy shores on the north and east coasts and heavily fissured limestone cliffs on the remaining coastline. A small pond lies behind the northern beach, and the centre of the island contains areas of scrub. The island has no permanent habitation although two restaurants operate along its northern coast. The island is accessed primarily by water. Prickly Pear West is separated from its smaller sister island by a narrow channel. Apart from one small beach, the island is a low limestone outcrop with low cliffs. It is the more rockier and rugged of the two islands. The islands cover an area of 98 ha.

■ Birds

This IBA is significant for its breeding seabirds. The population of Laughing Gull *Larus atricilla* are globally important, while those of Brown Booby *Sula leucogaster*, Red-billed Tropicbird *Phaethon aethereus*, Brown Pelican *Pelecanus occidentalis* and Least Tern *Sterna antillarum* are regionally so. The majority of *Sula leucogaster* (495 pairs in 2004) nest on Prickly Pear West (along with Brown Noddy *Anous stolidus* and small numbers of Red-footed Booby *S. sula*, while most of the *Larus atricilla* (2,500 pairs in 2004) are on Prickly Pear East (along with Bridled Tern *Sterna anaethetus*).

■ Other biodiversity

The beaches on Prickly Pear are important for nesting Critically Endangered hawksbill *Eretmochelys imbricata* and Endangered green *Chelonia mydas* turtles.

■ Conservation

Prickly Pear East is privately owned while Prickly Pear West is crown owned. Neither cay is protected, but the surrounding waters fall within the Seal Island and Prickly Pear Cay East Marine Park. Prickly Pear East's inshore coral reef system and two peak-season restaurants are popular with visitors from the mainland and from neighbouring St Martin. It represents Anguilla's most accessible and most visited small island. Threats to the IBA have not been documented although it seems likely that there is some disturbance to the seabird colonies from visitors to the islands. It is unknown if rats *Rattus* spp. are present and/or impacting the seabird populations.

■ Site description

Cove Pond IBA is at the south-west end of Anguilla. It is part of a larger coastal lagoon complex that includes Gull Pond to the west and Merrywing Pond to the east. Anguilla's highest dune system borders the pond's southern shoreline. Cove Pond is divided by a causeway built as an access road to the Cap Juluca Resort. The IBA is recovering from hurricane damage (Luis in 1995 and Lenny in 1999) which swept sand from the dune into the pond making it relatively shallow. High-end resort development borders the south-west and northern shores and a road runs along its eastern end. A golf course, constructed in 2005–2006, is located east of the road.

■ Birds

This IBA supports regionally significant populations of Least Tern *Sterna antillarum* (up to 34 pairs breeding). Two (of the four) Lesser Antilles EBA restricted-range birds occur at this IBA, namely Green-throated Carib *Eulampis holosericeus* and Lesser Antillean Bullfinch *Loxigilla noctis*. A vagrant Antillean Euphonia *Euphonia musica* (also a Lesser Antilles restricted-range bird) was found at Gull Pond in 2000, and both Caribbean Elaenia *Elaenia martinica* and Pearly-eyed Thrasher *Margarops fuscatus* occur, but are not resident. Snowy Plovers *Charadrius alexandrinus* and Wilson's Plover *C. wilsoni* breed in the IBA, and the Near Threatened Piping Plover *C. melodus* has been recorded.

■ Other biodiversity

Nothing recorded.

■ Conservation

Cove Pond IBA is a mix of land ownership, but is unprotected. Pockets of crown-owned property line the crown-owned pond and sand dunes, and the entire Cap Juluca Resort was constructed on crown-owned, though privately leased land. The Government of Anguilla has indicated that the pond, from its most eastern edge to the Cap Juluca causeway, and the protective sand dunes, will be legally designated a "protected area". There has been some discussion of opening the pond to the sea using horizontal channels through the dune. Construction is underway to develop Gull Pond into a mega-yacht marina. As both Gull Pond and Cove Pond are a connected system, the impacts of the marina and the accompanying built development may have significant negative impacts on the pond, despite its soon-to-be protected status.

■ Site description

Long Pond IBA is a large brackish-water pond located on the south-east coast of Anguilla. It is separated from the sea to the east by low sand-dunes and an open sandy area. The pond is bordered to both the east and the west by dirt roads lined with seagrape. The pond's south shore is c.100 m from the sea from which it is separated by wind-swept shrubland (growing on coralline limestone). Its northern shore is lined by villas and smaller homes. A limited buffer area of pioneering buttonwood bushes separates the buildings from the pond. Power lines were recently installed along the eastern dirt road, presumably to provide electricity for a planned built development.

■ Birds

This IBA supports regionally significant populations of Least Tern *Sterna antillarum* (130 pairs breeding in 2007). The *S. antillarum* colony is the largest in Anguilla and represents 40% of the territory's population. Snowy Plover *Charadrius alexandrinus* and Wilson's Plover *C. wilsoni* breed in the IBA and the Near Threatened Piping Plover *C. melodus* has been recorded. Two (of the four) Lesser Antilles EBA restricted-range birds occur at this IBA, namely Green-throated Carib *Eulampis holosericeus* and Caribbean Elaenia *Elaenia martinica*.

■ Other biodiversity

Nothing recorded.

■ Conservation

Long Pond IBA (or at least the land around it) is privately owned by a number of families, and the pond is not protected by any form of legislation. Built development (villas and houses) has occurred along the northern shore and it is now expanding to the eastern and southern sides. Dirt roads and tracks have been cut through the vegetation on both of these sides to allow for such expansion. While the sea rarely breaches the dune, the pond can overflow into the sea after heavy rains. In dry periods the pond begins to dry out but rarely to more than half its full extent.

■ Site description
Grey Pond IBA is a large shallow, brackish lagoon on the south-eastern coast of the mainland. It has relatively steep-sided limestone slopes on its southern and eastern shores and has formed behind a belt of sand-dunes on its north-eastern corner at Savannah Bay. The rocky limestone pavement is covered by low, wind-swept scrub vegetation. Sile Bay is located c.375 m to the south while the tip of Gibbon Point lies approximately 500 m to the east. There is limited residential and villa development on the sloping hills south of the pond.

■ Birds
This IBA supports a regionally important breeding population of Least Tern *Sterna antillarum*. Up to 150 terns have bred at this site, although more recently this has declined to c.45 birds. A diversity of waterbirds frequent the site, including breeding Snowy Plover *Charadrius alexandrinus*, Willet *Catoptrophorus semipalmatus* and White-cheeked Pintail *Anas bahamensis*. Two (of the four) Lesser Antilles EBA restricted-range birds occur at this IBA, namely Green-throated Carib *Eulampis holosericeus* and Caribbean Elaenia *Elaenia martinica*.

■ Other biodiversity
The ground lizard *Ameiva plei* (endemic to Anguilla, St Martin and St Barthélemy) occurs throughout the limestone pavement.

■ Conservation
Grey Pond IBA is not protected by any form of legislation. It is privately owned by a land development company, along with surrounding property in Sile Bay and Savannah Bay. Permission to develop the area was granted in 2006. A development comprising of 825 rooms on 100 ha of land has been approved and construction is expected to commence in 2011 and continue through to 2026. Initially, a marina had been proposed for the area. The creation of a marina would have involved opening Grey Pond to the sea and dredging the coastal and pond area. Following the completion of a series of assessments, it was decided that Grey Pond was not a suitable site for a marina. The pond remains earmarked for development and depending on the type of alteration, resting and nesting habitat for *Sterna antillarum* and other wetland birds may be affected.

■ Site description
Scrub Island IBA is the largest of Anguilla's outer islands (348 ha) and is separated from the north-easternmost point of the mainland by a 500 m wide channel. The island is low lying with a rocky, fractured limestone coast punctuated by four sandy beaches. There is a large pond on the west side (at Scrub Bay) and a complex of lagoons at Deadman's Bay. The ponds are lined in places by mangroves and low trees. The centre of the island is largely scrub stretching to the coastline of heavily fissured limestone and low, rocky cliffs. The island is uninhabited although the windblown remains exist of a former tourism development in the east and a wide, grassy disused airstrip in the centre.

■ Birds
This IBA supports significant seabird colonies. Populations of breeding Roseate Tern *Sterna dougallii* and Laughing Gull *Larus atricilla* are globally important, while those of Royal Tern *S. maxima* and Least Tern *S. antillarum* are regionally so. Eight species of seabird breed on the island, and the ponds and lagoons attract a diversity of waterbirds. Two (of the four) Lesser Antilles EBA restricted-range birds occur at this IBA, namely Caribbean Elaenia *Elaenia martinica* and Pearly-eyed Thrasher *Margarops fuscatus*.

■ Other biodiversity
Both the Critically Endangered leatherback *Dermochelys coriacea* and Endangered green *Chelonia mydas* turtles are thought to nest. The Endangered Anguillan racer *Alsophis rijgersmaeri* (Anguilla's only native snake) occurs, as do the ground lizard *Ameiva plei* (endemic to Anguilla, St Martin and St Barthélemy), little dwarf gecko *Sphaerodactylus parvus* (an Anguilla Bank endemic) and island dwarf gecko *Sphaerodactylus sputator* (northern Lesser Antilles endemic).

■ Conservation
Scrub Island IBA is privately owned and is unprotected. There are large numbers of goats present on the island (presumably impacting the vegetation). An unsurfaced airstrip still remains in the centre of the island although scrub is slowly regenerating across this area. Rats *Rattus* spp. are present and must be suppressing the seabird populations. There are occasional proposals for hotel or resort developments on the island, a venture which has been tried before without success as the ruins testify on the south-east end of the island.

ANTIGUA & BARBUDA

LAND AREA **441 km^2** ALTITUDE **0–402 m**
HUMAN POPULATION **83,000** CAPITAL **St John's**
IMPORTANT BIRD AREAS **12, totalling 196 km^2**
IMPORTANT BIRD AREA PROTECTION **20%**
BIRD SPECIES **182**
THREATENED BIRDS **4** RESTRICTED-RANGE BIRDS **11**

JOSEPH PROSPER, VICTOR JOSEPH, ANDREA OTTO AND SHANEE PROSPER
(ENVIRONMENTAL AWARENESS GROUP[1])

Magnificent Frigatebirds at Codrington Lagoon. (PHOTO: JIM KUSHLAN)

INTRODUCTION

The nation of Antigua and Barbuda comprises three main islands—Antigua, Barbuda and Redonda—all at the northern end of the Lesser Antilles. Antigua is the largest island (280 km^2) and lies 40 km north-east of Montserrat (to the UK), 67 km east of Nevis (St Kitts and Nevis) and 60 km north of Guadeloupe (to France). Barbuda is smaller (161 km^2) and located 42 km north of Antigua. Redonda is the smallest island (just 2.1 km^2) and is situated 45 km west-south-west of Antigua, between Nevis (St Kitts and Nevis, 25 km to the north-west) and Montserrat (20 km south-east).

Antigua and Barbuda are emergent parts of a 3,400 km^2 sub-marine platform. The depth of water between the two islands is just 27–33 m. Antigua is 19 km across (east to west) and 15 km from north to south. It has an intricate, deeply indented coastline with numerous islands, creeks, inlets and associated sand-bars (behind which wetlands have developed). A large portion of the east, north and south coasts are protected by fringing reefs. A flat, low-lying dry central plain gives rise to gently rolling limestone hills and valleys (vegetated with xeric scrub) in the north and east. The higher, volcanic mountains of the south-west support moist evergreen forest. Barbuda is a low limestone island with a less varied coastline although it supports extensive reef systems, especially off the east coast. Codrington Lagoon in the north-west is a large, almost enclosed saltwater lagoon bordered by mangroves and sand ridges. An area of highlands (c.35 m high) on the eastern side of the island is an escarpment with a scarp slope on the north and west, a gentle slope to the south, and sea cliffs on the east. Redonda is the cone of a volcano that rises abruptly from the sea in steep cliffs that surround the island. Antigua and Barbuda support some of the most extensive mangrove woodlands in the eastern Caribbean.

Antigua and Barbuda's climate is a tropical maritime one. Average annual precipitation is 1,050 mm but severe droughts are experienced every few years (Barbuda is drier than Antigua). Rainfall is concentrated between September and November, while January to April is considered the driest period. Most of Antigua's land area (up to 92%) was under sugarcane cultivation for 300 years (the industry closed in the 1960s). The abandonment of sugar (and cotton) has resulted in a large increase of livestock production (especially cattle, but also free-roaming sheep and goats). However, tourism is the island's major industry now.

1 The Environmental Awareness Group is a national NGO and is referred to throughout this chapter under the acronym EAG.

■ Conservation

Legislation for the establishment of formal terrestrial protected areas in Antigua and Barbuda is lacking. In the 1990s, a draft of a new Forestry Act was written, making provision for the designation of protected areas. However this draft has not yet been enacted. The Government's Environment Division is currently engaged in the (GEF-funded) Ridges-to-Reef project that may declare specific areas as nature reserves or other protected areas, but currently, very little formal protection is afforded the terrestrial (and land-based wetland) biodiversity in Antigua and Barbuda. The marine environment is better catered for legislatively. The Fisheries Act allows for the designation of marine reserves, and in 2005 the North-East Marine Management Area (NEMMA) was declared—a 3,100-ha area that extends from the northern tip of Antigua to its eastern extremity, and includes 30 offshore islands. This protected area receives more than 50,000 visitors each year, yet tourism contributes nothing to protected area management or conservation. The Fisheries Division (of the Ministry of Lands and Fisheries) has formally invited the Environmental Awareness Group (EAG) to participate in a quasi-autonomous Site Management Entity for the area's management. There is a keen desire to ensure that NEMMA becomes financially self-sustaining, and contributes tangibly to both biodiversity conservation and local livelihoods. Unfortunately, at the moment, this piece of legislation provides legal authority only for the management of the marine and not terrestrial habitats within the protected area. The Wild Bird Protection Act is a specific piece of legislation for bird protection. However, the act currently allows for an open hunting season (July 16–January 31), and thus provides little protection to e.g. the Vulnerable West Indian Whistling-duck *Dendrocygna arborea* or the Near Threatened Caribbean Coot *Fulica caribaea*, both of which are listed as game species. Enforcement of this act and specific provisions for threatened species likely to be hunted are needed if this legislation is to assist bird conservation in the country.

The EAG and its partners in the Antiguan Racer Conservation Project (ARCP) and the subsequent Off-shore Islands Conservation Project (OICP) have been leading an international effort to eradicate black rats *Rattus rattus* from 10–15 islands, and to raise awareness of the need for an effective off-shore island management system. The ARCP has, to date, reintroduced the endemic (and Critically Endangered) Antiguan racer *Alsophis antiguae* to three rat-free islands, resulting in an increase in the population from 50 snakes in 1995 to 300 in January 2008. Interest that the ARCP generated, coupled with the realisation that many jobs in tourism and fisheries depend on a healthy and biodiverse environment, recently culminated in proposals for a new protected area system to promote conservation and sustainable use. Proposed areas encompass the ranges of Antigua and Barbuda's most threatened species and habitats, from coral reefs to dry forests, and will afford an unprecedented opportunity for biodiversity restoration and protection in the country. Other than protected area designation, Antigua and Barbuda urgently needs greater enforcement and awareness of species protection laws, and an improvement in hunters' abilities to distinguish between protected and game species. There is also a need for training professionals involved in developing and administering conservation programs, and technicians involved in field implementation. At the same time, basic research is needed to assess population size, habitat use and threats to the country's key bird species to ensure that major conservation initiatives are based on sound science, and their efficacy monitored.

Effective biodiversity conservation is severely constrained by the dearth of trained staff in both the government and NGO sectors, a lack of biological knowledge, and inadequate investment by a debt-ridden government. These shortcomings are highlighted in the National Biodiversity Strategy and Action Plan which emphasises the need for protected area and critical species management plans and for strengthened capacities in government and NGOs to manage biodiversity. Habitat loss, degradation and disturbance are serious conservation concerns. Wetland habitats in particular are under pressure on Antigua. They are converted to other uses (e.g. human settlement, tourism, and agriculture); degraded through clear-cutting of mangroves and swamp forest for lumber and agriculture;

Valley Church Bay. (PHOTO: LISA SORENSON)

polluted with sewage, industrial water and pesticides; and impacted by natural catastrophes such as drought and hurricanes whose local effects are exacerbated by the other pressures. Severe overgrazing has resulted in large areas denuded of vegetation. Antigua's garbage landfill is within a coastal ravine. During storm surges, hurricanes and high winds, this trash is dispersed over the coastal-zone, possibly resulting in entanglement or consumption of debris (e.g. seabirds or other marine-life ingesting plastics). Introduced predators include small Indian mongoose *Herpestes auropunctatus*, dogs, cats, rats and mice which are impacting breeding seabirds, ducks and shorebirds on the main islands and offshore islands alike (including Redonda). Excessive and under-regulated sport and subsistence hunting is an issue, especially for *Dendrocygna arborea*. Significant progress has been made with regards to raising knowledge and awareness on the threatened status of *D. arborea* and the importance of local wetlands through participation in the West Indian Whistling-duck and Wetlands Conservation Project. Joseph "Junior" Prosper, EAG colleagues, and volunteers have implemented a monitoring program, counting ducks simultaneously on all wetlands four times per year. Prosper has also led efforts to educate hunters, landowners, decision makers and schoolchildren through talks, field trips, school projects and advocacy. Although serious threats to wetlands remain, this work has had a positive impact: many hunters have stopped shooting whistling-ducks and instead report their sightings to EAG, several wetlands have been saved from destruction, and nesting and local populations of whistling-ducks have increased.

■ Birds

At least 182 bird species have been recorded in Antigua and Barbuda, c.65% of which are Neotropical migratory birds. Eleven (of the 38) Lesser Antilles EBA restricted-range birds occur on the islands. These include relatively uncommon (wet forest dependent) species such as Bridled Quail-dove *Geotrygon mystacea*, Antillean Euphonia *Euphonia musica*, Scaly-breasted Thrasher *Margarops fuscus* and Pearly-eyed Thrasher *M. fuscatus*. However, the most significant species is the Barbuda Warbler *Dendroica subita* which is endemic to Barbuda. Although it occurs in the mangroves and the dry shrubland around Codrington Lagoon, little is known of this species' island distribution, population or ecological requirements. *Dendroica subita* is one of the country's four globally threatened birds, the threat category and national population sizes of which are listed in Table 1. The Near Threatened Buff-breasted Sandpiper *Tryngites subruficollis* and Piping Plover *Charadrius melodus* are only known as a vagrants/ rare migrants to the island and are not considered as national conservation priorities. Both *Dendrocygna arborea* and Caribbean Coot *Fulica caribaea* are at risk due to the deterioration of wetland habitats across the islands (something that is impacting other waterbirds such as Masked Duck *Nomonyx dominicus*), especially as they probably rely on a "network" of sites during their annual cycle. Hunting is also affecting the whistling-duck, and presumably the Near Threatened White-crowned Pigeon *Patagioenas leucocephala*.

Barbuda supports the Caribbean's largest colony of Magnificent Frigatebird *Fregata magnificens* with the population estimated at 5,300 individuals (1,743 occupied nests) in March 2008. Less is known about the current status (or population sizes) of the country's other breeding seabirds although surveys by EAG are being undertaken. An interesting recent addition to Antigua's breeding avifauna is the Little Egret *Egretta garzetta*. Previously known to breed only on Barbados within the Western Hemisphere, three nests (and an island population of 12 birds) were documented on Antigua in March 2008 (Kushlan in press).

IMPORTANT BIRD AREAS

Antigua and Barbuda's 12 IBAs—the island's international site priorities for bird conservation—cover 196 km² (including marine areas), and about 20% of the country's land area. The IBAs have been identified on the basis of 23 key bird species (listed in Table 1) that variously trigger the IBA criteria. These 23 species include all four regularly occurring globally threatened birds, all 11 restricted-range species, and 10 congregatory waterbirds/seabirds.

Magnificent Frigatebird chick at Codrington Lagoon—the Caribbean's largest frigatebird colony. (PHOTO: JIM KUSHLAN)

Most of Antigua and Barbuda's IBAs meet two or more of the criteria categories, quite often combining significance for congregatory waterbirds/seabirds with restricted-range birds and/or globally threatened birds. However, Redonda IBA (AG001) qualifies solely on the basis of its congregatory species. For many of the congregatory species, significant (i.e. >1% of the global or Caribbean population of the species) populations are only found in one IBA. Codrington Lagoon and Creek IBA (AG002) is significant not just as the largest *F. magnificens* colony in the Caribbean, but also as the only IBA where the globally threatened (and endemic) *Dendroica subita* occurs, and also the only IBA in the country where the restricted-range Lesser Antillean Flycatcher *Myiarchus oberi* is found. Walling's Forest IBA (AG008) and Christian Valley IBA (AG009) together represent Antigua's wet forest ecosystem and are the only IBAs in the country where the restricted-range *G. mystacea*, *E. musica*, *Margarops fuscus* and *M. fuscatus* are found, highlighting their critical importance in maintaining Antigua's biodiversity.

State, pressure and response scores at each IBA should be monitored annually to provide an objective status assessment and highlight management interventions that might be required to maintain these internationally important biodiversity sites. Any such basic site status monitoring would be best informed by species-specific status information for the key species listed in Table 1.

KEY REFERENCES

D'Auvergne, E. C. (1999) A use analysis of the Codrington Lagoon, Barbuda. St Lucia: OECS Natural Resources Management Unit. (Unpublished report).

Bacon, P. R. (1991) The status of mangrove conversation in the CARICOM islands of the Eastern Caribbean. (Unpublished report to the European Commission as part of the Tropical Action Plan for the Caribbean Region).

Table 1. Key bird species at Important Bird Areas in Antigua and Barbuc

Key bird species	Criteria			National populat
West Indian Whistling-duck *Dendrocygna arborea*	VU ■		■	
Little Egret *Egretta garzetta*			■	
Magnificent Frigatebird *Fregata magnificens*			■	6,
Brown Pelican *Pelecanus occidentalis*			■	
Masked Booby *Sula dactylatra*			■	
Red-footed Booby *Sula sula*			■	
Brown Booby *Sula leucogaster*			■	
Caribbean Coot *Fulica caribaea*	NT ■			
Laughing Gull *Larus atricilla*			■	4,
Royal Tern *Sterna maxima*			■	
Least Tern *Sterna antillarum*			■	650–
White-crowned Pigeon *Patagioenas leucocephala*	NT ■			
Bridled Quail-dove *Geotrygon mystacea*		■		
Purple-throated Carib *Eulampis jugularis*		■		
Green-throated Carib *Eulampis holosericeus*		■		
Antillean Crested Hummingbird *Orthorhyncus cristatus*		■		
Caribbean Elaenia *Elaenia martinica*		■		
Lesser Antillean Flycatcher *Myiarchus oberi*		■		
Scaly-breasted Thrasher *Margarops fuscus*		■		
Pearly-eyed Thrasher *Margarops fuscatus*		■		
Barbuda Warbler *Dendroica subita*	NT ■	■		
Lesser Antillean Bullfinch *Loxigilla noctis*		■		
Antillean Euphonia *Euphonia musica*		■		

All population figures = numbers of individuals.
Threatened birds: Vulnerable ■; Near Threatened ■.
Restricted-range birds ■. **Congregatory birds** ■.

Egret colony at McKinnon's Saltpond IBA—one of only two New World breeding sites for Little Egret. (PHOTO: JIM KUSHLAN)

	Antigua and Barbuda IBAs											
	AG001	AG002	AG003	AG004	AG005	AG006	AG007	AG008	AG009	AG010	AG011	AG012
Criteria		■	■	■	■	■	■	■	■	■	■	■
		■			■		■	■	■	■		■
	■	■	■	■	■	■	■			■	■	■
		200	74	80	134	65	71			30	38	134
			8									
	500	5,300										
			110		70	40				50		
	170											
	300											
	300											
					40							110
		1,000	800	400	200	1,000	500			300	400	
						20						
			500		100	50						
								50	50			
								✓	✓			
					✓		✓	✓	✓	✓		✓
		✓			✓		✓	✓	✓	✓		✓
		✓			✓		✓	✓	✓	✓		✓
		✓			✓		60	✓	✓	✓		✓
		✓										
								100	✓			
								100	✓			
		30										
		✓						150	350			
								✓	✓			

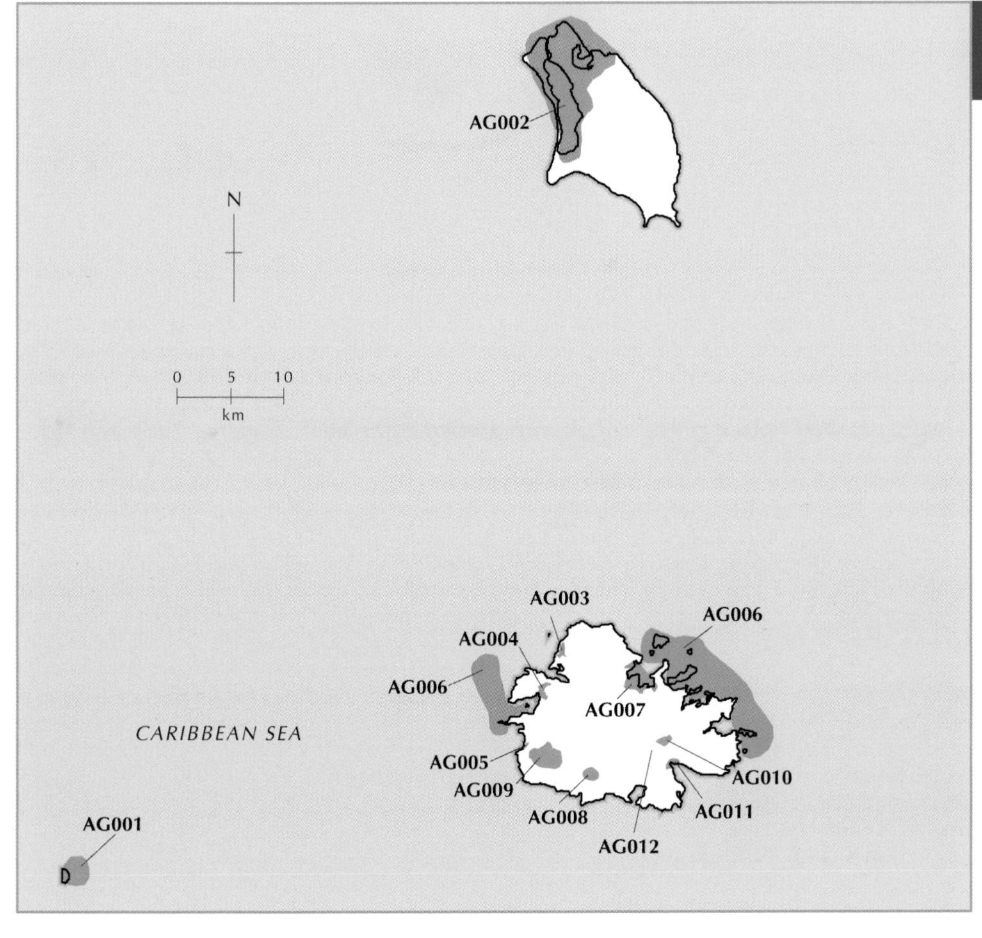

Figure 1. Location of Important Bird Areas in Antigua and Barbuda.

Bunce, L. (1993) Coral reef monitoring and baseline data of Antigua's coral reefs. Antigua: Island resources Foundation. (Unpublished report).

Cooper B. R. and Bowen, V. (2001) Integrating management of watersheds and coastal areas in small island developing states: national report for Antigua and Barbuda. (Unpublished report for Environment Division, Ministry of Tourism and Environment).

Daltry, J. C. (2007) An introduction to the herpetofauna of Antigua, Barbuda and Redonda, with some conservation recommendations. *Applied Herpetology* 4: 97–134.

Danforth, S. T. (1934) The birds of Antigua. *Auk* 51: 350–364.

Gricks, N. (1987) *Revised list of birds of Antigua and Barbuda and Redonda*. Antigua: privately published.

Gricks, N. B., Horwith, B. and Lindsay, K. (1997) Birds of Antigua, Barbuda and Redonda. (Unpublished report).

Holland, C. S. and Williams, J. M. (1978) Observations on the birds of Antigua. *American Birds* 32: 1095–1105.

Horwith, B. and Lindsay, K. (1997) *A biodiversity profile: Antigua, Barbuda, Redonda*. St Thomas, USVI: Island Resources Foundation (Eastern Caribbean Biodiversity Programme, Biodiversity Publication 3).

Howell, C. A. (1991) *Antigua and Barbuda: country environmental profile*. St Michael, Barbados: Caribbean Conservation Association, Island Resources Foundation and Environmental Awareness Group.

Jarecki, L. (1999) A proposed biological monitoring plan for Cades Bay Estuary, Fitches Creek Estuary, and Fish Pond, Carlisle Bay Antigua. (Unpublished report for the Environmental Awareness Group).

Kushlan, J. A. (in press) Status of the Magnificent Frigatebird (*Fregrata magnificens*) on Barbuda, West Indies. *J. Carib. Orn.* (2008).

Kushlan, J. A. and Prosper, J. W. (2008) Little Egret (*Egretta garzetta*) nesting on Antigua: a second nesting site in the Western Hemisphere. Submitted to *J. Carib. Orn.*

MacPherson, J. (1973) *Caribbean lands: a geography of the West Indies*. Port-of-Spain and Kingston: Longman Caribbean.

Martin-Kaye, P. (1959) *Reports on the geology of the Leeward and British Virgin Islands*. St Lucia: Voice Publishing Company.

Martin-Kaye, P. (1969) A summary of the geology of the Lesser Antilles. *Overseas Geology and Mineral Resources* 10: 263–286.

Pregill, G. K., Steadman, D. W. and Watters, D. R. (1994) Late Quaternary vertebrate faunas of the Lesser Antilles: historical components of Caribbean biogeography. *Bull. Carnegie Mus. Nat. Hist.* 30: 1–51.

Raffaele, H. Wiley J., Garrido, O., Keith, A. and Raffaele, J. (1998) *A guide to the birds of the West Indies*. Princeton, New Jersey: Princeton University Press.

Spencer, W. (1981) *A guide to birds of Antigua*. Antigua: privately published.

Sorenson, L. G., Bradley, P. E. and Haynes Sutton, A. (2004) The West Indian Whistling-Duck and Wetlands Conservation Project: a model for species and wetlands conservation and education. *J. Carib. Orn.* Special Issue: 72–80.

Sorenson, L., Bradley, P., Mugica, L. and Wallace, K. (2005) West Indian Whistling-Duck and Wetlands Conservation Project: symposium report and project news. *J. Carib. Orn.* 18: 102–105.

Sorenson, L. G. (2006) The West Indian Whistling-Duck and Wetlands Conservation Project. *Rainforest Alliance's Eco-Index*. http://www.eco-index.org/search/results.cfm?projectID=979

ACKNOWLEDGEMENTS

The authors would like to thank Brian Cooper, Carole McCauley, Jim Kushlan and Lisa Sorenson (Society for the Conservation and Study of Caribbean Birds) for their contributions to and reviews of this chapter.

AG001 Redonda

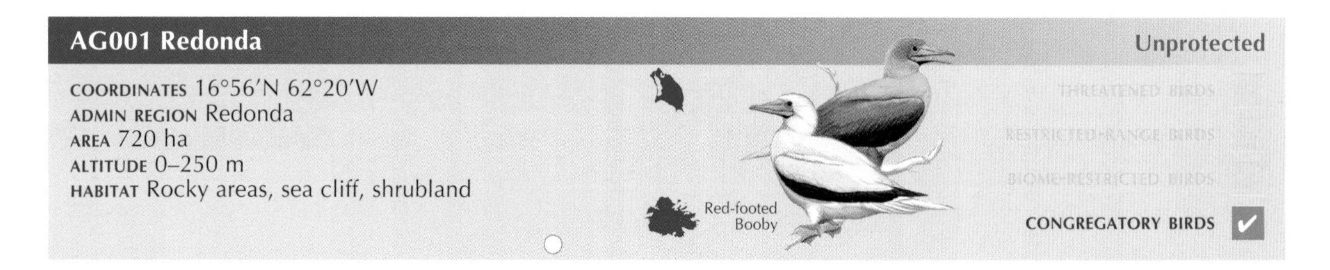

■ Site description
Redonda IBA is a small, uninhabited island situated 56 km west-south-west of Antigua, in the waters between the islands of Nevis (St Kitts and Nevis) and Montserrat (to the UK). It lies 22 km north-west of Montserrat, and 32 km south-east of Nevis. The IBA is a remnant of an extinct volcano, and protrudes steeply from the sea, mostly as sheer cliffs. It is 1.6 long and 0.5 km wide, with an area calculated at somewhere between 160 and 260 ha. Redonda has one beach (accessible by boat), but is otherwise completely surrounded by steep cliffs. The summit slopes at c.39° and the only "flat" land is a saddle of 0.4 ha at the southern end of the island, about 230 m up. This is accessible only by helicopter. Vegetation comprises coarse grasses and *Opuntia* cacti. A few individual short-leafed fig *Ficus citrifolia* cling to the western cliff-face.

■ Birds
This IBA is notable for its breeding seabirds. Breeding colonies of Magnificent Frigatebird *Fregata magnificens*, Masked Booby *Sula dactylatra*, Red-footed Booby *S. sula* and Brown Booby *S. leucogaster* are regionally significant. Other species breed, including over 140 Brown Noddy *Anous stolidus*. Burrowing Owl *Athene cunicularia* is thought to be resident.

■ Other biodiversity
Of the six reptile species recorded from Redonda, three (lizards) are endemic to the island: *Ameiva atrata, Anolis nubilus,* and a potentially new *Sphaerodactylus* sp. The Endangered green *Chelonia mydas* and Critically Endangered hawksbill *Eretmochelys imbricata* turtles have both been seen in the waters around the island.

■ Conservation
Redonda is state owned. From the 1860s until c.1920 the island was mined for its guano (phosphates), yielding 7,000 tons per year. In 1901, the human population was 120, but the island has been uninhabited since the First World War when mining operations were abandoned. The island supports a healthy population of goats which may indirectly impact the seabird populations. Rats are present (as would be expected after at least 40 years of human habitation) and in the last few years brown rats *Rattus norvegicus* have become established and will no doubt be having a serious impact on the seabird (and endemic reptile) populations.

AG002 Codrington Lagoon and Creek

Site description

Codrington Lagoon and Creek IBA is on the north-western side of Barbuda. It is a large (c.12 km long and 2 km wide), almost enclosed salt-water lagoon that is bordered by mangroves and sand ridges. The best developed mangroves border the lagoon along the eastern and north-western sides, and also along Codrington Creek which connects the lagoon to the sea in the north. Dry scrub vegetation borders the lagoon on the north-east side.

Birds

This IBA supports the largest breeding colony of Magnificent Frigatebird *Fregata magnificens* in the Caribbean, estimated at 1,743 nests and 5,300 birds in March 2008. A significant population (c.200) of Vulnerable West Indian Whistling-ducks *Dendrocygna arborea* occurs in the lagoon, as do four (of Antigua and Barbuda's 11) Lesser Antilles EBA restricted-range birds. The Near Threatened endemic Barbuda Warbler *Dendroica subita* occurs and the presence of up to 1,000 Laughing Gulls *Larus atricilla* is globally significant.

Other biodiversity

This is a major nesting site for the Critically Endangered leatherback *Dermochelys coriacea* and hawksbill *Eretmochelys imbricata* turtles. The lagoon is also of considerable importance as a major breeding ground for (economically important) fish and lobster.

Conservation

Codrington Lagoon and Creek IBA is publicly owned by the people of Barbuda (as is all land in Barbuda) and administered by the Barbuda Council. The frigatebird colony is within a "Wildlife Reserve" (on the north-west side of the creek), and the lagoon has been declared a Ramsar site. The government Environment Division in conjunction with the Barbuda Council has embarked on a project to help set up proper management systems for the lagoon and associated wetland areas. The EAG carried out a survey of the lagoon (late-1990s) and developed a monitoring plan for the area. Resources were not available to implement the plan, but it is being used to inform management decisions for the area. The government and EAG are educating tour boat operators in an attempt to minimise the disturbance from visitors being taken to see the frigatebird colony. Tour boats (and even helicopters) do approach too close to the colony on occasions. The lagoon's biodiversity needs more study to reinforce the case for minimal development within the IBA.

AG003 McKinnon's Saltpond

Site description

McKinnon's Saltpond IBA is on the west facing coast of north-west Antigua, 3 km north of St John's and just inland from Runaway Bay. It was once a mangrove-lined lagoon (the largest natural pond on the island), but a road along the western and northern seaward side has cut it off from the sea (except for a small culvert that allows limited exchange of water). The resultant increased water levels (now c.1 m deep) have killed the majority of the mangroves so that most of the shoreline is open (albeit fringed by the "skeletons" of the dead mangroves). Small stands of degraded mangroves remain on the southern and western (seaward) edges. A man-made causeway (accommodating fuel lines from a facility 3 km offshore to the storage area of an oil refinery) runs through the pond, isolating the southernmost section from the main body of water. There is extensive resort development around and adjacent to the pond, as well as private homes.

Birds

This IBA is notable for its waterbird populations. Numbers of Brown Pelican *Pelecanus occidentalis* and Least Tern *Sterna antillarum* are regionally significant, while those of Laughing Gull *Larus atricilla* are globally so. The Vulnerable West Indian Whistling-duck *Dendrocygna arborea* is often present at the site (although not all year) with 56 adults and 18 ducklings found in 2006. Little Egret *Egretta garzetta* breeds in this IBA: three nests found in 2008 represent a significant percentage of the New World breeding population. Numerous Neotropical migratory shorebirds use this site.

Other biodiversity

The Critically Endangered hawksbill turtle *Eretmochelys imbricata* nests on the nearby beaches.

Conservation

McKinnon's Saltpond is under mixed ownership, and is not protected in any way. There are claims on different parts of the pond by various landowners who have individual ideas about the development of the pond. The Government's Environment Division is currently dredging the pond to reduce flooding and increase the aesthetics of the area. *Dendrocygna arborea* has not been seen in the IBA since the dredging started. The sand flats surrounding the area open to dogs, mongoose and other domestic animals, and are used as parking areas for visitors to the pond or beach nearby. These activities adversely impact the nesting terns and shorebirds.

AG004 Hanson's Bay–Flashes

White-cheeked Pintail

■ Site description

Hanson's Bay–Flashes IBA is on the west coast of Antigua, just west of St John's. More specifically it lies south-east of Five Island Village, and south-west of Gray's Farm/Green Bay. The area comprises a complex wetland system (with salinas and mangroves) which forms the seaward outfall of the Body Ponds–Big Creek watershed (Antigua's largest). The dense stand of mangroves (the largest mangrove woodland on Antigua) borders Hanson's Bay, and this grades to salina and wet pasture (and bordering shrubland) on the landward side.

■ Birds

This IBA is significant as a roosting and feeding area for the Vulnerable West Indian Whistling-duck *Dendrocygna arborea*, with up to 80 occurring. Globally significant counts of 400 Laughing Gulls *Larus atricilla* have been reported. Good numbers of shorebirds, herons and egrets occur along with up to 100 White-cheeked Pintail *Anas bahamensis*, and the three species of restricted-range hummingbirds.

■ Other biodiversity

Nothing recorded.

■ Conservation

The area is state-owned, but totally unprotected. There are no conservation actions being undertaken at this site. Hunting is the largest threat to the birds within this IBA: the Bird Protection Law of 1913 permits an open-season for hunting. Tourism related development on the shores at the mouth of the Flashes is very likely in the future, and it is possible that the garbage dump at Cook's is affecting water quality in the Flashes.

AG005 Valley Church Bay

West Indian Whistling-duck

■ Site description

Valley Church Bay IBA is on the west coast of Antigua, close to Jolly Harbour and its associated tourist developments. This area was once a large swamp. However, a road has cut it off from the sea, preventing any exchange of water, turning the area into a shallow coastal salt pond encircled by a narrow fringe of mangroves. The area is bordered on the east by a major road and on the west by a narrow beach and sea-grass beds. On the seaward side of this IBA are a number of restaurants and a large hotel (which occupies part of the original mangrove). A road divides the pond to give access to the restaurants from the main road.

■ Birds

This IBA supports populations of four (of the 11) Lesser Antilles EBA restricted-range birds that occur in Antigua. However, it is primarily notable for its waterbirds including important populations of the Vulnerable West Indian Whistling-duck *Dendrocygna arborea* which nests underneath vegetation close to the swamp (96 adults and 38 ducklings in 2006 is a maximum count) and Near Threatened Caribbean Coot *Fulica caribaea*. Regionally significant populations of Brown Pelican *Pelecanus occidentalis*, Laughing Gull *Larus atricilla* and Least Tern *Sterna antillarum* occur. Good numbers of Neotropical migratory shorebirds use this IBA. Masked Duck *Nomonyx dominica* has been recorded.

■ Other biodiversity

The Critically Endangered hawksbill *Eretmochelys imbricata* and leatherback *Dermochelys coriacea* turtles are known to occur in the bay.

■ Conservation

Valley Church Bay is a mix of private and state ownership. However, the wetland is not protected and the surrounding habitat is presently being destroyed by hotel and residential development. Construction rubble and other solid waste have been dumped at the southern end of the pond, and the nesting and feeding areas of *Dendrocygna arborea* are being encroached upon. There are land reclamation plans by the private owners which would result in the complete destruction of the ponds. EAG has tried working with landowners (including one of the restaurateurs) to encourage greater awareness, conservation and potentially even the development of birdwatching infrastructure.

■ Site description

Offshore Islands IBA comprises many of Antigua's 51 offshore islands. The majority are concentrated off the north-east coast of the mainland in the North Sound area. These include Redhead, Rabbit, Galley, Lobster, "Jenny", Great Bird, Hellsgate and the Exchange islands. The islands of York and Green are located off the eastern most tips of the mainland, and the Five-Island islets to the west of the mainland. The islands range in size from c.40 ha (Green Island) to 0.25 ha (most of the Five-Island islets), and are characterised by limestone cliffs, xeric dry scrub and cactus vegetation, and surrounding mangroves and coral reef systems.

■ Birds

This IBA is notable for waterbirds and seabirds. The Vulnerable West Indian Whistling-duck *Dendrocygna arborea* breeds on some of the islands, and the numbers of Laughing Gull *Larus atricilla* are globally significant. Populations of Brown Pelican *Pelecanus occidentalis*, Royal Tern *Sterna maxima* and Least Tern *S. antillarum* are important regionally.

■ Other biodiversity

The Critically Endangered Antiguan racer *Alsophis antiguae* is found on Great Bird Island and, as a result of reintroductions since 1999, Rabbit, Green and York islands. Significant numbers of the Critically Endangered hawksbill turtle *Eretmochelys imbricata* nests on the beaches of a number of the offshore islands (e.g. 60 nesting females on Long Island).

■ Conservation

The Offshore Islands exhibit a range of ownership. Some are privately-owned or leased (e.g. Green and York islands are leased by the Mill Reef Club), some are state-owned, and some are disputed (e.g. Great Bird Island and Rabbit Island). There is currently no legislation that provides for the protection or management of the terrestrial biodiversity of these islands, although the North East Marine Management Area embraces the marine components of the east coast islands. EAG has been leading an international effort to eradicate black rats from 10–15 islands, and to raise awareness of the need for an effective island management system. The Antiguan Racer Conservation Project has been reintroducing the endemic *Alsophis antiguae* to three rat-free islands (and numbers have increased from 50 in 1995 to 300 in January 2008). Breeding bird numbers are being monitored and are increasing on the islands cleared of rats. Unregulated tourism and recreational use of the islands is causing direct and indirect impact and disturbance to the biodiversity.

AG007 Fitches Creek Bay–Parham Harbour

Unprotected

COORDINATES 17°07'N 61°47'W
ADMIN REGION Antigua
AREA 730 ha
ALTITUDE 0–3 m
HABITAT Inland wetland, mangrove, salina

West Indian Whistling-duck

THREATENED BIRDS 1
RESTRICTED-RANGE BIRDS 4
BIOME-RESTRICTED BIRDS
CONGREGATORY BIRDS ✓

■ Site description

Fitches Creek Bay–Parham Harbour IBA is in north-east Antigua. They represent two large bays between Barnacle Point (and the VC Bird International Airport) to the north and Old Fort Point at the elbow of Crabb's Peninsula to the south. The bays are separated by Blackman's Point, and support an almost unbroken stretch of mangroves. At Fitches Creek Bay a deep, mangrove-lined channel extends 1.5 km inland, and mangroves fringe the coast for c.1 km. Parham Harbour supports two broad areas on either side of the town. The western section consists of Vernon's Island and Byam's Wharf; the eastern section starts east of Parham town and follows the coast to Old Fort Point on Crabb's Peninsula. South and west of Vernon's Island is a salina and saltpond, and in the area east of Parham town are three small mangrove creeks, a salina and saltmarsh.

■ Birds

This IBA is significant for supporting populations of four (of Antigua's 11) Lesser Antilles EBA restricted-range birds, but primarily this is a wetland site with good numbers of egrets, herons, and wintering shorebirds. Up to 500 Laughing Gulls *Larus atricilla* have been reported (a globally significant number), and the Vulnerable West Indian Whistling-duck *Dendrocygna arborea* occurs (and nests) in significant numbers.

■ Other biodiversity

Nothing recorded.

■ Conservation

This area is privately-owned and is not protected. EAG has implemented education efforts in an attempt to save these wetlands. The IBA is used as a dumping site for of solid waste/garbage by members of the public. There is a road built across the swamp to gain access to the sea. Hurricanes Luis and Marilyn caused (directly and indirectly) significant damage to the taller mangroves.

■ Site description

Walling's Forest IBA is in south-west Antigua, in the volcanic Shekerly Mountains. The forest is on the north-west slopes of Signal Hill (Antigua's second highest mountain). Boggy Peak (and Sage Hill) lies c.4 km to the west, beyond which is the Christian Valley IBA (AG009). The slopes of Signal Hill are traversed by contour drainage ditches so that all most run off feeds into Walling's reservoir, created by a dam built in 1900 with a view to supplying neighbouring villages with potable water (something it no longer does). The IBA supports the largest and best remaining tract of moist evergreen forest on the island. A popular trail to the top of Signal Hill starts near the reservoir which is quite heavily used by tourists, and locals at weekends.

■ Birds

This IBA supports populations of nine (of the 11) Lesser Antilles EBA restricted-range birds. Within Antigua, some of these species (Bridled Quail-dove *Geotrygon mystacea*, Scaly-breasted Thrasher *Margarops fuscus*, Pearly-eyed Thrasher *M. fuscatus* and Antillean Euphonia *Euphonia musica*) are entirely confined to the Walling's Forest and Christian Valley IBA (AG009) ecosystem. A significant population of the Near Threatened White-crowned Pigeon *Patagioenas leucocephala* also occurs.

■ Other biodiversity

Seven species of bat occur including the Near Threatened insular single leaf bat *Monophylus plethodon* and Brazilian free-tailed bat *Tadarida brasilensis*.

■ Conservation

Walling's Forest and reservoir are partially state-owned. About 5 ha of Wallings was reforested in the early 1900s to protect the watershed. The Forestry Department (Ministry of Agriculture) offers some protection from logging for building and charcoal—the primary threat to this area. However, the draft Forestry Act (of the 1990s) which would enable formal protection to this area has not been enacted. Forestry Department maintain some areas, some trails and offer guided tours. A management plan has been written but not implemented. Forest conversion and disturbance for residential development, tourism (e.g. the recent construction of an aerial ropeway) and agriculture pose a real threat to the integrity of this ecosystem.

■ Site description

Christian Valley IBA is in south-west Antigua, in the volcanic Shekerly Mountains. The valley lies to the north of Boggy Peak (Antigua's highest mountain). It is surrounded to the south and east by this peak and Sage Hill, with Mount McNish to the north-east. Walling's Forest IBA (AG008) is c.4 km to the east of Boggy Peak. The entrance to Christian Valley is from the road between Jennings and Bolans. Along the entrance road is the ruin of the colonial Blubber Valley sugar estate house (now surrounded by a chicken farm). The Christian Valley estate house lies at the end of the road. During the seventeenth and eighteenth centuries, the surrounding hills served as a refuge for escaped African slaves ("maroons"). Maroon trails are still used by foresters and farmers.

■ Birds

This IBA is significant for supporting populations of nine (of the 11) Lesser Antilles EBA restricted-range birds that occur in Antigua. Within Antigua, some of these species (Bridled Quail-dove *Geotrygon mystacea*, Scaly-breasted Thrasher *Margarops fuscus*, Pearly-eyed Thrasher *M. fuscatus* and Antillean Euphonia *Euphonia musica*) are entirely confined to the Christian Valley and Walling's Forest IBA (AG008) ecosystem. A significant population of the Near Threatened White-crowned Pigeon *Patagioenas leucocephala* also occurs at this IBA.

■ Other biodiversity

No globally threatened or endemic species have been recorded, but the area has a diverse flora and insect fauna. Seven bat species (including Antillean fruit-eating bat *Brachyphylla cavernarum* and greater bulldog bat *Noctilio leporinus*) have been documented.

■ Conservation

Christian Valley is state owned. It is the subject of a watershed management program administered by the Forestry Division (Ministry of Agriculture) which affords the area some protection. The draft Forestry Act (of the 1990s) which would enable formal protection to this area has not been enacted. A government agricultural station in the valley oversees tree fruit production. Logging for building and charcoal is a real threat, as is the conversion of forest as a result of the expansion of human settlement, tourism, and agriculture. Water from the valley flows into the sea at Lignumvitae Bay, the site of the major tourist developments of Jolly Beach and Jolly Harbour.

Site description

Potworks Dam IBA is in south-east Antigua, c.2 km north of the village of Bethesda and Bethesda Dam IBA (AG012). The reservoir dam was under construction in 1969 when the area flooded as a result of exceptionally heavy rains. The dam site filled with water before preparation of the site was finished, but is considered a municipal reservoir and is an important source of water for domestic and agricultural use. Due to the very shallow sides to the reservoir and frequent droughts, the shoreline fluctuates greatly over time, reducing opportunities for the establishment of aquatic vegetation. Land surrounding most of the reservoir is former sugar cane estate land that is now farmed or grazed (primarily by free-roaming cattle).

Birds

The planted woodland and shrubland around parts of this IBA support populations of four (of the 11) Lesser Antilles EBA restricted-range birds that occur in Antigua. However, Potworks Dam is most notable for its waterbirds and numerous Neotropical migratory shorebirds. Over 100 Ruddy Duck *Oxyura jamaicensis* and 50 Pied-billed Grebe *Podilymbus podiceps* can be found, but the numbers of Laughing Gull *Larus atricilla* are globally significant, and Brown Pelican *Pelecanus occidentalis* are regionally so. Significant numbers of the Vulnerable West Indian Whistling-duck *Dendrocygna arborea* have been recorded.

Other biodiversity

Nothing recorded.

Conservation

The dam and surrounding lands were bought by the government in the late-1950s, but most of these lands are now leased to farmers. Now, due to intense crop cultivation in the area the reservoir is becoming highly polluted by pesticides and fertiliser run-off. Potworks Reservoir is a proposed wildlife reserve, but currently enjoys no formal protection. The Forestry Division has planted trees (e.g. neem, whitewood, *Lignum vitae*) on a small area on the northern side of the reservoir to protect the reservoir edges. The government-owned Public Utilities Company also tries to restrict use of the water for recreational purposes, but cattle still drink the water and feed on any marginal vegetation.

Site description

Christian Cove IBA is at the head of Willoughby Bay on the south-east coast of Antigua, just north of Christian Point. The IBA is south-east of Bethesda Dam IBA (AG012). The road to Bethesda forms the western boundary of the IBA. Christian Cove is a wetland basin comprising fringing mangroves (shorter mangroves to the north and taller stands on the south-east side), salt-marsh, and a small freshwater marsh. The freshwater marsh is in the north-western corner of the IBA, and is fed by a stream that is bisected by the Bethesda road. Dry woodland of manchineel *Hippomane mancinella* and *Acacia* spp. surrounds the wetland.

Birds

This IBA is globally significant for its population of 400 Laughing Gulls *Larus atricilla*, and Vulnerable West Indian Whistling-duck *Dendrocygna arborea*. The wetland supports a wide diversity of shorebirds and waterbirds (including ducks, herons, egrets and terns). The Near Threatened Piping Plover *Charadrius melodus* and Buff-breasted Sandpiper *Tryngites subruficollis* have all been recorded, although not in significant numbers.

Other biodiversity

Nothing recorded.

Conservation

Christian Cove is state owned but is not protected in any way, and there are currently no conservation efforts being implemented at the site. Development pressures are a constant threat to these wetlands.

■ Site description

Bethesda Dam IBA is a man-made reservoir in the south-eastern part of the island, just north-west of Bethesda village. The reservoir is surrounded by grass and scrub covered hills to the west and south, and an agricultural area to the north. Bethesda Dam was constructed (by the government) in the 1970s to provide irrigation water for the agricultural area. Water levels vary, but during the rainy season the reservoir covers c.5 ha and provides good feeding and cover for waterbirds.

■ Birds

This IBA is significant for its populations of the Near Threatened Caribbean Coot *Fulica caribaea* (up to 110 have been recorded) and Vulnerable West Indian Whistling-ducks *Dendrocygna arborea*. During 2006–2007, 204 whistling-ducks were seen at the reservoir. The species can be highly mobile, but is usually present and does breed at this site. Four (of the 11) Lesser Antilles EBA restricted-range birds occur in the shrubland around the reservoir.

■ Other biodiversity

This reservoir was one of the first places where an invasive alien species of reed was observed. The reed has since spread rapidly to other freshwater sites.

■ Conservation

Bethesda Dam is state-owned (the dam was a government project, built on government land), but it not protected in any way. There is no conservation management or action on-going at this site. However, the EAG does undertake some informal monitoring of the waterbirds. Hunting takes place within the IBA, posing a threat to the whistling-ducks and other waterbirds. Also, villa construction and landscaping associated with a bar/restaurant has impacted the preferred roosting area of *D. arborea*. Bethesda village is close to the eastern side of the reservoir. Villagers do frequent the IBA, often resulting in disturbance to the birds.

ARUBA

LAND AREA 193 km² ALTITUDE 0–188 m
HUMAN POPULATION 103,500 CAPITAL Oranjestad
IMPORTANT BIRD AREAS 4, totalling 6.1 km²
IMPORTANT BIRD AREA PROTECTION 9%
BIRD SPECIES 207
THREATENED BIRDS 1 RESTRICTED-RANGE BIRDS 0 BIOME-RESTRICTED BIRDS 1

ADRIAN DEL NEVO
(APPLIED ECOLOGICAL SOLUTIONS INC.)

The islands within St Nicolas Bay are globally important for their seabird colonies. (PHOTO: ADRIAN DEL NEVO)

INTRODUCTION

Aruba is a country within the Kingdom of the Netherlands[1], and one of the three Netherlands Antilles islands (Aruba, Bonaire and Curaçao) that lie off the north-west coast of Venezuela. Aruba is the westernmost of the three islands (c.65 km west of Curaçao) and the island nearest to mainland Venezuela. It lies just 27 km north of the Paraguaná peninsula in Falcón state, and (unlike Bonaire and Curaçao) its geological origins are as a former part of South America. The island is 33 km long and 8–9 km wide. It is a generally flat, barren and riverless island although the interior features some rolling (volcanic) hills including Hooiberg (165 m) and Mount Jamanota, the highest on the island at 188 m. The leeward southern and western coasts (sheltered from ocean currents) are shallow, with long sandy beaches (the main attraction for ever growing numbers of tourists), sheltered bays and lagoons and offshore islets. Aruba's climate is warm (moderated by the constant trade winds from the Atlantic) and dry, with an average annual precipitation of c.500 mm falling primarily between October and January. The resultant vegetation is sparse and dominated by cacti with some thorn scrub. The resident human population is greatly supplemented by a large and expanding tourist industry.

Conservation

The main provision for wildlife conservation in Aruba is the 3,400-ha Parke Nacional Arikok (administered by the Fundacion Parke Nacional Arikok) which covers Mount Jamanota and significant areas of cactus scrub. A number of other "protected areas" such as the Het Spaans Lagoen (Spanish Lagoon) Ramsar site exist, but none of these have qualified as Important Bird Areas in the current analysis. With no single government ministry responsible for environmental policy and implementation, wildlife conservation issues are not dealt with in a coordinated, efficient way. An island-wide Coastal Zone Management Plan is currently being developed and aims to identify areas of importance and those requiring protection and conservation. It will hopefully also address some of the administrative and practical issues associated with nature conservation within Aruba. The formal protection afforded the Important Bird Areas is difficult to assess. Bubali

1 At some point in the near future the "Netherlands Antilles" will be dissolved. St Maarten and Curaçao will become separate countries within the Kingdom of the Netherlands (similar to the status currently enjoyed by Aruba). The islands of Bonaire, Saba and St Eustatius will be linked directly to the Netherlands as overseas territories.

Bubali Wetlands IBA is a well recognised "nature reserve". (PHOTO: ADRIAN DEL NEVO)

Wetlands IBA (AW001), Oranjestad Reef Islands IBA (AW003), and San Nicolas Bay Reef Islands IBA (AW004) are all state owned but their formal protection status is unknown. However, they do enjoy informal, de facto or non-regulatory protection to varying degrees.

A number of different non-governmental organisations (NGOs) have been established to address different environmental concerns and these have focused upon marine turtles, cetaceans, endemic and introduced snakes, and seabirds. The projects these institutions are implementing are helping to address the critical need for an island-wide inventory of both terrestrial and marine resources to feed into a prioritised and coordinated conservation plan for Aruba. For example, the Aruba Tern Project (ATP) was started in 1999 and aims to provide a quantitative description of the population ecology of the island's seabirds. The ATP has also documented the status and distribution of other bird families (such as shorebirds and waterfowl) on Aruba. In collaboration with the government of Aruba the ATP also conducts field surveys (point counts) and bird banding to assess the status, distribution, and habitat condition of terrestrial (thorn scrub) bird species. An island-wide bird monitoring plan is currently being developed by the ATP in collaboration with the Dutch Caribbean Nature Alliance.

The main threat faced by Aruba's biodiversity is increased and seemingly unchecked urbanisation which is leading to the loss of the island's unique desert habitat, and is also resulting in the expansion of unplanned beach houses around the coast. In particular, the human encroachment and development along the coast, mostly associated with tourism and recreational activities, pose a serious threat to nesting seabirds within San Nicolas Bay and on the Oranjestad Reef Islands. Disturbance associated with recreational activities (e.g. off-road vehicles) pose a threat to fragile habitat, particularly the already limited areas of sand-dunes. Unauthorised visits to seabird nesting islands are also a serious threat.

■ Birds

Aruba's 207 recorded bird species comprise up to 70 resident (current or former) breeding species and 164 migrants. Most of the migrants (some of which winter on the island) are Neotropical migrants from breeding grounds in North America, although many are vagrants (recorded on average less than once a year). A smaller number of species are of South American origin, representing either dispersing individuals or austral migrants overshooting their northern South American wintering grounds having originated from breeding grounds further south or west. However, the island is seriously under-recorded in terms of its bird life. Three species (Collared Plover *Charadrius collaris*, American Oystercatcher *Haematopus palliatus* and Black Noddy *Anous minutus*) are new or recently confirmed breeders, demonstrating how much there is still to discover on the island.

Aruba supports just one Northern South America biome-restricted bird, namely Bare-eyed Pigeon *Patagioenas corensis*, and no restricted-range species. However, two subspecies are endemic to the island—Burrowing Owl *Athene cunicularia arubensis* and Brown-throated Parakeet *Aratinga pertinax arubensis*—both of which warrant further taxonomic research. Whereas the terns (*Sterna* spp.) have shown a dramatic increase in their breeding populations over the last 50 years (see below), some parts of the island are experiencing a decline in some bird species such as Yellow Oriole *Icterus nigrogularis*, Burrowing Owl *A. cunicularia*, and Crested Bobwhite *Colinus cristatus*. It would appear that Scaly-naped Pigeon *Columba squamosa* has been extirpated and the White-tailed Hawk *Buteo albicaudatus* probably so.

Aruba's endemic subspecies of Brown-throated Parakeet. (PHOTO: JEFF WELLS)

BAHAMAS

LAND AREA **13,940 km²** ALTITUDE **0–63 m**
HUMAN POPULATION **330,550** CAPITAL **Nassau**
IMPORTANT BIRD AREAS **39, totalling 4,700 km²**
IMPORTANT BIRD AREA PROTECTION **23%**
BIRD SPECIES **300**
THREATENED BIRDS **6** RESTRICTED-RANGE BIRDS **7**

PREDENSA MOORE AND LYNN GAPE
(BAHAMAS NATIONAL TRUST)

Brown Noddy nesting in Graham's Harbour IBA, San Salvador: the Bahamas islands support significant populations of many seabird species. (PHOTO: WILLIAM HAYES)

INTRODUCTION

The Commonwealth of The Bahamas is an archipelago of c.700 islands and c.2,000 cays and rocks extending over 1,100 km. The archipelago, which lies north and east of Cuba, runs from east of the southern end of Florida (USA), south-east until it terminates at the Turks and Caicos Islands (to the UK) which are geologically a continuation of the islands. The Bahamas are exposed parts of a limestone platform that is divided into several shallow banks. Little Bahama Bank is located along the northern coasts of Grand Bahama and encompasses all of Abaco and its North Atlantic offshore rocks and cays. The Great Bahama Bank (which is rich in marine life) stretches from north of the Biminis and Berry Islands, southward to hug the southern shoreline of New Providence and the western shores of Andros, Eleuthera, Cat Island, the Exumas, Long Island and the Ragged Islands. The Cay Sal Bank (which is biologically impoverished) is located at the extreme western sea border of The Bahamas, very close to Cuba. The islands of the Bahamas are low and flat with ridges that usually rise to no more than 15–20 m. However, there are precipitous slopes under water, between and within the convoluted banks. The Tongue of the Ocean is a 30-km wide trench between New Providence and Andros which drops to depths of 2,000 m. The islands have no rivers or streams and the soil is fertile but thin, and often lodged in shallows and "banana holes" within the harsh limestone rock. A freshwater lens exists close to the surface, resting on the underlying salt-water.

The Bahamas are often divided, ecologically, into three regions: Northern Bahamas (Grand Bahama, Biminis, Berry Islands, Abacos, North Andros, and New Providence) where all the larger islands are covered primarily by Caribbean pine *Pinus caribaea* woodland (with a broadleaf shrub and palm understorey), although much of this woodland was logged in the mid-twentieth century; Central Bahamas (South Andros, Eleuthera, Cat Island, the Exumas, Ragged Islands, Long Island, Rum Cay, Conception Island and San Salvador), in which the islands are covered primarily in broadleaf "coppice"—a dense, low semi-evergreen forest; and Southern Bahamas (Crooked Island, Acklins Island, Samana Cay, Mayaguana, Little and Great Inagua), where the islands are drier and support dry shrubland. New Providence, in spite of being one of the smaller islands, is home to c.69% of the Bahamian population and the nation's capital. Grand Bahama is second only to New Providence in terms of development, and it supports 16% of the population. It is also home to the longest underwater cave system in the world. The rest of the

The Exuma Cays Land and Sea Park IBA in the northern Exumas, Central Bahamas. (PHOTO: OLGA STOKES)

Bahamas islands are called the "Family Islands" which are sparsely populated and retain their natural beauty. Of these Family Islands, Great and Little Abaco (and its cays) are considered "the sailing capital of the world", and the islands have a booming tourist trade. Andros is the largest island in the Bahamas, with extensive creeks, interlacing channels, bays, bights and inlets. It is also home to many blue holes and as a result is renowned for its cave-diving. Inagua is the southernmost island in the Bahamas with the nation's only Ramsar site—Inagua National Park—which is home to over 40,000 Caribbean Flamingo *Phoenicopterus ruber* (and many other waterbirds). The company Morton Bahamas Ltd. produces salt from the salinas at one end of Lake Rosa (which occupies c.30% of the island). Morton is one of the largest salt producers in North America.

The Bahamas has the third highest per capita income in the western hemisphere (after the USA and Canada). Tourism is the primary economic activity, accounting for c.65% of the gross domestic product (GDP). The government's current economic thrust is to put an anchor resort on each of the major Family Islands which will have huge implications for the biodiversity of these otherwise relatively untouched islands. Offshore finance is the nation's second largest industry, accounting for c.15% of GDP. The settlement history of the Bahamas is convoluted and often different on each island. Plantations were established on some of the islands during the late eighteenth century, and large-scale agriculture was trialed in the mid-twentieth century when much of remaining virgin pine forests in the Northern Bahamas were logged. Subsequent development (especially on New Providence and Grand Bahama, but also locally on the other inhabited islands) has had a profound negative impact on the surrounding habitats.

The climate of the Bahamas is subtropical to tropical, and is moderated significantly by the waters of the Gulf Stream which keeps the islands warmer than Florida in the winter and cooler in the summer. Summer is the rainy season with June and October the wettest months. However, the Southern Bahamas only get half the rainfall that the northern Bahamas receive. The islands are frequently hit by hurricanes; for example, Hurricane Andrew in 1992, Floyd in 1999, Francis and Jeanne in 2004, and Wilma in 2005. Low-pressure systems associated with tropical waves and resulting in strong winds and drenching rain are a regular feature in the Bahamas.

■ Conservation

In the Bahamas, the Ministry of Environment is currently the principal government department involved in conservation and the environment. Within this ministry is the Bahamas Environment Science and Technology Commission, also known as the BEST Commission, which was established in 1994. The BEST Commission manages the implementation of multilateral environmental agreements and reviews environmental impact assessments and environmental management plans for development projects within the Bahamas. The Bahamas National Trust (BNT)[1] was established in 1959 under the Bahamas National Trust Act. It is a non-profit organisation, funded by private donations, an endowment fund and a significant subvention from the Government of the Bahamas. BNT advises the government on conservation policies and is charged with safeguarding the nation's environmental heritage. One of its statutory roles is to hold environmentally important lands in trust for the country. BNT also has the responsibility for managing the national park system. The park system now consists of 25 parks and protected areas (10 parks were designated in 2002), covering 283,400 ha throughout the archipelago. Many of these extraordinary and often innovatively managed parks are also IBAs and are mentioned in more detail within the individual IBA profiles below. BNT works in partnership with the Bahamas government, local business, national and international conservation organisations, schools and the community.

In the Bahamas, there is a constant quest for economic advancement, but without the necessary knowledge and appreciation that the nation's environment has limitations, this could have catastrophic long-term consequences. In the past, valuable timber (pine and coppice) were cut, monoculture agriculture was practiced, and introduced livestock (goats) and slash-and-burn agriculture expanded to less arable areas. At the same time, subsistence, commercial and recreational hunting and fishing, introduction of alien species, urban sprawl, road works, careless tapping of the freshwater lens, interference with natural drainage, dredging and reclamation of wetlands and tidal mangroves, pesticide spraying to eradicate mosquitoes, malaria, yellow fever, crop pests, problems of sewage and solid waste disposal and many other human intrusions have all taken a huge toll on local biodiversity, and thus threaten the essence of the nation's valuable tourism product.

In order to promote appropriate development for the Family Islands (which have previously been little impacted by development), there is an urgent need for a national land management or development plan. This would help identify sensitive areas (such as the IBAs) which should be subject to limited exploitation and/or should be placed in the protected area system. As an island archipelago, the Bahamas needs to be particularly sensitive to the tourism carrying capacity, water resource use and wetland destruction. Strategic planning for

1 The Bahamas National Trust (BirdLife in the Bahamas) is referred to throughout this chapter by the acronym BNT.

the marina needs for the entire archipelago could effectively limit destruction of mangrove wetlands and tidal creeks. However, for such planning to be adopted there needs to be a clear appreciation and understanding of the need to limit or mitigate the effects of development on the biodiversity of the islands.

Lack of environmental legislation and, more importantly, the lack of enforcement of environmental legislation continue to be an obstacle for conservation in the Bahamas. The very nature of the archipelagic nation creates enforcement problems compounded by insufficient human resources in both the Royal Bahamas Police Force and the Royal Bahamas Defense Force. Draft enabling legislation for the environment has recently been developed by the BEST Commission, and includes Environmental Impact Assessment Final Draft Regulations; Pollution Control and Waste Management Final Draft Regulations; Draft National Environmental Policy; and Environmental Management Final Draft Legislation. Enactment of such legislation will provide the basic framework for the coherent management of the nation's unique environment.

All conservation partners in the Bahamas agree that a stronger environmental ethic needs to be established. This can only be accomplished through a major public outreach campaign targeting both school-age and adult citizens as well. In particular, decision-makers need to be made aware of our environmental responsibilities so that collectively the threats outlined below can be addressed. Government agencies and the BNT are faced with a paucity of trained environmental staff. Many of those that are trained seek employment in unrelated but higher salaried professions in the financial or legal sectors. Even in-country field research capacity is minimal but vital to inform regulations for marine and terrestrial natural resource management. However, there is growing awareness that visiting researchers and international projects have a responsibility to help with this training and capacity issue. The Kirtland's Warbler Training and Research Program, a collaboration between BNT, U.S. Forest Service, The Nature Conservancy and the College of the Bahamas has been exemplary in providing opportunities for Bahamian students to gain expert field and academic training.

Habitat destruction and degradation caused by human population growth and extensive changes in land use practices is impacting on the birdlife and other biodiversity. Local species extinctions are happening, e.g. the Great Lizard-cuckoo *Saurothera merlini* has been extirpated from New Providence over the last 10 years. While the habitat loss that leads to such extinctions is best addressed through improved planning, legislation, protection and enforcement, the BNT is working to engage local communities in the protection of critical areas. For example, local Site Support Groups in Abaco, New Providence and Inagua are working with the BNT to develop native tree nurseries and to re-plant areas with native vegetation. BNT is also working with local nurseries to promote the propagation of native trees and vegetation by these private-sector businesses. In the Bahamas, it is common practice to treat the wetlands as wastelands to be filled in to provide more land or to be dredged for canals and marinas. The work of the BNT through the West Indian Whistling-duck and Wetlands Conservation Program (a program of the Society for the Conservation and Study of Caribbean Birds) has gone some way to raising awareness of the critical importance of wetlands for biodiversity, as nursery grounds for economically important fisheries, and for coastal zone protection (including flood and hurricane damage mitigation). BNT has recently partnered with RARE Conservation to implement a Pride Campaign, a social marketing campaign to educate Bahamians about the value of wetlands and change the perception of them as "wastelands" or dumping grounds. The site focus for this Pride Campaign is Harrold and Wilsons Ponds National Park, one of the IBAs described below.

Biodiversity in the Bahamas is facing a constant threat by introduced or invasive species, both plants (e.g. Brazilian pepper *Schinus terebinthefolius* and casuarina *Casuarina equisetifolia*) and animals (e.g. feral cats *Felis catus*, raccoons *Procyon lotor* and wild hogs *Sus scrofa*) alike. The historic and cultural practice of using small islands as natural corrals for goats has impacted the vegetation on many remote cays. The BNT is working with a Site Support Group to manage invasive plants at Harrold and Wilsons Ponds National Park, and with Friends of the Environments (another Site Support Group) in Abaco to manage the feral cat population. In the last two years, feral cats predated 50% of the "Bahama Parrot" *Amazona leucocephala bahamensis* nests on the island. Conservation agencies in conjunction with the BEST Commission have adopted, and are promoting, a National Invasive Species Policy.

■ Birds

Over 300 species of bird have been recorded from the Bahamas, 109 of which breed on the islands, 169 are migrants that pass through the islands or winter, and 45 are vagrants that have occurred only a few times each. Only three breeding landbirds are summer visitors: Antillean Nighthawk *Chordeiles gundlachii*, Grey Kingbird *Tyrannus dominicensis* and Black-whiskered Vireo *Vireo altiloquus*. However, many of the seabirds are only present during their spring and summer breeding seasons. Neotropical migrants (that breed in North America) comprise c.50% of the total land bird population in the northern islands from November through March. The number and diversity of migrants declines from north to south through the islands. Bahamas Endemic Bird Area (EBA) restricted-range birds total seven extant species (see Table 1).

Infrastructure for visitation (and education and awareness) is being put in place by BNT in a number of IBAs such as this viewing platform and boardwalk in the Blue Holes National Park, Central Andros. (PHOTO: SHELLEY CANT)

The "Bahama Parrot" has been the focus of research and conservation project actions such as the management of the predatory feral cat population on Abaco. (PHOTO: HENRY NIXON)

Table 1. Key bird species at Important Bird Areas in the Bahamas.

Key bird species	Criteria	National population		BS001	BS002	BS003	BS004	BS005	BS006	
			Criteria	■ (VU)		■ (NT)	■ (NT)		■ (NT)	
				■ (RR)			■ (RR)		■ (RR)	
					■ (CB)	■ (CB)		■ (CB)	■ (CB)	
West Indian Whistling-duck *Dendrocygna arborea*	VU ■ ■									
Audubon's Shearwater *Puffinus lherminieri*	■							315		
Caribbean Flamingo *Phoenicopterus ruber*	■									
Reddish Egret *Egretta rufescens*	■									
White-tailed Tropicbird *Phaethon lepturus*	■							258	50–249	
Magnificent Frigatebird *Fregata magnificens*	■							50–249		
Brown Pelican *Pelecanus occidentalis*	■							<50		
Masked Booby *Sula dactylatra*	■									
Brown Booby *Sula leucogaster*	■									
Piping Plover *Charadrius melodus*	NT ■ ■					70				
Laughing Gull *Larus atricilla*	■							1,923		25
Gull-billed Tern *Sterna nilotica*	■									
Royal Tern *Sterna maxima*	■									5
Sandwich Tern *Sterna sandvicensis*	■									
Roseate Tern *Sterna dougallii*	■							990		
Common Tern *Sterna hirundo*	■									
Least Tern *Sterna antillarum*	■							654		
Bridled Tern *Sterna anaethetus*	■				480			10,665		
Sooty Tern *Sterna fuscata*	■							10,665		
Brown Noddy *Anous stolidus*	■							1,281		
White-crowned Pigeon *Patagioenas leucocephala*	NT ■						250–999			25
Cuban Amazon *Amazona leucocephala*	NT ■								3, 600	
Bahama Woodstar *Calliphlox evelynae*	■						<50		50–249	
Thick-billed Vireo *Vireo crassirostris*	■			<50			<50			5
Bahama Swallow *Tachycineta cyaneoviridis*	VU ■ ■			50–249						5
Bahama Mockingbird *Mimus gundlachii*	■								<50	5
Pearly-eyed Thrasher *Margarops fuscatus*	■									
Olive-capped Warbler *Dendroica pityophila*	■			50–249			<50		✓	
Kirtland's Warbler *Dendroica kirtlandii*	NT ■								<50	
Bahama Yellowthroat *Geothlypis rostrata*	■						<50		50–249	

Key bird species	Criteria	National population		BS021	BS022	BS023	BS024	BS025	BS026	I
			Criteria	■ (VU)	■ (VU)	■ (NT)	■ (VU)	■ (NT)	■ (VU)	
							■ (RR)			
				■ (CB)	■ (CB)	■ (CB)	■ (CB)	■ (CB)	■ (CB)	
West Indian Whistling-duck *Dendrocygna arborea*	VU ■ ■			<50	50–249		50–249		250–999	
Audubon's Shearwater *Puffinus lherminieri*	■							750–2,997		5
Caribbean Flamingo *Phoenicopterus ruber*	■									
Reddish Egret *Egretta rufescens*	■									
White-tailed Tropicbird *Phaethon lepturus*	■									750–
Magnificent Frigatebird *Fregata magnificens*	■									
Brown Pelican *Pelecanus occidentalis*	■							<50		
Masked Booby *Sula dactylatra*	■									
Brown Booby *Sula leucogaster*	■									
Piping Plover *Charadrius melodus*	NT ■ ■									
Laughing Gull *Larus atricilla*	■				50–249		250–999		50–249	
Gull-billed Tern *Sterna nilotica*	■				50–249		50–249			
Royal Tern *Sterna maxima*	■							50–249		
Sandwich Tern *Sterna sandvicensis*	■							50–249		
Roseate Tern *Sterna dougallii*	■					50–249		50–249		
Common Tern *Sterna hirundo*	■									
Least Tern *Sterna antillarum*	■				50–249		50–249			
Bridled Tern *Sterna anaethetus*	■					50–249		250–999		5
Sooty Tern *Sterna fuscata*	■							7,500–29,997		
Brown Noddy *Anous stolidus*	■									25
White-crowned Pigeon *Patagioenas leucocephala*	NT ■			<50	50–249	250–999		50–249		
Cuban Amazon *Amazona leucocephala*	NT ■									
Bahama Woodstar *Calliphlox evelynae*	■						✓			
Thick-billed Vireo *Vireo crassirostris*	■						50–249			
Bahama Swallow *Tachycineta cyaneoviridis*	VU ■ ■									
Bahama Mockingbird *Mimus gundlachii*	■						50–249			
Pearly-eyed Thrasher *Margarops fuscatus*	■									
Olive-capped Warbler *Dendroica pityophila*	■									
Kirtland's Warbler *Dendroica kirtlandii*	NT ■									
Bahama Yellowthroat *Geothlypis rostrata*	■						<50			

All population figures = numbers of individuals.
Threatened birds: Vulnerable ■; Near Threatened ■. **Restricted-range birds** ■. **Congregatory birds** ■.

		Bahamas IBAs										
008	BS009	BS010	BS011	BS012	BS013	BS014	BS015	BS016	BS017	BS018	BS019	BS020
■	■		■	■		■	■			■		
■	■	■	■	■		■				■		
	■			■	■	■	■	■	■		■	■
<50												
											150–747	250–999
												750–2,997
	85			38			<50					
						250–999	50–249	50–249	50–249			
						<50						
						<50	<50					
									50–249			
									50–249			50–249
					50–249				50–249			
					250–999							
			50–249	250–999		<50				250–999		
✓	<50	<50	<50	<50		<50				<50		
			<50	<50		<50				250–999		
-249	<50					50–249						
	<50	50–249	<50	<50						50–249		
		<50										
										60		
<50	<50	<50	<50	<50						✓		

028	BS029	BS030	BS031	BS032	BS033	BS034	BS035	BS036	BS037	BS038	BS039
										■	■
	■	■									
■	■	■	■	■	■	■	■	■	■	■	■
											50–249
279			150–747				150–747				
											20,000–49,999
											250–999
60										50–249	
						297		50–249			
										<50	250–999
										<50	
150		50–249				1,650	150–747		50–249	250–999	
	50–249										50–248
	15										50–249
30				<50						<50	50–249
								750–2,997			50–249
					150–747			50–249			50–249
											50–249
		150–747	150–747		150–747			50–249			
			50–249	150–747	50–249			50–249			
900											
										50–249	50–249
											2,500–9,999
	✓	✓									
	✓	<50									
	✓	<50									
	✓	<50									

The Bahama Yellowthroat is endemic to the Bahamas while the Bahama Woodstar occurs also in the Turks and Caicos Islands.
(PHOTOS: ANTHONY HEPBURN)

The Bahamas EBA includes the Turks and Caicos Islands (to the UK) with which the Bahamas share four of the restricted-range birds, namely Bahama Woodstar *Calliphlox evelynae*, Bahama Mockingbird *Mimus gundlachii*, Pearly-eyed Thrasher *Margarops fuscatus* and Thick-billed Vireo *Vireo crassirostris*. Of the remainder, Olive-capped Warbler *Dendroica pityophila* occurs also in Cuba, but Bahama Yellowthroat *Geothlypis rostrata* and Bahama Swallow *Tachycineta cyaneoviridis* are endemic to the islands. The yellowthroat is common on Grand Bahama and Abaco, less common on Andros and Cat Island, uncommon on New Providence and non-existent on the other islands. The swallow is locally common and breeds on Grand Bahama, Abaco and Andros, less common on New Providence, and uncommon to non-existent in the central and southern Bahama Islands. An eighth restricted-range bird (and third national endemic) was the Brace's Emerald *Cholorostilbon bracei* which is now extinct. It was known only from a single specimen collected in 1877. A subspecies of the Greater Antillean Oriole, *Icterus dominicensis northropi* is found only on Andros (where it is threatened), having been extirpated from Abaco.

Globally threatened birds in the Bahamas include the Vulnerable West Indian Whistling-duck *Dendrocygna arborea* and *Tachycineta cyaneoviridis*, and the Near Threatened White-crowned Pigeon *Patagioenas leucocephala*, Cuban Amazon *Amazona leucocephala bahamensis*, Piping Plover *Charadrius melodus* and Kirtland's Warbler *Dendroica kirtlandii*. *Dendrocygna arborea* only occurs on Andros, Inagua, Cat Island, Long Island and Exuma where significant numbers occur in a few areas (such as Hog Cay off Long Island). The species is protected by law under the Bahamas Wild Birds (Protection) Act. *Tachycineta cyaneoviridis* relies on pine forests for breeding, but the movements of the species outside the breeding season are poorly known although it appears that significant numbers over-winter in the country. *Patagioenas leucocephala* is a target for recreational hunting, but poaching and excessive hunting is common because although laws exist for the species' protection, enforcement is inadequate. *Charadrius melodus* is an uncommon winter resident in the Bahamas although some specific beaches and tidal flat areas (which need to be designated as protected areas) do support significant numbers. Eleuthera supports the largest population of wintering *Dendroica kirtlandii* currently known, and is the focus of a multi-institutional initiative, the Kirtland's Warbler Training and Research Program.

Over 14 species of seabirds breed in the Bahamas, but their preferred habitats of isolated cays with steep cliffs or rocky shorelines, and with low vegetation near to deep water, are being lost due to increased human uses of coastal areas through resort developments, disturbance, and increased pollution of near-shore waters. Seabird eggs (and adults) are also collected. Recent (2002–2006) surveys in the Northern Bahamas identified over 60 seabird breeding locations in Grand Bahama, Biminis, Berry Islands and Abacos showing just how important these northern islands are for their seabird populations.

The Bahama islands are of great importance to wetland birds, but their usage of individual wetland sites varies seasonally and between years depending on weather and local conditions. This suggests that a network of protected wetland sites is critical to the long-term viability of the nation's waterbird populations. Large numbers of migratory shorebirds use these wetlands as stop-over sites and as wintering grounds, as do ducks and significant numbers of resident egrets and herons and other species. However, these waterbirds face many threats including draining and infilling of wetlands, contamination of food supplies, oil spills, introduced mammalian predators, disturbance, and hunting. However, conservation efforts can have a profound impact. In 1905, the National Audubon Society (BirdLife in the US) requested the Government of the Bahamas to provide legal protection for the Caribbean Flamingo *Phoenicopterus ruber*. The government responded by passing the Wild Birds (Protection) Act. An initial attempt to save the flamingo breeding colonies on Andros failed in the 1950s, but a research program was established and a colony was discovered on Great Inagua. A 99-year lease was agreed, the Inagua National Park was established, and the flamingo colony (over the next 40 years) increased from less than 10,000 birds to over 40,000. Conservation of birdlife in the Bahamas has been concentrated on a few high-profile species such as the Caribbean Flamingo *Phoenicopterus ruber*, West Indian Whistling-duck *Dendrocygna arborea*, "Bahama Parrot" *Amazona leucocephala bahamensis*, White-crowned Pigeon *Patagioenas*

Over 40,000 Caribbean Flamingos breed in Great Inagua as a result of successful, long-term conservation action on the island. (PHOTO: OLGA STOKES)

leucocephala and Kirtland's Warbler *Dendroica kirtlandii*. However, more attention is now being paid to critical sites (such as IBAs) and habitats (such as the dry forests) as well as the species themselves.

IMPORTANT BIRD AREAS

The Bahamas' 39 IBAs—the nation's international site priorities for bird conservation—cover 4,700 km² (including extensive marine areas). The IBAs include nine of the BNT-managed national parks and protected areas. However, just two IBAs are protected in their entirety. Seven are part protected, part unprotected, while for 30 of the IBAs there is currently no legal protection.

The IBAs have been identified on the basis of 30 key bird species (listed in Table 1) that variously trigger the IBA criteria. These 30 species include six globally threatened birds (two Vulnerable and four Near Threatened), all seven restricted-range species, and 20 congregatory waterbirds/seabirds.

Significant populations of the Bahamas' key bird species are found in two or more IBAs. Also, as the IBAs are almost evenly split between the Northern, Central and Southern Bahamas, there is good geographic representation for most species (where this is possible) throughout the archipelago. For shear numbers, both the North Atlantic Abaco Cays IBA (BS005) and Cay Sal IBA (BS025) stand out as supporting the largest numbers of seabirds, while Great Inagua IBA (BS039) is home to the largest congregation of waterbirds.

Great Inagua IBA supports huge numbers of waterbirds. (PHOTO: LYNN GAPE)

Figure 1. Location of Important Bird Areas in the Bahamas.

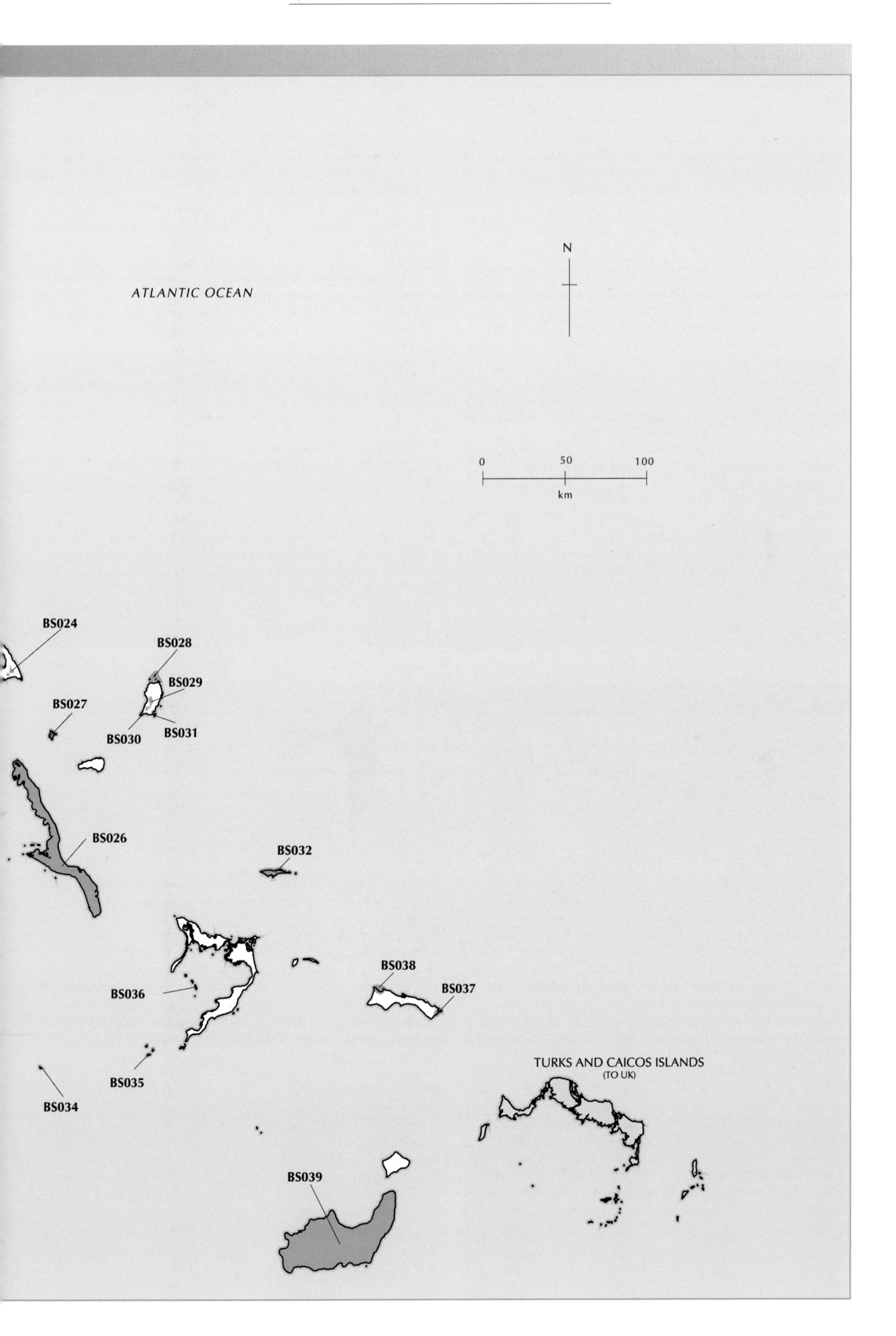
ATLANTIC OCEAN
N
0
50
100
km
BS024
BS028
BS029
BS027
BS030
BS031
BS026
BS032
BS038
BS037
BS036
BS035
BS034
TURKS AND CAICOS ISLANDS
(TO UK)
BS039

Monitoring currently being undertaken by local Site Support Groups, and also on some of the high profile species should be used to feed into the annual assessment of state, pressure and response scores at each of the Bahamas' IBAs in order to provide an objective status assessment, and highlight management interventions that might be required to maintain these internationally important biodiversity sites. With over 75% of IBAs unprotected, key species monitoring and status assessments will be critical to lobby for protection and develop conservation strategies.

KEY REFERENCES

Bainton, A. M. and White, A. W. (2006) *A bibliography of birds, ornithology and birding in The Bahamas and Turks and Caicos Islands*. Nassau, Bahamas: Media Enterprises Ltd.

Bendon, J. (1997) Moon over Mayaguana: return to Booby Cay. *Iguana Times* 6(4): 81–88.

Etheridge, H. (2001) *Yachtman's guide to the Bahamas (including Turks and Caicos)*. Coral Gables: Tropic Island Publishers.

Hallett, B. (2006) *Birds of The Bahamas and the Turks and Caicos Islands*. Oxford, U.K.: Macmillan Caribbean.

Kushlan, J. A. (2006) Seabird nesting and conservation in the northern Bahamas. (Unpublished report for the Bahamas National Trust).

Raffaele, H. Wiley J., Garrido, O., Keith, A. and Raffaele, J. (1998) *A guide to the birds of the West Indies*. Princeton, New Jersey: Princeton University Press.

Sealey, N. E. (1994) *Bahamian landscapes: an introduction to the geography of the Bahamas*. Nassau, Bahamas: Media Publishing Ltd.

Sprunt, A. (1984) The status and conservation of seabirds of the Bahama Islands. Pp157–168 in J. P. Croxall, P. G. H. Evans and R. W. Schreiber eds. *Status and conservation of the world's seabirds*. Cambridge, U.K.: International Council for Bird Preservation (ICBP Techn. Publ. 2).

Sutton, A. H., Sorenson, L. G. and Keeley, M. A. (2001) *Wondrous West Indian wetlands: teachers' resource book*. Boston, Mass.: West Indian Whistling-Duck Working Group of the Society of Caribbean Ornithology.

Wardle, C. and Moore, P. (2006) Important Bird Area program field trip to Mangrove Cay and South Andros. Nassau: Bahamas National Trust. (Unpublished field trip report).

Wardle, C. and Moore, P. (2007) Important Bird Area program field trip to Cat Island. Nassau: Bahamas National Trust. (Unpublished field trip report).

Wardle, C. and Moore, P. (2007) Important Bird Area program field trip to South Andros. Nassau: Bahamas National Trust. (Unpublished field trip report).

White, A. W. (1998) *A birder's guide to the Bahama Islands (including Turks and Caicos)*. Colorado Springs, Colorado: American Birding Association.

ACKNOWLEDGEMENTS

The authors would like to thank the BNT Ornithology Group, The Sam Nixon Bird Club (Inagua), Carolyn Wardle, Henry Nixon, Anthony White, Herb Rafaelle, Bruce Hallett, Rosemarie Gnam, William Mackin, Lisa Sorenson, Jim Kushlan, William Hayes, David Ewert, Joe Wunderle, Kim Thurlow, The Nature Conservancy Bahamas Programme, Morton Bahamas Ltd., Friends of the Environment, Anita Knowles, David Knowles, Caroline Stahala and Frank Rivera for their invaluable contributions to bird conservation in the Bahamas, and thus to the development of the Important Bird Area program.

BS001 Lucayan National Park

■ Site description

Lucayan National Park IBA encompasses a section of south-central Grand Bahama including the tidal Gold Rock Creek and adjacent beach. The IBA supports a wide diversity of habitats including a tall dune system, mixed scrub, wet coppice, pine forest, mangrove swamp and beach. Within the park, Ben's Cave and Burial Ground Cave are entrances to one of the longest underwater cave system in the world. Explorers have found pre-Columbian human skeletons and artefacts in Burial Mound Cave. The Grand Bahama South Shore IBA (BS003) adjoins the park to the west.

■ Birds

This IBA is significant for supporting three (of the 7) Bahamas EBA restricted-range birds, namely Thick-billed Vireo *Vireo crassirostris*, Bahama Swallow *Tachycineta cyaneoviridis* and Olive-capped Warbler *Dendroica pityophila*. *Tachycineta cyaneoviridis* is Vulnerable, and is regularly seen in the Lucayan National Park during the breeding season. The key bird species are all confined to the coppice and pine forest north of the east–west Queen's Highway. Waterbirds frequent the mangrove swamps, and shorebirds and terns occur along the beach.

■ Other biodiversity

The recently discovered Lucayan oar-foot "shrimp" *Spelionectes lucayensis* is endemic to the caves in this IBA. The Bahamas blind cave fish *Lucifuga (Stygicola) spelaeotes* (a Bahamian endemic) occurs. Buffy flower bat *Erophylla sezekorni* occurs in Ben's Cave during the summer. Two endemic orchids *Encyclia fucata* and *Cattleyopsis lendenii* flourish in the park.

■ Conservation

Lucayan National Park IBA is managed by BNT, and there is a boardwalk through the mangroves at Gold Rock Creek. Speculative proposals to develop a resort in eastern Grand Bahama could impact on the borders of this IBA, and developments are occurring all the time outside of the national park. Natural forest fires within the pine forest are a threat that needs management.

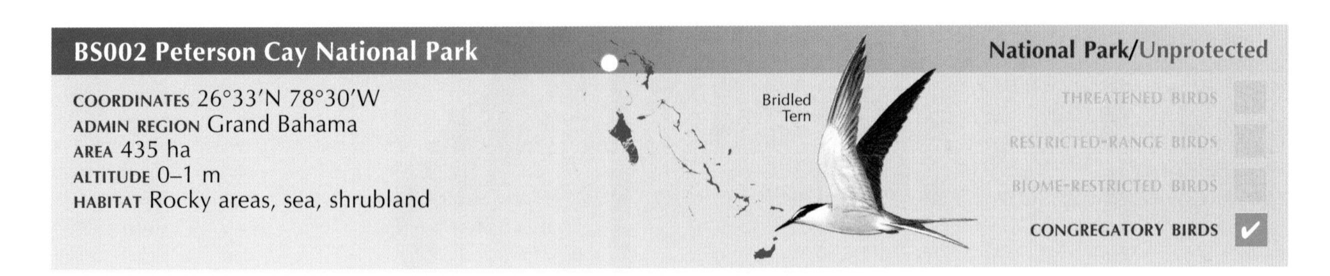

■ Site description

Peterson Cay National Park IBA lies c.2 km offshore on the south (leeward) side of Grand Bahama, c.2 km east of the entrance to the Grand Lucayan Waterway. It is a windswept and sparsely vegetated limestone island, and the only cay on the south side of Grand Bahama. The cay has a rocky shoreline with a sandy beach on the north side, and shrubland on the top of the cay. Shallow sand bars and coral reefs extend to the west of the cay. The IBA includes all marine areas up to 1 km from the cay.

■ Birds

This IBA supports a globally significant nesting colony of Bridled Tern *Sterna anaethetus*, with 160 pairs found in 2005.

■ Other biodiversity

No globally threatened or restricted-range species have been recorded.

■ Conservation

Peterson Cay National Park IBA is crown owned, and managed as a national park (the smallest in the Bahamas) by BNT. Marine areas up to 500 m from the cay are protected as part of the national park, leaving some of the marine areas of the IBA unprotected. The cay is uninhabited, but is actively used for ecotourism by resident kayak tour guides and resident and visiting boaters. There is potential for uncontrolled tourism to introduce predators such as rats *Rattus* spp. (or indeed other animals) to the cay. It is not know if this has already happened. Disturbance to the tern colony is also a threat, and the extent to which this already happens is unknown.

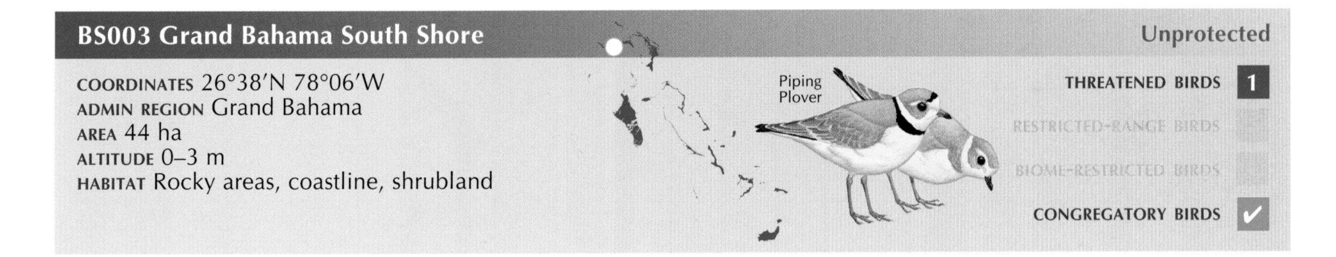

■ Site description

The Grand Bahama South Shore IBA extends along the south coast of Grand Bahama from the Grand Lucayan Waterway's south entrance eastward for c.11 km through the settlement of Ole Freetown and on to the western boundary of the Lucayan National Park (IBA BS001). It comprises a long stretch of uninterrupted sandy beach, beach flats and dunes including Barbary beach

■ Birds

This IBA is significant for the Near Threatened Piping Plover *Charadrius melodus* which winters on the beach along with a range of other shorebirds, and also herons and egrets. During the 2006 census 70 *Charadrius melodus* were recorded at this site.

■ Other biodiversity

Nothing recorded.

■ Conservation

Grand Bahamas South Shore IBA is crown land but is unprotected. It is a popular beach for recreational activity, attracting hundreds of residents and tourist alike. There are small restaurant and bar developments along the beach (outside of the IBA), but currently no large developments (as yet). The heavy recreational use of the beach causes disturbance to *C. melodus* and other wintering shorebirds. Invasive alien *Casuarina* trees threaten the stability of the beach, and the native vegetation behind the beach.

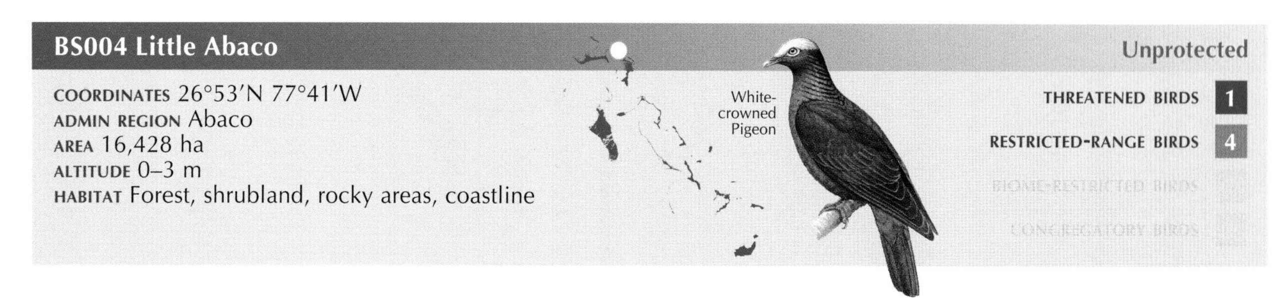

■ Site description

Little Abaco IBA is at the northernmost end of Abaco where it is just 15 km north of Grand Bahama. Little Abaco extends for about 30 km west of the northern point of Great Abaco Island (at Angel Fish Point) to which it is joined by a short causeway ("the bridge"). The island supports extensive tracts of virgin Caribbean pine *Pinus caribaea* forest, and has long stretches of sandy beach. There are five settlements: Crown Haven (at the westernmost tip), Fox Town, Wood cay, Mount Hope and Cedar Harbour.

■ Birds

The pine forests in this IBA support four (of the seven) Bahamas EBA restricted-range birds, namely Bahama Woodstar *Calliphlox evelynae*, Thick-billed Vireo *Vireo crassirostris*, Olive-capped Warbler *Dendroica pityophila* and Bahama Yellowthroat *Geothlypis rostrata*. The resident endemic race of Yellow-throated Warbler *Dendroica dominica* also occurs. A sizeable population of the Near Threatened White-crowned Pigeon *Patagioenas leucocephala* breeds. Little Abaco can be the first landfall for many Neotropical migrants in the fall.

■ Other biodiversity

No globally threatened or restricted-range species have been recorded.

■ Conservation

Little Abaco IBA is a mixture of crown and private lands, but it is currently unprotected. The pine forest in the IBA is thought to be the oldest, and only remaining virgin stand in the Bahamas. However, it is being degraded through illegal clearance, bulldozing and other human activities. The Government is currently building a trash transfer station within the pine forest, and as the human population of Abaco increases, the pressure on the forest for development and lumber will intensify.

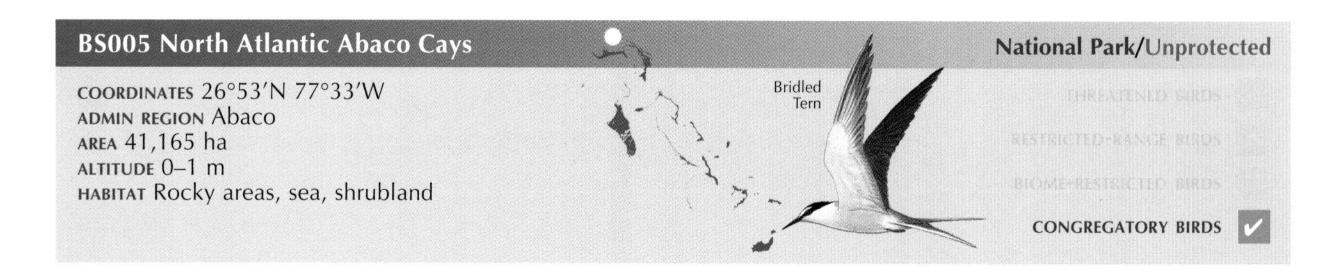

■ Site description

North Atlantic Abaco Cays IBA embraces the cays along the northern and north-eastern edge of the Little Bahama Bank. It runs from the 1,555-ha Walker's Cay National Park in the north (the northernmost point of the Bahamas), east and south-east to Scotland Cay (north of Marsh Harbour) including Pensacola, Spanish, Powell, Manjack, Green Turtle, Whale and Great Guana cays, and many isolated rocks. The vegetation on many of the cays comprises fringing mangroves and scrub. Gilliam Bay, at the south-east point of Green Turtle Cay, has extensive sand and mudflats at low tide.

■ Birds

This IBA is significant for its breeding seabirds. The breeding populations of Laughing Gull *Larus atricilla*, Roseate Tern *Sterna dougallii*, Least Tern *S. antillarum* and Bridled Tern *S. anaethetus* are globally important. Those of Audubon's Shearwater *Puffinus lherminieri*, White-tailed Tropicbird *Phaethon lepturus*, Magnificent Frigatebird *Fregata magnificens*, Brown Pelican *Pelecanus occidentalis*, Sooty Tern *S. fuscata* and Brown Noddy *Anous stolidus* are regionally so. Brown Booby *Sula leucogaster* also breeds and the flats at Gilliam Bay support many shorebirds.

■ Other biodiversity

Nothing recorded.

■ Conservation

North Atlantic Abaco Cays IBA is a mixture of private and crown ownership. Walker's Cay is protected as a national park and managed by BNT. This includes a large marine area as well as the cay. The rest of the IBA is unprotected. Many of the cays are uninhabited. Others are sparsely populated all or part of the year. Game and commercial fishing and tourism related activities are the primary occupation of the residents within the IBA. Threats include illegal egg collecting and hunting, clearance for development, pollution (from urban developments and visiting boaters), disturbance, and introduced alien predators.

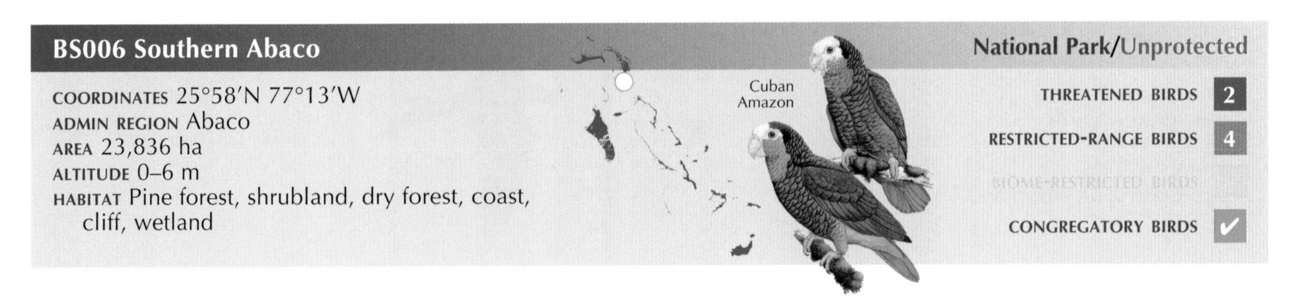

BS006 Southern Abaco

COORDINATES 25°58′N 77°13′W
ADMIN REGION Abaco
AREA 23,836 ha
ALTITUDE 0–6 m
HABITAT Pine forest, shrubland, dry forest, coast, cliff, wetland

■ Site description

Southern Abaco IBA embraces a large swathe of southern Abaco including the 8,296-ha Abaco National Park, and areas up to the east coast of southern Abaco, Hole-in-the-Wall at the southernmost tip of the island (where there are some low, coastal cliffs), and areas to the west of the park. Most of the IBA comprise undeveloped Caribbean pine *Pinus caribaea* forest and black land coppice.

■ Birds

The IBA is significant for supporting the majority of the Abaco population of the Near Threatened Cuban Amazon ("Bahama Parrot") *Amazona leucocephala*. Surveys in 2006 estimated c.3,600 individuals. Based on a number of recent sightings, small numbers of the Near Threatened Kirtland's Warbler *Dendroica kirtlandii* are thought to winter in the IBA. The pine forests support good populations of four (of the seven) Bahamas EBA restricted-range birds, namely Bahama Woodstar *Calliphlox evelynae*, Bahama Mockingbird *Mimus gundlachii*, Olive-capped Warbler *Dendroica pityophila* and Bahama Yellowthroat *Geothlypis rostrata*. The resident race of Yellow-throated Warbler *Dendroica dominica* also occurs, as do Cuban Emerald *Chlorostilbon ricordii* and Key West Quail-dove *Geotrygon chrysia*. A regionally significant population of White-tailed Tropicbird *Phaethon lepturus* breed at Hole-in-the-Wall.

■ Other biodiversity

The Atala hairstreak butterfly *Eumaeus atala* (confined to southern Florida, Cuba and Bahamas) is abundant in the pine forests of southern Abaco.

■ Conservation

Over 40% of this IBA is protected within the Abaco National Park which was established in 1994, primarily to protect the "Bahama Parrot". The parrot nests in limestone sinkholes within the pine forest areas, but uses the coppice extensively for feeding. BNT has developed a management plan for the park, and for the parrot which is vulnerable to predation by feral cats, introduced racoons and other predators. Game and pig hunting takes place in the park and surrounding areas which are primarily privately owned and unprotected. Fire is a significant threat and has been the focus of significant conservation efforts. The parrot has also been the focus of much conservation and research attention.

BS007 Red Bays

COORDINATES 25°13′N 78°11′W
ADMIN REGION Andros
AREA 1,369 ha
ALTITUDE 0–7 m
HABITAT Coastlne, forest, shrubland

■ Site description

Red Bays IBA it at the northernmost end of Andros island, on the west coast. It is centred on the settlement of Red Bay, the only settlement on the west coast of Andros. It was founded in the 1800s by Seminole Indians and escaped slaves from Florida. Sponge fishing is an active occupation as is the unique woven straw work produced by the residents. The IBA embraces a diverse area of Caribbean pine *Pinus caribaea* forest, broadleaf coppice, mangroves, shoreline scrub and beach. There is some small-scale agriculture (mostly slash-and-burn agriculture) with second-growth vegetation taking over abandoned areas.

■ Birds

This IBA supports regionally significant numbers of wintering Laughing Gull *Larus atricilla* and breeding Royal Tern *Sterna maxima*. Six (of the seven) Bahamas EBA restricted-range birds occur, including the Vulnerable Bahama Swallow *Tachycineta cyaneoviridis*. The Near Threatened White-crowned Pigeon *Patagioenas leucocephala* is found in significant numbers. Other species such as Cuban Emerald *Chlorostilbon ricordii*, Western Spindalis *Spindalis zena* and Great Lizard-cuckoo *Coccyzus merlini* are present, and the endemic subspecies of Greater Antillean Oriole *Icterus dominicensis northropi* occurs in the coconut palm trees within the Red Bay settlement. A diversity of waterbirds frequents the coast.

■ Other biodiversity

Nothing recorded.

■ Conservation

Red Bays IBA is a mixture of crown and privately owned land and is unprotected. Local development (in the form of slash-and-burn to cultivate and build) is causing some habitat destruction, and there is disturbance to breeding seabirds.

■ Site description

San Andros Pond IBA is located in northern North Andros where it is situated within the security boundary at the San Andros Airport. It comprises a small freshwater pond with associated shrubland and coppice (but also cultivated fields and verges associated with the airport). The pond is immediately surrounded by overgrown vegetation. The IBA is situated within the airport security boundary.

■ Birds

This IBA supports a significant number of Vulnerable West Indian Whistling-duck *Dendrocygna arborea*. The current status of the ducks is unknown since the sides of the pond became overgrown thus precluding easy observation. The Vulnerable Bahama Swallow *Tachycineta cyaneoviridis* also occurs at the IBA in good numbers. Three (of the seven) Bahamas EBA restricted-range birds occur, namely Bahama Woodstar *Calliphlox evelynae*, *T. cyaneoviridis* and Bahama Yellowthroat *Geothlypis rostrata*. The pond attracts a range of waterbirds while the coppice is important for wintering Neotropical migrant landbirds.

■ Other biodiversity

Nothing recorded.

■ Conservation

San Andros Pond IBA is on a mix of crown and private lands, and is within the airport security boundary. Access is restricted within the airport boundary due to increased security measures, and the pond can only be visited by special permission. This provides the pond and the associated birds some degree of *de facto* protection. However, any expansion of the airport could easily destroy this IBA.

■ Site description

Stafford Creek to Andros Town IBA embraces a large tract of land extending along the north-east coast of Central Andros from the settlements of Stafford Creek in the north, through Staniard Creek and Coakley Town to Andros Town (also known as Fresh Creek) in the south. It encompasses the sandy beach flats, Caribbean pine *Pinus caribaea* forest, broadleaf coppice, wetland, and inland blue holes. In extends inland to include the Blue Hole National Park. The area is used for large scale domestic and commercial agriculture; fly, sport, and commercial fishing; and ecotourism, general tourism and research.

■ Birds

This IBA supports important populations of four (of the seven) Bahamas EBA restricted-range birds, namely Bahama Woodstar *Calliphlox evelynae*, Bahama Swallow *Tachycineta cyaneoviridis*, Bahama Mockingbird *Mimus gundlachii* and Bahama Yellowthroat *Geothlypis rostrata*. *Tachycineta cyaneoviridis* is a Vulnerable species. Up to 85 Near Threatened Piping Plover *Charadrius melodus* have been recorded wintering along this stretch of coast. Other characteristic birds within the IBA include Great Lizard-cuckoo *Coccyzus merlini*, Key West Quail-dove *Geotrygon chrysia*, Cuban Emerald *Chlorostilbon ricordii* and many others.

■ Other biodiversity

The Vulnerable rock iguana *Cyclura cychlura cychlura* occurs throughout the pine and coppice areas.

■ Conservation

Stafford Creek to Andros Town IBA is a mix of crown and private lands, most of which is unprotected. However, the western portion of the IBA (including areas of pine forest and coppice) is protected within the Blue Holes National Park (managed and being developed for visitation by the BNT). Offshore from the beaches (and just outside) of this IBA is the Andros Barrier Reef National Park. There are two research centres within the IBA: Forfar Field Station, midway between Stafford and Staniard Creeks, which is a field site of the International Field Studies Program; and the Bahamas Environmental and Research Centre located at Staniard Creek, a joint project of George Mason University and College of the Bahamas. Development and agriculture threaten vital habitats and hunting causes disturbance to the birds.

■ Site description

Owenstown IBA is in northern Central Andros, inland from the northern end of the Stafford Creek to Andros Town IBA (BS009). It comprises the former commercial lumber settlement of Owenstown, on the north bank of Stafford Creek, and includes the western portion of the creek. The town was abandoned after major deforestation of the native Caribbean pine *Pinus caribaea* forest in the 1970s and is now overgrown with landscaping vegetation and weeds. Some native trees have returned. The habitat immediately surrounding the town consists of pine forest, broadleaf coppice and coconut palms. The area remains uninhabited.

■ Birds

This IBA supports important populations of four (of the seven) Bahamas EBA restricted-range birds, namely Bahama Woodstar *Calliphlox evelynae*, Bahama Mockingbird *Mimus gundlachii*, Olive-capped Warbler *Dendroica pityophila* and Bahama Yellowthroat *Geothlypis rostrata*. Other characteristic birds within the IBA include the endemic subspecies of Greater Antillean Oriole *Icterus dominicensis northropi*, Greater Antillean Bullfinch *Loxigilla violacea*, Western Spindalis *Spindalis zena*, Northern Bobwhite *Colinus virginianus* and many Neotropical migrant warblers. Stafford Creek supports many waterbirds, including Black Rail *Laterallus jamaicensis*.

■ Other biodiversity

Nothing recorded.

■ Conservation

Owenstown IBA is on crown land but is currently unprotected. As an abandoned town there are seemingly no threats although there has been little research in this area to support this assumption.

■ Site description

Mangrove Cay IBA lies between Middle Bight and South Bight in the middle of Andros, with Bog Wood Cay to the north, and South Andros Island to the south. Settlements are confined to the east coast of the island where a mangrove creek runs parallel to the seashore behind a sand dune. The west side of the island is uninhabited. The island consists of Caribbean pine *Pinus caribaea* forest, broadleaf coppice, freshwater blue holes, inland wetlands and mangroves and beaches. There is some agriculture practiced around the settlements.

■ Birds

This IBA supports populations of four (of the seven) Bahamas EBA restricted-range birds, namely Bahama Woodstar *Calliphlox evelynae*, Thick-billed Vireo *Vireo crassirostris*, Bahama Mockingbird *Mimus gundlachii* and Bahama Yellowthroat *Geothlypis rostrata*. The IBA is also significant for the Near Threatened White-crowned Pigeon *Patagioenas leucocephala*. The mangrove creek supports wintering shorebirds and other waterbirds.

■ Other biodiversity

Nothing recorded.

■ Conservation

Mangrove Cay is a mix of private and crown land, but is currently unprotected. The communities of Moxey Town, Bastian Point and Lisbon Creek are expanding and the gradual development is resulting in habitat loss and fragmentation. The mangrove creek has been severely degraded in places by causeways and other obstacles cutting off the flow of water. There has been little research of conservation activity on Mangrove Cay.

BS012 Driggs Hill to Mars Bay

■ Site description

Driggs Hill to Mars Bay IBA is on the eastern side of South Andros Island. From Driggs Hill at the northernmost tip of South Andros it runs south (following the road) for c.48 km through Congo Town, The Bluff, Kemp's Bay, over deep creek and Little Creek to Mars Bay in the south. The IBA extends c.5 km inland from the east coast, and embraces a number of blue holes including Rat Bat Lake and Twins, north of Congo Town airport, and Nine Tasks Blue Hole and Evelyn Green Blue Hole south of The Bluff. The IBA supports impenetrable shrubland coppice and unexplored wetlands, numerous creeks and a shallow shoreline with tidal flats. The human population is small and focus on low key agriculture, fishing and tourism activities

■ Birds

This IBA supports populations of four (of the seven) Bahamas EBA restricted-range birds, namely Bahama Woodstar *Calliphlox evelynae*, Thick-billed Vireo *Vireo crassirostris*, Bahama Mockingbird *Mimus gundlachii* and Bahama Yellowthroat *Geothlypis rostrata*. The IBA is also significant for the Near Threatened White-crowned Pigeon *Patagioenas leucocephala*. Mars Bay is important for wintering Near Threatened Piping Plover *Charadrius melodus*. Other species present in the IBA include Great Lizard-cuckoo *Coccyzus merlini*, the endemic subspecies of Greater Antillean Oriole *Icterus dominicensis northropi* and the Bahamas only known nesting site for Cave Swallow *Petrochelidon fulva* (in limestone cavities in Nine Tasks Blue Hole and at Twins Blue Hole). The IBA also supports many waterbirds.

■ Other biodiversity

The Vulnerable Andros rock iguana *Cyclura cychlura cychlura* occurs in this IBA.

■ Conservation

Driggs Hill to Mars Bay IBA is a mix of private and crown land, but is unprotected. The ecosystem is currently relatively intact, although development is an ever-present threat while this IBA remains unprotected. The IBA is one of the premier *Patagioenas leucocephala* hunting sites in the Bahamas, and this should be monitored in relation to annual population estimates for this Near Threatened bird. Disturbance (by people and dogs) of shorebirds (especially *Charadrius melodus*) on the tidal beach flats is a problem that needs to be monitored and managed.

BS013 Goulding Cay Wild Bird Reserve

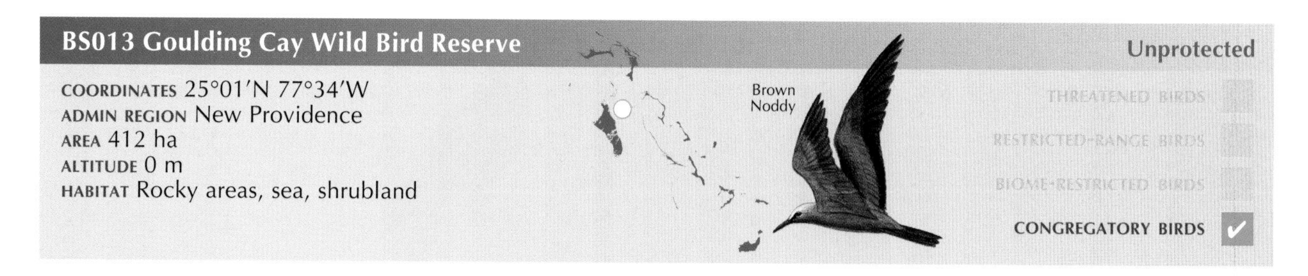

■ Site description

Goulding Cay IBA lies 3 km off the westernmost end of New Providence. It is directly offshore of Jaws Beach, near Lyford Cay. Goulding Cay is a 4-ha uninhabited offshore rocky cay with low coastline vegetation such as bay marigolds, bay lavender, bay cedar, sea purslane and railroad vines. The IBA includes all marine areas within 1 km of the cay.

■ Birds

This IBA is seabird colony. Regionally significant numbers of Bridled Tern *Sterna anaethetus* and Brown Noddy *Anous stolidus* nest on the cay each summer (May–August). Sooty Tern *S. fuscata* also breed on the cay.

■ Other biodiversity

Nothing recorded.

■ Conservation

Goulding Cay Wild Bird Reserve is crown owned land and legally recognised reserve, making hunting on the island illegal. However, formal protected status has yet to be granted by the government. The BNT Ornithology Group has been monitoring the seabirds since 2004 with little apparent change in the populations being counted each year. There is no evidence of egg collecting or indeed of the presence of rats *Rattus* spp. on the island, and disturbance from tourist and diving boats (the cay is a popular dive site, but landing is difficult) appears to be minimal.

■ **Site description**
Harrold and Wilson Ponds National Park IBA is in central New Providence, south-west of Nassau. It encompasses a large area of freshwater ponds with areas of mud, and fringing vegetation of reeds, sedge, broadleaf coppice and some pine lands. Being so close to the nation's capital agriculture and commercial and residential development had encroached on the site before it was designated a national park in 2002.

■ **Birds**
This IBA supports a diversity of species and is particularly important for its waterbirds. The population of Laughing Gull *Larus atricilla* is globally significant, while those of Gull-billed Tern *Sterna nilotica* and Royal Tern *S. maxima* are regionally so. Large numbers of cormorants, herons, egrets, ibises, ducks and shorebirds frequent the IBA. The Near Threatened White-crowned Pigeon *Patagioenas leucocephala* occurs, as does the Vulnerable Bahama Swallow *Tachycineta cyaneoviridis* which is one of three Bahamas EBA restricted-range birds to be found around the ponds.

■ **Other biodiversity**
Nothing recorded.

■ **Conservation**
Harrold and Wilson Ponds National Park IBA is a mix of crown and private lands. However, the area is now designated a national park under the management of the BNT. Being so close to Nassau, this IBA is an ideal educational and ecotourism site, and an interpretation and public use plan has been developed. Implementation of this plan has started (2007) and boardwalks, observation platforms and educational signage have been installed. The IBA still faces the threat of pollution from adjacent housing developments (and squatters) and dumping and infill to "reclaim" land. A commercial chicken farm has been closed down and the land will be annexed to the park. Invasive plants such as *Casuarina* and Brazilian pepper crowd out native species, but are the focus of a BNT invasive species management project in the park.

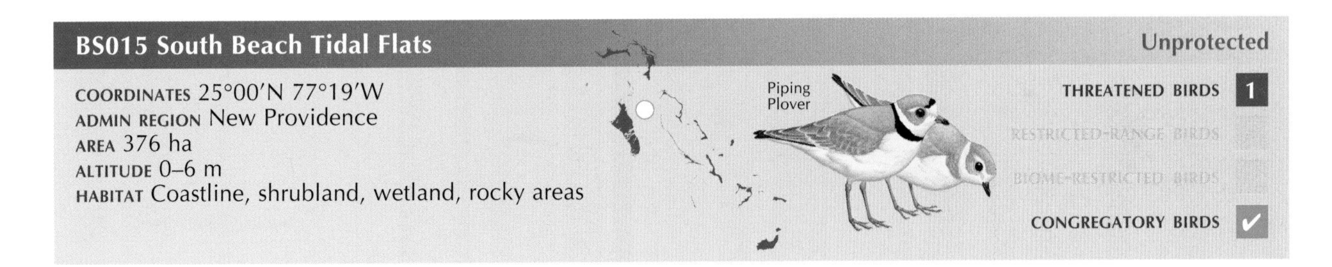

■ **Site description**
South Beach Tidal Flats IBA extends along c.3 km of New Providence's south-eastern coastline. It follows the line of Marshall Road, from Blue Hill Road, south-west towards Cay Point. The IBA is characterised by sand and limestone tidal flats, with rocky banks supporting low mangroves. It also includes some freshwater wetlands just inland of the beach, and shrublands adjacent to the beach.

■ **Birds**
This IBA is significant for its wintering population of the Near Threatened Piping Plover *Charadrius melodus*. The numbers of Laughing Gull *Larus atricilla* and Royal Tern *Sterna maxima* are regionally important. Large numbers of a wide diversity of shorebirds use this IBA as a stop-over site and as wintering habitat. Least Tern *S. antillarum* are common in the nesting season, migrant warblers and resident land birds can be found in the shrubland along the shoreline, and the freshwater wetlands Marshall Road support a wide range of waterbirds.

■ **Other biodiversity**
Nothing recorded.

■ **Conservation**
South Beach Tidal Flats IBA comprises crown lands (the tidal zone) and private lands inland, none of which is currently protected. The population in this area is expanding rapidly leading to habitat destruction from development and disturbance of birds by people and dogs. The shoreline is a popular beach and picnic area, and it is also a favoured launching point for resident fishermen. Pollution from adjacent developments and illegal dumping are additional threats to this IBA.

BS016 Salt Cay

COORDINATES 25°09′N 77°03′W
ADMIN REGION New Providence
AREA 968 ha
ALTITUDE 0–2 m
HABITAT Sea, rocky areas, shrubland

■ Site description

Salt Cay IBA is an island c.5 km north-east of Nassau, and c.1.5 km north of the eastern end of Paradise Island. Also known as Blue Lagoon Island, it is the easternmost island in a chain of cays that extends towards Eleuthera. The island has been much altered over time. Originally supporting a salt marsh, this was dredged out in the 1900s and connected to the sea to make the lagoon. Over 5,000 palm trees were planted at this time. The eastern end of the island is a popular tourism and recreation destination. The western end is very narrow and rocky. The island, which is c.3 km long, supports shrubland and has a mix of sandy and rocky shoreline. The IBA includes marine areas up to 1 km from the cay.

■ Birds

This IBA is regionally significant for its population of Laughing Gull *Larus atricilla*. Many wintering shorebirds occur, and Roseate Tern *Sterna dougallii* is reported to nest although numbers are unknown. White-cheeked Pintail *Anas bahamensis* nest on Salt Cay, but move their young to Paradise Island once they have fledged.

■ Other biodiversity

Nothing recorded.

■ Conservation

Salt Cay IBA is privately owned and unprotected. The eastern end of the island is heavily used by day visitors (taking boat trips from Nassau). Dolphin Encounters—a natural seawater dolphin experience facility—is based around the lagoon and is one of the Bahamas' premier tourist attractions. Further development will impact three breeding seabirds. It is unknown what introduced predators are present on the island, although it is likely that rats *Rattus* spp. occur and are predating gull and tern eggs and chicks.

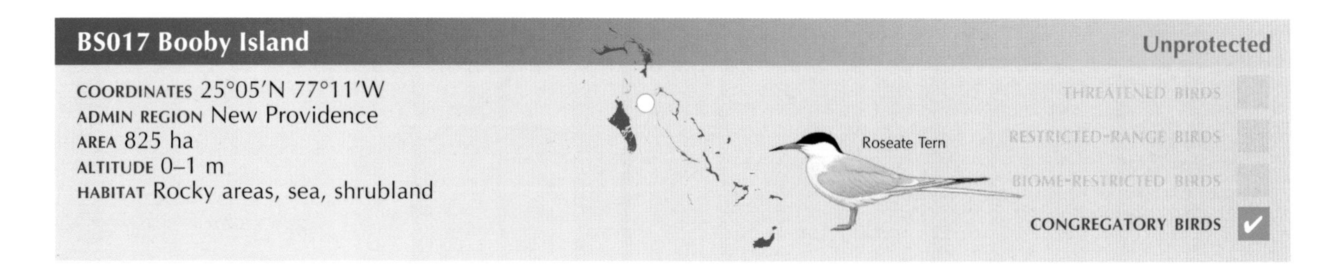

BS017 Booby Island

COORDINATES 25°05′N 77°11′W
ADMIN REGION New Providence
AREA 825 ha
ALTITUDE 0–1 m
HABITAT Rocky areas, sea, shrubland

■ Site description

Booby Island IBA lies 22 km north-east of the eastern end of New Providence, towards the western end of the chain of cays that extends towards Eleuthera. It is north-east of Rose island. Booby Island is 3 km long and less than 100 m wide, and has a low, rocky coralline shoreline that makes access difficult. It supports minimal vegetation such as sea purslane, bay cedar and other salt resistant plants. The IBA includes marine areas up to 1 km from the island.

■ Birds

This IBA supports a number of breeding seabirds. The population of Roseate Tern *Sterna dougallii* is globally significant, while those of Laughing Gull *Larus atricilla*, Least Tern *S. antillarum* and Bridled Tern *S. anaethetus* are regionally so. Brown Noddy *Anous stolidus*, Sooty Tern *S. fuscata* and Brown Booby *Sula leucogaster* also breed in the IBA. A range of shorebirds have been recorded.

■ Other biodiversity

Nothing recorded.

■ Conservation

Booby Island IBA is crown land but is unprotected. The BNT Ornithology Group visited the island to count breeding seabirds in September 2007 which could form the baseline for monitoring this important seabird island. Rats *Rattus* spp. and illegal egg collecting are potential but unconfirmed problems.

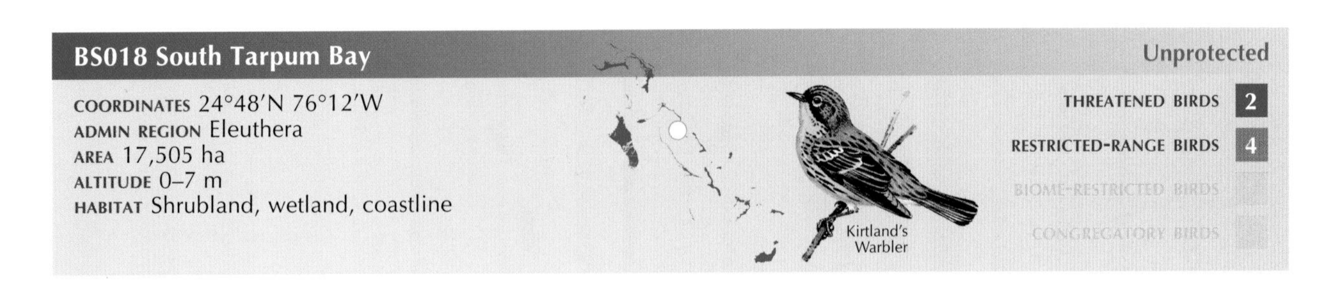

■ Site description

South Tarpum Bay IBA embraces the southern third of Eleuthera Island. It extends from Tarpum Bay and Winding Bay in the north for c.35 km through Rock Sounds to Bannerman Town at the southernmost end of the island. The IBA is a mosaic of small agricultural and fishing settlements, small agricultural plots, mature broadleaf coppice of varying heights, abandoned plantation, shrubland, coastal coppice and beach habitats.

■ Birds

This IBA supports the largest known concentration of wintering Near Threatened Kirtland's Warbler *Dendroica kirtlandii* which was discovered in the IBA in 2002. During the winter 2003–2004 at least 60 birds were recorded at 15 different locations in southern Eleuthera. The Near Threatened White-crowned Pigeon *Patagioenas leucocephala* also occurs in significant numbers, and four (of the seven) Bahamas EBA restricted-range birds, namely Bahama Woodstar *Calliphlox evelynae*, Bahama Yellowthroat *Geothlypis rostrata*, Thick-billed Vireo *Vireo crassirostris* and Bahama Mockingbird *Mimus gundlachii* are present. Great Lizard-cuckoo *Coccyzus merlini* and Greater Antillean Bullfinch *Loxigilla violacea* also occur.

■ Other biodiversity

No globally threatened or endemic terrestrial species have been recorded.

■ Conservation

South Tarpum Bay IBA is a mix of crown and privately owned land, but none of it is protected. Habitat is being lost as a result of increased residential and resort development, and slash-and-burn land clearance is common. The BNT Ornithology Group discovered *D. kirtlandii* in this IBA in 2002 since when the species has been the focus of an intensive, multi-institutional research program (the Kirtland's Warbler Research and Training Program). The species' winter habitat preferences are for early successional fruiting scrub and low coppice. Wild sage (*Lantana involucrata* and *L. bahamensis*), West Indian snowberry (*Chiococca alba*), and black torch (*Erithalis fruticosa*) appear to be especially important and this should be considered in relation to any conservation management interventions.

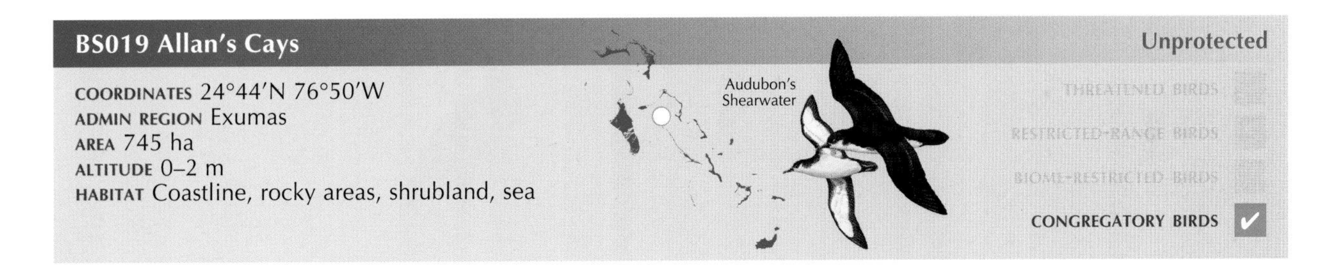

■ Site description

Allan's Cays IBA is at the northern end of the Exuma Cays between Ship Channel Cay and Highborne Cay. It comprises three small, uninhabited cays, namely Allan's Cay, Southwest Allan's Cay and Leaf Cay. The shoreline of Allan's and Southwest Allan's Cays is comprised of mainly honeycomb limestone rock (including cliffs on Allan's Cay) and Leaf Cay has sandy soil and beaches. The cays support some areas of shrubland. The IBA includes marine areas up to 1 km from the cays.

■ Birds

The rocky cliffs on Allan's Cay support a regionally significant colony of Audubon's Shearwater *Puffinus lherminieri*.

■ Other biodiversity

The Vulnerable rock iguanas *Cyclura cychlura inornata* and *C. c. figginisi* occur on Leaf Cay and Southwest Allan's Cay. All iguanas are protected by law in the Bahamas.

■ Conservation

Allan's Cay IBA is crown owned but unprotected. The cays are a popular scuba-diving and snorkelling destination and there are daily powerboat trips to the cays from Nassau. There is a constant threat of disturbance to the birds and the iguanas by commercial and private boating activity, including from dogs taken ashore for exercise. *Puffinus lherminieri* faces natural threats from resident Barn Owl *Tyto alba* and wintering Peregrine Falcon *Falco peregrinus*, but more worryingly rats *Rattus* spp. were confirmed as present in 2007.

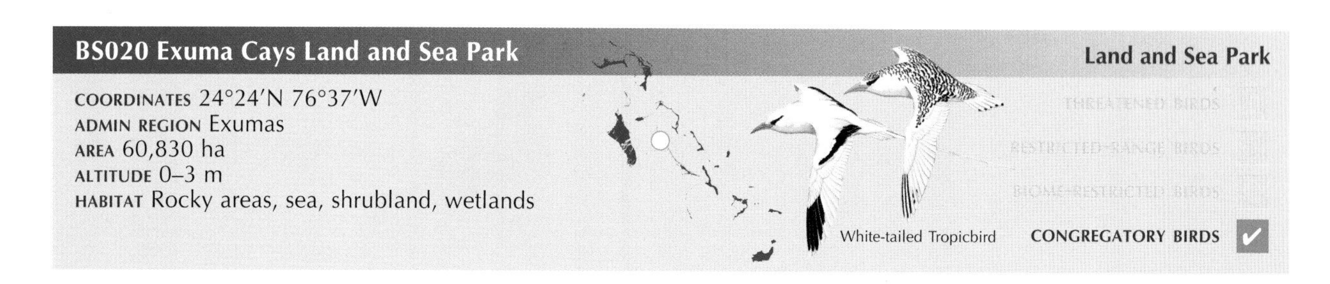

Site description

Exuma Cays Land and Sea Park IBA embraces a large section of the northern Exumas. It stretches for 35 km from Wax Cay Cut in the north to Conch Cut in the south and includes Little Wax Cay, Shroud Cay, Hawksbill Cay, Cistern Cay, Warderick Wells, Halls Pond Cay, Bells Cay, Little Bells Cay and many others. The IBA boundary is the same as the land and sea park, and thus extends about 7.5 km either side of the cays. The cays support a variety of habitats including shrubland and low coppice, wetlands, mangroves, sandy and rocky beaches, tidal flats, low cliffs and coral reef. The park headquarters building and visitors centre is located on Warderick Wells.

Birds

This IBA supports a globally significant population of White-tailed Tropicbird *Phaethon lepturus* (primarily on the eastern cliffs of Shroud Cay, and the northern cliffs of Warderick Wells). The breeding population of Audubon's Shearwater *Puffinus lherminieri* on Long Rock (also called Long Cay) is regionally important, as are the breeding Least Terns *Sterna antillarum* (primarily on Warderick Wells). The mangroves support a range of waterbirds, and the restricted-range Bahama Mockingbird *Mimus gundlachii* and Thick-billed Vireo *Vireo crassirostris* occur in the shrubland.

Other biodiversity

The Vulnerable Bahamian hutia *Geocapromys ingrahami* has been introduced on Little Wax Cay (where they have devastated the cay's vegetation) and Waderwick Wells (where the population is c.25,000). Critically Endangered hawksbill *Eretmochelys imbricata* and Endangered green *Chelonia mydas* and loggerhead *Caretta caretta* turtles forage in the park. The Endangered rock iguana *Cyclura riley rileyi* is (introduced) on Bush Hill Cay, and the Vulnerable *C. cychlura inornata* and *C. cychlura figginisi* are also present (introduced) on a number of cays.

Conservation

Exuma Cays Land and Sea Park IBA includes some privately owned islands, but all cays are covered by the regulations of the land and sea park which is managed by the BNT. It is the oldest land and sea park in the world (established in 1958) and since 1986 it has been managed as a strict no-take zone—nothing living or dead, can be removed from the park, which is essentially pristine. The IBA is a popular yachting (and tourist) destination resulting in some disturbance of nesting seabirds, although this threat is being actively managed by the BNT. Predation of nests and adult birds by rats *Rattus* spp. and other introduced predators is a problem.

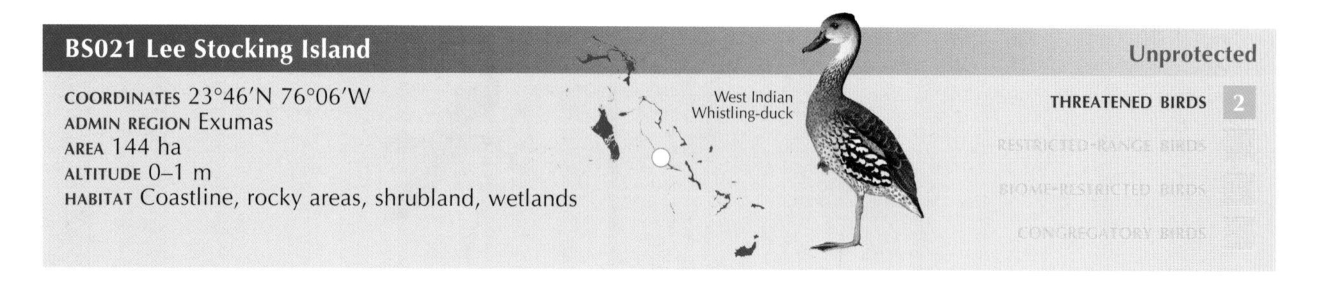

Site description

Lee Stocking Island is in the southern Exumas, just north of Great Exuma Island. The island is c.7 km long. There are no roads on the island, but there is some settlement. The Hotel Higgins eco-resort is in the IBA, as are a marine research centre, and an airstrip. The island comprises sandy beaches, rocky areas, tidal flats, lagoons, wetlands, coral reefs and shrubland. There are two small freshwater ponds at the north end of the airstrip. The IBA includes marine areas up to 1 km from the island.

Birds

This IBA is significant for supporting a population of the Vulnerable West Indian Whistling-duck *Dendrocygna arborea* (which frequent the airstrip ponds), and the Near Threatened White-crowned Pigeon *Patagioenas leucocephala*. The restricted-range Bahama Mockingbird *Mimus gundlachii* and Thick-billed Vireo *Vireo crassirostris* occur along with other characteristic birds including Burrowing Owl *Athene cunicularia*, Greater Antillean Bullfinch *Loxigilla violacea* and a range of waterbirds.

Other biodiversity

The marine environment surrounding this IBA supports the Endangered Nassau grouper *Epinephelus striatus* and queen conch *Strombus gigas*, both of which are commercially valuable and are being studied by researchers based on the island. Critically Endangered hawksbill *Eretmochelys imbricata* and Endangered green *Chelonia mydas* and loggerhead *Caretta caretta* turtles forage in the IBA.

Conservation

Lee Stocking Island IBA is a mix of crown and privately owned lands, but is unprotected. The Caribbean Marine Research Centre is on the island and serves marine scientist from the USA and the Bahamas. Tourists from yachts can visit the centre. There is currently minimal development on the island and as long as it remains ecologically sensitive the threats to the IBA and its key species will be minimal. It is unknown whether rats *Rattus* spp. (or other predators) are a problem and this should be investigated.

■ Site description

Grog Pond IBA is situated c.16 km north-west of George Town on Great Exuma. It is bounded on the north by the Queen's Highway, and on the east, south and west by Bahama Sound Development. Grog Pond is an inland wetland. Grog Pond is a shallow, brackish water lake with clumps of black mangroves and fringing saltmarsh, buttonwood and coppice.

■ Birds

This IBA is significant for supporting a population of the Vulnerable West Indian Whistling-duck *Dendrocygna arborea*, and the Near Threatened White-crowned Pigeon *Patagioenas leucocephala*. The numbers of Laughing Gull *Larus atricilla*, Gull-billed Tern *Sterna nilotica* and Least Tern *S. antillarum* present in the IBA are regionally significant. The restricted-range Bahama Mockingbird *Mimus gundlachii* and Thick-billed Vireo *Vireo crassirostris* occur along with Greater Antillean Bullfinch *Loxigilla violacae* and a range of waterbirds including duck, herons, egrets, ibises and shorebirds

■ Other biodiversity

Nothing recorded.

■ Conservation

Grog Pond IBA is privately owned and unprotected. It has the potential to become a community-led eco-tourism site, recreation area and a centre for students and adults to learn about the environment, and the BNT has been pursuing this concept. However, the surrounding coppice has been divided into residential plots and it appears that development is imminent. The area has been used as an illegal garbage dump (despite the "no dumping" signs). Hunting is also prevalent at this site, as is the collection of pond-stone by local builders for patios and walkways.

■ Site description

Tee Cay, Goat Cay and Long Rocks IBA is located between northern Cat Island and (to the west) Little San Salvador. The islands are physically nearer to (1–3 km from) Little San Salvador. Goat Cay lies north-east of Little San Salvador, Long Rocks lies due east, Tee Cay south-east. The cays are uninhabited limestone ridges partially covered with scrubland vegetation such as seagrape, cacti, haulback and other native plants. There is a sandy cove on Goat Cay. The IBA includes marine areas up to 1 km from the cays.

■ Birds

This IBA is significant for its breeding seabirds. The population of Roseate Tern *Sterna dougallii* is thought to be globally significant and that of Bridled Tern *Sterna anaethetus* regionally so. Sooty Tern *S. fuscata*, Brown Noddy *Anous stolidus*, Magnificent Frigatebird *Fregata magnificens* and Brown Booby *Sula leucogaster* are all thought to breed on the cays. The Near Threatened White-crowned Pigeon *Patagioenas leucocephala* has been reported nesting on Goat Cay.

■ Other biodiversity

Nothing recorded.

■ Conservation

Tee Cay, Goat Cay and Long Rocks IBA is poorly known and there is little direct information available except from boaters. Breeding season surveys of the seabirds are a clear priority. The cays are unprotected. The seabirds are prone to predation from introduced species (e.g. rats *Rattus* spp.) from visiting boats, and from refugees that are occasionally landed in the IBA.

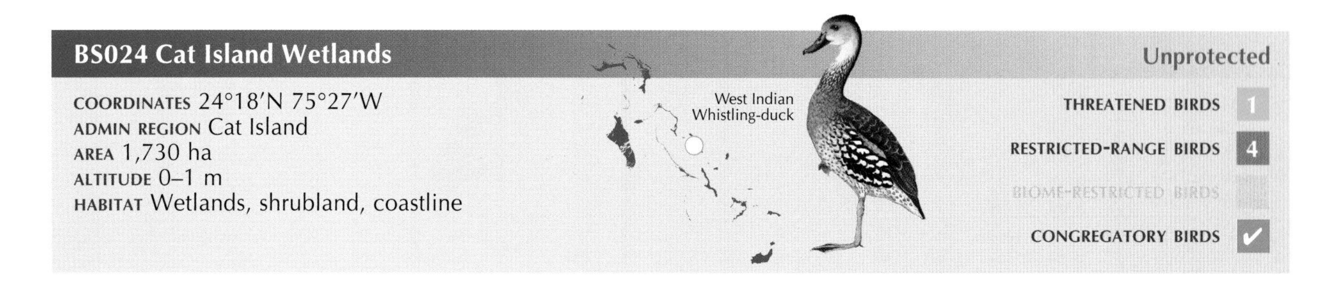

■ Site description

Cat Island IBA is south-east of Eleuthera on the Atlantic edge of the Great Bahama Bank. The island is c.80 km long and just a few kilometres wide except at the southern end which broadens out to embrace the large, brackish Gambier Lake. A paved road runs the length of the island with a series of dirt roads crossing the island to the ocean side (locally called the "north shore"). There are a number of settlements along the road on the western shore. The 63-m Mount Alvernia is towards the south of the island and is the highest point in the Bahamas. The island supports a range of freshwater and saltwater wetlands, tidal flats, beach and adjacent broadleaf coppice.

■ Birds

This IBA is significant for its population of the Vulnerable West Indian Whistling-duck *Dendrocygna arborea*. The population of Laughing Gull *Larus atricilla* is globally important while those of Gull-billed Tern *Sterna nilotica* and Least Tern *Sterna antillarum* are regionally so. The terns breed at Gambier Lake which is also a nesting site for other terns, Reddish Egret *Egretta rufescens* and a range of waterbirds. Four (of the seven) Bahamas EBA restricted-range birds, namely Bahama Woodstar *Calliphlox evelynae*, Bahama Yellowthroat *Geothlypis rostrata*, Thick-billed Vireo *Vireo crassirostris* and Bahama Mockingbird *Mimus gundlachii* are present.

■ Other biodiversity

The Bahamian endemic Bahama pygmy boa *Tropidophis canus* occurs, as do a number of other snakes, lizards, frogs and freshwater turtles.

■ Conservation

The Cat Island Wetlands IBA is a mixture of crown and privately owned land, but is unprotected. Small scale farming (including corn, which *D. arborea* feeds on) and fishing supports most of the local population. However, local and international tourism has begun to grow on the island resulting in habitat destruction from urban development. Illegal hunting of birds is a problem, as are introduced predators.

■ Site description

Cay Sal IBA is located due south of Miami, midway between Florida and Cuba. It is closer to Florida and Cuba than to Andros. The IBA comprises Double Headed Shot Cays, Elbow Cay, Damas and Anguilla Cays and Cay Sal that are situated along the northern and eastern edges of the Cay Sal Bank. These cays are presently uninhabited, except as a harbour for yachts sailing between Cuba and Florida. The cays are rocky, with some sandy beaches, a saltwater lagoon on Cay Sal, and some low shrubland. The IBA includes marine areas up to 1 km from the cays.

■ Birds

This IBA supports significant numbers of seabirds. The populations of Roseate Tern *Sterna dougallii* and Bridled Tern *S. anaethetus* are globally important, and those of Audubon's Shearwater *Puffinus lherminieri*, Brown Pelican *Pelecanus occidentalis*, Royal Tern *S. maxima*, Sandwich Tern *S. sandvicensis* and Sooty Tern *S. fuscata* are regionally so. Other seabirds frequent the IBA as non-breeding residents. Elbow Cay is the main nesting cay for the seabirds. The Near Threatened White-crowned Pigeon *Patagioenas leucocephala* breeds along with a small number of other resident landbirds. The IBA is an important stop-over site for Neotropical migratory landbirds and shorebirds.

■ Other biodiversity

Critically Endangered hawksbill *Eretmochelys imbricata*, and Endangered green *Chelonia mydas* and loggerhead *Caretta caretta* turtles nest in the IBA. The Cay Sal anole *Anolis fairchildi* is endemic to the IBA, and the Bahama pygmy boa *Tropidophis canus* occurs.

■ Conservation

Cay Sal IBA is crown land, but is currently unprotected. There are apparently plans to build a marina on Cay Sal which will inevitably lead to habitat loss and disturbance of the nesting seabirds. Elbow Cay has a fresh water cistern and refugees from Cuba frequently stop there, and have decimated the *Puffinus lherminieri* colony to obtain fresh meat. Introduced predators such as rats *Rattus* spp. are a potential threat, although it is not know whether they are present on the islands. The seabirds and the threats to them are seldom monitored.

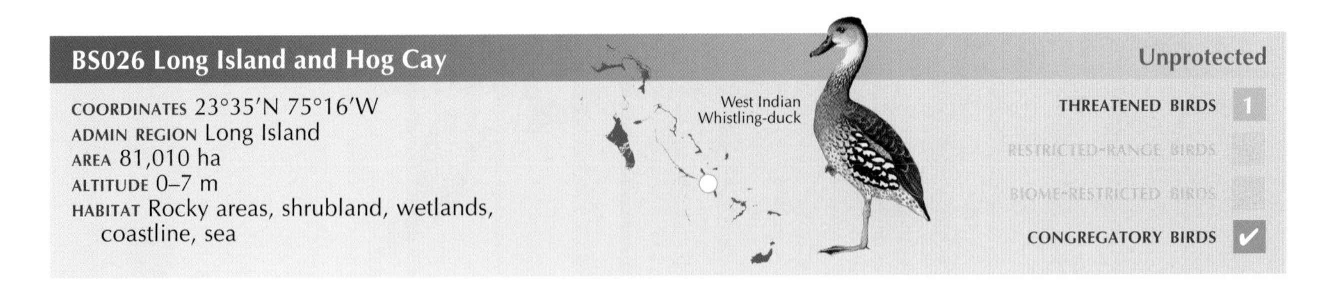

■ Site description

Long Island and Hog Cay IBA lies south of Cat Island and south-east of the southern end of the Exumas. The island is about 128 km long and a maximum of 6.5 km wide. Hog Cay is a privately-owned island on the leeward side of northern Long Island. Long Island supports a variety of habitats including shrubland, coppice, freshwater and saltwater wetlands, mangroves swamps and tidal flats. Wetlands are scattered throughout the interior of the island and there are frequent roadside ponds. Fishing and farming are the main occupations of the local population.

■ Birds

This IBA is significant for supporting a large population of the Vulnerable West Indian Whistling-duck *Dendrocygna arborea* which roost on Hog Cay each night. The island's wetlands are also home to a diversity of waterbirds including ducks, herons, egrets and migratory shorebirds. Sandwich Tern *Sterna sandvicensis* and Roseate Tern *S. dougallii* breed on Hog and Galliott Cays. The breeding population of Laughing Gull *Larus atricilla* is regionally important. The restricted-range Bahamas Mockingbird *Mimus gundlachii* and Thick-billed Vireo *Vireo crassirostris* also occur. A population of the Near Threatened White-crowned Pigeon *Patagioenas leucocephala* occurs, but the numbers involved are unknown.

■ Other biodiversity

The Near Threatened Gervais's funnel-eared bat *Nyctiellus lepidus* and Brazilian free-tailed bat *Tadarida brasiliensis* occur (along with a number of other bat species).

■ Conservation

Long Island and Hog Cay IBA is a mixture of crown and privately owned land, but none of it is protected. The owner of Hog Cay provided daily feed for the large flock of *D. arborea* which roost on the cay at night. Residential and urban development is leading to habitat destruction, and illegal hunting is a problem. Feral cats, wild goats and pigs are all common and are impacting the vegetation and nesting birds.

■ Site description

Conception Island IBA lies c.40 km south-west of San Salvador, midway between Cat Island and Rum Cay. It is c.5 km by 2.5 km and it encircles an interior lagoon. The island is uninhabited and comprises coral reefs, sandy beaches, rocky and low coralline cliff shores, mangrove, low scrub and coppice. Offshore to the east lies Booby Cay, and to the south-west is South Rocks. The island is an attractive destination for yachts. The IBA includes marine areas up to 1 km from the islands.

■ Birds

This IBA is characterised by its breeding seabirds. The population of White-tailed Tropicbird *Phaethon lepturus* is globally significant while those of Audubon's Shearwater *Puffinus lherminieri*, Bridled Tern *Sterna anaethetus* and Brown Noddy *Anous stolidus* are regionally so. Booby Cay has one of the largest colonies of Sooty Tern *S. fuscata* in the Bahamas (and is also where the *A. stolidus* nests). The restricted-range Bahama Mockingbird *Mimus gundlachii* and Bahama Woodstar *Calliphlox evelynae* are present and ducks, herons and shorebirds are common in the interior lagoon. A population of the Near Threatened White-crowned Pigeon *Patagioenas leucocephala* occurs, but the numbers involved are unknown.

■ Other biodiversity

Critically Endangered hawksbill *Eretmochelys imbricata* and Endangered green *Chelonia mydas* turtles are common in the interior lagoon.

■ Conservation

Conception Island is owned by the crown and is protected as a national park under the management of the BNT. However, Booby Cay and South Rocks and the surrounding shallow water are not included in the protected area. Hunting and illegal egg collecting by boaters and fishermen stopping over on the island are significant threats to the breeding seabirds. The mouth of the lagoon is sometimes illegally blocked by fishermen in order to catch fish and turtles trapped in the interior.

■ Site description
Graham's Harbour IBA lies off the north coast of San Salvador where several pristine cays are found in the "harbour's" shallow waters. The area is characterised by shallow reefs and rock throughout the bay, with White Cay nearest the reef edge, Green Cay on the north-western side, and Gaulin, Cato and Cut cays near to the north shore of San Salvador. The cays are uninhabited with rocky shorelines and some sandy beaches, and supporting low scrub.

■ Birds
The cays in Graham's Harbour are important seabird colonies. Regionally significant populations of Brown Booby *Sula leucogaster* and Bridled Tern *Sterna anaethetus* nest on Green Cay, which also supports some breeding Magnificent Frigatebird *Fregata magnificensis*. Frigatebirds and Brown Booby *Sula leucogaster* nest on White Cay, and Brown Noddy *Anous stolidus*, *Sterna anaethetus* and Sooty Tern *S. fuscata* nest on Gaulin and Cato Cays, albeit not in significant numbers.

■ Other biodiversity
About 250 Endangered rock iguanas *Cyclura rileyi rileyi* were living on Green Cay in 1997.

■ Conservation
Graham's Harbour IBA is a mix of crown and privately owned land and is currently unprotected. However, the BNT has targeted this area as a potential national park and a managed ecotourism site. Invasive plants (that crowd out native flora) are a potential threat that needs monitoring as it would ruin the pristine state of the cays. Similarly, invasive predators such as rats *Rattus* spp. could deplete the seabird populations. The arrival of such alien invasives should be monitored for, along with the seabird populations. Visitation by tourists needs to be well controlled to avoid disturbance to nesting seabirds.

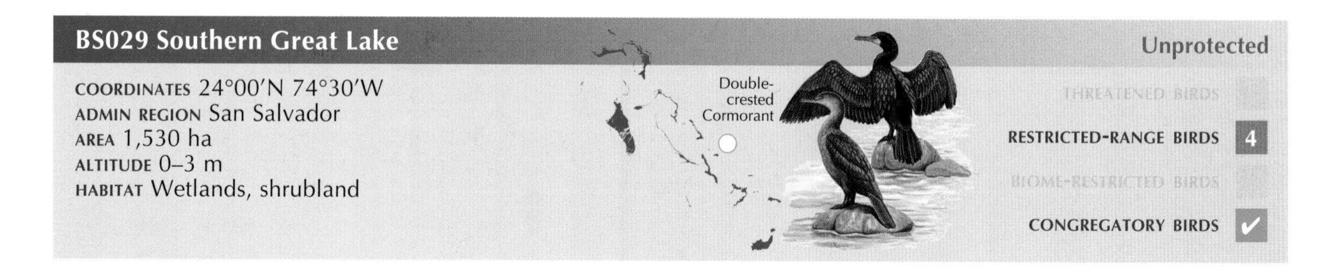

■ Site description
Southern Great Lake IBA embraces the saline wetland that occupies a large proportion of the interior of southern San Salvador. The wetlands are extensive and largely unobserved or explored due to the difficulty of access. The wetland is surrounded by dry shrubland and there are fringing mangroves. San Salvador is a small island (8 km by 19 km) with less than 1,000 people resident. The southern wetlands are therefore little disturbed.

■ Birds
This IBA supports globally significant breeding populations of Gull-billed Tern *Sterna nilotica* and Laughing Gull *Larus atricilla*. The Great Lake is home to a wide diversity of waterbirds including the endemic diminutive race of Double-crested Cormorant *Phalacrocorax auritus*, egrets and herons. Four (of the 7) Bahamas EBA restricted-range birds, namely Bahama Woodstar *Calliphlox evelynae*, Pearly-eyed Thrasher *Margarops fuscatus*, Thick-billed Vireo *Vireo crassirostris* and Bahama Mockingbird *Mimus gundlachii* are present. The endemic race of West Indian Woodpecker *Melanerpes superciliaris* is present in the IBA.

■ Other biodiversity
The Endangered rock iguanas *Cyclura rileyi rileyi* is found in the interior lake areas. An endemic blind snake *Leptotypholops columbi* is present.

■ Conservation
Southern Great Lake IBA is crown land, but is currently unprotected. An observation platform overlooking the northern end of the lake (near Cockburn Town) is the only easily accessible viewing point and thus the wetland and the populations of its waterbirds are poorly known. Resort development is an ever present (but as yet unrealised) threat. The expanded airport at Cockburn Town has recently caused considerable habitat destruction although this has not impinged on the lake system.

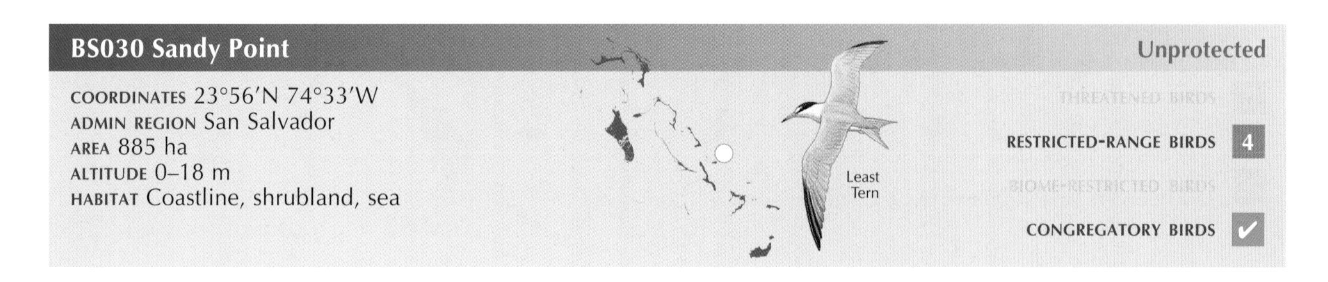

■ **Site description**

Sandy Point IBA is located at the south-western tip of San Salvador. The IBA includes residential areas (a subdivision of an urban development called "Columbus Landing"), the ruins known as Watling's Castle and surrounding shrubland. However, the primary interest is the sandy beaches.

■ **Birds**

The beaches in this IBA support a regionally important population of Least Tern *Sterna antillarum*. Many *Charadrius* spp. plovers use the beaches too. The surrounding shrubland is home to four (of the 7) Bahamas EBA restricted-range birds, namely Bahama Mockingbird *Mimus gundlachii*, Bahama Woodstar *Calliphlox evelynae*, Pearly-eyed Thrasher *Margarops fuscatus* and Thick-billed Vireo *Vireo crassirostris*. The endemic race of West Indian Woodpecker *Melanerpes superciliaris* also occurs.

■ **Other biodiversity**

Nothing recorded.

■ **Conservation**

Sandy Point IBA is a mix of crown and privately owned land and is unprotected. The beaches are public and are a popular destination for tourists and locals from the residential community within the IBA. The recreational traffic on the beaches poses a serious threat to the nesting *S. antillarum*. Predation from household pet cats and dogs, and also from introduced predators such as rats *Rattus* spp. is also a problem.

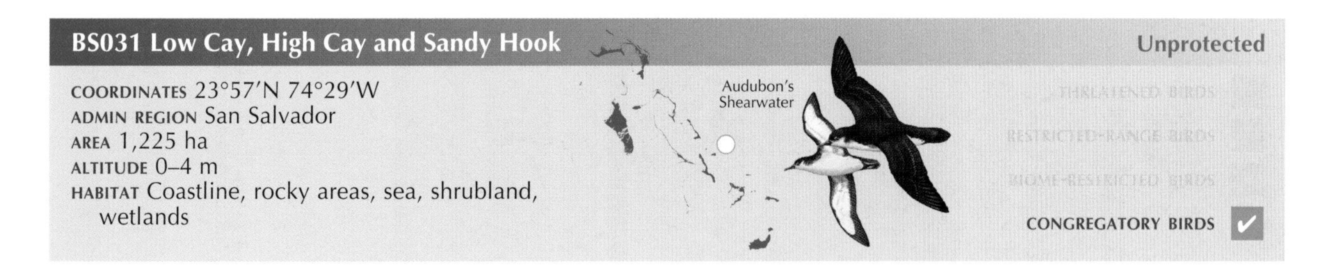

■ **Site description**

Low Cay, High Cay and Sandy Hook IBA is at the south-eastern end of San Salvador. Sandy Hook is a subdivision of an urban development called "Columbus Landing". It is a peninsula with the sandy Snow Bay Beach on it eastern side, and Pigeon Creek (a tidal lagoon) to its north and west. Low and High Cays are small rocky cays located 0.5–1 km offshore to the south-east of Sandy Hook. Pigeon Creek supports mangrove and there is some shrubland on Sandy Hook, but the primary habitats of importance are the sandy beaches and rocky cays. The IBA includes marine areas 1 km from the shore and from the cays.

■ **Birds**

This IBA is home to regionally significant populations of breeding Least Tern *Sterna antillarum* (on the beaches at Sandy Hook), and Audubon's Shearwater *Puffinus lherminieri* and Bridled Tern *S. anaethetus* (on the offshore cays). Mixed flocks of seabirds, including (additionally) Roseate Tern *S. dougallii,* Sooty Tern *S. fuscata* and Brown Noddy *Anous stolidus* are seen feeding close to the mouth of Pigeon Creek in the fall.

■ **Other biodiversity**

Nothing recorded.

■ **Conservation**

Low Cay, High Cay and Sandy Hook IBA is a mix of crown and privately-owned land. The residential development at Sandy Hook is likely to be further expanded which will cause inevitable habitat destruction and increase the disturbance to birds on Snow Bay Beach. The cays are visited by tourists (on jet skis) from the Club Med Resort. Local "guides" use the cays for commercial purposes. This visitation is unregulated and will cause inevitable disturbance to the nesting seabirds. It is unknown whether rats *Rattus* spp. are present on the cays.

BS032 Samana Cay

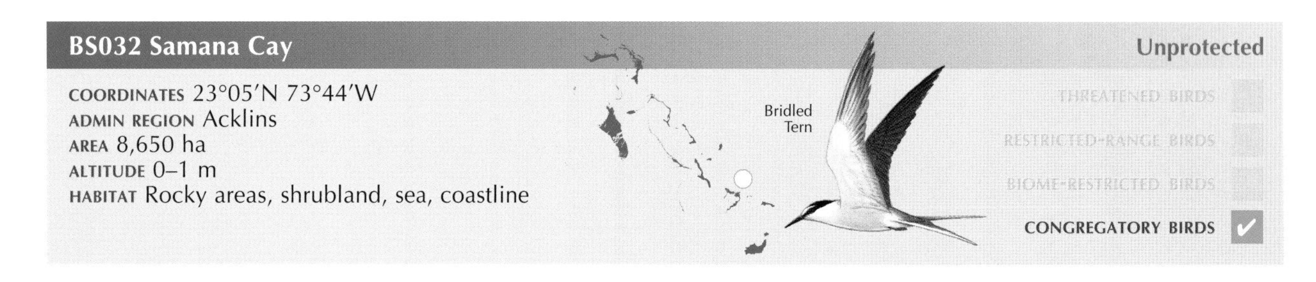

■ Site description

Samana Cay is located 32 km north-east of Crooked Island (Crooked and Acklins islands). It is a small island (c.16 km by 3 km) completely surrounded by coral reefs and with a small offshore islet—Propeller Cay—situated off its eastern end. The cascarilla tree *Croton elutaria* grows profusely on the island. Samana Cay is uninhabited (although there is evidence of Lucayan inhabitants present up until the 1500s). However, it is visited frequently by locals for fishing and collecting cascarilla bark. There is no fresh water on the cay. The IBA includes marine areas up to 1 km from the cay.

■ Birds

A globally significant population of Bridled Tern *Sterna anaethetus* nests on Propeller Cay, along with a range of other seabirds including Audubon's Shearwater *Puffinus lherminieri*, Sooty Tern *S. fuscata* and Brown Noddy *Anous stolidus*. A regionally important population of Royal Tern *S. maxima* breeds in the IBA. Brown Booby *Sula leucogaster* has been found roosting on Propeller Cay although there is no evidence of breeding.

■ Other biodiversity

Nothing recorded.

■ Conservation

Samana Cay is crown land but is unprotected. Disturbance to seabirds is likely to be caused by visitors (locals collecting bark and fishing and boaters). There is some local, small scale farming practiced on the island, and refugees land on the island seeking food and shelter. It is unknown whether rats *Rattus* spp. are present on the cays.

BS033 Cay Lobos

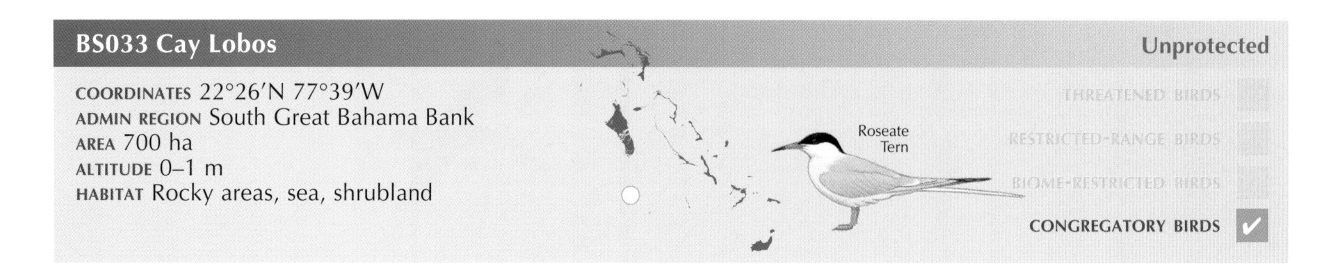

■ Site description

Cay Lobos IBA is a minute cay (c.250 long) located on the southern edge of the Great Bahama Bank, c.32 km north of Cuba's Cayo Romano. The cay (which is in Bahamian territorial waters) is uninhabited. The Cay Lobos lighthouse was built on the island in 1860, and this is the dominant feature. A small area of low shrubland surrounds the lighthouse, but the rest of the cay comprises sandy beach and surrounding reef.

■ Birds

This IBA supports a globally significant breeding population of Roseate Tern *Sterna dougallii*, and regionally important populations of Least Tern *S. antillarum* and Bridled Tern *S. anaethetus*. Many Neotropical migratory birds were collected on the island (attracted by the light of the lighthouse) between 1899 and 1901, but there is little subsequent information on the landbirds using this site.

■ Other biodiversity

Nothing recorded.

■ Conservation

Cay Lobos is crown land but is unprotected. There is no threat of development on the island which is, however, a stopping point for fishermen (both Bahamian and Cuban) who inevitably disturb the breeding seabirds. Illegal egg collecting and killing of the birds by refugees and fishermen is thought to be a threat to the seabird populations. It is unknown if rats *Rattus* spp. are present on the cay. Scuba-divers visit the cay to dive on the surrounding reefs.

■ Site description

Cay Verde IBA is an isolated cay at the south-easternmost edge of the Grand Bahama Bank, 48 km east of Greater Ragged Island and 110 km west of the southern tip of Acklins Island. It covers about 16 ha and supports extensive growth of sea grape *Coccoloba uvifera*, prickly pear *Opuntia sp.* and sea lavender. However, there is no fresh water on the cay, and it is uninhabited. The IBA includes marine areas up to 1 km from the cay.

■ Birds

This IBA supports a large seabird colony. The breeding populations of Magnificent Frigatebird *Fregata magnificens* (99 pairs) and Brown Booby *Sula leucogaster* (550 pairs) are regionally significant although these estimates (made in 1979) were 60% lower than counts done in 1907. Other seabirds nest on the island including Sooty Tern *Sterna fuscata*, Bridled Tern *S. anaethetus*, Brown Noddy *Anous stolidus* and Audubon's Shearwater *Puffinus lherminieri* although there is no recent data.

■ Other biodiversity

Nothing recorded.

■ Conservation

Cay Verde IBA is crown owned but is unprotected. Illegal egg collecting and killing of the birds by refugees and fishermen is thought to be a threat to the seabird populations. It is unknown if rats *Rattus* spp. are present on the cay. With little recent information concerning the status of the island and its seabirds, this IBA should be a target for monitoring expedition.

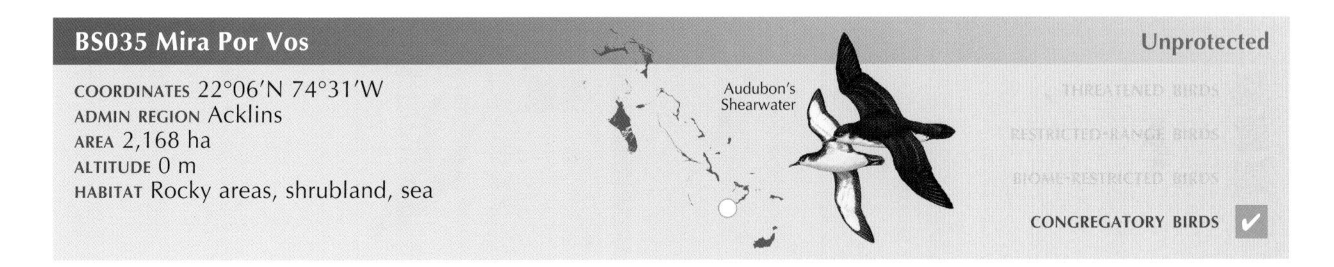

■ Site description

Mira Por Vos IBA comprises a series of uninhabited rocky islands and shoals spread across c.100 km^2. It is located c.14 km south-west of Salina Point, Acklins. South Cay supports a pond and North Rock one of the main seabird colonies.

■ Birds

This IBA is home to many seabirds. The breeding populations of Brown Booby *Sula leucogaster* (on North Rock) and Audubon's Shearwater *Puffinus lherminieri* are regionally significant. Sooty Tern *Sterna fuscata*, Bridled Tern *S. anaethetus* and Brown Noddy *Anous stolidus* also breed in the IBA. Caribbean Flamingo *Phoenicopterus ruber* and Reddish Egret *Egretta rufescens* have been seen at a pond on South Cay.

■ Other biodiversity

Nothing recorded.

■ Conservation

Mira Por Vos IBA is crown land but is currently unprotected. Little is known about the threats to the islands and their seabirds, but it is possible that refugees and fishermen land on the rocks to take eggs and birds for food. It is unknown if rats *Rattus* spp. are present on any of the islands.

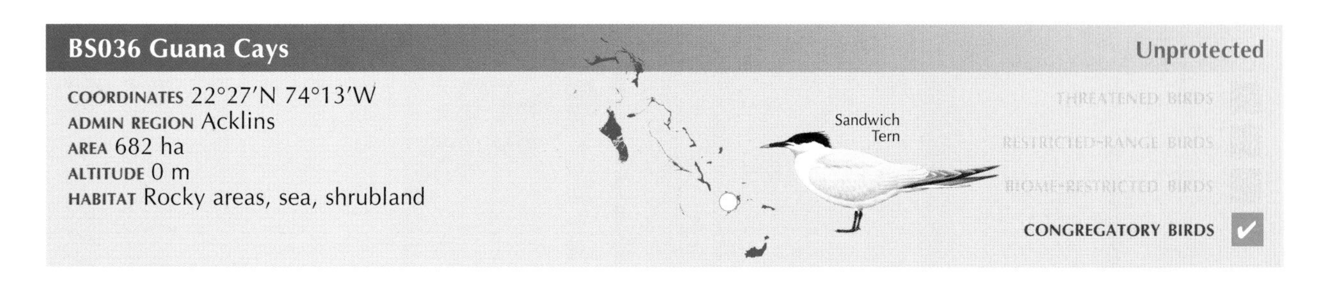

■ Site description

Guana Cays IBA comprises a group of small cays (including North Cay, Fish Cay and Guana Cay) and associated rocks and reefs. The cays are aligned in a loose chain across the south-western reef edge and entrance to the Bight of Acklins, lying between Crooked Island's Long Cay to the north and Binnacle Hill on Acklins Island to the south. They are assumed to be uninhabited, have rocky (coralline) coastlines, and support some scrub vegetation. The IBA extends to include marine areas up to 1 km from the cays.

■ Birds

This IBA supports a range of seabirds. The breeding colonies of Sandwich Tern *Sterna sandvicensis* and Roseate Tern *S. dougallii* are globally significant, while those of Least Tern *S. antillarum*, Bridled Tern *S. anaethetus* and Magnificent Frigatebird *Fregata magnificens* are regionally so.

■ Other biodiversity

Nothing recorded, although it is likely that globally threatened sea-turtles are present.

■ Conservation

Guana Cays IBA is crown land but is unprotected. The cays are poorly known in terms of their biodiversity, and an up-to-date assessment of their seabird populations is needed. Any such visit should assess current threats to the IBA such as disturbance from tourists (scuba divers, bone-fishers) and fishermen, or indeed the potential presence of predators such as rats *Rattus* spp.

■ Site description

Booby Cay IBA lies less than 500 m offshore from the easternmost end of the isolated Mayaguana Island. This uninhabited cay covers only c.75 ha, and dips in the centre of the island have formed two ponds which shrink and grow in water level and salinity according to rainfall (although they occupy c.30% of the island). There is a sandy beach along the north-western shore, and the south-east portion of the cay supports impenetrable shrubland coppice vegetation. Buttonwood, cacti and other plants grow around the central ponds.

■ Birds

This IBA supports a regionally significant population of Brown Booby *Sula leucogaster*. It is unknown if other seabirds breed on the island. There are reports of up to 80 non-breeding Caribbean Flamingo *Phoenicopterus ruber* on the cay (presumably part of the resident non-breeding flock on Mayaguana).

■ Other biodiversity

A subspecies of the Critically Endangered Bahamas rock iguana *Cyclura carinata bartschi* is endemic to Booby Cay (although in 1998 a colony was established on Mayaguana).

■ Conservation

Booby Cay IBA has been leased by the crown to a private individual who established the goats on the island that have significantly impacted the vegetation. There is no protection afforded this cay, although the BNT has proposed that it be included in the national parks system on the basis of the presence of the iguana and the breeding seabirds. Some goats have been removed, but this action needs to be completed to safeguard the island's biodiversity. Local conch fishermen occasionally overnight on the island. The status of introduced predators including cats and rats *Rattus* spp. is unknown.

BS038 Booby Rocks and Pirates Bay

COORDINATES 22°19′N 72°44′W
ADMIN REGION Mayaguana
AREA 3,620 ha
ALTITUDE 0–3 m
HABITAT Rocky areas, sea, shrubland, wetlands

■ Site description
Booby Rocks and Pirates Bay IBA is located on the north-western tip of the isolated Mayaguana Island. Booby Rocks are a cluster of rocks c.400 m offshore from the rocky shore of Northwest Point, at the head of the wide shallow Pirates Bay. Sandy beaches extend along Pirates Bay to Blackwood Point at the north-eastern tip of the bay. The IBA includes marine areas up to 1 km from the shore and Booby Rocks, and also the shallow mangrove wetlands (with adjacent coppice) lie on the landward side of the bay.

■ Birds
This IBA supports a range of seabirds, all in regionally significant populations. There is a Brown Booby *Sula leucogaster* nesting colony on Booby Rocks, and White-tailed Tropicbird *Phaethon lepturus* nest on the cliffs at Northwest Point. Non-breeding numbers of Brown Pelican *Pelecanus occidentalis*, Masked Booby *Sula dactylatra* and Royal Tern *Sterna maxima* are also regionally important. Magnificent Frigatebird *Fregata magnificens* nest in the IBA. The wetlands support shorebirds, ducks, herons and egrets. Reddish Egret *Egretta rufescens* is apparently common, and up to 200 Caribbean Flamingo *Phoenicopterus ruber* frequent the wetlands at Blackwood Point. The Near Threatened White-crowned Pigeon *Patagioenas leucocephala* breeds in the IBA.

■ Other biodiversity
Nothing recorded.

■ Conservation
Booby Rocks and Pirates Bay IBA is unprotected crown and private land. A mega resort development started to be built in 2006. It includes plans to connect the wetlands to the sea for a commercial marina in the north-western corner of the island (within the IBA) which will have a serious impact on the natural vegetation and these currently undisturbed fresh and salt-water habitats. It is unknown if predators are present on Booby Rocks, or indeed if there are other threats impinging on the seabird populations within the IBA.

BS039 Great Inagua

COORDINATES 21°04′N 73°21′W
ADMIN REGION Inagua
AREA 178,140 ha
ALTITUDE 0–5 m
HABITAT Wetlands

■ Site description
Great Inagua IBA embraces the entire island of Great Inagua—the southernmost (and third largest) island in the Bahamas, lying just 90 km north-east of the easternmost tip of Cuba. The island is c.90 km by 30 km, and Lake Rosa occupies c.30% of the western end. Lake Rosa is a permanent shallow brackish lake, up to 1.5 m deep with small islands scattered throughout. It is fringed with brackish marshes, and dense mangrove swamps on the northern and eastern borders. The rest of the island comprises seasonal marshes, open shrubland and broadleaf coppice on the higher ground. The western portion of Lake Rosa is managed for commercial salt production.

■ Birds
This IBA is home to a wide diversity and large numbers of waterbirds. Over 40,000 Caribbean Flamingo *Phoenicopterus ruber* occur (the largest colony outside of Cuba), and populations of Reddish Egret *Egretta rufescens*, Roseate Tern *Sterna dougallii*, Common Tern *Sterna hirundo* and the Vulnerable West Indian Whistling-duck *Dendrocygna arborea* are globally significant. A number of other waterbirds are present in regionally important numbers. Over 6,000 Near Threatened Cuban Amazon ("Bahama Parrot") *Amazona leucocephala bahamensis* occur on the island, and there are records of the Vulnerable Bahama Swallow *Tachycineta cyaneoviridis* although numbers involved are unknown.

■ Other biodiversity
Critically Endangered hawksbill *Eretmochelys imbricata* and Endangered green *Chelonia mydas* turtles are present, and the endemic Inagua freshwater turtle *Chrysemys malonei* occurs.

■ Conservation
Great Inagua is a mix of crown and private land. About 50% of it is protected within the Inagua National Park (which is also designated a Ramsar site), although the park has just one warden to manage and monitor it. Recognising these issues, the BNT has been working with the local Sam Nixon Bird Club (a Site Support Group) to monitor the IBA and its birds and to develop micro-enterprises to assist in the establishment of ecotourism on the island. Wild pigs, donkeys and cats all represent a threat to the natural vegetation and nesting waterbirds. Occasional unauthorised hunting occurs within portions of the national park.

BARBADOS

LAND AREA **431 km²** ALTITUDE **0–343 m**
HUMAN POPULATION **279,000** CAPITAL **Bridgetown**
IMPORTANT BIRD AREAS **7, totalling 2.06 km²**
IMPORTANT BIRD AREA PROTECTION **74%**
BIRD SPECIES **230**
THREATENED BIRDS **0** RESTRICTED-RANGE BIRDS **4**

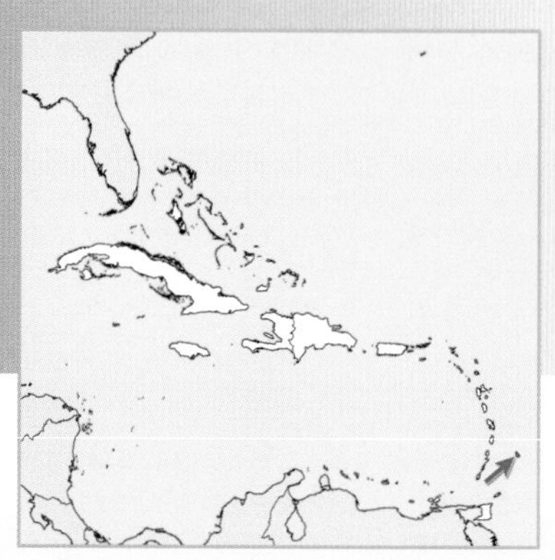

WAYNE BURKE

Chancery Lane. (PHOTO: RYAN CHENERY)

INTRODUCTION

Barbados is situated in the Atlantic, east of the Windward Islands towards the southern end of the Lesser Antilles. It is the most easterly of the Lesser Antilles, lying 165 km east of St Vincent and 220 km north-north-east of Tobago. The island is teardrop shaped, 31 km long from north to south, and 22 km east to west at its widest (in the south). Barbados is divided into 11 parishes with Bridgetown, the capital, in the south-western parish of St Michael. It is densely populated. In contrast to the older, mountainous volcanic islands of the Lesser Antilles, Barbados is a geologically-recent, low-lying, coral island. The majority of the island (85%) rises in a series of raised coral limestone terraces. Two of these old sea cliffs run north–south inland of the west coast, and another runs east–west inland of the south coast. The oldest and highest ridge (over 300 m) overlooks the island's east coast. The limestone in this part of the island is fractured in places by an extensive network of gullies up to 30 m deep. The remainder of the island, on the north-east coast (north and east of Hackleton's Cliff, in the parishes of St Andrew, St Joseph and St John) is eroded into irregular topography with sharp ridges and steep valleys exposing older sedimentary rock. This rolling, rugged landscape is known as the Scotland District and is the least populated part of the island. The coast of Barbados is primarily sandy beach, but sea cliffs (some up to 35 m high) dominate in the south-east and at the northern end of the island.

Barbados' climate is tropical marine with a dry season characterised by north-east trade winds between November and May. Average annual precipitation is 1,400–1,500 mm, falling mainly during June to October. The deciduous and semi-deciduous forest that once covered the island was almost entirely removed for cultivation (especially for sugar cane) within c.60 years of British settlement in 1627. Relict woodlands persist on steep slopes at Turner's Hall Woods (c.20 ha), and the under-cliff woodland in the Scotland District is regenerating. Mature woodland also survives in the gullies that cross the coral surface of the island—c.250 km of the total gully system is wooded. Surface water is scarce on the limestone island. The wetland of greatest significance is the 33-ha Graeme Hall Swamp on the south coast. This wetland supports areas of fresh water with sedges, and others of brackish water with red *Rhizophora mangle* and white *Laguncularia racemosa* mangroves. The smaller, seasonal Chancery Lane Swamp is also a very important wetland. Much of the former sugar cane land has been, or is being converted to golf courses. Though these are mostly "green deserts", their irrigation ponds do offer some habitat for waterbirds.

Conservation

The 1907 Wild Birds Protection Act provides a measure of protection to most resident birds and some migrants. It is currently being reviewed, and proposals have been made to include the Vulnerable West Indian Whistling-duck *Dendrocygna arborea* and the island endemic Barbados Bullfinch *Loxigilla barbadensis*. Unfortunately, the only migratory shorebirds listed are Upland Sandpiper *Bartramia longicauda*, Buff-breasted Sandpiper *Tryngites subruficollis*, Hudsonian Godwit *Limosa haemastica* and Ruff *Philomachus pugnax*. Habitat protection is afforded under various designations such as the Graeme Hall Swamp Nature Preserve and Bird Sanctuary (given legal protection after it was designated a Ramsar site); Chancery Lane Swamp "special study area" (a poor second-best to the stricter protection suggested for this area by Captain Maurice Hutt); the Scotland District "protected landscape" (a poor second-best to the proposal by Hutt to make this area a national park); and the under-cliff woodlands of St John and St Joseph at Hackelton's Cliff and Joe's River, and Turner's Hall Wood which are now listed as "national forest candidates". If the endemic Barbados racer *Liophis perfuscus* still survives it will most likely to be found in these under-cliff woodlands.

Initiatives focused on Graeme Hall Swamp have been at the forefront of conservation in Barbados. Captain Hutt started efforts to protect the swamp in the 1970s, and Dr Karl Watson (University of the West Indies, UWI), continued these efforts. However, Peter Allard finally seized the opportunity to purchase the western portion of the swamp in 1994, and opened the Graeme Hall Nature Sanctuary (GHNS) in May 2004 with a mandate for conservation, environmental education and nature tourism. The whole 33-ha Graeme Hall Swamp was designated a Ramsar site in 2005 (and a proposal to government to make the area a national park is pending). Over 11,000 school children have now visited GHNS on discounted tours from the island's secondary and primary schools. GHNS has also sponsored celebrations of World Wetlands Day to raise awareness (among invited school children and UWI students) of the numerous benefits of wetland conservation. Karl Watson also works tirelessly to educate the public and students about the need for conservation measures. A regular water sampling regime is conducted for GHNS by the Centre for Resource Management and Environmental Studies (CERMES) of UWI. CERMES, through its affiliation with Caribbean Coastal Marine Productivity (CARICOMP), also conducts regular monitoring of the health of the swamp's mangrove ecosystem. GHNS staff record daily observations of bird life in the sanctuary, but throughout the rest of the island there is no formal bird conservation work being implemented, although a network of dedicated birdwatchers keeps detailed records of the island's avifauna.

Threats to birds and habitats on the island stem from a single root cause: the pressures of commercial and residential development on a small, densely populated island. The proposed construction of a "wind farm" in St Lucy by Barbados Light and Power is just one of many such development projects, which in this case could add to the mortality of migrant shorebirds in the parish. There is an urgent need for wise management of the inevitable encroachment of new developments (e.g. housing estates, golf courses etc.) on existing woodlands or indeed agricultural land. The pressure is enormous and the resistance limited. The threat to woodlands could be mitigated by a low-cost tree planting scheme (using native tree species) implemented on tracts of government-owned land in the Scotland District. Secondary school environmental groups could play an active part in such a scheme, keeping costs low and returns high.

The complex issue of bird shooting on the island also needs to be addressed. This is not to argue for elimination of the "sport" but for regulation based on accurate data concerning numbers of each species being shot. With greater transparency and a more accurate picture of numbers, informed decisions about species-specific bag limits could be set, thus protecting the most vulnerable birds (such as American Golden Plover *Pluvialis dominica* which has a global population estimated at just 200,000 individuals). The artificially maintained shooting swamps provide habitat for many non-target waterbirds for at least part of the year and some that are maintained throughout the year provide year-round habitat. These swamps are important components of the island's wetland network, and they exist solely as a result of shooting-specific management actions. Even with regulated hunting, the ideal would be for the maintenance of two "no-shooting" wetlands (one in the north and one in the east) to offer sanctuary for migratory shorebirds. This, combined with the preservation of Chancery Lane Swamp IBA (BB005), would provide necessary refuge. Unfortunately, such projects are severely constrained by lack of funding.

Birds

Typical of oceanic islands the breeding avifauna of Barbados is depauperate with just 28 species breeding. However, Barbados' position east of the main island chain makes it a first landfall for wandering migrants (e.g. trans-Atlantic species) resulting in over 33% of the island's recorded species (75 of 230) being considered vagrants. This is reflected in the occurrence of globally threatened birds on the island. The Near Threatened Piping Plover *Charadrius melodus* is known from a single record (of a bird shot) in 1957; four West Indian Whistling-ducks *Dendrocygna arborea* (Vulnerable) were present at Graeme Hall Swamp in 1961; two Black-capped Petrels *Pterodroma hasitata* (Endangered) 10–12 km south-west of Barbados in April 2003 were the first island records; and Caribbean Coot *Fulica caribaea* (Near Threatened) started breeding on the island in 1999. The Critically Endangered Eskimo Curlew *Numenius borealis* was once a regular autumn passage migrant (into the late nineteenth century) but the last certain record on Barbados (and indeed anywhere) was of a single bird shot in September 1963.

Congo Road Swamp—one of four shooting swamps in St Philip. (PHOTO: JIM KUSHLAN)

The island-endemic Barbados Bullfinch. (PHOTO: STEVE MLODINOW)

Table 1. Key bird species at Important Bird Areas in Barbados.

			Barbados IBAs						
			BB001	BB002	BB003	BB004	BB005	BB006	BB007
		Criteria		■	■	■	■	■	■
Key bird species	Criteria	National population	■	■	■	■	■	■	■
Masked Duck *Nomonyx dominicus*	■	30							30
Audubon's Shearwater *Puffinus lherminieri*	■	150–300	150						
Little Egret *Egretta garzetta*	■	24			24	21	✓		
American Golden Plover *Pluvialis dominica*	■	10,000–25,000		4,050–9,000				4,050–9,000	
Greater Yellowlegs *Tringa melanoleuca*	■	1,000–5,000		2,550				2,550	
Lesser Yellowlegs *Tringa flavipes*	■	25,000–100,000		13,200				13,200	
Pectoral Sandpiper *Calidris melanotos*	■	10,000–25,000		6,600				6,600	
Green-throated Carib *Eulampis holosericeus*	■			✓	✓	✓	✓	✓	✓
Antillean Crested Hummingbird *Orthorhyncus cristatus*	■			✓	✓	✓	✓	✓	✓
Caribbean Elaenia *Elaenia martinica*	■				✓				
Barbados Bullfinch *Loxigilla barbadensis*	■			✓	✓	✓	✓	✓	✓

All population figures = numbers of individuals.
Restricted-range birds ■. Congregatory birds ■.

Little Egret colony at Graeme Hall Swamp—the first New World breeding site for the species. (PHOTO: JIM KUSHLAN)

Barbados is a geologically young island compared with its neighbouring Antillean islands and, as a result, levels of endemism are low. The Barbados Bullfinch *Loxigilla barbadensis*, which is common in all habitats, has recently been recognised as a species distinct from the Lesser Antillean Bullfinch *L. noctis*. It is the only island endemic, and it occurs alongside three other Lesser Antilles Endemic Bird Area restricted-range birds, namely Green-throated Carib *Eulampis holosericeus*, Antillean Crested Hummingbird *Orthorhyncus cristatus* and Caribbean Elaenia *Elaenia martinica*. The two hummingbirds are common in woodlands and gardens. Two more restricted-range species are known from the island, but are not considered to sustain viable populations. The Scaly-breasted Thrasher *Margarops fuscus* was last recorded in the 1920s and is now almost certainly extirpated from the island, and the Pearly-eyed Thrasher *M. fuscatus* is known from just a few sightings (possibly of vagrants) over the last two years.

Barbados is most important for its waterbirds. The network of natural and (in the case of the shooting swamps) artificially maintained wetlands provides critical habitat for an increasing number of waterbird species. In recent decades, a range of waterbirds have been added as breeding species to the Barbados avifauna. For example, Little Egret *Egretta garzetta* first nested in 1994 (representing the first breeding record of this species in the New World); Snowy Egret *E. thula* also nested for the first time in 1994; Pied-billed Grebe *Podilymbus podiceps* in 2004; Black-bellied Whistling-duck *Dendrocygna autumnalis* in 2002; and Masked Duck *Nomonyx dominicus* in 1990. These, and other waterbirds, rely on a functioning network of wetlands to provide their various feeding and breeding requirements throughout the year.

It is as a staging post for Arctic-nesting Neotropical migratory shorebirds that Barbados stands out as of global importance. Adverse weather conditions in the Atlantic can cause large flights of shorebirds to put down in Barbados' wetlands. Unfortunately, many of these birds are shot at privately-owned wetlands designed and maintained specifically to attract the flocks. Though exact data are not available, the number of birds killed by the 10 active shooting swamps on the island ranges between 15,000 and 30,000 each July-to-October shooting season. Information from hunters in the five shooting swamps in the northern parish of St Lucy suggests that between 2,400–3,000 birds are shot at each swamp, and that a combined total of 12,000–15,000 birds are shot each year in this parish. The same hunters have suggested that the number of birds shot represents "just" 10% of the total number of birds passing (although this estimate comes with obvious biases), indicating that 150,000–300,000 shorebirds could be using the island's wetlands each autumn. Pectoral Sandpiper *Calidris melanotos* and Lesser Yellowlegs *Tringa flavipes* make up 70–75% of the birds shot, and 10% are American Golden Plovers *Pluvialis dominica*. It is worth noting that the last confirmed record of *Numenius borealis* was of a bird killed at a shooting swamp in St Lucy in 1963.

IMPORTANT BIRD AREAS

Barbados' seven IBAs—the island's international priority sites for bird conservation—cover 185 ha (including marine areas), but only 0.1% of the island's land area. Three of the IBAs have some form of protective designation, representing 74% of the area covered by the IBAs, although some of these designations do not alleviate the threat of development. The IBAs have been identified on the basis of 11 key bird species (listed in Table 1) that variously trigger the IBA criteria. These species include all four restricted-range species, and seven congregatory waterbirds/ seabirds. The IBAs are wetland focused, and together they form an important national network of sites for the waterbird species that rely on them. The occurrence of the various restricted-range birds is mostly incidental with three of these species being widespread and common in most habitats. However, in terms of the IBA network, the Caribbean Elaenia *Elaenia martinica* is found only in Graeme Hall Swamp IBA (BB003). If either of the restricted-range *Margarops* thrashers were found to be breeding on the island again, it would be appropriate to identify woodland IBA supporting populations of all of the restricted-range birds. No globally threatened species occur on the island in numbers significant for IBA identifications. However, the small, recently established population of the Near Threatened Caribbean Coot *Fulica caribaea* should be monitored as it grows.

Threats to the IBAs are essentially those that are outlined above (see Conservation), namely pressure from developments. Ironically, Graeme Hall Swamp IBA (BB003), Barbados' best protected area, faces probably the greatest threats, especially related to the maintenance of water quality. There is measureable mild biocide runoff into the swamp from surrounding agricultural land, including from government-owned agricultural land inside the swamp watershed. More concerning though is the South Coast Sewerage Project (SCSP) treatment plant which is situated in the government-owned eastern section of the wetland. In the event of a plant failure, there is a plan is to discharge raw sewerage directly into the wetland. This occurred in July 2005 with a small discharge that resulted in a limited fish-kill and blue-green algal bloom. However, a major discharge of untreated sewerage would result in serious eutrophication.

State, pressure and response scores at each IBA should be monitored annually to provide an objective status assessment and highlight management interventions that might be required to maintain these internationally important biodiversity sites. Monitoring the status of the key bird species (listed in Table 1) at each IBA will be an important component of this broader site-monitoring process. Constant vigilance is needed to ensure that Bird Rock IBA (BB001) remains rat free, and that the Audubon's Shearwater *Puffinus lherminieri* colony is allowed to thrive.

Figure 1. Location of Important Bird Areas in Barbados.

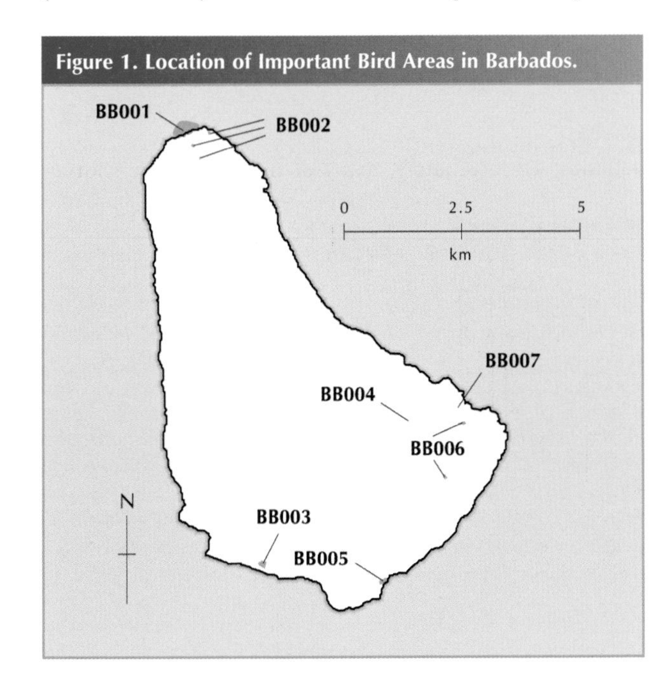

KEY REFERENCES

BRODKORB, P. (1964) Fossil birds from Barbados, West Indies. *J. Barbados Mus. and Hist. Soc.* 31: 1964.

BUCKLEY, P. A. AND BUCKLEY, F. G. (2004) Rapid speciation by a Lesser Antillean endemic, Barbados Bullfinch *Loxigilla barbadensis*. *Bull. Brit. Orn. Club* 124: 108–123.

BURKE, W. (2005) Another first for Barbados at Graeme Hall Swamp. *J. Barbados Mus. and Hist. Soc.* 51.

COPPIN, L. AND JONES, L. (2005) An assessment of the Cattle Egret (*Bubulcus ibis*) population in Barbados. *J. Barbados Mus. and Hist. Soc.* 51.

FIELDEN, H. W (1889) On the breeding of *Puffinus auduboni* in the island of Barbados. *Ibis* (6)1: 60–63.

FROST, M. AND MASSIAH, E. (2001) Caribbean Coot (*Fulica caribaea*) – the return of a former breeding resident bird. *J. Barbados Mus. and Hist. Soc.* 47.

FROST, M. AND MASSIAH, E (2001) The Recent colonisation of Masked Duck (*Nomonyx dominicus*) on Barbados. *J. Barbados Mus. and Hist. Soc.* 47.

FROST, M. AND MASSIAH, E (2001) Four new and rare Nearctic–Neotropical landbird migrants during autumn from Barbados. *Carib. J. Sci.* 37: 3–4.

FROST, M. AND MASSIAH, E. (2001) An annotated list of recent bird species new to Barbados. *J. Barbados Mus. and Hist. Soc.* 47.

HUGHES, G. (1750) *The natural history of Barbados*. London: printed for the author (in 10 books). (Reprinted in 1972, New York: Arno Press).

HUTT, M. B. (1985) The case for the preservation of the Graeme Hall Swamp area. (Unpublished report).

HUTT, M. B. (1991) Shooting of migrating shorebirds in Barbados. Pp.77–91 in T. Salathé, ed. *Conserving migratory birds*. Cambridge, U. K.: International Council for Bird Preservation (Techn. Publ. 12).

HUTT, M. B. AND HUTT, H. (1992) The birds of Barbados, West Indies: an annotated checklist. (Unpublished manuscript).

LOVETTE, I., SEUTIN, G., RICKLEFS, R. AND BERMINGHAM, E. (1999) The assembly of an island fauna by natural invasions: sources and temporal patterns in the avian colonization of Barbados. *Biological Invasions* 1: 33–41.

MASSIAH, E. (1996) Identification of Snowy Egret and Little Egret. *Bird News* 9: 11.

MCNAIR, D., MASSIAH, E. AND FROST, M. (1999) New and rare species of Nearctic landbird migrants during autumn for Barbados and the Lesser Antilles. *Carib. J. Sci.* 35: 1–2.

MCNAIR, D. B., SIBLEY, F., MASSIAH, E. B. AND FROST, M. D. (2002) Ground-based Nearctic–Neotropic landbird migration during autumn in the eastern Caribbean. Pp.86-1-3 in F. E. Hayes and S. A. Temple, eds. *Studies in Trinidad and Tobago ornithology honouring Richard ffrench*. St Augustine, Trinidad: Dept. Life Sci., Univ. West Indies (Occ. Pap. 11).

RAFFAELE, H. WILEY J., GARRIDO, O., KEITH, A. AND RAFFAELE, J. (1998) *A guide to the birds of the West Indies*. Princeton, New Jersey: Princeton University Press.

WATTS, D. (1987) *The West Indies: patterns of development, culture and environmental change since 1492*. Cambridge, U.K.: Cambridge University Press

ACKNOWLEDGEMENTS

The author would like to thank Kerryann Branford (GHNS), Ryan Chenery (GHNS), Martin Frost, Renata Goodridge (CERMES and CARICOMP), Lorna Inniss (Coastal Zone Management Unit), Ed Massiah and Shontelle Wellington (Ministry of Energy and Environment) for their generous contributions to and reviews of this chapter. Also, Captain Hutt (1919–1998), on whose observations and records modern field ornithology in Barbados has been built.

■ Site description

Bird Rock IBA lies offshore at the northernmost tip of Barbados. The coastline comprises cliffs and boulder-strewn shores, and the 0.5-ha Bird Rock is c.30 m from this shoreline. A narrow footpath leads to a small sandy beach almost opposite Bird Rock, but at the present, no human settlement encroaches on the rock or the immediately adjacent coast. The IBA includes all marine areas up to 1 km from the island.

■ Birds

This IBA is significant for its breeding population of Audubon's Shearwater *Puffinus lherminieri*. "Considerable numbers" were recorded in the late nineteenth century. More recently (1996) counts of birds coming to the site after sunset suggested a population of 50–100 pairs. No other seabirds are known to breed on Bird Rock.

■ Other biodiversity

Nothing recorded.

■ Conservation

Bird Rock IBA is within a Natural Heritage Conservation Area that extends down the east coast of Barbados. It is also within the study area for a Coastal Zone Management Plan. Building is restricted within 400 m of the cliff along this section of coastline. However, any coastal development that permitted security lighting near the cliff would likely result in disturbance and disorientation of the shearwaters breeding on Bird Rock. No rats are present, but Bird Rock should be monitored for this and other potential predators.

■ Site description

The St Lucy Shooting Swamps IBA comprises five separate wetlands, each less than 2 ha in extent, in the northernmost parish of St Lucy. The wetlands are all on private lands and are artificially created and maintained for the express purpose of providing habitat to lure migrating Neotropical shorebirds down so they can be shot. Generally, the immediate environs of these wetlands are farmed for cattle pasture or sugarcane.

■ Birds

This IBA is critical for Neotropical migratory shorebirds migrating south between July and October. It is reported that a collective total of 10,000–15,000 Nearctic nesting shorebirds (70–75% of which are Pectoral Sandpipers *Calidris melanotos* and Lesser Yellowlegs *Tringa flavipes*) are shot each year in the swamps of this IBA. The number of birds shot is an unknown percentage of the total numbers using the wetlands, but could be c.10% or more. Numbers of birds stopping (and thus the numbers shot) vary from year to year as a result of weather conditions. More birds stop when affected by adverse weather associated with tropical Atlantic depressions and storms. These wetlands also provide useful year-round habitat for non-target wetland birds, and three (of the four) Lesser Antilles EBA restricted-range birds occur. The last confirmed record of Eskimo Curlew *Numenius borealis* was shot at Fosters Swamp in this IBA.

■ Other biodiversity

The endemic lizard *Anolis extremus* occurs.

■ Conservation

Each of the individual shooting swamps is privately-owned and managed for shooting. Swamp management by the shooting clubs includes water being pumped onto them in the wet (shooting) season. A ban or restriction on hunting would result in the cessation of swamp management and the gradual drying out of the wetlands. Of particular concern are the numbers of American Golden Plovers *Pluvialis dominica* being shot (10% of the total bag from this IBA)—depending on the year this can be up to 0.6% of the global population shot solely within this IBA. Setting bag limits for this species and establishing "no-shooting" swamps within St Lucy would be desirable conservation goals, as would the pumping of water during the dry season to maintain habitat for other waterbirds.

■ **Site description**
Graeme Hall Swamp IBA is in south-west Barbados, 5 km east of Bridgetown, just inland from the coast. Residential and commercial tourism development surrounds the swamp along the southern, eastern, western and north-western boundaries. The coast road (Highway 7) runs between the swamp and the sea to the south, and agricultural lands border the swamp in the north-east. The IBA supports the largest body of inland water on the island. The swamp is divided into a freshwater marsh (eastern section) and a brackish lake (western section) by a north–south, man-made roadway and drainage canal (the swamp's only connection to the sea).

■ **Birds**
This IBA is one of only two documented breeding sites for Little Egret *Egretta garzetta* in the Western Hemisphere, with up to 24 birds counted on their favoured mangrove island in the brackish lake. The wetlands support a wide diversity of waterbirds (residents, migrants and vagrants) although not the numbers of shorebirds that used to occur when it was managed as a shooting swamp. Populations of all four Lesser Antilles EBA restricted-range birds occur in this IBA. The mangroves support the highest density of resident Yellow Warblers *Dendroica petechia* on the island.

■ **Other biodiversity**
Graeme Hall Swamp IBA supports the largest remaining stands of red and white mangrove woodland on the island. Introduced green monkeys and mongooses occur.

■ **Conservation**
The brackish western sector of this IBA is privately-owned and managed by the Graeme Hall Nature Sanctuary, and has been the focus of restoration activities. The eastern sector of the IBA is government-controlled and has not undergone any significant restorative effort, being managed by the Ministry of Health for mosquito control. This control consists of clearing vegetation, removing obstructing mangroves and intensive thermal fogging with Malathion. Increased protection is required for this IBA although a proposed plan for a national park describes the significant reduction of suitable waterbird habitat in preference to human recreational use. Invasive alien species are presumably impacting the avifauna. Increasing noise and light pollution from events at Graeme Hall Nature Sanctuary is a concern for nesting egrets. Further manipulation or modification of this already heavily modified wetland must be careful not to disrupt or disturb its attractiveness for birds in favour of human recreation.

■ **Site description**
East Point Pond IBA is in south-eastern Barbados, c.3 km inland (south-west) of the coast at St Marks, and south of Highway 4. This small pond is artificially maintained and surrounded by grazing land. It forms part of a complex of wetland and contiguous pastureland that exists throughout the St Philip Shooting Swamps IBA (BB006).

■ **Birds**
This IBA is significant as a feeding area for the Little Egrets *Egretta garzetta* (21 were counted in 2007) that breed at Graeme Hall Swamp IBA (BB003). It is also the only documented nesting site for the Near Threatened Caribbean Coot *Fulica caribaea* on the island, with about 3–4 pairs breeding. Populations of three (of the four) Lesser Antilles EBA restricted-range birds occur, namely Green-throated Carib *Eulampis holosericeus*, Antillean Crested Hummingbird *Orthorhyncus cristatus* and the island-endemic Barbados Bullfinch *Loxigilla barbadensis*.

■ **Other biodiversity**
Nothing recorded.

■ **Conservation**
East Point Pond IBA is privately owned (on the lands of East Point House) and not formally protected. It is artificially maintained with water pumped into it (although as at June 2008 pumping appears to have stopped). It is not currently threatened, although any change in land use or management regime (such as the cessation of pumping) would have a profound and damaging impact on the waterbirds, many of which rely on this IBA as part of an island-wide network of wetlands that is needed to satisfy their annual feeding and breeding requirements.

■ Site description

Chancery Lane Swamp IBA is on the south coast of Barbados. It is a seasonal, coastal wetland comprising an irregular mosaic of shallow open water, mudflats and grassy areas. This natural wetland IBA is behind a well developed coralline sand-dune system bound by a vegetated berm to seaward, pasture and an inland cliff. Residential development is encroaching towards the western end of the marsh from Fairy Valley Rock and Chancery Lane.

■ Birds

This IBA is a critical refuge for Neotropical migratory (and vagrant) shorebirds but is significant as a feeding area for the Little Egrets *Egretta garzetta* (>20) that breed at Graeme Hall Swamp IBA (BB003). Populations of three (of the four) Lesser Antilles EBA restricted-range birds occur, namely Green-throated Carib *Eulampis holosericeus*, Antillean Crested Hummingbird *Orthorhyncus cristatus* and the island-endemic Barbados Bullfinch *Loxigilla barbadensis*.

■ Other biodiversity

Chancery Lane is the only place in Barbados known to support stands of buttonwood *Conocarpus erectus*. The coastal wetlands are rare in terms of the island's ecosystems. The beach is an important nesting site for globally threatened sea-turtles (especially the Critically Endangered hawksbill *Eretmochelys imbricata* and [possibly] leatherback *Dermochelys coriacea* turtles).

■ Conservation

Chancery Lane Swamp IBA is privately owned but has been designated a Natural Heritage Conservation Area and a Special Study Area. It also embraces an important archaeological site. The Chancery Lane beach and surroundings are a popular recreation area for local residents and tourists who stay in hotels and guesthouses nearby. A large portion of the land in the Special Study Area has been approved for development. Any commercial or residential development that does not leave it intact or does not allow an adequate buffer zone will degrade the wetland, and severely damage the landscape value. A permanent protective status preventing development encroachment and enabling appropriate water management are essential for the long-term survival of Chancery Lane Swamp IBA. This IBA forms part of an island-wide network of wetlands that the waterbirds of Barbados rely on to satisfy their annual feeding and breeding requirements.

■ Site description

The St Philip Shooting Swamps IBA comprises four shooting swamp wetlands, each less than 2 ha in extent, in the south-eastern parish of St Philip. Two of the swamps are contiguous, while Congo Road and Hampton Swamp are separate. The wetlands are all on private lands and are artificially created and maintained for the express purpose of providing habitat to lure migrating Neotropical shorebirds down so they can be shot. Generally, the immediate environs of these wetlands are cattle pasture.

■ Birds

This IBA is critical for Neotropical migratory shorebirds migrating south between July and October. It is reported that a collective total of 10,000–15,000 Nearctic nesting shorebirds (70–75% of which are Pectoral Sandpipers *Calidris melanotos* and Lesser Yellowlegs *Tringa flavipes*) are shot each year in the swamps of this IBA. The number of birds shot is an unknown percentage of the total numbers using the wetlands, but it could be c.10% or more. Numbers of birds stopping (and thus the numbers shot) vary from year to year as a result of weather conditions. More birds stop when affected by adverse weather associated with tropical Atlantic depressions and storms. These wetlands also provide useful year-round habitat for non-target wetland birds, and for populations of three (of the four) Lesser Antilles EBA restricted-range birds.

■ Other biodiversity

The endemic lizard *Anolis extremus* occurs.

■ Conservation

These four shooting swamps are privately-owned and managed for shooting. Swamp management by the shooting clubs includes water being pumped onto them during the wet (shooting) season. A ban or restriction on hunting would likely result in the cessation of swamp management and the gradual drying out of the wetlands. Of particular concern are the numbers of American Golden Plovers *Pluvialis dominica* being shot (10% of the total bag from this IBA)—depending on the year this can be up to 0.6% of the global population shot solely within this IBA. Setting bag limits for this species and establishing "no-shooting" swamps within St Philip would be desirable conservation goals, as would pumping water onto the swamps in the dry season to maintain habitat for waterbirds. Pressure from residential development is a constant and real threat in this parish—any such developments adjacent to the swamps would likely result in the abandonment of management.

BB007 Bayfield Pond

Unprotected

COORDINATES 13°10′N 59°27′W
ADMIN REGION St Philip
AREA 1 ha
ALTITUDE 37–39 m
HABITAT Freshwater wetland

Masked Duck

THREATENED BIRDS
RESTRICTED-RANGE BIRDS 3
BIOME-RESTRICTED BIRDS
CONGREGATORY BIRDS ✓

■ Site description

Bayfield Pond IBA is a small, permanent pond situated in the village of Bayfield, on the south-east coast of Barbados. The near circular pond is surrounded on all sides by houses with just a narrow strip of herbaceous vegetation and some trees between the waters edge and the residential roads. Floating aquatic plants cover a significant portion of the pond's surface.

■ Birds

This IBA is important for Masked Duck *Nomonyx dominicus* which is resident although numbers increase up to 30 individuals during the dry season suggesting this pond is of national importance for the species, and emphasising the reliance of some waterbirds on an island-wide network of wetlands to satisfy their annual feeding and breeding requirements. Populations of three (of the four) Lesser Antilles EBA restricted-range birds occur, namely Green-throated Carib *Eulampis holosericeus*, Antillean Crested Hummingbird *Orthorhyncus cristatus* and the island-endemic Barbados Bullfinch *Loxigilla barbadensis*.

■ Other biodiversity

Nothing recorded.

■ Conservation

Bayfield Pond IBA is privately owned and not formally protected. Being surrounded by housing it is subject to human disturbance, the affects of pollution, changes in aquatic vegetation and exotic introductions (such as invasive plants and also exotic fish).

BERMUDA

LAND AREA **53 km^2** ALTITUDE **0–76 m**
HUMAN POPULATION **66,160** CAPITAL **Hamilton**
IMPORTANT BIRD AREAS **1, totalling 7.6 km^2**
IMPORTANT BIRD AREA PROTECTION **100%**
BIRD SPECIES **375**
THREATENED BIRDS **3** RESTRICTED-RANGE BIRDS **0**

ANDREW DOBSON (BERMUDA AUDUBON SOCIETY) AND JEREMY MADEIROS
(DEPARTMENT OF CONSERVATION SERVICES, BERMUDA GOVERNMENT)

Castle Harbour Islands, at the eastern end of Bermuda.
(PHOTO: ANDREW DOBSON)

INTRODUCTION

Bermuda, a UK Overseas Territory, is situated in the western North Atlantic, 917 km from Cape Hatteras, the nearest landfall in the USA. It is made up of a mini-archipelago of approximately 150 islands, of which the eight largest are joined by bridges or causeways. Bermuda is volcanic in origin and is the largest of three volcanic seamounts, which rise from mid-oceanic depths of over 3,500 m. Over time, and due to ideal conditions for coral growth, limestone deposits built up over the eroded volcanic base such that the present visible islands are entirely formed of limestone. The present-day islands comprise a hilly, rolling landscape with relatively steep, parallel hills and ridges separated by inter-dune lows or valleys, some of which extend below water level to form wetland areas. There are very few natural level areas of any size, the largest being the inland peat-marsh basins in Devonshire and Pembroke Parishes. Extensive land reclamation by the US military in the early 1940s to build naval and air force operating bases added over 1.5 km^2 to Bermuda's area and greatly increased the amount of relatively low, level area on the otherwise hilly island.

Bermuda's climate is considered sub-tropical, mainly due to the influence of the Gulf Stream, which passes to the west and north-west of the island. Annual rainfall is 1,400 mm, distributed fairly evenly throughout the year. Monthly average temperatures range from 18°C in February to 27°C in August. Bermuda was discovered in about 1505 but has only been settled since 1609. Today, the country is largely suburban in nature and is one of the most densely populated countries in the world. The impact of human development on native fauna and flora has been significant. About 94% of present flora is introduced, much of it now naturalised.

Conservation

Bermuda's first conservation legislation was passed in 1616, when the Governor issued a proclamation against "the spoyle and havocke of the cahows". A number of current acts cover the protection of species and habitats. These include: the Coral Reef Preserves Act 1966; the Fisheries Act 1972; the Protection of Birds Act 1975; the Endangered Animals and Plants Act 1976; the Bermuda National Parks Act 1986; and the Protected Species Act 2003. Although birds are covered there are other terrestrial species—for example, the Critically Endangered Bermuda skink *Eumeces longirostris*—that have no protection unless they are found within a national park. There are a number of government bodies responsible for the management of natural resources on Bermuda, which all fall under the umbrella of the Ministry of Environment and Sport. These include the Department of Environmental Protection, the Department of Conservation Services, the Department of Planning and the Department of Parks. There are also 13 NGOs which play an important role in assisting the government in meeting its conservation objectives. In 2003, Bermuda produced a five-year Biodiversity Strategy and Action Plan. This outlines 12 objectives and a series of prioritised actions for achieving them. A total of 9% or c.500 ha of Bermuda's land area is designated as park and nature reserves. National parks are areas that are protected under the National Parks Act 1986 for the enjoyment of present and future generations. Nature reserves are areas of special scientific interest, and all forms of development are prohibited. There are 177 ha of privately-owned nature reserve, of which the Bermuda National Trust and Bermuda Audubon Society own about half.

At the time of its discovery, Bermuda was almost entirely covered in a dense, mostly evergreen forest dominated by several endemic tree species, notably Bermuda cedar *Juniperus bermudiana,* Bermuda palmetto palm *Sabal bermudiana* and Bermuda olivewood *Cassine laneanum.* Massive disturbance of Bermuda's vegetative cover was carried out by the early settlers, through clearing for agriculture and lumber, wholesale burning of large areas of forest to control a plague of rats, and the introduction of mammals such as pigs, rats, goats and cattle. By the late 1800s, only the Bermuda cedar remained common, becoming Bermuda's main forest cover, with an understorey of other, less common, native and endemic plants. Following an infestation of accidentally introduced juniper scale insect in the late 1940s, up to 96% of the Bermuda cedar forest died, leaving the land essentially defoliated. Over 1,000 species of exotic plants have been introduced since that time, with many becoming invasive and completely dominating the inland vegetation cover. In 2003, only some coastal cliff and dune areas were still dominated by native plant communities, along with some of the more exposed and isolated offshore islands. There are a number of native reforestation projects carried out by the Bermuda Department of Conservation Services, notably the Nonsuch Island Living Museum, the Walsingham Trust Property, Paget Marsh Nature Reserve and Morgan's Island Nature Reserve. The Department of Conservation Services manages the Nonsuch Island Living Museum project which has (since 1960) focused on the restoration of island habitats and their floral and faunal communities to pre-colonial status. Much of the island is now covered with a dense, 40-year old replanted forest comprising native and endemic tree and shrub species. This unique project has created a perfect environment for the establishment of a more secure Bermuda Petrel *Pterodroma cahow* population (see below).

Bermuda's small size and geographical isolation have made it very difficult for mammals, amphibians and reptiles to become naturally established. Prior to human settlement, only one species of terrestrial reptile was resident, the endemic rock lizard or Bermuda skink *Eumeces longirostris,* which is now only common on some of the offshore islands, in particular the Castle Harbour Islands. Because of accidental and deliberate introduction by man, there are two species of rat—*Rattus norvegicus* and *Rattus rattus*—now common on Bermuda, as well as the house mouse *Mus musculus.* There is also a large population of domestic and feral house cats, which impact on bird populations in some areas. Dogs are much better controlled, with a dog authority enforcing annual licensing and compulsory microchip tagging, ensuring that there are very few, if any, feral animals. Other mammals sold primarily as pets, such as guinea pigs and hamsters, are occasionally released in parks or nature reserves by irresponsible owners. Attempts by pet stores to introduce other exotic mammals, such as African pygmy hedgehogs and de-scented skunks, were recently prevented by the introduction of new legislation.

The coast of Nonsuch Island Living Museum, the focus of a restoration project since 1960. (PHOTO: ANDREW DOBSON)

■ Birds

About 375 bird species have been recorded in Bermuda, most of which are Neotropical migrants. Although most of these migrants stay only for a matter of days or weeks to rebuild energy reserves before continuing on their migration, a number will stay on in varying numbers for the entire winter period (October to April). For example, up to 25 species of North American wood warbler regularly winter on Bermuda, in addition to orioles and various raptors.

Three globally threatened birds occur on Bermuda. The Near Threatened Piping Plover *Charadrius melodus* and Buff-breasted Sandpiper *Tryngites subruficollis* are both regular to occasional migrants, visiting mainly in autumn. The third species is the Endangered Bermuda Petrel (or Cahow) *Pterodroma cahow*. The Cahow is Bermuda's national bird. It was super-abundant before human settlement, but was nearly driven to extinction by earlier settlers. It was thought extinct for almost 300 years but was rediscovered breeding on the rocky islets in Castle Harbour in 1951. The population in 1960 was just 18 pairs, producing eight fledged chicks annually. In the 2007–2008 breeding season the population had reached a record high of 85 established nests (with an additional six nest burrows being investigated) with a total of 40 chicks successfully fledging. The birds nest on four rocky islets near Cooper's and Nonsuch Islands. A recovery program for Bermuda's national bird has been under way since 1951 and is managed by the Department of Conservation Services under the direction of the Terrestrial Conservation Officer. As part of this recovery program, (pre-fledging) birds have been translocated to artificial nest burrows on Nonsuch Island in order to establish a new breeding colony. Over five years, 101 chicks have successfully fledged from Nonsuch. Cahows spend the first 3–4 years of life out at sea, but the first Nonsuch Island fledglings were seen prospecting nest burrows on Nonsuch in February–April 2008, providing hope that a new colony could be established as early as 2009. Nonsuch Island represents a more elevated and less vulnerable nesting island for the Cahows than the rocky islets that are frequently over-washed and eroded by hurricanes (compounded by sea-level rise). Nonsuch Island has the capacity to hold a significantly larger breeding population that the other islets.

Cahow chick being translocated to an artificial nest burrow on Nonsuch Island. (PHOTO: ANDREW DOBSON)

Bermuda supports just 18 resident (regular) breeding birds, including one endemic sub-species, the Bermuda White-eyed Vireo *Vireo griseus bermudianus*, and several native species including Mourning Dove *Zenaida macroura*, Common Ground-dove *Columbina passerina*, Grey Catbird *Dumetella carolinensis*, Eastern Bluebird *Sialia sialis* and Barn Owl *Tyto alba*. The Yellow-crowned Night-heron *Nyctanassa violacea* was successfully reintroduced to Bermuda for natural control of the abundant red land crab *Gecarcinus lateralis*, after the discovery of sub-fossil remains in limestone caves and the descriptions of early settlers confirmed that the bird once nested there. Breeding of the Green Heron *Butorides virescens*, suspected for years, was confirmed in 2003 around the edge of the mangrove-fringed Trott's Pond and Mangrove Lake; it extended its breeding range in 2004 and by 2008 was nesting in at least seven locations around Bermuda. Three seabird species visit Bermuda to breed: the Cahow *P. cahow* (as discussed above); White-tailed Tropicbird *Phaethon lepturus*, with c.2,000 nesting pairs found on coastal cliffs and offshore islands around Bermuda; and Common Tern *Sterna hirundo* nests on small rocks and islets in several of the larger inshore harbours, maintaining a small population of 18–25 nesting

The Bermuda Petrel or Cahow, the islands' national bird. (PHOTO: JEREMY MADEIROS)

Bermuda's endemic subspecies of White-eyed Vireo. (PHOTO: ANDREW DOBSON)

White-tailed Tropicbird, Bermuda's commonest seabird. (PHOTO: ANDREW DOBSON)

pairs. Hurricane Fabian (2003) had a disastrous effect on *S. hirundo*, with only six pairs returning to breed in 2004, and 9–10 pairs breeding in 2008.

The Department of Conservation Services has management programs in place for all of the seabird species. Efforts for the White-tailed Tropicbird *Phaethon lepturus*, Bermuda's commonest seabird, include studies of breeding success and chick growth-rates at a number of nesting locations, and installation of artificial nests by the department's Conservation Unit at managed nature reserve areas such as the Castle Harbour Islands. Recent hurricane activity has resulted in the loss of many of the natural nest sites for this species through cliff collapse and erosion. Artificial nests provide an alternative that has been readily accepted by the birds (which otherwise rely on natural cavities as they are unable to dig their own). These efforts are augmented by the Bermuda Audubon Society which provides artificial nests to private land-owners and has installed nests at several of its own nature reserves. Bermuda has by far the largest nesting population of this species in the North Atlantic and hosts nearly half of the breeding population for the *P. lepturus catsbyii* subspecies. Research is also ongoing on the greatly diminished population of *S. hirundo* on the island and has indicated a shortage of male birds, which, coupled with the threat posed by hurricane activity to their tiny nesting islands, seriously threatens the future of this species on Bermuda. Several other species of seabird, some of which may have nested on the island at the time of human colonization, visit Bermuda as vagrants.

Figure 1. Location of the Important Bird Area in Bermuda.

ATLANTIC OCEAN
N
BM001
0 5 10
km

IMPORTANT BIRD AREAS

Cooper's Island and Castle Islands is Bermuda's only IBA—the territory's international priority site for bird conservation. It covers 7.6 km² (including marine areas), and about 1% of the islands' land area and has been identified on the basis of two key bird species (listed in Table 1) that trigger the IBA criteria. With the recent protection afforded Cooper's Island (Cooper's Island National Park) almost all of the IBA is protected to some extent. The monitoring work undertaken for the White-tailed Tropicbird *Phaethon lepturus* and Bermuda Petrel *Pterodroma cahow* should be used to inform the annual assessment of state, pressure and response scores at the IBA to help provide an objective status assessment and highlight any additional management interventions that might be required to maintain this internationally important biodiversity site.

Table 1. Key bird species at the Important Bird Area in Bermuda.

Key bird species	Criteria	National population	Bermuda IBA BM001
Bermuda Petrel *Pterodroma cahow*	■ ■	255	255
White-tailed Tropicbird *Phaethon lepturus*	■	6,000	2,100

All population figures = numbers of individuals.
Threatened birds: Endangered ■. **Congregatory birds** ■.

KEY REFERENCES

ANDERSON, A., DE SILVA, H., FURBERT, J., GLASSPOOL, A., RODRIGUES, L., STERRER, W. AND WARD, J. (2001) *Bermuda biodiversity country study.* Bermuda: Bermuda Aquarium Museum and Zoo and the Bermuda Zoological Society, Bermuda.

ANON. (2008) Bermuda Petrel returns to Nonsuch Island (Bermuda) after 400 years. *World Birdwatch* 30(2): 7.

BEEBE, W. (1932) *Nonsuch: land of water.* New York: Brewer, Warren and Putnam.

BEEBE, W. (1932) *Half mile down.* New York: Harcourt, Brace and Company.

BRADLEE, T. S. AND MOWBRAY, L. L. (1931) *A list of birds recorded from the Bermudas.* Compiled by Warren F. Eaton. Boston: Boston Society of Natural History.

BRITTON, N. L. (1918) *Flora of Bermuda.* New York: S. Scribner's Son.

DOBSON, A. (2002) *A birdwatching guide to Bermuda.* Chelmsford, U.K.: Arlequin Press.

DOBSON, A. AND MADEIROS, J. (2006) Bermuda. Pp.29–36 in S. M. Sanders, ed. *Important Bird Areas in the United Kingdom Overseas Territories.* Sandy, U.K.: Royal Society for the Protection of Birds.

HAYWARD, S. J., GOMEZ, V. H. AND STERRER, W., EDS. (1981) *Bermuda's delicate balance.* Bermuda: Bermuda National Trust.

LEFROY, J. H. (1877–79; 1981) *Memorials of the discovery and early settlement of the Bermudas or Somers Islands, 1515–1685.* Bermuda: Bermuda Historical Society and Bermuda National Trust.

MADEIROS, J. (2008) Cahow nesting season update April 2008. *Bermuda Audubon Soc. Newsletter* 19(1): 2–5.

PHILLIPS-WATLINGTON, C. (1996) *Bermuda's botanical wonderland.* London: Macmillan Education.

ROWE, M. P. (1998) *An explanation of the geology of Bermuda.* Bermuda: Bermuda Government, Ministry of the Environment.

STERRER, W., ED. (1986) *Marine fauna and flora of Bermuda.* New York: John Wiley and Sons.

STERRER, W., ED. (1992) *Bermuda's marine life.* Bermuda: Bermuda Natural History Museum and Bermuda Zoological Society.

STERRER, W. AND CAVALIERE, A. R. (1998) *Bermuda's seashore plants and seaweeds.* Bermuda: Bermuda Natural History Museum and Bermuda Zoological Society.

THOMAS, M. L. (2004) *The natural history of Bermuda.* Bermuda: Bermuda Zoological Society.

VERRILL, A. E. (1902) *The Bermuda Islands.* Reprinted from the Transactions of the Connecticut Acad. Sci. 11, with some changes. New Haven, Connecticut.

WINGATE, D. B. (1982) Successful reintroduction of the Yellow-crowned Night-heron as a nesting resident on Bermuda. *Colonial Waterbirds* 5: 104–15.

WINGATE, D. B. (1982) The restoration of Nonsuch Island as a living museum of Bermuda's pre-colonial terrestrial biome. Pp.225–238 in P. J. Moors, ed. *Conservation of island birds: case studies for the management of threatened island species.* Cambridge, U.K.: International Council for Bird Preservation (Techn. Publ. 3).

ACKNOWLEDGEMENTS

The authors would like to thank Mr Jack Ward, Director of Conservation Services, Bermuda Government, and Dr David Wingate, Bermuda Audubon Society (former Bermuda Government Conservation Officer, retired).

BM001 Cooper's Island and Castle Islands

National Nature Reserve/National Park/World Heritage Site

COORDINATES 32°21'N 64°39'W
ADMIN REGION St George's
AREA 760 ha
ALTITUDE 0–21 m
HABITAT Sea, reef, coastline, shrubland

White-tailed Tropicbird

THREATENED BIRDS ■
RESTRICTED-RANGE BIRDS
BIOME-RESTRICTED BIRDS
CONGREGATORY BIRDS ■

■ Site description

Cooper's Island and Castle Islands IBA (43 ha of which is terrestrial) is situated at the east end of Bermuda. The 31-ha Cooper's Island was connected to St David's Island during the (1943) construction of the US Air Force base (now the international airport). It is on the eastern side of Castle Harbour and juts out into the centre of the Castle Harbour Islands which form a chain across the south-eastern edge of the harbour. The Castle Islands Nature Reserve includes Nonsuch (the largest at 6.9 ha, and most ecologically diverse), Castle Island (1.9 ha), Charles Island (1.8 ha) and about 15 other smaller islands that are surrounded by extensive coral reefs and sea grass beds. The former NASA tracking station on the diverse, and ecologically important Cooper's Island peninsula has recently been incorporated into a new national park.

■ Birds

This IBA supports the world population (85 pairs in 2008) of the Endangered Bermuda Petrel (or Cahow) *Pterodroma cahow* which breeds on four of the islets. Birds have been translocated to Nonsuch Island to establish a new breeding colony. The breeding population (700 pairs) of White-tailed Tropicbird *Phaethon lepturus* is globally significant. The marine portion of this IBA covers areas where *P. cahow* congregates before flying to the nesting islets, and is a concentration point for shearwaters each spring. The endemic subspecies of White-eyed Vireo *Vireo griseus bermudianus* occurs in the IBA.

■ Other biodiversity

The Critically Endangered Bermuda cedar *Juniperus bermudiana* and Bermuda sedge *Carex bermudiana* and Endangered Bermuda palmetto palm *Sabal bermudana* and Bermuda olivewood *Cassine laneanum* grow in the IBA. The Critically Endangered Bermuda skink *Eumeces longirostris* is found on the Castle Harbour islands, and may occur on Cooper's Island. The Critically Endangered hawksbill *Eretmochelys imbricata* and Endangered green *Chelonia mydas* and loggerhead *Caretta caretta* are present, with the latter recently nesting on Coopers Island.

■ Conservation

The recently designated Cooper's Island National Park embraces the majority of the peninsula, including the former NASA tracking station. There is some pedestrian access, but some of the most environmentally sensitive sections of Coopers Island may remain off-limits except by special permit or on guided tours. The park provides a valuable buffer to the Castle Islands Nature Reserve (part of which, due to the presence of early colonial fortifications, is a world heritage site). Access to Castle and Charles islands is strictly regulated (but landing is allowed). The other islands east of Castle Roads Channel (including Nonsuch and the Cahow nesting islands) are all restricted access nature reserves with landing by special permit only. There is occasional illegal landing on the Cahow nesting islands.

BONAIRE

LAND AREA **288 km²** ALTITUDE **0–240 m**
HUMAN POPULATION **14,000** CAPITAL **Kralendijk**
IMPORTANT BIRD AREAS **6, totalling 238 km²**
IMPORTANT BIRD AREA PROTECTION **50%**
BIRD SPECIES **214**
THREATENED BIRDS **2** RESTRICTED-RANGE BIRDS **3** BIOME-RESTRICTED BIRDS **2**

JEFF WELLS (BOREAL SONGBIRD INITIATIVE) AND
ADOLPHE DEBROT (CARMABI FOUNDATION)

Washington-Slagbaai National Park. (PHOTO: ROWAN O. MARTIN)

INTRODUCTION

Bonaire, which is politically part of the Kingdom of the Netherlands[1], is one of the three Netherlands Antilles islands (Aruba, Bonaire and Curaçao) that lie off the north-west coast of Venezuela. Bonaire is the easternmost of the three islands (c.50 km east of Curaçao) and the island furthest (c.85 km) from mainland Venezuela. It is 35 km long, 8–15 km wide, and consists of a volcanic core, surrounded by limestone formations. The northern end of the island, within Washington-Slagbaai National Park (IBA AN009), is dominated by hills including Mount Brandaris, the island's highest point. The flat, low-elevation southern end of the island contains the Pekelmeer (IBA AN014), once a series of natural shallow lagoons that have been modified over hundreds of years for salt production. Bonaire has jurisdiction over an offshore island—Klein Bonaire (IBA AN012)—situated c.1 km from the central west coast. Klein Bonaire is a low coral-limestone island fringed with sandy beaches.

1 At some point in the near future the "Netherlands Antilles" will be dissolved. St Maarten and Curaçao will become separate countries within the Kingdom of the Netherlands (similar to the status currently enjoyed by Aruba). The islands of Bonaire, Saba and St Eustatius will be linked directly to the Netherlands as overseas territories.

Bonaire (as Aruba and Curaçao) is very dry with an average annual rainfall of 450 mm falling mostly in the period October–January. As a result, the island's vegetation is generally xerophytic with many areas dominated by columnar cactus intermixed with low scrub and large expanses of land largely devoid of vegetation, especially along the eastern shoreline which receives slightly less rainfall on average than the western side of the island. Virtually all trees on the island were removed by the early nineteenth century and woody vegetation continued to be cut for charcoal production into the twentieth century. Grazing animals were introduced by 1700 and have significantly altered the vegetation. Free-roaming goats and donkeys have continued to have an impact in many areas even to the present day. In some regions, notably within Washington-Slagbaai National Park, there are patches of thicker and taller (3–4 m) thorn scrub forest supporting some epiphytic growth. Lac Bay (IBA AN013) on the south-eastern side of the island supports Bonaire's only significant mangrove woodland.

Bonaire's human population is significantly less than that of neighbouring Aruba (100,000) and Curaçao (138,000). The island's economy is largely dependent on ecotourism centered on scuba diving within the marine park. Apart from tourism, the salt production industry, a small oil transfer facility,

Free-ranging donkeys are a major threat to Bonaire's vegetation. (PHOTO: BERT DENNEMAN)

banking, and fishing provide much of the remaining employment outside of the service and support sectors.

Conservation

Bonaire has a relatively long history of natural resources protection and legislation. The Bonaire Nature Management Plan 1999–2004 (ratified by the Island Council in 1999) defines protected zones and recommends a number of other portions of the island to be designated with varying levels of conservation protection (see "Important Bird Areas" below). Areas currently protected include: Washington-Slagbaai National Park (IBA AN009) that encompasses 17% of the island's land area; Bonaire National Marine Park that extends from the high-water mark to the 60-m depth contour around the coast of Bonaire and Klein Bonaire, covering an area of c.2,700 ha; and Klein Bonaire (IBA AN012) that was designated a protected area in 2000. These three areas are under the management authority of the local NGO STINAPA Bonaire, with the marine park primarily financed through scuba-diver user fees.

Conservation action (research, monitoring and education) on Bonaire is currently focused largely on marine issues, and is implemented through STINAPA Bonaire. However, CARMABI Foundation has a long legacy of carrying out and supporting a variety of ecological research (including on birds) throughout the Netherlands Antilles. CARMABI has recently completed a vegetation mapping and analysis project that provides baseline information and recommendations for protecting Bonaire's habitats. The foundation has also undertaken or facilitated a number of surveys of nesting terns and plovers. Other research and outreach and education campaigns implemented by University of Sheffield (UK) have focused on the Yellow-shouldered Amazon *Amazona barbadensis*. National Audubon Society (BirdLife in the USA) has facilitated some ornithological/birdwatching training of staff from Washington-Slagbaai National Park. A long-term monitoring program for bird populations in the park was initiated by STINAPA in 2007.

Biodiversity faces a wide range of threats on Bonaire including disturbance from recreational activities, pollution of (and increasing run-off into) the marine environment and wetlands, and unplanned development. However, the greatest impact on the environment can be attributed to the direct destruction of vegetation by free-ranging goats and donkeys, predation of the native fauna by cats, and the capture (and keeping) of parrots and other birds for pets. These issues will only be addressed through an integrated program aimed at changing local attitudes towards traditional land and biodiversity use, legal enforcement and wardening, invasive species control and eradication, and securing of long-term funding for the management of the island's protected areas.

Birds

Over 210 species of bird have been recorded from Bonaire. Only 55 of these species are resident (current or former) breeding species, the vast majority being migrants, winterers, and occasional vagrants. Most of the migrants are Neotropical migrants from breeding grounds in North America, although many (50–60 species) are vagrants (recorded on average less than once a year). A smaller number of species are of South American origin, representing either dispersing individuals or austral migrants overshooting their northern South American wintering grounds having originated from breeding grounds further south or west.

Bonaire's resident avifauna is a rather unique, with species of West Indian origin mixed with those originating in South America. This is demonstrated by the presence of two Northern South America biome-restricted birds, namely Bare-eyed Pigeon *Patagioenas corensis* and Yellow-shouldered Amazon *A. barbadensis*, as well as a number of more wide-ranging South American species. *Amazona barbadensis* is also one of three restricted-range species that constitute the Netherlands Antilles secondary Endemic Bird Area (EBA), the other two being West Indian birds, namely Caribbean Elaenia *Elaenia martinica* and Pearly-eyed Thrasher *Margarops fuscatus*. At least 16 subspecies have been described from Aruba, Bonaire and Curaçao; four exclusively from Bonaire. The Bonaire form of Brown-throated Parakeet *Aratinga pertinax xanthogenius* is particularly well-differentiated (based on plumage and vocalisations), and a subspecies of Grasshopper Sparrow *Ammodramus savannarum caribaeus* found only on Bonaire and Curaçao appears vocally distinct from other North American and Caribbean forms. Similarly, vocalisations of White-tailed Nightjar *Caprimulgus cayennensis insularis* from Bonaire and Curaçao are distinct from mainland forms. Further taxonomic research is needed on all of these subspecies.

The threat category and national population sizes of the globally threatened birds are listed in Table 1. The Vulnerable *A. barbadensis* has a disjunct range in northern coastal Venezuela (Falcón, Lara, Anzoátegui and Sucre) and the islands of Margarita and La Blanquilla. The Bonaire birds are clearly important in the context of a global population estimated at 2,500–9,999 individuals. However, over 300 birds were illegally caught for the local pet trade between 1998 and 2002 which has presumably halted any potential population growth on the island. In 2008 several broods were poached from nests on Bonaire, some of which have reportedly

The Vulnerable Yellow-shouldered Amazon. (PHOTO: ROWAN O. MARTIN)

Table 1. Key bird species at Important Bird Areas in Bonaire.

Key bird species	Criteria	National population		AN009	AN010	AN011	AN012	AN013	AN014
			Criteria	■ ■ ■ ■	■ ■ ■	■ ■ ■ ■	■ ■ ■	■ ■ ■	■
Caribbean Flamingo *Phoenicopterus ruber*	■	1,500–7,000		500					5,000
Caribbean Coot *Fulica caribaea*	NT ■	250				246			
Royal Tern *Sterna maxima*	■	255							170
Sandwich "Cayenne" Tern *Sterna sandvicensis*	■	540		360					340
Common Tern *Sterna hirundo*	■	115		20					60
Least Tern *Sterna antillarum*	■	2,375		412		452	100		582
Bare-eyed Pigeon *Patagioenas corensis*	■	500–1,000		✓	✓	✓	✓	✓	
Yellow-shouldered Amazon *Amazona barbadensis*	VU ■ ■ ■	650		250	267	100		100	
Caribbean Elaenia *Elaenia martinica*	■	250–500		✓	✓	✓	✓	✓	
Pearly-eyed Thrasher *Margarops fuscatus*	■			✓	✓	✓			

All population figures = numbers of individuals.
Threatened birds: Vulnerable ■; Near Threatened ■. **Restricted-range birds** ■. **Biome-restricted birds** ■. **Congregatory birds** ■.

Bonaire's endemic subspecies of Brown-throated Parakeet. (PHOTO: BERT DENNEMAN)

Figure 1. Location of Important Bird Areas in Bonaire.

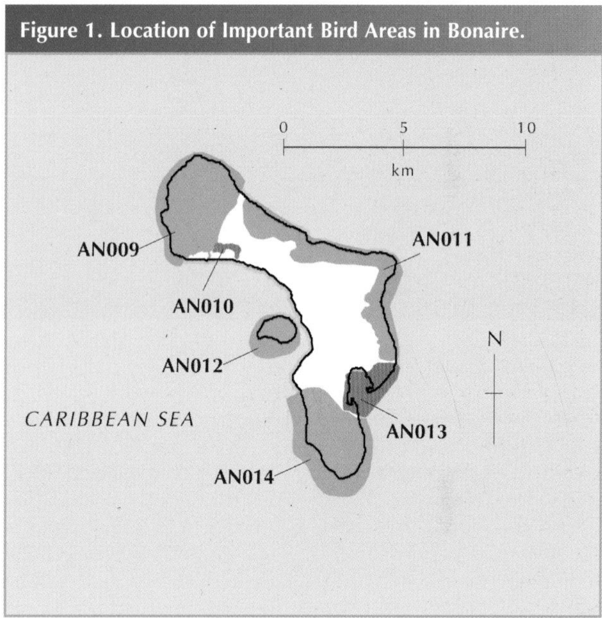

been seen as pets in homes in Bonaire and neighbouring Curaçao.

Bonaire is of global importance for its waterbird populations including Caribbean Flamingo *Phoenicopterus ruber* whose numbers, over the last 10 years have fluctuated between c.1,500 and 7,000 breeding individuals (though most normally averaging c.5,000). The flamingos fly to mainland Venezuela to feed in lagoons along the coast of the state of Falcón where hundreds are regularly seen but are not known to breed. The movements of the flamingos within the island and to-and-from mainland Venezuela are poorly known and warrant further research. Bonaire supports significant populations of breeding terns (*Sterna* spp.), including the "Cayenne" form of Sandwich Tern *S. sandvicensis eurygnatha*, primarily within Washington-Slagbaai National Park IBA (AN009) and Pekelmeer Saltworks IBA (AN014).

IMPORTANT BIRD AREAS

Bonaire's six IBAs—the island's international site priorities for bird conservation—cover 23,830 ha (including their marine extensions). They embrace c.55% of the island's land area. Washington-Slagbaai National Park IBA (AN009) and Klein Bonaire IBA (AN012) are formally protected within the national system. Parts of Washikemba–Fontein–Onima IBA (AN011), Pekelmeer Saltworks IBA (AN014) and Lac Bay IBA (AN013) have been identified as proposed protected areas within the Bonaire Nature Management Plan, but these recommendations have not been acted upon. However, the latter two IBAs are designated Ramsar sites, offering them formal recognition of their importance.

The IBAs have been identified on the basis of 10 key bird species that variously trigger the IBA criteria (see Table 1). The majority of these birds occur in two or more IBAs. However, Royal Tern *Sterna maxima* only nests in Pekelmeer Saltworks IBA (AN014), and the Near Threatened Caribbean Coot *Fulica caribaea* only occurs on the freshwater reservoirs in Washikemba–Fontein–Onima IBA (AN011). Perhaps of greater concern is the fact that c.60% of the Vulnerable Yellow-shouldered Amazon *Amazona barbadensis* population occurs outside of formal protected areas, leaving the species totally exposed to capture for the local pet trade. For example, Dos Pos IBA (AN010) contains some of the most important breeding and roosting sites for the species on Bonaire but receives no protection from future development (although there are no immediate threats to this area), or poaching.

Yellow-shouldered Amazon nesting cliffs at Fontein.
(PHOTO: ROWAN O. MARTIN)

There is an urgent need to establish secure, protected areas for breeding terns (*Sterna* spp.) on Klein Bonaire IBA (AN012), the islands in Goto Lake (IBA AN009) and in the Pekelmeer Saltworks IBA (AN014) through the eradication of cats and rats where possible (e.g. on Klein Bonaire), signage, fencing, and regular patrols. Such proactive management would likely see a dramatic increase in the breeding tern (and plover *Charadrius* spp.) populations. More attention should also be given to balancing the management of Pekelmeer Saltworks IBA for its ecological values in addition to its economic value. Washington-Slagbaai National Park IBA would benefit from a concerted program of removing goats, donkeys and pigs that are so dramatically impacting the vegetation. The landbird (and vegetation) monitoring program started in 2007 should help to determine the impact these grazing animals have had.

Amazona barbadensis would benefit from increased patrolling of the Washington-Slagbaai National Park IBA in an effort to stop poaching, although this would be difficult and costly. More practical would be a public awareness campaign to raise local pride in combination with enforcement of the laws prohibiting the possession of unregistered birds, thereby reducing local demand for wild-caught birds. Ideally this would reach beyond Bonaire to the neighbouring island of Curaçao as a (currently unknown) proportion of parrots poached on the island are exported to Curaçao. *Amazona barbadensis* on Bonaire is perceived by many as an agricultural pest. A detailed study to determine the extent of agricultural damage caused by the parrot, accompanied by measures to address this conflict with humans is also needed. Further research to determine the factors limiting the parrot population on Bonaire is required to inform management decisions within the IBAs.

State, pressure and response scores at each IBA should be monitored annually to provide an objective status assessment and highlight management interventions that might be required to maintain these internationally important biodiversity sites.

KEY REFERENCES

De Boer, B. A. (1979) Flamingos on Bonaire and in Venezuela. Willemstad, Curaçao: CARMABI Foundation. (STINAPA Documentation Series 3)

De Freitas, J. A., Nijhof, B. S. J., Rojer, A. C. and Debrot, A. O. (2005) *Landscape ecological vegetation map of the island of Bonaire (Southern Caribbean)*. Amsterdam: Royal Netherlands Academy of Arts and Sciences.

De Freitas, J. A., Debrot, A. O. and Criens, S. (2006) Natuur- en cultuurhistorische waarden van plantage Onima, Bonaire. Willemstad, Curaçao: CARMABI Foundation. (Unpublished report).

Debrot, A. O. (1997) Klein Bonaire: a brief biological inventory. Willemstad, Curaçao: CARMABI Foundation. (Unpublished report).

Debrot, A. O., Boogerd, C. and Van den Broeck, D. (in press) Dutch Antilles III: Curaçao and Bonaire. In Bradley P. E. and Norton, R. L. eds. *Breeding seabirds of the Caribbean*. Gainesville, Florida: Univ. Florida Press.

Hilty, S. L. (2003) *Birds of Venezuela*. Princeton: Princeton University Press.

Jackson, J. (2000) Distribution, population changes and threats to Least Terns in the Caribbean and adjacent waters of the Atlantic and Gulf of Mexico. Pp 109–117 in E. A. Schreiber and D. S. Lee, eds. *Status and conservation of West Indian Seabirds*. Ruston, USA: Society of Caribbean Ornithology (Spec. Publ. 1).

Kress, S. (1998) Applying research for effective management: case studies in seabird restoration. Pp 141–154 in J. M. Marzluff and R. Sallabanks, eds. *Avian Conservation*. Washington, D.C.: Island Press.

Ligon, J. (2006) Annotated checklist of birds of Bonaire. Downloaded from http://www.bonairediveandadventure.com (23 March 2008).

Miller, J. Y., Debrot, A. O. and Miller, L. D. (2003) Butterflies of Aruba and Bonaire with new species records for Curaçao. *Car. J. Sci.* 39: 170–175.

Nagelkerken, I., van der Velde, G., Gorissen, M. W., Meijer, G. J., vant Hof, T. and den Hartog, C. (2000) Importance of mangroves, seagrass beds and the shallow coral reef as a nursery for important coral reef fishes, using a visual census technique. *Estuar. Coastal Shelf Sci.* 51: 31–44.

Nijman, V., Aliabadian, M., Debrot, A. O., de Freitas, J. A., Gomes, L. G. L., Prins, T. G. and Vonk, R. (2008) Conservation status of Caribbean Coot *Fulica caribaea* in the Netherlands Antilles, and other parts of the Caribbean. *Endangered Species Research* 4: 241–246.

Prins, T. G., De Freitas, J. A. and Roselaar, K. C. S. (2003) First specimen record of the Barn Owl *Tyto alba* in Bonaire, Netherlands Antilles. *Carib. J. Science* 39: 144–147.

Prins, T. G. and Nijman, V. (2007) Checklist of the birds of Bonaire. Amsterdam: Zoological Museum Amsterdam. (Unpublished report).

Prins, T. G., Roselaar, K. C. S and Nijman, V. (2005) Status and breeding of Caribbean Coot in the Netherlands Antilles. *Waterbirds* 28:146–149.

Raffaele, H. Wiley J., Garrido, O., Keith, A. and Raffaele, J. (1998) *A guide to the birds of the West Indies*. Princeton, New Jersey: Princeton University Press.

Schreiber, E. A. (2000) Action plan for conservation of West Indian seabirds. Pp 182–191 in E. A. Schreiber and D. S. Lee, eds. *Status and conservation of West Indian Seabirds*. Ruston, USA: Society of Caribbean Ornithology (Spec. Publ. 1).

Voous, K. H. (1965) Nesting and nest sites of Common Tern and Dougall's Tern in the Netherlands Antilles. *Ibis* 107: 430–431.

Voous, K. H. (1983) *Birds of the Netherlands Antilles*. Zutphen, The Netherlands: De Walburg Pers.

Wagenaar-Hummelinck, P. and Roos, P. J. (1969) Een natuurwetenschappelijk onderzoek gericht op het behoud van het Lac op Bonaire [A scientific survey of Lac on Bonaire]. Willemstad, Curaçao: CARMABI Foundation. (Unpublished report of the Natural Science Working Group).

Wells, J. V. and Childs Wells, A. (2006) The significance of Bonaire, Netherlands Antilles, as a breeding site for terns and plovers. *J. Carib. Orn.* 19: 21–26.

ACKNOWLEDGEMENTS

The authors would like to thank Bert Denneman (Vogelbescherming Nederlands), Eric Newton (Netherlands Antilles Min. Public Health and Social Development), Adrian del Nevo (Applied Ecological Solutions Inc.), Rowan Martin (University of Sheffield), Kalli de Meyer (Dutch Caribbean Nature Alliance), Fernando Simal (STINAPA Bonaire), and Sam Williams (University of Sheffield) for their contributions to and reviews of this chapter.

AN009 Washington-Slagbaai National Park

National Park/Ramsar Site

COORDINATES 12°17′N 68°24′W
ADMIN REGION Bonaire
AREA 6,900 ha
ALTITUDE 0–243 m
HABITAT Shrubland, salina, coast

Yellow-shouldered Amazon

THREATENED BIRDS	1
RESTRICTED-RANGE BIRDS	3
BIOME-RESTRICTED BIRDS	2
CONGREGATORY BIRDS	✓

■ Site description

Washington-Slagbaai National Park IBA encompasses c.25% of Bonaire at the northern end of the island. It is the area of greatest geographic relief, including the island's highest point, Mount Brandaris. The park has a generally well-maintained road network for visitor access, and supports some of the island's most extensive areas of vegetation (xerophytic shrublands with columnar cacti). There are a number of water holes in the park, and salinas/lagoons along the coast. An oil storage facility is located on the south-eastern border. Dos Pos IBA (AN010) is adjacent to Goto Lake, and the Washikemba–Fontein–Onima IBA (AN011) abuts the park's north-eastern corner.

■ Birds

This IBA is a significant nesting, roosting, and foraging area for c.300–400 Vulnerable Yellow-shouldered Amazon *Amazona barbadensis*. The shrublands support all three Netherlands Antilles secondary EBA restricted-range birds, the two Northern South America biome-restricted birds, and a number of endemic subspecies. Globally significant numbers of Common Tern *Sterna hirundo*, and regionally important numbers of Sandwich *S. sandvicensis* and Least *S. antillarum* terns nest. A regionally important concentration of 500 Caribbean Flamingo *Phoenicopterus ruber* occurs.

■ Other biodiversity

The endemic fish *Poecilia vandepolli* occurs in the freshwater streams and ponds. One Bonaire and Curaçao endemic plant, seven endemic lizards, and a number of endemic land-snails are found inside the park.

■ Conservation

This IBA is a state-owned protected area that includes two Ramsar sites—Goto Lake and Boca Slagbaai. The park is managed by STINAPA Bonaire National Park Foundation. Researchers from CARMABI Foundation have conducted ecological studies; those from Sheffield University (UK) are working on the parrot; from University of Amsterdam on raptors; and from Zoological Museum Amsterdam on waterbirds and fish. Once a privately-owned ranch (exporting animals, and producing charcoal and aloe resin), the park still has free-roaming goats and donkeys (and now pigs) which continue to negatively impact the vegetation. Illegal trapping of *A. barbadensis* is a major threat. Feral cats and human disturbance are problems for nesting terns and shorebirds, and oil spills represent a potential threat due to the park's close proximity to a major oil shipping lane and oil storage facility.

■ Site description

Dos Pos IBA is in northern Bonaire, on the leeward (western) side of the island and immediately east of Goto Lake. It is south-east of Washington-Slagbaai National Park IBA (AN009) and is characterised by numerous small hills and sheltered valleys. Roi Sangu—a 100-m wide canyon with 25-m high cliffs—is within the eastern portion of the IBA. The area also embraces a former fruit plantation (mainly mangoes) and a small, ephemeral pond.

■ Birds

This IBA is significant for its population of the Vulnerable Yellow-shouldered Amazon *Amazona barbadensis*. In winter, c.40% of the island's parrots roost (in three separate roosts) in the IBA. Roi Sangu is an important nesting area for the species. All three of the Netherlands Antilles secondary EBA restricted-range birds occur at this IBA, as do both the Northern South America biome-restricted bird species. Dos Pos IBA appears to be important for Neotropical migrant passerines, and the small pond supports some waterbirds (including the Near Threatened Caribbean Coot *Fulica caribaea*).

■ Other biodiversity

As one of the few freshwater sites of Bonaire, Dos Pos has received extensive natural history attention, and a recent study documented the butterfly fauna of the site. Several endemic lizards and land-snail species occur.

■ Conservation

Dos Pos is adjacent to the Washington-Slagbaai National Park but is itself unprotected. It is a mix of state and private ownership. Feral donkeys and high densities of goats have had a profound effect on the area's habitat structure and composition. Their continued presence combined with an apparent increase of feral pigs is preventing natural regeneration of the flora. *Amazona barbadensis* is threatened by habitat loss and degradation, introduced mammalian predators including cats and rats and by poaching for the local pet trade. Although natural lands are being developed at an alarming rate on Bonaire, this area is not likely to be under immediate threat.

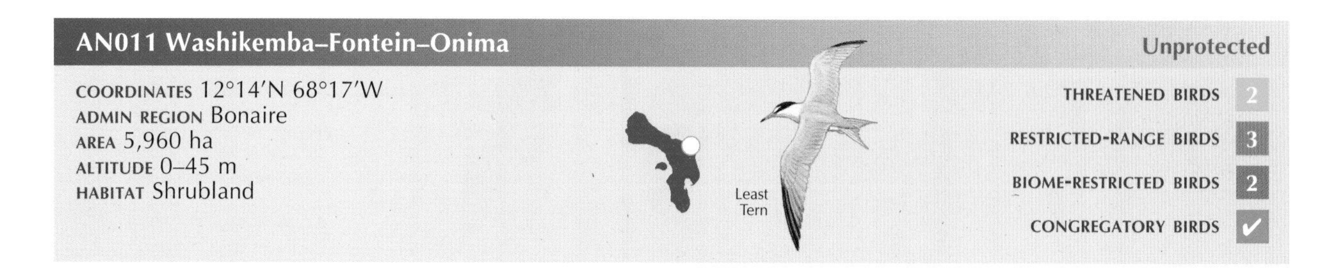

■ Site description

Washikemba–Fontein–Onima IBA is situated along the windward (north and east) coast of Bonaire, between the Washington-Slagbaai National Park IBA (AN009) in the north, and Lac Bay IBA (AN013) in the south-east. It extends inland to the escarpments from Washikemba in the south to Playa Grandi in the north, and includes the natural springs and cliffs of Fontein and the cliffs and intermittent ponds at Washikemba, Onima dam, and inland from Playa Grandi. There is very limited human settlement in this barren area but it does embrace a number of sites of cultural and historic significance with ancient inscriptions of Bonaire's original (aboriginal) inhabitants. The vegetation is sparse, with cacti, small shrubs, *Euphorbia* species and low-growing salt-tolerant plants.

■ Birds

This IBA is important for the Vulnerable Yellow-shouldered Amazon *Amazona barbadensis* which breeds in the Onima–Fontein cliffs. Over 100 birds form a winter roost at Fontein. More than 240 Near Threatened Caribbean Coot *Fulica caribaea* occur and breed at reservoirs in Onima, Playa Grandi and Washikemba. The coast from Boca Onima to Washikemba is regionally important for breeding Least Tern *Sterna antillarum*. All three Netherlands Antilles secondary EBA restricted-range birds occur at this IBA, as do the two Northern South America biome-restricted species.

■ Other biodiversity

The fauna of this area includes at least five endemic land-snails, two endemic lizards, and six endemic arthropods.

■ Conservation

This IBA is under a mix of private and state ownership, and none of it is currently protected. However, about half of the area is within regions recommended (in the 1999–2004 Bonaire Nature Management Plan) for status as "Island Park" or "Protected Landscape". Ecological research, waterbird and freshwater fish surveys have been carried out (variously) by the CARMABI Foundation, Zoological Museum of Amsterdam and Jeff Wells. Threats to the area and its birds include: the continued illegal trapping of wild *A. barbadensis*; destructive foraging of free-ranging goats and donkeys; potential depredation by feral cats on nesting terns and shorebirds; and human disturbance of tern nesting colonies.

AN012 Klein Bonaire

■ Site description

Klein Bonaire IBA is a low, coral-limestone island, situated c.1 km offshore from the central west coast of Bonaire, opposite Bonaire's capital Kralendijk. The island is dominated by low shrubby vegetation that has been severely impacted from a long history of the felling of trees and overgrazing by introduced goats. The island's shoreline includes three salinas (c.36 ha), five freshwater springs or wells, sandy beach areas and coral rubble strands with low shrubby vegetation. The island is uninhabited but has been used in the past for camping by residents, and as a quarantine facility (as evidenced by a ruin of a small building).

■ Birds

This IBA is significant for Caribbean Elaenia *Elaenia martinica*—a Netherlands Antilles secondary EBA restricted-range species—and the Bare-eyed Pigeon *Patagioenas corensis*—one of the two Northern South America biome species on the island. The breeding population of 100 Least Terns *Sterna antillarum* is regionally important. The terns and a number of shorebirds (especially plovers) nest along the shoreline and in the salinas. Ruby-topaz Hummingbird *Chrysolampis mosquitus* occurs at relatively high densities.

■ Other biodiversity

Klein Bonaire IBA is the most important sea-turtle nesting area in Bonaire, supporting good numbers of the Critically Endangered hawksbill *Eretmochelys imbricata* and the Endangered loggerhead *Caretta caretta* turtles. Bonaire endemics present on the island include three land-snails and one lizard. The island harbours the only major population of the Vulnerable kalabari tree *Zanthoxylum flavum*.

■ Conservation

Klein Bonaire is state-owned, and is protected (along with its surrounding reef) within the Bonaire National Marine Park. The island is also a designated Ramsar site. The government has delegated management of the marine park to the non-profit organisation STINAPA Bonaire (that also manages the Washington-Slagbaai National Park IBA, AN009). Goats were eradicated in the 1980s. Further assessment is required to determine if rats and mice still occur, but feral cats were confirmed as present in 2006. Predation of nests and eggs by cats (and possibly rats and mice) is likely the greatest threat to the nesting terns. Disturbance from occasional visitors may also be a problem. Biological inventories and some rare plant reintroduction work have been carried out by the CARMABI Foundation. The most recent bird survey was in 2006.

AN013 Lac Bay

■ Site description

Lac Bay IBA is on the south-east side of Bonaire and comprises a shallow bay (with sea-grass beds) protected from the open ocean by a fringing reef at its mouth. The island's only significant mangrove woodland (c.100 ha) is within the IBA. On the north side of the bay are large expanses of saltflats and small salinas. A small resort and two windsurfing centres are located on the south side and a small harbour for fishing vessels (that fish outside of the bay) and associated buildings (including a restaurant) are on the northern side of the bay mouth. Scattered farms and homes abut the edge of the IBA in its north-west corner.

■ Birds

This IBA is significant as a (sporadic) roost site for c.100 Vulnerable Yellow-shouldered Amazon *Amazona barbadensis*. Also, two (of the 3) Netherlands Antilles secondary EBA restricted-range birds occur at this IBA, as do both of Bonaire's Northern South America biome-restricted species. Numbers of Caribbean Flamingo *Phoenicopterus ruber* occasionally exceed 200, and the IBA supports good numbers of breeding and wintering shorebirds, breeding herons, breeding Least Terns *Sterna antillarum*, and a roost (historically over 100) of Magnificent Frigatebird *Fregata magnificens*.

■ Other biodiversity

Lac Bay lagoon is important as nursery habitat for reef fishes and queen conch *Strombus gigas*, and contains sea-grass beds used by globally threatened sea-turtles.

■ Conservation

Lac Bay IBA is state owned. The marine environment is protected (activities in the area, and use of natural resources are regulated) within the Bonaire National Marine Park. Lac Bay is also a Ramsar site and has been proposed for national park designation. A multi-year management and education program for Lac Bay was undertaken by STINAPA and the Marine Park (with funding from WWF Netherlands). The bay has become popular for a variety of recreational water-sports (jet-skis and kite-surfing are banned) which may cause disturbance of foraging flamingos and other birds. There are also naturalist-guided kayak trips among the mangroves. However, the greatest potential threats are probably from pollution and increased nutrient loading, and also sedimentation from adjacent development-related land clearing. The roost sites for *A. barbadensis* should be mapped and protected from clearance/cutting.

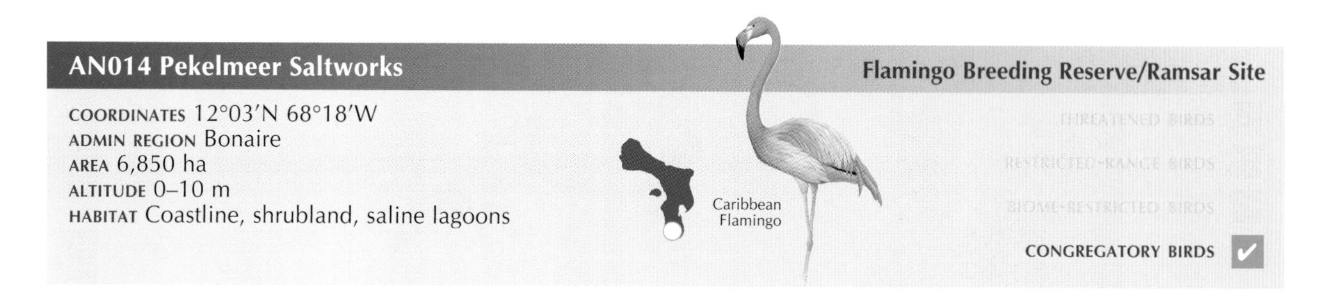

AN014 Pekelmeer Saltworks

Flamingo Breeding Reserve/Ramsar Site

COORDINATES 12°03′N 68°18′W
ADMIN REGION Bonaire
AREA 6,850 ha
ALTITUDE 0–10 m
HABITAT Coastline, shrubland, saline lagoons

■ Site description

Pekelmeer Saltworks IBA is the flat, low-elevation southern end of the island. It was once a series of natural shallow lagoons, but these have been modified over hundreds of years for salt production. Since the 1950s the area has been further modified for large-scale industrial salt production, with water levels tightly controlled within a series of condenser lagoons for maximum salt production. Most of the area (over 90%) is under active management for salt production. Low bushes (mostly buttonwood *Conocarpus erecta*) grow along the shores of some of the lagoons but the area is largely open. Lac Bay IBA (AN013) lies on the south-east coast, just to the north of Pekelmeer.

■ Birds

This IBA is globally significant for Caribbean Flamingos *Phoenicopterus ruber*—numbers fluctuate between 1,500 and 7,000 (most often c.5,000) individuals, with a maximum of 1,300 pairs nesting in 1996. The IBA is also an important nesting area for terns, including: 60 Common Terns *Sterna hirundo* breed (globally important); 340 Sandwich Tern *S. sandvicensis* (regionally important); 170+ Royal Tern *Sterna maxima* (regionally important); and 582 Least Tern *Sterna antillarum* (regionally important). Hundreds (and sometimes thousands) of migratory and wintering shorebirds use the site, as do numerous herons. The Northern South America biome-restricted Bare-eyed Pigeon *Patagioenas corensis* occurs.

■ Other biodiversity

No threatened or endemic species have been recorded although there are numerous studies documenting various aspects of the area's biodiversity, such as arthropods, lizards, land-snails and euryhaline fishes.

■ Conservation

Much or most of the IBA is government owned but is leased to the commercial salt works company. However, a 55-ha area (including an island) has been set-aside since 1969 as a Flamingo Breeding Reserve, which is where most of the birds nest. Pekelmeer (and the flamingo reserve) are designated as a Ramsar site. A section along the south-west side has been proposed as a "Strict Reserve" and the remainder (excluding the crystalliser basins on the western side) as an "Island Park" in the 1999–2004 Bonaire Nature Management Plan. The CARMABI Foundation has carried out research within the IBA. The bird populations within the IBA co-exist with the industrial activities of the salt production area. While there may be some disturbance to the birds, the character and extent of any disturbance has not been assessed.

BRITISH VIRGIN ISLANDS

LAND AREA **153 km²** ALTITUDE **0–521 m**
HUMAN POPULATION **23,550** CAPITAL **Road Town**
IMPORTANT BIRD AREAS **3, totalling 53 km²**
IMPORTANT BIRD AREA PROTECTION **45%**
BIRD SPECIES **210**
THREATENED BIRDS **3** RESTRICTED-RANGE BIRDS **8**

CLIVE PETROVIC (BVI ENVIRONMENTAL CONSULTANT), ESTHER GEORGES AND NANCY WOODFIELD PASCOE (BVI NATIONAL PARKS TRUST)

The Magnificent Frigatebird colony on Great Tobago IBA is one of the top five for the species in the insular Caribbean. (PHOTO: ANDY MCGOWAN)

INTRODUCTION

The British Virgin Islands (BVI), a UK Overseas Territory, is at the eastern end of the Greater Antillean chain of islands in the northern Caribbean Sea, and comprise more than 60 islands, cays and rocks. As an archipelago, the Virgin Islands are politically divided between BVI (which stretch out to the north-east) and the United States Virgin Islands (USVI, the south-western group of islands), and are located on the Puerto Rican Bank. The archipelago once formed a continuous landmass with Puerto Rico and was only isolated in relatively recent geologic time. With the exception of the isolated limestone island of Anegada, the islands are volcanic in origin and are mostly steep-sided with rugged topographic features and little flat land, and are surrounded by coral reefs. The main islands are Tortola, Virgin Gorda, Jost Van Dyke and Anegada, although 15 islands are inhabited. Tortola is the largest island (c.20 km long and 5 km wide) and supports over 80% of the population. The capital, Road Town, is on Tortola.

BVI experiences a dry subtropical climate with average annual rainfall of c.700 mm (in coastal areas) and 1,150 mm on higher ground falling mostly September–November. The most abundant vegetation types on the islands are cactus scrub and dry woodland, although much of this has been modified. Habitat alteration during the plantation era and the introduction of invasive alien species has had major impacts on populations of native flora and fauna. Economic development and population expansion have placed great stresses on the natural environment, which is especially true on the main island of Tortola where coastal development pressures have resulted in the degradation of most wetland ecosystems. The impact of human activities on the fauna has been substantial and many invertebrate species have decreased in abundance or been extirpated by human induced habitat changes, especially within native freshwater habitats. Whilst these freshwater ecosystems have always been ephemeral, they are now frequently dry and lack their former species richness. Vertebrate fauna have been similarly impacted by the loss of freshwater habitats such as ghuts, and freshwater fish have been eliminated or greatly reduced in numbers.

■ Conservation

The conservation of natural resources in the BVI is provided through local legislation such as the National Parks Act (2006), the Fisheries Act (2003), and the Endangered Animals and Plants Ordinance (1987). The Wild Birds Protection Ordinance (1959) protects 21 species of rare or threatened wild birds within the BVI and also the nests, eggs and young of all bird

The British Virgin Islands National Parks Trust runs regular birdwatching tours as part of its environmental education and public awareness activities. (PHOTO: BVINPT)

species. The Physical Planning Act (2004) addresses land development issues, environmental assessments and historical preservation. Responsibility for conservation in the BVI falls to the Ministry of Natural Resources and Labour. The Conservation and Fisheries Department (CFD) and the BVI National Parks Trust (BVINPT) are the two main conservation implementing agencies within the ministry. CFD is responsible for biodiversity conservation including environmental monitoring and fisheries management. The National Parks Ordinance established the BVINPT in 1961 (replaced by the National Parks Act 2006) as a statutory body with responsibility for the territory's terrestrial and marine national parks. The BVINPT is a non-profit organisation that receives an annual subvention from the BVI government and raises the remainder of its budget from park entrance fees and mooring permit fees. The Trust manages 21 national parks and protected areas, five of which are bird sanctuaries, including Dead Chest, Fallen Jerusalem, Great Tobago, Little Tobago and Prickly Pear. Overall there are 20 designated bird sanctuaries, some of which are also proposed protected areas. Environmental education courses are part of the Community College curriculum, with additional public awareness environmental programmes managed by the BVINPT and CFD.

Birds

More than 210 species of birds have been recorded from the BVI. A small core of permanent resident breeding species is augmented by seabirds in the summer months and numerous Neotropical migrants from North America during the winter months. Puerto Rico and Virgin Islands EBA restricted-range species (of which there are 27) are represented by eight species (see Table 1), none of which is endemic to BVI. Of these eight species, the Antillean Mango *Anthracothorax dominicus* is a rare permanent resident and there are few recent confirmed reports although it is suspected to be more common on Anegada and perhaps Guana Island. Puerto Rican Flycatcher *Myiarchus antillarum* may be a rare nesting species but breeding has not been confirmed. The Puerto Rican Screech-owl *Megascops nudiceps* has been considered extirpated from the territory for this IBA analysis. While it was previously recorded with regularity, there are no reliable recent sightings although there have been reports of the characteristic calls and possible pellets both on Tortola and Guana Island.

Three globally threatened (all Near Threatened) birds have been recorded in BVI but their status is such that they have not been considered in the IBA analysis. Piping Plover *Charadrius melodus* is a rare migrant and the Caribbean Coot *Fulica caribaea* is now a rare visitor to the islands (although previously it was a rare permanent resident). However, a total of 33 adult coots, with 10 young, were counted on Josiah's Bay pond on Tortola in March 2006, so a population could re-establish itself if suitable wetland conditions persist. White-crowned Pigeon *Patagioenas leucocephala* was historically abundant throughout the islands but was nearly extirpated by hunting and is now considered rare and largely confined

Caribbean Elaenia, one of eight restricted-range species on the islands. (PHOTO: VINCENT LEMOINE)

Globally significant populations of Roseate Tern breed on the islands.
(PHOTOS: VINCENT LEMOINE)

to mangrove habitats. It may still nest in small numbers in isolated areas, particularly on Anegada. In the mid-1990s, the Guana Island Wildlife Sanctuary embarked on an ambitious project to reintroduce this species and re-establish a breeding colony on Guana. The effort has been successful and young birds from Guana may be responsible for an apparent increase in recent sightings, especially on Beef Island. More research is required on these threatened (and indeed the rarer restricted-range) species to determine their current status and populations.

During the summer months, BVI hosts a range of seabirds that nest on the offshore cays and islets. At least 15 species of seabird breed on the islands and the populations of Roseate Tern *Sterna dougallii* and Magnificent Frigatebird *Fregata magnificens* are globally significant. These nesting seabirds face a range of threats including human disturbance, and predation from a range of invasive mammalian predators.

IMPORTANT BIRD AREAS

BVI's three IBAs—the territory's international priority sites for bird conservation—cover 53 km² (including marine areas), and about 10% of the islands' land area. The IBAs have been identified on the basis of 16 key bird species (listed in Table 1) that variously trigger the IBA criteria. These 16 species comprise all eight extant restricted-range species, and eight congregatory seabirds. The Near Threatened Caribbean Coot *Fulica caribaea*, Piping Plover *Charadrius melodus* and White-crowned Pigeon *Patagioenas leucocephala* have not been considered in the analysis due to the nature of their small populations on the islands (see above). Great Tobago IBA (VG001) is protected as a national park and a bird sanctuary, and Anegada wetlands IBA (VG003) is partly (the western end) protected as an island-level nature reserve and Ramsar site. However, the eastern end wetlands are unprotected, as is Green

Wetlands on eastern Anegada, important for a wide diversity of waterbirds
(PHOTO: ANDY MCGOWAN)

Table 1. Key bird species at Important Bird Areas in the British Virgin Islands.

Key bird species	Criteria	National population	Criteria	VG001	VG002	VG003
				British Virgin Islands IBAs		
			■	■	■	■
			■	■	■	■
Brown Pelican *Pelecanus occidentalis*	■					27
Masked Booby *Sula dactylatra*	■					18
Magnificent Frigatebird *Fregata magnificens*	■			1,500–3,000		
Laughing Gull *Larus atricilla*	■			200		150
Royal Tern *Sterna maxima*	■					21
Sandwich Tern *Sterna sandvicensis*	■					275
Roseate Tern *Sterna dougallii*	■				1,755	157
Least Tern *Sterna antillarum*	■					180
Bridled Quail-dove *Geotrygon mystacea*	■			✓		
Antillean Mango *Anthracothorax dominicus*	■					✓
Green-throated Carib *Eulampis holosericeus*	■			✓	✓	✓
Antillean Crested Hummingbird *Orthorhyncus cristatus*	■			✓		✓
Caribbean Elaenia *Elaenia martinica*	■			✓	✓	✓
Puerto Rican Flycatcher *Myiarchus antillarum*	■			✓	✓	✓
Pearly-eyed Thrasher *Margarops fuscatus*	■			✓	✓	✓
Lesser Antillean Bullfinch *Loxigilla noctis*	■				✓	✓

All population figures = numbers of individuals.
Restricted-range birds ■. Congregatory birds ■.

Cay IBA (VG002), and all three IBAs face a range of threats (see individual IBA profiles below). The system of salt ponds on Anegada is most urgently in need of national level protection. Great Tobago IBA represents one of the five main Magnificent Frigatebird *Fregata magnificens* colonies in the insular Caribbean. With the control of goats on the island, the frigatebird colony (and indeed the vegetation) should be monitored to assess the impact that the goats had, and to inform future management actions.

The IBA process has highlighted the need for more systematic surveys and monitoring to be undertaken. In the mid-1990s Green Cay IBA supported a large Roseate Tern *Sterna dougallii* colony. However, numbers appear to have declined at this site. At the same time, the numbers of this species on other islands (e.g. Cockroach Island, Cistern Rock, Necker Island, Virgin Gorda) have increased, but fluctuate dramatically. This suggests that the BVI population of this bird is highly mobile and sensitive to disturbance. The network of breeding sites needs to be monitored on an annual basis to determine the conservation actions to be taken. There are also several potential additional IBAs in the territory, but further data are required. For example, the network of ponds and mangroves on Tortola (including Bar Bay), and Guana and Norman islands need more quantitative information, but may all qualify as IBAs. Guana and Norman islands appear to support significant breeding populations of Brown Pelican *Pelecanus occidentalis*. State, pressure and response scores at each IBA should be monitored annually to provide an objective status assessment and highlight management interventions that might be required to maintain these internationally important biodiversity sites. Site status monitoring would be helpfully informed by regular survey results focused on the key bird species (both restricted-range species and seabirds alike) listed in Table 1.

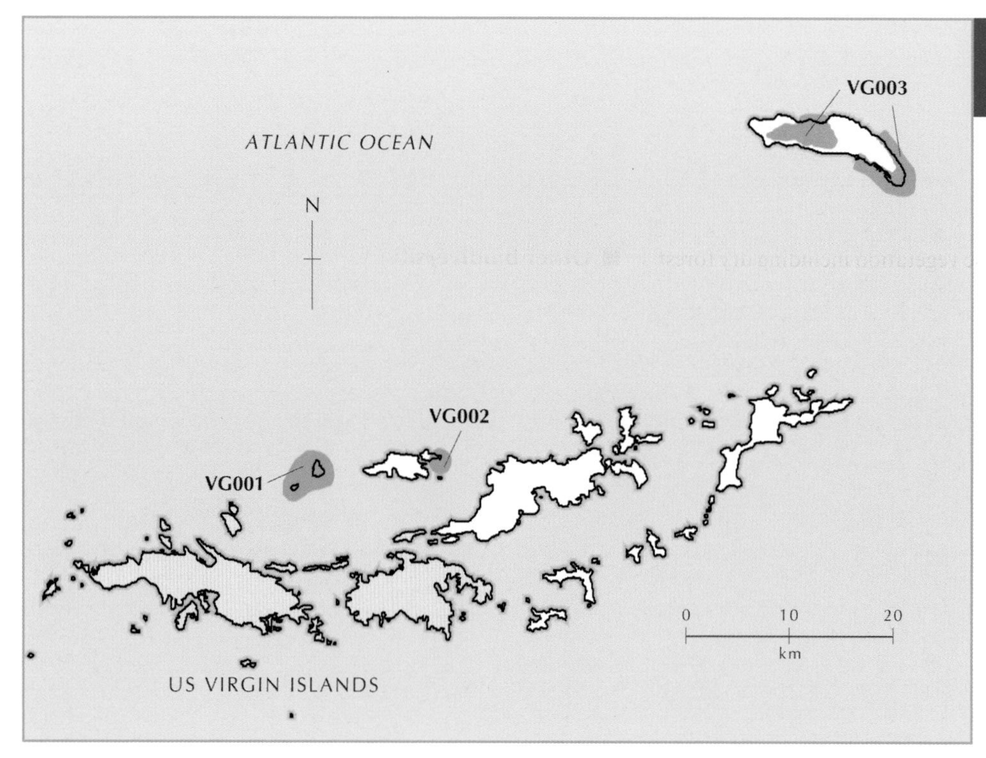

Figure 1. Location of Important Bird Areas in the British Virgin Islands.

KEY REFERENCES

BARNES, J. A. (1992) Flamingos return to BVI. *Forum News* (NGO Forum for the UK Dependent Territories) 7: 2.

CLUBBE, C., GILLMAN, M., ACEVEDO-RODRÍGUEZ, P. AND WALKER, R. (2004) Abundance, distribution and conservation significance of regionally endemic plant species on Anegada, British Virgin Islands. *Oryx* 38: 342–246.

COLLAZO, J. A., SALIVA, J. E. AND PIERCE, J. (2000) Conservation of the Brown Pelican in the West Indies. Pp 39–45 in E. A. Schreiber and D. S. Lee, eds. *Status and conservation of West Indian Seabirds*. Ruston, USA: Society of Caribbean Ornithology (Spec. Publ. 1).

CONYERS, J. (1996) The BVI flamingo restoration project. *Critter Talk* (Bermuda Zoological Society) 19(2): 1–2.

GOODYEAR, N. C. (1992) Flamingos return to Anegada: status update. *National Parks Trust News* (British Virgin Islands), August: 1.

LABASTILLE, A. AND RICHMOND, M. (1973) Birds and mammals of Anegada Island, British Virgin Islands. *Carib. J. Sci.* 13: 91–109.

LINDSAY, K., HORWITH, B. AND SCHREIBER, E. A. (2000) Status of the Magnificent frigatebird in the Caribbean. Pp 58–64 in E. A. Schreiber and D. S. Lee, eds. *Status and conservation of West Indian Seabirds*. Ruston, USA: Society of Caribbean Ornithology (Spec. Publ. 1).

MCGOWAN A., BRODERICK, A. C., CLUBBE, C., GORE, S., GODLEY, B. J., HAMILTON, M., LETTSOME, B., SMITH-ABBOTT, J. AND WOODFIELD, N. K. (2006) Darwin Initiative action plan for the coastal biodiversity of Anegada, British Virgin Islands. Downloaded from: http://www.seaturtle.org/mtrg/projects/anegada/ (July 2008).

MCGOWAN, A., BRODERICK, A. C., GORE, S., HILTON, S., WOODFIELD, N. K. AND GODLEY, B. J. (2006) Breeding Seabirds in the British Virgin Islands. *Endangered Species Research* 3: 1–6. Downloaded from: http://www.int-res.com/journals/esr/ (July 2008).

MCGOWAN, A., WOODFIELD, N., HILTON, G., BRODERICK, A. AND GODLEY, B. (2007) A Rigorous Assessment of the Avifauna of a Small Caribbean Island: A Case Study in Anegada, British Virgin Islands; Caribbean Journal of Science, Vol. 43, No. 1, 99–116, 2007

NORTON, R., CHIPLEY, R. AND LAZELL, J. (1989) A contribution to the ornithology of the British Virgin Islands. *Carib. J. Sci.* 25: 115–118

PETROVIC, C., GEORGES, E. AND WOODFIELD, N. (2006) British Virgin Islands. Pp.55–64 in S. M. Sanders, ed. *Important Bird Areas in the United Kingdom Overseas Territories*. Sandy, U.K.: Royal Society for the Protection of Birds.

RAFFAELE, H. WILEY J., GARRIDO, O., KEITH, A. AND RAFFAELE, J. (1998) *A guide to the birds of the West Indies*. Princeton, New Jersey: Princeton University Press.

VEITCH, C. R. (1998) Survival of the Anegada Rock Iguana: an assessment of threats and possible remedial actions. Papakura, New Zealand: Unpublished report to BVI National Parks Trust.

ACKNOWLEDGEMENTS

The authors would like to thank the following for their invaluable contributions to bird conservation in the BVI: Dr Andy McGowan (Exeter University and the Anegada Darwin Initiative Project), Sarah Sanders (RSPB), Dr Lisa Sorenson (West Indian Whistling Duck Working Group of the Society for the Conservation and Study of Caribbean Birds),Trish Baily, Liana Jarecki, James Lazell, the late Rowan Roy, and all the BVINPT staff and volunteers.

VG001 Great Tobago

■ Site description

Great Tobago IBA is one of the most westerly of islands in BVI, just 4 km west of Jost Van Dyke Island and c.11 km north-west of St John (USVI). It is a small (104 ha), uninhabited steep island of volcanic origin with exposed sea cliffs and mixed but sparse vegetation including dry forest and coastal scrub. The IBA includes all marine areas up to 1 km from the island. Recent removal of most of the feral goats by the BVINPT has resulted in the regeneration of vegetation particularly within the bird colony at Man O' War Bay (on the eastern side).

■ Birds

This IBA supports the BVI's only colony of Magnificent Frigatebird *Fregata magnificens*. With c.3,000 breeding birds this is one of the five largest colonies in the insular Caribbean and is globally significant. The main nesting areas are in the trees at Man O' War Bay and on the steep eastern hillsides. Other seabirds nesting on the island (albeit not in significant numbers) include Brown Booby *Sula leucogaster*, Brown Pelican *Pelecanus occidentalis*, Laughing Gull *Larus atricilla*, Roseate Tern *Sterna dougalli* and both tropicbird species *Phaethon* spp. Six (of the eight) Puerto Rico and the Virgin Islands EBA restricted-range birds have been recorded from this IBA, including Bridled Quail-dove *Geotrygon mystacea*. However, Puerto Rican Flycatcher *Myiarchus antillarum* might only be a visitor to the island.

■ Other biodiversity

No globally threatened or endemic species have been recorded.

■ Conservation

Great Tobago Island is a designated national park and bird sanctuary under the management of the BVINPT. The Trust has established a culling program for the goats that are present on the island and have been severely impacting the vegetation. About 95% of the goats have been removed and vegetation is starting to regenerate. Loss of trees caused by hurricanes has posed a threat to the frigatebird colony (especially when compounded by the grazing pressure of goats that have prevented post-hurricane regeneration). As a short-term measure, artificial nesting platforms have been installed within the frigatebird colony to compensate for the lost trees. The island is relatively undisturbed, but entanglement of seabirds in monofilament fishing line appears to be a substantial threat.

■ Site description

Green Cay IBA lies just 1 km east of the eastern end of Jost Van Dyke Island. It is a small (6 ha) volcanic island comprised of igneous bedrock reaching a maximum elevation of 30 m. The shore is rocky on the eastern end, but this grades to a broad sandy beach on the western (sheltered) end of the island. The IBA includes all marine areas up to 1 km from the island. Green Cay is covered in xerophytic coastal scrub.

■ Birds

This IBA is globally significant for its Roseate Tern *Sterna dougallii* colony. In 1996 a total of 1,755 were counted making Green Cay the largest breeding colony for this species in the US and British Virgin Islands. However, more recently numbers appear to have declined. A range of other seabirds breed in small numbers. Five (of the eight) Puerto Rico and the Virgin Islands EBA restricted-range birds have been recorded from this IBA, namely Green-throated Carib *Eulampis holosericeus*, Caribbean Elaenia *Elaenia martinica*, Pearly-eyed Thrasher *Margarops fuscatus* and Puerto Rican Flycatcher *Myiarchus antillarum* (although this latter species might only be a visitor to the island).

■ Other biodiversity

No globally threatened or endemic species have been recorded.

■ Conservation

Green Cay IBA has been purchased by the BVI Government and is indentified as a proposed protected area. However, it has been proposed as a protected area within the system of national parks. High rates of visitation by yachts to the adjacent island of Sandy Spit which is 160 m to the south) will result in increasing disturbance to the nesting terns unless mitigated.

■ Site description

Anegada is 25 km north of Virgin Gorda, at the eastern end of the BVI. Unlike the other islands in the territory, it is a flat limestone island with an arid landscape. The soils are shallow and alkaline, and with low rainfall the vegetation is mainly xerophytic coastal scrub and dry woodland. Mangroves fringe much of the south and east coast, and the interior wetlands. The western sandy plain supports edaphic and xeric vegetation. The entire north coast is sandy beach. The IBA covers the networks of salt ponds and mangrove at the eastern and western ends of the island, and the adjacent marine areas.

■ Birds

This IBA supports a wide diversity of shorebirds, waterbirds and seabirds. The breeding population of Roseate Tern *Sterna dougallii* is globally significant, while those of Sandwich Tern *S. sandvicensis*, Royal Tern *S. maxima*, Least Tern *S. antillarum*, Laughing Gull *Larus atricilla* and Brown Pelican *Pelecanus occidentalis* are regionally so. Seven (of the eight) Puerto Rico and the Virgin Islands EBA restricted-range birds occur in the mangroves and coastal scrub. Antillean Mango *Anthracothorax dominicus* is rare on the island, and the Puerto Rican Flycatcher *Myiarchus antillarum* may only be a visitor.

■ Other biodiversity

The Critically Endangered Anegada rock iguana *Cyclura pinguis* occurs and is the focus of recovery efforts led by BVINPT. Critically Endangered hawksbill *Eretmocheyls imbricata* and Endangered green *Chelonia mydas* turtles nest on the island. Plants include 16 that are globally threatened including three island endemics (*Acacia anegadensis*, *Metastelma anegadense* and *Senna polyphylla*) that are Critically Endangered.

■ Conservation

Anegada is government owned. The Anegada Nature Reserve (also a Ramsar site) embraces the 445-ha Flamingo Pond—a salt pond at the western end of the island. The extensive ponds and stands of mangrove on the eastern end of the island are presently unprotected. Conservation activities have included the reintroduction of the Caribbean Flamingo *Phoenicopterus ruber* (18 birds introduced in 1992 started breeding in 1995, and the population in 2007 numbered over 140), *Cyclura pinguis* recovery activities and the development of a coastal biodiversity action plan. The IBA is threatened from development and grazing animals, and invasive mammalian predators (cats, rats, mongoose, dogs and pigs). Historically, hunting was responsible for the local extirpation of the flamingo, West Indian Whistling-duck *Dendrocygna arborea* and White-crowned Pigeon *Patagioenas leucocephala*.

CAYMAN ISLANDS

LAND AREA **262 km²** ALTITUDE **0–43 m**
HUMAN POPULATION **46,600** CAPITAL **George Town**
IMPORTANT BIRD AREAS **10, totalling 67 km²**
IMPORTANT BIRD AREA PROTECTION **22%**
BIRD SPECIES **222**
THREATENED BIRDS **4** RESTRICTED-RANGE BIRDS **4**

PATRICIA E. BRADLEY, MAT COTTAM (CAYMAN ISLANDS DEPARTMENT OF ENVIRONMENT, DOE), GINA EBANKS-PETRIE (DOE) AND JONI SOLOMON (DOE)

Malportas Pond, part of the Central Mangrove Wetland IBA on Grand Cayman.
(PHOTO: KRISTAN D. GODBEER/DOE)

INTRODUCTION

The Cayman Islands, a UK Overseas Territory, comprises three low-lying islands: Grand Cayman (197 km²), Little Cayman (28 km²) and Cayman Brac (38 km²). These emergent limestone bluffs are situated along the submerged Cayman Ridge, which is continuous with the Sierra Maestra mountains of south-eastern Cuba. Cayman Brac is the closest to Cuba being c.250 km south, and is also the easternmost of the Cayman Islands. Little Cayman is separated from Cayman Brac by a 7-km wide channel, with Grand Cayman located some 140 km to the south-west. It was not until 1741 that permanent settlements were established on Grand Cayman. The territory's economy changed and grew dramatically in the 1960s with the advent of international banking and tourism. The Cayman Islands are one of the largest banking centres in the world, and the tourism industry now caters to over two million visitors each year. This recent economic boom and population increase has placed a burgeoning pressure on the natural environment. Historically, forests were exploited for timber and fuel. Agriculture is also well established, especially in those areas supporting pockets of wind-blown soil deposits. More recently, clearance of vegetation for the extraction of aggregate and urban development has accounted for significant losses of terrestrial and wetland habitat.

The islands are sheltered by the Greater and Lesser Antilles from the north-east trade winds and have a tropical marine climate with two distinct seasons: a wet season from May to November and a relatively dry season from December to April. Damaging storms strike the islands at a frequency of approximately once every 10 years. The habitats of the Cayman Islands may be broadly categorized into four vegetation types: wetlands, coastal, dry forest, woodland and shrubland, and man-modified areas. Wetland habitats include mangrove swamps, saline lagoons and ponds, wet grassland, freshwater ponds, brackish sedge and *Typha* swamps. The marine mangrove ecosystem includes black mangrove *Avicennia germinans*, white mangrove *Laguncularia racemosa*, red mangrove *Rhizophora mangle*, and buttonwood *Conocarpus erectus*. Coastal habitats include fringing reefs, shoreline, littoral woodland and shrubland and marine bluffs (cliffs). The shoreline includes red mangrove, sandy coral beach, cobble and Ironshore, a Pleistocene conglomerate that overlaps the bluff limestone as a low elevation coastal terrace. While most of the shoreline is characteristically of extremely low elevation, the high bluffs of Grand Cayman and Cayman Brac provide nesting habitat for seabirds. Dry forest and shrubland dominate the karstic limestone which comprises the interior of the islands. Extensive hardwood forests once covered the drier eastern regions of Grand Cayman and Cayman Brac bluff. However, by the beginning of the twentieth century, most of the mature trees had been felled.

The three islands differ in a number of features. Grand Cayman supports c.96% of the total population of the country, the majority of whom live on the western half of the island. North Sound (in western Grand Cayman) is a north-facing shallow marine bay extending for c.100 km^2 and is perhaps the most significant topographical feature. Originally bordered by mangrove swamps to the west, south and east, development has severely reduced fringing vegetation along the southern and western portions. The bed of the Sound has also been extensively dredged, to facilitate the passage of marine craft and to provide fill for development. Grand Cayman reaches its maximum elevation, 18 m, in the Mastic Forest of North Side.

Little Cayman is the least populated of the islands and South Town (Blossom Village) is the main settlement. New roads, small hotels and condominiums have been constructed to meet the growing demands of the tourism industry, the main attractions for which are scuba diving and nature tourism. The south coast comprises fringing reefs, with a rubble ridge forming shallow, protected sounds. Coastal areas, along with the western half of the island, consist of a platform of Ironshore formation rock overlain by beach ridge vegetation, mangrove swamps and saline coastal lagoons. The limestone central bluff rises to a height of 14 m in the area known as Sparrowhawk Hill, the maximum elevation on the island.

Cayman Brac supports a population of c.1,600. The majority are centred along the comparatively low-lying northern coastal shelf, a situation that has, in the past, exposed people to the devastating impacts of high seas, most notably during the 1932 storm. The dominant topographic feature of Cayman Brac is a dolostone plateau, known locally as "the Bluff". Outcropping in the west, the Bluff rises slowly to the north-east, attaining a height of 46 m before terminating in vertical marine cliffs.

Dry forest within the Bluff Forest IBA on Cayman Brac. (PHOTO: KRISTAN D. GODBEER/DOE)

Forest clearance on the edge of the Mastic Reserve IBA, Grand Cayman, shows the need for conservation zoning within the island's development plan. (PHOTO: KRISTAN D. GODBEER/DOE)

■ Conservation

The Department of Environment (DOE) of the Cayman Islands Government and the National Trust for the Cayman Islands, the only environmental NGO in the territory, are the two organisations responsible for advocacy and conservation provisions. The Crown (the Cayman Islands Government) and the Trust, jointly or separately, are owners of all protected land in the islands. The DOE has established, and monitors, several marine parks around the islands, and has drafted legislation (currently pending) to enable the establishment of a series of terrestrial national parks. This draft conservation legislation, the National Conservation Law, was first proposed in 2000, but remains "on the table". It urgently needs to be enacted as it would provide the framework for profound changes in conservation management, including introducing regulations and enforcement across the spectrum, from the creation of national parks to control of the importation of exotic species.

The Development Plan for Grand Cayman 2002 is in the appellate process after the Government removed the proposed conservation and nature tourism zones from the Plan. The DOE, the National Trust and many private individuals are appealing this decision. The Sister Islands (Little Cayman and Cayman Brac) have decided not to implement a Development Plan, but instead have proposed a Sustainable Management Plan, which is currently in first draft. This Plan does not make provision for any system of protected areas. Together, these gaps make conservation planning difficult and leave costly land purchase as the only route to conserve environmentally important sites.

The Cayman Islands Government Environmental Protection Fund (EPF) was established in 1997 through a levy of US$2–4 tax on every person departing the country. One of the main purposes of the fund is the purchase of conservation land and the government has recently confirmed its intent to use the EPF to purchase land in the Barkers area on Grand Cayman, as a move towards establishing the country's first national park. This would set a welcome precedent for further acquisitions of land for conservation on the three Islands. Combined with effective use of the EPF, enactment of the National Conservation Law would help provide the effective conservation management which the Cayman Islands' unique biodiversity urgently needs.

■ Birds

A total of 222 species of birds have been recorded from the Cayman Islands, of which 49 are breeding birds, and 173 occur as non-breeding (primarily Neotropical) migrants. Of the migrants, 97 species are recorded annually. The Cayman Islands are a secondary Endemic Bird Area (EBA). One species was historically endemic to the islands, namely the Grand Cayman Thrush *Turdus ravidus*, but the five other restricted-range birds are found also in other Caribbean Endemic Bird Areas. *Turdus ravidus* was last recorded in 1938. Habitat loss and storms may have contributed to its extinction, though the exact cause of its disappearance remains unknown. The other restricted-range birds include: Jamaican Oriole *Icterus leucopteryx*, extirpated from the Cayman Islands having been last recorded in 1968; Vitelline Warbler *Dendroica vitellina*; Thick-billed Vireo *Vireo crassirostris alleni* (endemic subspecies), extirpated from Little Cayman but still present on the other two islands; Yucatan Vireo *Vireo magister caymanensis* (endemic subspecies), confined to Grand Cayman; and Caribbean Elaenia *Elaenia martinica caymanensis* (endemic subspecies) which is common on all three islands. The Cayman Islands supports 17 endemic subspecies of birds, one of which—the Cuban Bullfinch *Melopyrrha nigra taylori* (endemic to and scarce on Grand Cayman)—has recently been proposed as a separate species, Taylor's Bullfinch.

Four of the Cayman Islands' resident birds are considered globally threatened (see Table 1). The Vulnerable West Indian Whistling-duck *Dendrocygna arborea* breeds on Grand Cayman, Little Cayman and intermittently on Cayman Brac. Birds regularly fly over from Little Cayman to forage on Cayman Brac. In the mid-1980s, total maximum counts fell to 180–220 in Grand Cayman and Little Cayman, but numbers have increased dramatically over the last 20 years to 2,156 birds in 2003, the majority of which (1,500–1,800) occur on Grand Cayman. The increase is attributed to the active enforcement of legal protection causing a reduction in illegal hunting, and the start up of an artificial feeding programme on Grand Cayman.

The Near Threatened Cuban Amazon *Amazona leucocephala* occurs as two endemic races on the Cayman Islands, *A. l. caymanensis* ("Cayman Parrot") on Grand Cayman, with a population estimated at 1,408–1,935 birds in 1995, and *A. l. hesterna* ("Brac Parrot") on Cayman Brac, with a population estimated at 350–430. The latter previously occurred on Little Cayman but was extirpated sometime before 1944. The Grand Cayman population ranges throughout much of the island with subpopulations in the forests and individual pairs breeding close to urban areas in the West Bay peninsula. Its greatest threat is the ongoing fragmentation and clearance of dry forest, 95% of which is privately owned and unprotected. The Near Threatened Vitelline Warbler *D. vitellina* occurs on the Swan Islands (to Honduras) and the Cayman Islands, although 97% of its population is thought to be on the Cayman Islands where it is common. There are three races, two on the Cayman Islands, *D. v. vitellina* on Grand Cayman and *D. v. crawfordi* on Little Cayman and Cayman Brac, and *D. v. nelsoni* on Greater Swan Island. The Near Threatened White-crowned Pigeon *Patagioenas leucocephala* is uncommon to locally common in Cayman, and has been recorded breeding on all three islands. It is also a common passage migrant. Numbers have declined sharply since the mid-1980s, and the DOE has recommended its protection. However, the species remains on the game bird list (under current legislation), and is still hunted.

Six species of seabird breed in the Cayman Islands, three of which are resident: Red-footed Booby *Sula sula* and Magnificent Frigatebird *Fregata magnificens* are resident and breed on Little Cayman; Brown Booby *Sula leucogaster* (the third resident species with 100–120 pairs in 2003) and White-tailed Tropicbird *Phaethon lepturus* (110 pairs in 2003) breed

West Indian Whistling-duck numbers in the Cayman Islands have increased dramatically since the early 1980s to over 2,000 birds. (PHOTO: KRISTAN D. GODBEER/DOE)

The White-crowned Pigeon population has declined sharply since the mid-1980s but is still legally hunted. (PHOTO: KRISTAN D. GODBEER/DOE)

on Cayman Brac; c.10 pairs of Bridled Tern *Sterna anaethetus* nest on Grand Cayman; and a declining number (85 pairs in 2004) of summer breeding Least Tern *S. antillarum* nest throughout the islands.

IMPORTANT BIRD AREAS

The Cayman Islands' 10 IBAs—the territory's international priority sites for bird conservation—cover 6,700 ha (including marine areas). The IBAs have been identified on the basis of 10 key bird species (listed in Table 1) that variously trigger the IBA criteria. These 10 species comprise all four (extant) restricted-range species, and the four globally threatened birds. Of the 10 IBAs in the Cayman Islands, only two are protected in their entirety—namely Booby Pond Nature Reserve (KY007) and Botanic Park and Salina Reserve (KY003). Five IBAs have no protection whatsoever (including three dry forest IBAs in eastern Grand Cayman), and three have partial (between 20 and 70% of their areas) protection.

The three IBAs on Little Cayman cover 50% of the island's area. These include Booby Pond Reserve (KY007)—a Ramsar site protecting the largest *Sula sula* colony in the insular Caribbean; the entire Crown Wetlands (KY009) which is important as breeding habitat for the Vulnerable West Indian Whistling-duck *Dendrocygna arborea*; and dry forest at Sparrowhawk Hill (KY008) which supports the Near Threatened Vitelline Warbler *Dendroica vitellina vitellina*. The single IBA on Cayman Brac covers 9% of the island's area. It is an important breeding site for the Near Threatened "Brac Parrot" *Amazona leucocephala hesterna*, which possibly merits full species status. It has the smallest population of any *Amazona* parrot and the most limited range at 38 km².

Six IBAs on Grand Cayman cover 20% of the island's area. Five sites hold the Near Threatened "Cayman Parrot" *Amazona leucocephala caymanensis* and include 80–85% of its forest breeding habitat. However, only a small proportion

Table 1. Key bird species at Important Bird Areas in the Cayman Islands.

				Cayman Islands IBAs									
				KY001	KY002	KY003	KY004	KY005	KY006	KY007	KY008	KY009	KY010
			Criteria	■	■	■	■	■	■	■	■	■	■
					■	■	■	■	■	■	■		■
Key bird species		**Criteria**	**National population**	■						■		■	
West Indian Whistling-duck *Dendrocygna arborea*	VU ■	■	2,156	1,500		30				60		405	
Magnificent Frigatebird *Fregata magnificens*		■	600							460–600			
Red-footed Booby *Sula sula*		■	20,000							20,000			
Least Tern *Sterna antillarum*		■	255	165								180	
White-crowned Pigeon *Patagioenas leucocephala*	NT ■			✓	✓	✓		✓		✓	✓	✓	✓
Cuban Amazon *Amazona leucocephala*	NT ■		1,755–2,365	253–379	25–350	30	150	170	150				180–210
Caribbean Elaenia *Elaenia martinica*		■			✓	✓	✓	✓	✓	✓	✓		✓
Thick-billed Vireo *Vireo crassirostris*		■			✓	✓	✓	✓	✓				✓
Yucatan Vireo *Vireo magister*		■			✓	✓	✓	✓	✓				
Vitelline Warbler *Dendroica vitellina*	NT ■	■			✓	✓	✓	✓	✓	60	✓		✓

All population figures = numbers of individuals.
Threatened birds: Vulnerable ■; Near Threatened ■. **Restricted-range birds** ■. **Congregatory birds** ■.

Figure 1. Location of Important Bird Areas in the Cayman Islands.

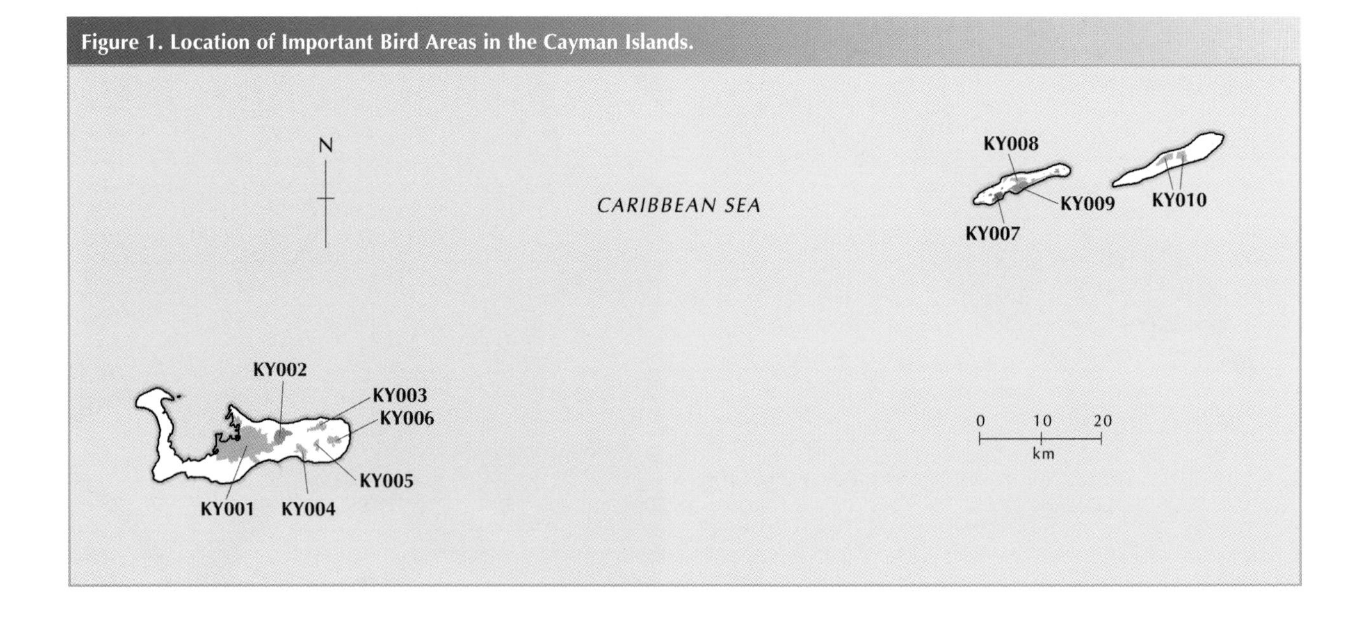

The "Cayman Parrot" occurs in five Grand Cayman IBAs that support 80–85% of its breeding habitat. (PHOTO: MARK F. ORR)

of this habitat is protected. The Central Mangrove Wetland (KY001) supports the largest breeding population of *Dendrocygna arborea* outside of Cuba and is important for *A. l. caymanensis*. The Mastic Reserve (KY002) and three dry forest IBAs in the eastern districts embrace much of the remaining breeding habitat of *A. l. caymanensis* as well as that of *Dendroica vitellina*, Thick-billed Vireo *V. crassirostris alleni,* Yucatan Vireo *V. magister caymanensis* and Cuban Bullfinch *Melopyrrha nigra taylori.*

It is intended that all of the IBAs will be incorporated into the Cayman Islands DOE Protected Areas Management Programme. To be effective, all of the IBAs would require legal protection (which may include land purchase using the EPF), advocacy and planning. While there is sufficient information available on the globally threatened and restricted range species to identify IBAs, the process of identification has made clear that further research and monitoring and urgent sustainable management strategies are required. The results from monitoring currently undertaken for the parrot, the whistling-duck and the seabirds (and any additional species monitoring undertaken in the future) should be used to inform the annual assessment of state, pressure and response scores at each IBA in order to provide an objective status assessment and highlight management interventions that might be required to maintain these internationally important biodiversity sites.

KEY REFERENCES

BANGS, O. (1919) The races of *Dendroica vitellina* Cory. *Bull. Mus. Comp. Zool.* 62: 493–495.

BARLOW, J. C. AND WALKER, M. (1997) A study of inter-island song variation in the Thick-billed Vireo (*Vireo crassirostris*). *El Pitirre* 10: 35–36.

BEARD, J. S. (1955) The classification of tropical American vegetation-types. *Ecology* 36: 89–100.

BRADLEY, P. E. (1986) A Census of *Amazona leucocephala caymanensis*, Grand Cayman and *Amazona leucocephala hesterna*, Cayman Brac. George Town, Grand Cayman: Cayman Islands Government (C.I. Gov. Tech. Pub. 1).

BRADLEY, P. E. (1995) *Birds of the Cayman Islands.* Photographs by Y-J. Rey-Millet. 2nd edition. Italy: Caerulea Press.

BRADLEY, P. E. (2000) *The birds of the Cayman Islands: an annotated check-list*. Tring, UK: British Ornithologists' Union (BOU Check-list 19).

BRADLEY, P.E. (2002) Management Plan to conserve the brown booby colony and its habitat on Cayman Brac, 2002–2006. George Town, Grand Cayman: Cayman Islands Government.

BRADLEY, P. E. (in press) The Cayman Islands. In P. E. Bradley and R. L. Norton eds. *Breeding seabirds of the Caribbean*. Gainesville, Florida: Univ. Florida Press.

BRADLEY, P. E., COTTAM, M., EBANKS-PETRIE, G. AND SOLOMON, J. (2006) Cayman Islands. Pp.65–98 in S. M. Sanders, ed. *Important Bird Areas in the United Kingdom Overseas Territories*. Sandy, U.K.: Royal Society for the Protection of Birds.

BRUNT, M. A. AND DAVIS, J. EDS. (1994) *The Cayman Islands: natural history and biogeography*. The Netherlands: Kluwer Acad. Publ.

BUDEN, D. W. (1985) New subspecies of Thick-billed Vireo (Aves: Vireonidae) from the Caicos Islands, with remarks on taxonomic status of other populations. *Proc. Biol. Soc. Wash.* 98: 591–597.

BURTON, F. J., BRADLEY, P. E., SCHREIBER, E. A., SCHENK, G. A. AND BURTON, R. W. (1999) Status of Red-footed Boobies *Sula sula* on Little Cayman, British West Indies. *Bird Conserv. Internat.* 9: 227–233.

DIAMOND, A. W. (1980) Ecology and species turnover of the birds of Little Cayman. *Atoll Res. Bull.* 241: 141–164.

GARRIDO, O. H., PARKES, K. C., REYNARD, G. B., KIRKCONNELL, A. AND SUTTON, R. L. (1991) Taxonomy of the Stripe-headed Tanager, Genus *Spindalis* (Aves: Thraupidae), in the West Indies. *Wilson Bull.* 109: 561–594.

JOHNSTON, D. W. (1969) The thrushes of Grand Cayman Island, B.W.I. *Condor* 71: 120–128.

NTCI (1992) Census of *Amazona leucocephala caymanensis* on Grand Cayman. George Town, Grand Cayman: Cayman Islands National Trust.

NTCI (1995) Census of *Amazona leucocephala caymanensis* on Grand Cayman. George Town, Grand Cayman: Cayman Islands National Trust.

PAYNTER, R. A. (1956) Birds of the Swan Islands. *Wilson Bull.* 68: 103–110.

PETERS, J. L. (1926) A review of the races of *Elaenia martinica* (Linne). *Occas. Pap. Boston Soc. Nat. Hist.* 5: 197–202.

RAFFAELE, H. WILEY J., GARRIDO, O., KEITH, A. AND RAFFAELE, J. (1998) *A guide to the birds of the West Indies*. Princeton, New Jersey: Princeton University Press.

STEADMAN, D. W. AND MORGAN, G. S. (1985) A new species of Bullfinch (AVES: EMBERIZIDAE) from a late Quaternary cave deposit on Cayman Brac, West Indies. *Proc. Biol. Soc. Wash.* 98: 544–553.

WALKER, M. R. (1996) The Thick-billed Vireo: a conservation perspective for a West Indian endemic. *El Pitirre* 10: 112.

WILEY, J. W. (1991) Status and conservation of parrots and parakeets in the Greater Antilles, Bahama Islands, and Cayman Islands. *Bird Conserv. Internat.* 1: 187–214.

WILEY, J. W. AND WUNDERLE, J. M. (1993) The effects of hurricanes on birds, with special reference to Caribbean islands. *Bird Conserv. Internat.* 3: 319–349.

ACKNOWLEDGEMENTS

The authors would like to thank Kristan D. Godbeer (DOE), Frank Roulstone (National Trust for the Cayman Islands), and Sarah Sanders (Royal Society for the Protection of Birds) for their help in reviewing this chapter.

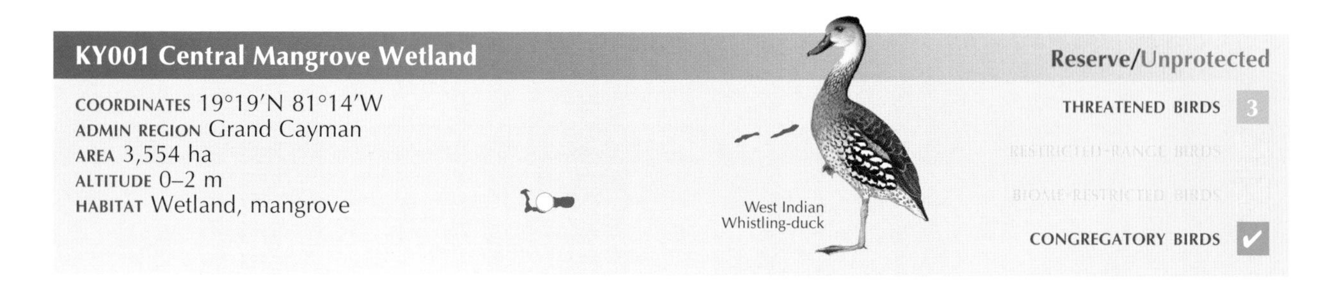

Site description

The Central Mangrove Wetland IBA is a large (30% of the island area), almost totally pristine wetland occupying the centre of Grand Cayman. It includes Meagre Bay Pond (40 ha) and Pease Bay Pond (6 ha) on the south coast, Malportas Pond (44 ha) on the north coast, and a mangrove islet, Booby Cay. The wetland's eastern boundary adjoins the Mastic Reserve IBA (KY002). Vegetation is a mix of the four mangrove species with some areas of monospecific stands. There are many seasonal areas of open water and, interspersed throughout, "dry cays" supporting dry forest. The wetland is important for rainfall generation in the western areas, groundwater replenishment, nature tourism, agriculture, fisheries, the dive industry and hurricane protection.

Birds

This IBA holds at least 1,500 Vulnerable West Indian Whistling-duck *Dendrocygna arborea*—83% of the Cayman Islands population. The Near Threatened "Cayman Parrot" *Amazona leucocephala caymanensis* (over 250 birds) and White-crowned Pigeon *Patagioenas leucocephala* occur. The breeding population (55 pairs) of Least Tern *Sterna antillarum* is regionally important. Large numbers of waterbirds are present with a mixed heronry of 500 pairs of egrets and herons and at least 1,000 wintering ducks occurring alongside shorebirds and marshbirds.

Other biodiversity

The plant (herb) *Agalinis kingsii* is endemic to this IBA and the Salina Reserve (IBA KY003).

Conservation

The Central Mangrove Wetland IBA is 19% protected under marine conservation law, 7% owned and protected by the National Trust for the Cayman Islands, 9% owned by the Crown and unprotected, and 75% privately owned and unprotected. Both *A. leucocephala caymanensis* and *D. arborea* rely on the outer *Avicennia* forest zone for breeding, but at least four areas of this forest have been cleared for marl-mining pits. Large areas of *Avicennia* were badly damaged during Hurricane Ivan in 2004. Other threats include destruction of parrot nest sites during illegal trapping, shooting of parrots as a crop pest, and feral dogs and cats predating the young ducks. Proposed road and urban developments also threaten the site. The government removed all conservation zones from the 2001 Development Plan for Grand Cayman, and efforts to have the IBA designated a Ramsar site have so far failed. Malportas Pond is the site of a government-funded feeding station for *D. arborea*.

Site description

The Mastic Reserve IBA is an area of dry forest that lies inland on North Side, central-east Grand Cayman. It is bounded on the west and south by the Central Mangrove Wetland IBA (KY001) and to the north by agricultural land. The IBA, which includes the highest point on Grand Cayman ("The Mountain"), comprises a series of east–west karstic ridges. The forest has both primary- and second-growth trees with areas of grassland, once used for agriculture. It is the largest area of contiguous dry forest in the Cayman Islands, and has retained its pristine nature because there is no road access; only two footpaths bisect the site.

Birds

This IBA supports three Near Threatened birds: "Cayman Parrot" *Amazona leucocephala* (up to 350 of the endemic subspecies *caymanensis*); Vitelline Warbler *Dendroica vitellina vitellina* (up to 1% of the global population); and White-crowned Pigeon *Patagioenas leucocephala*. All four Cayman Islands secondary EBA restricted-range birds occur in the IBA, as does the Cuban Bullfinch *Melopyrrha nigra taylori*. The extinct Grand Cayman Thrush *Turdus ravidus* and the extirpated Jamaican Oriole *Icterus leucopteryx* were known from this IBA.

Other biodiversity

This IBA has the highest degree of endemism and biodiversity in the Cayman Islands. Other endemic species, including some endemic to Grand Cayman, include four reptiles, five butterflies and 10 plants.

Conservation

The Mastic Reserve IBA is c.70% (314 ha) owned and managed by the National Trust for the Cayman Islands. The remainder is unprotected private land, although purchase of this is the focus of a Trust campaign. The long-term conservation plan aims to create a larger protected area, by combining the Mastic Reserve and the adjoining Central Mangrove Wetland IBA. The Mastic trail, a 4-km traditional footpath that runs north–south, was opened for visitors in 1994 (following a grant from RARE Conservation). There are plans for urban development on the north, south and eastern boundaries of the IBA. A proposed north–south road would enable access for the development of private lands. Specific threats to the parrots include illegal felling of nesting trees and removal of young for the illegal pet trade; predation by rats *Rattus* spp. and feral cats; and illegal shooting as a crop pest.

KY003 Botanic Park and Salina Reserve

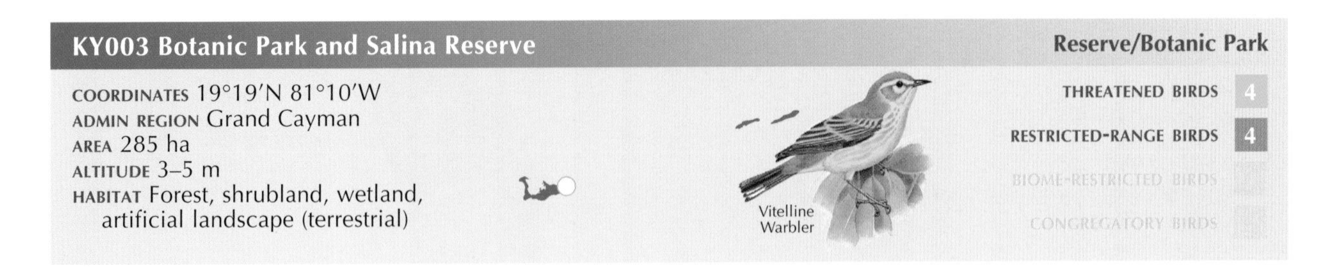

■ Site description

The Botanic Park and Salina Reserve IBA comprises two separate areas in eastern Grand Cayman. The 50-ha Botanic Park lies immediately east of the Frank Sound Road (and west of Frank Sound Forest IBA KY004) in the centre of the island and supports dry forest fragments interspersed with shrubland, *Conocarpus* wetlands, a lake and horticultural areas. The 235-ha Salina Reserve is inland on the north-east coast and is a large temporary, freshwater herbaceous wetland bounded by an intricate mosaic of sedges, reeds and *Conocarpus* shrubland with dry forest on the northern boundary, where *Swietenia mahagoni* is dominant. There are no clear trails through the reserve which means that it has been left fairly isolated and is still very much undisturbed.

■ Birds

This IBA supports small breeding populations of the Vulnerable West Indian Whistling-duck *Dendrocygna arborea*, and the Near Threatened "Cayman Parrot" *Amazona leucocephala*, White-crowned Pigeon *Patagioenas leucocephala* and Vitelline Warbler *Dendroica vitellina vitellina*. All four Cayman Islands secondary EBA restricted-range birds occur in the IBA (especially in the Botanic Park), namely *D. vitellina*, Caribbean Elaenia *Elaenia martinica*, Yucatan Vireo *Vireo magister* and Thick-billed Vireo *V. crassirostris*. The Cuban Bullfinch *Melopyrrha nigra taylori* also occurs.

■ Other biodiversity

The IBA is a centre for the captive breeding and release of the Critically Endangered Grand Cayman blue iguana *Cyclura lewisi*. The Near Threatened white-shouldered bat *Phyllops falcatus* and Brazilian free-tailed bat *Tadarida brasiliensis* occur. The plant *Agalinis kingsii* is endemic to this IBA and the Central Mangrove Wetland (IBA KY001). Many other Cayman Islands endemic plants, insects and reptiles occur.

■ Conservation

The Queen Elizabeth II Botanic Park is jointly owned and protected by the National Trust for the Cayman Islands and the Crown. About half of the reserve is open to the public. However, this area is likely to become an island surrounded by urban development, which is a potential threat to the free-roaming iguanas (*C. lewisi*) released in the area. The Salina Reserve is 100% owned and protected by the Trust. Its inaccessibility and sharp, rocky terrain helps maintain the reserve in a relatively pristine state, and provides some deterrent to feral dogs and cats that might otherwise impact the released iguanas.

KY004 Frank Sound Forest

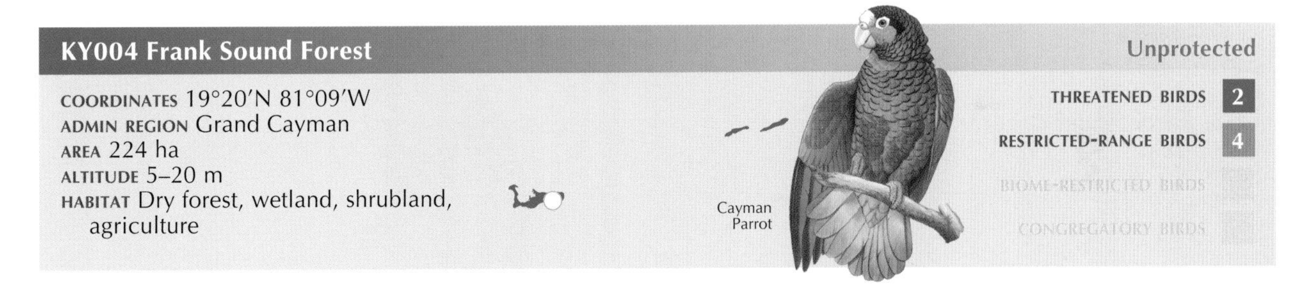

■ Site description

Frank Sound Forest IBA lies east of east of Frank Sound Road at the eastern end of Grand Cayman. The forest surrounds but is not connected to the Botanic Park (IBA KY003) that is on its north-western boundary. It comprises dry forest bordered by *Conocarpus* wetlands, dry shrubland and agricultural land.

■ Birds

This IBA supports two Near Threatened birds: "Cayman Parrot" *Amazona leucocephala* (endemic subspecies *caymanensis*) and Vitelline Warbler *Dendroica vitellina vitellina*. All four Cayman Islands secondary EBA restricted-range birds occur in the IBA, namely *D. vitellina*, Caribbean Elaenia *Elaenia martinica*, Yucatan Vireo *Vireo magister* and Thick-billed Vireo *V. crassirostris*. The Cuban Bullfinch *Melopyrrha nigra taylori* also occurs.

■ Other biodiversity

A number of reptiles (four species) and plants (six species) that are endemic to Grand Cayman occur in this IBA.

■ Conservation

Frank Sound Forest IBA is privately owned and unprotected. Parts are being cleared for agriculture and urban development causing forest loss and fragmentation which in turn will impact the bird populations reliant on this and other eastern Grand Cayman forests. Young parrots are collected from the nest for the illegal pet trade, and the nest trees are usually destroyed. Illegal shooting of parrots as a crop pest continues.

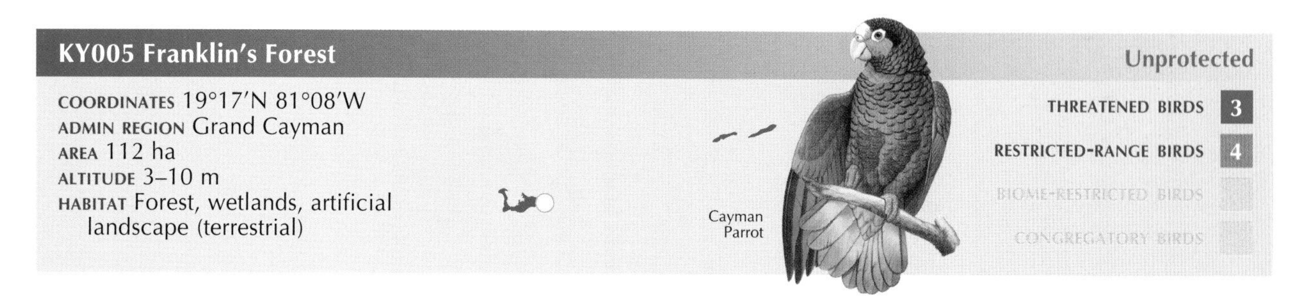

■ Site description

Franklin's Forest IBA lies in the centre of Grand Cayman's eastern districts. It is between Frank Sound Forest IBA (KY004, to the west) and Eastern Dry Forest IBA (KY006, to the east) and comprises dry forest bordered by *Conocarpus* wetlands and agricultural plantations.

■ Birds

This IBA supports three Near Threatened birds: "Cayman Parrot" *Amazona leucocephala* (150+ of the endemic subspecies *caymanensis*); Vitelline Warbler *Dendroica vitellina vitellina* (more than 1% of the global population); and White-crowned Pigeon *Patagioenas leucocephala*. All four Cayman Islands secondary EBA restricted-range birds occur in the IBA, namely *D. vitellina*, Caribbean Elaenia *Elaenia martinica*, Yucatan Vireo *V. magister* and Thick-billed Vireo *V. crassirostris*. The Cuban Bullfinch *Melopyrrha nigra taylori* also occurs. The extinct Grand Cayman Thrush *Turdus ravidus* was last seen in this IBA in 1938.

■ Other biodiversity

The Critically Endangered Grand Cayman blue iguana *Cyclura lewisi* occurs in adjacent shrubland. Grand Cayman endemics include four reptiles and six plants.

■ Conservation

Part of Franklin's Forest IBA has recently been purchased by the Crown. However, the remainder is privately owned and all of it is unprotected. As such, it is threatened by clearance and fragmentation. Young parrots are illegally collected from the nest for the illegal pet trade, and the nest trees are usually destroyed. Illegal shooting of parrots as a crop pest continues with over 200 birds shot in 2000.

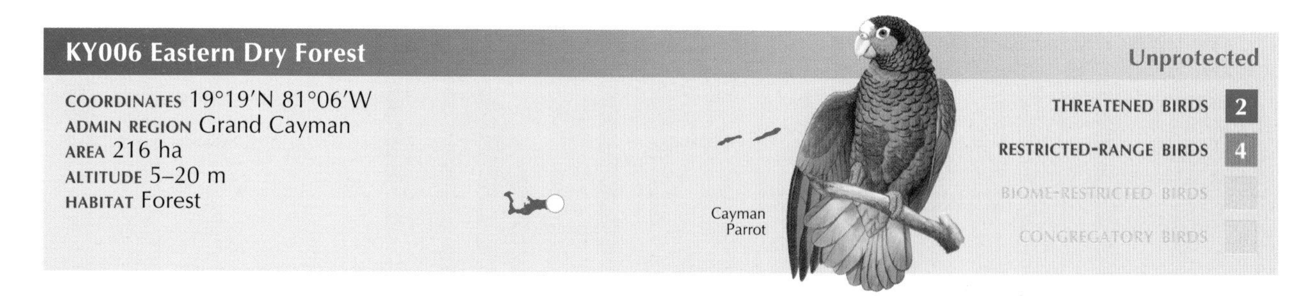

■ Site description

The Eastern Dry Forest IBA lies within Grand Cayman's eastern districts. It is north of East End town and comprises dry forest that is rapidly being cleared and fragmented. During the 1980s, this site contained some of the largest trees on Grand Cayman and was a major parrot breeding site.

■ Birds

This IBA supports two Near Threatened birds: "Cayman Parrot" *Amazona leucocephala* (endemic subspecies *caymanensis*) and Vitelline Warbler *Dendroica vitellina vitellina*. All four Cayman Islands secondary EBA restricted-range birds occur in the IBA, namely *D. vitellina*, Caribbean Elaenia *Elaenia martinica*, Yucatan Vireo *Vireo magister* and Thick-billed Vireo *V. crassirostris*. The Cuban Bullfinch *Melopyrrha nigra taylori* also occurs.

■ Other biodiversity

The Critically Endangered Grand Cayman blue iguana *Cyclura lewisi* occurs in adjacent shrubland. A number of reptiles (four species) and plants (six species) that are endemic to Grand Cayman occur in this IBA.

■ Conservation

Eastern Dry Forest IBA is privately owned and unprotected. This is a very fragmented forest site on Grand Cayman and, being close to East End town, has traditionally been a source of young parrots for the illegal pet trade. This illegal practice continues, causing loss of nesting trees when the trunk is cut open to reach into the deep nest-cavity. Predation by feral cats is a further threat. The DOE is currently establishing an annual parrot survey of the islands.

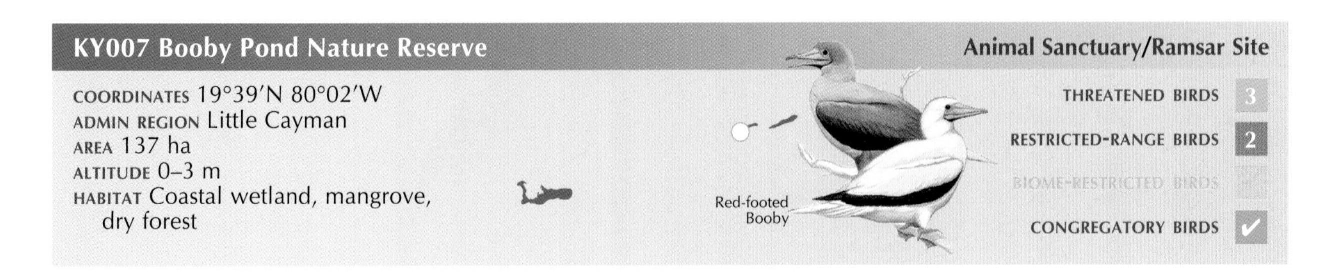

Site description

Booby Pond Nature Reserve IBA lies on the south coast of Little Cayman behind the beach ridge at South Hole Sound. The IBA comprises the 43-ha Booby Pond—a seasonally flooded, enclosed hyper-saline lagoon with a broken mangrove fringe—and on the northern, inland side of the pond, an area of dry forest. A large seabird rookery is situated on this northern side of the pond. The southern shore of the Booby Pond is being encroached upon by hotel buildings.

Birds

This IBA is named after the globally significant (indeed the insular Caribbean's largest) colony of Red-footed Booby *Sula sula* that nest on the north side of the pond in the mangrove, shrubland and inland dry forest. The 2000–2004 population was estimated at 3,824–5,854 pairs (and c.20,000 birds). A regionally important colony of Magnificent Frigatebird *Fregata magnificens* nest alongside the *S. sula*. Up to 20 pairs of Vulnerable West Indian Whistling-duck *Dendrocygna arborea* also breed in the IBA. The Near Threatened Vitelline Warbler *Dendroica vitellina crawfordi* and White-crowned Pigeon *Patagioenas leucocephala* are present. Large numbers of egrets, duck and shorebirds frequent the IBA.

Other biodiversity

The Vulnerable Lesser Cayman Islands rock iguana *Cyclura nubila caymanensis* occurs here, along with the lizard *Anolis maynardi*, which is endemic to Little Cayman. The mollusc *Cerion nanus* is Critically Endangered and endemic to Little Cayman. A number of other Cayman endemics occur including three reptiles, three molluscs, two insects, two fish and seven plants.

Conservation

This IBA has the greatest protection of any site in the Cayman Islands. It is protected as the Booby Pond and Rookery Animal Sanctuary, is designated as a Ramsar site, and 135 ha of the IBA is owned by the National Trust for the Cayman Islands (which has a visitor centre on site). In spite of this protection, illegal development has encroached (without legal sanction) on the south shore of the Booby Pond, and is thought to contribute to pollution in the pond. A major concern for the booby colony is the potential impact of a proposed new airport, in the vicinity of the pond. Disturbance from night lights, aircraft, and access roads may have significant detrimental effects. Predation from rats *Rattus* spp., feral cats and domestic dogs has not been quantified, though control measures against feral cats have been initiated by DOE.

KY008 Sparrowhawk Hill

Unprotected

COORDINATES 19°41′N 80°02′W
ADMIN REGION Little Cayman
AREA 153 ha
ALTITUDE 10–14 m
HABITAT Dry forest

Vitelline Warbler

THREATENED BIRDS 2
RESTRICTED-RANGE BIRDS 2
BIOME-RESTRICTED BIRDS
CONGREGATORY BIRDS

Site description

Sparrowhawk Hill IBA is an area of pristine dry forest in the centre of Little Cayman. It lies on the Central Bluff—an area of raised limestone outcrop that reaches an elevation of 14 m, the island's highest point. At present, there is no public road into the island's interior near the forest.

Birds

This IBA supports a significant population of the Near Threatened Vitelline Warbler *Dendroica vitellina crawfordi*. The forest represents c.3% of the suitable habitat within the species' range. The Near Threatened White-crowned Pigeon *Patagioenas leucocephala* breeds in the IBA (although they are not resident). *Dendroica vitellina* is one of the two Cayman Islands secondary EBA restricted-range birds to occur in the IBA, the other one being Caribbean Elaenia *Elaenia martinica*. The dry forest is home to a range of Neotropical migratory birds during the winter.

Other biodiversity

The Vulnerable Lesser Cayman Islands rock iguana *Cyclura nubila caymanensis* occurs in the IBA, along with the lizard *Anolis maynardi*, which is endemic to Little Cayman. A number of other Cayman endemics occur including four reptiles, two butterflies and five plant species.

Conservation

Sparrowhawk Hill IBA is privately owned, and with no public roads into the area it remains relatively pristine. However, the land owners have built survey tracks on the west and south edges of the IBA. The threat of development is ever present.

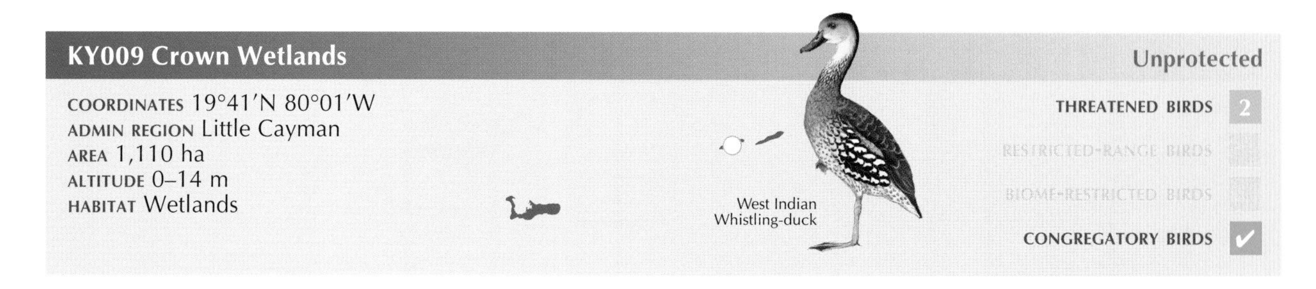

■ Site description

The Crown Wetlands IBA embraces the pristine wetlands of Little Cayman which cover 40% of the island's land area. There are four main types of wetlands in the IBA. On the north and south coasts, the mangrove wetlands—each associated with a hyper-saline lagoon—are Easterly Pond (3 ha), Rosetta Flats Pond (2 ha), Sandy Point Pond (3.5 ha), Tarpon Lake complex (236 ha), Spot Bay Pond (5 ha), Jackson's Pond (9 ha) and Grape Tree Pond (10 ha). In the south-west, Preston Bay westerly ponds (8.4 ha) are brackish herbaceous wetlands with buttonwood. Charles Bight Pond (8.5 ha) is an inland buttonwood wetland on the eastern bluff, and the 0.1-ha Coot Pond is a temporary freshwater wetland (grassland and buttonwood) on the south-east coast. All of the wetlands dry out seasonally, except Tarpon Lake.

■ Birds

This IBA is significant for its breeding population of up to 135 pairs of Vulnerable West Indian Whistling-duck *Dendrocygna arborea*. The largest concentrations of the duck are at Jackson's Pond, Grape Tree Pond and Charles Bight Pond. The Near Threatened White-crowned Pigeon *Patagioenas leucocephala* also occurs. The breeding population of 60 pairs of Least Tern *Sterna antillarum* is regionally important. Large numbers of egrets and herons (at least 1,000), ducks (over 1,650) and shorebirds (1,400) frequent the IBA.

■ Other biodiversity

The Vulnerable Lesser Cayman Islands rock iguana *Cyclura nubila caymanensis* occurs in the IBA, along with the lizard *Anolis maynardi*, which is endemic to Little Cayman. Three other reptiles are Cayman Island endemics.

■ Conservation

The Crown Wetlands IBA is almost entirely (1,094 ha) crown land, but has no formal protection. The wetlands are almost pristine, except for the development of a circum-island road in 1994. In 1999, ten raised wildlife observation platforms were constructed on the major ponds (part of a Foreign and Commonwealth Office-funded avitourism project). While the government allowed the development of these avitourism facilities, the wetlands remain unprotected, and the government is currently looking at pending land claims that would place many hectares of the wetlands into private hands.

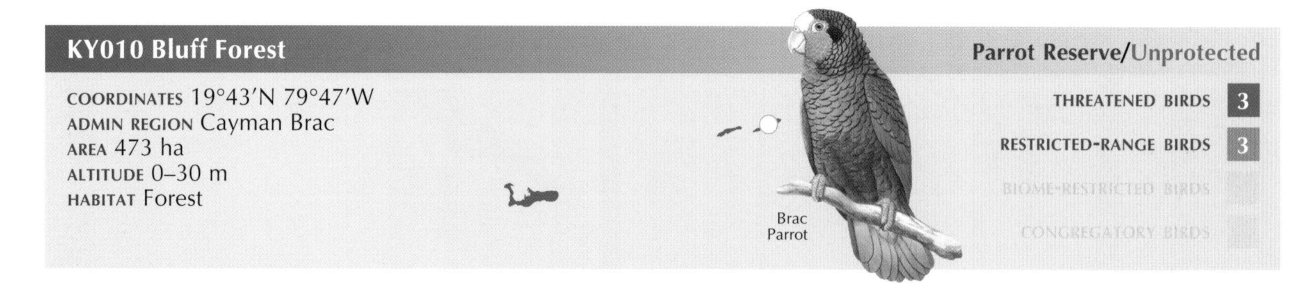

■ Site description

Bluff Forest IBA comprises dry forest on jagged karstic limestone, of which almost 20% is a protected parrot reserve. The forest is highly diverse with cedar *Cedrela odorata* cavities being particularly important for nesting parrots. These trees are common within the *Bursera-Exothea-Chionanthus* community, but there is little or no recruitment of seedlings or young trees, due to infestation from the mahogany shoot-borer *Hypsipyla grandella*. There has been a long history of disturbance by logging, and the forest is a mosaic of primary and second-growth trees.

■ Birds

This IBA supports 60–70 pairs of the Near Threatened "Brac Parrot" *Amazona leucocephala* (the endemic subspecies *hesterna*), and about 9% of the global population of the Near Threatened Vitelline Warbler *Dendroica vitellina crawfordi*. The Near Threatened White-crowned Pigeon *Patagioenas leucocephala* breeds in the IBA. Three (of the four) Cayman Islands secondary EBA restricted-range birds to occur in the IBA, namely *D. vitellina*, Caribbean Elaenia *Elaenia martinica* and Thick-billed Vireo *Vireo crassirostris*. The IBA is home to a range of Neotropical migratory birds during the winter.

■ Other biodiversity

The Vulnerable Lesser Cayman Islands rock iguana *Cyclura nubila caymanensis* occurs here. A number of other Cayman endemics occur including five reptiles, two butterflies and eight plants. The cedar *Cedrela odorata*, so important for the parrot, is Vulnerable.

■ Conservation

The Bluff Forest is c.80% (345 ha) privately owned, with 113 ha protected as the Brac Parrot Reserve which is owned by the National Trust for the Cayman Islands. However, primary forest immediately bordering this area has recently been cleared for residential development. Most active parrot nests have been found in dead or dying *Cedrela odorata* cavities. With little or no recruitment of this tree species, there will be insufficient protected breeding habitat to sustain the parrot population in the long-term. Additional forest habitat needs to be conserved. Other threats include illegal shooting and human disturbance; predation by escalating populations of rats *Rattus* spp. and feral cats; illegal capture of parrots as pets; potential inter-breeding with illegally imported Grand Cayman parrot; and the establishment of competing feral populations of exotic parrots. *Patagioenas leucocephala* is hunted during the summer breeding season—a threat to the pigeon, and a disturbance to other landbirds at a critical time.

CUBA

LAND AREA **109,886 km²** ALTITUDE **0–1,974 m**
HUMAN POPULATION **11,239,043** CAPITAL **Havana**
IMPORTANT BIRD AREAS **28, totalling 231,657 km²**
IMPORTANT BIRD AREA PROTECTION **76%**
BIRD SPECIES **371**
THREATENED BIRDS **29** RESTRICTED-RANGE BIRDS **11**

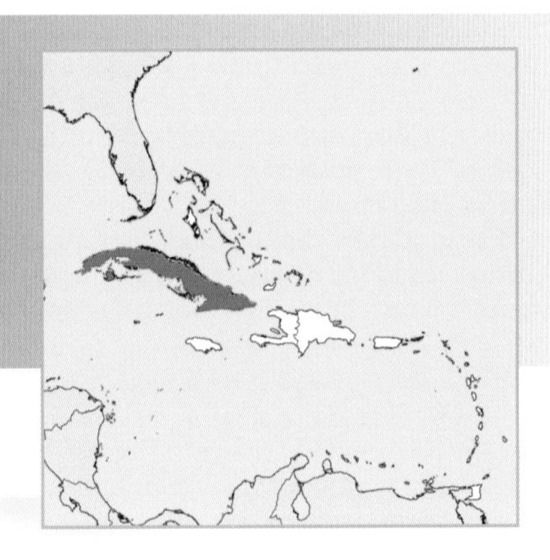

COORDINATOR: SUSANA AGUILAR MUGICA
(CENTRO NACIONAL DE ÁREAS PROTEGIDAS, CNAP)

Alejandro de Humboldt IBA represents the largest and best preserved tract of tropical and montane forest in the insular Caribbean. (PHOTO: JULIO LARRAMENDI)

INTRODUCTION

Cuba is the largest and most westerly island of the insular Caribbean, accounting for more than 50% of the region's land area. It is bordered by the Bahamas and the Florida peninsula to the north, Turks and Caicos Islands to the north-east, Hispaniola to the east, Cayman Islands to the south, Jamaica to the south-east, and the waters of the Gulf of Mexico to the west. Mainland Cuba is surrounded by four archipelagos: Sabana Camagüey (off the north coast of central Cuba); Los Canarreos (including the Isle of Pines of the south coast of western Cuba); Jardines de la Reina (south coast of eastern Cuba); and Los Colorados north coast of westernmost Cuba). Combined, these archipelagos comprise 4,195 islets and cays which cover c.3,715 km².

Mainland Cuba is 1,250 km long and averages 150 km wide. Vast plains occupy 79% of the land area but are interrupted by four mountain systems: Guaniguanico Mountain Range (western Cuba), Guamuhaya Massif (on the southern side of central Cuba), the Nipe-Sagua-Baracoa Massif (easternmost Cuba), and the Sierra Maestra (the south coastal range of south-east Cuba). Pico Turquino in the Sierra Maestra, is the highest point on the island at 1,974 m. The climate is tropical–subtropical, with an average annual rainfall of 1,375 mm and temperatures ranging from 21 to 27 °C. Annual rainfall varies between less than 200 mm on the south coast (around Guantánamo) to 3,400 mm in the Nipe-Sagua-Baracoa mountain range.

Climate, geography and topography have combined to produce a wide diversity of ecosystems on the island (including five terrestrial ecoregions, three biogeographical areas, and 39 floristic districts). The flora of the island is particularly rich, with 921 species of bryophytes, 500 pteridophytes and 6,519 higher plant species. Cuba is the most biologically diverse island in the West Indies and exhibits exceptional levels of endemism, particularly at higher elevations and in the east of the country. More than 50% of flora and 32% of vertebrate fauna are endemic to Cuba, with these proportions especially high among vascular plants (52%) and herpetofauna (86%). Many of these endemic species are locally restricted. Mammals are represented by 42 species, including hutias (*Capromys pilorides*, *Mysateles melanurus*, *M. prehensilis*, and *Mesocapromys auritus*), the Cuban solenodon *Solenodon cubanus*, and various species of bats, including *Nyctiellus lepidus*, the smallest bat in the world. Among the 204 species of herpetofauna, *Eleutherodactylus iberia* is notable for being the smallest frog in the Northern Hemisphere. At least 13,000 invertebrate species have been reported, and among these, molluscs and in particular land snails, have significant levels of endemism. There are also unique species of arachnids, such as the scorpion *Microfityus fundorai*. Marine life is also extremely diverse (although endemism is not high) and in a healthy state, with 963 fish species, 58 corals, 160 sponges, and 68 gorgonids.

The growth of the sugarcane industry in the early twentieth century led to the destruction and alteration of habitats across the plains of Cuba. This was followed by deforestation for urban development and livestock farming and as a result all Cuban habitats have been affected, either by fragmentation, pollution, degradation, modification or introduction of exotic species. Fortunately, reforestation efforts that started in 1960 have borne fruit, and forest now covers 21% of the country. Tourist development has also caused major disturbances, especially in the northern cays which have been severely impacted. Finally, Cuba is frequently affected by hurricanes and storms, and recent climate changes have contributed to increased cyclonic activity, drought periods, and fires.

■ Conservation

Conservation efforts began on the island in the 1930s, with the creation of the Sierra Cristal National Park. This was followed by the designation of other areas, albeit as a formal exercise rather than on-the-ground management action. By 1959, nine national parks, and the first natural reserves and national monuments had been created, and the first proposal for a national system of protected areas (Sistema Nacional de Áreas Protegidas, SNAP) made. However, these early protected areas had no structured protection categories, adequate personnel nor infrastructure. In 1980, the Empresa para la Protección de la Flora y la Fauna (National Enterprise for the Protection of Flora and Fauna, ENPFF) of the Ministry of Agriculture was created to manage c.30 protected areas. Installations were built and technical and administrative personnel were assigned to individual areas, and a proposal was made to expand this capacity development to a further 73 protected areas. Four biosphere reserves were declared at this time through UNESCO.

The creation of the Ministerio de Ciencia Tecnología y Medio Ambiente (Ministry of Science, Technology, and Environment, CITMA) and within it the Centro Nacional de Áreas Protegidas (National Centre for Protected Areas, CNAP) in 1995 marked a new era in the realisation of the national protected area system. CNAP is responsible for planning and management of the overall national system of protected areas in Cuba. ENPFF is also responsible for the administration of the majority of them. CITMA manages a smaller portfolio of sites, primarily those with fragile ecosystems or in need of strict protection. The national system of protected areas follows the 1994 IUCN protected area management classification scheme (with some Cuba-specific modifications).

Significant progress has been made in the development and institutionalisation of Cuba's protected areas in recent years. The protected area system currently includes 263 proposed areas (80 of which are considered nationally significant and 183 locally significant), covering c.22% of the country's area. There are 96 protected areas currently functioning at a practical level within Cuba, of which 45 areas have been legally approved by the Comité Ejecutivo del Consejo de Ministros (Executive Committee of the Ministerial Council), whilst 14 are waiting for formal recognition.

The protected areas provide a framework for quite considerable research, conservation, education and awareness efforts of CNAP, ENPFF and other institutions including Instituto de Ecología y Sistemática (IES), the Museo Nacional de Historia Natural, the Facultad de Biología de la Universidad de la Habana, and the Centro Oriental de Ecosistemas y Biodiversidad (BIOECO). However, in spite of these efforts, the protected areas (and the land surrounding them) still face multiple threats from habitat destruction and degradation, hunting, alien invasive species, illegal trade in species and pollution. Specific threats to the Important Bird Areas (IBAs) are detailed under each individual IBA profile.

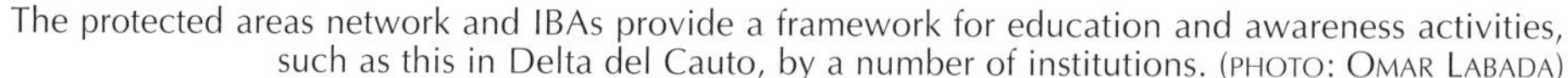

The protected areas network and IBAs provide a framework for education and awareness activities, such as this in Delta del Cauto, by a number of institutions. (PHOTO: OMAR LABADA)

Bee Hummingbird—the world's smallest bird—is endemic to Cuba. (PHOTO: TIM STEWART)

The Endangered Cuban Sparrow represents one of six bird genera endemic to Cuba, and occurs in three localised populations (such as this one from Cayo Coco) across the island. (PHOTO: PETE MORRIS)

■ Birds

Of the 371 bird species reported for Cuba about 42% breed on the island, and 70% are migratory (c.114 species are regular winter residents). Of the breeding species, 135 are resident (including eight introduced birds) and 14 arrive from South America to breed on the island in the summer. The island is home to six endemic genera (namely *Cyanolimnus*, *Starnoenas*, *Xiphidiopicus*, *Ferminia*, *Teretistris* and *Torreornis*), 28 endemic birds (including the world's smallest bird, the Bee Hummingbird *Mellisuga helenae*) and 60 endemic subspecies. Cuba is an Endemic Bird Area (EBA), but only 11 species have ranges of less than 50,000 km² (which is the threshold that defines a restricted-range bird). However, there are 48 species in Cuba confined to the Greater Antilles biome (which includes all the island endemics and the restricted-range birds), and these have been used to define IBAs in Cuba (see below). Some of the restricted-range species are shared with neighbouring islands (e.g. Thick-billed Vireo *Vireo crassirostris* and Olive-capped Warbler *Dendroica pityophila* with Bahamas, Palm Crow *Corvus palmarum* with Hispaniola, Bahama Mockingbird *Mimus gundlachii* with Bahamas and Jamaica).

The Vulnerable Cuban Parakeet is one of a number of species threatened by the illegal pet trade in Cuba. (PHOTO: PETE MORRIS)

There are 29 threatened birds known from Cuba, three of which are Critically Endangered (namely, Cuban Kite *Chondrohierax wilsonii*, Ivory-billed Woodpecker *Campephilus principalis* and Bachman's Warbler *Vermivora bachmanii*), eight Endangered, seven Vulnerable, and 11 Near Threatened. However, five of these threatened birds have been excluded from the IBA analysis, as the island is not currently known or suspected to support viable populations. The five species are the Critically Endangered *V. bachmanii*, the Vulnerable Bahama Swallow *Tachycineta cyaneoviridis* and Cerulean Warbler *Dendroica cerulean*, and the Near Threatened Caribbean Coot *Fulica caribaea* and Golden-winged Warbler *Vermivora chrysoptera*. The threat category and national population sizes of the threatened birds are listed in Table 1.

The endemic Cuban Macaw *Ara tricolor* became extinct in the nineteenth century due to heavy hunting. Other species heading in the same direction as the macaw are: the Cuban Kite *Chondrohierax wilsonii* which was once widespread, but habitat destruction has caused its decline, and it is presently confined to the Alejandro de Humboldt IBA (CU027); and Ivory-billed Woodpecker *Campephilus principalis*, a small population of which may survive in south-eastern Cuba, although it has not been observed since 1987. A number of the threatened species have very localised populations which contributes to their endangerment, as is the case with the Endangered Zapata Rail *Cyanolimnas cerverai* and Zapata Wren *Ferminia cerverai* (both confined to Ciénaga Zapata, CU006) and Cuban Sparrow *Torreornis inexpectata* (confined to three small, disjunct areas in Cuba).

Cuba's location within the Caribbean makes the island an important migratory corridor and wintering site for a large number of Neotropical migratory birds. It is located on both the Mississippi and East Atlantic migratory flyways, and thousands of raptors, ducks, shorebirds and landbirds are recorded each year across the country. Cuba is extremely important for waterbirds, and the country's network of natural and artificial wetlands (which includes the two largest wetlands in the Caribbean, namely Ciénaga Zapata and Delta del Cauto) provides critical habitat for the largest recorded concentrations in the Caribbean of a number of species (including flamingos, ibises, cormorants, egrets, anhingas and spoonbills). Major seabird breeding colonies are concentrated on the offshore cays of the Sabana-Camagüey Archipelago.

Apart from the widespread loss and degradation of habitats, Cuba's birds are threatened by hunting (both sport hunting and illegal poaching), collecting eggs, and illegal capture and trade of Cuban psittacids (i.e. Cuban Amazon *Amazona*

Table 1. Key bird species at Important Bird Areas in Cuba.

Key bird species	Criteria	National population	CU001	CU002	CU003	CU004	CU005	CU006
Criteria			■ ■ ■	■ ■ ■	■ ■	■ ■ ■	■ ■ ■ ■	■ ■ ■ ■
Northern Bobwhite *Colinus virginianus*	NT ■				✓	✓	24,000	✓
West Indian Whistling-duck *Dendrocygna arborea*	VU ■ ■ ■	5,000–9,999	✓		50–100		✓	✓
Blue-winged Teal *Anas discors*	■	100,000–999,999						
Black-capped Petrel *Pterodroma hasitata*	EN ■ ■	250–499						
Caribbean Flamingo *Phoenicopterus ruber*	■	100,000–999,999						3,000
White Ibis *Eudocimus albus*	■	10,000–24,999						
Glossy Ibis *Plegadis falcinellus*	■	25,000–99,999			20,000			
Magnificent Frigatebird *Fregata magnificens*	■	1,000–4,999						
Brown Pelican *Pelecanus occidentalis*	■	1,000–4,999			146			
Double-crested Cormorant *Phalacrocorax auritus*	■	10,000–24,999			820			
Osprey *Pandion haliaetus*	■							
Cuban Kite *Chondrohierax wilsonii*	CR ■ ■	50–249						
Gundlach's Hawk *Accipiter gundlachi*	EN ■ ■	300–400	✓	✓		6		6
Black Rail *Laterallus jamaicensis*	NT ■	500–999						✓
Zapata Rail *Cyanolimnas cerverai*	EN ■ ■ ■ ■	<50						✓
Sandhill Crane *Grus canadensis*	■	500–999					171	120
Piping Plover *Charadrius melodus*	NT ■ ■	100–249						
Short-billed Dowitcher *Limnodromus griseus*	■	10,000–24,999			10,000			
Lesser Yellowlegs *Tringa flavipes*	■	10,000–24,999						
Least Sandpiper *Calidris minutilla*	■	10,000–24,999						
Laughing Gull *Larus atricilla*	■	10,000–24,999						
Gull-billed Tern *Sterna nilotica*	■	1,000–4,999						
Royal Tern *Sterna maxima*	■	5,000–9,999						
Sandwich Tern *Sterna sandvicensis*	■	500–999						
Common Tern *Sterna hirundo*	■	100–249						
Least Tern *Sterna antillarum*	■	1,000–4,999						
Bridled Tern *Sterna anaethetus*	■	1,000–4,999						
White-crowned Pigeon *Patagioenas leucocephala*	NT ■		✓		✓	✓	800,000	✓
Plain Pigeon *Patagioenas inornata*	NT ■ ■		✓	✓				✓
Grey-headed Quail-dove *Geotrygon caniceps*	VU ■ ■	2,500–9,999		✓		30		✓
Key West Quail-dove *Geotrygon chrysia*	■		✓	✓			✓	✓
Blue-headed Quail-dove *Starnoenas cyanocephala*	EN ■ ■	1,000–2,499	✓	✓		✓		✓
Cuban Parakeet *Aratinga euops*	VU ■ ■	2,500–9,999		30				8
Cuban Amazon *Amazona leucocephala*	NT ■ ■		✓	✓			580	100
Great Lizard-cuckoo *Saurothera merlini*	■		✓	✓		✓	✓	✓
Bare-legged Owl *Gymnoglaux lawrencii*	■		✓	✓		✓	✓	✓
Cuban Pygmy-owl *Glaucidium siju*	■		✓	✓		✓	✓	✓
Antillean Nighthawk *Chordeiles gundlachii*	■		✓			✓		✓
Cuban Nightjar *Caprimulgus cubanensis*	■		✓			✓	✓	✓
Antillean Palm-swift *Tachornis phoenicobia*	■		✓			✓	✓	✓
Cuban Emerald *Chlorostilbon ricordii*	■		✓	✓		✓	✓	✓
Bee Hummingbird *Mellisuga helenae*	NT ■ ■		✓	20		✓		✓
Cuban Trogon *Priotelus temnurus*	■		✓	✓		✓	✓	✓
Cuban Tody *Todus multicolor*	■		✓	✓		✓	✓	✓
West Indian Woodpecker *Melanerpes superciliaris*	■		✓	✓		✓	✓	✓
Cuban Green Woodpecker *Xiphidiopicus percussus*	■		✓	✓		✓	✓	✓
Fernandina's Flicker *Colaptes fernandinae*	VU ■ ■	600–800		✓		8		450
Ivory-billed Woodpecker *Campephilus principalis*	CR ■	<50						
Greater Antillean Pewee *Contopus caribaeus*	■		✓	✓		✓	✓	✓
Loggerhead Kingbird *Tyrannus caudifasciatus*	■		✓	✓		✓		✓
Giant Kingbird *Tyrannus cubensis*	EN ■ ■	250–999	✓	✓		✓	✓	✓
La Sagra's Flycatcher *Myiarchus sagrae*	■		✓	✓		✓	✓	✓
Cuban Vireo *Vireo gundlachii*	■		✓	✓		✓	✓	✓
Thick-billed Vireo *Vireo crassirostris*	■ ■							
Cuban Palm Crow *Corvus minutus*	EN ■ ■ ■	2,500–9,999						
Cuban Crow *Corvus nasicus*	■		✓	✓			✓	✓
Cuban Martin *Progne cryptoleuca*	■					✓	✓	✓
Zapata Wren *Ferminia cerverai*	EN ■ ■ ■	1,000–2,499						✓
Cuban Gnatcatcher *Polioptila lembeyei*	■ ■							
Bahama Mockingbird *Mimus gundlachii*	■ ■							
Cuban Solitaire *Myadestes elisabeth*	NT ■ ■			40		50		
Bicknell's Thrush *Catharus bicknelli*	VU ■							
Olive-capped Warbler *Dendroica pityophila*	■ ■			✓		100		
Yellow-headed Warbler *Teretistris fernandinae*	■ ■		✓	✓		✓	✓	✓
Oriente Warbler *Teretistris fornsi*	■ ■							
Greater Antillean Oriole *Icterus dominicensis*	■		✓	✓		✓	✓	✓
Cuban Blackbird *Dives atroviolaceus*	■		✓	✓		✓	✓	✓
Red-shouldered Blackbird *Agelaius assimilis*	■ ■		✓				✓	4
Tawny-shouldered Blackbird *Agelaius humeralis*	■							✓
Greater Antillean Grackle *Quiscalus niger*	■		✓	✓		✓	✓	✓
Cuban Sparrow *Torreornis inexpectata*	EN ■ ■ ■	250–999						✓
Cuban Bullfinch *Melopyrrha nigra*	■		✓	✓		✓	✓	✓
Cuban Grassquit *Tiaris canorus*	■			✓		✓		✓
Painted Bunting *Passerina ciris*	NT ■		✓	✓				✓
Western Spindalis *Spindalis zena*	■		✓	✓		✓	✓	✓

All population figures = numbers of individuals.
Threatened birds: Critically Endangered ■; Endangered ■; Vulnerable ■; Near Threatened ■. **Restricted-range birds** ■. **Biome-restricted birds** ■. **Congregatory birds** ■.

		Cuba IBAs										
008	CU009	CU010	CU011	CU012	CU013	CU014	CU015	CU016	CU017	CU018	CU019	CU020
■	■	■	■	■	■	■	■	■	■	■	■	■
				■	■	■	■		■		■	
		■	■	■	■	■	■	■	■	■	■	■
■	■			■	■	■	■			■		■
	✓	✓	✓	✓		✓		✓	✓		✓	✓
100	100			200	✓	1,000	✓		✓		✓	450
	120,000											
										46		
				12,000	15,000	100,000						20,000
	1,000											
	20,000											10,000
				216								
					30	800						
						40,000						
		✓	✓	✓		✓	✓	✓	✓	✓	✓	✓
				102								
✓				32	11		98					
	5,000											3,000
	5,000											5,000
	10,000											10,000
,000	1,000			820	184							
	500				140							
100				36								
96												
86												
162					108							50
				578								
350	✓	✓		✓	✓	✓	✓		✓	✓	✓	✓
					✓	✓	✓		✓		✓	
		✓	✓			✓		✓		✓	✓	
		✓	✓	✓	✓	✓	✓	✓		✓		
				✓				✓	✓	✓	✓	
		✓	53	✓				14	✓			126
		✓	✓	✓				38	48		✓	
		✓	✓	✓	✓	✓	✓	✓	✓	✓	✓	✓
		✓	✓		✓	✓	✓	✓	✓	✓	✓	✓
		✓	✓		✓	✓	✓	✓	✓	✓	✓	✓
				✓		✓					✓	
		✓		✓	✓	✓	✓	✓	✓			✓
						✓		✓	✓		✓	✓
		✓	✓	✓	✓	✓	✓	✓	✓	✓	✓	✓
											✓	
		✓	✓		✓	✓	✓	✓	✓	✓	✓	✓
		✓	✓	✓	✓	✓	✓	✓	✓	✓	✓	✓
		✓	✓	✓	✓	✓	✓	✓	✓	✓	✓	✓
		✓	✓	✓	✓	✓	✓	✓	✓	✓	✓	✓
					✓	✓		✓	✓	✓	✓	13
		✓	✓	✓	✓	✓	✓		✓	✓	✓	✓
		✓	✓	✓	✓	✓	✓	✓	✓	✓	✓	✓
									105	✓	✓	
		✓	✓	✓	✓	✓	✓		✓	✓	✓	✓
		✓	✓	✓	✓	✓	✓	✓	✓	✓	✓	✓
				✓								
									✓			
		✓	✓	✓	✓	✓	✓	✓	✓	✓		
		✓		✓	✓	✓	✓	✓	✓			
				✓	✓	✓	✓				✓	
				✓			✓					
										✓		
										✓		
				✓	✓	2–132	✓	✓	✓	✓	✓	
		✓	✓	✓	✓	✓	✓			✓	✓	✓
		✓	✓		✓	✓	✓	✓	✓	✓	✓	✓
		✓	✓	✓	✓	✓	✓	✓	✓			✓
		✓	✓	✓	✓	✓	✓	✓	✓	✓	✓	✓
				✓								
		✓	✓	✓	✓	✓	✓	✓	✓	✓	✓	✓
			✓		✓	✓	✓		✓	✓	✓	✓
✓			✓	✓		✓	✓		✓			✓
		✓	✓	✓	✓	✓	✓	✓	✓	✓	✓	✓

Table 1 ... continued. Key bird species at Important Bird Areas in Cuba.

Key bird species	Criteria				National population	CU021	CU022	CU023	CU024	CU025	CU026	CU027	CU028
					Criteria	■	■	■	■	■	■	■	■
						■	■		■		■	■	■
						■	■	■	■	■	■	■	■
								■			■		
Northern Bobwhite *Colinus virginianus*	NT ■					✓	✓	✓			✓	✓	✓
West Indian Whistling-duck *Dendrocygna arborea*	VU ■		■	■	5,000–9,999	✓		35					
Blue-winged Teal *Anas discors*				■	100,000–999,999								
Black-capped Petrel *Pterodroma hasitata*	EN ■			■	250–499								
Caribbean Flamingo *Phoenicopterus ruber*				■	100,000–999,999								
White Ibis *Eudocimus albus*				■	10,000–24,999								
Glossy Ibis *Plegadis falcinellus*				■	25,000–99,999								
Magnificent Frigatebird *Fregata magnificens*				■	1,000–4,999								
Brown Pelican *Pelecanus occidentalis*				■	1,000–4,999			300					
Double-crested Cormorant *Phalacrocorax auritus*				■	10,000–24,999								
Osprey *Pandion haliaetus*				■						6,143			
Cuban Kite *Chondrohierax wilsonii*	CR ■		■		50–249							✓	
Gundlach's Hawk *Accipiter gundlachi*	EN ■		■		300–400	✓	✓	25	✓	✓	✓	✓	✓
Black Rail *Laterallus jamaicensis*	NT ■				500–999								
Zapata Rail *Cyanolimnas cerverai*	EN ■	■	■	■	<50								
Sandhill Crane *Grus canadensis*				■	500–999								
Piping Plover *Charadrius melodus*	NT ■			■	100–249								
Short-billed Dowitcher *Limnodromus griseus*				■	10,000–24,999								
Lesser Yellowlegs *Tringa flavipes*				■	10,000–24,999								
Least Sandpiper *Calidris minutilla*				■	10,000–24,999								
Laughing Gull *Larus atricilla*				■	10,000–24,999								
Gull-billed Tern *Sterna nilotica*				■	1,000–4,999								
Royal Tern *Sterna maxima*				■	5,000–9,999								
Sandwich Tern *Sterna sandvicensis*				■	500–999								
Common Tern *Sterna hirundo*				■	100–249								
Least Tern *Sterna antillarum*				■	1,000–4,999								
Bridled Tern *Sterna anaethetus*				■	1,000–4,999								
White-crowned Pigeon *Patagioenas leucocephala*	NT ■					✓		300	✓		✓	✓	✓
Plain Pigeon *Patagioenas inornata*	NT ■		■									✓	
Grey-headed Quail-dove *Geotrygon caniceps*	VU ■		■		2,500–9,999	✓	✓		✓	✓		✓	✓
Key West Quail-dove *Geotrygon chrysia*			■			✓			✓	✓			
Blue-headed Quail-dove *Starnoenas cyanocephala*	EN ■		■		1,000–2,499							✓	
Cuban Parakeet *Aratinga euops*	VU ■		■		2,500–9,999		✓		✓			250	
Cuban Amazon *Amazona leucocephala*	NT ■		■				✓		✓			150	
Great Lizard-cuckoo *Saurothera merlini*			■			✓	✓	✓	✓	✓	✓	✓	✓
Bare-legged Owl *Gymnoglaux lawrencii*			■			✓	✓	✓	✓	✓		✓	✓
Cuban Pygmy-owl *Glaucidium siju*			■			✓	✓	✓	✓	✓	✓	✓	✓
Antillean Nighthawk *Chordeiles gundlachii*			■					✓	✓			✓	✓
Cuban Nightjar *Caprimulgus cubanensis*			■			✓	✓	✓	✓	✓	✓	✓	✓
Antillean Palm-swift *Tachornis phoenicobia*			■			✓	✓	✓				✓	
Cuban Emerald *Chlorostilbon ricordii*			■			✓	✓	✓	✓	✓	✓	✓	✓
Bee Hummingbird *Mellisuga helenae*	NT ■		■				✓		✓		✓	40	34
Cuban Trogon *Priotelus temnurus*			■			✓	✓		✓	✓	✓	✓	✓
Cuban Tody *Todus multicolor*			■			✓	✓	✓	✓	✓	✓	✓	✓
West Indian Woodpecker *Melanerpes superciliaris*			■			✓	✓	✓	✓	✓	✓	✓	✓
Cuban Green Woodpecker *Xiphidiopicus percussus*			■			✓	✓	✓	✓	✓	✓	✓	✓
Fernandina's Flicker *Colaptes fernandinae*	VU ■		■		600–800								✓
Ivory-billed Woodpecker *Campephilus principalis*	CR ■				<50							✓	
Greater Antillean Pewee *Contopus caribaeus*			■			✓	✓	✓	✓	✓	✓	✓	✓
Loggerhead Kingbird *Tyrannus caudifasciatus*			■			✓	✓	✓	✓	✓	✓	✓	✓
Giant Kingbird *Tyrannus cubensis*	EN ■		■		250–999				✓	✓		✓	
La Sagra's Flycatcher *Myiarchus sagrae*			■			✓	✓	✓	✓	✓	✓	✓	✓
Cuban Vireo *Vireo gundlachii*			■			✓	✓	✓	✓	✓	✓	✓	✓
Thick-billed Vireo *Vireo crassirostris*		■	■										
Cuban Palm Crow *Corvus minutus*	EN ■	■	■		2,500–9,999								
Cuban Crow *Corvus nasicus*			■			✓	✓		✓			✓	
Cuban Martin *Progne cryptoleuca*			■			✓		✓	✓				✓
Zapata Wren *Ferminia cerverai*	EN ■	■	■		1,000–2,499								
Cuban Gnatcatcher *Polioptila lembeyei*		■	■			✓			✓		✓	✓	✓
Bahama Mockingbird *Mimus gundlachii*		■	■										
Cuban Solitaire *Myadestes elisabeth*	NT ■		■				✓		✓			✓	
Bicknell's Thrush *Catharus bicknelli*	VU ■												
Olive-capped Warbler *Dendroica pityophila*		■	■				✓		✓				
Yellow-headed Warbler *Teretistris fernandinae*		■	■										
Oriente Warbler *Teretistris fornsi*		■	■			✓	✓	✓	✓	✓	✓	✓	✓
Greater Antillean Oriole *Icterus dominicensis*			■			✓	✓	✓	✓	✓	✓	✓	✓
Cuban Blackbird *Dives atroviolaceus*			■			✓	✓	✓	✓	✓	✓	✓	✓
Red-shouldered Blackbird *Agelaius assimilis*		■	■										
Tawny-shouldered Blackbird *Agelaius humeralis*			■			✓	✓	✓		✓	✓	✓	✓
Greater Antillean Grackle *Quiscalus niger*			■			✓	✓	✓	✓	✓	✓	✓	✓
Cuban Sparrow *Torreornis inexpectata*	EN ■	■	■		250–999								700
Cuban Bullfinch *Melopyrrha nigra*			■			✓	✓	✓	✓	✓	✓	✓	✓
Cuban Grassquit *Tiaris canorus*			■			✓	✓	✓	✓	✓	✓	✓	✓
Painted Bunting *Passerina ciris*	NT ■					✓		100	✓				
Western Spindalis *Spindalis zena*			■			✓	✓	✓	✓	✓	✓	✓	✓

All population figures = numbers of individuals.

Threatened birds: Critically Endangered ■; Endangered ■; Vulnerable ■; Near Threatened ■. **Restricted-range birds** ■. **Biome-restricted birds** ■. **Congregatory birds** ■.

leucocephala and Cuban Parakeet *Aratinga euops*) and other taxa such as Bee Hummingbird *Mellisuga helenae*, Cuban Bullfinch *Melopyrrha nigra*, and grassquits *Tiaris* spp. Also, the impacts of introduced species such as cats, pigs *Sus scrofa*, mongoose *Herpestes auropunctatus*, rats *Rattus* spp., and more recently, the catfish *Claria* sp. have not been evaluated yet, but are thought to be significant in some areas and for some species. The potential impact of catfish (which are present in the Ciénaga Zapata) on the Endangered *Cyanolimnas cerverai* is of particular concern.

The Cuban IBA program has facilitated the collection of baseline data and some population estimates for a number of IBAs. (PHOTO: ARTURO KIRKCONNELL)

IMPORTANT BIRD AREAS

Cuba's network of 28 IBAs—the island's international site priorities for bird conservation—cover c.231,657 km². The IBAs have been identified on the basis of 75 key bird species (listed in Table 1) that variously meet the IBA criteria. These species include 24 threatened birds, 11 restricted-range species, 48 biome-restricted species and 23 congregatory waterbird and seabird species. All 28 IBAs support populations of globally threatened birds; 18 contain Cuba EBA restricted-range species and 24 IBAs support a significant proportion of the Greater Antilles biome-restricted species; 12 IBAs are home to globally significant congregations of seabirds or waterbirds, and seven have regionally significant congregations.

Most IBAs are partially or wholly included in the national system of protected areas, with some level of legal protection and management. However, there are several environmentally sensitive areas that lack any form of protection, such as Delta del Mayarí (CU023) and the Humedal Sur de Pinar del Río (CU003), both of which support important habitat for resident and migratory waterbirds. Five of Cuba's six Ramsar sites are part of the IBA network: Ciénaga de Lanier y Sur de la Isla de la Juventud (CU005), Río Máximo (CU014), Gran Humedal del Norte Ciego de Ávila (CU012), Delta del Cauto (CU020), and Ciénaga de Zapata (CU006). Additionally, six of the IBAs are designated biosphere reserves and the Desembarco del Granma National Park (CU019) and the Alejandro de Humboldt National Park (CU027) have been declared World Heritage Sites.

The Centro Nacional de Áreas Protegidas (CNAP) is responsible for coordinating the Cuban Important Bird Area (IBA) program which started in 2000 through the project "Eastern Cuba: saving a unique Caribbean wilderness" with funds from the British Birdwatching Fair and BirdLife International. This project focused on an extensive fieldwork program in eastern Cuba, implemented by CNAP, Instituto de Ecología y Sistemática, Museo Nacional de Historia Natural, Facultad de Biología de la Universidad de la Habana,

Río Máximo IBA is one of six Ramsar sites in Cuba, and supports the largest breeding colony of Caribbean Flamingos in the Caribbean, with more than 50,000 pairs. (PHOTO: ANIET VENEREO)

Figure 1. Location of Important Bird Areas in Cuba.

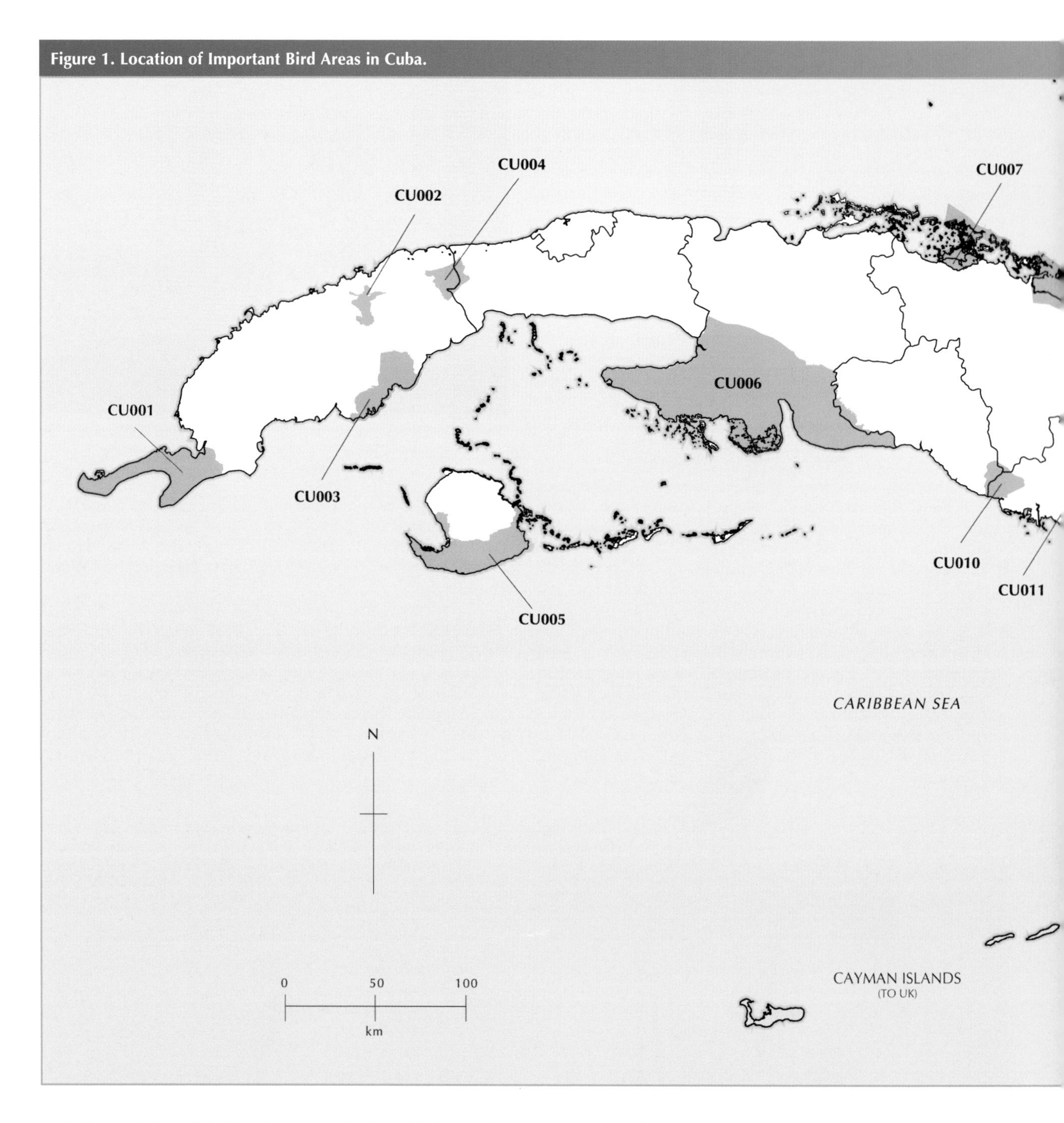

and Centro Oriental de Ecosistemas y Biodiversidad. Results from this fieldwork have contributed to the information presented in this chapter (which was otherwise an output of three participatory national IBA workshops), and many of the proposals for the establishment of new protected areas that are detailed within the various IBA profiles below. These same institutions (and others) have been involved in public education and awareness initiatives, campaigns and events, including the Society for the Conservation and Study of Caribbean Birds' annual Caribbean Endemic Bird Festival and West Indian Whistling-Duck and Wetlands Conservation Program. Increasing awareness at a local, community level is critical to the development of effective IBA conservation actions.

The IBA program and other initiatives have facilitated the collection of baseline data (including some population estimates) for many of the IBAs (especially in eastern Cuba, and also for waterbirds). These data need to be built on to monitor the status of key bird species listed in Table 1 (especially the globally threatened birds) at each IBA. Information concerning the status of these key species can be used to inform the annual assessment of state, pressure and response scores at each of Cuba's IBAs in order to provide an objective status assessment and highlight management interventions that might be required to maintain these internationally important biodiversity sites.

KEY REFERENCES

Acosta, M., Mugica, L. and Denis, D. (2002) Dinámica de los gremios de aves que habitan la arrocera Sur del Jíbaro, Sancti Spiritus, Cuba. *El Pitirre* 15: 25–30.

Blanco, P. (2006) Distribución y áreas de importancia para las aves del orden Charadriiformes en Cuba. Ciudad de la Habana: Universidad de la Habana. (Unpublished Doctoral thesis).

CIGEA (2000) = Centro de Información, Gestión y Educación Ambiental (2000) *Panorama ambiental de Cuba*. Ciudad de la Habana, Cuba: Ministerio de Ciencia Tecnología y Medio Ambiente.

Fong, A., Maceira, D., Alverson, W. S. and Shopland, J. M. eds. (2005) *Cuba: Siboney-Juticí*. Chicago: The Field Museum. (Rapid Biological Inventories Report 10).

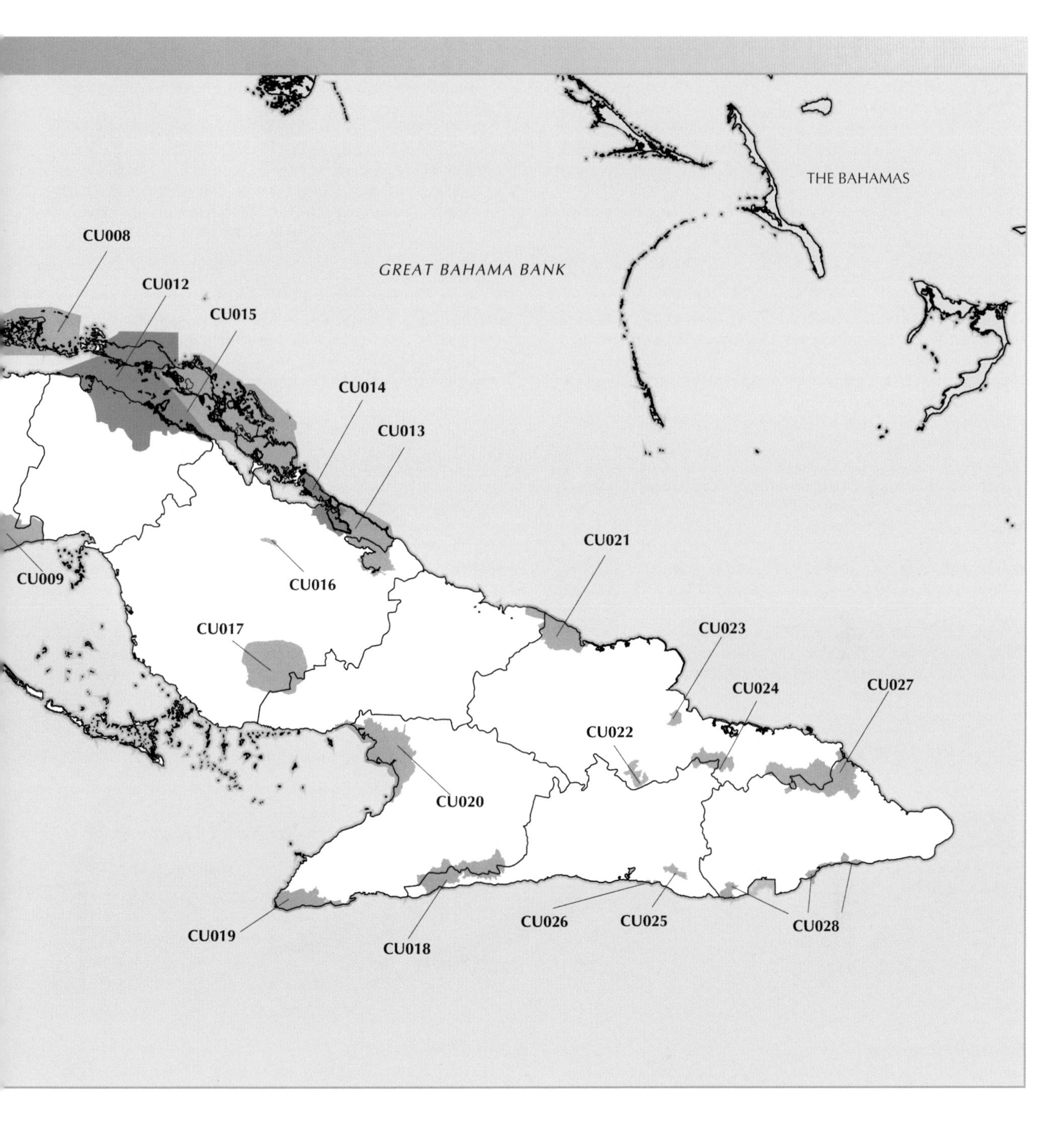

Garrido, O. H. and Kirkconnell, A. (2000) *Field guide of the birds in Cuba*. Ithaca, New York: Cornell Univ. Press.

González, H., Llanes, A., Sánchez, B., Rodríguez, D., Pérez, E., Blanco, P. and Pérez, A. (1999) Estado de las comunidades de aves residentes y migratorias en ecosistemas cubanos en relación con el impacto provocado por los cambios globales 1989–1999. Ciudad de la Habana: Instituto de Ecología y Sistemática. (Unpublished final report).

González, H. (2002) *Aves de Cuba*. Vaasa, Finland: UPC Print.

González, H., Pérez, E., Rodríguez, D., Rodríguez, P., Llanes, A., Begué, G. and Hernández, A. (2007) Distribución, diversidad y abundancia de las comunidades de aves del Parque Nacional Alejandro de Humboldt y la Reserva Ecológica Baitiquirí. Ciudad de la Habana: Instituto de Ecología y Sistemática. (Unpublished final report to BirdLife International for the project "Eastern Cuba: saving a unique Caribbean wilderness").

González, H., Pérez, E., Rodríguez, D. and Barrio Valdés, O. (2005) Adiciones a la avifaunaterrestre de Cayo Sabinal, Cuba. *J. Carib. Orn.* 18: 24–28.

González, H., Álvarez, M., Hernández, J. and Blanco, P. (2001) Composición, abundancia y subnicho estructural de las comunidades de aves en diferentes habitats de la Sierra del Rosario, Pinar del Río, Cuba. *Poeyana* 481–483: 6–19.

Hechavarria, G., Triay, O., Almeida de la Cruz, M., Segovia, Y., Torres, M., García, Z., García, A., Cala de la Hera, Y., Galindo, A. and Pérez, J. (2005) Avifauna asociada al Parque Nacional Desembarco del Granma, Cuba. (Unpublished report).

Kirkconnell, A. (2007) Results from nine field surveys in Granma, Holguin and Guantánamo provinces between 2004 and 2007. Ciudad de la Habana: Museo Nacional de Historia Natural. (Unpublished final report to BirdLife International for the project "Eastern Cuba: saving a unique Caribbean wilderness").

Kirkconnell, A., Stotz, D. F. and Shopland, J. M. eds. (2005) *Cuba: Península de Zapata*. Chicago: The Field Museum. (Rapid Biological Inventories Report 14).

Kirkconnell, A. (1998) Aves de Cayo Coco, Archipiélago de Sabana-Camagüey, Cuba. *Torreia* 43: 22–39.

Labrada, O. and Cisneros, G. (2005) Aves de Cayo Carenas, Ciénaga de Birama, Cuba. *J. Carib. Orn.* 18: 16–17.

Morales, M., Castillo, U. and Collazo, U. L. (2004) Dinámica poblacional de las áreas tróficas del Refugio de Fauna Las Picuas Cayo del Cristo. (Unpublished report).

Morales Leal, J. (1996) El Flamenco Rosado Caribeño. *Flora Fauna* 0: 14–17.
Mugica, L., Acosta, M., Denis, D., Jiménez, A., Rodríguez, A. and Ruiz, X. (2006) Rice culture in Cuba as an important wintering site for migrant waterbirds from North America. Pp.172–176 in G. C. Boere, C. A. Galbraith and D. A. Stroud, eds. *Waterbirds around the world*. Edinburgh, U.K.: The Stationary Office.
Navarro, N., Llamacho, J. and Peña, C. (1997) Listado preliminar de la avifauna de Sierra de Nipe, Mayarí, Holguín, Cuba. *El Pitirre* 10: 65.
Ocaña, F., Cigarreta, S., Peña, C., Fernández, A., Lambert, D., González, P., Monteagudo, S. and Vega, A. (2004) Informe de los resultados del estudio de cuatro humedales en la costa norte de la provincia de Holguín, Cuba. (Unpublished report to the Ramsar Convention).
Perera, S. (2004) Dinámica de la comunidad de aves acuáticas del Refugio de Fauna Río Máximo, Camagüey. Ciudad de la Habana: Facultad de Biología, Universidad de la Habana. (Unpublished Diploma thesis).
Primelles J. and Barrio, O. (2005) Lista preliminar de las aves del Refugio de Fauna Cayos Ballenatos y Manglares de la Bahía de Nuevitas, Cuba. (Unpublished report).
Rodríguez, F., (2002) Highest Osprey flight for Cuba. *El Pitirre* 15: 127–128.
Rodríguez, F., Martell, M., Nye, P. and Bildstein, K. L. (2001) Osprey migration through Cuba. Pp.107–117 in K. L. Bildstein and D. Klem eds. *Hawkwatching in the Americas*. Kempton, Philadelphia: Hawk Migration Association of North America.
Rodríguez-Batista, D. (2000) Composición y estructura de las comunidades de aves en tres formaciones vegetales de Cayo Coco, Archipiélago de Sabana-Camagüey, Cuba. Ciudad de la Habana: Instituto de Ecología y Sistemática. (Unpublished Doctoral thesis).
Ruiz, E., Rodríguez, D., Llanes, A., Rodríguez, P., Pérez, E., González, H., Blanco, P., Arias, A. and Parada, A. (in prep.) Avifauna de los cayos Santa María y Las Brujas, del Archipiélago Sabana-Camagüey, noreste de Villa Clara, Cuba. *J. Carib. Orn.*
Sánchez, B. (2005) Inventario de la avifauna de Topes de Collantes, Sancti Spíritus, Cuba. *J. Carib. Orn.* 18:7–12.
Sánchez, B. N., Navarro, N., Oviedo, R., Peña, C., Hernández, A., Reyes, E., Blanco, P., Sánchez, R. and Herrera, A. (2003) Composición y abundancia de las aves en tres formaciones vegetales de la altiplanicie de Nipe, Holguín, Cuba. *Orn. Neotrop.* 14: 215–231.
Sánchez, B., Rodríguez, D., Torres, A., Rams, A. and Ortega, A. (1992). Nuevos reportes de aves para el corredor migratorio de Gibara, Holguín, Cuba. Pp. 22–23 in *Comunicaciones breves de Zoología* (Instituto de Ecología y Sistemática).
Torres, A. (1994) Listado de las aves observadas dentro del corredor migratorio de Gibara, provincia Holguín, Cuba. *Garciana* 22: 1–4.

ACKNOWLEDGEMENTS

The coordinator and individual IBA profile authors would like to thank the following institutions for their assistance and support: Centro Nacional de Áreas Protegidas, Centro Oriental de Ecosistemas y Biodiversidad, Empresa Nacional para la Protección de la Flora y la Fauna, Universidad de la Habana, Instituto de Ecología y Sistemática, Museo Nacional de Historia Natural, Ecovida, Unidad del CITMA de Holguín, Unidad Presupuestada de Servicios Ambientales Alejandro de Humboldt, Estación de Monitoreo de Caibarién, Villa Clara, Museo de Historia Natural de Holguín, Centro de Ecosistemas Costeros, Centro de Investigaciones Medioambientales de Camagüey and Centro de Estudios y Servicios Ambientales, Villa Clara. Denis Dennis provided support as the national World Bird Database coordinator.

CU001 Guanahacabibes

National Park/Managed Resources Protected Area/Biosphere Reserve

COORDINATES 21°55′N 84°30′W
ADMIN REGION Pinar del Rio
AREA 101,116 ha
ALTITUDE 0–19 m
HABITAT Mangroves, marsh, coast, semi-deciduous and evergreen forests

Blue-headed Quail-dove

THREATENED BIRDS 9
RESTRICTED-RANGE BIRDS 2
BIOME-RESTRICTED BIRDS 31
CONGREGATORY BIRDS

■ Site description

Guanahacabibes IBA is located in the municipality of Sandino, Pinar del Río province, in westernmost Cuba. It comprises the forested, flat limestone plain of the Guanahacabibes peninsula, itself formed by the peninsulas of Cabo de San Antonio and Corrientes. Cliffs rise to 19 m on the south coast from where the land slopes gently across the peninsula down to sea level on the north coast. A range of limestone formations, including caves, are found within the IBA. The Cueva la Barca, contains a rich, nationally important cave biota. The town of La Bajada (1,146 inhabitants) is located within the IBA.

■ Birds

This IBA supports 190 bird species (31 of which are biome-restricted species), including 11 Cuban endemics and nine globally threatened species of which the Blue-headed Quail Dove *Starnoenas cyanocephala*, Giant Kingbird *Tyrannus cubensis* and Gundlach's Hawk *Accipiter gundlachi* are all Endangered. The Cuban EBA restricted-range Yellow-headed Warbler *Teretistris fernandinae* and Red-shouldered Blackbird *Agelaius assimilis* both occur. Guanahacabibes forms part of the migratory corridor of the Mississippi flyway, and is a bottleneck site during fall migration. Bird capture rates during migration mist-netting studies have been higher in this IBA than anywhere else in Cuba.

■ Other biodiversity

The Endangered (and endemic) frog *Eleutherodactylus guanahacabibes* occurs, as do other endemic reptiles including *Anolis quadriocellifer* and *Antillophis andreai peninsulae*. A number of endemic rodents and bats occur, and 14 plant species are confined to the IBA.

■ Conservation

In 1959, the areas of El Veral and Cabo Corrientes within the Guanahacabibes IBA were designated as natural reserves, and as strict conservation areas in 1963. The whole IBA was declared a biosphere reserve in 1987, within which the core zone, Guanahacabibes National Park, was approved by the government in 2001. Residents of La Bajada work mainly in forestry, apiculture, cattle farming, and cultivation of tobacco and other crops. Some are employed in a nearby scuba-diving centre. Other land uses include selective logging and pig foraging. Scientific research is the only activity conducted in the core zone. Threats to the IBA include ecosystem degradation, invasive species, tourism-related development and disturbance. Fishing, hunting and harvesting of natural resources also exert pressure on the ecosystem.

Hiram González, Alina Pérez, Alejandro Llanes, Eneider Pérez

■ Site description

Mil Cumbres IBA embraces a complex landscape within the municipalities of Los Palacios, La Palma and Bahía Honda, in Pinar del Río province. The landscape comprises agricultural plains, karst valleys, mojotes, slate cliffs, sinkholes, sulphur springs and hills, and includes the Sierra de los Órganos, Sierra del Pan de Guajaibón, and the Cajálbana Plateau. The IBA protects an important aquifer and the catchment areas of the San Marcos and San Diego rivers.

■ Birds

This IBA supports 32 biome-restricted species, 15 of which are Cuba endemics and 11 globally threatened birds. The Endangered Blue-headed Quail-dove *Starnoenas cyanocephala*, Giant Kingbird *Tyrannus cubensis*, and Gundlach's Hawk *Accipiter gundlachi* occur and the area is particularly important for the Vulnerable Fernandina's Flicker *Colaptes fernandinae*, and Near Threatened Painted Bunting *Passerina ciris*, Cuban Solitaire *Myadestes elisabeth*, and Plain Pigeon *Patagioenas inornata*.

■ Other biodiversity

The herpetofauna includes the Critically Endangered frog *Eleutherodactylus symingtoni* and other Pinar del Río endemics *Anolis vermiculatus* and *A. bartschi*. Mammals include the hutias *Mysateles prehensilis* and *Capromys pilorides*, and 10 bat species. Two locally endemic freshwater fish are also present. Of a flora with 1,143 species, 52 are endemic to Cajálbana and 24 to Sierra de la Güira.

■ Conservation

Mil Cumbres IBA is a Managed Resources Protected Area created in 1976. It is currently awaiting approval to be included as a site of national significance in the National System of Protected Areas. It is administered by the Empresa para la Protección de la Flora y la Fauna (EPFF). Since the late nineteenth century, this area has suffered from excessive forest exploitation (for timber) and clearance of lands for agriculture and livestock farming. Only the highest (least accessible) areas have retained their forest cover. In spite of this, a species-rich flora, diverse fauna, and attractive landscapes still remain in the area. Among the main threats are illegal hunting and logging, uncontrolled grazing, and the use of agrochemicals in the tobacco plantations. The presence of nearby mineral reserves and an active timber industry will maintain pressure on the area.

Hiram González, Arturo Kirkconnell

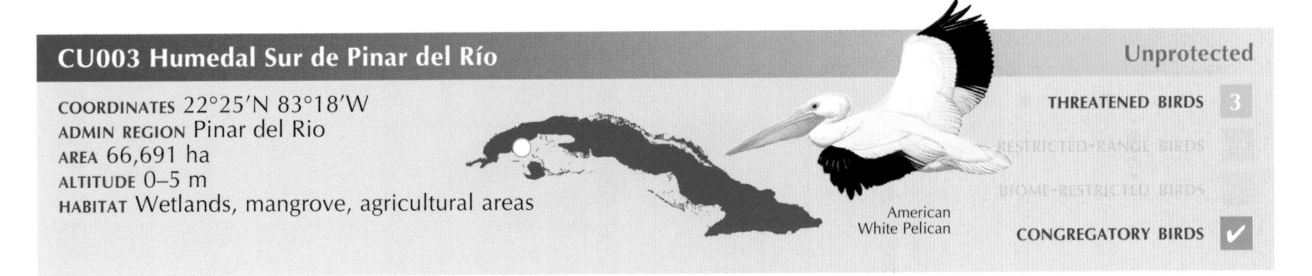

■ Site description

Humedal Sur de Pinar del Río IBA includes the coastal zone between the municipalities of Los Palacios and Consolación del Sur, in the south of Pinar del Río province. It comprises a fringe of natural coastal wetlands associated with rice cultivation and cattle farming. The coastal wetlands, which include several lagoons such as Maspotón and Casa Media, are bordered by mangroves. More than 90% of the IBA consists of wetland habitat, 30% of it is natural (mangroves, coastal lagoons, salt marshes, intertidal mudflats) and the rest is artificial (i.e. seasonal rice paddies).

■ Birds

This IBA supports more than 101 bird species and holds significant numbers of resident and migratory waterbirds, particularly shorebirds and ducks. More than 10,000 dowitchers *Limnodromus* spp. have been observed. Other common waterbirds include Little Blue Heron *Egretta caerulea*, Glossy Ibis *Plegadis falcinellus*, West Indian Whistling-duck *Dendrocygna arborea* (Vulnerable), Blue-winged Teal *Anas discors*, and Brown Pelican *Pelecanus occidentalis*. The largest concentration of American White Pelican *Pelecanus erythrorhynchos* recorded in the Caribbean (c.400 individuals) was in this IBA where it is a common winter resident.

■ Other biodiversity

The coastal lagoons support populations of the Vulnerable American crocodile *Crocodylus acutus* and the hutia *Capromys pilorides*.

■ Conservation

Humedal Sur de Pinar del Río IBA has no legal protection status, but conservation projects and environmental education campaigns are being implemented in the area. The coastal strip is managed by the Empresa para la Protección de la Flora y la Fauna (EPFF), and the rice paddies by Los Palacios Agro-industrial Complex. Although this area has traditionally been used for hunting, it has valuable coastal wetland remnants that are critical for sustaining waterbird populations. These wetlands are partly affected by residents from Los Palacios, la Cubana, Paso Real de San Diego, Alonso de Rojas, and Consolación del Sur, who use them to gain access to the rice paddies. Despite the agricultural uses of the IBA, the application of fertilisers has decreased by 50% in the last 15 years, due to the economic crisis in Cuba.

Lourdes Mugica, Martín Acosta

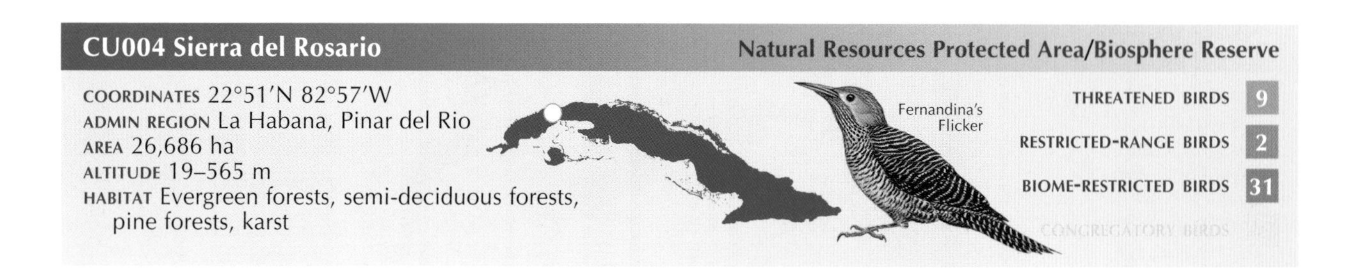

■ Site description

Sierra del Rosario IBA is located to the east of Cordillera de Guaniguanico, in the municipalities of Artemisa, Candelaria, and Bahía Honda, straddling the border between Pinar del Río and Havana provinces. Las Peladas Natural Reserve and El Salón Ecological Reserve comprise the core zone of the biosphere reserve. The IBA boundaries are the same as those of the biosphere reserve. The reserve's eastern entrance is located 50 km south-west of Havana. The IBA supports 4,800 people in eight communities, one of which, Las Terrazas has developed a sustainable rural economy and is also an ecotourism centre. There are remains of seventeenth century French coffee plantations around Las Terrazas.

■ Birds

This IBA is home to 93 bird species (32 of which are biome-restricted species), including 16 Cuba endemics and 10 globally threatened birds. These threatened species include the Endangered Blue-headed Quail-dove *Starnoenas cyanocephala*, Giant Kingbird *Tyrannus cubensis* and Gundlach's Hawk *Accipiter gundlachi*, the Vulnerable Fernandina's Flicker *Colaptes fernandinae* and the Near Threatened Northern Bobwhite *Colinus virginianus* and White-crowned Pigeon *Patagioenas leucocephala*.

■ Other biodiversity

Mammals include the endemic hutias *Capromys pilorides* and *Mysateles prehensilis*, and 11 species of bats. The lizards *Anolis vermiculatus*, *A. bartschi*, and *A. mestrei* are endemic to the Pinar del Río karst, and five amphibians are also local endemics.

■ Conservation

Sierra del Rosario IBA was the first biosphere reserve to be declared in Cuba (in 1985). The core zones of Las Peladas and El Salón were approved by the government in 2008. Farming is fundamental to the livelihoods of the reserve's residents. Activities in the transition zone include cattle ranching, forestry, mixed crops, tourism (around Las Terrazas and Soroa), and conservation. However, the core zones are limited to research, monitoring, and environmental education. Sustainable livelihood practices are implemented in this IBA, in particular by the Las Terrazas community. Threats include illegal hunting and logging, forest fires, erosion, and pollution caused by untreated discharges from pig and poultry farms. Access to the zone is still regulated, but very large numbers of visitors have been reported at camp sites. The reserve has also been affected by the construction of two large dams in the transition zone.

Alina Pérez, Hiram González, Fidel Hernández

■ Site description

Ciénaga de Lanier y Sur de la Isla de la Juventud IBA is in the southern part of Isla de la Juventud, in the Canarreos Archipelago, off the south coast of western Cuba. It includes the Ciénaga de Lanier, which stretches from Ensenada de Siguanea to the eastern Ensenada of San Juan, and divides the north of the island from the karstic plains and coastal zones to the south. The IBA supports a wide diversity of wetland and coastal habitats. Cocodrilo and Cayo Piedras with 300 and 20 inhabitants, respectively, are located within the IBA, whereas the communities of Santa Fé, Mella, and La Reforma are in surrounding areas.

■ Birds

This IBA is notable for its resident and migratory waterbirds, including globally significant populations of the Vulnerable West Indian Whistling-duck *Dendrocygna arborea* and the resident subspecies of Sandhill Crane *Grus canadensis nesiotes*. Terrestrial species include the Endangered Giant Kingbird *Tyrannus cubensis* and a huge population of Near Threatened White-crowned Pigeon *Patagioenas leucocephala*.

■ Other biodiversity

Reptiles include the Endangered Cuban crocodile *Crocodylus rhombifer*, and the Vulnerable American crocodile *C. acutus* and Cuban ground iguana *Cyclura nubila*. The hutia *Mysateles meridionalis* is endemic to Isla de la Juventud, and the fish *Atractosteus tristoechus* is restricted to Zapata and Lanier swamps.

■ Conservation

Ciénaga de Lanier y Sur de la Isla de la Juventud IBA was declared a managed resource protected area in 1990 and in 1998 the southern part of the island and the eastern cays were declared a special region of sustainable development. Its core zones, Punta Francés National Park and Punta del Este Ecological Reserve are awaiting approval as nationally significant areas. The Los Indios Ecological Reserve is also awaiting approval. The IBA was declared a Ramsar site in 2002. Local uses of the area include small-scale fishing (by a cooperative in Cocodrilo) and small-scale forest exploitation through selective logging. Limestone was quarried near Cayo Piedras, and could be resumed in the future. This IBA is threatened by a large-scale tourism development plan as well as the impacts of introduced invasive species. Dams constructed to the north of the marsh are restricting the freshwater input to the wetland.

Susana Aguilar, Juan Pedro Soy, Aryanne Cerrano

■ Site description

Embracing the entire Zapata Peninsula in southern Matanzas province, this IBA is the largest wetland in the Caribbean. It has extensive cave lake systems with spectacular blue holes, flooded caves and important water resources. The IBA also provides critical habitat in the form of forest, flooded palm savannas, open waterand salinas, reefs and mangroves. It has unique submerged marine terraces and coral reefs, valuable archaeological and paleontological sites, and a history of traditional use of natural resources in the surrounding rural communities.

■ Birds

This IBA is home to 40 biome-restricted birds, including 21 Cuba endemics and 17 globally threatened species. The Endangered Zapata Wren *Ferminia cerverai* and Zapata Rail *Cyanolimnas cerverai* are endemic to the IBA. The Endangered Cuban Sparrow *Torreornis inexpectata inexpectata*, Gundlach's Hawk *Accipiter gundlachi*, Blue-headed Quail-dove *Starnoenas cyanocephala* and Giant Kingbird *Tyrannus cubensis* also occur. The area supports large concentrations of waterbirds, including breeding populations of Sandhill Crane *Grus canadensis nesiotes* and Wood Stork *Mycteria americana*. This is the last site in Cuba where the Critically Endangered Bachman's Warbler *Vermivora bachmanii* was observed, in 1964.

■ Other biodiversity

The Endangered Cuban crocodile *Crocodylus rhombifer*, Vulnerable American crocodile *C. acutus* and Critically Endangered dwarf hutia *Mesocapromys nanus* occur. Globally threatened sea-turtles nest on the beaches, and the IBA supports the largest population of the endemic Endangered fish *Atractosteus tristoechus*, considered a living fossil.

■ Conservation

Ciénaga de Zapata was declared a biosphere reserve in 2000 and currently awaits approval as a managed resource protected area. The proposed core zones are Ciénaga de Zapata National Park (approved by the Comité Ejecutivo del Consejo de Ministros in 2008), Bermejas Wildlife Refuge, and Los Sábalos Wildlife Refuge. Main activities are forestry, tourism and fishing. Tourism is concentrated in La Boca, Guamá, Playa Larga and Playa Girón, where wildlife watching, recreational fishing, hiking and beach tourism take place. The IBA has been affected by forest fires, storms, poor water management and restoration, as well as hunting, fishing, and illegal logging. Other threats include invasive plants and fauna (e.g. the carnivorous catfish *Clarias gariepinus*).

Hiram González, Alejandro Llanes, Arturo Kirkconnell

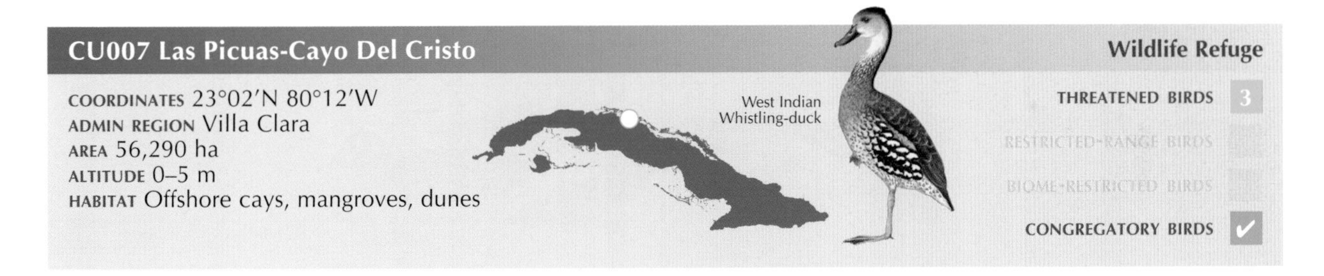

■ Site description

Las Picuas-Cayo Del Cristo IBA is located in the Sabana-Camagüey Archipelago, a group of cays off the north-east coast of Villa Clara province. The IBA extends from Cayo Blanquizal to Cayo Cristo. These cays and islets are characterised by a muddy-sandy substrate, and are dominated by mangrove forests and dunes with associated shrubby vegetation. The administrative centre of the IBA is in the coastal town of Carahatas, with around 666 residents who mostly work in fishing through a state fisheries cooperative, but also in agriculture and cattle farming.

■ Birds

This IBA supports globally significant breeding colonies of Caribbean Flamingo *Phoenicopterus ruber*, Glossy Ibis *Plegadis falcinellus* and the Vulnerable West Indian Whistling-duck *Dendrocygna arborea*. It represents an important waterbird site with 46 species nesting including herons, ibises, spoonbills, cormorants, and pelicans. More than 17 waterbird species roost communally on several of cays and islets. The Near Threatened White-crowned Pigeon *Patagioenas leucocephala* breeds in the IBA.

■ Other biodiversity

Marine life includes the Vulnerable West Indian manatee *Trichechus manatus* and globally threatened sea-turtles. The Vulnerable Cuban ground iguana *Cyclura nubila* occurs on more than 20 cays. Two plants are Cuban endemics, one of which is the Endangered *Pilosocereus robinii*.

■ Conservation

Las Picuas-Cayo del Cristo IBA was approved as a wildlife refuge in 2001 and is currently managed by the Empresa para la Protección de la Flora y la Fauna (ENPFF). The most important threat facing this IBA is overfishing and excessive use of trawl nets. More than 15 trawlers operate year-round; an activity which is unregulated by local fishing authorities and significantly disturbs marine habitats and fauna. Illegal hunting of manatees occurs year-round and of sea-turtles during the nesting season. The queen conch *Strombus gigas* is also unsustainably harvested and used as bait. Other threats include dumping of construction waste, oil and lubricant spills from boats, exotic flora, hurricanes and other natural phenomena, as well as the cutting of *Coccotrinax litoralis* and *Eugenia* spp. to make fishing tools.

María Morales, Vicente Berovides, Susana Aguilar

CU008 Cayería Centro-Oriental de Villa Clara

Wildlife Refuge/National Park/Biosphere Reserve

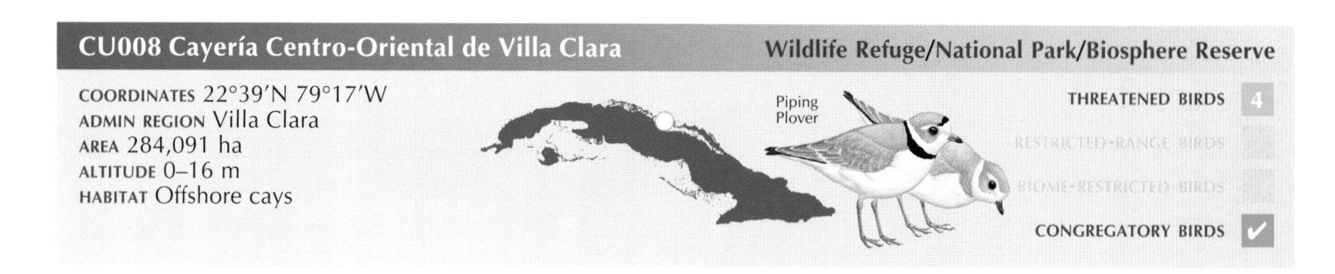

■ Site description
Cayería Centro-Oriental de Villa Clara IBA is located in the Sabana-Camagüey Archipelago, and represents a group of cays off the north coast of Villa Clara province. The IBA embraces several cays including Cobos, Francés, Las Brujas, Ensenachos, Fragoso, and Santa María. These are flat cays, with diverse natural features including coastal evergreen forest, coastal scrub, mangroves, coral reefs, and sandy beaches. The towns of Caibarién, Camajuaní, and Encrucijada are close to but outside of the IBA.

■ Birds
This IBA supports globally significant breeding colonies of Laughing Gull *Larus atricilla*, Royal Tern *Sterna maxima*, Sandwich Tern *S. sandvicensis*, Common Tern *S. hirundo* and Least Tern *S. antillarum*. The area is a wintering site for the Near Threatened Piping Plover *Charadrius melodus* and there are also unconfirmed reports of Brown Noddy *Anous stolidus* breeding. A wide diversity of waterbirds use this area including the Vulnerable West Indian Whistling-duck *Dendrocygna arborea*.

■ Other biodiversity
The Critically Endangered eared hutia *Mesocapromys auritus* is endemic to Cayo Fragoso. The lizard *Anolis pigmaequestris* is endemic to Cayo Francés. The most diverse coral reef of the Sabana-Camagüey Archipelago is within the IBA. Of the 248 plant species present, 29 are endemic.

■ Conservation
The Lanzanillo-Pajonal-Fragoso Wildlife Refuge (the only protected area approved in the IBA) and the Las Loras Wildlife Refuge are managed by the Empresa para la Protección de la Flora y la Fauna (ENPFF). The Maja-Español Inner Cays and Francés-Español Outer Cays wildlife refuges have only been proposed and lack management. Portions of the Santa María-Los Caimanes National Park as well as the Buenavista Biosphere Reserve are within the IBA and are managed by the Ministry of Science Technology and Environment. Threats to the IBA include habitat destruction, fragmentation, disturbance, gas emissions, dust, and introduced species. Beach tourists cause disturbance to *C. melodus* and construction-related habitat loss might affect species such as Bahama Mockingbird *Mimus gundlachii* and Great Lizard-cuckoo *Saurothera merlini santamariae*. Furthermore, local fishermen harvest eggs and chicks of gulls, cormorants and flamingos for food.

Edwin Ruiz, Angel Arias, Daysi Rodríguez, Hiram González, Alejandro Llanes, Eneider Pérez, Patricia Rodríguez, Pedro Blanco

CU009 Humedal del Sur de Sancti Spiritus

Wildlife Refuge

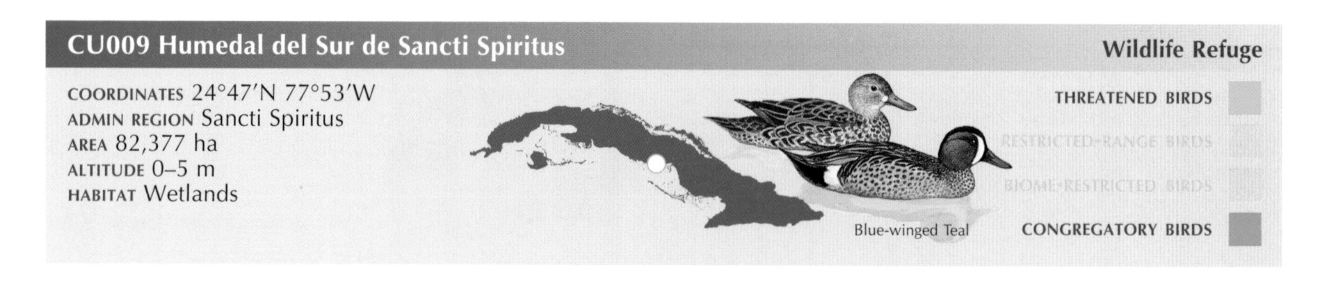

■ Site description
Humedal del Sur de Sancti Spiritus IBA is a wetland and coastal area between Las Nuevas, Mapo and Tunas de Zaza, on the south coast in Sancti Spiritus province. It comprises a group of natural coastal wetlands, rice paddies, and pastures for livestock farming. The rice paddies act as seasonal wetland ecosystems, with El Jíbaro being the most important in the country and a critical site for congregatory waterbirds. To the south the IBA is bordered with a fringe of mangrove forests and coastal wetlands including the El Basto and La Limeta lagoons. Access to this area is facilitated by the rice paddy embankments, although the lagoons can be reached by land. The rice paddies are managed by the Sur del Jíbaro Agroindustrial Complex.

■ Birds
This IBA is notable for resident and migratory waterbirds, with congregations of migratory shorebirds exceeding 10,000 individuals, and over 20,000 Glossy Ibis *Plegadis falcinellus* observed on occasions. The migratory Blue-winged Teal *Anas discors* is particularly abundant (>100,000 individuals) in the coastal lagoons. Up to 100 Vulnerable West Indian Whistling-duck *Dendrocygna arborea* use the remote coastal wetlands as diurnal resting areas and the rice paddies as nocturnal feeding sites.

■ Other biodiversity
The endemic hutia *Capromys pilorides* is present in the mangrove area, as are various endemic reptiles belonging to the genera *Anolis* and *Epicrates*.

■ Conservation
Humedal del Sur de Sancti Spiritus IBA includes the locally significant Tunas de Zaza Wildlife Refuge approved by the Comite Ejecutivo del Consejo de Ministros in 2001. Conservation projects and environmental education campaigns are currently being implemented in the IBA. These wetlands are affected by activities of the residents of Las Nuevas, Jibaro, Mapo, Tunas de Zaza (and other areas) who traditionally use the IBA as a hunting site. Local residents use hand-made traps and nets to capture egrets, ibises, flamingos and ducks. Although use of fertilisers in the artificial wetlands has decreased in the last 15 years, natural wetlands, like those close to Tunas de Zaza, are being affected by sewage discharges, causing the death of mangroves.

Martín Acosta, Lourdes Mugica, Ariam Jiménez

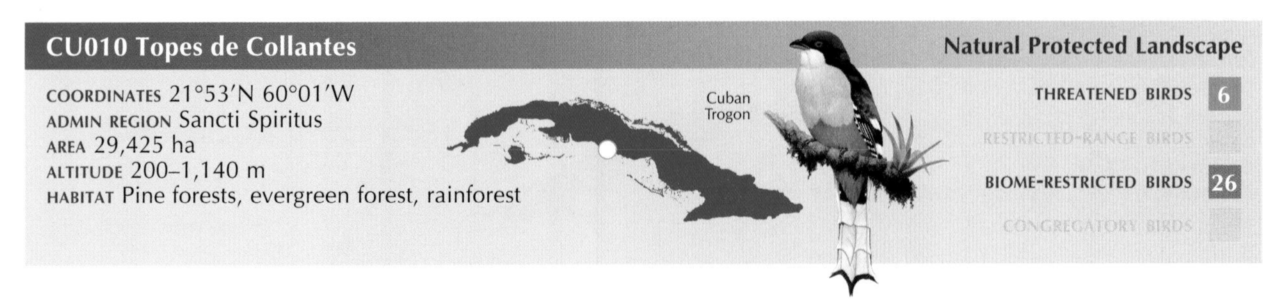

■ Site description

The Topes de Collantes IBA is an area of hills within the Sierra del Escambray (part of the Guamuhaya massif) in south-western Sancti Spiritus province. These hills are characterised by their structural and tectonic complexity (including areas of karst limestone) as well as their cool climate. This has resulted in a diverse range of vegetation types including evergreen forest, pine forest, riparian forest and grasslands, semi-deciduous forest, and secondary scrub containing *Dichrostachys cinerea* in deforested areas.

■ Birds

This IBA supports populations of 26 (of the 48) Greater Antilles biome-restricted including the Endangered Gundlach's Hawk *Accipiter gundlachi*, Vulnerable Cuban Parakeet *Aratinga euops* and Grey-headed Quail-dove *Geotrygon caniceps*, and Near Threatened Cuban Amazon *Amazona leucocephala*. Other species of interest include Cuban Green Woodpecker *Xiphidiopicus percussus*, Limpkin *Aramus guarauna*, Cuban Martin *Progne cryptoleuca*, Tree Swallow *Tachycineta bicolor*, Wood Thrush *Hylocichla mustelina*, Louisiana Waterthrush *Seiurus motacilla*, Ovenbird *Seiurus aurocapilla*, and Red-legged Honeycreeper *Cyanerpes cyaneus*.

■ Other biodiversity

Of the 488 higher plant species present, 53 are endemic, 22 of which with some category of global threat. Twenty-five endemics of the Guamuhaya sector have been reported in Pico Potrerillo, including local endemics such as *Vernonia potrerilloana*, *Rondeletia potrerilloana* and *Psychotria martii*.

■ Conservation

Topes de Collantes IBA has been proposed as a natural protected landscape within the National System of Protected Areas. It is managed by the Topes de Collantes Touristic Complex of the Gaviota Touristic Enterprise. Land is mostly used for agriculture, forestry and tourism. The principal threat to the IBA is the high demand for nature tourism and related activities such as the commercialisation and extraction of species, mostly orchids, by tourists. Natural areas suffer from heavy human pressure and a high incidence of erosion, as well as water pollution (especially in the Río Vegas Grandes). Other threats include dry season forest fires and invasive plants such as *Dichrostachys cinerea*.

Daysi Rodríguez

■ Site description

Alturas de Banao IBA is in the Guamuhaya massif of the Cordillera de Banao, central Cuba. It comprises an area of low, karstic hills in the municipalities of Sancti Spiritus, Fomento and Trinidad, and includes the uppermost part of the Banao and Higuanojo river basins. The landscape is one of canyons, valleys and sinkholes. The IBA is surrounded by the communities of Banao, El Pinto, Cacahual, Cuarto Congreso, Santa Rosa, Gavilanes, La Veintitrés, La Güira, and Los Limpios.

■ Birds

This IBA supports 72% (77 species) of Cuba's breeding resident birds including 25 (of the 48) Greater Antilles biome-restricted species. Of particular note is the presence of the Endangered Gundlach's Hawk *Accipiter gundlachi*, Vulnerable Cuban Parakeet *Aratinga euops* and Grey-headed Quail-dove *Geotrygon caniceps*, and the Near Threatened Cuban Amazon *Amazona leucocephala*, Northern Bobwhite *Colinus virginianus* and Painted Bunting *Passerina ciris*. The Near Threatened Cerulean Warbler *Dendroica cerulean* has also been recorded in the IBA on passage.

■ Other biodiversity

The herpetofauna in this IBA is particularly rich. Amphibians include the Endangered frogs *Eleutherodactylus emiliae* and *E. caspari*, and the Vulnerable *E. limbatus*. The Endangered Cuban long-nosed toad *Bufo longinasus dunni* is restricted to Caja de Agua. Of the 900 species of plant recorded, 225 are endemic.

■ Conservation

The IBA follows the boundaries of the Alturas de Banao Ecological Reserve which is awaiting approval by the Comité Ejecutivo del Consejo de Ministros. The reserve is divided into the following administrative areas: Jarico, María Antonia, El Regalo, Hoyo del Naranjal, Caja de Agua, and La Sabina. There are pasture areas for grazing animals, including horses, cattle and buffalo, the latter in the Hoyo del Naranjal management area. The IBA is primarily used for research and nature tourism by students, visitors, and residents of the surrounding communities. Activities such as illegal logging and hunting, firewood extraction and recreation are threatening the area.

Maikel Cañizares, Susana Aguilar

CU012 Gran Humedal del Norte Ciego de Ávila

Ecological Reserve/Wildlife Refuge/Ramsar Site

COORDINATES 22°30′N 78°26′W
ADMIN REGION Ciego de Ávila
AREA 268,728 ha
ALTITUDE 0–13 m
HABITAT Wetlands, offshore cays, forest, mangroves, coastal dry scrub

Cuban Sparrow

THREATENED BIRDS 10
RESTRICTED-RANGE BIRDS 5
BIOME-RESTRICTED BIRDS 29
CONGREGATORY BIRDS ✓

■ Site description

Gran Humedal del Norte Ciego de Ávila IBA is in the north-eastern portion of Ciego de Avila province, covering nearly all of its coastal zone as well as the Cayo Guillermo and Cayo Coco. It also includes part of La Yana catchment area and limestone formations located in the northern part of the province. Wetlands include Laguna de la Leche and Laguna la Redonda which feed the underground rivers of the area. Communities in the area include La 21, Marbella, La Loma, Rinconada, Malezal, Santa Bárbara, and Delia, among others.

■ Birds

This IBA supports globally significant populations of the Vulnerable West Indian Whistling-duck *Dendrocygna arborea*, Near Threatened Piping Plover *Charadrius melodus*, and Sandhill Crane *Grus canadensis nesiotes*, and regionally important populations of Royal Tern *Sterna maxima* and Magnificent Frigatebird *Fregata magnificens*. Cayo Guillermo has Cuba's largest gull and tern colonies, and Cuba's only breeding site for "Cayenne" Tern *S. sandvicensis eurygnatha* and Audubon's Shearwater *Puffinus lherminieri*. Endangered Cuban Sparrow *Torreornis inexpectata varonai* and Great Lizard-cuckoo *Saurothera merlini santamariae* subspecies are endemic to Cayo Coco and the Sabana-Camagüey Archipelago, respectively.

■ Other biodiversity

Fauna includes the hutia *Capromys pilorides*, the Vulnerable Cuban ground iguana *Cyclura nubila* and the Vulnerable American crocodile *Crocodylus acutus*. On Cayo Paredón, the endemic bat *Phyllops falcatus* can be found.

■ Conservation

Administered within the Gran Humedal del Norte Ciego de Ávila IBA are the Centro Oeste de Cayo Coco Ecological Reserve, El Venero Wildlife Refuge and the (approved) Sierra de Judas de La Cunagua Wildlife Refuge. Other protected areas have been proposed (but are not yet administered) within the IBA, namely Elemento Natural Destacado Dunas de Cayo Guillermo, Reserva Florística Manejada Monte El Coy and Refugio de Fauna Cayo Alto. The IBA is a Ramsar site. Tourism and its related activities (e.g. hotel and road construction, pest fumigation) are major threats. Both Cayo Guillermo and Cayo Coco have tourist infrastructure and activities that are incompatible with sustainable management, and there are mid-term plans for to develop Cayo Paredón Grande. Other threats include invasive species, and disturbance of nesting seabirds.

Daysi Rodríguez, Yarelys Ferrer, Alain Parada, Patricia Rodríguez, Pedro Blanco, Raúl Inguanzo, Idael Ruiz, Oscar Ortiz

■ Site description

Cayos Sabinal y Ballenatos, y Manglares de la Bahía de Nuevitas IBA is off the north coast of Camagüey province, between the Gran Humedal del Norte Ciego de Ávila IBA (CU012) and Río Máximo IBA (CU014). Surrounded by fringing mangroves, and extending from Ensenada del Gremio to Punta Júcaro, Bahía de Nuevitas contains a group of limestone cays that, together with Cayo Libertad, form the Cayos Ballenatos. Low, permanently flooded marshy areas formed by numerous islets, inner lagoons, channels and brackish lagoons are abundant. Cayo Sabinal belongs to the Sabana-Camagüey Archipelago and is connected to the mainland by a 2 km causeway.

■ Birds

This IBA supports globally significant populations of Caribbean Flamingo *Phoenicopterus ruber* (>15,000 individuals), the Near Threatened Piping Plover *Charadrius melodus* and Vulnerable West Indian Whistling-duck *Dendrocygna arborea* and a number of other waterbirds. Seabirds include regionally important numbers of Laughing Gull *Larus atricilla*, Gull-billed Tern *Sterna nilotica*, and Least Tern *S. antillarum*. Terrestrial species include the restricted-range Cuban Gnatcatcher *Polioptila lembeyei* and Oriente Warbler *Teretistris fornsi* and a number of globally threatened birds.

■ Other biodiversity

The Vulnerable Cuban ground iguana *Cyclura nubila*, West Indian manatee *Trichechus manatus* and American crocodile occur. The endemic hutia *Capromys pilorides* is present on the Cayos Ballenatos.

■ Conservation

Cayo Sabinal has been proposed as a managed resources protected area and includes parts of the protected areas of Maternillo-Tortuguilla Ecological Reserve and Laguna Larga Managed Floristic Reserve. The Empresa para la Protección de la Flora y la Fauna (ENPFF) manages the cay, albeit without a remit or guidance to ensure its sustainable development. Land use includes conservation, apiculture, extraction of forest and non-forest products, recreational fishing, and beach and nature tourism. Threats include illegal hunting and logging, exotic species, and different degrees of habitat modification and pollution. Fires on the cays and on the coastal strip are also a risk. This site is close to the tourist region of Camagüey (Santa Lucia Beach). A proposal exists to develop Cayo Sabinal for sustainable tourism.

Jarenton Primelles, Omilcar Barrios, Susana Aguilar, Hiram González, Eneider Pérez

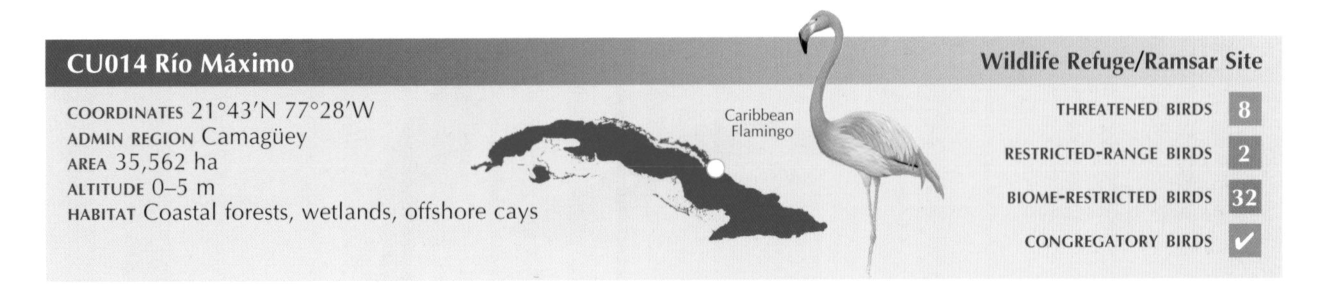

■ Site description

Río Máximo IBA is a wildlife refuge in north-eastern Camagüey province and is bordered by the deltas of the Máximo and Cagüey rivers in the municipality of Minas. The area is characterised by coastal and marine ecosystems with numerous cays, the largest of which are Cayo Güajaba and Cayo Sifonte. Río Máximo also has biodiverse terrestrial ecosystems. There are four human settlements with c.2,170 residents near the wildlife refuge, but closest to the site is Mola.

■ Birds

This IBA is home to some of the largest concentrations of waterbirds (including shorebiirds) in the Caribbean. It supports the world's largest breeding population of Caribbean Flamingo *Phoenicopterus ruber* (c.100,000 birds), 40,000 Double-crested Cormorant *Phalacrocorax auritus*, significant numbers (c.1,000) of the Vulnerable West Indian Whistling-duck *Dendrocygna* arborea and the largest population of Snowy Plover *Charadrius alexandrinus* in the Caribbean.

■ Other biodiversity

The Vulnerable West Indian manatee *Trichechus manatus* and American crocodile occur, as does the endemic fish *Cichlasoma tetracantha.* The Cagüey River delta is important for plants and supports a population of the threatened *Copernicia rigida* along with other rare species such as *C. vespertilionum* and *Trichillia pungens*.

■ Conservation

The Río Máximo IBA is a nationally significant protected area that has been managed by the Empresa para la Protección de la Flora y la Fauna (ENPFF) since 1986. It was approved as a wildlife refuge in 2001 and as a Ramsar site in 2002. Activities in the area are concentrated on habitat and species conservation, with an emphasis on birds. Management and protection measures have contributed to an increase in the flamingo population, from 5,000 nests in 1987 to more than 50,000 in 2007. The canalisation of Río Máximo has modified the ecosystem and increased the impact of droughts. During 2003–2005 a prolonged drought caused the salinisation of a large portion of the river and the reduction of freshwater habitats. Other problems relate to illegal fishing, and inappropriate fishing methods which represent a threat to manatees and dolphins, whilst also disturbing nesting and feeding areas for the waterbirds.

José Morales, Loydi Vázques, Ariam Jiménez

■ Site description

Cayos Romano-Cruz-Mégano Grande IBA is at the eastern end of the Sabana-Camagüey Archipelago off the north coast of Cuba. It comprises the large cays of Romano and Cruz and many small cays and islets in a landscape of mangrove forests, beaches, sinkholes, waterholes, marshes, and isolated areas of higher ground, such as Alto del Ají and Silla de Romano. There is a coral barrier reef bordering the northern side of the cays. Mangrove forests predominate on lowland marshy zones. Other vegetation types include evergreen forests and semi-deciduous forests, bog forests, coastal dry scrub, and salt marsh.

■ Birds

This IBA is important for its great diversity and density of waterbirds, including Roseate Spoonbill *Platalea ajaja*, White Ibis *Eudocimus albus*, Vulnerable West Indian Whistling-duck *Dendrocygna arborea*, Wood Stork *Mycteria americana*, and a globally important population of the Near Threatened Piping Plover *Charadrius melodus*. The woodland areas support 29 (of the 48) Greater Antilles biome-restricted species including the Endangered Gundlach's Hawk *Accipiter gundlachi*. The IBA is also an important wintering site for species such as the Near Threatened Painted Bunting *Passerina ciris.*

■ Other biodiversity

The snail *Liguus fasciatus romanoensis* is a local endemic. There are 16 species of introduced mammals present on Cayo Romano. Plants include 151 Cuban endemics, 12 of which are local endemics and 23 are threatened.

■ Conservation

Cayo Romano has been proposed as a managed resources protected area and the Empresa para la Protección de la Flora y la Fauna (ENPFF) has an administrative unit within the area. The Cayo Cruz Wildlife Refuge and the Refugio de Fauna Centro y Oeste de Cayo Paredón Grande are also proposed protected areas, but they lack any kind of protection or administration at the present time. Land uses include natural resource conservation, extraction of forest and non-forest products, sport and commercial fishing, and beach tourism. Among the main threats are modification and pollution of the cay, as well as invasive alien species. Introduced fauna includes horses, pigs, deer, antelope and ankole-watusi cattle.

Daysi Rodríguez, Susana Aguilar

CU016 Limones-Tuabaquey

■ Site description

Limones-Tuabaquey IBA is an ecological reserve situated in the east-central part of the Sierra de Cubitas, northern Camagüey province. It is an area of complex karst limestone landforms including closed sinkholes (dolines), deep eroded channels (karren) and more than 50 caves. There are important archaeological and historic sites in the IBA, as well as aboriginal pictographs in the Matías, Las Mercedes, María Teresa, and Pichardo caves. The towns of Lesca and Vilató lie just outside the IBA, the latter having the largest population in Sierra de Cubitas and the best preserved indigenous traditions of the region.

■ Birds

This IBA supports populations of 27 (of the 48) Greater Antilles biome-restricted species, 13 of which are endemic to Cuba. Globally threatened birds include the Endangered Blue-headed Quail-dove *Starnoenas cyanocephala* and Gundlach's Hawk *Accipiter gundlachi* The Vulnerable Fernandina's Flicker *Colaptes fernandinae* also occurs. Each year, the Near Threatened White-crowned Pigeon *Patagioenas leucocephala* migrates to the region of Cubitas from adjacent areas to breed.

■ Other biodiversity

The fauna of this IBA includes 46 Cuban endemic species (with Mirador de Limones being particularly important). Mammals are represented by the hutias *Capromys pilorides* and *Mysateles prehensilis*, and 12 species of bats.

■ Conservation

Limones-Tuabaquey IBA is a nationally significant protected area that was approved as an ecological reserve in 2008. The area is owned by the Military Farm Enterprise of Camagüey and is jointly managed by the Ministry of Science, Technology and Environment and the Integral Military Farm La Cuba. Livestock farming, mining, and agriculture occur in the buffer zone and surrounding areas. A nature tourism project is currently being implemented. About 60% of the IBA is selectively logged, but the proliferation of roads and trails that has facilitated deforestation and burning, resulting in structural changes to the forest. Approximately 11% of the IBA is untouched and the potential for recovery in the remaining forest is good if the correct management is put in place. Other problems are mining related dust emissions and sound pollution.

Juan Carlos Reyes, Luis Ramos, Susana Aguilar

CU017 Sierra del Chorrillo

■ Site description

Sierra del Chorrillo IBA is a large area of hills in the centre of the municipality of Najasa, south-eastern Camagüey province. It embraces the higher elevation points of Sierra del Chorrillo, Sierra de Najasa, and Guaicanámar, as well as the Monte Quemado Mirador and Sierra El Martillo. Local rivers include Najasa, Yáquimo and Sevilla. The area comprises a mosaic forest, shrubland, rivers and agricultural land. The communities of Arrollo Hondo and La Belén are located within the IBA.

■ Birds

This IBA supports populations of 32 (of the 48) Greater Antilles biome-restricted species, 11 of which are endemic to Cuba. Globally threatened birds include the Endangered Blue-headed Quail-dove *Starnoenas cyanocephala*, Gundlach's Hawk *Accipiter gundlachi*, Giant Kingbird *Tyrannus cubensis* and Cuban Palm Crow *Corvus minutus*. The IBA is home to the largest known populations for both *T. cubensis* (with 35 pairs) and C. palmarum. It also supports Cuba's most important population of the Near Threatened Plain Pigeon *Patagioenas inornata*, and significant breeding populations of the Vulnerable Cuban Parakeet *Aratinga euops* and Near Threatened Cuban Amazon *Amazona leucocephala*.

■ Other biodiversity

Seven reptile species are Cuban endemics including the Near Threatened snake Maja de Santa Maria *Epicrates angulifer*. Frogs include the endemic Endangered *Eleutherodactylus thomasi* and Vulnerable *E. limbatus*, and the mammals include one species of hutia and seven bat species (including the Near Threatened *Phyllonycteris poeyi* and *Brachyophylla nana*.

■ Conservation

Sierra del Chorrillo IBA has been managed by the La Belén unit of the Empresa para la Protección de la Flora y la Fauna (ENPFF) since 1998. In 2001 it was declared a managed resources protected area. The site has been affected by forestry, agriculture, road and house construction, and the introduction of exotic species. There are areas with different levels of human-induced modification. Another important impact is livestock farming, which disturbs vegetation, destroys part of the understory, compacts soil, and pollutes waters. Additionally, cattle compete with native fauna for shelter, space, and food, and can destroy nests of ground-nesting birds. Some of the 11 introduced mammals include blackbuck *Antilocapra cervicapra*, nilgai *Boselaphus tragocamelus*, water buffalo *Bubalus bubalis*, ankole-watusi *Bos* sp. and zebra *Equus zebra*.

Susana Aguilar, Arturo Kirkconnell

CU018 Turquino-Bayamesa

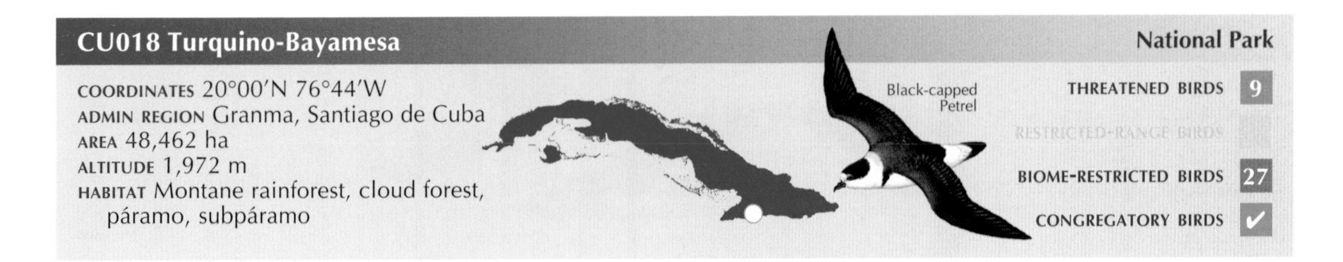

■ Site description

Turquino-Bayamesa IBA is within the Sierra Maestra massif. The landscape is one of rugged mountains including the highest peaks in the country, namely Pico Real del Turquino (1,972 m), Pico Cuba (1,872 m) and Pico La Bayamesa (1,752 m). The area has its own characteristic microclimate, with some of the lowest temperatures and the highest rainfall in Cuba, resulting in montane rainforest, cloud-forest and páramo-type vegetation, but with notable differences between northern and southern slopes. La Mula, Palma Mocha, and Potrerillo rivers originate within the IBA.

■ Birds

This IBA supports populations of 27 (of the 48) Greater Antilles biome-restricted species including the Endangered Blue-headed Quail-dove *Starnoenas cyanocephala*, Gundlach's Hawk *Accipiter gundlachi* and Giant Kingbird *Tyrannus cubensis*. The Endangered Black-capped Petrel *Pterodroma hasitata* has been recorded flying inland into these mountains and almost certainly breeds. The IBA is an eastern Cuba stronghold for the Vulnerable Fernandina's Flicker *Colaptes fernandinae*, and Cuba's only known wintering area for the Vulnerable Bicknell's Thrush *Catharus bicknelli*. An endemic subspecies of Oriente Warbler *Teretistris fornsi turquinensis* occurs.

■ Other biodiversity

Local endemics include the Critically Endangered frogs *Eleutherodactylus albipes* and *E. turquinensis*, the Endangered *E. melacara*, and the lizard *Anolis guazuma*. Flora includes the Endangered *Magnolia cubensis*, the Vulnerable *Juniperus saxicola* and *Tabebuia oligolephis*, and *Pinus maestrensis*, as well as 26 local and 100 Sierra Maestra endemics.

■ Conservation

The IBA comprises Turquino National Park (created in 1991 and approved in 2001) and La Bayamesa National Park (recently created and awaiting approval), both managed by the Empresa para la Protección de la Flora y la Fauna (ENPFF). Residents of the surrounding communities work with conservation, as well as in agriculture and silviculture. Ecotourism has also been widely implemented in Turquino National Park. Threats include deforestation and habitat degradation (e.g. through illegal logging), forest fires, soil erosion due to run-off, road construction, illegal hunting, abandoned agricultural land encouraging invasive flora, and introduced fauna. Some landowners within the IBA do not follow land management protocols appropriate to the protected area.

Eneider Pérez, Alejandro Llanes

CU019 Desembarco del Granma

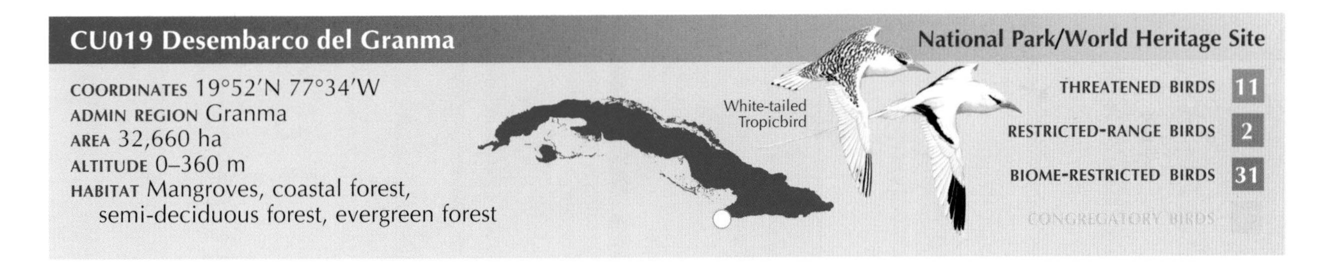

■ Site description

Desembarco del Granma IBA is in the Sierra Maestra district of south-eastern Granma province. It forms part of the Cabo Cruz Marine Terraces Complex, the second largest and best preserved system of emergent and submerged marine terraces in the world. This limestone area is one of forested slopes, coastal cliffs, lagoons, mangroves, sinkholes, and cave systems and includes scenic sites such as Hoyo de Morlotte and Cueva del Fustete. Within the area are three communities, one of which (Cabo Cruz) originated from a community fishing cooperative. The communities of Las Coloradas, Río Nuevo, and Las Palmonas are located near the site.

■ Birds

This IBA supports populations of 31 (of the 48) Greater Antilles biome-restricted species, 14 of which are endemic to Cuba, including the Endangered Blue-headed Quail-dove *Starnoenas cyanocephala*, Gundlach's Hawk *Accipiter gundlachi* and Giant Kingbird *Tyrannus cubensis*. The Vulnerable West Indian Whistling-duck *Dendrocygna arborea*, Grey-headed Quail-dove *Geotrygon caniceps* and Fernandina's Flicker *Colaptes fernandinae*, and Near Threatened Cuban Amazon *Amazona leucocephala leucocephala* and White-crowned Pigeon *Patagioenas leucocephala* (which nests in Cabo Cruz) also occur. This IBA is one of the few known breeding sites for White-tailed Tropicbird *Phaethon lepturus* in Cuba, albeit in small numbers.

■ Other biodiversity

Endemics include the snails *Polymita venusta* and *Liguus vittatus*, the lizards *Anolis guafe* and *A. confuses*, and the primitive Cuban night lizard *Cricosaura typical* (a monotypic genus). Flora includes 12 local endemics. The Vulnerable tree cactus *Dendrocereus nudiflorus* is present, with some individuals exceeding 500 years old.

■ Conservation

The Desembarco del Granma National Park was created in 1985 and approved in 2001—the first area within this category to be declared in Cuba. It is managed by the Empresa para la Protección de la Flora y la Fauna (ENPFF). It was also the first Cuban site to be declared a World Heritage Site in 1999. Threats invasive flora and fauna (particularly fish species escaped from fish farms), soil erosion and compacting, as well as habitat fragmentation and degradation associated with deforestation. The Cabo Cruz lagoons are threatened by pollution and hyper-salinisation.

Ernesto Reyes, Yuself Cala de la Hera

CU020 Delta del Cauto

■ Site description
Delta del Cauto IBA is in south-eastern Cuba and embraces the most extensive, complex and best preserved deltaic system in the insular Caribbean. It is also the second largest wetland in the region after Ciénaga de Zapata (IBA CU006). It is a difficult area to access which has left the mosaic of diverse estuaries, marshes, saltmarshes, lagoons (such as Biramas and Leonero) and open areas relatively undisturbed. Well preserved mangrove predominates, reaching up to 30 m in height. The communities of Cabezada, Manajuana, El Yarey, and Guasimilla are located within the wildlife refuge.

■ Birds
Numbers of duck and shorebirds in this IBA exceed 100,000 during the winter. Principal species include Blue-winged Teal *Anas discors*, Fulvous Whistling-duck *Dendrocygna bicolor*, Northern Shoveler *A. clypeata*, Least Sandpiper *Calidris minutilla*, and Short-billed Dowitcher *Limnodromus griseus*. Globally significant numbers of Caribbean Flamingo *Phoenicopterus ruber* and Glossy Ibis *Plegadis falcinellus* breed. The Vulnerable West Indian Whistling-duck *Dendrocygna arborea*, Gundlach's Hawk *Accipiter gundlachi*, Cuban Parakeet *Aratinga euops* and Fernandina's Flicker *Colaptes fernandinae* occur.

■ Other biodiversity
The herpetofauna includes the local endemic lizard *Anolis birama* and the endemic (and Vulnerable) Cuban small-eared toad *Bufo empusus*. The IBA supports the largest breeding population of the Vulnerable American crocodile *Crocodylus acutus*. Mammals include the endemic hutia *Capromys pilorides* and bat *Noctilio leporinus*.

■ Conservation
Delta del Cauto Wildlife Refuge was established in 1991 and approved as a nationally significant protected area in 2001. It became a Ramsar site in 2002. The Empresa para la Protección de la Flora y la Fauna (ENPFF) manages its two administrative units of Monte Cabaniguan Wildlife Refuge (Las Tunas province) and Delta del Cauto Wildlife Refuge (Granma). Main threats include illegal hunting, fishing and logging, forest fires, and dumping waste. Problems have also resulted from the damming of the Río Cauto, leading to changes in river dynamics, saltwater intrusion during storms, land salinisation and changes to the mangrove and swamp forests. Commercial fishing in Biramas and Leonero lagoons has a negative impact on the waterbird breeding colonies.

Martín Acosta, Omar Labrada, Lourdes Mugica, Susana Aguilar

CU021 Gibara

■ Site description
Gibara IBA is on the coast of north-eastern Cuba. It comprises a flat coastal plain along a stretch of c.35 km of shore that becomes more rugged inland with karstic limestone caverns, sinkholes and elevated formations such as at Cupeicillo la Candelaria. The Gibara and Cacoyoguin rivers reach the sea at Bahía de Gibara. Gibara itself is a town of c.17,000 inhabitants. The coast has areas of mangroves and lagoons whilst inland are forested areas of varying types.

■ Birds
This IBA supports populations of 30 (of the 48) Greater Antilles biome-restricted species, including the Endangered Giant Kingbird *Tyrannus cubensis* and Gundlach's Hawk *Accipiter gundlachi* , and the Vulnerable Grey-headed Quail-dove *Geotrygon caniceps*. The Vulnerable West Indian Whistling-duck *Dendrocygna arborea* and Near Threatened White-crowned Pigeon *Patagioenas leucocephala* also occur. This area is one of the main migratory corridors in the Cuban Archipelago, with numerous vagrant and "occasional" bird records coming from the IBA. Over 200 resident and migratory bird species have been recorded.

■ Other biodiversity
The area is home to many local and regional endemic species and subspecies including lizards such as *Leiocephallus stictigaster gibarensis*, *L. raviceps delavarai* and *Anolis jubar gibarensis*.

■ Conservation
The Gibara IBA includes the Balsas de Gibara Wildlife Refuge and the Caletones Ecological Reserve both of which have been proposed as protected areas but are awaiting approval and currently lack any management. Programs related to land use, local ecosystem management on the coastal plain, as well as environmental education campaigns have been implemented in the area. Until recently, migratory birds were banded at Playa Caletones. Threats include cutting coastal forest for charcoal, shifting agriculture in the coastal forest, overfishing, and the removal of mangrove bark for tanning leather. Birds have long been captured in the mangroves of Gibara for the national pet trade, including migrants such as Piping Plover *Charadrius melodus*. This 100-year-old practice is still very much alive within the IBA and at Santa Fé de Havana, the final destination for most birds captured in Gibara. The construction of a wind farm is a potential threat for birds during their spring migration.

Nils Navarro

■ Site description

La Mensura IBA is in the Nipe-Cristal region of the Nipe-Sagua-Baracoa massif in eastern Cuba. It is an area of complex limestone peaks that surround the Sierra de Nipe. The prominent peaks include La Mensura, La Mensurita, and El Gurugú. The area has a great diversity of ecosystems such as rainforest, Cuban pine *Pinus cubensis* forests, charrascal, and karst vegetation. Close to the IBA are the communities of Biran Uno and Guamutas (in Cueto municipality), and Pueblo Nuevo de Pinares, Mensura Uno and Mensura Dos on the plateau of Mayarí.

■ Birds

This IBA supports populations of 30 (of the 48) Greater Antilles biome-restricted species, 16 of which are endemic to Cuba, including the Endangered Gundlach's Hawk *Accipiter gundlachi*, the Vulnerable Cuban Parakeet *Aratinga euops* and Grey-headed Quail-dove *Geotrygon caniceps*, the Near Threatened Bee Hummingbird *Mellisuga helenae* (occurring at low densities), the rare Bare-legged Owl *Gymnoglaux lawrencii*, and pine-forest specialist Olive-capped Warbler *Dendroica pityophila*. The subspecies of Stygian Owl *Asio stygius siguapa* is endemic. The IBA is home to numerous wintering Neotropical migrants.

■ Other biodiversity

Land molluscs include 60 endemic species (nine of which are local endemics) including the snails *Alcadia euglypta*, *Polydontes sobrina*, *Polymita venusta*, and *Xenopoma hendersoni*. The endemic lizards *Anolis anfiloquioi*, *A. cupeyalensis* and *A. isolepis*, and snake *Tropidophis wrighti* are present. La Mensura is home to 268 endemic plants, 90 of which are locally endemic.

■ Conservation

La Mensura National Park was created in 1992 and was approved by Comité Ejecutivo del Consejo de Ministros in 2008. It is managed by the Empresa para la Protección de la Flora y la Fauna (ENPFF). Residents from surrounding communities have historically used timber and other resources (guano, yarey and royal palm) for building houses and fuel. Subsistence farming is the main activity, but cattle grazing and nature tourism also occur in the area. There are guided interpretative trails close to the administrative and service areas. Some portions of the IBA, specifically in Casimba are considered as a mining reserve. Some threats to the area include fires, soil erosion and intense exploitation of forest resources around the periphery of the park.

Ernesto Reyes, Bárbara Sánchez

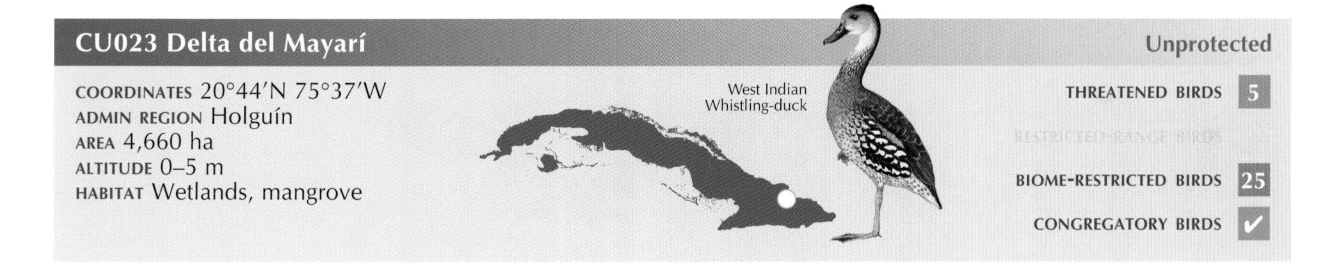

■ Site description

Delta del Mayarí IBA comprises a significant wetland and mangrove areas within the Bahía de Nipe near the north coast of south-eastern Cuba. The Bahía de Nipe is the largest semi-closed bay in the Caribbean and one of the largest in the world. Almost half of it is covered by mangrove forests of which the Delta del Mayarí, with its sub-tidal estuaries and brackish lagoons forms and important part. The Río Mayarí is the main river that drains towards the bay, the others being the Nipe and Tacajó rivers. The northern and eastern parts of this IBA are surrounded by the waters of Bahía de Nipe, and the south and west parts by land, with the nearest community being Guatemala.

■ Birds

This IBA supports populations of 25 (of the 48) Greater Antilles biome-restricted species, nine of which are Cuban endemics. A wide diversity of waterbirds (79 species) occur, including regionally significant numbers of Brown Pelican *Pelecanus occidentalis*. Globally threatened birds include the Endangered Gundlach's Hawk *Accipiter gundlachi*, Vulnerable West Indian Whistling-duck *Dendrocygna arborea*, and Near Threatened White-crowned Pigeon *Patagioenas leucocephala*. The Near Threatened Painted Bunting *Passerina ciris* is one of many Neotropical migrants that use this IBA on passage or over winter.

■ Other biodiversity

The endemic bat *Phyllonycteris poeyi* and regional and national endemic species of snail, such as *Chondropoma confertum*, *Obeliscus bacillus*, and *Oleacina solidula* all occur, as do several endemic *Anolis* lizard species and the turtle *Trachemys decussata*.

■ Conservation

Delta del Mayarí IBA has no protection status at present, but a proposal to declare it a wildlife refuge is underway. The main socio-economic activity in Guatemala was related to the sugarcane industry, but this ceased after a nationwide restructuring of agro-industrial complexes. The local economy currently relies on the fishing industry, including platform fishing, aquaculture (for mangrove oysters), as well as blue crab *Callinectes sapidus* and clam harvesting. Important threats to the area include the unsustainable extraction of mangrove wood (for charcoal) and bark (for tanning leather), and forest clearance for agriculture. These activities have reduced the mangrove area from 1,630 ha in 1983 to 994 ha in 2003.

Carlos M. Peña

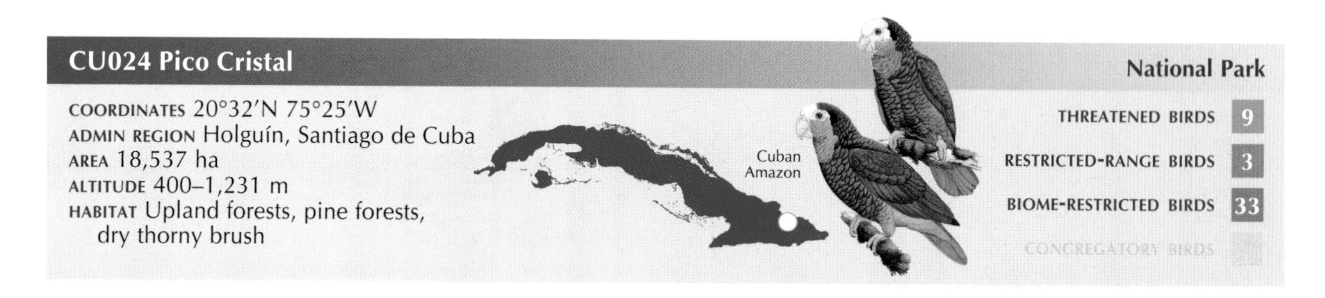

■ Site description

Pico Cristal IBA is located in the Nipe-Cristal mountain region of the Nipe-Sagua-Baracoa massif in eastern Cuba. The area comprises a mosaic of upland forests, pine forests, dry thorny brush over serpentine soils, and upland scrub. Within the IBA are two communities: La Zanja (in Frank País municipality), and Batista (in Segundo Frente municipality). The communities of Baconal, El Palenque, and El Culebro are located in the areas surrounding the IBA.

■ Birds

This IBA supports populations of 33 (of the 48) Greater Antilles biome-restricted species. Globally threatened species include the Endangered Gundlach's Hawk *Accipiter gundlachi* and Giant Kingbird *Tyrannus cubensis* (EN), the Vulnerable Cuban Parakeet *Aratinga euops* and Grey-headed Quail-dove *Geotrygon caniceps*, and the Near Threatened Cuban Amazon *Amazona leucocephala* and Cuban Solitaire *Myadestes elisabeth*.

■ Other biodiversity

Vulnerable Cuban solenodon *Solenodon cubanus* (confined to eastern Cuba) occurs along with the hutias *Capromys pilorides*, *Mysateles melanurus* and *M. prehensilis*. Endemic reptiles (of which there are 35, with nine confined to eastern Cuba) include *Sphaerodactylus celicara*, *Chamaeleolis porcus* and *Epicrates angulifer*. This IBA is part of an important speciation centre for Cuban serpentine flora, with more than 300 endemics, 50 of which are locally endemic.

■ Conservation

Pico Cristal was Cuba's first national park. It was declared in 1930, but its legal status was not implemented until it was approved by the Comité Ejecutivo del Consejo de Ministros in 2001. The IBA follows the national park boundaries. Activities in areas surrounding the park include conservation and coffee cultivation. Constant threats include logging, hunting, the capture of parrot chicks, and invasive species such as rats *Rattus* spp. and feral dogs, cats, and pigs. Forestry companies are very interested in exploiting the area's pine forests. Additionally, 10% of the area (1,500 ha) is within an area zoned for mining. Intense mining and forestry activities have provoked soil and road erosion, and have led to the establishment of new human settlements, habitat fragmentation, and deforestation caused by the logging itself as well as forest fires. These economic activities have proven to be incompatible with conservation efforts.

Juan Pedro Soy, Ernesto Reyes

CU025 Gran Piedra-Pico Mogote

Ecological Reserve/Natural Protected Landscape/Biosphere Reserve

COORDINATES 20°00′N 75°37′W
ADMIN REGION Santiago de Cuba
AREA 4,532 ha
ALTITUDE 600–1,224 m
HABITAT Forest

Osprey

THREATENED BIRDS 3
RESTRICTED-RANGE BIRDS
BIOME-RESTRICTED BIRDS 25
CONGREGATORY BIRDS

■ Site description

Gran Piedra-Pico Mogote IBA is located in Sierra de la Gran Piedra, on the south-eastern coast of Cuba, between the Santiago de Cuba tectonic basin to the west and the Río Baconao River to the north and east. The coastal terraces of the karstic limestone plateau of Santiago de Cuba lie to the south. The city of Santiago de Cuba is 25 km to the west of the western edge of the IBA. There are several dispersed settlements within the area.

■ Birds

This IBA supports populations of 25 (of the 48) Greater Antilles biome-restricted species, 10 of which are Cuban endemics. The Endangered Gundlach's Hawk *Accipiter gundlachi* and Giant Kingbird *Tyrannus cubensis*, and Vulnerable Grey-headed Quail-dove *Geotrygon caniceps* all occur. Studies at this IBA have shown it to be an important migratory corridor for raptors, with 3,793 Osprey *Pandion haliaetus* recorded on passage in 2006 (one of the largest concentrations of this species in the world). A new migratory route for American Swallow-tailed Kite *Elanoides forficatus* was discovered in the IBA.

■ Other biodiversity

The dragonfly *Hypolestes trinitatis*, frog *Eleutherodactylus guantanamera*, and several species of reptiles are Vulnerable. Endemism is high within the flora, herpetofauna and snails.

■ Conservation

Gran Piedra-Pico Mogote IBA embraces the Gran Piedra Natural Protected Landscape (approved in 2008) and the Pico Mogote Ecological Reserve (awaiting approval as a locally significant protected area), both part of the core zone of the Baconao Biosphere Reserve. There is an ecological station at Gran Piedra at which diverse studies have been undertaken. Some of the threats to this IBA include cutting trees for charcoal, illegal logging, shifting agriculture on steep slopes, introduced animals and plants. The primary activity of the local community is agriculture, mostly centred in the Gran Piedra Integral Forestry Enterprise and to a lesser degree at a hotel installation and other service centres in the community. In Gran Piedra an intensive community environmental education program is being implemented.

Freddy Rodríguez

CU026 Siboney-Juticí

Ecological Reserve

COORDINATES 19°57′N 75°46′W
ADMIN REGION Santiago de Cuba
AREA 1,857 ha
ALTITUDE 0–120 m
HABITAT Coastline, shrubland, limestone hills

Cuban Gnatcatcher

THREATENED BIRDS 4
RESTRICTED-RANGE BIRDS 2
BIOME-RESTRICTED BIRDS 23
CONGREGATORY BIRDS ✓

■ Site description

Siboney-Juticí IBA is an area of marine terraces and limestone karst just to the east of the city of Santiago de Cuba. The landscape is one of gorges, canyons, closed sinkholes (dolines), deep eroded channels (karren), caves and grottos. The topography, climate and location of the area have resulted in this IBA being home to a diverse and unique biota. The seasonal creeks of Sardinero and Juticí only run after intense downpours. Outside the IBA, the coastal community of Siboney (1,023 residents) receives a large number of visitors each year.

■ Birds

This IBA supports populations of 23 (of the 48) Greater Antilles biome-restricted species, 10 of which are Cuban endemics. The Endangered Gundlach's Hawk *Accipiter gundlachi* and the restricted-range Cuban Gnatcatcher *Polioptila lembeyei* and Oriente Warbler *Teretistris fornsi* all occur. The IBA is very important for Neotropical migrants, primarily passerines and raptors. An unprecedented 6,143 Osprey *Pandion haliaetus* have been recorded on passage, as have the largest numbers of migrant American Swallow-tailed Kite *Elanoides forficatus*, Merlin *Falco columbarius*, and Peregrine Falcon *Falco peregrinus* noted in the insular Caribbean.

■ Other biodiversity

There are 21 endemic land molluscs species within the IBA (including the locally endemic snail *Macroceramus jeannereti*), and also three endemic amphibians, 13 endemic reptiles and 167 endemic plants (12 of which are Endangered and two Vulnerable).

■ Conservation

Siboney-Juticí IBA is an Ecological Reserve which was approved in 2001 by the Comité Ejecutivo del Consejo de Ministros and is now managed by the Centro Oriental de Ecosistemas y Biodiversidad (BIOECO). In 1970, a 67-ha area west of Siboney beach was declared a speleological natural reserve (one of the first protected areas in the country). There is an ecological station in the IBA that has served as an important centre for research on the area's forests. The dry coastal scrub on karren is threatened by fire (both natural and human-caused). Wood extraction by the community has been a significant threat, mostly during periods of economic crisis, although this has decreased considerably in recent years. However, illegal sand extraction is affecting certain areas in the IBA.

FREDDY RODRÍGUEZ

CU027 Alejandro de Humboldt

National Park/Biosphere Reserve/World Heritage Site

■ Site description

Alejandro de Humboldt IBA is in the Nipe-Sagua-Baracoa massif, stretching inland from the coast between Moa and Baracoa. This area represents the largest and best preserved tract of tropical and montane forest in the insular Caribbean, and supports the highest species diversity and levels of endemism in the country. It has a unique landscape of non-calcareous karst, pristine rivers, pools and waterfalls, extensive tracts of pine and broadleaf tropical forest, and stunning panoramic vistas. Eleven communities (totalling more than 5,000 people) are within the IBA, and another nine are within the buffer zone.

■ Birds

This IBA supports populations of 35 (of the 48) Greater Antilles biome-restricted species, 10 of which are Cuban endemics, and what might be the last populations of the Critically Endangered Cuban Kite *Chondrohierax wilsonii* and Ivory-billed Woodpecker *Campephilus principalis*. The Endangered Gundlach's Hawk *Accipiter gundlachi*, Blue-headed Quail-dove *Starnoenas cyanocephala* and Giant Kingbird *Tyrannus cubensis*, the Vulnerable Cuban Parakeet *Aratinga euops* and Grey-headed Quail-dove *Geotrygon caniceps* also occur. Oriente Warbler *Teretistris fornsi* is the most abundant endemic and the central bird of mixed species flocks.

■ Other biodiversity

Fauna includes the Endangered Cuban solenodon *Solenodon cubanus*, the Critically Endangered frog *Eleutherodactylus iberia* and Endangered *E. principalis*, the endemic snail *Polymita picta* and the lizard *Anolis rubribarbus*. Funnel-eared bat *Nyctiellus lepidus* also occurs. Important plant species include *Podocarpus ekmani* and *Dracaena cubensis* (a primarily Old World genus).

■ Conservation

Alejandro de Humboldt IBA comprises what used to be the Jaguaní and Cupeyal del Norte natural reserves and the Ojito de Agua Wildlife Refuge. In 2001 these were merged into a national park which was declared a World Natural Heritage site the same year. It is managed by the Alejandro de Humboldt Environmental Services Unit and is a core zone of the Cuchillas del Toa Biosphere Reserve. Land use includes forestry, mining, and farming. There are nickel, chrome, iron, and cobalt deposits with mining potential. The coastal zone of Baracoa–Moa has good potential for ecotourism. The main threats are soil erosion, forest fires, unsustainable forest management, illegal logging, hunting and trafficking, invasive species, and unsustainable land use in the buffer zone.

ENEIDER E. PÉREZ, HIRAM GONZÁLEZ, GERARDO BEGUE, BÁRBARA SÁNCHEZ, ALEJANDRO LLANES

CU028 Hatibonico-Baitiquirí-Imías

Ecological Reserve/Natural Reserve/Unprotected

COORDINATES 19°59′N 74°59′W
ADMIN REGION Guantánamo
AREA 16,746 ha
ALTITUDE 0–401 m
HABITAT Dry forest, semi-desert shrubland, evergreen forest

Cuban Sparrow

THREATENED BIRDS 7
RESTRICTED-RANGE BIRDS 3
BIOME-RESTRICTED BIRDS 29
CONGREGATORY BIRDS

■ Site description

Hatibonico-Baitiquirí-Imías IBA is on the southern coast of Guantánamo province, in the Sagua-Baracoa sub-region of easternmost Cuba. The IBA is split into two areas either side of Guantánamo bay: Hatibonico (a small town in Caimanera municipality) just to the west of the bay, and the small towns of Baitiquirí (San Antonio del Sur municipality) and Imías to the east. This is the driest area of the Cuba resulting in characteristic semi-desert (shrubland and dry forest) and fauna, with numerous regional and local endemics. The area of Hatibonico (which also supports evergreen tropical forest) has naturally sculpted rock formations known as "monitongos"

■ Birds

This IBA supports populations of 29 (of the 48) Greater Antilles biome-restricted species, 12 of which are Cuban endemics including the Endangered Gundlach's Hawk *Accipiter gundlachi*, the Vulnerable Grey-headed Quail-dove *Geotrygon caniceps* and Fernandina's Flicker *Colaptes fernandinae* and the Near Threatened Bee Hummingbird *Mellisuga helenae*. A large population of the Endangered Cuban Sparrow *Torreornis inexpectata sigmani* is in the dry forest—one of only three areas for the bird. Large numbers of migrating Ospreys *Pandion haliaetus* take advantage of the wind currents along the ridges.

■ Other biodiversity

The Vulnerable Cuban ground iguana *Cyclura nubila* occurs along with the endemic hutia *Epicrates angulifer*. The flora includes 90 endemic species including three that are Endangered, namely *Apassalus parvulus*, *Melocactus harlowii*, and the fern *Notholaena ekmanii* which is confined to the monitongos of Hatibonico.

■ Conservation

Hatibonico-Baitiquirí-Imías IBA includes the Hatibonico Ecological Reserve which was approved as a protected area in 2001 and is managed by the Alejandro de Humboldt Environmental Services Unit (of the Ministry of Science Technology and Environment). Baitiquiri Ecological Reserve (soon to be administered) and the Imías Natural Reserve have been proposed as protected areas, but as yet lack approved status or any management. The main economic activities in this IBA are agroforestry (coffee, cacao and other crops) and salt production in Caimanera. The Hatibonico forest is threatened by the expansion of *Dichrostachys cinerea* which occupies c.15% of the area, as well as by forest fires, erosion, and silvicultural practices of a local farming enterprise.

Freddy Rodríguez

CURAÇAO

LAND AREA **444 km²** ALTITUDE **0–375 m**
HUMAN POPULATION **138,000** CAPITAL **Willemstad**
IMPORTANT BIRD AREAS **5, totalling 163 km²**
IMPORTANT BIRD AREA PROTECTION **86%**
BIRD SPECIES **215**
THREATENED BIRDS **1** RESTRICTED-RANGE BIRDS **1** BIOME-RESTRICTED BIRDS **1**

ADOLPHE DEBROT (CARMABI FOUNDATION) AND
JEFF WELLS (BOREAL SONGBIRD INITIATIVE)

Mount Christoffel, the highest point in Curaçao. (PHOTO: JEFF WELLS)

INTRODUCTION

Curaçao, which is politically part of the Kingdom of the Netherlands[1], is one of the three Netherlands Antillean islands (Aruba, Bonaire and Curaçao) that lie off the north-west coast of Venezuela. Curaçao is the middle one of the three islands (c.80 km east of Aruba, and 50 km west of Bonaire), and lies c.70 km off the coast of mainland Venezuela. It is the largest of the three "Leeward" Netherlands Antillean islands and c.56 km long by 4–10 km wide. The island consists of a basaltic centre of volcanic origin, rimmed in the coastal zones by marine limestone terraces. The landscape is rugged and rocky with prominent cliffs. Christoffelberg, at the north-west end of the island, is the highest point in Curaçao. The island has a dry, windy climate with an annual average rainfall of c.550 mm. The resultant vegetation comprises xeric shrubland characterised by columnar cacti. Curaçao possesses several semi-enclosed inland bays, which are densely fringed by mangroves. There are also several enclosed shallow, hyper-saline lagoons that are important for waterbirds, as are the numerous man-made freshwater catchment dams, some of which retain significant amounts of fresh water well into the dry season. The island is surrounded by coral reefs and a variety of rocky, sandy or rubble shores.

Curaçao has the largest human population of the Leeward Netherlands Antilles, and about 30% of the island has been occupied by housing and industry. In bygone eras, the vegetation of the island was heavily impacted by agricultural activities such as the cultivation of seasonal crops, charcoal production and livestock grazing. However, traditional agricultural activities have all but ceased due to economic factors, rampant theft of livestock and produce, and an increase in speculative private land ownership. As a consequence, dense secondary woodlands are gradually recuperating from the impacts of man and beast, and are not uncommon, especially on the western half of the island.

■ Conservation

In Curaçao about 30% of the island surface area has been legally designated as conservation habitat since 1997 by means of a land-use ordinance, the Curaçao Island Development Plan. Environmental policy and implementation is overseen

1 At some point in the near future the "Netherlands Antilles" will be dissolved. St Maarten and Curaçao will become separate countries within the Kingdom of the Netherlands (similar to the status currently enjoyed by Aruba). The islands of Bonaire, Saba and St Eustatius will be linked directly to the Netherlands as overseas territories.

by the Department of Environment and Nature (Ministry of Public Health and Social Development), the Curaçao Urban Planning and Public Housing Service, the Curaçao Environmental Service, and the Curaçao Agricultural and Fishery Service. However, the CARMABI Foundation is the island's park service and formally manages nine conservation areas distributed around the island amounting to c.3,000 ha of terrestrial and lagoonal habitat (primarily within North-east Curaçao parks and coastal IBA, AN015) and 600 ha of coral reefs (in the Curaçao Underwater Park on the leeward side of the island). The 2001 draft nature management plan for Curaçao identifies the need to update and expand all island level protective legislation. However, the plan still awaits ratification by the island council. Government funding for environmental protection and management remains sparse and often non-structural in spite of the recognised importance of the environment to sustainable tourism. In 2006, funding to the island park service (CARMABI Foundation) was cut by 70%. Limited institutional capacity due to insufficient funding has been identified as a key bottleneck to environmental management and conservation in the Netherlands Antilles.

Since 1988 when Defensa Ambiental (Environmental Defense) was founded, the local environmental NGO movement has grown significantly. A number of organisations are currently active, including Amigu di Tera (Friends of the Earth), Reef Care and Kids for Corals (with a focus on coral reefs), Uniek Curaçao (focusing on awareness and guided tours), Korsou Limpi i Bunita (clean ups), and the Stichting Dierenbescherming (animal cruelty and welfare). All of these organisations contribute to general environmental awareness, but none have a significant focus on the island's avifauna. The CARMABI Foundation is the only organisation that has conducted or supported (albeit intermittent) studies on priority birds, and organised birding excursions and workshops.

At present, urbanisation through real estate and tourist development, coastal development and disturbance (especially due to tourist-oriented recreational activity) are the principal threats to remaining habitat areas on Curaçao. These factors pose a major threat to the future integrity of the legally designated conservation areas, and the resultant habitat fragmentation (and disturbance) threatens a number of sensitive species such as nesting terns (*Sterna* spp.), Scaly-naped Pigeon *Patagioenas squamosa*, Curaçao Barn Owl *Tyto alba bargei* and White-tailed Hawk *Buteo albicaudatus*. An antiquated ordinance dating from 1926 provides limited legal protection to some native bird species (including some terns), but enforcement remains totally lacking.

■ Birds

Of the 215 species of bird that have been recorded from Curaçao only 57 are resident (current or former) breeding species, the vast majority being migrants, winterers, and occasional vagrants. Most of the migrants are Neotropical migrants (especially warblers) from breeding grounds in North America, although many are vagrants (recorded on average less than once a year). Little is known about the significance of the island habitats to these migrant birds. The rainy season usually begins (September–October) just in time to yield peak insect swarms for migrants such as swallows and warblers. A small number of Curaçao's species are of South American origin, representing either dispersing individuals or austral migrants overshooting their northern South American wintering grounds having originated from breeding grounds further south or west.

As with Aruba and Bonaire, due to the proximity of the islands to the mainland of South America and the general mobility of birds, there are no island endemic bird species. Nevertheless, the islands have been sufficiently separated from other sources of interbreeding such that at least 16 subspecies have been described from Aruba, Bonaire and Curaçao. Eleven of these species breed on Curaçao, and two are totally restricted to the island, namely the Brown-throated Parakeet *Aratinga pertinax pertinax* and the Barn Owl *Tyto alba bargei*. A subspecies of Grasshopper Sparrow *Ammodramus savannarum caribaeus* found only on Curaçao and Bonaire appears vocally distinct from other North American and Caribbean forms. Similarly, vocalisations of White-tailed Nightjar *Caprimulgus cayennensis insularis* from Curaçao and Bonaire are distinct from mainland forms, but further taxonomic research is needed on all of these subspecies. Curaçao's resident avifauna is a rather unique, with species of West Indian origin mixed with those of South American origin. This is demonstrated by the presence of a Northern South America biome-restricted bird, the Bare-eyed Pigeon *Patagioenas corensis* as well as a number of more wide ranging South American species. Caribbean Elaenia *Elaenia martinica*, the only restricted-range bird on the island (part of the Netherlands Antilles secondary Endemic Bird Area), is one of the West Indian birds found on the island.

Klein Curaçao IBA, globally significant for nesting Least Terns. (PHOTO: ADOLPHE DEBROT)

Curaçao is globally important for its nesting terns (Common Tern *Sterna hirundo* and Least Tern *S. antillarum*) and for the Near Threatened Caribbean Coot *Fulica caribaea*, the national population sizes of which are listed in Table 1. Just two species of tern (listed above) currently nest on Curaçao, but Royal Tern *S. maxima*, "Cayenne" Tern *S. sandvicensis eurygnatha* and Roseate Tern *S. dougalli* all bred during the twentieth century. The relatively large size of Curaçao, the wide variety of habitats, and the more pristine and diverse state of its vegetation compared to the smaller and still more pastoral neighbouring islands of Aruba and Bonaire, mean that Curaçao provides some of the best opportunities for protecting native bird diversity in the Leeward Netherlands Antilles.

IMPORTANT BIRD AREAS

Curaçao's five IBAs—the island's international priority sites for bird conservation—cover 16,280 ha (including marine areas) and c.24% of the land area. At 13,555 ha the North-east Curaçao parks and coast IBA (AN015) makes up c.83% of this IBA coverage. The North-east Curaçao parks and coast IBA embraces Curaçao's two terrestrial national parks (totalling c.2,300 ha). The remainder of the IBA is a Protected Conservation Area, a designation common to part of the Malpais-Sint Michiel IBA (AN016) and the Jan Thiel Lagoon IBA (AN018) although in none of these areas is there active management for conservation. Muizenberg IBA (AN017) is designated as protected parkland, but also suffers from a lack of active management. Klein Curaçao IBA (AN019) is not protected in any way. The IBAs have been identified on the basis of five key bird species (see Table 1) that variously trigger the IBA criteria. Each of these birds occurs in two or more IBAs although the majority of the Near Threatened Caribbean Coot *Fulica caribaea* occur in the threatened and unmanaged Muizenberg IBA (AN017).

Caribbean Coot and Caribbean Flamingos at Muizenberg IBA. (PHOTO: ADOLPHE DEBROT)

Table 1. Key bird species at Important Bird Areas in Curaçao.

Key bird species	Criteria		National population	Curaçao IBAs AN015	AN016	AN017	AN018	AN019
Criteria					■	■		
				■	■			
				■	■			■
				■	■	■	■	■
Caribbean Coot *Fulica caribaea*	NT ■	■	1,000		100	800		
Common Tern *Sterna hirundo*		■	400		45		225	
Least Tern *Sterna antillarum*		■	1,860	514			10–60	858
Bare-eyed Pigeon *Patagioenas corensis*	■			✓	600			✓
Caribbean Elaenia *Elaenia martinica*	■			✓	✓			

All population figures = numbers of individuals.
Threatened birds: Near Threatened ■. **Restricted-range birds** ■. **Biome-restricted birds** ■. **Congregatory birds** ■.

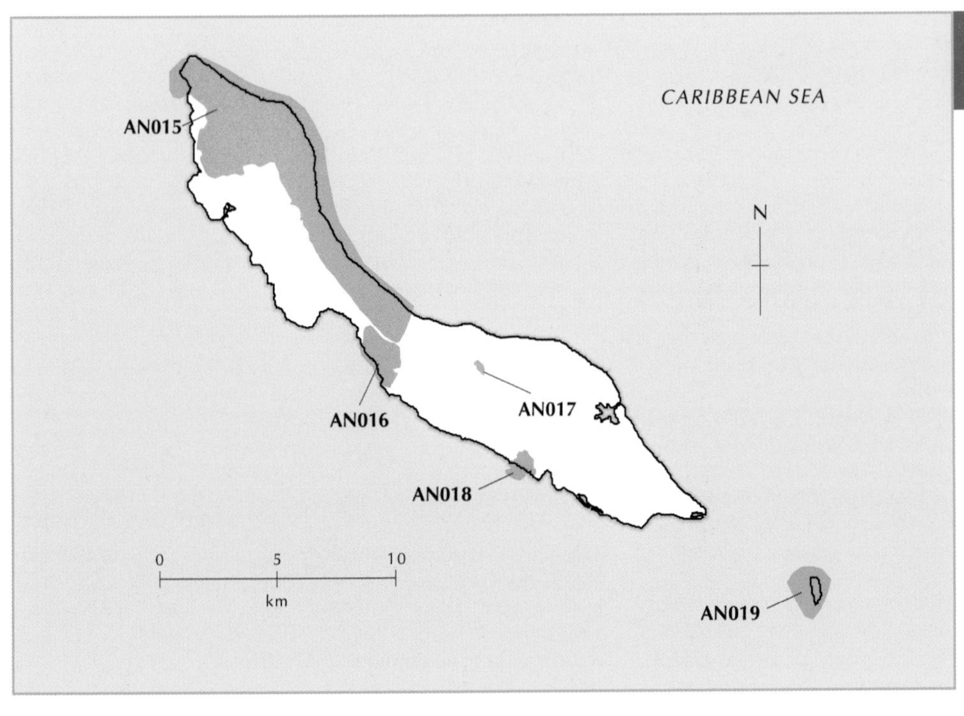

Figure 1. Location of Important Bird Areas in Curaçao.

Jan Thiel Lagoon IBA. (PHOTO: ADOLPHE DEBROT)

All of the IBAs have urgent management requirements if the populations of the birds for which they are internationally important are to thrive. However, securing disturbance free zones around the tern nesting colonies appears to be one of the greatest needs. If implemented effectively, the tern populations would increase dramatically (as seen at the protected colonies on Aruba) and perhaps some of the 1,200 pairs of "Cayenne" Tern *S. sandvicensis eurygnatha* that used to breed (pre-1962) at Jan Thiel Lagoon IBA might return.

Monitoring the populations of the terns and waterbirds should be used for the assessment of state, pressure and response scores at each of Curaçao's IBAs in order to provide an objective status assessment as well as to highlight management interventions that might be required to maintain these internationally important biodiversity sites.

KEY REFERENCES

ANON. (1989) Reactie op het eilandelijk ontwikkelingsplan Curaçao, in 't bijzonder v.w.b. Malpais. Willemstad: CARMABI/ STINAPA. (Unpublished report).

BEERS, C. E., DE FREITAS, J. A. AND KETNER, P. (1997) Landscape ecological vegetation map of the island of Curaçao, Netherlands Antilles. Publication Foundation for Scientific Research in the Caribbean Region: 138. 54 pp.

BOKMA, W. (1972) Malpais, toekomstig vogelreservaat op Curacao? *STINAPA report* 6: 31–36.

CUPPENS, M. AND VOGELS, J. (2004) Characterization of foraging areas of the Caribbean Flamingo *Phoenicopterus ruber* on Curaçao (Netherlands Antilles): the relationship between abiotic factors, food abundance and flamingo density. Nijmegen: Radboud University/ CARMABI. (Unpublished thesis).

DEBROT, A. O. (2000) A review of records of the extinct West Indian monk seal *Monachus tropicalis* (Carnivora: Phocidae), for the Netherlands Antilles. *Mar. Mamm. Sci.* 16: 834–837.

DEBROT, A. O. (2003) A review of the freshwater fishes of Curacao, with comments on those of Aruba and Bonaire. *Carib. J. Science* 39: 100–108.

DEBROT, A. O., ESTEBAN, N., LE SCAO, R., CABALLERO, A. AND HOETJES, P. C. (2005) New sea-turtle nesting records for the Netherlands Antilles provide impetus to conservation action. *Car. J. Sci.* 41: 334–339.

DEBROT, A. O., BOOGERD, C. AND VAN DEN BROECK, D. (in press) Dutch Antilles III: Curaçao and Bonaire. In Bradley P. E. and Norton, R. L. eds. *Breeding seabirds of the Caribbean.* Gainesville, Florida: Univ. Florida Press.

DEBROT, A. O. AND DE FREITAS, J. A. (1991) Wilderness areas of exceptional conservation value in Curaçao, Netherlands Antilles. *Nederlandse Commissie voor Internationale Natuurbescherming, Meded.* 26: 1–25.

DEBROT, A. O. AND DE FREITAS, J. A. (1999) Avifaunal and botanical survey of the Jan Thiel Lagoon Conservation Area, Curaçao. Willemstad: CARMABI Foundation (Unpublished report).

DEBROT, A. O. AND PORS, L. P. J. J. (2001) Beheers- en inrichtingsplan conserveringsgebied Jan Thiel. Willemstad: CARMABI Foundation (Unpublished report 2).

HULSMAN, H., VONK, R., ALIABADIAN, M., DEBROT, A. O. AND NIJMAN, V. (2008) Effect of introduced species and habitat alteration on the occurrence and distribution of euryhaline fishes in fresh- and brackish-water habitats on Aruba, Bonaire and Curaçao (South Caribbean). *Contrib. Zool.* 77: 45–52.

NIJMAN, V., ALIABADIAN, M., DEBROT, A. O., DE FREITAS, J. A., GOMES, L. G. L., PRINS, T. G. AND VONK, R. (2008) Conservation status of Caribbean Coot *Fulica caribaea* in the Netherlands Antilles, and other parts of the Caribbean. *Endangered Species Research* 4: 241–246.

PRINS, T. G. AND NIJMAN, V. (2007) Checklist of the birds of Curaçao. Amsterdam: Zoological Museum Amsterdam. (Unpublished report).

PRINS, T. G. AND NIJMAN, V. (2005) Historic changes in status of Caribbean Coot in the Netherlands Antilles. *Oryx* 39: 125–126.

PRINS, T. G., ROSELAAR, K. C. S AND NIJMAN, V. (2005) Status and breeding of Caribbean Coot in the Netherlands Antilles. *Waterbirds* 28:146–149.

SMELTER, M. (2005) Food-web structure, and dispersion of food items of flamingos, in hyper-saline lakes in Curaçao, Netherlands Antilles. Nijmegen: Radboud University/ CARMABI. (Unpublished thesis).

VOOUS, K. H. (1983) *Birds of the Netherlands Antilles.* Zutphen, The Netherlands: De Walburg Pers.

WELLS, J. V. AND CHILDS WELLS, A. (2006) The significance of Bonaire, Netherlands Antilles, as a breeding site for terns and plovers. *J. Carib. Orn.* 19: 21–26.

ACKNOWLEDGEMENTS

The authors would like to thank Bert Denneman (Vogelbescherming Nederlands), Adrian del Nevo (Applied Ecological Solutions Inc.), Kalli de Meyer (Dutch Caribbean Nature Alliance), Vincent Nijman and Tineke Prins (Zoological Museum Amsterdam) for their contributions to and reviews of this chapter.

Site description

North-east Curaçao parks and coastal IBA extends along coast from Westpunt and Noordpunt at the northern end of island around the northern tip and east along coast through Playa Grandi and ending near Hato airport. At the northern end of the island it extends inland from the 100-ha coastal Shete Boka National Park to include the 2,000-ha Christoffel National Park. The area comprises coastal limestone terraces and inland hills supporting evergreen woodland, coastal lagoons with sea grass beds and mangroves, and dry deciduous shrubland on volcanic soils. Christoffel National Park supports one of largest contiguous blocks of shrubland remaining on island.

Birds

This IBA is regionally significant for its breeding colony of 500+ Least Terns *Sterna antillarum*. There is a Brown Booby *Sula leucogaster* roosts of c.12 birds within the IBA, and just offshore are important feeding areas for this species and the terns. The Netherlands Antilles secondary EBA restricted-range Caribbean Elaenia *Elaenia martinica*, and the Northern South America biome-restricted Bare-eyed Pigeon *Patagioenas corensis* are both numerous. The shrubland is important for the endemic subspecies of Brown-throated Parakeet *Aratinga pertinax pertinax*. In total, 10 (of the 11) Leeward Netherlands Antilles endemic subspecies of birds breed in this IBA which is also important for large numbers of Neotropical migrants.

Other biodiversity

The Shete Boka National Park is a nesting area for three globally threatened sea-turtle species. The Christoffel National Park supports c.50 rare plant species (that are found almost entirely within the park boundaries) and seven rare mammal species. The Hato area is home to c.55 Leeward Netherlands Antilles endemics or near-endemics.

Conservation

This IBA is largely state owned. The Shete Boka and Christoffel national parks are managed by the CARMABI Foundation, and the rest is legally designated as Protected Conservation Area but is unmanaged. Extensive research on all aspects of the ecology, biology and geology of this area has been carried out by CARMABI and visiting scientists. The tern colonies are threatened by disturbance from humans, and disturbance and predation from feral dogs and cats. Free-ranging goats have impacted the shrubland in Christoffel National Park, but eradication measures have lead to a recovery in vegetation. Elsewhere, illegal livestock rearing is increasing, as is illegal dumping, and effective management is urgently needed.

Site description

Malpais–St Michiel IBA is on the southern side of central Curaçao. It is an area of basalt rock over-capped by coralline limestone hills. Malpais is a former plantation, just north of St Michiel Bay. The IBA is diverse, including: two freshwater lakes (created by a dam) which retain at least some water during most dry seasons; a hyper-saline lagoon at St Michiel (on the south coast c.5 km north-west of Willemstad, lying inland from the coral reef-fringed bay); dry deciduous vegetation (on the volcanic soils); and a well developed *Coccoloba swartzii–Erithal fruticosa* woodland habitat on the limestone. The island's increasingly large dumpsite is just north and upstream of the Malpais freshwater ponds.

Birds

This IBA is significant for the Near Threatened Caribbean Coot *Fulica caribaea* which breeds at Malpais (numbers regularly exceed 100). The shrubland is important for the Netherlands Antilles secondary EBA restricted-range Caribbean Elaenia *Elaenia martinica*, and the Northern South America biome-restricted Bare-eyed Pigeon *Patagioenas corensis* (up to 600 roost below Malpais dam, along with c.165 Brown-throated Parakeets *Aratinga pertinax*). The St Michiel lagoon supports a globally important population (15 pairs) of Common Tern *Sterna hirundo*, and is one of a network of sites supporting Curaçao's Caribbean Flamingos *Phoenicopterus ruber*.

Other biodiversity

The endemic freshwater fish *Poecilia vandepolli* is present in the lakes. Endemic land-snails are present, as are a number of endemic plant species. The IBA is floristically important.

Conservation

Malpais–St Michiel IBA is under mixed (private and state) ownership. The Malpais area is managed as part of a conservation area and has excellent well-signed and designated hiking trails which are maintained by the local conservation organisation Uniek Curaçao. The dumpsite (and pig farm) near the Malpais ponds pose a potential threat of wetland contamination. Uncontrolled recreational access by hikers (and dog walkers) is a threat to birds and other fauna. Poor maintenance of the dam may result in one of the two lakes drying out during prolonged dry seasons, significantly decreasing the area's value for waterbirds and the endemic fish. The main threat at St Michiel is uncontrolled public access that disturbs the flamingos (the nesting terns are located on the less disturbed west side of the lagoon). The CARMABI Foundation and Zoological Museum of Amsterdam have carried out research in this IBA.

■ Site description

Muizenberg IBA comprises an intermittent shallow lake/ wetland in the northern suburbs of Willemstad, central Curaçao. The wetland has been created by the damming of a stream that drains the surrounding low hills. It is bounded on all sides by busy roads (on the west side the roads abut the wetland). On the north-east side the area is flanked by agricultural lands with small farms. The wetland typically retains some water for more than six months each year (and in wetter years water can be present year-round). Periodically inundated grassland and shrubland surround the wetland. A separate small pond (Kaya Fortuna) is situated 200 m west of Muizenberg.

■ Birds

This IBA is significant for its population of Near Threatened Caribbean Coot *Fulica caribaea*. The species is a resident breeder, with congregations of up to 800 birds during the wet season.

Caribbean Flamingos *Phoenicopterus ruber* occasionally feed in the wetlands, with flocks of up to 170 birds recorded. Many other waterbirds (both residents and migrants) are supported within this IBA.

■ Other biodiversity

The endemic freshwater fish *Poecilia vandepolli* is present in this wetland.

■ Conservation

Muizenberg IBA is state owned and legally designated as protected parkland, although it is not being actively managed. Although it is one of Curaçao's two most important and rare freshwater areas, no biological conservation or research projects are known to have been implemented at the site apart from the Zoological Museum of Amsterdam's 2006 inventory of waterbirds and freshwater fish. The IBA is threatened by the unregulated dumping of garbage, pollution, drainage of surrounding wetlands and recreational disturbance.

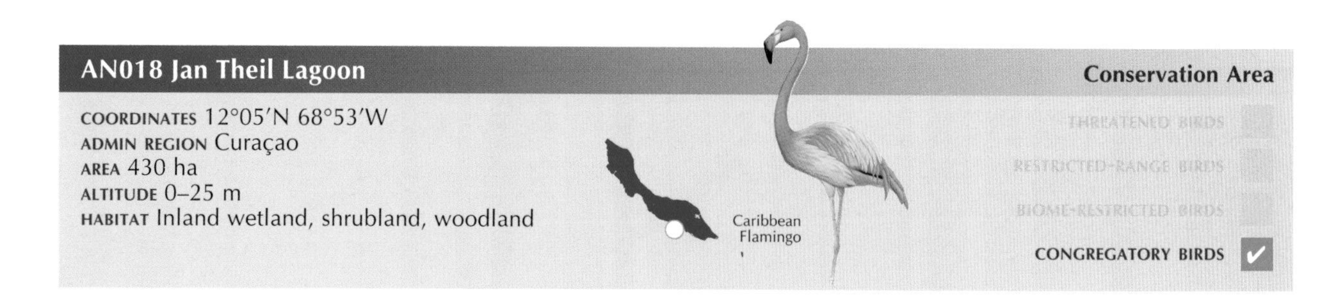

■ Site description

Jan Thiel Lagoon IBA is on the south coast of southern Curaçao, east of Willemstad. The site comprises 80 ha of hyper-saline lagoons (with islands) surrounded by c.228 hectares of dry deciduous woodlands and (on the limestone) evergreen shrubland. The vegetation near the margins of the lagoon is characterised by water and salt-tolerant tree species. There is also one spring and several abandoned dams within the IBA which support fresh water during the rainy season, significantly increasing the diversity of habitats. Hotels and resorts surround the lagoon.

■ Birds

This IBAs population of Common Tern *Sterna hirundo* (75 pairs) is globally significant, while the fluctuating population of 10–60 pairs of Least Tern *S. antillarum* is regionally so. Up to 1,200 pairs of Sandwich Tern *S. sandvicensis* used to breed (pre-1962) but disturbance has reduced numbers to insignificant levels. The lagoon is an important foraging area for the 200–300 Caribbean Flamingo *Phoenicopterus ruber* that occur on the island. The IBA is also important for migratory shorebirds, other migrants, and for resident populations of at least 10 (of the 11) Leeward Dutch Antilles endemic subspecies of birds that breed on Curaçao.

■ Other biodiversity

Reptiles and land snails endemic to the Leeward Dutch Antilles are well represented in the IBA. Plants include the extremely rare *Vitex cymosa*—known from less than 10 trees on the island, and rare evergreen species such as *Croton niveus, Mayteneus tetragona, Schoephia schreberi* and *Adelia ricinella.*

■ Conservation

Jan Thiel Lagoon is a part state- and part privately-owned conservation area. A management plan developed by the CARMABI Foundation has been approved by the government, but implementation has not been financed. Uncontrolled public recreational access has resulted in a c.90% reduction of breeding terns. With adequate protection, tern (and flamingo) abundance will certainly recover. In 1999, the government proposed to build a road through the conservation area, but the plans were successfully contested in court (by the local environmental group Defensa Ambiental). An abandoned, unsealed landfill bordering the lagoon is an unquantified threat.

■ Site description

Klein Curaçao IBA is a small (c.170 ha), flat, offshore reef island about 10 km south-east of the south-easternmost point of Curaçao. The island was originally well-vegetated but was extensively mined for phosphate in the late nineteenth and early twentieth centuries and overgrazed by livestock since the 1800s. As a consequence, the island has been devoid of all trees and bushes for more than 100 years. Up until 2000 the vegetation consisted of a few small herbs and grasses but since a highly successful restoration project, the native flora has grown and landscape changes are taking place rapidly.

■ Birds

This IBA is significant for its breeding population of Least Tern *Sterna antillarum*, 286 of which nested in 2002, making this globally important. Historically, far greater numbers of terns bred at this IBA. Bare-eyed Pigeon *Patagioenas corensis*—the only Northern South America biome-restricted species occurring in Curaçao—has been recorded on this island. House Sparrow *Passer domesticus* colonised in the 1990s.

■ Other biodiversity

Klein Curaçao is the most important sea turtle nesting beach within Curaçao's jurisdiction, with Endangered loggerhead *Caretta caretta* and green *Chelonia mydas*, and Critically Endangered hawksbill *Eretmochelys imbricata* turtles nesting. The island was historically important for the now extinct Caribbean monk seal *Monachus tropicalis*. The island is surrounded by a luxuriant reef system.

■ Conservation

The island is state owned, legally designated as "open land", and thus has no protected status (in spite of requests for such by the CARMABI Foundation). Feral goats were eradicated from the island in 1996 by the Curaçao Agriculture and Animal Husbandry and Fishery Service. Feral cats were eliminated by CARMABI in 2004. These two successful eradications have paved the way for ecological recovery which has been assisted (since 2000) by CARMABI through the planting of drought and salt resistant native trees, shrubs, herbs and grasses. These plants are now spreading naturally over the island. Based on the vegetation recovery attained, the first native land bird (Bananaquit *Coereba flaveola*) was reintroduced from Curaçao in 2005 and has since been breeding. The island's main threat is now disturbance from uncontrolled recreational access by over 600 visitors per week (in 2006).

DOMINICA

LAND AREA **754 km²** ALTITUDE **0–1,447 m**
HUMAN POPULATION **72,400** CAPITAL **Roseau**
IMPORTANT BIRD AREAS **4, totalling 106 km²**
IMPORTANT BIRD AREA PROTECTION **93%**
BIRD SPECIES **176**
THREATENED BIRDS **3** RESTRICTED-RANGE BIRDS **19**

STEPHEN DURAND AND BERTRAND JNO. BAPTISTE
(FORESTRY, WILDLIFE AND PARKS DIVISION)

Morne Trois Pitons National Park, looking north-east towards Delices. (PHOTO: PAUL REILLO, RSCF/FWP DIVISION)

INTRODUCTION

The Commonwealth of Dominica is the most northerly of the Windward Islands, and at the mid-point of the Lesser Antillean chain. It lies between the French islands of Guadeloupe (c.28 km to the north) and Martinique (c.40 km to the south). The island is c.47 km long by 26 km wide, and is divided into 10 administrative parishes. Dominica is one of the youngest islands in the Lesser Antilles. Its volcanic origins have created an island characterised by very rugged and steep terrain. The volcanic cone of Morne Diablotin (1,447 m), along with Morne Au Diable on the northern peninsula, dominates the topography of the northern half of the island, while a chain of mountains (including Morne Trois Pitons, Morne Micotrin, Morne Watt, Morne Anglais, and Morne Plat Pays) extends through the south of the island.

Dominica's climate is classified as humid tropical marine, characterised by little seasonal or diurnal variation and strong, steady trade winds. The island is among the wettest in the Caribbean, a factor which gives rise to its lush vegetation. Rainfall is higher in the interior which receives >10,000 mm annually, and drops off substantially to 1,200 mm per year on the leeward (western) side of the island. Dominica's vegetation comprises more than 1,000 species of flowering plants with about 68 rainforest tree species. Major terrestrial ecosystems include mature tropical rainforest, montane thicket and cloud forest (elfin woodland), and littoral woodland along the windward coast. Dry scrub and xeric woodland occupy much of the leeward coast. Dominica has, with justification, been referred to as the "nature island of the Caribbean". With its mostly unspoiled mountainous landscape, perennial streams, rivers and numerous waterfalls, and its great diversity of flora and fauna, the island is considered to be among the most beautiful and pristine countries in the world. The country's undisturbed forests are undoubtedly more extensive than on any other island in the Lesser Antilles.

Dominica's economy is heavily dependent upon tourism and agriculture, with c.20% of the workforce employed in the agricultural sector, particularly banana crops which have traditionally formed the backbone of the economy. Other primary agricultural exports include vegetables (e.g. dasheen, hot peppers and pumpkins), herbs, plantains, citrus, coconut oil and other essential oils. The island's lack of sandy beaches means that Dominica's tourism industry is dominated by scuba-diving and nature-tourism niche markets. Dominica's population is primarily concentrated along the flatter, coastal areas although in recent times there has been some limited residential development in the interior.

■ Conservation

The Forestry, Wildlife and National Parks Division (FWP Division, within the Ministry of Agriculture, Fisheries and Forestry) is the governmental agency responsible for the management of forest reserves and national parks, and for the conservation and protection of wild flora and fauna. Approximately 60% of Dominica is still under natural vegetative cover, albeit mostly on privately-owned lands. However, 20% of the island is under some form of protected-area status. The legally established protected area system comprises: two forest reserves (totalling 5,688 ha), namely the Northern Forest Reserve and Central Forest Reserve; the three national parks (totalling c.10,746 ha) of Morne Trois Pitons National Park (a World Heritage Site), Morne Diablotin National Park, and Cabrits National Park; and the Soufriere–Scotts Head Marine Reserve that surrounds the Scotts Head Peninsula. It is important to note that the Cabrits National Park includes a 426-ha marine component. There are also large tracts of "Unallocated State Lands" in the Governor, La Guerre, Upper Layou, Morne Plaisance, and Fond Figues areas. Conservation in Dominica is implemented within the context of a number of statutes and pieces of legislation including the Forest Act, the National Parks and Protected Areas Act, and the Forestry and Wildlife Act. This latter act was amended in 1988 to make the Imperial Amazon *Amazona imperialis* and Red-necked Amazon *A. arausiaca* "specially protected birds". However, this existing legislation needs strengthening to maximise its impact, and the FWP Division is in need of capacity building and professional training in natural resource management to more effectively execute the legislation. The lack of a proper land-use policy for the country creates its own challenges for the land management and conservation programs being implemented.

In 1980, World Wildlife Fund funded the Forestry Division's research project to determine the status of the populations of the island's two parrot species following the devastation of the birds' habitats during the passage of Hurricane David in 1979. In 1989 RARE Centre assisted the Division with "Project Sisserou" which was designed to raise public awareness about the importance of the island's parrot species. The parrots have provided a focus for conservation action. There is currently an ongoing initiative between the Rare Species Conservatory Foundation (RSCF) and the Government of Dominica to support the Forestry Division's parrot research and conservation program. This initiative started in 1997 and has broadened its scope significantly since then, bringing new technology and techniques to the Dominica Parrot Research Program; support and assistance in the establishment of the Morne Diablotin National Park; management and care for the Dominica Parrot Captive Breeding Program at the Parrot Conservation and Research Centre; and continuity in the FWP Division's environmental protection program. RSCF has also been supporting the Division's efforts to celebrate the annual month-long Caribbean Endemic Birds Festival—an important regional environmental awareness initiative (of the Society for the Conservation and Study of Caribbean Birds).

In spite of the conservation legislation and protective measures in place, biodiversity is under pressure in Dominica. Habitat is being lost due to agricultural expansion, housing development and proliferation of quarrying activities, but it is the illegal clearance of forest for agricultural activities (including marijuana) in the island's interior and areas used by the parrots that is of particular concern. Mature gommier trees (*Dacryodes excelsa*) are slashed for illegal gum harvesting. This is one of the key tree species used by both parrots for food and nesting. Natural disasters (e.g. tropical storms, hurricanes and volcanic activity) are also a significant threat to habitat. Hurricanes in particular are a major threat to the parrot populations, their nest trees, and foraging areas. Hurricane David devastated Dominica in 1979, nearly extirpating *A. imperialis*, and reducing *A. arausiaca* to a fragment of its former range.

Imperial Amazon (left) and Red-necked Amazon (right). (PHOTOS: PAUL REILLO, RSCF/FWP DIVISION)

Female Blue-headed Hummingbird, endemic to Dominica and Martinique. (PHOTO: PAUL REILLO, RSCF/FWP DIVISION)

Black-capped Petrel found at Trafalgar in May 2007. (PHOTO: ARLINGTON JAMES, FWP DIVISION)

Table 1. Key bird species at Important Bird Areas in Dominica.

Key bird species	Criteria	National population	Dominica IBAs: DM001	DM002	DM003	DM004
Criteria			■ ■	■ ■	■ ■	■
White-tailed Tropicbird *Phaethon lepturus*	■	200			80	100
Magnificent Frigatebird *Fregata magnificens*	■	300				300
Brown Booby *Sula leucogaster*	■	400			250	150
Bridled Tern *Sterna anaethetus*	■	500			250	250
Brown Noddy *Anous stolidus*	■	650			300	350
Black Noddy *Anous minutus*	■	350				350
Red-necked Amazon *Amazona arausiaca*	VU ■ ■	850	500	200–500		
Imperial Amazon *Amazona imperialis*	EN ■ ■	100	50–80	<50		
Lesser Antillean Swift *Chaetura martinica*	■		✓	✓		
Purple-throated Carib *Eulampis jugularis*	■		✓	✓		
Green-throated Carib *Eulampis holosericeus*	■		✓	✓		
Antillean Crested Hummingbird *Orthorhyncus cristatus*	■		✓	✓	✓	
Blue-headed Hummingbird *Cyanophaia bicolor*	■		✓	✓		
Caribbean Elaenia *Elaenia martinica*	■		✓	✓	✓	
Lesser Antillean Pewee *Contopus latirostris*	■		✓	✓		
Lesser Antillean Flycatcher *Myiarchus oberi*	■		✓	✓	✓	
Scaly-breasted Thrasher *Margarops fuscus*	■		✓	✓	✓	
Pearly-eyed Thrasher *Margarops fuscatus*	■		✓	✓		
Brown Trembler *Cinclocerthia ruficauda*	■		✓	✓		
Rufous-throated Solitaire *Myadestes genibarbis*	■		✓	✓		
Forest Thrush *Cichlherminia lherminieri*	VU ■ ■	200	100	100		
Plumbeous Warbler *Dendroica plumbea*	■		✓	✓	✓	
Lesser Antillean Bullfinch *Loxigilla noctis*	■		✓	✓	✓	
Antillean Euphonia *Euphonia musica*	■		✓	✓	✓	
Lesser Antillean Saltator *Saltator albicollis*	■		✓	✓	✓	

All population figures = numbers of individuals.
Threatened birds: Endangered ■; Vulnerable ■. **Restricted-range birds** ■. **Congregatory birds** ■.

Birds

A total of 176 species of birds have been recorded for Dominica, of which about 66% are Neotropical migrants and 34% (62 species) are resident species. Nineteen of the 38 Lesser Antilles EBA restricted-range birds occur on the island, including the Blue-headed Hummingbird *Cyanophaia bicolor* which occurs just on Dominica and neighbouring Martinique, and Plumbeous Warbler *Dendroica plumbea* which is shared only with Guadeloupe. Most significantly though, Dominica supports two single-island endemic *Amazona* parrots—the only small island in the Caribbean to do so. The Imperial Amazon (or "Sisserou") *Amazona imperialis* is Dominica's national bird and is featured on the country's flag and Coat-of-Arms. It is also the largest of all *Amazona* parrots. It shares the island with the Red-necked Amazon (or "Jaco") *A. arausiaca*.

Both of Dominica's parrots are globally threatened (the Sisserou being Endangered and the Jaco Vulnerable) and promoted as conservation flagship species. In the past, hunting for food and the pet trade, loss of habitat through natural disasters (particularly hurricanes) and increasing agricultural expansion were the main threats to both species. In recent times *A. arausiaca* has been depredating on farmers' citrus and passion fruit crops in several localities on the island. As

A. imperialis is globally threatened and at risk from both environmental and anthropogenic factors, there is a need for regular monitoring and further research on its breeding biology, and to extend parrot research activities into remote areas within the Northern Forest Reserve, Morne Diablotin National Park and Morne Trois Pitons National Park. Dominica supports a third globally threatened bird species—the Vulnerable Forest Thrush *Cichlherminia lherminieri* (see Table 1 for threat categories and population sizes of the island's globally threatened birds), and may yet prove to be a breeding location for the Endangered Black-capped Petrel *Pterodroma hasitata*. Breeding of this species has not been proven, but a small flock was seen at sea in April 1984 (off the south-east coast) and there are reports of birds calling at night in the south of the island from the 1980s. Most recently (in May 2007) a bird was found just outside the Morne Trois Pitons National Park.

IMPORTANT BIRD AREAS

Dominica's four IBAs—the island's international priority sites for bird conservation—cover 106 km² (including marine areas), and about 13% of the islands' land area. The two forest IBAs are national parks (totalling c.9,845 ha), and thus 93% of the area covered by the island's IBAs is under protection. The two seabird/ marine IBAs are not formally protected.

The IBAs have been identified on the basis of 25 key bird species (listed in Table 1) that variously trigger the IBA criteria. These 25 species include three globally threatened birds, all 19 restricted-range species, and six congregatory seabirds. The two forested national park IBAs (Morne Diablotin and Morne Trois Pitons, IBAs DM001 and DM002) embrace populations of all the restricted-range species and the majority of the populations of all three globally threatened species (the two *Amazona* parrots and the Forest Thrush

Figure 1. Location of Important Bird Areas in Dominica.

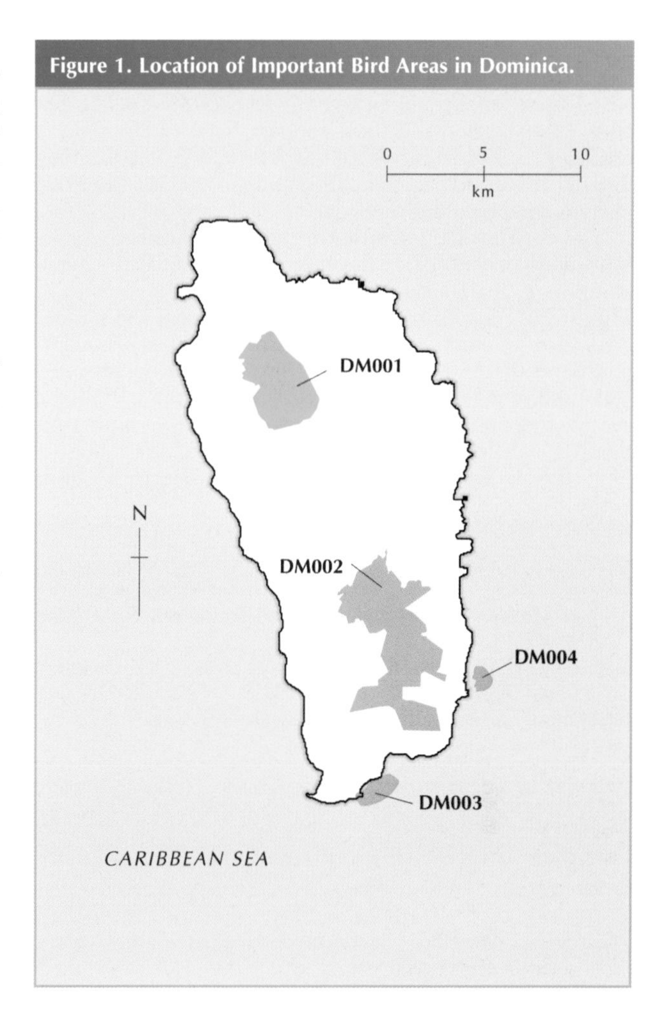

Morne Diablotin.
(PHOTO: PAUL REILLO, RSCF/FWP DIVISION)

Cichlherminia lherminieri), emphasising how critically important these two IBAs are for the maintenance of the island's biodiversity. Point Des Foux IBA (DM003) and L'Ilet IBA (DM004) between them support most of Dominica's breeding seabirds, but without any formal protection they remain vulnerable to poaching and potentially invasive mammalian predators.

The existing parrot monitoring program needs to be expanded to include field assessments (surveys and subsequent monitoring) for *C. lherminieri* and the seabird populations. Monitoring results should be used to inform the annual assessment of state, pressure and response scores at each of the island's IBAs to provide an objective status assessment and highlight management interventions that might be required to maintain these internationally important biodiversity sites.

KEY REFERENCES

Caribbean Conservation Association (1991) *Dominica: country environmental profile*. St Michael, Barbados: Caribbean Conservation Association

James, A. (2004) *Flora and fauna of Cabrits National Park, Dominica*. Roseau, Dominica: Forestry, Wildlife and Parks Division (Ministry of Agriculture and the Environment).

James, A., Durand, S., and Jno. Baptiste, B. (2005) *Dominica's birds*. Roseau, Dominica: Forestry, Wildlife and Parks Division (Ministry of Agriculture and the Environment).

Lack, A. J., Whitefoord, C., Evans, P. G. H and James, A. (1997) *Dominica, nature island of the Caribbean: illustrated flora*. Roseau, Dominica: Ministry of Tourism.

Raffaele, H. Wiley J., Garrido, O., Keith, A. and Raffaele, J. (1998) *A guide to the birds of the West Indies*. Princeton, New Jersey: Princeton University Press.

Wiley, J. W., Gnam, R., Koenig, S. E., Dornelly, A., Galvez, X., Bradley, P. E., White, T., Zamore, M., Reillo, P. R. and Anthony, D. (2004) Status and conservation of the family Psittacidae in the West Indies. *J. Carib. Orn.* 17: 94–154.

Zamore, M. P. (2000) *The wildlife of Dominica*. Roseau, Dominica: Forestry, Wildlife and Parks Division (Ministry of Agriculture and the Environment).

ACKNOWLEDGEMENTS

The authors would like to thank Arlington James and David Williams (Forestry, Wildlife and Parks Division), and Paul Reillo (Rare Species Conservatory Foundation) for their input in reviewing this chapter.

DM001 Morne Diablotin National Park

National Park

COORDINATES 15°30'N 61°23'W
ADMIN REGION St Andrew, St John, St Joseph, St Peter
AREA 3,360 ha
ALTITUDE 579–1,447 m
HABITAT Tropical forest

Imperial Amazon

THREATENED BIRDS 3
RESTRICTED-RANGE BIRDS 19
BIOME-RESTRICTED BIRDS
CONGREGATORY BIRDS

■ Site description

Morne Diablotin National Park IBA is located within one of the northern mountain ranges and boasts the highest mountain peak—Morne Diablotin—on Dominica. The IBA is 95% forested, and bounded on the north-east, east and south sides by the Northern Forest Reserve. Privately-owned agricultural lands are found on the north- and south-western boundaries of the park. The nearest residential area is 10 km distant on the west coast.

■ Birds

This IBA is important, and indeed the national park was established for the Endangered Imperial Amazon *Amazona imperialis* and the Vulnerable Red-necked Amazon *A. arausiaca*. The majority of the world population of *A. imperialis* is found within this IBA with 50–80 birds currently present, and it also supports the largest population of *A. arausiaca*. A healthy population of the Vulnerable Forest Thrush *Cichlherminia lherminieri* also occurs. Populations of all 19 Lesser Antilles EBA restricted-range birds are supported within this IBA.

■ Other biodiversity

The Endangered endemic Dominican tink frog *Eleutherodactylus amplinympha* occurs, as does the Vulnerable Lesser Antillean iguana *Iguana delacatissima*, the endemic Dominica anole *Anolis oculatus* and the regionally endemic least gecko *Sphaerodactylus vincenti*. The IBA also supports populations of the endemic subspecies of agouti *Dasyprocta leporina* and opossum *Didelphys marsupialis insularis*. The endemic boa *Boa* (*constrictor*) *nebulosa* may warrant full species status.

■ Conservation

Morne Diablotin National Park was established in 2000 to protect the critically important populations of the two endemic *Amazona* parrots. The IBA is mostly state owned, with some private land holdings (used for agriculture) on the western side. The Northern Forest Reserve abuts the IBA, and the Central Forest Reserve is in close proximity, both providing a buffer to some potential threats. Unauthorised occupation of some areas in the interior for illegal agricultural farming activities leads to some deforestation, including the felling of tree species utilised by both parrot species for nesting and foraging. These activities are also occurring on private lands in close proximity to the IBA. Illegal hunting for feral pigs *Sus scrofa* can lead to the discovery of parrot nests and subsequent poaching activities, and is thus a major concern. Hurricanes and tropical storms can impact nest trees/ cavities and food sources for both parrot species.

DM002 Morne Trois Pitons National Park

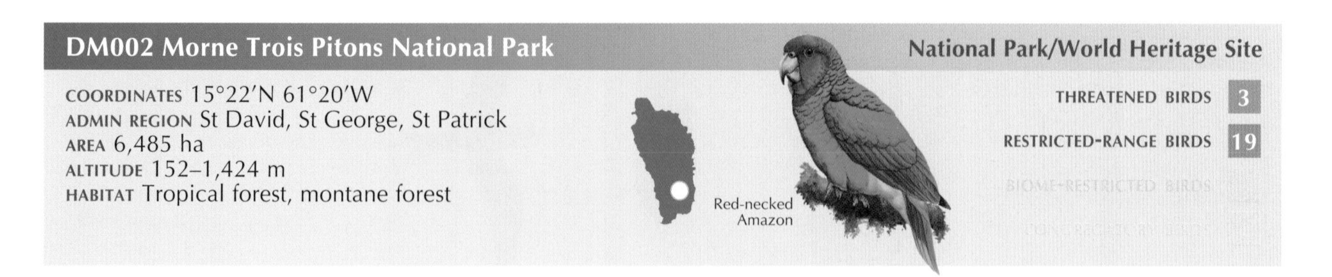

■ Site description

The Morne Trois Pitons National Park IBA is in the highlands of southern Dominica. The IBA rises from the lowlands in the south-east of the island towards the three prominent peaks of Morne Trois Pitons. The IBA, Dominica's largest national park, comprises some of the best remaining examples of volcanic island ecosystems in the Caribbean with active volcanic and geothermal areas (e.g. Boiling Lake and the Valley of Desolation), mountains, lakes and clear streams (Emerald Pool, Freshwater Lake, Middleham falls), tropical forest and elfin forest on the highest peaks. Communities close to the west, south-west and south-east boundaries of the IBA include Eggleston, Cochrane, Grand Fond, Laudat, Giraudel, Petite Savanne and Bagatelle.

■ Birds

This IBA is important for the Endangered Imperial Amazon *Amazona imperialis* and the Vulnerable Red-necked Amazon *A. arausiaca*. It supports the smaller of the two known populations of these parrots, with <50 *A. imperialis* and 200–500 *A. arausiaca*, but is nevertheless critical the survival of both. A healthy population of the Vulnerable Forest Thrush *Cichlherminia lherminieri* also occurs. Populations of all 19 Lesser Antilles EBA restricted-range birds are supported within this IBA. In May 2007, an Endangered Black-capped Petrel *Pterodroma hasitata* was found in the village of Trafalgar, just 3 km from the boundaries of the park.

■ Other biodiversity

The Endangered endemic Dominican tink frog *Eleutherodactylus amplinympha* occurs, as do the endemic Dominica anolis *Anolis oculatus* and bromeliad *Pitcairnia micotrinensis*. The IBA also supports populations of the endemic subspecies of agouti *Dasyprocta leporina* and opossum *Didelphys marsupialis insularis*. The endemic boa *Boa* (*constrictor*) *nebulosa* may warrant full species status.

■ Conservation

Morne Trois Pitons National Park IBA is the largest of Dominica's three national parks. It is state owned and was established as a UNESCO World Heritage Site in 1997. It includes the 375-ha Archbold Preserve (donated by John D. Archbold). Unauthorised occupation of some areas for illegal agricultural farming activities leads to some deforestation, and if allowed to continue, will impact the populations of both species of globally threatened parrot. Agricultural expansion is of great concern for the long-term viability of the parrots. Volcanic activity is an ever present threat to the IBA and the birds it supports.

DM003 Point Des Foux

■ Site description

Point Des Fous IBA is at the southernmost tip of Dominica, on the Atlantic coast between the villages of Grand Bay and Scotts Head. The IBA embraces the rocky sea cliffs at Point Des Foux, the coastline to the north-east, some of the wooded areas inland from the coast on the cliff tops, and marine areas up to 1 km from Point Des Foux. It is an area of steep and rugged terrain, and is in close proximity to Petite Coulibri Estate and Morne Fous Estate. Residential areas are c.2 km to the north and west of the site.

■ Birds

This IBA is important as a breeding site for six species of seabird including regionally important populations of Brown Booby *Sula leucogaster*, Brown Noddy *Anous stolidus*, Bridled Tern *Sterna anaethetus* and White-tailed Tropicbird *Phaethon lepturus*. This site and L'Ilet IBA (DM004) represent the two main seabird breeding colonies in the country. Populations of eight (of the 19) Lesser Antilles EBA restricted-range birds occur at this IBA including Plumbeus Warbler *Dendroica plumbea*.

■ Other biodiversity

The Vulnerable Lesser Antillean iguana *Iguana delacatissima* occurs, as do the endemic Dominica anolis *Anolis oculatus*, Dominica ameiva *Ameiva fuscata* and boa *Boa* (*constrictor*) *nebulosa* (which may warrant full species status).

■ Conservation

Point Des Foux IBA is privately owned and not formally protected. The Soufriere–Scotts Head Marine Reserve is c.3 km to the east of the IBA. The steep terrain has restricted human activity to a large extent. However, illegal hunting for some bird species and their eggs is a major concern. Facing the Atlantic, the breeding seabirds in this IBA are vulnerable to the impacts of tropical storms and hurricanes.

■ Site description

L'Ilet IBA is a small (c.200 m^2) islet, c.200 m from shore to the south of the village of Boetica on the south-east, Atlantic coast of Dominica. The islet is rocky with vegetation covering the top. The IBA includes all marine areas up to 1 km from the islet. The Morne Trois Pitons National Park IBA (DM002) is just 5 km from L'Ilet.

■ Birds

This IBA is globally significant for its population of Black Noddy *Anous minutus*, and regionally important as a breeding site for six other species of seabird, namely Magnificent Frigatebird *Fregata magnificens*, Brown Booby *Sula leucogaster*, Brown Noddy *A. stolidus*, Bridled Tern *Sterna anaethetus* and White-tailed Tropicbird *Phaethon lepturus*. This site and Point Des Foux IBA (DM003) represent the two main seabird breeding colonies in the country. A number of land birds also occur including the restricted-range Antillean Crested Hummingbird *Orthorhyncus cristatus*, Green -throated Carib *Eulampis holosericeus* and Lesser Antillean Bullfinch *Loxigilla noctis*.

■ Other biodiversity

The flora and fauna recorded for the island does not include any endemic or threatened species.

■ Conservation

L'Ilet IBA is state owned but unprotected. The nearest residential area is c.3 km from the islet. This, and the fact that access is only by swimming or boat, has limited the human activity at the site and therefore disturbance to the breeding seabirds. However, illegal hunting for some bird species and their eggs is a major concern. Being situated on the exposed Atlantic coast, L'Ilet is vulnerable to high winds associated with hurricanes and tropical storms, and to being over-washed by waves and storm-surges.

DOMINICAN REPUBLIC

LAND AREA 48,730 km² ALTITUDE -40–3,087 m
HUMAN POPULATION 9,365,800 CAPITAL Santo Domingo
IMPORTANT BIRD AREAS 21, totalling 7,416 km²
IMPORTANT BIRD AREA PROTECTION 14%
BIRD SPECIES 306
THREATENED BIRDS 23 RESTRICTED-RANGE BIRDS 34

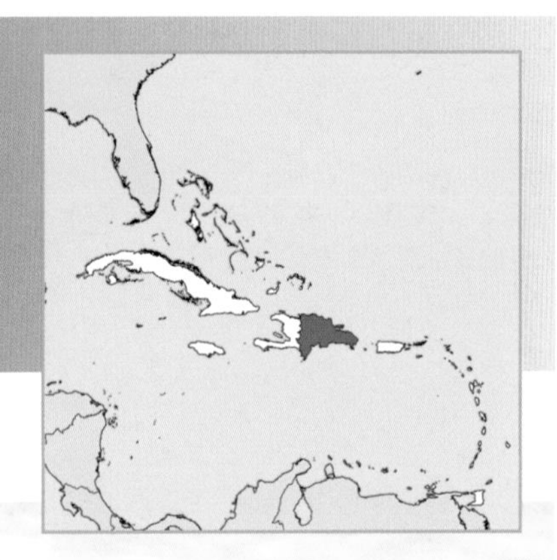

LAURA PERDOMO AND YVONNE ARIAS
(GRUPO JARAGUA)

Sierra Martín García IBA—a poorly studied area in need of conservation action.
(PHOTO: RICARDO BRIONES)

INTRODUCTION

The Dominican Republic occupies the eastern two-thirds of the island of Hispaniola which, at 77,900 km², is the second largest island in the Caribbean. The Republic of Haiti, with which the Dominican Republic shares a 360-km border, occupies the rest of Hispaniola. The island lies 80 km east of Cuba, 90 km west of Puerto Rico, and 150 km from north-eastern Jamaica. The topography of the Dominican Republic is dominated by four principal mountain systems that run from north-west to south-east, namely the Cordillera Septentrional (Northern Mountain Range); the Cordillera Central (Central Mountain Range), which extends into Haiti where it is called the Massif du Nord; the Sierra de Neiba, which extends into Haiti as the Montagnes du Trou d´Eau; and Sierra de Bahoruco, which in Haiti continues as the Massif de la Selle. These parallel mountain ranges are responsible for the longest and most voluminous rivers in the Caribbean: Yaque del Norte, Yaque del Sur, Yuna-Camú, and Nizao. The country also contains the largest number of lakes and lagoons in the insular Caribbean, including Lago Enriquillo, the largest body of still-water in the region. The diverse habitats (five distinct ecoregions are recognised) include 1,500 km of coastline, freshwater and brackish wetlands, dry forest, broadleaf forest and pine forest, xeric areas, savannas, and dunes. The climate is warm, with a mean annual temperature of 27ºC.

Hispaniola is considered to have the highest biodiversity in the Caribbean, distributed across an intricate mosaic of environments and microclimates that have formed as a result of a complex geological history. This has produced sites which range from 40 m below sea- level (e.g. Hoya de Enriquillo), to those at more than 3,000 m above (within the Cordillera Central), as well as sites such as Isla Alto Velo that supports unique species confined to only 1 km². Rates of endemism across most taxonomic groups are high. However, most habitats (but especially cloud forest and moist broadleaf forest) that support these endemic species have been (and continue to be) severely affected by deforestation and other human pressure. With the growing human population concentrated in coastal regions, habitats in these areas (e.g. beaches, coastal wetlands and mangroves) are suffering from multiple threats. Not only are the habitats being destroyed directly by cutting forests, draining or polluting wetlands and urban and agricultural expansion, but invasive alien species (including plants, predatory and grazing animals) are impacting what habitat remains and the species that rely upon them.

The Vulnerable Hispaniolan Amazon and Hispaniolan Parakeet are both captured for the illegal pet trade. (PHOTO: LANCE WOOLAVER)

■ Conservation

The Environment and Natural Resources General Law (No.64-00) is the legal framework that protects wild areas and biodiversity in the Dominican Republic. This law allows for the creation of sector-specific laws such as the Law of Protected Areas (No.202-04) which regulates the National System of Protected Areas (NSPA). However, attempts in 2004–2005 by the president and the legislature to eviscerate the national parks system by selling-off protected lands for tourism and development activities shows how fragile the parks are from a legal standpoint. In response, conservation and academic groups are working towards modifying the nation's constitution to declare the National System of Protected Areas as inalienable, non-sequestrable national treasures, and not subject to statutory limitations.

The National System of Protected Areas has improved in terms of the quantity of protected areas and their management categories over the last 20 years. In 1980, only nine areas (4.2% of the country's land area) were legally protected, but this number increased to 19 (11.2%) from 1981 to 1990, and to 86 areas (25.4%) between 2002 and 2008. The Jaragua-Bahoruco-Enriquillo Biosphere Reserve is unique in the country, embracing a number of protected areas and the Lago Enriquillo Ramsar Site. The Directorate of Protected Areas administers the management of protected area system, although a number of NGOs collaborate with or have been assigned to protected areas (e.g. Grupo Jaragua to Jaragua National Park, and Fundación Moscoso Puello to Valle Nuevo) under co-management agreements with the Directorate. However, in spite of this enlightened approach to management, only 10 of the Dominican Republic's national parks have management plans, and for only six is there some level of implementation.

The protected areas of the Dominican Republic face multiple threats to the effective conservation of their biodiversity. These include uncertainties in land ownership, the lack of an appropriate system of compensation for the expropriation of land for conservation purposes, lack of clear policies for the administration and management of funds generated by protected areas, inadequate management of the areas, as well as delayed local development as a result of centralised policies. Knowledge of regulations and permitted uses of the protected areas is lacking, as is a general awareness of their importance, value and the ecological services they provide. Together with imprecise boundaries, these deficiencies lead to disturbances such as expansion of agricultural activities (including cattle grazing), as well as forest fires, deforestation, illegal hunting, fishing, and trafficking of endangered species. Other threats relate to the expansion of unsustainable tourism,

Slash-and-burn agriculture in Los Haitises IBA is threatening the integrity of this critical ecosystem. (PHOTO: LANCE WOOLAVER)

mining, and hydro-electric projects. Finally, poverty levels in communities adjacent to the parks have led to unsustainable land-use practices and illegal human settlement both within the protected areas and their buffer zones. The threats faced by the nation's protected areas (many of which are IBAs) are indicative of what is happening to biodiversity across the country.

■ Birds

Of the 306 bird species reported for the Dominican Republic c.140 are breeding residents. Hispaniola is also an important over-wintering area for Neotropical migrants, with 136 species recorded. The Hispaniolan avifauna exhibits exceptional levels of endemism. The island is an Endemic Bird Area (EBA) with 36 restricted-range species, 34 of which are known from the Dominican Republic. The remaining two species, Grey-crowned Palm-tanager *Phaenicophilus poliocephalus* and Thick-billed Vireo *Vireo crassirostris*, have only been recorded in Haiti. A total of 28 of the restricted-range birds are endemic to the island, the others being shared with adjacent EBAs. For example, Vervain Hummingbird *Mellisuga minina*, Stolid Flycatcher *Myiarchus stolidus*, Greater Antillean Elaenia *Elaenia fallax* and Golden Swallow *Tachycineta euchrysea* are all shared with Jamaica. Six of the restricted-range species represent genera endemic to Hispaniola, namely *Calyptophilus*, *Dulus* (also a monotypic family), *Microligea*, *Nesoctites*, *Phaenicophilus* and *Xenoligea*. Endemism is also high at the sub-specific level with over 35 subspecies described from Dominican Republic.

There are records of 23 threatened species in the Dominican Republic, including one Critically Endangered, four Endangered, nine Vulnerable, and nine Near Threatened species. However, three of the Near Threatened species were excluded from the IBA analysis since they are not considered to sustain significant populations in the country, namely Back Rail *Laterallus jamaicensis*, Piping Plover *Charadrius melodus* and Golden-winged Warbler *Vermivora chrysoptera*. The threat category and national population sizes (where known) of the globally threatened birds are listed in Table 1. The Critically Endangered Ridgway's Hawk *Buteo ridgwayi* is now confined to Los Haitises IBA (DO018) where the small population is declining. The Endangered Black-capped Petrel *Pterodroma hasitata* maintains a small breeding colony in the Sierra de Bahoruco IBA (DO006), and this IBA also supports critical populations of the other Endangered species, namely Hispaniolan Crossbill *Loxia megaplaga*, La Selle Thrush *Turdus swalesi* and Bay-breasted Cuckoo *Coccyzus rufigularis*. Many of the globally threatened birds are restricted to the high altitude broadleaf and pine forests.

The Dominican Republic is also important for large breeding and wintering populations of waterbirds and seabirds. Laguna Limón (DO019) and Laguna Cabral (DO008) support the largest reported population of the Near Threatened Caribbean Coot *Fulica caribaea*, with up to 6,000 and 3,000 individuals, respectively. Laguna Cabral is also home to some of the largest wintering concentrations of ducks in the Caribbean with up to 160,000 individuals (of various species) reported. Seabirds are primarily concentrated on the satellite islands around the Dominican Republic's coast. They are relatively poorly known in terms of colony status and size, but the Sooty Tern *Sterna fuscata* colony (of 80,000 pairs) on Isla Alto Velo is one of the largest in the Caribbean. Monitoring of the other known breeding islands would provide valuable information that may result in new IBAs being defined.

IMPORTANT BIRD AREAS

Dominican Republic contains 21 IBAs—the country's international site priorities for bird conservation—cover c.14% of the land surface of the country. The IBAs have been identified on the basis of 49 key bird species (listed in Table 1) that variously meet the IBA criteria. These species include

The Critically Endangered Ridgway's Hawk is confined to Los Haitises IBA where the population is declining. (PHOTO: ELADIO FERNÁNDEZ)

A critical population of the Endangered Hispaniolan Crossbill occurs within the Sierra de Bahoruco IBA. (PHOTO: ELADIO FERNÁNDEZ)

Table 1. Key bird species at Important Bird Areas in the Dominican Republic.

Key bird species	Criteria			National population	DO001	DO002	DO003	DO004	DO005	DO006	DO007
Criteria						■	■	■	■	■	■
						■	■	■	■	■	■
					■					■	■
West Indian Whistling-duck *Dendrocygna arborea*	VU ■								✓		✓
Lesser Scaup *Aythya affinis*			■	90,000							
Ruddy Duck *Oxyura jamaicensis*			■	10,000							
Black-capped Petrel *Pterodroma hasitata*	EN ■		■	60–120						60–120	
Magnificent Frigatebird *Fregata magnificens*			■	250–999							
Brown Pelican *Pelecanus occidentalis*			■	250–999							250
Ridgway's Hawk *Buteo ridgwayi*	CR ■	■		240–360							
Caribbean Coot *Fulica caribaea*	NT ■		■	2,500–9,999							
Least Tern *Sterna antillarum*			■	250–999							300
Bridled Tern *Sterna anaethetus*			■	1,000–2,499	1,000–2,499						
Sooty Tern *Sterna fuscata*			■	130,000							130,000
Brown Noddy *Anous stolidus*			■		50–380						
White-crowned Pigeon *Patagioenas leucocephala*	NT ■			2,500–9,999							✓
Plain Pigeon *Patagioenas inornata*	NT ■										
Grey-headed Quail-dove *Geotrygon caniceps*	VU ■										
Hispaniolan Parakeet *Aratinga chloroptera*	VU ■	■				✓	30			✓	30
Hispaniolan Amazon *Amazona ventralis*	VU ■	■				✓	30	30	✓	✓	30
Hispaniolan Lizard-cuckoo *Saurothera longirostris*		■				✓	✓	✓	✓	✓	✓
Bay-breasted Cuckoo *Coccyzus rufigularis*	EN ■	■		2,500–9,999		✓	✓	✓		33	
Ashy-faced Owl *Tyto glaucops*		■				✓	✓		✓	✓	✓
Least Pauraque *Siphonorhis brewsteri*	NT ■	■		10,000–19,999			✓	✓	✓	✓	✓
Hispaniola Nightjar *Caprimulgus ekmani*		■					✓	✓		✓	
Antillean Mango *Anthracothorax dominicus*		■				✓	✓	✓	✓	✓	✓
Hispaniolan Emerald *Chlorostilbon swainsonii*		■				✓	✓	✓		✓	
Vervain Hummingbird *Mellisuga minima*		■				✓	✓	✓	✓	✓	✓
Hispaniolan Trogon *Priotelus roseigaster*	NT ■	■				✓	✓	✓		30	
Narrow-billed Tody *Todus angustirostris*		■				✓	✓	✓		✓	
Broad-billed Tody *Todus subulatus*		■				✓	✓	✓		✓	✓
Antillean Piculet *Nesoctites micromegas*		■				✓	✓	✓		✓	✓
Hispaniolan Woodpecker *Melanerpes striatus*		■				✓	✓	✓	✓	✓	✓
Greater Antillean Elaenia *Elaenia fallax*		■					✓	✓		✓	
Hispaniolan Pewee *Contopus hispaniolensis*		■				✓	✓	✓	✓	✓	✓
Stolid Flycatcher *Myiarchus stolidus*		■				✓	✓	✓	✓	✓	✓
Flat-billed Vireo *Vireo nanus*		■						✓	✓	✓	✓
Hispaniolan Palm Crow *Corvus palmarum*	NT ■	■				50–249	90	✓	90	90	
White-necked Crow *Corvus leucognaphalus*	VU ■	■				✓	✓	✓	30	30	30
Palmchat *Dulus dominicus*		■				✓	✓	✓	✓	✓	✓
Golden Swallow *Tachycineta euchrysea*	VU ■	■				✓	✓	36		30	
Pearly-eyed Thrasher *Margarops fuscatus*		■									✓
Rufous-throated Solitaire *Myadestes genibarbis*		■				✓	✓	✓		✓	
Bicknell's Thrush *Catharus bicknelli*	VU ■					✓	30			✓	
La Selle Thrush *Turdus swalesi*	EN ■	■					✓	✓		✓	
Antillean Siskin *Carduelis dominicensis*		■					✓	✓		✓	
Hispaniolan Crossbill *Loxia megaplaga*	EN ■	■		3,100–3,500			50–249			3,000	
Green-tailed Warbler *Microligea palustris*		■				✓	✓	✓		✓	✓
White-winged Warbler *Xenoligea montana*	VU ■	■				✓	✓	30		30	
Black-crowned Palm-tanager *Phaenicophilus palmarum*		■				✓	✓	✓	✓	✓	✓
Chat Tanager *Calyptophilus frugivorus*	VU ■	■				✓	✓	✓		30	
Antillean Euphonia *Euphonia musica*		■				✓	✓	✓		✓	

All population figures = numbers of individuals.
Threatened birds: Critically Endangered ■; Endangered ■; Vulnerable ■; Near Threatened ■. **Restricted-range birds** ■. **Congregatory birds** ■.

Dominican Republic IBAs												
009	DO010	DO011	DO012	DO013	DO014	DO015	DO016	DO017	DO018	DO019	DO020	DO021
■	■	■	■	■	■	■	■	■	■	■	■	■
■	■	■	■	■		■	■		■		■	■
										■	■	
								✓				
											540–600	
									240–360			
										1,000–6,000		
											2,500–9,999	
									30			
30			30									
30		✓		✓		✓						
30	30	✓		✓		✓			30		30	✓
✓		✓	✓	✓		✓	✓		✓		✓	✓
✓	✓	✓	✓	✓		✓	✓		✓		✓	✓
✓	✓								✓		✓	
							✓		✓		✓	
✓	✓	✓	✓	✓		✓	✓		✓		✓	✓
✓	✓	✓	✓	✓		✓	✓					
✓	✓	✓	✓	✓		✓	✓		✓		✓	✓
30	✓	✓	30	✓		✓						
✓	✓	✓	✓	✓		✓	✓		✓			✓
✓	✓	✓	✓	✓		✓	✓		✓		✓	
✓	✓	✓	✓	✓		✓	✓		✓		✓	
✓	✓	✓	✓	✓		✓					✓	✓
✓	✓	✓	✓	✓								
✓	✓	✓	✓	✓		✓	✓		✓		✓	✓
✓	✓	✓	✓	✓		✓	✓		✓		✓	✓
✓	✓		✓						✓		✓	✓
✓	✓	✓	90			✓		90				
✓									250–999		✓	
✓	✓	✓	✓	✓		✓	✓		✓		✓	✓
✓		30	✓									
											✓	✓
✓	✓	✓	✓	✓		✓			✓			
	30	30		✓	250–999	✓			✓		✓	
		✓				✓						
✓		✓	✓									
		50–249										
	✓	✓	✓			✓	✓		✓		✓	
	✓	✓	✓									
✓	✓	✓	✓	✓		✓	✓		✓		✓	✓
30	30	30	30				✓					
✓	✓	✓	✓	✓			✓		✓			

Cayos Siete Hermanos IBA supports globally significant numbers of breeding seabirds. (PHOTO: RICARDO BRIONES)

20 threatened birds (see "Birds" above), all 34 restricted-range species, and 10 congregatory species. Of the 21 IBA identified, 20 support critical populations of globally threatened birds; 17 are home to important assemblages of restricted-range species; five support globally significant populations of congregatory waterbirds or seabirds; and four are important for congregatory birds at a regional level.

All but two IBAs belong partially or totally to the National System of Protected Areas, and thus are formally protected under a recognised management category. IBAs at Punta Cana (DO021) and Honduras (DO016) lack any type of formal protection, whilst Loma Nalga de Maco–Río Limpio IBA (DO002) protected in part as a national park. The majority of the country's life zones, habitats and vegetative associations are represented within the IBA network. Some of the IBA are recognised under other international designations, such as the Lago Enriquillo Ramsar Site, the Jaragua-Bahoruco-Enriquillo Biosphere Reserve, and Los Haitises and Sierra de Bahoruco as Alliance for Zero Extinction (AZE) sites.

Grupo Jaragua has been coordinating the IBA program in the Dominican Republic since 2002. Grupo Jaragua is a non-governmental, non-profit organisation established in 1987 and whose mission is to bring about the effective management of Dominican Republic's natural resources and biodiversity through research and projects aimed at solving local conservation problems. Despite most of their efforts being concentrated in the Jaragua National Park and surrounding communities, Grupo Jaragua pays special attention to the development of the Jaragua-Bahoruco-Enriquillo Biosphere Reserve (which embraces three IBAs) through community participation projects. The participation of multiple key actors from government institutions, the private sector, non-governmental organisations, community organisations, international cooperation agencies, and interested individuals has enabled the successful development and implementation of the IBA program in the country, as well as the achievement of local and national capacity building. The documentation of the Dominican Republic's IBAs represents a significant step in the program and will allow the development of more complete conservation agendas for these sites.

State, pressure and response scores have been collated for 11 (DO001 Cayos Siete Hermanos, DO003 Armando Bermudez National Park, DO005 Lago Enriquillo, DO006 Sierra de Bahoruco, DO007 Jaragua National Park, DO008 Laguna Cabral, DO011 Valle Nuevo, DO013 Loma Quita Espuela, DO018 Los Haitises, DO019 Laguna Limón and DO020 Del Este National Park) of the Dominican Republic's IBAs, but should be monitored annually at all IBAs to provide an objective status assessment and highlight management interventions that might be required to maintain these

Valle Nuevo IBA is one of 11 sites monitored for threats to IBAs ("Pressure"), the condition of IBAs ("State") and conservation actions taken at IBAs ("Response"). (PHOTO: RICARDO BRIONES)

Figure 1. Location of Important Bird Areas in the Dominican Republic.

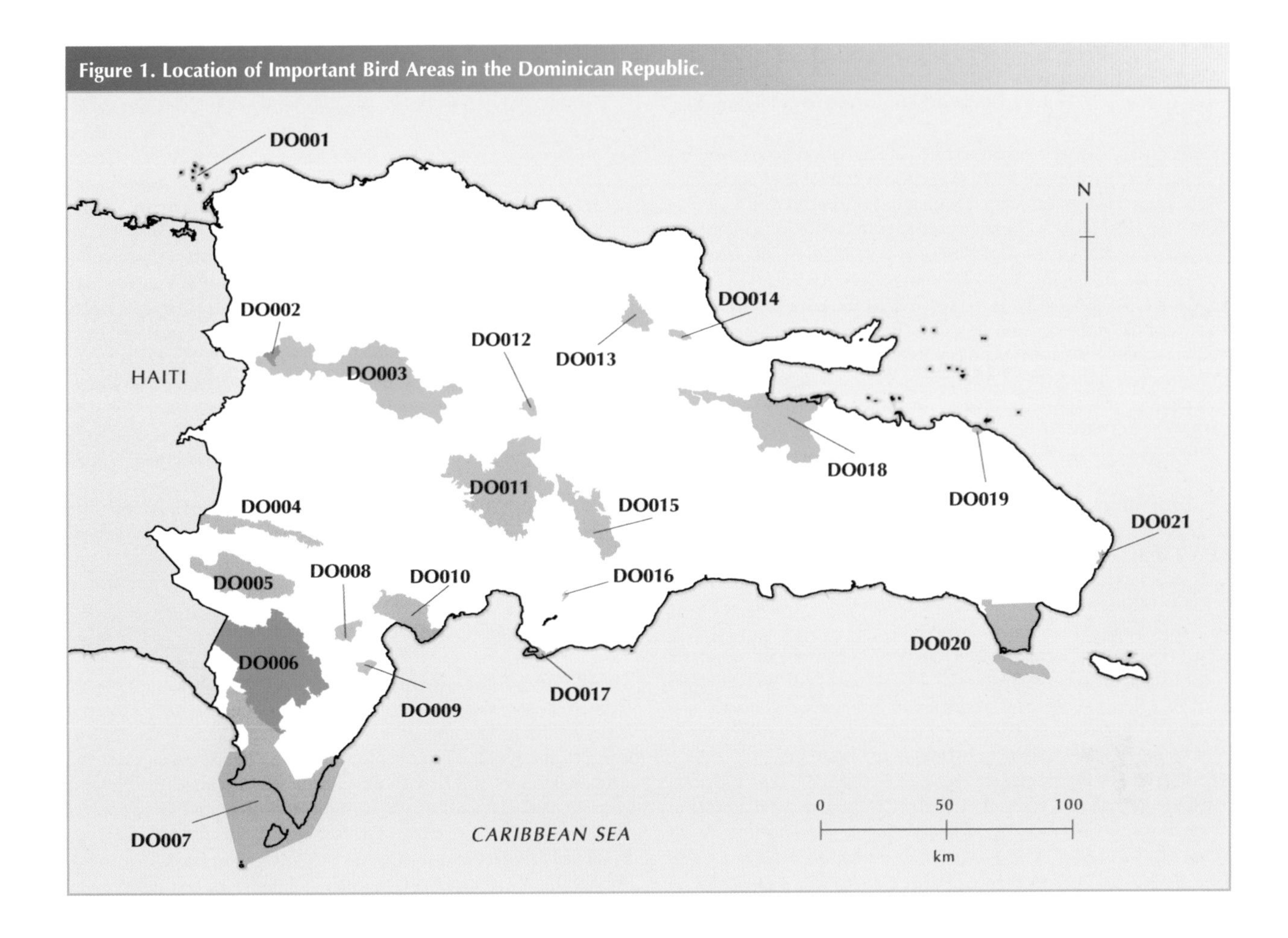

internationally important biodiversity sites. Monitoring of the country's globally threatened birds (especially the Critically Endangered and Endangered species), waterbirds and seabirds is urgently needed, and can usefully inform the annual status assessment of the IBA network.

KEY REFERENCES

Abreu, D. and Guerrero, K. eds. (1997) Evaluación ecológica integral: Parque Nacional del Este, República Dominicana, Tomo 1: *Recursos Terrestres*. Nassau, Bahamas: Media Publishing Ltd. and The Nature Conservancy.

Álvarez, V. (1998) Reconocimiento ecológico y diversidad biológica de los manglares de la Provincia Montecristi. Informe Final. In: Informe Final Subcontrato Biología Marina Parque Nacional Montecristi. Proyecto Biodiversidad CIBIMA/GEF-PNUD/ONAPLAN. Conservación y Manejo de la Biodiversidad en la Zona Costera de la República Dominicana.

BirdLife International and Grupo Jaragua (2005) Informe del II Taller Nacional de Identificación de Áreas Importantes para la Conservación de las Aves. 24 de febrero del 2005. Programa Áreas Importantes para la Conservación de las Aves en República Dominicana (Unpublished report).

BirdLife International and Grupo Jaragua (2006) Listado de aves registradas en las AICAs según la Base de Datos de las Aves del Mundo. (Unpublished report).

Bolay, E. (1997) *The Dominican Republic: a country between rain forest and desert; contributions to the ecology of a Caribbean island.* Weikersheim, Germany: Margraf Verlag.

CIBIMA (1998) = Centro de Investigaciones de Biología Marina (1998) Informe Final Subcontrato Biología Marina Parque Nacional Montecristi. Doc 1/4 Síntesis. Proyecto Biodiversidad CIBIMA/GEF-PNUD/ONAPLAN. Conservación y Manejo de la Biodiversidad en la Zona Costera de la República Dominicana. República Dominicana.

Ducks Unlimited, Inc. (2007) Duck count report for Dominican Republic. Data retrieved for: 2000–2006 from http://www.ducks.org/Conservation/International/1345/DominicanRepublicWaterfowlSurveys.htmlUCN.

García, G. G. (2007) Especie nueva Pozonia (Araneae: Araneidae) para República Dominicana. Museo Nacional de Historia Natural de Cuba. *Revista Soledonón.* 6: 41–44.

González, J. (2001) Informe de la vegetación Laguna Limón, Miches. Santo Domingo, República Dominicana: Departamento de Horticultura, Jardín Botánico Nacional. (Unpublished Report).

Hispaniolan Ornithological Society (2006) Listado de las aves observadas en Áreas Importantes para la Conservación de las Aves de República Dominicana. (Unpublished data compiled by S. Brauning and J. Brocca).

Hoppe, J. (1989) *Los parques nacionales de la República Dominicana/the national parks of the Dominican Republic.* Santo Domingo, Dominican Republic: Editora Corripio. Santo Domingo, República Dominicana. (Colección Barceló 1).

Keith, A. (in press) Hispaniola: Haiti and Dominican Republic, and including Navassa Island (U.S.). In Bradley P. E. and Norton, R. L. eds. *Breeding seabirds of the Caribbean.* Gainesville, Florida: Univ. Florida Press.

Keith, A. R., Wiley, J. W., Latta, S. and Ottenwalder, J. (2003) *The birds of Hispaniola: Haiti and the Dominican Republic.* Tring, U.K.: British Ornithologists' Union (BOU Check-list No 21).

Latta, C., Rimmer, C. C. and McFarland, K. P. (2003) Winter bird communities in four habitats along an elevational gradient on Hispaniola. *Condor* 105: 179–197.

Latta, S., Rimmer, C., Keith, A., Wiley, J., Raffaele, H., McFarland, K. and Fernandez, E. (2006) *Birds of the Dominican Republic and Haiti.* Princeton, New Jersey: Princeton University Press.

Nuñez, F., Ramírez, N., McPherson, M. and Portorreal, F. (2006) *Plan de conservación del Parque Nacional Juan Bautista Pérez Rancier (Valle Nuevo).* Santo Domingo, República Dominicana: Editora Amigo del Hogar.

SCOLUM, M., MITCHELL AIDE, T., ZIMMERMAN, J. K. AND NAVARRO, L. (2000) La vegetación leñosa en helechales y bosques de ribera en la Reserva Científica Ébano Verde, República Dominicana. *Moscosoa* 11: 38–56.

SEMARENA (2004) = Secretaría de Estado de Medio Ambiente y Recursos Naturales (2004) *Programa Nacional de Valorización de Áreas Protegidas*. Santo Domingo: Editora Búho.

SEMARENA (2004) = Secretaría de Estado de Medio Ambiente y Recursos Naturales (2004) *Reserva de la Biosfera Jaragua-Bahoruco-Enriquillo*. Santo Domingo: SEMARENA.

SEMARENA/SAP (2005) = Secretaría de Estado de Medio Ambiente, Subsecretaría de Áreas Protegidas y Biodiversidad, Dirección de Áreas Protegidas y Recursos Naturales (2005) *Plan de manejo del Parque Nacional Sierra de Bahoruco*. Santo Domingo: SEMARENA/SAP.

SEA/DVS (1990) = Secretaría de Estado de Agricultura, Subsecretaría de Estado de Recursos Naturales, Departamento de Vida Silvestre (1990) *Evaluación de los recursos naturales en la Sierra Martín García y Bahía de Neiba*. Santo Domingo: SEA/DVS.

SEA/DVS (1992) = Secretaría de Estado de Agricultura, Subsecretaría de Estado de Recursos Naturales, Departamento de Vida Silvestre (1992) *Estudio preliminar sobre la fauna de Ébano Verde*. Santo Domingo: SEA/DVS.

SEA/DVS (1992) = Secretaría de Estado de Agricultura, Subsecretaría de Estado de Recursos Naturales, Departamento de Vida Silvestre (1992) *Reconocimiento y evaluación de los recursos naturales en el Bahoruco Oriental: Proyecto La Diversidad Biológica en la República Dominicana*. Santo Domingo: SEA/DVS.

SEA/DVS (1994) = Secretaría de Estado de Agricultura, Subsecretaría de Estado de Recursos Naturales, Departamento de Vida Silvestre (1994) *Reconocimiento y evaluación de los recursos naturales de la Loma La Humeadora: Proyecto Estudio y Conservación de la Biodiversidad en la República Dominicana*. Santo Domingo: SEA/DVS.

SEA/DVS (1995) = Secretaría de Estado de Agricultura, Subsecretaría de Estado de Recursos Naturales, Departamento de Vida Silvestre (1995) *Reconocimiento y evaluación de los recursos naturales de la Sierra de Bahoruco: Proyecto Estudios Biológicos y Socioeconómicos del Suroeste (Sierra de Neiba, Lago Enriquillo y Sierra de Bahoruco para elaborar Estrategias de un Manejo Sostenible a través de una Reserva Biosfera)*. Santo Domingo: SEA/DVS.

SEA/DVS (1995) = Secretaría de Estado de Agricultura, Subsecretaría de Estado de Recursos Naturales, Departamento de Vida Silvestre (1995) *Reconocimiento y evaluación de los recursos naturales de la Sierra de Neyba: Proyecto Estudios Biológicos y Socioeconómicos del Suroeste (Sierra de Neiba, Lago Enriquillo y Sierra de Bahoruco para elaborar Estrategias de un Manejo Sostenible a través de una Reserva Biosfera)*. Santo Domingo: SEA/DVS.

TOLENTINO, L. AND PEÑA, M. (1998) Inventario de la vegetación y uso de la tierra en la República Dominicana. *Moscosoa* 10:179–203.

WOOLAVER, L. (2005) Conserving Ridgway's Hawk in the Dominican Republic. *BirdLifeCaribbean Regional Newsletter*. December 2005, No. 3.

ACKNOWLEDGEMENTS

The authors would like to thank the staff of Grupo Jaragua, Yolanda León, Esteban Garrido, Ernst Rupp, Rafael Lorenzo, Eduardo Vasquez, Jesús Almonte, José Manuel Mateo, Juana Peña, Bolívar Cabrera and José Ramón Martínez for their continuous support and strengthening of the IBA framework in the Dominican Republic; also the technical staff of the Natural Resources and Environment Secretariat, the Hispaniolan Ornithological Society, the National Botanical Garden, and the Dominican Environmental Consortium. Special thanks to Jaragua Community Volunteers, our pillar in the community of Oviedo, Pedernales.

DO001 Cayos Siete Hermanos

Wildlife Refuge

COORDINATES 19°57′N 71°47′W
ADMIN REGION North Region (Cibao)
AREA 3,084 ha
ALTITUDE 0–3 m
HABITAT Offshore cays

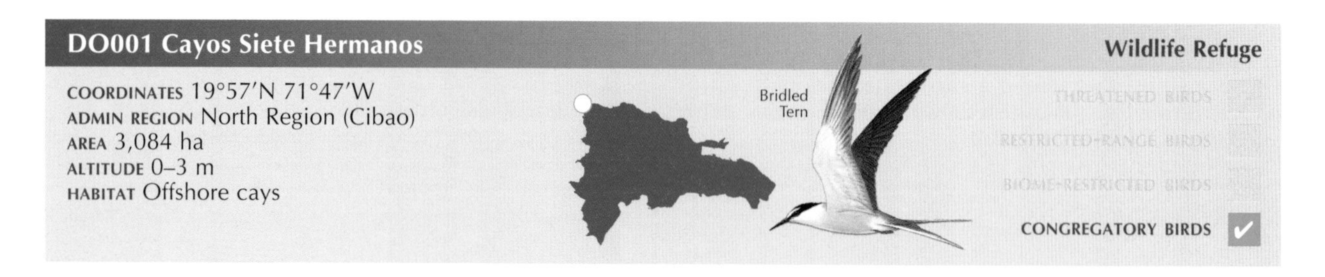

■ Site description

Cayos Siete Hermanos IBA is in Montecristi province and comprises a group of small, barren islands on the Montecristi Bank, stretching 5–15 km from the north-westernmost coast of the Dominican Republic. The islands of Torurú, Monte Chico and Terrero form the closest group to the mainland while Monte Grande, Ratas, Muerto and Arenas make up the most distant, westward group. Low thorny shrubs, grasses, herbs and cacti make up the scarce vegetation present on these sandy islets. Marine areas up to 1 km from each island are included within the IBA. The coastal region of Montecristi Province is an important fishing area associated with the Yaque del Norte River.

■ Birds

This IBA is significant for its breeding seabirds, with the population of Bridled Tern *Sterna anaethetus* being globally important and that of Brown Noddy *Anous stolidus* regionally so. Other species such as Sooty Tern *S. fuscata* also nest on the islands. The most significant seabird colonies are reported from Monte Chico and Ratas islands, with nesting primarily concentrated between May and August.

■ Other biodiversity

The Critically Endangered hawksbill turtle *Eretmochelys imbricata* is present, along with the commercially-important queen conch *Strombus gigas*. Cayos Siete Hermanos (and their coral reefs) support a diverse marine fauna and are important marine nursery grounds for the Montecristi Bank.

■ Conservation

Cayos Siete Hermanos IBA is part of Cayos Siete Hermanos Wildlife Refuge and supports activities such as fishing, marine research, ecotourism, birdwatching and traditional tourism. However, the cays have been subject to significant human disturbance, mainly by fishermen from both the Dominican Republic and Haiti. Disturbances include: cutting trees for firewood, establishment of camps (with their associated refuse), indiscriminate and inappropriate fishing practices (e.g. use of chemicals and harpoons), overfishing, and unsustainable collection of the eggs of *Sterna anaethetus*, *S. fuscata* and *Anous stolidus* (which are thought to have aphrodisiac properties) for food. Other threats include the removal of sea-turtle eggs and individuals, the presence of rats *Rattus* sp., unsustainable tourism, sand extraction, and an increasing sediment load in marine waters. Most human activity is concentrated on Torurú, Terrero, Rata, Muerto and Arenas islands. Guards from the refuge and the navy have, in the last few years, have been protecting the seabird colonies during the breeding season.

DO002 Loma Nalga de Maco–Río Limpio

National Park/Unprotected

COORDINATES 19°14′N 71°29′W
ADMIN REGION Northern Region (Cibao)
AREA 20,349 ha
ALTITUDE 600–1,990 m
HABITAT Evergreen broadleaf forest, cloud forest, open forest, riparian forest

Bay-breasted Cuckoo

THREATENED BIRDS 10
RESTRICTED-RANGE BIRDS 25
BIOME-RESTRICTED BIRDS
CONGREGATORY BIRDS

■ Site description

Loma Nalga de Maco–Río Limpio IBA is located in the northern region of the Dominican Republic, towards the westernmost end of the Cordillera Central, close to the border with the Republic of Haiti. Nalga de Maco National Park belongs to the municipality of Pedro Santana, province of Elías Piña. To the north it borders Los Almácigos municipality, Santiago Rodríguez province, to the east with the Armando Bermúdez National Park (DO003), to the west Restauración municipality, Dajabón province, and to the north-west with Río Limpio, a local coffee growing community. Communities surrounding the protected area are rural and generally lack basic services. A unique dwarf cloud forest survives in this IBA.

■ Birds

This IBA supports populations of 25 (of the 34) Hispaniolan EBA restricted-range birds, 10 of which are threatened including the Endangered Bay-breasted Cuckoo *Coccyzus rufigularis* (reported from the Río Limpio-Carrizal area) and the Vulnerable White-necked Crow *Corvus leucognaphalus*, Golden Swallow *Tachycineta euchrysea*, Bicknell's Thrush *Catharus bicknelli* (wintering), White-winged Warbler *Xenoligea montana*, and (Eastern) Chat Tanager *Calyptophilus frugivorus* (*frugivorus*).

■ Other biodiversity

Seven amphibian species are present (primarily in the riparian vegetation of the Río Limpio), representing 36% of those reported for the Cordillera Central. The flora is diverse with a high degree of endemism, particularly so in the summit area of the IBA.

■ Conservation

Loma Nalga de Maco–Río Limpio IBA is under mixed ownership. Nalga de Maco is a national park created in 1995 and ratified by law in 2000 and 2004. Río Limpio borders Nalga de Maco, but contains private lands and is not legally protected. Río Limpio offers visitor accommodation, and access to the national park via a two-day long hiking trail. Among the threats to this IBA are agriculture (including slash-and-burn practices), cattle ranching, forest fires, land invasions and human settlements, and disturbances caused by illegal migratory movements and scientific research. A visitor centre has been built at the end of what will become the "Hispaniola Trail", which will facilitate enjoyment of the national parks throughout the Cordillera Central.

DO003 Armando Bermudez National Park

National Park

COORDINATES 19°10′N 71°05′W
ADMIN REGION Northern Region (Cibao)
AREA 78,957 ha
ALTITUDE 900–3,080 m
HABITAT Pine forests, moist broadleaf forests, montane moist scrub, transtion forests

Hispaniolan Trogon

THREATENED BIRDS 13
RESTRICTED-RANGE BIRDS 31
BIOME-RESTRICTED BIRDS
CONGREGATORY BIRDS

■ Site description

Armando Bermudez National Park IBA is on the northern slope of the Cordillera Central, extending from Ciénaga de Manabao to Nalga de Maco National Park (IBA DO002). To the north it is bounded by the communities of Mata Grande, La Diferencia, Los Ramones, Lomita, and La Cidra. To the south is the José del Carmen Ramírez National Park with which it shares the highest peaks in the Cordillera Central, namely Pico Duarte, La Pelona and Pico Yaque. The most important rivers in the country originate from this IBA, including Yaque del Norte, Jagua, Bao, Amina, Guayubín, Mao, and Cenovi. The park is adjacent to small communities in Jarabacoa, San José de Las Matas, and Santiago Rodríguez districts.

■ Birds

This IBA supports populations of 31 (of the 34) Hispaniolan EBA restricted-range species. It is particularly significant for threatened species associated with montane and pine forests such as the Endangered Bay-breasted Cuckoo *Coccyzus rufigularis*, La Selle Thrush *Turdus swalesi* and Hispaniolan Crossbill *Loxia megaplaga*, and the Vulnerable White-winged Warbler *Xenoligea montana*, Golden Swallow *Tachycineta euchrysea* and (Eastern) Chat Tanager *Calyptophilus frugivorus* (*frugivorus*). The Vulnerable Hispaniolan Parakeet *Aratinga chloroptera* and Hispaniolan Amazon *Amazona ventralis* also occur, and the IBA is a winter refuge for migratory species such as Vulnerable Bicknell's Thrush *Catharus bicknelli.*

■ Other biodiversity

Mammals include the Endangered Hispaniolan solenodon *Solenodon paradoxus* and Vulnerable Hispaniolan hutia *Plagiodontia aedium*. This is also one of the few areas where the endemic pine *Pinus occidentalis* occurs.

■ Conservation

The Armando Bermúdez National Park was created in 1956 and ratified in 2000 and 2004. The park is mostly used for conservation and research, although there is some agriculture. Pico Duarte is the primary "ecological" destination in the country. Popular visitor activities include hiking, camping, rafting and birdwatching. Local communities are actively involved in the area's management and conservation and generate income through ecotourism activities such as donkey rides and guided tours. The IBA has five visitor centres and a small eco-lodge. Main threats include agricultural expansion, cattle grazing, invasive alien species, fuelwood and timber extraction, dove hunting, fires, and rural infrastructure development. There are proposals to develop hydro-electric plants in the area.

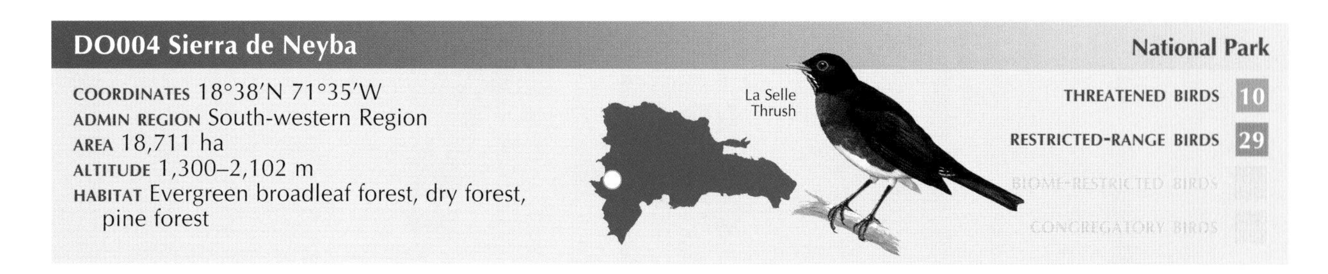

■ Site description

Sierra de Neyba IBA is north of Lago Enriquillo (IBA DO005) and from the town of Galván it crosses into Haiti under the name of "Montagnes du Trou d'Eau". It extends south-east among the valleys of El Cercado, Hondo Valle, and Hoya de Enriquillo, descending gradually to the valley of the Río Yaque del Sur. The San Juan and Neyba valleys divide this IBA from the Cordillera Central and Sierra de Bahoruco respectively. The Sierra de Neyba is composed of limestone and now supports little primary forest. What does remain includes open pine forest (c.1% of the area), evergreen broadleaf forest, and dry forest (c.26% of the area). Nearly all forest below 1,600 m has been cut.

■ Birds

This IBA supports populations of 29 (of the 34) Hispaniolan EBA restricted-range species. It is particularly significant for the Endangered Bay-breasted Cuckoo *Coccyzus rufigularis* and La Selle Thrush *Turdus swalesi*, and the Vulnerable White-winged Warbler *Xenoligea montana*, Golden Swallow *Tachycineta euchrysea* and (Eastern) Chat Tanager *Calyptophilus frugivorus* (*frugivorus*). The Vulnerable Hispaniolan Parakeet *Aratinga chloroptera* was abundant but may have been extirpated due to poaching. The Critically Endangered Ridgway's Hawk *Buteo ridgwayi* may possibly still occur.

■ Other biodiversity

The Critically Endangered *Eleutherodactylus parabates* and the locally endemic *E. notitode* occur along with nine other endemic amphibians. Reptiles are represented by 39 island endemics including the locally endemic lizard *Anolis placidus*. Mammals include the Endangered Hispaniolan solenodon *Solenodon paradoxus* and Vulnerable Hispaniolan hutia *Plagiodontia aedium*. The flora includes over 170 endemics.

■ Conservation

The Sierra de Neyba National Park was created in 1995 (with boundaries set in 2004). Little conservation action has been undertaken, and there are numerous information gaps. Nevertheless, there are 24 park staff and a number of local conservation committees. Recreational activities include hiking, horse riding, mountain bikes, camping, agro-ecotourism, and birdwatching. Threats include slash-and-burn agriculture and agricultural expansion, livestock farming, charcoal production and logging. Landslides and floods are common. Illegal hunting of doves and trafficking of the parrot and parakeet are traditional local practices. Almost 40% of the park's dry forest area has been affected by shifting agriculture and other activities, resulting in erosion and habitat degradation. Uncontrolled immigration from Haiti is a serious problem.

■ Site description

Lago Enriquillo IBA is in the Neyba Valley between Independencia and Bahoruco provinces, south-western Dominican Republic. It comprises a closed system of hyper-saline wetlands in Hoya de Enriquillo (40 m below sea-level), and receives waters from the Sierra de Neyba and Bahoruco mountain ranges, respectively to the north and south of the lake. Lago Enriquillo is flanked by marshy areas such as at Caño Boca de Cachón and Villa Jaragua. This is the largest lake in the insular Caribbean, with a surface area of 256 km^2 and a maximum depth of 24 m. It contains three islands, the largest being Isla Cabritos (24 km^2), and Islita and Barbarita which connect to the lake shore when water levels drop.

■ Birds

This IBA is an important wetland site, supporting large numbers of waterbirds including hundreds of Caribbean Flamingos *Phoenicopterus ruber*, and ibises, egrets, herons and shorebirds. The Vulnerable West Indian Whistling-duck *Dendrocygna arborea* occurs. The areas adjacent to the lake support 14 (of the 34) Hispaniolan EBA restricted-range species, including the Vulnerable Hispaniolan Amazon *Amazona ventralis* and White-necked Crow *Corvus leucognaphalus*.

■ Other biodiversity

This IBA supports the country's only remaining viable population of the Vulnerable American crocodile *Crocodylus acutus*. The Critically Endangered Ricord's iguana *Cyclura ricordi* and Vulnerable rhinoceros iguana *C. cornuta* both occur. A rich ichthyofauna including *Limia sulphurophila* which is endemic to the lake.

■ Conservation

Isla Cabritos was declared as a national park in 1974, but it was not until 1996 that Lago Enriquillo and the surrounding marshy areas were incorporated into it. Lago Enriquillo and Isla Cabritos National Park is one of the core zones of the Jaragua-Bahoruco-Enriquillo Biosphere Reserve and was also Hispaniola's first Ramsar site. It is primarily used for fishing, aquaculture, and agriculture. However, the IBAs ecological integrity is threatened by cattle ranching, unsustainable fishing practices, hunting and capture of flamingos and crocodiles, destruction of vegetation, and the canalisation and deviation of water for irrigation and associated activities. In addition, the use of pesticides pollutes both the soil and water, and is impacting the preferred habitats of waterbirds and crocodiles. This IBA features key places for recreation and wildlife observation with great potential for sustainable tourism, conservation, research, and environmental education.

Site description

The Sierra de Bahoruco IBA is in south-west of the Dominican Republic, between the provinces of Pedernales, Independencia and Barahona. It is bordered to the north by Hoya de Enriquillo, to the south by Jaragua National Park (IBA DO007) and Barahona, and to the east by Jimaní. To the west it connects with Haiti's Massif de la Selle. Sierra de Bahoruco comprises an eastern section (represented in IBA DO009) and a western section embraced by this IBA. The highest peak is Loma del Toro, and the vegetation is a diverse range of forest types across a wide variety of life zones.

Birds

This IBA supports 32 (of the 34) Hispaniolan EBA restricted-range species. Threatened birds include the Endangered Black-capped Petrel *Pterodroma hasitata* (small numbers breed in the IBA), Bay-breasted Cuckoo *Coccyzus rufigularis*, La Selle Thrush *Turdus swalesi* and the largest known population of Hispaniolan Crossbill *Loxia megaplaga*. The IBA provides vital wintering habitat for 21 Neotropical migratory species including the Vulnerable Bicknell's Thrush *Catharus bicknelli*. The Critically Endangered Ridgway's Hawk *Buteo ridgwayi* occurred until 1994.

Other biodiversity

Threatened mammals include the Endangered Hispaniolan solenodon *Solenedon paradoxus* and Vulnerable Hispaniolan hutia *Plagiodontia aedium*. Many *Eleutherodactylus* frogs occur, all of which are Critically Endangered or Endangered, including the locally endemic *E. rufifemoralis* (Critically Endangered). Reptiles include the Vulnerable rhinoceros iguana *Cyclura cornuta* and Hispaniolan slider *Trachemys decorata*.

Conservation

Sierra de Bahoruco National Park was created in 1983 and its boundaries ratified by the Laws 64-00 and 202-04. It is one of the three core zones of the Jaragua-Bahoruco-Enriquillo Biosphere Reserve. The IBA contains important but unprotected sections in La Placa and Puerto Escondido. Activities include research, conservation, recreation and ecotourism. Main threats include agricultural expansion, introduced animals, forest fires, illegal logging, capture of parrot chicks and illegal hunting. Temporary settlement of illegal immigrants from Haiti moving through this area results in habitat damage and disturbances. Tree-nesting species are affected by the unsustainable removal of dead and diseased trees. Most of these problems are a result of weak park management and enforcement.

Site description

Jaragua National Park IBA is in the south-western corner of the Dominican Republic on the Barahona peninsula. It borders the Sierra Bahoruco IBA (DO006), and lies across the municipalities of Pedernales and Oviedo, close to the border with Haiti. Within the IBA are the Beata and Alto Velo islands, the Los Frailes and Piedra Negra cays, and Laguna Oviedo—a "Watchable Wildlife Pond" and a proposed Ramsar site. Surrounding communities are Juancho, La Colonia de Juancho, Oviedo, Los Tres Charcos, Manuel Goya, and Pedernales. About 4 km from Tres Charcos is Fondo Paradí, a popular birdwatching area and "Ecotourism Pilot Site".

Birds

This IBA is important for its wetlands and islands, although the forest harbours 18 (of the 34) Hispaniolan EBA restricted-range species including the Vulnerable Hispaniolan Amazon *Amazona ventralis*, Hispaniolan Parakeet *Aratinga chloroptera* and White-necked Crow *Corvus leucognaphalus*. The wetlands support more than 20,000 waterbirds including the Vulnerable West Indian Whistling-duck *Dendrocygna arborea* and regionally important populations of species including Caribbean Flamingo *Phoenicopterus ruber*. The Sooty Tern *Sterna fuscata* colony (80,000 pairs) on Isla Alto Velo is one of the largest in the Caribbean.

Other biodiversity

Reptiles include the Critically Endangered Ricord's iguana *Cyclura ricordi* and hawksbill turtle *Eretmochelys imbricata* and *Sphaerodactylus ariasae*—the smallest amniote vertebrate in the world. *Anolis altavelensis* is endemic to Isla Alto Velo. The IBA is an invertebrate hotspot including newly discovered species such as *Beatadesmus ivonneae*. The plants *Pseudophoenix ekmanii* (Critically Endangered), *Pimenta haitensis*, and *Coccothrinax ekmanii* are almost endemic to the IBA.

Conservation

Jaragua National Park was declared in 1983 and its boundaries set in 1986. There have been attempts to modify the legal framework that protects it. Since 1989, Grupo Jaragua and SEMARENA have jointly managed the IBA, facilitating and coordinating local community conservation actions. Lands are used for conservation, research, recreation, nature tourism and education, but small-scale fishing, agriculture, and livestock farming is practised in neighbouring communities. Main threats to this IBA include the development of tourist projects (although the beaches are unsuitable for large-scale tourism), agricultural expansion, introduced species, land invasions, mining, hunting, fishing, and extraction of eggs and chicks of the parrots, parakeets and seabirds.

DO008 Laguna Cabral

Wildlife Refuge

COORDINATES 18°17′N 71°14′W
ADMIN REGION South-western Region
AREA 5,615 ha
ALTITUDE 0–92 m
HABITAT Wetlands, dry forest, dry scrub

Lesser Scaup

THREATENED BIRDS 3
RESTRICTED-RANGE BIRDS 12
BIOME-RESTRICTED BIRDS
CONGREGATORY BIRDS ✓

■ Site description

Laguna Cabral IBA is located in the south-western region between the provinces of Barahona and Independencia, c.20 km inland from the Bahía de Neyba. Laguna Cabral is the largest freshwater wetland in the Dominican Republic, and the IBA includes the nearby wetlands of Laguneta Seca, and the Cristóbal and Peñón Viejo hills to the north. The IBA is surrounded by the communities of Cabral, Peñón, Cristóbal and La Lista. In the flat southern section of the IBA, plantains and coconuts are cultivated alongside pastures and other crops.

■ Birds

The IBA supports 12 (of the 34) Hispaniolan EBA restricted-range species, including the Vulnerable Hispaniolan Parakeet *Aratinga chloroptera* and Near Threatened Least Pauraque *Siphonorhis brewsteri*. However, it is for the waterbirds that this IBA is primarily significant. Huge (globally important) concentrations of duck have included counts of up to 90,000 Lesser Scaup *Aythya affinis* and 10,000 Ruddy Duck *Oxyura jamaicensis*. Up to 3,000 Near Threatened Caribbean Coot *Fulica caribaea* have also been recorded. Other duck present are American Wigeon *Anas americana* (up to 10,000), Blue-winged Teal *A. discors* (up to 25,000) and White-cheeked Pintail *A. bahamensis* (up to 22,000).

■ Other biodiversity

The Vulnerable toad *Bufo guentheri*, Hispaniolan slider *Trachemys decorata* and rhinoceros iguana *Cyclura cornuta* occur. The aquatic fauna of the lake in includes crustaceans such as *Palaemon pandaliformis*, the endemic fish including *Nandopsis haitienensis* and various species of the genera *Limia* and *Gambusia*. Eight plants are endemics, such as *Justicia abeggii*, *Tournefortia sufruticosa*, *Neoabbottia paniculata*, and *Malpighia micropetala*.

■ Conservation

Laguna Cabral IBA was declared a Scientific Reserve in 1983 and a Wildlife Refuge in 1996. It has been proposed as a Ramsar site. Habitat loss has resulted from agricultural activities such as cattle ranching; the planting of non-timber species, fires, and felling trees for charcoal production is reported. Fish stocks have also diminished because of overfishing, the introduction of exotic species, and pesticide pollution. Additionally, natural aquatic systems have been altered through canalisation. Other threats include land invasion, illegal constructions, the hunting of turtles, iguanas, coots and the persecution of Caribbean Flamingo *Phoenicopterus ruber*. Among the proposed initiatives for the sustainable management and conservation of this site are ecotourism, environmental education, and monitoring.

DO009 Bahoruco Oriental

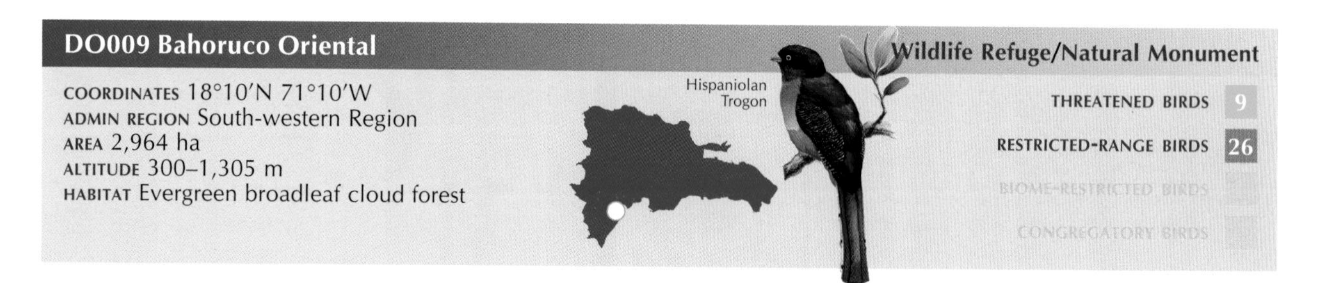

■ Site description

Bahoruco Oriental IBA is in south-west of the Dominican Republic, between the provinces of Pedernales, Independencia and Barahona. The Sierra de Bahoruco is the country's southernmost mountain range, and comprises an eastern section (this IBA) and a western section (represented in IBA DO006). Bahoruco Oriental IBA embraces the mountainous part of the province of Barahona, bounded to the north by the Valle de Neyba, to the east by the Caribbean Sea, to the south by the Nizaíto river valley, and to the west by the valleys of Polo and La Cueva. The landscape is a mosaic of primary forest (including Hispaniola's only magnolia *Magnolia hamori* forest, and the largest *Prestoea montana* forest), secondary forests, vast coffee plantations, farming and secondary vegetation areas.

■ Birds

This IBA supports populations of 26 (of the 34) Hispaniolan EBA restricted-range species including the Vulnerable Grey-headed Quail-dove *Geotrygon caniceps leucometopia*, Hispaniolan Amazon *Amazona ventralis*, Hispaniolan Parakeet *Aratinga chloroptera*, White-necked Crow *Corvus leucognaphalus*, Golden Swallow *Tachycineta euchrysea* and (Western) Chat Tanager *Calyptophilus frugivorus* (*tertius*). The Vulnerable Bicknell's Thrush *Catharus bicknelli* has been recorded, but not in significant numbers.

■ Other biodiversity

Frogs include the Critically Endangered *Eleutherodactylus rufifemoralis*, the Endangered *E. armstrongi* and Vulnerable *E. audanti*. Endemic reptiles include *Anolis bahorucoensis*, *Chamaelinorops barbouri* and *Wetmorena haetiana*. Threatened mammals are represented by the Endangered Hispaniolan solenodon S*olenodon paradoxus*. Rare *Lephantes* orchids are also present, some restricted to microhabitats within the IBA.

■ Conservation

Bahoruco Oriental IBA was declared the Biological Reserve Padre Miguel Domingo Fuerte (Bahoruco Oriental) in 1996 (with boundaries ratified in 2000). The management category was changed into Wildlife Refuge/Natural Monument in 2004. Land use is mainly for agriculture. However, research and rural tourism activities are also carried out in this area, as well as projects and initiatives aimed at social, economic, and environmental sustainability. This IBA has suffered multiple impacts since the 1930s and 1940s from agriculture, cattle ranching, deforestation, slash-and-burn practices, mining (silica and "Larimar" or blue pectolite), and road construction. Other threats include bird hunting, extraction and illegal trade in flora (e.g. ferns) and fauna (e.g. parrot and parakeet chicks), introduced and invasive flora and fauna, as well as natural and intentionally lit forest fires.

■ **Site description**

Sierra Martín García IBA is in south-west Dominican Republic between the provinces of Azua and Barahona, next to Puerto Viejo. This limestone massif emerges from the sea opposite the city of Barahona, at the north-east end of the Bahía of Neyba, and runs from Puerto Alejandro to Punta Martín García. There is evidence of Taino (the original island settlers) presence in local caves, especially in the town of Barreral, where plant fossils and the oldest Taino settlements have been found.

■ **Birds**

This IBA supports 23 (of the 34) Hispaniolan EBA restricted-range species including the Vulnerable (Eastern) Chat Tanager *Calyptophilus frugivorus* (*frugivorous*), White-winged Warbler *Xenoligea montana*, Hispaniolan Amazon *Amazona ventralis* and (wintering) Bicknell's Thrush *Catharus bicknelli*. The Near Threatened Least Pauraque *Siphonorhis brewsteri* is present. Poor management, regulation and enforcement resulted in the loss of the Critically Endangered Ridgway's Hawk *Buteo ridgwayi* from this IBA.

■ **Other biodiversity**

This is a key site for the conservation of rare endemic plants such as *Arcooa gonavensis*, *Cnidosculus acrandus*, and *Fuertesia domingensis*. The palm *Coccothrinax boschiana* is endemic to the sierra and forms stands known locally as "guanales".

■ **Conservation**

Sierra Martín García IBA has been legally protected at a national park since 1996. Its boundaries were defined in 2004 and an administrator was assigned for the first time in 2007. Local residents use this area for subsistence activities such as slash-and-burn agriculture, charcoal production and fishing. Other activities include limited scientific research, birdwatching and ecotourism. Threats include limestone extraction, intentional fires, excessive cattle and goat grazing, extraction of sand and gravel, as well as hunting of parrots and iguanas. Feral dogs, cats, and mongoose *Herpestes auropunctatus* prey on iguana eggs and juveniles and presumably birds. Human activities followed by the effects of rain and wind have removed vegetation and ground cover, resulting in a deteriorating, eroded and degraded landscape in which desertification is a real possibility. This IBA has been poorly studied and there are few conservation projects or actions being implemented.

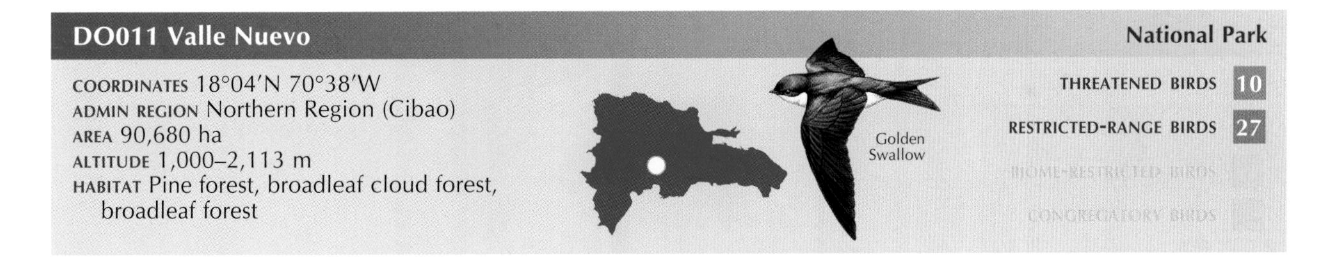

■ **Site description**

Valle Nuevo IBA embraces the highlands at Alto de la Bandera hill, located in Constanza La Vega province, north-central Dominican Republic. It is surrounded by the provinces of Monseñor Nouel, La Vega, Azua and San José de Ocoa and is bordered to the east by Loma La Humeadora (DO015). With five river basins and more than 700 rivers, Valle Nuevo is an important catchment area supplying water for the provinces of the northern and southern regions, and Santo Domingo. The area supports a range of forest types including pine, broadleaf, cloud and *Prestoea montana* forest. Approximately 20 communities with a total population of 3,500 inhabitants live within this IBA.

■ **Birds**

This IBA supports populations of 27 (of the 34) Hispaniolan EBA restricted-range species including the Endangered Hispaniolan Crossbill *Loxia megaplaga* and La Selle Thrush *Turdus swalesi*, and the Vulnerable Hispaniolan Amazon *Amazona ventralis*, Hispaniolan Parakeet *Aratinga chloroptera*, Golden Swallow *Tachycineta euchrysea*, (Eastern) Chat Tanager *Calyptophilus frugivorus* (*frugivorous*), White-winged Warbler *Xenoligea montana*. It also provides wintering habitat for Neotropical migrants such as the Vulnerable Bicknell's Thrush *Catharus bicknelli*.

■ **Other biodiversity**

Mammals such as the Endangered Hispaniolan solenodon *Solenedon paradoxus* and Vulnerable Hispaniolan hutia *Plagiodontia aedium* occur, and there are 29 endemic species of reptiles including *Anolis aliniger* and *Celestus darlingtoni*. There are 138 endemic plant species, including *Magnolia pallescens*. This is a critical area for *Vegaea pungens* and is important for the ferns *Cyathea insignis* and *C. harrissi*.

■ **Conservation**

Valle Nuevo IBA was declared as a restricted area (zona vedada) in 1961, a scientific reserve in 1983, and a national park in 1996. Lands have traditionally been used for forest exploitation, agriculture, and cattle ranching. Intentional fires have altered the natural fire regime of this IBA, which in 1983 suffered the worst forest fire in the history of the country. In recent years, nearly 5,000 ha have been lost to agricultural expansion and an additional 4,500 ha have been affected by forest fires. Other threats include agrochemical pollution, inadequate waste and waterway management, erosion, landslides, and road construction. All these threats affect biodiversity, ecological dynamics and succession, as well as water quality. Programs to develop the site's ecotourism potential are being implemented.

DO012 Ebano Verde Scientific Reserve

Scientific Reserve

COORDINATES 19°04′N 70°32′W
ADMIN REGION Northern Region (Cibao)
AREA 2,993 ha
ALTITUDE 900–1,565 m
HABITAT Broadleaf cloud forest, riparian forest, secondary forest

White-winged Warbler

THREATENED BIRDS 6
RESTRICTED-RANGE BIRDS 24
BIOME-RESTRICTED BIRDS
CONGREGATORY BIRDS

■ Site description

Ebano Verde Scientific Reserve IBA is in north-central Dominican Republic on the north-eastern slope of the Cordillera Central, La Vega province, and municipality of Jarabacoa. The IBA contains the catchment areas of the Jimenoa, Camú, Jatubey and Jayaco rivers. It is named after the local name (Ebano Verde) for the species *Magnolia pallescens*, characteristic of the area's broadleaf cloud forests. *Prestoea montana* forest also occurs in the IBA. Ebano Verde Scientific Reserve has a long history of forest exploitation, with forest lands and products still being used by people from surrounding communities.

■ Birds

This IBA provides habitat for 24 (of the 34) Hispaniolan EBA restricted-range species, and is critical for globally threatened birds including the Vulnerable Grey-headed Quail-dove *Geotrygon caniceps leucometopia*, Golden Swallow *Tachycineta euchrysea*, White-winged Warbler *Xenoligea montana* and (Eastern) Chat Tanager *Calyptophilus frugivorus* (*frugivorous*). The Hispaniolan endemic race of Rufous-collared Sparrow *Zonotricia capensis antillarum* occurs in the IBA.

■ Other biodiversity

Globally threatened frogs include the Endangered Hispaniolan giant tree-frog *Osteopilus vastus*, *Eleutherodactylus auriculatoides* and *E. pituinus*. The rare lizard *Anolis insolitus* and the endemic fish *Poecilia dominicensis* are present in this IBA. It also supports 156 species of endemic spermatophytes, and orchids are both prominent and highly endemic, with 81 species.

■ Conservation

Ebano Verde Scientific Reserve was created in 1989, ratified in 2000 and validated in 2004. Major threats to this IBA include introduced flora and fauna, livestock farming, intentional forest fires, timber extraction and trafficking, the capture of bird chicks, natural phenomena, habitat destruction and modification due to agricultural and urban expansion. Despite the IBAs restricted public use, it has huge potential for ecotourism. Activities such as birdwatching, hiking, enjoyment of landscapes and panoramic views are all activities enjoyed in the IBA. The Arroyazo Biological Station has acted as a central office for research conducted at this IBA.

■ Site description

Loma Quita Espuela IBA is c.15 km from the north-east of the city of San Francisco de Macorís in northern Dominican Republic. It lies on the eastern slope of the Cordillera Septentrional in the provinces Maria Trinidad Sanchez and Duarte, and it includes the areas of Quita Espuela and La Canela. This IBA includes five hills: Quita Espuela is the highest and at the centre of the IBA (985 m), Vieja (730 m), El Quemao (565 m), La Canela (560 m) and Firme Los Sabrosos (510 m). The forested slopes of these hills protect the sources of several streams that supply water to a number of nearby towns.

■ Birds

This IBA supports populations of 19 (of the 34) Hispaniolan EBA restricted-range species including the Vulnerable Hispaniolan Amazon *Amazona ventralis* and Hispaniolan Parakeet *Aratinga chloroptera*. The IBA is important for Neotropical migrants including the Vulnerable Bicknell's Thrush *Catharus bicknelli*.

■ Other biodiversity

Mammals such as the Endangered Hispaniolan solenodon *Solenedon paradoxus* and Vulnerable Hispaniolan hutia *Plagiodontia aedium* have been reported. The tree *Mora abbotti* is endemic to the Cordillera Septentrional and occurs within the IBA.

■ Conservation

Loma Quita Espuela is scientific reserve that was established in 1992. It is managed by the Office of the Subsecretary of Protected Areas and Biodiversity and the Loma Quita Espuela Foundation. Main uses include farming, charcoal production, research, conservation, ecotourism, local pilgrimages and recreation. Among the threats to this IBA are habitat loss related to forest fires, slash-and-burn farming, agricultural expansion, charcoal production, cattle grazing, and road and path construction. Other threats are related to extraction of flora and fauna, bird hunting, invasive alien fauna, and pollution of water sources. This IBA faces problems with land invasions, human settlements and management conflicts with neighbouring communities regarding land tenure and protected area boundaries.

DO014 Loma Guaconejo

Scientific Reserve

COORDINATES 19°19′N 69°59′W
ADMIN REGION Northern Region (Cibao)
AREA 2,329 ha
ALTITUDE 6–606 m
HABITAT Broadleaf forest, broadleaf scrub, pasturelands

Bicknell's Thrush

THREATENED BIRDS 1
RESTRICTED-RANGE BIRDS
BIOME-RESTRICTED BIRDS
CONGREGATORY BIRDS

■ Site description
Loma Guaconejo IBA is at the eastern end of the Cordillera Septentrional in the provinces of María Trinidad Sánchez and Duarte. It lies east of Loma Quita Espuela IBA (DO013) and is named after the local name (Guaconejo) of the plant *Stevensia ebracteata*. This IBA embraces the Río Helechal catchment area, which in turns feeds the Boba and Nagua rivers that finally supply water to the municipalities of Nagua and El Factor. Loma Guaconejo retains one of the best preserved moist broadleaf forests in the Cordillera Septentrional. Around 2,000 inhabitants from 16 communities live around this IBA.

■ Birds
This IBA is significant for supporting a large wintering population of the Vulnerable Bicknell's Thrush *Catharus bicknelli*. A range of other Neotropical migratory birds (including various *Dendroica* warbler species) winter in these forests. The avifauna has been poorly studied and other key bird species probably occur.

■ Other biodiversity
Four endemic frogs include the Critically Endangered *Eleutherodactylus parabates*, and *E. inoptatus*. The forests include a large population of *Calyptronoma plumeriana*, intermixed with the endemic *Tabebuia ricardii*, *Plumeria magna* and other trees considered exclusive to Loma Quita Espuela and Los Haitises (IBA DO018).

■ Conservation
Loma Guaconejo was designated as a scientific reserve in 1996 and its boundaries were defined in 2004. The area is jointly managed by the State Secretariat of Environment and Natural Resources and the Society for the Integral Development of the Northeast (SODIN). The land is used for conservation, agriculture, and recreation. The Cuesta Colorada Ecotourism and Environmental Capacity Centre is also located in this area. SODIN has facilitated self-management and participatory processes, especially in María Trinidad Sánchez province. Since 1995 it has worked toward sustainable management of the IBAs natural resources. Currently the society is collaborating with local guides and Peace Corps Volunteers in developing an environmental education program for the buffer zone. Some threats to the area include water pollution, sand extraction, erosion, shifting agriculture, fires, trafficking of timber, livestock farming, deficient environmental sanitation, unemployment, low levels of education, and solid waste accumulation.

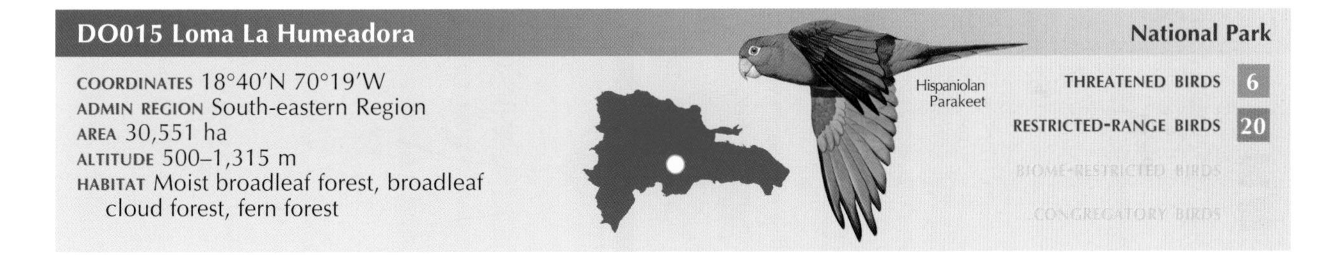

DO015 Loma La Humeadora

National Park

COORDINATES 18°40′N 70°19′W
ADMIN REGION South-eastern Region
AREA 30,551 ha
ALTITUDE 500–1,315 m
HABITAT Moist broadleaf forest, broadleaf cloud forest, fern forest

THREATENED BIRDS 6
RESTRICTED-RANGE BIRDS 20
BIOME-RESTRICTED BIRDS
CONGREGATORY BIRDS

■ Site description
Loma La Humeadora IBA is on the south-easternmost slope of the Cordillera Central, 10 km to the south-west of the municipality of Villa Altagracia and to the north-east of San Cristóbal city. This IBA comprises Loma La Humeadora, with an area of c.84 km^2 and a surrounding group of lower-elevation hills which support forest remnants. Average annual rainfall is 2,300 mm and numerous rivers and streams originate within the area, representing an important hydrological resource for the surrounding area.

■ Birds
This IBA is home to 20 (of the 34) Hispaniolan EBA restricted-range species including the Endangered La Selle Thrush *Turdus swalesi*, and the Vulnerable Hispaniolan Amazon *Amazona ventralis* and Hispaniolan Parakeet *Aratinga chloroptera*. The IBA is important for Neotropical migrants including the Vulnerable Bicknell's Thrush *Catharus bicknelli*, and also for species with declining distributions on Hispaniola such as Limpkin *Aramus guarauna* and Scaly-naped Pigeon *Patagioenas squamosa*.

■ Other biodiversity
Restricted-range Hispaniolan plant endemics include *Pricramnia dyctioneura*, *Podocarpus hispaniolensis*, *Urera domingensis*, *Omphalea ekmanii*, and *Piper luteobaccum* (20% of the plant species are Hispaniolan endemics). Species previously considered exclusive to Lomas La Sal and La Golondrina (to the north of the IBA), such as *Chaetocarpus domingensis*, *Cinnamomum alainii*, and *Gonocalyx tetrapterus* have been found in Loma La Humeadora.

■ Conservation
In 1992 Loma La Humeadora IBA was declared as a restricted area (zona vedada) to protect the streams and rivers that originate in this area. In 1996 it was declared as national park (which was ratified in 2000, with boundaries set in 2004). Loma La Humeadora has been seriously disturbed by farming activities, slash-and-burn practices, firewood extraction and charcoal production. Only remnants of primary forest remain. This affects the landscape, the fauna and flora, the environmental services such as water production, and as a result, the general well-being of the human populations that depend on this resource. Apart from scheduled visits to fulfil management tasks and sporadic visits by birdwatchers, conservation actions within the IBA are scarce.

DO016 Honduras

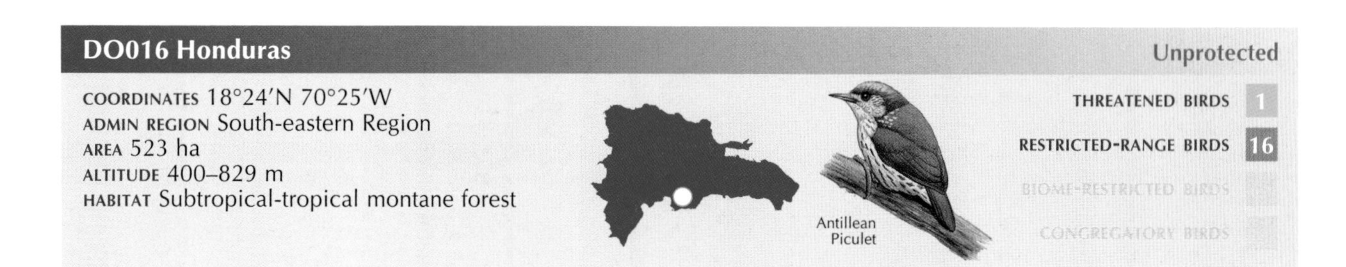

Site description
Honduras IBA is in the south-eastern region of the Dominican Republic, c.15–20 km north of Baní (in Peravia province, and the municipality of Matadero). About 90% of this IBA supports montane forest, although little has been documented concerning this IBA so details are unclear. There are also cultivated areas within the IBA.

Birds
This IBA supports populations of 16 (of the 34) Hispaniolan EBA restricted-range species including the Vulnerable (Eastern) Chat Tanager *Calyptophilus frugivorus* (*frugivorus*), and others such as Hispaniolan Lizard-cuckoo *Saurothera longirostris*, Ashy-faced Owl *Tyto glaucops*, Hispaniolan Nightjar *Caprimulgus ekmani*, Antillean Piculet *Nesoctites micromegas*, Palmchat *Dulus dominicus*, Green-tailed Warbler *Microligea palustris* and Black-crowned Palm-tanager *Phaenicophilus palmarum*. The Vulnerable Bicknell's Thrush *Catharus bicknelli* has been recorded, but numbers are unknown.

Other biodiversity
New species, such as the spider *Pozonia andujari* have been reported, but the flora and fauna are poorly known and need further work.

Conservation
Honduras IBA is a mixture of private and state owned lands, and currently (2008) has no legal protection. Detailed information regarding threats to the IBA is lacking, but during recent exploratory visits human settlements, introduced animals, and the evidence of agriculture (over c.10% of the area) were observed in the area. Threats to the key species in the IBA are unknown and this requires further research. As well as the area's agricultural tradition; some lands (35% of the area) are used for research, recreation, and bird tourism, whilst others are unexploited (55% of the area). The Dominican Republic IBA Program is exploring the possibility of joint initiatives with the private sector to protect the remaining critical forests and perform more thorough research in this IBA. It is also facilitating discussions with the State Secretariat of Environment and Natural Resources about potential protection mechanisms for the site (e.g. Private Reserve, Reserve or Protected Municipal Area).

DO017 Bahía de las Calderas

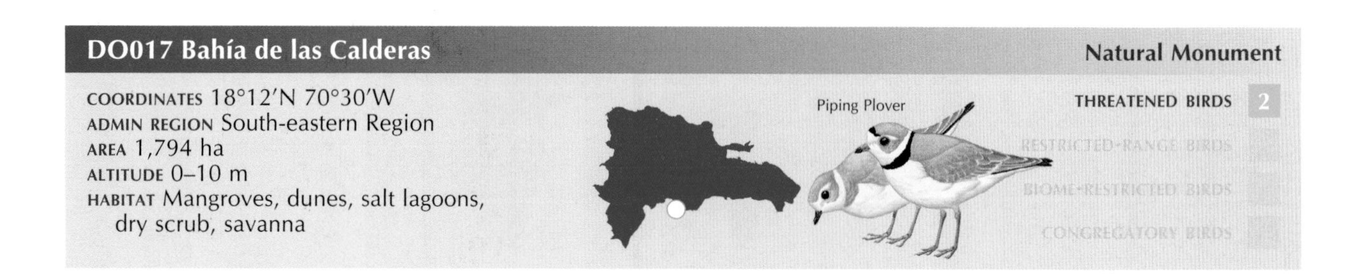

Site description
Bahía de las Calderas IBA is located in the Peravia province, in the south-eastern region of the Dominican Republic, about 115 km south-west of Santo Domingo. It is on the peninsula of Las Calderas, which is primarily covered by a hyper-saline pond—Salado del Muerto—used for salt extraction. Several towns are present in Las Calderas as well as a naval base. Bahía de las Calderas IBA also includes the dunes of Baní, that stretch 15 km from the town of Matanzas to Puerto Hermoso.

Birds
This IBA is significant for its populations of the Vulnerable West Indian Whistling-duck *Dendrocygna arborea* and Near Threatened Hispaniolan Palm Crow *Corvus palmarum*. A wide diversity of waterbirds use this IBA and nesting species include Snowy Plover *Charadrius alexandrinus*, Wilson's Plover *Charadrius wilsonia*, Least Tern *Sterna antillarum* and Willet *Catoptrophorus semipalmatus*. The Near Threatened Piping Plover *Charadrius melodus* has been recorded in small numbers. Many rare and vagrant species records come from this IBA.

Other biodiversity
There have been reports of the Vulnerable rhinoceros iguana *Cyclura cornuta* within the IBA. Many other reptile species frequent the area. The ichthyofauna is represented by the endemic species *Limia perugiae* and *Cyprinodon* spp., and others such as *Elops saurus*, *Megalops atlanticus*, *Gerres cinereus* and *Centropomus undecimalis*. *Simarouba berteroana* is a dune-stabilising tree endemic to this region.

Conservation
Bahía de Las Calderas IBA has been legally protected as Dunas de las Calderas Natural Monument since 1996. Activities in this area include small-scale agriculture, aquaculture, fishing, scientific research, birdwatching, recreation, and traditional tourism. Mangroves along the lagoon have been used for charcoal production and firewood. Threats to this IBA include fires, introduced and feral animals, extraction of non-timber vegetation, and sand extraction for commercial and construction purposes. Other threats include water drainage and canalisation; housing, commercial and industrial development; land invasions, and human settlements. Furthermore, recreational activities such as the use of 4x4 vehicles and beach tourism result in solid waste pollution, disturbance and habitat damage. Ecotourism is promoted through the construction of trails and student visits.

DO018 Los Haitises

National Park

COORDINATES 19°01′N 69°37′W
ADMIN REGION Northern Region (Cibao)
AREA 63,416 ha
ALTITUDE 0–287 m
HABITAT Tropical moist forest, karst forest, mangroves, wetlands, coastal forest

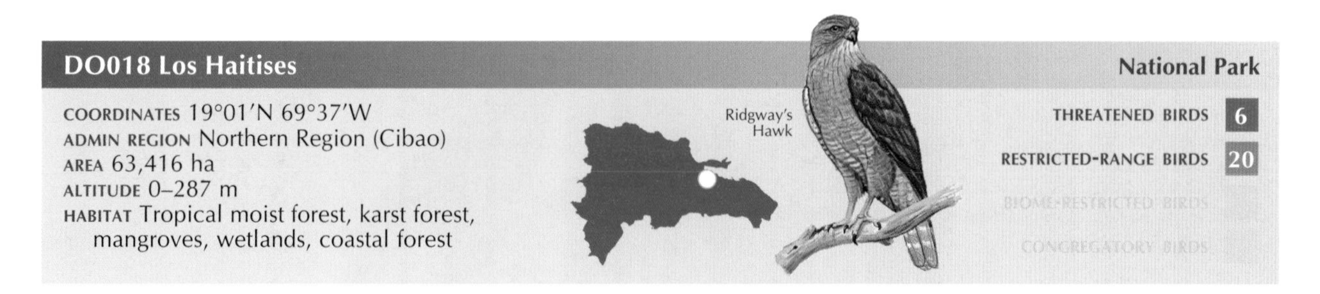

■ Site description
Los Haitises IBA is located in the north-eastern region of the country, and extends from the southern portion of the Cibao Oriental Valley to the town of Sabana de la Mar, south-west of the Bahía de Samaná. The IBA is a national park, and lies across the provinces of Monte Plata, Hato Mayor and Samaná. Due to its size, the park has been divided into two administrative sectors, an eastern sector that includes Sabana de la Mar and adjacent areas, and a southern sector that includes Monte Plata and San Francisco de Macorís. The IBA embraces areas of limestone karst (supporting moist broadleaf forest), secondary forest, agricultural areas, and the Dominican Republic's largest area of mangroves. The park offers a vast cave system with pictograms and petroglyphs.

■ Birds
This IBA supports populations of 20 (of the 34) Hispaniolan EBA restricted-range species. However, it is as the last known refuge for the Critically Endangered Ridgway's Hawk *Buteo ridgwayi* that this IBA is significant. There are about 50 pairs within the IBA, but productivity is low and the species is becoming scarce, even within Los Haitises. Other threatened birds include the Vulnerable Hispaniolan Amazon *Amazona ventralis*, White-necked Crow *Corvus leucognaphalus* and Bicknell's Thrush *Catharus bicknelli*, and the Near Threatened Least Pauraque *Siphonorhis brewsteri*.

■ Other biodiversity
The Endangered Hispaniolan solenodon *Solenodon paradoxus*, and Vulnerable Hispaniolan hutia *Plagiodontia aedium* and West Indian manatee *Trichechus manatus* occur within the IBA.

■ Conservation
Los Haitises IBA was first created as a forest reserve in 1968 and later as national park in 1976. Then, in 1996 its boundaries were extended and ratified in 2000 and 2004. Factors responsible for the current dire status of *Buteo ridgwayi* include habitat destruction and degradation (including the felling of trees used as nest sites by the species), agricultural expansion, fire-induced habitat fragmentation, nest destruction and reduction of feeding resources, hunting by poultry farmers, and lack of knowledge about the species. Other threats to this IBA include indiscriminate fishing, illegal hunting, cattle ranching, land invasions and human settlements, pollution and agrochemical use, forest fires, development projects, inadequate tourism, introduced animals, extraction of animal derivates (e.g. bat manure), and cave vandalism.

DO019 Laguna Limón

Wildlife Refuge

COORDINATES 18°55′N 68°50′W
ADMIN REGION South-eastern Region
AREA 1,083 ha
ALTITUDE 0 m
HABITAT Wetlands

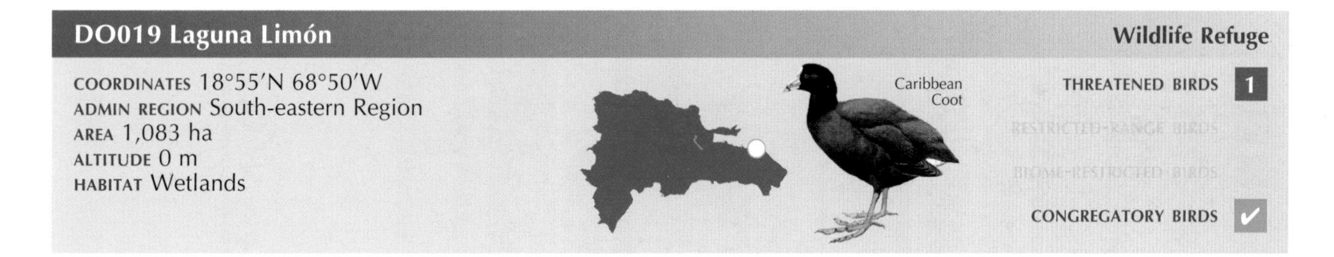

■ Site description
Laguna Limón IBA is located in the north-east of the Dominican Republic in the province of El Seibo, 27 km from the town of Miches. It is part of a system of coastal lagoons located on the coastal plain between the Cordillera Oriental and the Atlantic Ocean. It has a maximum depth of 2 m and receives waters from several streams such as Caño El Negro and Río Lisas, although it is still connected to the ocean. The vegetation in the area is dominated by *Typha domingensis* and the genus *Machaerium*.

■ Birds
This IBA supports a globally significant population of the Near Threatened Caribbean Coot *Fulica caribaea*. Up to 6,000 coots have been recorded from the lagoon. Many other waterbirds use this IBA including large (but not significant) numbers of Blue-winged Teal *Anas discors*, White-cheeked Pintail *A. bahamensis*, and Ruddy Duck *Oxyura jamaicensis*.

■ Other biodiversity
The Vulnerable fresh water turtle *Trachemys decorata* is present. More studies on the flora and the fauna of this site are needed.

■ Conservation
Laguna Limón IBA was declared a scientific natural reserve in 1983, a strict natural reserve in 1995, and a wildlife refuge in 2004. The surrounding lands are mostly privately owned. Activities around and within the lagoon include fishing, aquaculture, and farming, as well as coconut and rice cultivation. Some threats to this region include agrochemical pollution, disturbance by cattle, extraction of sand and gravel for construction, unsustainable hunting and fishing, disturbance from (tourist) motor boats, and natural and intentional fires. Another threat is the presence of the invasive aquatic plant *Hydrilla verticillata*. Changes in water levels within the lagoon appear to be the result of deforestation in the surrounding uplands leading to increased runoff and sedimentation. Conservation and research efforts are being performed through waterbird (duck) counts in the area.

■ Site description

Del Este National Park IBA is on the south coast at the easternmost end of the Dominican Republic. The park covers most of the peninsula that extends from the towns of Boca de Yuma and Bayahibe in the municipality of San Rafael del Yuma. It also includes the 110-km² Isla Saona (and the nearby 22-ha Isla Catalinita) from which it is separated by the Catuano channel. The park embraces coastal habitats, lagoons, mangroves, scrub forest and extensive woodlands. The park offers several tourist attractions such as its excellent beaches and evidence of the country's pre-Columbian heritage. It is the most visited protected area in the country.

■ Birds

This IBA supports 18 (of the 34) Hispaniolan EBA restricted-range species including the Vulnerable Hispaniolan Amazon *Amazona ventralis* and White-necked Crow *Corvus leucognaphalus*. Other globally threatened species include the Vulnerable Bicknell's Thrush *Catharus bicknelli*, and the Near Threatened Least Pauraque *Siphonorhis brewsteri* and White-crowned Pigeon *Patagioenas leucocephala*. Isla Saona supports the country's larges breeding colony (200 pairs) of Magnificent Frigatebird *Fregata magnificens*. Other seabirds nest on Isla Saona (and nearby Isla Catalina), but numbers are unknown.

■ Other biodiversity

The Endangered Hispaniolan solenodon *Solenodon paradoxus* and Vulnerable Hispaniolan hutia *Plagiodontia aedium* occur, as does the Critically Endangered hawksbill turtle *Eretmochelys imbricata* and the Endangered frog *Eleutherodactylus probolaeus*. The endemic freshwater fish *Limia perugiae* is present, as are many endemic (and threatened plants.

■ Conservation

Del Este National Park was declared a protected area in 1975. Its boundaries have been considerably altered by law. Activities in the area include conservation, research and birdwatching. Approximately 8% of this IBA has been affected by agriculture (e.g. coconut crops), mostly on Isla Saona. The main threat to this IBA relates to the 260,000 tourists that visit the park (mostly Isla Saona) each year. The modification of the park boundaries has rendered the coastal area vulnerable to unsustainable tourist development. The IBA, it is still vulnerable to real estate speculation due to land tenure irregularities. Other threats include introduced species, indiscriminate and unsustainable fishing, hunting of pigeons, intentional forest fires, land invasions and illegal settlements inside the protected area, extraction and trafficking of parrot chicks.

■ Site description

Punta Cana IBA is at the easternmost tip of the Dominican Republic, in the province of Altagracia and north-east of Del Este National Park (DO020). The town of Bávaro lies to the north of the IBA, and to the east are the Mona channel and the Caribbean Sea. Little information has been documented concerning the vegetation or habitats of this IBA. However, it appears to comprise lowland moist forests, shrubland, pastureland and urban areas. Within the site is the Ojos Indigenous Reserve and Ecological Park, which is mostly used for recreation and tourism.

■ Birds

This IBA supports 13 (of the 34) Hispaniolan EBA restricted-range species including Vulnerable Hispaniolan Amazon *Amazona ventralis*. The area supports many other native and migratory birds (116 species have been recorded) such as the rare Double-striped Thick-knee *Burhinus bistriatus* which in the Caribbean is confined to Hispaniola. It is present in the open pasturelands of this IBA.

■ Other biodiversity

Nothing recorded.

■ Conservation

Punta Cana IBA currently (2008) has no official protection status. Lands are privately owned, including the 610-ha Ojos Indigenous Reserve and Ecological Park, which was donated in 1994 by the Punta Cana Resort and Club to the Punta Cana Ecological Foundation. In general, main uses include tourism, recreation, and agriculture. There is no detailed information of threats to this IBA, but since it is located in one of the most popular tourist zones in the country, loss of habitat for tourist development could be considered as one of its main threats.

GRENADA

LAND AREA **344 km²** ALTITUDE **0–840 m**
HUMAN POPULATION **96,000** CAPITAL **Saint George's**
IMPORTANT BIRD AREAS **6, totalling 21.5 km²**
IMPORTANT BIRD AREA PROTECTION **85%**
BIRD SPECIES **164**
THREATENED BIRDS **1** RESTRICTED-RANGE BIRDS **7**

BONNIE L. RUSK
(DIRECTOR, GRENADA DOVE CONSERVATION PROGRAMME)

Critically Endangered Grenada Dove, less than 140 survive in Grenada's dry forests.
(PHOTO: GREG R. HOMEL)

INTRODUCTION

The island nation of Grenada lies at the southernmost end of the Lesser Antilles, just c.160 km north of the Venezuelan coast. It comprises three main islands: Grenada (311 km²) in the south; and the southern Grenadines islands of Carriacou (32 km², and 37 km north of Grenada) and Petit Martinique (0.7 km², and 6 km east of Carriacou) along with their associated offshore islets. Grenada is the highest remaining part of the submarine Grenada Bank that extends 180 km from Bequia (in the St Vincent Grenadines), south past Grenada. An active submarine volcano—"Kick-'em-Jenny"—sits 3 km north of Grenada. Its first recorded explosion into the atmosphere was in 1936. Carriacou and Petit Martinique are also exposed summits of the Grenada Bank.

The main island of Grenada is volcanic. The island's highest point is Mount St Catherine, part of a central chain of rugged mountains (that also includes Mount Qua Qua and Mount Sinai) that run north–south through the centre of the island. South of this mountain chain is a system of curving ridges (the Southern Mountains) that run toward the south and then bend to the east and north-east and make up the Grand Etang Forest Reserve. These mountains descend gradually to an extensive lowland coastal plain in the east. Grenada supports a wide diversity of forest types with (in 1982) cloud-forest (including elfin woodlands, palm brake and montane thickets) on the highest peaks (covering c.1,700 ha), then rainforests and lower montane rainforest (2,280 ha), evergreen and semi-evergreen seasonal forest, deciduous forest and dry woodlands (less than 1,750 ha), littoral woodland or dry coastal scrub (less than 1,230 ha), and mangrove forest (less than 190 ha). Mangroves have greatly diminished in numbers and area, and pockets are found primarily along the eastern coastline from True Blue to Requim, with some around Mount Hartman and Woburn Bays, and at Calivigny. One of the largest areas, at Levera, was recently (2003) mostly destroyed for development.

Like many Caribbean islands, Grenada was cleared of most of its forests to make way for sugarcane cultivation, but natural disasters paved the way for the introduction of other crops. In 1782, nutmeg *Myristica fragrans* was introduced to Grenada, and thrived in the island's ideal soils. The collapse of the sugar estates and the introduction of nutmeg and cacao encouraged the development of smaller landholdings, and the island developed a land-owning farmer class. Currently, Grenada relies on tourism as its main source of foreign exchange, especially since the construction of an international airport in 1985, located on the south-westernmost peninsula of Point Saline. Grenada's tropical climate is defined by a hot, humid rainy season from June to January (with maximum rainfall in November). Average annual rainfall varies between 4,000 mm in the mountainous interior to 1,500 mm in some coastal areas, and as little as 750 mm on the Point Saline peninsula. Grenada is divided into six parishes for administrative purposes.

The deep fertile soils on Carriacou and Petit Martinique resulted in extensive clearing for agriculture, primarily cotton and for a short period for sugarcane and fruit trees. By the late-1870s soil fertility had decreased and livestock were introduced. To this day, grazing continues to cause significant soil erosion, particularly during the dry season when the animals are released and graze freely. In spite of the clearances and livestock, recent (2004) estimates suggest that Carriacou is 65% forested (forest and woodland), albeit with just 135 ha in forest reserves (primarily High North Forest Reserve). The islands support seasonal evergreen forest, dry thorn scrub and deciduous forest, mangroves and fringing coral reefs. Offshore islets to the between Grenada and Carriacou are uninhabited.

■ Conservation

The protection of Grenada's natural resources (forests and wildlife) is directed by a series of policies and legislation, primarily under the management of the government's Forestry and National Parks Department (FNPD). The Ordinances for the Protection of Forests, Soil and Water Conservation (1954–1958, with an amendment in 1984) addresses the protection of areas to provide natural and undisturbed habitat for the flora and fauna of Grenada, and gives protection mainly to the Grand Etang Forest Reserve (the area around Grand Etang Lake, Grenada) and High North Forest Reserve (Carriacou). These forest reserves serve as sanctuaries for wildlife, and the legislation prohibits hunting, trapping, and carrying firearms within them. The Birds and other Wildlife Protection Ordinances (of 1956, 1964 and 1966) were intended to provide protection to wild birds and other wildlife, with all birds and their eggs being protected throughout the year except those listed in an annex for which there is an open season. The Critically Endangered Grenada Dove *Leptotila wellsi* is not specifically mentioned in this legislation, but it is protected within the context of "all wild birds".

In 1988, the Government of Grenada and the Organization of American States developed a Plan and Policy for a System of National Parks and Protected Areas for Grenada and Carriacou. However, this plan has not been implemented nor formally adopted. The National Parks and Protected Areas Act (1990) led to the establishment of both the Mount Hartman National Park and the Perseverance Protected Area (Dove Sanctuary). An amendment to this Act in 2007 enabled the re-designation of the Mount Hartman National Park boundaries (in 2008). The FNPD initiated a participatory, National Forest Policy review which resulted in a government-approved new Forest Policy (2002). The FNPD has also undertaken a review of all of Grenada's key forest and wildlife legislation, and has drafted revisions although (as of June 2008) these revisions have not yet been implemented. The revised legislation addresses issues such as endangered species, and grants specific mention to species of special concern, such as the *L. wellsi* and the Grenada Hook-billed Kite *Chondroheirax uncinatus mirus*.

About 23% of Grenada is currently forested: c.4,000 ha as higher elevation forests and c.3,000 ha of evergreen, semi-evergreen and deciduous seasonal forest and woodland. Almost 70% of this forest area is Crown land. However, most of Grenada's protected land (primarily the Grand Etang Forest Reserve and the proposed Mount St Catherine Forest Reserve) cover just the high elevation montane forest-types leaving the seasonal forest, deciduous forest, dry woodlands, dry coastal scrub and mangrove forest very poorly protected. Grenada lies south of the hurricane belt, and before Hurricane Ivan (September 2004) and Emily (July 2005), the last hurricane to hit the island was in 1955 (Janet). Hurricane Ivan had a profoundly devastating effect on the island's economy, agricultural sector and ecosystems (including those on which *L. wellsi* depends). Biodiversity on the island is being affected by a range of factors (mostly related to development and agricultural pressures) which are described under each of the IBA profiles below. However, the introduction of the mongoose *Herpestes auropunctatus* during the 1880s is particularly notable. This alien invasive predator is now abundant and is thought to be having a significant impact on Grenada's mammal, bird and herpetofauna.

■ Birds

Over 160 species of bird have been recorded from Grenada, with resident landbirds represented by just 35 species. The remainder is comprised of Neotropical migrants, waterbirds and seabirds. A number of birds of South American origin are

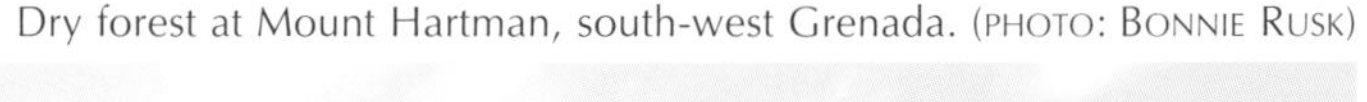

Dry forest at Mount Hartman, south-west Grenada. (PHOTO: BONNIE RUSK)

Table 1. Key bird species at Important Bird Areas in Grenada.

Key bird species	Criteria	National population	GD001	GD002	GD003	GD004	GD005	GD006
		Criteria	■ ■	■ ■	■ ■	■ ■	■ ■	■ ■
Grenada Dove *Leptotila wellsi*	■ ■	136	16	6	14		38	50
Green-throated Carib *Eulampis holosericeus*	■		✓	✓		✓		
Antillean Crested Hummingbird *Orthorhyncus cristatus*	■		✓	✓	✓	✓	✓	✓
Caribbean Elaenia *Elaenia martinica*	■		✓	✓	✓	✓	✓	✓
Grenada Flycatcher *Myiarchus nugator*	■		✓	✓	✓	✓	✓	✓
Lesser Antillean Bullfinch *Loxigilla noctis*	■		✓	✓		✓	✓	✓
Lesser Antillean Tanager *Tangara cucullata*	■		✓	✓	✓	✓	✓	✓

(Grenada IBAs: GD001–GD006)

All population figures = numbers of individuals. Dove population figures are based on numbers of calling males, and assume 1:1 sex ratio.
Threatened birds: Critically Endangered ■. **Restricted-range birds** ■.

present on the island. However, the zoogeographic boundary of the West Indian region separates Grenada from Trinidad and Tobago which has a predominantly South American avifauna. Lesser Antilles EBA restricted-range birds (of which there are 38) are represented by seven species (see Table 1), one of which, Grenada Dove *Leptotila wellsi*, is endemic to the island of Grenada. The Grenada Flycatcher *Myiarchus nugator* and Lesser Antillean Tanager *Tangara cucullata* are restricted to Grenada and St Vincent. Four other Lesser Antilles restricted-range birds have been recorded from Grenada but have not been included in the IBA analysis as their current status is unclear. The Purple-throated Carib *Eulampis jugularis* was seen several times in the 1960s although it has been suggested that these may have been vagrants. Scaly-breasted Thrasher *Margarops fuscus* was described as "not common" and then "very rare" prior to 1940 and its current status is unknown. The Antillean Euphonia *Euphonia musica* is similar—recorded as "not very common" in the 1900s and "very rare" by the 1940s, the species does not seem to have been recorded since. Last, the Brown Trembler *Cinclocerthia ruficauda* is unknown by current birdwatchers in Grenada (or visiting birding tours) although it is listed as present in the country. The presence of Lesser Antilles EBA restricted-range birds on Carriacou and Petit Martinique is poorly documented.

Leptotila wellsi is Grenada's national bird, but it is Critically Endangered and since its abundance and distribution were first documented in 1987, the species has been limited to two isolated patches of secondary seasonal dry forest in the south-west and west of the island. The total population declined by about 50% between 1987 and 1990. In 1998, the population numbered only c.100 individuals, increasing to an estimated 180 individuals by 2004. Surveys in 2007 found 68 calling males suggesting a post-hurricane recovering population of c.136 individuals. With so few individuals in the population, the dove features prominently in the IBA analysis for Grenada, and is the focus of a range of conservation efforts being implemented by the government and the international scientific community. The Near Threatened Caribbean Coot *Fulica caribaea* occurs on Grenada with 30 individuals recorded from Lake St Antoine in 1971, and an unspecified number again in 1987. However, there are no recent reports documenting the current status of the species on the island. The Near Threatened Buff-breasted Sandpiper *Tryngites subruficollis* is a very rare migrant on the island.

Grenada Flycatcher, endemic to Grenada, St Vincent and the Grenadines. (PHOTO: ALLAN SANDER)

Little has been documented concerning the status and distribution of Grenada's breeding and non-breeding seabirds (or in fact waterbirds and migrants in general). However, important areas for breeding seabirds are the unpopulated islets between Grenada and Carriacou, especially the islands close to Isle la Ronde. Boobies (presumably mostly Red-footed Booby *Sula sula* and Brown Booby *S. leucogaster*) are by far the most important species group and significant rookeries (of unknown size) are to be found at "gwizo" (near Isle la Ronde), Les Tantes, and "Upper Rock". Significant (but undocumented) numbers of Magnificent Frigatebird *Fregata magnificens* are resident at Sandy and Green Islands. Brown Noddy (*Anous stolidus*) and Red-billed Tropicbird *Phaethon aethereus* are also found on these islands. Roseate Tern *Sterna dougallii*, Bridled Tern *Sterna anaethetus*, and Sooty Tern *Sterna fuscata* were observed (2004) around the islands between Carriacou and Kick-'em-Jenny. All of these birds depend on the abundant fish (schools of anchovies and various fry or "pischet") in the Isle la Ronde zone. Fishermen and other poachers target the young (fat-chested) boobies and Scaly-naped Pigeon *Patagioenas squamosa* ("Ramier"), which, in 1987, were abundant on Sugarloaf and Sandy Islands.

IMPORTANT BIRD AREAS

Grenada's six IBAs—the country's international site priorities for bird conservation—cover 21.5 km², equivalent to 30% of remaining forested area, but only 6.25% of total land cover. They have been identified on the basis of seven key bird species (listed in Table 1) that variously trigger the IBA criteria. These species are all Lesser Antilles EBA restricted-range birds, one of which (Grenada Dove *Leptotila wellsi*) is Critically Endangered. There is very little data concerning the distribution and abundance of these key species (with the exception of *L. wellsi*), the other restricted-range birds that have been recorded historically, or other potential IBA

Mount Hartman National Park, home to 37% of the world's Grenada Dove population. (PHOTO: LISA SORENSON)

trigger species such as the waterbirds and seabirds that are known to occur but in unknown numbers. As a result, five of the IBAs have been identified primarily for the Critically Endangered *L. wellsi*, and together these IBAs support nearly the entire species' population. However, six males hold territories outside the IBAs: two on the west coast and four in the south-west.

Three of the IBAs (Perseverance GD002, Grand Etang GD004 and Mount Hartman GD006), are currently legally protected. The remainder include one IBA that is primarily Crown land (GD003), one that is owned (mostly) by a single land owner (GD001), and one that is held by many private land owners (GD005). These three unprotected IBAs all support critical habitat for *L. wellsi*. Indeed, Woodlands IBA (GD005) supports 27% of the world population of the dove and should therefore be the focus of special management measures. Grand Etang IBA is Grenada's only montane protected area. It represents a significant portion of Grenada's interior high mountain forest covering different vegetation types, and it supports populations of all the restricted-range birds other than *L. wellsi*. It might yet prove to support other restricted-range species (such as Purple-throated Carib *Eulampis jugularis*, Scaly-breasted Thrasher *Margarops fuscus*, Antillean Euphonia *Euphonia musica* and Brown Trembler

Red-footed Booby, one of a number of seabirds that urgently need surveying to assess the importance of Grenada's breeding colonies. (PHOTO: ALLAN SANDER)

Figure 1. Location of Important Bird Areas in Grenada.

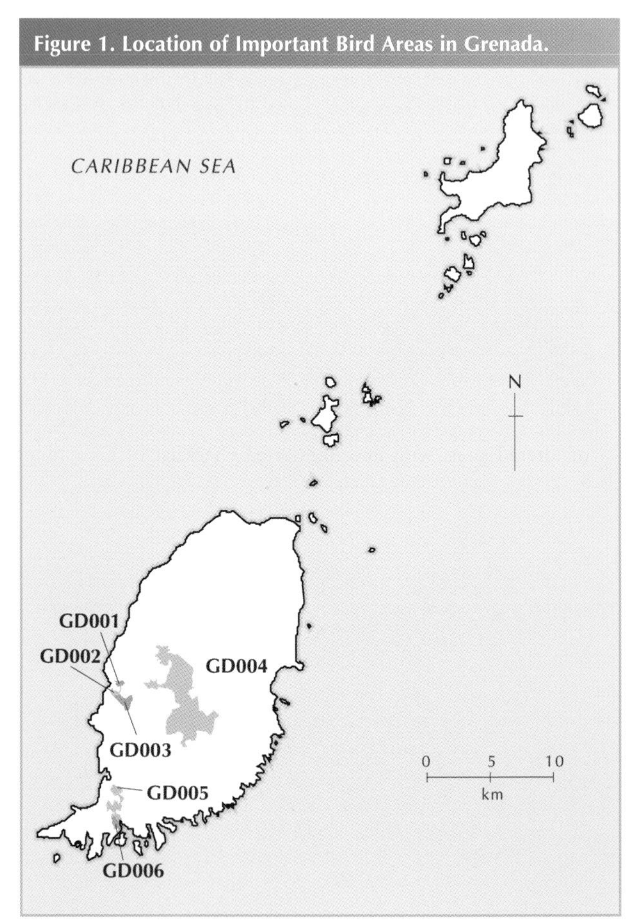

Cinclocerthia ruficauda; see above), and systematic surveys should be undertaken with these species in mind.

The paucity of data concerning the status and distribution of restricted-range species, waterbirds and seabirds suggests an urgent need for systematic surveys for these species groups throughout the islands that make up the nation of Grenada. With additional information it is likely that other IBAs will be identified. Mount St Catherines covers 573 ha of forest at 480–800 m in central Grenada. It probably supports the same key bird species that Grand Etang IBA supports, but this requires confirmation. Mount St Catherines is a proposed Forest Reserve currently being considered for designation by the Grenada Government. Surveys at Lake St Antoine may also highlight its continued importance for the Near Threatened Caribbean Coot *Fulica caribaea*, and surveys on/around the known seabird breeding colonies will help provide the necessary information to assess their international importance. Thus there is a clear and urgent need for surveys of wetlands and seabird colonies to establish a baseline against which to monitor and from which additional IBAs could possibly be described. The existing dove monitoring programme could be usefully expanded to include field assessments (surveys and subsequent monitoring) for other forested areas on Grenada and potentially the surrounding islands. All monitoring results should be used to inform the annual assessment of state, pressure and response scores at each of the country's IBAs to provide an objective status assessment and highlight management interventions that might be required to maintain these internationally important biodiversity sites.

KEY REFERENCES

Beard, J. S. (1949) The natural vegetation of the windward and leeward islands. *Oxford For. Mem.* 21: 136–158.

Blockstein, D. E. (1988) Two endangered birds of Grenada, West Indies: Grenada Dove and Grenada Hook-billed Kite. *Carib. J. Sci.* 24: 127–136.

Blockstein, D. E. (1991) Population declines of the endangered endemic birds of Grenada, West Indies. *Bird Conserv. Int.* 1:83–91.

Bond, J. (1993) *Birds of the West Indies* (5th ed.). Boston, Mass.: Houghton Mifflin Company.

Caribbean Conservation Association (1991) Grenada*: country environmental profile*. St George's, Grenada: Caribbean Conservation Association.

Devas, R. P. (1943) *Birds of Grenada, St. Vincent and the Grenadines*. St George's, Grenada; Carenage Press.

Evans, P. G. H. (1990) *Birds of the eastern Caribbean*. London: Macmillan Education Ltd.

Frost, M. D., Hayes, F. E. and Sutton, A. H. (in press) Saint Vincent, the Grenadines and Grenada. In P. E. Bradley and R. L. Norton eds. *Breeding seabirds of the Caribbean*. Gainesville, Florida: Univ. Florida Press.

Germano, J. M., Sander, J. M., Henderson, R. W. and Powell, R. (2003) Herptofaunal communities in Grenada: a comparison of altered sites, with and annotated checklist of Grenadian amphibians and reptiles. *Carib. J. Sci.* 39: 68–76.

Goodwin, D. (1983) *Pigeons and doves of the world*. Third edition. Ithaca, New York: Cornell University Press.

Groome, J. R. (1970) *A natural history of the island of Grenada, West Indies*. Arima, Trinidad: Caribbean Printers Ltd.

Howard, R. A. (1950) *Flora of the Lesser Antilles Leeward and Windward Islands*. Jamaica Plain, Mass.: Harvard University Press.

Johnson, M. S. (1985) *Forest inventory in Grenada*. Surbiton, UK: Land Resources Development Centre.

Lack, D. (1976) *Island biology: illustrated by the land birds of Jamaica*. Oxford: Blackwell Scientific Publications (Studies in Ecology 3).

Lack, D. and Lack, A. (1973) Birds on Grenada. *Ibis* 115: 53–59.

Lands and Surveys Division (1988) *Map of Grenada 1:25,000*. St George's, Grenada: Lands and Surveys Division, Government of Grenada.

Lugo, P. R. (2005) Composition and structure of Grenada Dove (*Leptotila wellsi*) habitat. St. George's, Grenada: Forestry and National Parks Dept. (Unpublished report).

Raffaele, H. Wiley J., Garrido, O., Keith, A. and Raffaele, J. (1998) *A guide to the birds of the West Indies*. Princeton, New Jersey: Princeton University Press.

Roberts, A. W. and Taylor, L. T. (1988) The Grenada Dove expedition. Reading, UK: University of Reading. (Unpublished report).

Rusk, B. L. (1992) The Grenada Dove and the Mt. Hartman Estate: management recommendations. St George's, Grenada: Forestry and National Parks Dept., Government of Grenada. (Unpublished report).

Rusk, B. L. (1998) Status of the endangered endemic Grenada Dove (*Leptotila wellsi*) on Grenada, West Indies. (Unpublished report).

Rusk, B. L. (2005) Draft post-Hurricane Ivan Grenada Dove evaluation. St George's, Grenada: Forestry and National Parks Dept., Government of Grenada. (Unpublished report).

Rusk, B. L. (2007). Grenada Dove and the Mt Hartman Estate. St. George's, Grenada: Government of Grenada and Cinnamon88. (Unpublished report for the "Developing the Four Seasons Resort, Grenada" project).

Rusk, B. L. (2008). Grenada Dove census 2007. St. George's, Grenada: Government of Grenada and Cinnamon88. (Unpublished report for the "Developing the Four Seasons Resort, Grenada" project).

Rusk, B. L., and Clouse, L. (2004) Status of the endangered endemic Grenada Dove (*Leptotila wellsi*). St George's, Grenada: Forestry and National Parks Dept., Government of Grenada. (Unpublished report).

Rusk, B. L. and Temple, S. A. (1995) Grenada Dove census, post habitat loss: west coast. (Unpublished report).

Rusk, B. L., Blockstein, D. E., Temple, S. A. and Collar, N. J. (1998) Draft recovery plan for the Grenada Dove. Washington, D.C.: Unpublished report for the World Bank and the Government of Grenada.

Smith, T. B., and Temple, S. A. (1982) Grenada Hook-billed Kites: recent status and life history notes. *Condor* 84:131.

Thomas, A. (2005) Overview of Biodiversity in Grenada. St George's, Grenada: Forestry and National Parks Dept., Government of Grenada. (Unpublished report).

Thornstrom, R., Massiah, E. and Hall, C. (2001) Nesting biology, distribution, and population estimate of the Grenada Hook-billed Kite *Chondroheirax uncinatus mirus*. *Carib. J.Sci.* 37: 278–281.

Toone, B., Ellis Joseph, S., Wirth, R. and Seal, U. (1994) Conservation assessment and management plan for pigeons and doves: report from a workshop held 10–13 March 1993, San Diego, USA. ICBP Pigeon and Dove Specialist Group and IUCN/SSC Captive Breeding Specialist Group.

Vincent, G. (1981) A report on the proposed Levera National Park, ECNAMP-CCA. St George's, Grenada: unpublished report.

Weaver, P. (1989) Forestry Department Technical Report: planning and management activities. (Unpublished FAO Tech. Coop. Prog. Rept. FO TCP/GRN/885).

Wells, J. G. (1887) A catalogue of the birds of Grenada, West Indies, with observations thereon. *Proc. U.S. Nat. Mus.* 9: 609–633.

Wunderle, J. M. (1985) An ecological comparison of the avifaunas of Grenada and Tobago, West Indies. *Wilson Bull.* 97: 356–365.

ACKNOWLEDGEMENTS

The author would like to thank Alan Joseph (Chief Forestry Officer, Forestry and National Parks Dept.), Anthony Jeremiah (Forestry and National Parks Dept.), other officers and staff of Grenada's Forestry and National Parks Dept., Jennifer Ellard (former Special Advisor to the Grenada Prime Minister and Minister of Finance), David Blockstein (Chair, Ornithological Council), Stanley A. Temple (University of Wisconsin-Madison and Aldo Leopold Foundation), George Ledec (World Bank), David Wege (BirdLife International), Lawrence Clouse, Verndal Phillip (research assistant), and B. Calliste (fisherman) for assistance in the field and/or in the drafting of this chapter.

■ Site description

The Woodford IBA is on the west coast of Grenada, north of Perseverance IBA (GD002). Its western border extends from the coast at Halifax Harbour to inland of the north–south coastal road along the Douce River. It is bordered to the north by the Douce River, except c.8 ha on the coastal side of the road, both north and south of the mouth of the Douce River. To the south, the IBA is bordered by a sanitary landfill and a proposed habitat corridor linking the site to the Perseverance Dove Sanctuary. Upriver is a gravel quarry, no longer in operation (2007). On the coastal side of the North–South road to the south are a mangrove swamp, Grenada's main garbage dump, and an asphalt plant.

■ Birds

This IBA is important for the Critically Endangered Grenada Dove *Leptotila wellsi*. The population was estimated at 12 pairs in 2003/2004 (pre-Hurricane Ivan) (Rusk & Clouse 2004), only four pairs 3–4 months after the hurricane in 2004 (a possible under estimate due to a change in calling behaviour post hurricane) (Rusk 2005), and eight pairs in 2007 (Rusk 2008). All seven of the Lesser Antilles EBA restricted-range birds occur at this IBA, as does the threatened endemic Grenada Hook-billed Kite *Chondrohierax uncinatus mirus* subspecies.

■ Other biodiversity

Nothing recorded.

■ Conservation

This site is private land owned primarily by a single landowner, and is unprotected. Due to its importance for *L. wellsi*, special management and/or its purchase for protection should be considered. Expanding industrial activity is the primary threat to the IBA, as well as clearing of land for agriculture. A proposed (2008) quarry on its southern border would negatively impact doves on adjacent sites. Hurricane Ivan (2004) caused extreme changes to vegetation structure and composition, and to the availability of resources used by the dove. Rats *Rattus* spp., mongoose *Herpestes auropunctatus* and other predators may affect the bird populations (including the dove).

■ Site description

The Perseverance IBA, a plantation destroyed by Hurricane Janet in 1955, is on the west coast of Grenada, inland from Halifax Harbour. Its western border is on the north–south coastal road, running inland c.400 m to its eastern border at the base of a steep cliff. To the north, the IBA is bordered by some houses and the Salle River. This IBA includes an unprotected habitat corridor north of the Salle River, inland of the sanitary landfill to the base of the ridge (behind which is a proposed quarry), linking this IBA with Woodford IBA (GD001). North of the river is a (non-operational) sanitary landfill. On the coastal side of the north–south road are a mangrove swamp, Grenada's main garbage dump, and an asphalt plant.

■ Birds

This IBA is important for the Critically Endangered Grenada Dove *Leptotila wellsi*. The population was estimated at six pairs in 2003/2004 (Rusk & Clouse 2004) (pre-Hurricane Ivan), only one pair 3–4 months after the hurricane in 2004 (Rusk 2005) (a possible under estimate due to a change in calling behaviour post hurricane), and three pairs in 2007 (Rusk 2008). It supports all of the Lesser Antilles EBA restricted-range birds and also the threatened endemic Grenada Hook-billed Kite *Chondrohierax uncinatus mirus* subspecies.

■ Other biodiversity

Nothing recorded.

■ Conservation

Perseverance IBA is a protected area (Dove Sanctuary) established in 1996 to protect *L. wellsi*. It comprises a c.32 ha (to be re-surveyed) portion of the 113-ha Government-owned Perseverance Estate (contrary to the government gazette which suggests it includes the entire estate). Protection measures are being explored for the nearby Beausejour–Grenville Vale IBA (GD003) that would link the two sites. Expanding industrial activity is the primary threat to the IBA. Toxic fumes (from landfill fires that started in 2004) and garbage are blown into the IBA. A proposed (2008) quarry would impact the habitat corridor to the Woodford IBA. Rats *Rattus* spp., mongoose *Herpestes auropunctatus* and cats may affect the bird populations (including *L. wellsi*). A metal recycling collection site borders the IBA to the north. Hurricane Ivan caused extreme changes to vegetation structure and composition, and to the availability of resources used by the dove. A fire (May 2008) burned the western-facing ridge down to the edge of the lower slopes, below which the doves are found.

■ Site description

Located near the west coast of Grenada, the Beausejour–Grenville Vale IBA is just inland from the north-south coastal road from Beausejour Bay. The bowl-shaped site is bordered to the west, north and south by a slope, rising to c.500 m in elevation, and includes the east-facing side of the eastern ridge (Grenville Vale). The slopes to the west, north and east are forested. The valley bottom comprises some fruit trees, an agricultural plot and cattle grazing. Immediately to the south is a housing development. The north-western side of this IBA is adjacent to the south-eastern border of the Perseverance IBA (GD002).

■ Birds

This IBA is important for the Critically Endangered Grenada Dove *Leptotila wellsi*. The population was estimated at c.15 pairs in 2003/2004 (pre-Hurricane Ivan) (Rusk & Clouse 2004) but only 1–6 pairs 3–4 months after the hurricane in 2004 (Rusk 2005) (a possible under/over estimate due to a change in calling behaviour post hurricane). Seven pairs were estimated in 2007 (Rusk 2008). Five (of the seven) Lesser Antilles EBA restricted-range birds occur at this IBA.

■ Other biodiversity

Nothing recorded.

■ Conservation

This site is both Crown land and under private ownership and is currently unprotected. Threats include the possible expansion of a housing development and cutting for agriculture. Rats *Rattus* spp., mongoose *Herpestes auropunctatus* and other predators may impact the *L. wellsi* population. A fire (May 2008) burned a large portion of the IBA, though remained upslope of most known dove territories. Hurricane Ivan caused extreme changes to vegetation structure and composition on the west coast. Protection measures are being explored by Government that could result in a contiguous protected area with the Perseverance IBA.

■ Site description

Grand Etang Forest Reserve is within the Southern Mountains of central Grenada. It encompasses mountains such as Mount Sinai (701 m), Mount Grandby (682 m) and Mount Qua Qua (735 m), and several crater lakes including the Grand Etang Lake near the centre of the reserve (at 530 m). Rainforests, lower montane rainforests and elfin woodlands characterise the steeper slopes throughout the high region. At lower altitudes, the trees are smaller and more thickly covered by epiphytes (ferns and mosses). Six plantations of exotics have been established on steep slopes since 1957.

■ Birds

Six (of the seven) Lesser Antilles EBA restricted-range birds occur at this IBA, and others may occur. Little survey or census work has been carried out. However, the site represents a major portion of Grenada's remaining high altitude forests that are so important for the restricted-range birds, and may yet prove to support populations of Purple-throated Carib *Eulampis jugularis*, Scaly-breasted Thrasher *Margarops fuscus*, Antillean Euphonia *Euphonia musica*, and Brown Trembler *Cinclocerthia ruficauda* (all of which were historically known to occur, but have not recently been recorded).

■ Other biodiversity

Eleutherodactylus euphronides is Endangered and confined to just 16 km^2 of Grenadian montane forest at elevations >300 m, including within this Forest Reserve. *Anolis aeneus and A. richardii* occur at this site along with other reptile and amphibian Grenada Bank endemics. Endemic plants include Grand Etang Fern *Danaea* sp., the Cabbage Palm *Oxeodoxa oleracea*, *Maythenus grenadensis*, *Rhytidophyllum caribaeum*, and *Lonchcarpus broadwayi*. The Mona monkey *Cercopithecus mona*, introduced from West Africa, is found in the upper montane forest of this IBA.

■ Conservation

This IBA was designated as a Forest Reserve in 1906. A visitor centre near the Grand Etang Lake is (in 2006) managed privately although nearby trails (and the forest) are maintained by the Forestry and National Parks Department. Illegal hunting and cutting of trees takes place within the forest reserve, though direct impacts on the forest birds are likely to be minimal. Regulations are enforced by the government.

■ Site description

Located in south-west Grenada, this IBA comprises discrete wooded areas within and adjacent to residential areas. Its southern border is the northern boundary of the Mt Hartman IBA (GD006). The site extends north across the Grand Anse Road through Petit Bouc to the eastern side to Café. The northernmost edge is a concrete road (with a concrete water tower). A golf course (with adjacent houses) is situated near the middle of the IBA.

■ Birds

This IBA is important for the Critically Endangered Grenada Dove *Leptotila wellsi*. The population was estimated at c.25 pairs in 2003/2004 (Rusk & Clouse 2004) (pre-Hurricane Ivan), c.12–15 pairs 3–4 months after the hurricane in 2004 (Rusk 2005) (a possible under/over estimate due to a change in calling behaviour post hurricane) and 19 pairs in 2007 (Rusk 2008). Six (of the 7) Lesser Antilles EBA restricted-range birds occur at this IBA, as does the Grenada Hook-billed Kite *Chondrohierax uncinatus mirus* subspecies.

■ Other biodiversity

Nothing recorded, although reptiles and amphibians endemic to the Grenada Bank have been recorded in the adjacent Mount Hartman Estate.

■ Conservation

This site is private land (with many land owners) and is unprotected. Due its importance for the Grenada Dove, special management of this site should be considered in partnership with private landowners. Purchasing portions of this site should also be considered. Expanding residential development (which is happening both along the edge of the IBA and within its boundaries) is the primary threat to the IBA. Roads through the IBA are also leading to habitat loss and are opening up new areas to increased disturbance and use. Hurricane Ivan (2004) severely affected the structure of the vegetation in this area.

■ Site description

Mount Hartman IBA comprises the (recently re-defined) national park on the coast in south-western Grenada. The park sits within the 187-ha Mount Hartman Estate; the 125 ha outside the park is privately owned. Originally three discrete parcels of woodland, the national park boundaries were re-designated (in 2008) to create one contiguous protected area of the same size. The IBA is bordered (except to the north) by a new resort development. To the north are residential houses, and to the north-east is a marina on Woburn Bay. Mount Hartman Bay lies to the south-west.

■ Birds

This IBA is the single most important site for the Critically Endangered Grenada Dove *Leptotila wellsi*. It supports a population estimated at c.20 pairs in 2003/2004 (pre-Hurricane Ivan) (Rusk & Clouse 2004), and c.11–20 pairs 3–4 months after the hurricane in 2004 (a possible under/over estimate due to a change in calling behaviour post hurricane) (Rusk 2005). Twenty-five pairs were estimated in 2007 (Rusk 2008) within the re-designated park boundaries. Six (of the 7) Lesser Antilles EBA restricted-range birds occur at this IBA, as does the threatened endemic Grenada Hook-billed Kite *Chondrohierax uncinatus mirus* subspecies.

■ Other biodiversity

The "Grenada Bank" endemic reptiles and amphibians *Corallus grenadensis*, *Anolis aeneus* and *A. richardii* occur.

■ Conservation

Mount Hartman was designated a national park in 1996 to protect *L. wellsi*. Disturbance from foot traffic and grazing are primarily confined to the lowlands outside the park, but grazing does sometimes occur within the park boundaries. Cutting for charcoal rarely occurs inside the park. Predators (rats *Rattus* spp., mongoose *Herpestes auropunctatus* and feral cats) may affect the bird populations (including *L. wellsi*). Hurricane Ivan caused extreme changes to vegetation structure and composition, and to the availability of resources used by the dove. The Mount Hartman Estate, previously government-owned, was sold (in 2008, excluding the national park) to a private developer for a large-scale tourism development involving a hotel, golf course and over a hundred villas. In conjunction with the development, the national park boundaries were re-designated to embrace the dove's centre of abundance and a contiguous area of suitable dove habitat. Mitigation measures have also been put in place to minimize the impact on critical dove habitat.

GUADELOUPE

LAND AREA 1,713 km^2 ALTITUDE 0–1,467 m
HUMAN POPULATION 453,000 CAPITAL Basse-Terre
IMPORTANT BIRD AREAS 9, totalling 505 km^2
IMPORTANT BIRD AREA PROTECTION 19%
BIRD SPECIES 251
THREATENED BIRDS 7 RESTRICTED-RANGE BIRDS 17

ANTHONY LEVESQUE AND ALAIN MATHURIN (AMAZONA)

Pointe des Châteaux IBA, south-easternmost Grande-Terre.
(PHOTO: FRANTZ DUZONT)

INTRODUCTION

Guadeloupe is an archipelago consisting of six groups of islands in the Lesser Antilles. Grande-Terre and Basse-Terre (1,520 km^2) are the two largest islands separated from each other by the narrow Salt River sea-channel, although they are connected by road. These are located 55 km south-east of Montserrat, 40 km north of Dominica and 100 km south of Antigua and Barbuda. The other five groups of islands consist of Marie-Galante (152 km^2), La Désirade (22 km^2), Petite-Terre (Terre de Haut and Terre de Bas, 2 km^2), Les Saintes (five main islands Terre de Bas, Terre de Haut, Cabrit Islet, Grand Islet and la Coche, 19 km^2) and the final group contains all the small islets. The islands of St Barthélemy (24 km^2) and the French part of Saint Martin (92 km^2) are situated 300 km to the north of Guadeloupe. These latter two islands were under the jurisdiction of Guadeloupe until February 2007 when they were officially detached (see the appropriate chapters for details of their current status). Guadeloupe is a *département d'outre-mer* (DOM, overseas department) of France (and an outermost region of the European Union).

Basse-Terre has a rugged volcanic relief while Grande-Terre features rolling hills and flat plains. The highest point of Guadeloupe is La Soufrière, an active volcano rising to 1,467 m in the south of Basse-Terre. Wet tropical forest has developed on its slopes, replaced by wet grasslands towards the summit, both benefiting from abundant rains. In contrast, Grande-Terre and its dependencies (Marie-Galante, Désirade, and Petite-Terre) are limestone islands influenced by the trade winds. The precipitation is significantly lower and vegetation is consequently xerophytic, with mangroves on the coast. Historically, large areas of land were cleared for sugarcane plantations and livestock (cattle and goats) grazing. Secondary dry woodland has developed in abandoned areas.

Hurricanes hit the islands periodically and can devastate the economy, which depends mostly on tourism and agriculture, as well as light industry and services. Guadeloupe is also highly dependant on large subsidies and imports from France and the European Union. Unemployment is high, particularly amongst the younger generation.

Conservation

The National Park of Guadeloupe covers 33,500 ha (20% of Guadeloupe's land area) of forested habitat on Basse-Terre (IBA GP002). It is the only national park in the archipelago. Combined with the Grand Cul-de-Sac Marin (IBA GP005) they have been designated as part of a 72,380-ha biosphere reserve. Grand Cul-de-Sac Marin (3,706 ha) and Petite-Terre

The Near Threatened and endemic Guadeloupe Woodpecker. (PHOTO: ANTHONY LEVESQUE)

Islets (990 ha) are the only two national nature reserves, and Grand Cul-de-Sac Marin is also recognised as a Ramsar site (for its internationally important coastal wetlands and coral reefs)

The Regional Department of Environment (the DIREN) is the representative in Guadeloupe of the Ministry of Ecology and Sustainable Development. Most of the forests are under the jurisdiction of and managed by the Office National des Forêt (French National Forest Office) and Conservatoire du littoral (Coastal Protection Agency). No institutions are currently focused on bird ecology or biology research work across the archipelago, although a 3-year study of the Guadeloupe Woodpecker *Melanerpes herminieri* was conducted in the 1990s, and in 2006, AMAZONA establish a bird research program including ringing (banding), specific terrestrial studies and shorebird monitoring in Grande-Terre.

Humans are the main cause of the decline of certain birds in Guadeloupe. Activities specifically threatening the birds include of poaching of adults, eggs and chicks, and habitat destruction and degradation as a result of urban and agricultural expansion. Introduced species such as rats *Rattus rattus*, mongoose *Herpestes auropunctatus* and domestic cats *Felis catus* are also a threat to birds. Hunting was introduced by law in 1953. It remains a priority to develop the effective regulation of hunting through rigorous monitoring of the activity. This is the responsibility of the rangers of the Office National de la Chasse et de la Faune Sauvage (the National Hunting and Wildlife Agency). Hunting is a particular issue for the globally threatened Forest Thrush *Cichlherminia lherminieri*, West Indian Whistling-duck *Dendrocygna arborea*, White-crowned Pigeon *Patagioenas leucocephala* and Caribbean Coot *Fulica caribaea* and their protection must be enforced if Guadeloupe is to retain these species in the long-term. The Near Threatened Buff-breasted Sandpiper *Tryngites subruficollis* is an irregular visitor to the island, but known to be hunted. Targetted research and monitoring programs for these threatened species urgently needed to complement the enforcement of hunting regulations. Environmental educational programs are required to highlight the main threats to the avifauna (including the impacts of alien invasive species), to raise awareness of the need for environmental protection, and to promote an understanding of the richness of Guadeloupe's natural environment.

■ Birds

Of Guadeloupe's 251 recorded bird species, 72 breed and over 180 species are Neotropical migrants occurring as winter visitors, transients or vagrants. There are 17 restricted-range birds (see Table 1) of the 38 occurring in the Lesser Antilles Endemic Bird Area (EBA), including the Vulnerable Forest Thrush *Cichlherminia lherminieri* and the island's only extant endemic bird, the Near Threatened Guadeloupe Woodpecker *Melanerpes herminieri*. Plumbeous Warbler *Dendroica plumbea* is restricted to Guadeloupe and neighbouring Dominica.

Brown Trembler—one of 17 restricted-range birds occurring in Guadeloupe. (PHOTO: ANTHONY LEVESQUE)

Plumbeous Warbler—endemic to Guadeloupe and Dominica. (PHOTO: FRANTZ DUZONT)

Most of these restricted-range species are found in Basse-Terre as they are forest-dependent birds. An endemic subspecies of House Wren *Troglodytes aedon guadelupensis* was last recorded in 1973, and may represent the latest of a number of island extinctions—three species of Psittacids (Guadeloupe Amazon *Amazona violacea,* Guadeloupe Parakeet *Aratinga labati* and Lesser Antillean Macaw *Ara guadeloupensis*) were extinct by the end of the eighteenth century. Black-capped Petrel *Pterodroma hasitata*, Burrowing Owl *Athene cunicularia,* Caribbean Flamingo *Phoenicopterus ruber* and Magnificent Frigatebird *Fregata magnificens* all used to breed in Guadeloupe.

Only two globally threatened species have been considered in the IBA analysis (see Table 1). However, seven species have been recorded: White-crowned Pigeon *Patagioenas leucocephala*, Caribbean Coot *Fulica caribaea*, Buff-breasted Sandpiper *Tryngites subruficollis* and Piping Plover *Charadrius melodus* (all Near Threatened) are all considered passage migrants or vagrants to the island. The Vulnerable West Indian Whistling-duck *Dendrocygna arborea* is trying to colonise but some are hunted when arriving from neighbouring islands. Due to the small numbers currently involved this species is not considered in the IBA analyses (although see IBA GP008 below). *Cichlherminia lherminieri,* which is restricted to Montserrat, Guadeloupe, Dominica and St Lucia, is principally found within woodland and swamp-forest. A new regulation forbids the hunting of *C. lherminieri* on Grande-Terre, and for Basse-Terre has set a bag limit of eight birds per day (only on Saturdays, Sundays and public holidays) between October and December. Surveys to establish a robust population estimate for the species throughout Guadeloupe will be essential to determine its status set appropriate conservation measures. The island-endemic *Melanerpes herminieri* is Guadeloupe's national and is confined to the same IBAs as *C. herminieri.* It inhabits semi-deciduous forest on igneous and clay soils, and evergreen forest, mangroves and swamp forests. It occurs from sea-level to the tree-line at 1,000 m, but is most common between 100–700 m. Clear-cutting and the removal of dead trees are the main threats, but damage from hurricanes, road construction, airport enlargement and land development are all threats. Introduced rats may also be a problem. The global population of *M. herminieri* has been estimated at 19,527 breeding pairs in 2007.

Guadeloupe supports a diverse range of waterbirds including 14 regular breeding species breeding and six that breed occasionally. There are 42 regular Neotropical migrant waterbirds and another 25 occurring less frequently). The main waterbird sites on the islands are: Grand Cul-de-Sac Marin (IBA GP005), Marais de Port-Louis, Pointe des Châteaux (IBA GP007), Petite-Terre Islets (IBA GP008) and Barrage de Gaschet (IBA GP004).

In February 2008, the first breeding report for Guadeloupe of the Vulnerable West Indian Whistling-duck *Dendrocygna arborea* (with nine ducklings) was documented from Petite-Terre Islets National Nature Reserve (IBA GP008). Seabirds are also an important component of Guadeloupe's avifauna. Seven species are abundant, regular breeders, namely: Red-billed Tropicbird *Phaethon aethereus* (245–445 pairs), White-tailed Tropicbird *P. lepturus* (50–90 pairs), Sooty Tern *Sterna fuscata* (2,450–3,300 pairs), Bridled Tern *S. anaethetus* (205–275 pairs), Brown Noddy *Anous stolidus* (435–525 pairs), Least Tern *S. antillarum* (50–75 pairs) and Roseate Tern *S. dougallii* (20–30 pairs). An additional three species breed rarely, namely: Audubon's Shearwater *Puffinus lherminieri*, Brown Booby *Sula leucogaster* and Brown Pelican *Pelecanus occidentalis*. The most important sites for breeding seabirds are Tête à l'Anglais Islet (IBA GP001), Pointe des Châteaux (IBA GP007), north cliffs of Grande-Terre (IBA GP003), la Désirade, Marie-Galante (IBA GP009) and Les Saintes. There is an important migratory corridor for Procellariidae located between Petite-Terre (IBA GP008), Désirade islets and Pointe

Barrage de Gaschet IBA, one of Guadeloupe's main waterbird sites. (PHOTO: ANTHONY LEVESQUE)

Sooty Terns at Tête à l'Anglais Islet IBA, one of the most important sites for seabirds in Guadeloupe. (PHOTO: PASCAL VILLARD)

des Châteaux with thousands of shearwaters (and some petrels) passing by.

IMPORTANT BIRD AREAS

Guadeloupe's nine IBAs—the country's international priorities sites for bird conservation—cover 505 km² (including marine areas) and about 19% of Guadeloupe's land area. Most of the IBAs lack formal protective designation. Only the central areas of the forest habitat of Basse-Terre (IBA GP002) are protected by the National Park of Guadeloupe. Portions of Grand Cul-de-Sac (IBA GP005) and Petite-Terre Islets (IBA GP008) are protected within the Grand Cul-de-Sac marin National Nature Reserve (Réserve Naturelle) and the îlets de la Petite-Terre National Nature Reserve, respectively. North cliffs of Grande-Terre (IBA GP003), Pointe des Châteaux (IBA GP007) and Marie-Galante North Cliffs and Îlet de Vieux-Fort (IBA GP009) are Littoral Conservation Areas belonging to the Conservatoire du littoral where lands up to 15 m from the coast (the littoral zone) are protected by law. The protection of most of these IBAs is part of the

Îlet de Vieux-Fort (off the coast of Marie-Galante) is important for seabirds, and is within a Littoral Conservation Area. (PHOTO: ANTHONY LEVESQUE)

Table 1. Key bird species at Important Bird Areas in Guadeloupe.

			Guadeloupe IBAs								
			GP001	GP002	GP003	GP004	GP005	GP006	GP007	GP008	GP009
Key bird species	**Criteria**	**National population**	■	■ ■ ■	■	■	■ ■ ■	■	■ ■	■ ■	■
Masked Duck *Nomonyx dominicus*	■	<50				10–80					
Red-billed Tropicbird *Phaethon aethereus*	■	735–3,000			120–210				75–105		450–900
Least Tern *Sterna antillarum*	■	150–750						60–150	30–72	300	
Bridled Tern *Sterna anaethetus*	■	615–825							150–180		180–210
Sooty Tern *Sterna fuscata*	■	7,350–9,900	18,468								
Brown Noddy *Anous stolidus*	■	1,305–1,575									600–660
Bridled Quail-dove *Geotrygon mystacea*	■			✓			✓				
Lesser Antillean Swift *Chaetura martinica*	■			✓							
Purple-throated Carib *Eulampis jugularis*	■			✓					✓	✓	
Green-throated Carib *Eulampis holosericeus*	■			✓			✓		✓	✓	
Antillean Crested Hummingbird *Orthorhyncus cristatus*	■			✓			✓		✓	✓	
Guadeloupe Woodpecker *Melanerpes herminieri*	NT ■ ■	58,581		22,500–30,990			2,310				
Caribbean Elaenia *Elaenia martinica*	■			✓			✓		✓	✓	
Lesser Antillean Pewee *Contopus latirostris*	■			✓			✓				
Lesser Antillean Flycatcher *Myiarchus oberi*	■			✓							
Scaly-breasted Thrasher *Margarops fuscus*	■			✓					✓	✓	
Pearly-eyed Thrasher *Margarops fuscatus*	■			✓			✓		✓	✓	
Brown Trembler *Cinclocerthia ruficauda*	■			✓			✓				
Forest Thrush *Cichlherminia lherminieri*	VU ■ ■			✓			✓				
Plumbeous Warbler *Dendroica plumbea*	■			✓			✓				
Lesser Antillean Bullfinch *Loxigilla noctis*	■			✓			✓		✓		
Antillean Euphonia *Euphonia musica*	■			✓							
Lesser Antillean Saltator *Saltator albicollis*	■			✓			✓		✓		

All population figures = numbers of individuals.
Threatened birds: Vulnerable ■; Near Threatened ■. **Restricted-range birds** ■. **Congregatory birds** ■.

Figure 1. Location of Important Bird Areas in Guadeloupe.

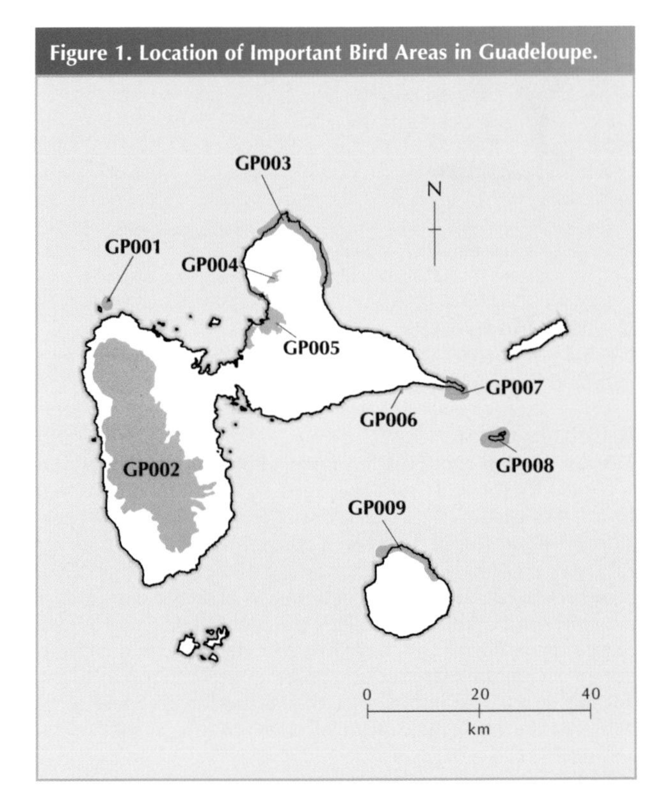

commitments and responsibilities of the French government under the 1976 Nature Protection Law, and to the international conventions (e.g. 1992 Convention on Biological Diversity, and the 1971 Ramsar Convention). These international commitments are particularly important since the European Union Birds (409/79/EC) and Habitats (43/92/EC) directives do not currently apply to the French Overseas Departments.

The IBAs have been identified on the basis of 23 key bird species (listed in Table 1) that variously meet the IBA criteria. These 23 species include two (of the seven) globally threatened birds (see "Birds" above), all 17 restricted-range bird species, and six congregatory waterbirds/ seabirds. The IBAs of Basse-Terre (GP002) and Grand Cul-de-Sac (GP005) are critical for supporting populations of the restricted-range birds, and are the only two IBAs for the globally threatened species. Seabirds are distributed a number of the remaining IBAs, showing that the protection of all the satellite islands will be essential to in maintaining the country's seabird populations. Marais de Port-Louis was proposed as an IBA, but less than 30% of the Lesser Antilles restricted-range species were present, and no globally threatened birds occur, so it has not met the criteria for international significance.

It is clear that a coherent monitoring program is urgently needed for some species (such as the Forest Thrush *Cichlherminia lherminieri*, other hunted species, the seabirds and waterbirds). The results from such future monitoring efforts (and any monitoring currently being undertaken) should be used to inform the annual assessment of state, pressure and response scores at each of Guadeloupe's IBAs which will provide an objective status assessment for these internationally important biodiversity sites and highlight management interventions required to maintain their integrity.

KEY REFERENCES

AEVA (2008) Statut de la population du Pic de la Guadeloupe (*Melanerpes herminieri*) en 2007. Petit-Bourg, Guadeloupe : (unpublished report: AEVA contribution 30).

Barlow, J. C. (1978) Another colony of the Guadeloupe House Wren. *Wilson Bull.* 90:635–637.

Breuil, M. (2002) Histoire naturelle des amphibiens et reptiles terrestres de l'archipel Guadeloupéen: Guadeloupe, Saint-Martin, Saint-Barthélemy. *Patrimoines Naturels* 54: 1–339.

DIREN (2001) *Atlas du patrimoine guadeloupéen: espaces naturels et paysages*. Sainte-Anne, Guadeloupe: DIREN.

Dumont, R. (2004) *Plan de gestion 2004–2008 de la Réserve Naturelle des îlets de la Petite-Terre.* Les Abymes, Guadeloupe: Office National des Forêts (Unpublished report).

Ibéné B., Leblanc, F. and Barataud, M. (2006) Complément d'inventaire des chauves-souris de la Guadeloupe. Sainte-Anne, Guadeloupe: L'ASFA, DIREN (Unpublished report).

Ibéné B., Leblanc. F and Pentier, C. (2007) Contribution à l'étude des chiroptères de la Guadeloupe. Sainte-Anne, Guadeloupe: L'ASFA, DIREN, *Groupe Chiroptères Guadeloupe* (Unpublished report).

Leblond, G. (2003) Les oiseaux marins nicheurs de Guadeloupe, de Saint-Martin et de Saint-Barthélemy. (Unpublished report: BIOS/DIREN).

Levesque, A. and Sorenson, L. (in prep.) First record of a West Indian Whistling-duck nesting in Guadeloupe.

Levesque, A. and Chevry, L. (2006) Suivi des limicoles à la Pointe des Châteaux (août à octobre 2006). Le Gosier, Guadeloupe: AMAZONA (Unpublished report: AMAZONA contribution 10).

Levesque, A. and Jaffard, M. E. (2002) Fifteen new bird species in Guadeloupe (F.W.I.). *El Pitirre* 15: 5–6.

Levesque, A. and Yésou, P. (2005) The abundance of shearwaters and petrels off the Lesser Antilles: results from a Guadeloupe-based study, 2001–2004. *North American Birds* 59: 672–677.

Levesque, A., Duzont, F. and Mathurin A. (2007) Liste des oiseaux de la Guadeloupe. Le Gosier, Guadeloupe: AMAZONA (Unpublished report: AMAZONA contribution 13).

Levesque, A., Duzont, F. and Ramsahai, A. (2005) Précisions sur cinq espèces d'oiseaux dont la nidification a été découverte en Guadeloupe (Antilles Françaises) depuis 1997. *J. Carib. Orn.* 18: 45–47.

Lurel, F. (1995) Inventaire et cartographie des groupements végétaux du site du barrage de Gaschet, communes de Petit-Canal et Port-Louis, et notes sur la faune en vue d'un programme de reboisement du secteur de Gaschet. Petit-Bourg, Guadeloupe. (Unpublished report).

Lurel, F. (1998) Végétation de l'étage des forêts semi-décidues Guadeloupe. Petit-Bourg, Guadeloupe. (Unpublished report).

Lurel, F. (1999) Végétation de la presqu'île de la Pointe des Châteaux de Guadeloupe. Etude préalable à l'Opération Grand Site. Petit-Bourg, Guadeloupe. (Unpublished report).

Lurel, F. (2000) Marais de Port-Louis entre terre et mer, diagnostic écologique et recommandations globales. Petit-Bourg, Guadeloupe. (Unpublished report : ACED contribution 55).

Lurel, F. (2003) Description végétation des falaises du nord est de Grande Terre Anse des Corps à Porte d'Enfer. Petit-Bourg, Guadeloupe. (Unpublished report).

Lurel, F., Fournet J., and Grandguillotte M. (2003) Falaises Est de Marie Galante. Petit-Bourg, Guadeloupe. (Unpublished report : Fiche inventaire ZNIEFF no. 0000–0031).

Lurel, F. and Warichi, A. (2002) Etude de la zone côtière inondable de Belle Plaine en vue d'un espace muséal de maison de la mangrove (Taonaba). Petit-Bourg, Guadeloupe. (Unpublished report).

Raffaele, H., Wiley J., Garrido, O., Keith, A. and Raffaele, J. (1998) *A guide to the birds of the West Indies*. Princeton, New Jersey: Princeton University Press.

Villard, P. (1999) *Le Pic de la Guadeloupel The Guadeloupe Woodpecker*. Brunoy, France : Société d'Etudes Ornithologiques de France (SOEF).

ACKNOWLEDGEMENTS

The authors would like to thank Maurice Anselme (CAR-SPAW), Bernard Deceuninck (LPO), Alison Duncan (LPO), Frantz Duzont (AMAZONA), Philippe Feldmann (AEVA), Béatrice Ibéné (L'ASFA), Luc Legendre (DIREN), Félix Lurel (ACED) and Pascal Villard (AEVA).

GP001 Tête à l'Anglais Islet

COORDINATES 16°22'N 61°45'W
ADMIN REGION Deshaies
AREA 339 ha
ALTITUDE 0–44 m
HABITAT Rocky area, coast, cliffs, sea

■ Site description
Tête à l'Anglais IBA is an oval shaped islet situated 3 km off the north-coast of Basse-Terre Island. The 1.5-ha islet is rocky with cliffs and steep slopes that reach 44 m in altitude. The vegetation is xerophytic and dominated by succulent plants including Royen's tree cactus *Pilosocereus royenii* and Spanish lady *Opuntia triancantha*. There is no habitation on the islet, but occasionally boats anchor by the island to fish, picnic or collect eggs.

■ Birds
This IBA is regionally significant for its breeding colony of Sooty Tern *Sterna fuscata*. Surveys in 2007 estimated a population of 6,156 (± 518) nesting pairs. The site is suitable for breeding and roosting by other seabirds, with 15 pairs of Roseate Tern *Sterna dougalli* and 50 individual Brown Noddy *Anous stolidus* known to breed. A small group of Brown Booby *Sula leucogaster* has been seen flying to roost on the islet.

■ Other biodiversity
Nothing recorded. A complete inventory of reptiles and mammals is needed.

■ Conservation
Tête à l'Anglais islet IBA has no legal protection. The island is owned by the state and managed by the French National Forest Office. In 1995, the site was reconised as a (Type 1) Zone Naturelles d'Intérêt Ecologique Faunistique et Floristique (ZNIEFF). Every year, until the end of the 1990s, poachers burnt the islet's vegetation to provider easy access for collecting seabird eggs. This practice is believed to have ceased. Rats *Rattus* spp. have not been detected on the island but an accidental introduction poses a potential threat to the nesting terns. The importance of the site for breeding terns deserves the implementation of research and conservation actions.

GP002 Basse-Terre Forest Massif

COORDINATES 16°10'N 61°40'W
ADMIN REGION Île de Basse-Terre
AREA 38,705 ha
ALTITUDE 28–1,467 m
HABITAT Forest, grassland

■ Site description
The Basse-Terre Forest Massif IBA encompasses almost all the central land mass of Basse-Terre island from nearly sea-level to the top of Soufrière volcano (the highest mountain in Guadeloupe). The boundary of this IBA follow those of the Departmental and State Forest as defined (and managed) by the French National Forests Office. The IBA embraces the Guadeloupe National Park which covers 60% of this IBA. Basse-Terre's cities and towns are located outside the IBA but are expanding towards the forest edge. Different types of vegetation exist due to diverse altitudes, wind exposure and volcanic soils.

■ Birds
This IBA supports populations of all 17 Lesser Antilles EBA restricted-range birds including c.50% of the known population of the Near Threatened (island-endemic) Guadeloupe Woodpecker *Melanerpes herminieri*. The Vulnerable Forest Thrush *Cichlherminia lherminieri* is also commonly found within the IBA. The Lesser Antilles subspecies of Ringed Kingfisher *Ceryle torquatus stictipennis* is found in Guadeloupe only this IBA (and has a population of less than 70 pairs). The endemic subspecies of House Wren *Troglodytes aedon guadelupensis* was last recorded in 1973, from this IBA.

■ Other biodiversity
The Endangered Guadeloupe big brown bat *Eptesicus guadeloupensis* and Thomas's yellow-shouldered bat *Sturnira thomasi*, and Vulnerable Dominican myotis *Myotis dominicensis* occur. The Endangered frogs *Eleutherodactylus pinchoni* and *E. barlagnei* are endemic to Basse-Terre. This IBA is rich in rare plants, 25 of which are endemic to Guadeloupe.

■ Conservation
The Basse-Terre Forest Massif IBA has mixed ownership. The total central zone includes all 17,300 ha of the National Park of Guadeloupe. This forest is protected by a 1948 ministerial decree that confers permission of usage to the French National Forests Office. Almost all the IBA is recognised as biosphere reserve. The lands managed by the French National Forests Office and the National Park of Guadeloupe are well protected. Threats are directly related to the presence of introduced mammals such as black rat *Rattus rattus* and mongoose *Herpestes auropunctatus*. Poaching is more common in areas surrounding the national park. The uncontrolled use of pesticides in banana plantations is also threat. A study of *Melanerpes lherminieri* was carried out in the National Park of Guadeloupe which represents the most studied site for this endemic species.

GP003 North Cliffs of Grande-Terre

Littoral Conservation Area

COORDINATES 16°29′N 61°27′W
ADMIN REGION Anse-Bertrand, Petit-Canal
AREA 3,960 ha
ALTITUDE 0–95 m
HABITAT Coastline, cliffs, sea

THREATENED BIRDS
RESTRICTED-RANGE BIRDS
BIOME-RESTRICTED BIRDS
CONGREGATORY BIRDS ✓

■ Site description
The North Cliffs of Grande-Terre IBA covers a 25-km stretch of cliffs from Pointe de la Petite Vigie (Anse-Bertrand) to Pointe Bellacaty (Petit-Canal). The cliffs rise abruptly from the sea and range in height from 25 to 75 m. Vegetation is limited to xerophytic grassland, low shrubs and cacti (ude to the lack of rain during 4 to 9 months of the year). The cliffs are important in providing nesting crevices for seabirds. The IBA extends 250 m inland from the cliff edge, and includes marine areas up to 1 km from the cliffs

■ Birds
This IBA is globally important for its breeding population of 40–70 pairs of Red-billed Tropicbird. Other seabirds breeding are White-tailed Tropicbird *P. lepturus* (5–10 pairs), Bridled Tern *Sterna anaethetus* (10–20 pairs), Brown Noddy *Anous stolidus* (30–60 pairs) and a group of Brown Booby *Sula leucogaster*. Seabirds are distributed all along the cliffs. However, the section with the largest number is situated between Piton and Pointe Grand Rempart. Audubon's Shearwater *Puffinus lherminieri* has not been confirmed breeding in Guadeloupe but this IBA presents suitable nesting habitat.

■ Other biodiversity
The cliffs support several rare, Lesser Antilles endemic herb species such as *Chamaesyce balbisii.*

■ Conservation
The North Cliffs of Grande-Terre IBA is owned by the state and is protected by the coastline law that prohibits construction within 15 m of the littoral zone. The site is recognised as a Zone Naturelles d'Intérêt Ecologique Faunistique et Floristique (ZNIEFF). The cliffs are threatened due to the proximity of agricultural activities (and heavy machinery) that are causing soil erosion and cliff degradation. The development of a wind farm has generated disturbance during the construction period and warrants monitoring after construction. Fishermen access the cliffs and disturb/ impact the seabird nesting areas. Introduce mammals such as black rat *Rattus rattus* and mongoose *Herpestes auropunctatus* prey on the seabird breeding colony. A survey of seabird colonies was undertaken in 2002.

GP004 Barrage de Gaschet

Unprotected

COORDINATES 16°25′N 61°29′W
ADMIN REGION Port-Louis
AREA 290 ha
ALTITUDE 4–11 m
HABITAT Artificial freshwater wetland, grassland, shrubland

Masked Duck

THREATENED BIRDS
RESTRICTED-RANGE BIRDS
BIOME-RESTRICTED BIRDS
CONGREGATORY BIRDS ✓

■ Site description
Barrage de Gaschet IBA is the largest artificial area of fresh water in Guadeloupe and is situated in northern Grand-Terre. It is the property of the Conseil Général (County Council) and was created at the beginning of the 1990s for agricultural irrigation. The reservoir is 4 km long and is surrounded by extensive cattle pastures. In certain areas, particularly the shallower parts, it has been colonised by water plants such as cattails *Typha* sp. Scattered shrubland also surround the area.

■ Birds
This IBA supports a wide diversity of waterbirds. It is regionally important for the population of Masked Duck *Nomonyx dominicus* (1–5 pairs), occasionally holding up to 80 individuals. Other waterbirds recorded are the Near Threatened Caribbean Coot *Fulica caribaea* (4–5 pairs) and 10–20 pairs of Pied-billed Grebe *Podilymbus podiceps*. In July 2006, a population of 129 *P. podiceps* was observed in the reservoir. Common Moorhen *Gallinula chloropus* has also been seen in high numbers, with up to 1,250 individuals noted. The Near Threatened Buff-breasted Sandpiper *Tryngites subruficollis* has also been recorded in the IBA as a vagrant. In 2007, seven Vulnerable West Indian Whistling-duck *Dendrocygna arborea* arrived on the reservoir—five of them were shot by hunters.

■ Other biodiversity
Twenty rare species and several threatened flowering plants such as *Byrsonina lucida, Sideroxylon salicifolium, Cordia collococca,* and *Bucida buceras* occur.

■ Conservation
Barrage de Gaschet IBA is state owned. The site was recognised as a (Type 1) Zone Naturelles d'Intérêt Ecologique Faunistique et Floristique (ZNIEFF) in 2000. Since its creation, the reservoir has been surrounded by fences prohibiting access, and therefore hunting. In recent years the fences have been damaged and opened, leading to increasing hunting pressure on the waterbirds. Hundreds of birds are present before mid-July but this does not compare with the large numbers arriving after the hunting season closes. *Fulica caribaea* is severely threatened by hunting even during the breeding season, despite the fact that hunting this species is prohibited by law. Mongoose *Herpestes auropunctatus* are abundant and prey on the eggs of waterbirds that nest on the ground. Traditional activities such as grazing and arable farming favour the expansion of invasive leguminous plant species such as aroma *Dichrostachys cinerea.* AMAZONA coordinates regular waterbird counts in this IBA.

GP005 East Coast of Grand Cul-de-Sac Marin

National Nature Reserve/Coastal State Forest/Biosphere Reserve/Ramsar Site

COORDINATES 16°21′N 61°29′W
ADMIN REGION Petit-Canal, Port-Louis, Abymes, Morne-à-l'Eau
AREA 2,785 ha
ALTITUDE 0–21 m
HABITAT Wetland, forest, grasslands

Guadeloupe Woodpecker

THREATENED BIRDS 2
RESTRICTED-RANGE BIRDS 12
BIOME-RESTRICTED BIRDS
CONGREGATORY BIRDS

■ Site description

East Coast of Grand Cul-de-Sac Marin IBA supports the most extensive mangrove and swamp forest of Guadeloupe. It lies on the west coast of Grand-Terre and is bounded to the north by Port-Louis Fishing Harbour and to the south by the Bridge of the Alliance. This massive forest is divided in two, separated by the town of Vieux-Bourg. The north-eastern section covers 1,675 ha and the south-east is 1,331 ha. It is bordered by the maritime division of Grand Cul-de-Sac Marin on the west and to the east by pastures and housing. The IBA embraces three herbaceous marshes (at Lambi, Choisy and Vieux-Bourg). All four species of mangrove are represented and the swamp forest is dominated by dragonsblood tree *Pterocarpus officinalis*.

■ Birds

This IBA is significant for supporting 12 (of the 17) of the Lesser Antilles EBA restricted-range birds including two globally threatened species, namely the Vulnerable Forest Thrush *Cichlherminia lherminieri* (which is very rare in the IBA) and Near Threatened Guadeloupe Woodpecker *Melanerpes herminieri* (which has a population of 770 pairs). The Near Threatened White-crowned Pigeon *Patagioenas leucocephala* is occasionally seen in the IBA, but breeding is uncertain and it is hunted without any control. Numerous Neotropical migratory warblers are recorded in the IBA.

■ Other biodiversity

The Near Threatened insular single leaf bat *Monophyllus plethodon* and tree bat *Ardops nichollsi annectens* occur.

■ Conservation

East Coast of Grand Cul-de-Sac Marin IBA is state owned. The southern part of the IBA is classified as a National Nature Reserve by a 1987 ministerial decree. This reserve is managed by the National Park of Guadeloupe. Bird ringing (banding) is conducted irregularly (notably along Lambi's Marsh trail). The whole IBA is classified as coastal state forest, a biosphere reserve and Ramsar site. AMAZONA regularly monitors (and occasionally rings) the birds in the northern section, and a regular ringing program will be established soon. The main threat is conversion of land (including forest) for cattle pasture or agriculture, and urban expansion. A proposed deep-water harbour at Port-Louis would have a serious impact on the mangroves. Other threats include introduced mammalian predators and the discharge of pollution into the Grand Cul-de-Sac Marin Nature Reserve. Hunting *C. lherminieri* is forbidden on Grande-Terre

■ Site description

The Saint-François Fishing Harbour IBA is on the south-east coast of Grand-Terre and comprises a sea wall of limestone, rocks and concrete built on a small coral reef. The sea wall, built in the late 1990s to protect the harbour, is 500 m long and 20 m wide, and connects a number of patches of emergent reef. There is very little vegetation on the sea wall (although seapurslane *Sesuvium portulacastrum*, seagrape *Coccoloba uvifera*, portia tree *Thespesia populnea*, goat's foot *Ipomoea pes-caprae* and various herbaceous species are present), and the land is not currently utilised. The IBA includes marine areas up to 1 km from the harbour.

■ Birds

This IBA supports a regionally significant breeding population of Least Tern *Sterna antillarum* with 20–50 pairs present along the sea wall. The population has been stable since the terns first colonised in 2002. Migratory shorebirds frequenting the IBA in small numbers include Ruddy Turnstone *Arenaria interpres*, Black-bellied Plover *Pluvialis squatarola*, Sanderling *Calidris alba* and Semipalmated Plover *Charadrius semipalmatus*.

■ Other biodiversity

No threatened or endemic species have been recorded.

■ Conservation

Saint-François Fishing Harbour IBA is state owned. AMAZONA is pursuing measures for the protection of this IBA through its designation as a Regional Nature Reserve. Since the discovery of the *S. antillarum* breeding colony, AMAZONA has monitored the population every year. During 2008 a bridge was built connecting the tern colony to the mainland. The resulting human disturbance prevented the terns from nesting. The status of this IBA will need to be reviewed if the terns fail to breed in subsequent years. The bridge will also expose the terns to the threat of predator from invasive mammals. The biggest threat to the colony prior to the construction of the bridge was the presence of Laughing Gulls *Larus atricilla* attracted by fish remnants thrown in the harbour by fishermen.

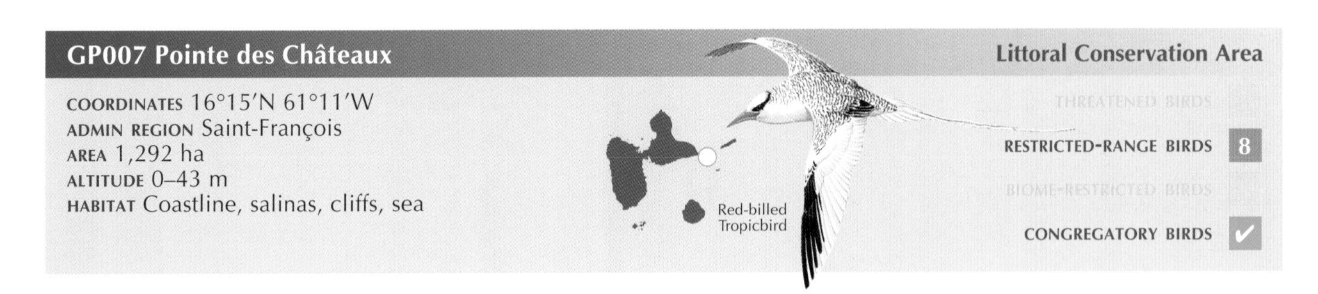

■ Site description

Pointe des Châteaux IBA comprises a peninsula at the south-easternmost tip of Grande-Terre. The peninsula is limestone and supports coastal dry zone (xerophytic) vegetation. It includes limestone cliffs (at the west of Anse to Plume), two rocky islets (La Roche and L'Eperon) and Pointe Colibri (at the extreme eastern tip). Six saline lagoons and several beaches separate these formations. Grande Saline is the largest lagoon at 15 ha. The others are all smaller than 1 ha. The IBA includes marine areas up to 1 km from the coast.

■ Birds

This IBA supports populations of eight (of the 17) Lesser Antilles EBA restricted-range birds. It is globally significant for Red-billed Tropicbird *Phaethon aethereus* (25–35 pairs breed) and regionally so for Least Tern *Sterna antillarum* (10–24 pairs) and Bridled Tern *S. anaethetus* (50–60 pairs). It is the second most important nesting site for Sooty Tern *Sterna fuscata* (1,500–2,000 pairs) in Guadeloupe. Other notable seabirds are Brown Noddy *Anous stolidus* (10–20 pairs) and White-tailed Tropicbird *Phaethon lepturus* (5–10 pairs). Birds like Wilson's Plover *Charadrius wilsonia* (3–5 pairs), the Near Threatened Piping Plover *Charadrius melodus*, Buff-breasted Sandpiper *Tryngites subruficollis* and White-crowned Pigeon *Patagioenas leucocephala* are observed occasionally in the salinas.

■ Other biodiversity

The restricted-range least gecko *Sphaerodactylus fantasticus* occurs. The xerophytic vegetation includes the Endangered guaiac tree *Guaiacum officinale*.

■ Conservation

Pointe des Châteaux IBA is state owned and classified as a Littoral Conservation Area, belonging to the Conservatoire du littoral. The site was recognised as a (Type 2) Zone Naturelles d'Intérêt Ecologique Faunistique et Floristique (ZNIEFF). It is the most visited touristic site in Guadeloupe receiving c.500,000 people per year. Regulations on tourism, access to the beaches and dunes, and off-road vehicle use are enforced to minimise the impact on the site. Information panels are also displayed throughout the IBA. However, disturbance is causing a decrease in breeding success in the salinas. Another significant threat is the predation of ground nesting birds by feral cats, dogs, mongoose *Herpestes auropunctatus* and black rat *Rattus rattus*. AMAZONA conducts monthly ringing (banding) of shorebirds and terrestrial birds, and Mr Leblond has conducted a seabird census.

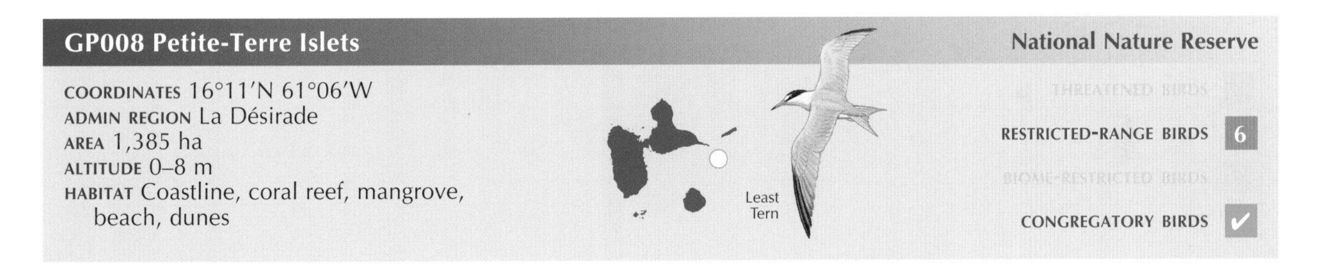

■ Site description

Petite-Terre Islets IBA is situated 12 km to the south of Désirade and 9.5 km to the south-east of Pointe des Châteaux (IBA GP007). The IBA comprises two uninhabited islets—the 31-ha Terre de Haut to the north and 118-ha Terre de Bas to the south—separated by a narrow channel but both surrounded by coral reefs. Terre de Haut is bordered by white sand beaches while terre de Bas has low rocky cliffs. The islets are limestone and support dune vegetation with xerophytic vegetation on the limestone, and mangroves in some sections. The IBA includes marine areas up to 1 km from the islets.

■ Birds

This IBA supports populations of six (of the 17) Lesser Antilles EBA restricted-range birds. A breeding population of 100 pairs of Least Tern *Sterna antillarum* is regionally significant. The first breeding record for Guadeloupe of the Vulnerable West Indian Whistling-duck *Dendrocygna arborea* was from the IBA. Near Threatened Piping Plover *Charadrius melodus* (observed in consecutive years), Buff-breasted Sandpiper *Tryngites subruficollis* and White-crowned Pigeon *Patagioenas leucocephala* (both seen "occasionally") have been recorded. American Oystercatcher *Haematopus palliatus* and Wilson's Plover *Charadrius wilsonia* breed. Large numbers of passage seabirds have been noted from this IBA including the Endangered Black-capped Petrel *Pterodroma hasitata*.

■ Other biodiversity

Petite-Terre holds 30–50% of the global population of the Vulnerable Lesser Antillean iguana *Iguana delicatissima*. The Endangered guaiac tree *Guaiacum officinale* and lesser Antillean endemic trumpet tree *Tabebuia pallida* occur. The Critically Endangered hawksbill *Eretmochelys imbricata* and Endangered green *Chelonia mydas* turtles nest on the beachs each year in the Nature Reserve.

■ Conservation

Petite-Terre Islets IBA is state owned and a designated National Nature Reserve (Réserve Naturelle) under a 1998 ministerial decree. The whole IBA is recognised as a Zone Naturelles d'Intérêt Ecologique Faunistique et Floristique (ZNIEFF). Rangers continuously monitor the national nature reserve. The main threats to the seabirds are disturbance from the numerous visitors and predation of *S. antillarum* by black rat *Rattus rattus*. A five-year management plan was prepared in 2003. A permanent bird ringing program gathers information on Bananaquit *Coereba flaveola*. Breeding shorebirds, terns and migratory seabirds are counted monthly.

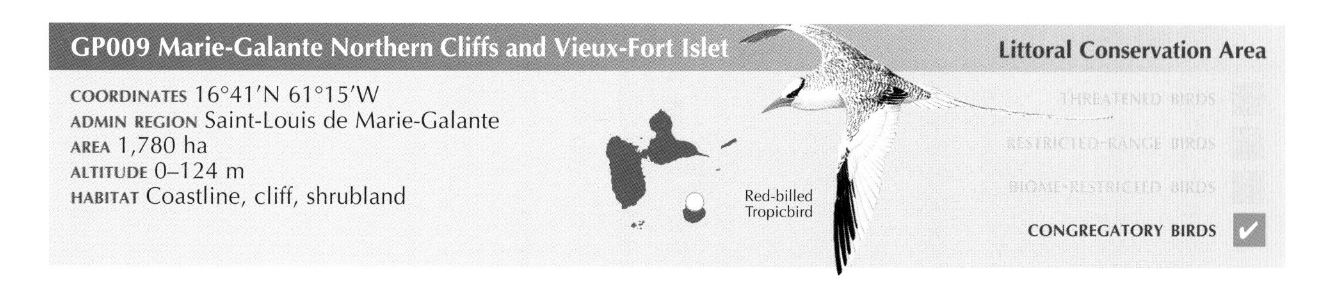

■ Site description

Marie-Galante Northern Cliffs and Vieux-Fort Islet IBA is located in the north-west coast of Marie-Galante Island. The cliffs extend (taking in 250 m of cliff-top land) for 14 km along the coast of Ménard Point to Anse Peak. The 0.29-ha Vieux-Fort islet is 600 m from Marie-Galante and has a maximum height of 6 m. Semi-deciduous vegetation predominates on the cliffs and is subjected to constraints of drought and wind, resulting in low stature "dwarf" shrubland. The Vieux-Fort Islet is covered primarily with herbaceous vegetation favoured by the nesting of Sooty Terns *Sterna fuscata.*

■ Birds

This IBA is important for its breeding seabirds. It supports a globally significant colony of Red-billed Tropicbird *Phaethon aethereus* (up to 300 pairs) and regionally important numbers of Brown Noddy *Anous stolidus* (200–220 pairs) and Bridled Tern *Sterna anaethetus* (60–70 pairs). The site holds the third largest colony of Sooty Tern *S. fuscata* (900–1,000 pairs) in Guadeloupe. Audubon's Shearwater *Puffinus lherminieri* are heard calling at night suggesting breeding behaviour but it has not been possible to estimate the size of the population.

■ Other biodiversity

The Endangered guaiac tree *Guaiacum officinale* occurs along with range of other trees and thorny shrubs.

■ Conservation

Marie-Galante Northern Cliffs and Vieux-Fort Islet IBA are state-owned. The area is a coastal state forest, and the cliffs are protected by the coastline law that prohibits construction within 15 m of the littoral zone. It also designated as a "Classified Site" and recognised as a Zone Naturelles d'Intérêt Ecologique Faunistique et Floristique (ZNIEFF). Vieux-Fort Islet has easy access along its flat shore. Landings and visits should be forbidden during the nesting season to minimise human disturbance which can harm the colony by over-exposing the eggs and chicks to the sun when the adults are forced to leave their nests. The cliffs are not visited often because access is difficult, but erosion could be caused by clearance of vegetation for agricultural activities (e.g. allottments). A seabird count was carried out in 2002.

HAITI

LAND AREA **27,750 km²** ALTITUDE **0–2,680 m**
HUMAN POPULATION **8,706,500** CAPITAL **Port-au-Prince**
IMPORTANT BIRD AREAS **10, totalling 232 km²**
IMPORTANT BIRD AREA PROTECTION **73%**
BIRD SPECIES **245**
THREATENED BIRDS **22** RESTRICTED-RANGE BIRDS **36**

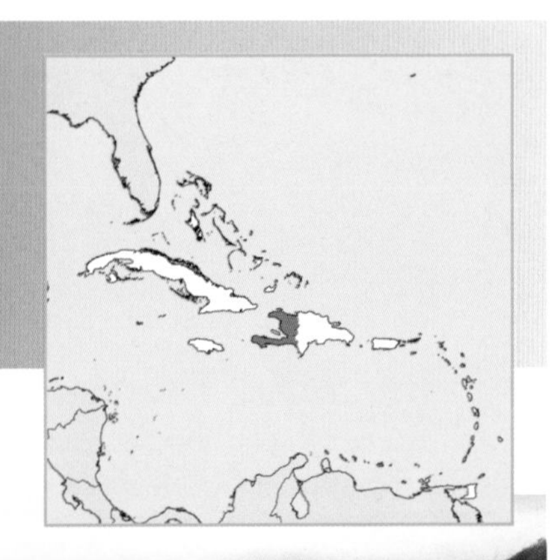

FLORENCE SERGILE
(SOCIÉTÉ AUDUBON HAITI/UNIVERSITY OF FLORIDA)

Small colonies of Endangered Black-capped Petrels breed along the La Selle escarpment which forms part of the Aux Diablotins IBA in the Massif de la Selle. (PHOTO: JIM GOETZ/CORNELL LAB OF ORNITHOLOGY)

INTRODUCTION

The Republic of Haiti is situated on the western third of the island of Hispaniola, the second largest island in the Caribbean. The Dominican Republic, with which Haiti shares a 360-km border, occupies the rest of Hispaniola. North-western Haiti is just 80 km east of Cuba. Haiti's landscape of rugged mountains interspersed with small coastal plains and river valleys has been divided politically into 10 "départements": Artibonite, Centre, Grande-Anse, Nippes, Nord, Nord-Est, Nord-Ouest, Ouest, Sud, Sud-Est. Haiti also has six satellite islands (totalling 954 km²), namely Île de la Tortue (off the north coast), La Gonâve, (north-west of Port-au-Prince), Île à Vache (off the southern tip of south-western Haiti), Les Cayemites (off the north coast of the Southern Peninsula) and the disputed island of Navassa (see separate Navassa chapter).

The northern region of Haiti consists of the Massif du Nord mountain range (an extension of the Dominican Republic's Cordillera Central) which extends from the border through the north-west peninsula. The Plaine du Nord lowlands lie along the northern border with the Dominican Republic, between the Massif du Nord and the Atlantic Ocean. Haiti's central region consists of the Plateau Central that runs south-east to north-west along both sides of the Guayamouc River, south of the Massif du Nord. South-west of this plateau are the Montagnes Noires, the north-western parts of which merge with the Massif du Nord. The southern region consists of the Plaine du Cul-de-Sac in the south-east, and the mountainous southern Tiburon Peninsula. The Plaine du Cul-de-Sac is a natural depression in which lies the lake of Trou Caïman and Haiti's largest lake, Lac Azuei. The Chaîne de la Selle mountain range is an extension of the Sierra de Bahoruco in the Dominican Republic. It extends from the Massif de la Selle in the east (Pic la Selle is Haiti's highest point) to the Massif de la Hotte in the west.

Haiti has a tropical climate with two main wet seasons: the north-east trade winds bring rain from April to June, and northerly winds bring drizzle from about September through November. However, the country's topography produces significant regional (and altitudinal) differences in temperature and rainfall. The resultant vegetation varies from subtropical very dry forest formations where cacti and scrub predominate, to tropical montane wet forest at the higher altitudes where Hispaniolan pines *Pinus occidentalis* and temperate vegetation thrive. Wetlands, lakes, lagoons, estuaries and a varied coastline provide additional diversity. In 1925, Haiti was lush, with 60% of its original forest cover. Since then, the population

Agriculture is mainly small-scale subsistence farming carried out by the country's largest, growing and economically impoverished population. (PHOTO: JAMIE RHODES)

(which is now at a density of c.300/km^2 and growing at a rate of 2.3% each year) has cut down all but c.2%, and in the process fertile farmland soils have been destroyed which in turn has contributed to desertification. Most Haitian logging is done to produce charcoal, the country's chief source of fuel. Deforestation has led to severe erosion in the mountainous areas, and also periodic (but often catastrophic) flooding. Droughts, earthquakes and hurricanes add to the human and environmental suffering. Haiti has remained the least-developed country in the Americas with c.80% of the population estimated to be living in poverty in 2003 (on an average income of <US$1 per day). About 66% of all Haitians work in the agricultural sector, which is mainly small-scale subsistence farming although mangos and coffee are Haiti's two most important exports.

■ Conservation

Since 1983, biodiversity protection in Haiti slowly turned into a reality resulting from a combination of government commitment, dedicated people and a national environmental awareness campaign. In spite of economic hardships the Ministry of the Environment (Ministère de l'Environnement, MDE) and Ministry of Agriculture, Natural Resources, and Rural Development (Ministère de l'Agriculture, des Ressources Naturelles et du Développement Rural, MARNDR) managed to establish a protected area system encompassing c.25,000 ha (c.1% of the country's land area) in four reserves. The reserves are: Macaya Biosphere Reserve in the Massif de la Hotte, the Parc National La Visite and the Forêt des Pins in the Massif de la Selle, and the Parc Historique La Citadelle, Sans Souci, les Ramiers in the north. Recognising that the full range of Haiti's ecosystems was not covered in these primarily montane reserves, 18 additional areas (totalling another 23,000 ha) were identified as potential protected areas (see Important Bird Areas below), although none of these have yet been officially designated.

Management of the protected areas started in 1992, initially with USAID funding targeting Macaya Biosphere Reserve, and then through a John D. and Catherine T. MacArthur Foundation project that ran until 1998. These projects facilitated the development of a World Bank financed park and forest technical assistance initiative (Appui technique à la Protection des Parcs et Forêt, ATPPF) that started in late 1998. Bridging the gap between the USAID and World Bank projects, the University of Florida focused activities on bird monitoring, species recovery plans and environmental education within Macaya Biosphere Reserve, La Visite and

Forest in La Visite National Park—Haiti's protected areas are primarily montane. (PHOTO: FONDATION SEGUIN)

La Citadelle parks, updating the bird work done in these areas during the 1980s. The World Bank ATPPF project aimed to develop the Mayaca, La Visite and Forêt des Pins protected areas and their buffer zones and train Haitian professionals. It led to the inclusion of Morne d'Enfer and Pic La Selle in the Parc La Visite boundary, and the Sapotille area into the northern side of the Macaya reserve. The MDE (which was created in 1995) and MARNDR are both chronically financially under-resourced. MARNDR (through its Service des Parcs et Sites Naturels) is responsible for managing the protected areas. However, natural resource management was moved down the agenda by the transitional government (2004–2006) who put a greater emphasis on agricultural production. With the country's high population density, poverty, and political instability, compounded by the small budgets for conservation, absence of trained staff, lack of clear policies, and shifting government priorities, sustained conservation efforts have been prevented from establishing. The protected areas in Haiti are essentially unprotected. There are personnel responsible for the parks, and basic offices do exist, but staff seem to be present only intermittently and access is entirely uncontrolled.

The conservation NGO sector includes Haiti-Net, created in 1992 to promote ecosystem management and environmental education in Haiti, Société Audubon Haïti (SAH) which was established in 2003 to conserve Haiti's natural ecosystems focusing on birds. SAH works in collaboration with the Vermont Center for Ecostudies, Sociedad Ornitológica de la Hispaniola and BirdLife, and is implementing conservation projects in both Macaya and La Visite. In 2005 SAH published Haiti's first book on birds as an educational tool to raise awareness about the country's unique biodiversity.

The pressures on Haiti's ecosystems and biodiversity are huge. They are primarily a result of the country's large, growing but economically impoverished population. Habitat destruction is leading to desertification, erosion and sedimentation, all of which negatively impact the human population. With so little forest left, its continued destruction will result in numerous species extinctions in the country. Habitat loss is compounded by unregulated, unsustainable hunting which is widespread, invasive mammalian predators, introduced exotic plants which are outcompeting the native flora, and the commercial export of plants and animals (e.g. for the pet trade) that has impoverished many life zones. To move forward with conservation in Haiti it will be essential to: focus on a few priority sites such as the Important Bird Areas (IBAs); address the livelihood needs of the people dependent on the resources (at both the site and species level) being conserved; involve these same stakeholders in the design and implementation of conservation actions; raise the level of awareness of biodiversity and conservation issues at the site level, but also within the government; conservation management training for local practitioners and national institutions; establish clear monitoring frameworks to determine the success or failure of particular management actions; and ensure projects are developed with long-term sustainability and commitment as prerequisites.

■ Birds

The Republic of Haiti supports over 245 species of bird, of which more than 73 are resident landbirds. The Hispaniolan avifauna exhibits exceptional levels of endemism. The island is an Endemic Bird Area (EBA), and 36 restricted-range species are known from Haiti, one of which, Grey-crowned Palm-tanager *Phaenicophilus poliocephalus* is endemic to Haiti. The majority of the restricted-range species are confined to, or occur in habitats above 1,000 m, emphasising the importance of mixed montane broadleaf–pine forest. A total of 28 of these restricted-range birds are endemic to the island, the others being shared with adjacent EBAs. For example, Vervain Hummingbird *Mellisuga minima*, Stolid Flycatcher *Myiarchus stolidus*, Greater Antillean Elaenia *Elaenia fallax* and Golden Swallow *Tachycineta euchrysea* are all shared with Jamaica. Six of the restricted-range species represent genera endemic to Hispaniola, namely *Calyptophilus*, *Dulus* (also a monotypic family), *Microligea*, *Nesoctites*, *Phaenicophilus* and *Xenoligea*. Endemism is also high at the sub-specific level with 47 subspecies described. All of the satellite islands support their own endemic subspecies, with seven found on Île de la Gonâve and three on Île de la Tortue. Our ornithological knowledge of Haiti is relatively poor, as a result of which there are a number of restricted-range birds whose current distribution and status in Haiti is unknown. These species include: Ashy-faced Owl *Tyto glaucops*, Ridgway's Hawk *Buteo ridgwayi*, Least Pauraque *Siphonorhis brewsteri*, Bay-breasted Cuckoo *Coccyzus rufigularis*, Flat-billed Vireo *Vireo nanus*, Hispaniolan Nightjar *Caprimulgus eckmani*. These species are not represented within the Important Bird Area (IBA) analysis (see Table 1), but further work in the IBAs may show them to be present, and their discovery in localities outside the IBAs may necessitate the definition of new IBAs.

There are significant populations of 18 globally threatened species currently known from Haiti. However, four additional species are listed from Haiti but have not been considered in the IBA analysis. These are the Near Threatened Buff-breasted Sandpiper *Tryngites subruficollis* which is only known as a vagrant; the Critically Endangered *Buteo ridgwayi* which has not been recorded for 20 years, but was known from Haiti's satellite islands and may yet occur; and the Endangered *Coccyzus rufigularis* and Near Threatened *Siphonorhis brewsteri*, the current status of which is unknown within the country. Most of the globally threatened birds (including three Endangered and eight Vulnerable birds: see Table 1) are concentrated in the remnant montane forests of the Massif de la Hotte (Macaya) and Massif de la Selle (La Visite). All of the globally threatened birds are poorly known within the country and population estimates at the site and national levels are not available.

The Grey-crowned Palm-tanager is endemic to Haiti. (PHOTO: ELADIO FERNÁNDEZ)

Table 1. Key bird species at Important Bird Areas in Haiti.

Key bird species	Criteria	National population	Haiti IBAs: HT001	HT002	HT003	HT004	HT005	HT006	HT007	HT008	HT009	HT010
		Criteria	■ ■	■ ■	■ ■	■ ■	■ ■	■	■ ■	■ ■	■ ■	■ ■
Northern Bobwhite *Colinus virginianus*	NT ■								✓			
West Indian Whistling-duck *Dendrocygna arborea*	VU ■		✓									
Black-capped Petrel *Pterodroma hasitata*	EN ■								✓			
Caribbean Coot *Fulica caribaea*	NT ■						✓	250–300				
White-crowned Pigeon *Patagioenas leucocephala*	NT ■				✓	✓						
Plain Pigeon *Patagioenas inornata*	NT ■				✓	✓				✓		✓
Hispaniolan Parakeet *Aratinga chloroptera*	VU ■ ■			✓			✓		✓	✓		
Hispaniolan Amazon *Amazona ventralis*	VU ■ ■			✓			✓		✓			
Hispaniolan Lizard-cuckoo *Saurothera longirostris*	■			✓	✓	✓			✓	✓		
Antillean Mango *Anthracothorax dominicus*	■		✓	✓	✓	✓	✓		✓	✓		✓
Hispaniolan Emerald *Chlorostilbon swainsonii*	■								✓	✓	✓	✓
Vervain Hummingbird *Mellisuga minima*	■			✓	✓	✓	✓		✓			
Hispaniolan Trogon *Priotelus roseigaster*	NT ■ ■								✓	✓		✓
Narrow-billed Tody *Todus angustirostris*	■			✓					✓	✓	✓	✓
Broad-billed Tody *Todus subulatus*	■		✓	✓			✓					
Antillean Piculet *Nesoctites micromegas*	■										✓	✓
Hispaniolan Woodpecker *Melanerpes striatus*	■			✓			✓		✓	✓	✓	✓
Greater Antillean Elaenia *Elaenia fallax*	■			✓					✓	✓	✓	✓
Hispaniolan Pewee *Contopus hispaniolensis*	■								✓	✓		
Stolid Flycatcher *Myiarchus stolidus*	■			✓		✓	✓					
Thick-billed Vireo *Vireo crassirostris*	■				✓	✓						
Hispaniolan Palm Crow *Corvus palmarum*	NT ■ ■			✓			✓		✓	✓		
White-necked Crow *Corvus leucognaphalus*	VU ■ ■						✓					
Palmchat *Dulus dominicus*	■			✓			✓					
Golden Swallow *Tachycineta euchrysea*	VU ■ ■								✓	✓	✓	✓
Rufous-throated Solitaire *Myadestes genibarbis*	■								✓	✓	✓	✓
Bicknell's Thrush *Catharus bicknelli*	VU ■								✓	✓	✓	
La Selle Thrush *Turdus swalesi*	EN ■ ■								✓	✓		
Antillean Siskin *Carduelis dominicensis*	■								✓	✓		
Hispaniolan Crossbill *Loxia megaplaga*	EN ■ ■								✓		✓	
Green-tailed Warbler *Microligea palustris*	■								✓	✓		
White-winged Warbler *Xenoligea montana*	VU ■ ■											✓
Black-crowned Palm-tanager *Phaenicophilus palmarum*	■		✓	✓								
Grey-crowned Palm-tanager *Phaenicophilus poliocephalus*	NT ■ ■										✓	✓
Chat Tanager *Calyptophilus frugivorus*	VU ■ ■								✓	✓	✓	✓
Hispaniolan Spindalis *Spindalis dominicensis*	■									✓	✓	✓
Antillean Euphonia *Euphonia musica*	■			✓								✓

All population figures = numbers of individuals.
Threatened birds: Endangered ■; Vulnerable ■; Near Threatened ■. **Restricted-range birds** ■.

The Endangered La Selle Thrush occurs in the remnants of montane forest in the Massif de la Selle. (PHOTO: ELADIO FERNÁNDEZ)

More than 155 waterbirds are found in Haiti. Although work has been done to survey and monitor the ducks (by Ducks Unlimited) there is very little documentation concerning the populations of shorebirds passing through or wintering in Haiti, or the numbers of resident waterbirds at the various wetlands. Seabirds are also poorly known in terms of colony sizes (or indeed distribution and species composition). Consequently, no congregatory species feature in the IBA analysis. Key waterbird sites that are known about (albeit with limited population data available) include Lagon-aux-Boeufs (IBA HT001); Acul Bay near Cap-Haitian; île de la Tortue in Basse-Terre and Coquillage (IBA HT003); Petit Paradis; Artibonite Delta and Étang Bois Neuf; Sources Puantes; Lac Azuéi (IBA HT006); Trou Caïman (IBA HT005); Étang de Miragoâne; Baradères—Cayemite mangroves; Étang Laborde-Lachaux near Camp-Perrin; île-à-Vache wetlands and mangroves; and the Île de la Gonâve mangroves.

IMPORTANT BIRD AREAS

Haiti's 10 IBAs—the country's international site priorities for bird conservation—cover 232 km², less than 1% of Haiti's land area. Five of the IBAs are within Haiti's embryonic protected areas system. In the Massif de la Hotte, Bois Musicien IBA (HT010) and Aux Bec-Croisés IBA (HT009) are both within the Macaya Biosphere Reserve. In the Massif du Nord, Les Todiers IBA (HT002) is within the Parc Historique la Citadelle, Sans-Souci, les Ramiers, and in the Massif de la Selle, Aux Diablotins IBA (HT007) and Aux Cornichons IBA (HT008) are within the La Visite National Park. However, effective conservation management within these parks is

The Macaya Biosphere Reserve embraces two of Haiti's IBAs. ((PHOTO: CHRIS RIMMER/VERMONT CENTER FOR ECOSTUDIES)

essentially non-existent and thus the protection afforded these critical sites minimal. The other five IBAs are not legally protected.

The IBAs have been identified on the basis of 37 key bird species (listed in Table 1) that variously meet the IBA criteria. These 37 species include 18 (of the 22) globally threatened birds, all 30 restricted-range species for which there are known populations, but no congregatory waterbirds/seabirds due to the lack of site-level population estimates for these species groups. Most of the globally threatened and restricted-range birds are confined to or occur in the IBAs within the La Visite and Macaya protected areas, emphasizing the importance of the montane forests in these two parks. However, there are some lowland, drier forest restricted-range birds (and the two globally threatened waterbirds) that are only present outside of these montane areas, showing that a network of sites is critical to conserve the full range of Haiti's unique biodiversity.

Trou Caïman IBA is one of a number of IBAs previously identified as potential additions to the protected area network in Haiti. (PHOTO: JEAN VILMOND HILAIRE/SAH)

Figure 1. Location of Important Bird Areas in Haiti.

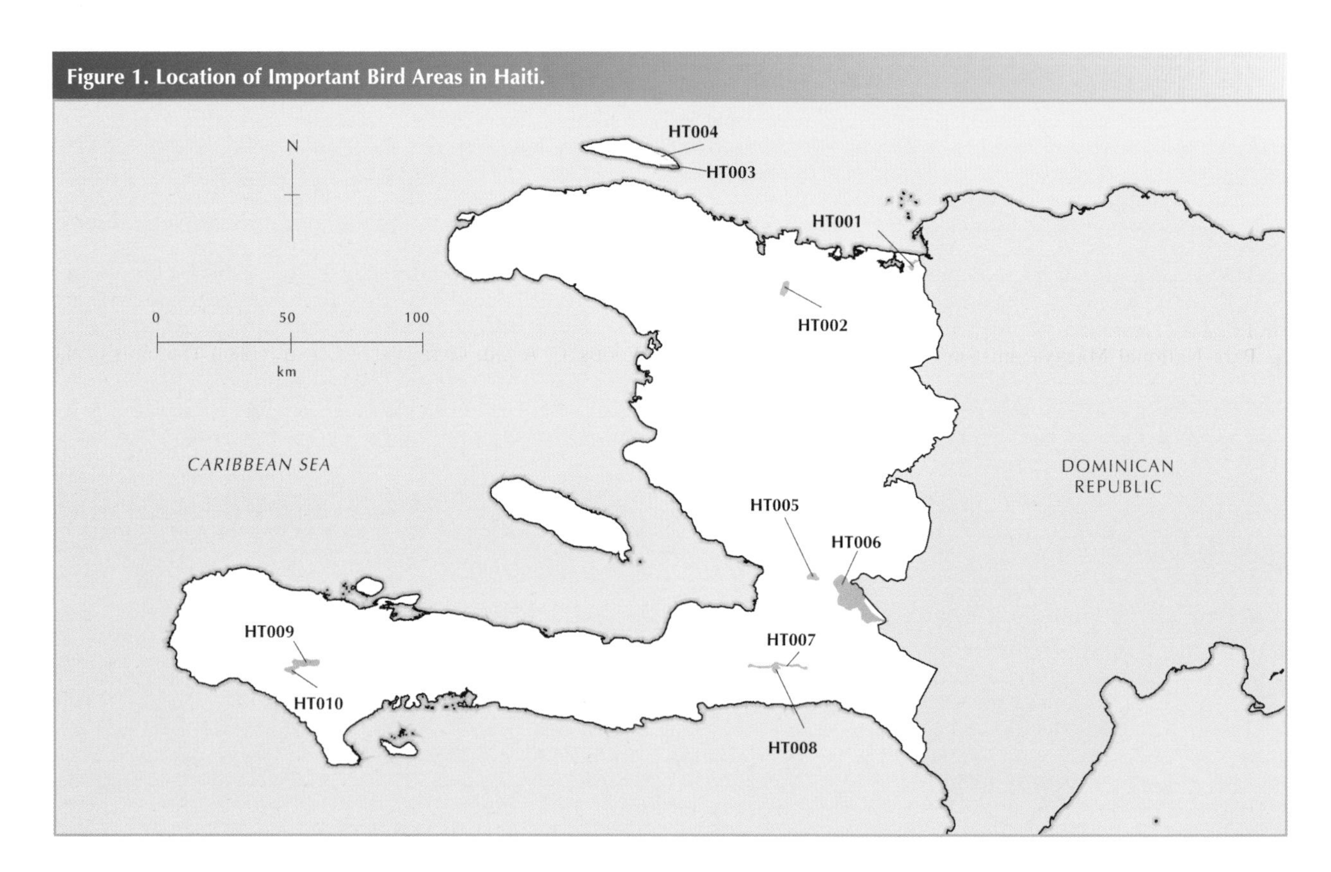

The IBA network as described in this chapter is not complete. With so many gaps in our knowledge of bird distributions, populations and abundance in Haiti, other sites will be identified in the future. Surveys could usefully focus on increasing our knowledge of the globally threatened and restricted-range birds whose current status and distribution in the country is unknown. Discoveries of any of these species may result in the definition of new IBAs. Similarly, fieldwork looking at waterbird populations and seabird colony size will almost certainly result in new IBAs being described—large numbers of shorebirds, waterbird and seabirds are present in the country, but without estimated population data, sites cannot be objectively described against the IBA criteria.

Four protected areas have been designated in Haiti. However, another 18 sites were identified as "areas to be protected". Of these 18 sites, one—Navassa Island—is described in a separate IBA chapter. Coquillage IBA (HT003), Lagon-aux-Boeufs IBA (HT001), Trou Caïman IBA (HT005) and Lac Azuéi IBA (HT006) represent another four of the potential protected areas. The remaining 13 sites are (from north to south): Baie de Fort-Liberté–Rivière du Massacre delta; Baie de l'Acul; Pointe Ouest; Petit Paradis; Artibonite Delta; Bassin Zim; Étang Bois-Neuf; Langue Blanche and Pointe Ouest; Les Arcadins; Étang de Miragoâne; Baie de St Louis du Sud/Grosse Cayes; Îles Cayemites and Baradères; and Pointe Diamant. The survey work that is so urgently required within Haiti could usefully focus on these potential protected areas to determine their current status and importance for the key globally threatened, restricted-range and congregatory bird species. Similar field assessments (surveys and subsequent monitoring) are needed for the key bird species in all 10 Haitian IBAs. The results should be used to help inform the assessment of state, pressure and response scores at each IBA to provide an objective status assessment and to highlight the management interventions that are required to maintain these internationally important biodiversity sites.

KEY REFERENCES

CEP (1996) *Status of protected area systems in the wider Caribbean region*. Kingston, Jamaica: UNEP-Caribbean Environment Programme (CEP Technical Report 36).

Keith, A. R., Wiley, J. W., Latta, S. and Ottenwalder, J. (2003) *The birds of Hispaniola: Haiti and the Dominican Republic*. Tring, U.K.: British Ornithologists' Union (BOU Check-list No 21).

Latta, S., Rimmer, C., Keith, A., Wiley, J., Raffaele, H., McFarland, K. and Fernandez, E. (2006) *Birds of the Dominican Republic and Haiti*. Princeton, New Jersey: Princeton University Press.

McPherson, H. and Graham, C. (1993) A survey of the birds of the Citadelle area, April 23–24, 1993. (Unpublished report).

McPherson, H., Fondots, C. and Graham, C. (1993) Birds of the Parc National Macaya, mist-netting January–April 1993. (Unpublished report).

MARNDR (1988) Pour une déclaration officielle de l'état d'urgence face à la dégradation de l'environnement national. Groupe de Travail pour le suivi du Colloque sur le reboisement tenu à Damien en avril 1987. Port-au-Prince : Ministère de l'Agriculture, des Ressources Naturelles et du Développement Rural. (Unpublished report).

Paryski, P. E., Woods, C. A. and Sergile, F. E. (1989) Conservation strategies and the preservation of biological diversity in Haïti. Pp.855–878 in C. A. Woods, ed. *Biogeography of the West Indies*. Gainesville, Florida: Sandhill Crane Press, Inc.

Raffaele, H. Wiley J., Garrido, O., Keith, A. and Raffaele, J. (1998) *A guide to the birds of the West Indies*. Princeton, New Jersey: Princeton University Press.

Rimmer, C. C., Garrido, E. and Brocca, J. L. (2005) Ornithological field investigations in La Visite National Park, Haiti, 26 January – 1 February 2005. Woodstock, Vermont: Vermont Institute of Natural Science. (Unpublished report).

Rimmer, C. C., Townsend, J. M., Townsend, A. K., Fernández, E. M. and Almonte, J. (2004) Ornithological field investigations in Macaya Biosphere Reserve, Haiti, 7–14 February 2004. Woodstock, Vermont: Vermont Institute of Natural Science. (Unpublished report).

Rimmer, C.C., Townsend, J. M., Townsend, A. K., Fernández, E. M. and Almonte, J. (2005) Avian diversity, abundance, and conservation status in the Macaya Biosphere Reserve of Haiti. *Orn. Neotrop.* 16: 219–230.

Rimmer, C. C., Klavins, J., Gerwin, J. A., Goetz, J. E. and Fernández, E. M. (2006) Ornithological field investigations in Macaya Biosphere Reserve, Haiti, 2–10 February 2006. Woodstock, Vermont: Vermont Institute of Natural Science. (Unpublished report).

Sergile, F. E. (1990) The Biosphere Reserve Henry Christophe: Potential for the management and conservation of natural resources in Haïti. Gainesville, Florida: University of Florida. (Unpublished MA Thesis).

Sergile, F. E. (2005) *A la découverte des oiseaux d'Haïti*. Petion-ville, Haiti: Société Audubon Haïti.

Sergile, F. E. and Mérisier, J. R. (1994) *Connaître et protéger la richesse naturelle d'Haïti*. Gainesville, Florida: Florida Museum of Natural History.

Sergile, F. E. and Woods, C. A. (1993) Haiti National Parks and Conservation Project: June 1992 – July 1993 report. Gainesville, Florida: University of Florida. (Unpublished report).

Sergile, F. E. and Woods, C. A. (1995) Guide de terrain des aires protégées en Haïti. Gainesville, Florida: Florida Museum of Natural History. (Unpublished report).

Sergile, F. E. and Woods, C. A. (1995) Veye richès peyi d'Ayiti. Gainesville, Florida: Florida Museum of Natural History. (Unpublished report).

Sergile, F. E. and Woods, C. A. (1998) Ile de la Tortue: profil de l'environnement. Haïti-NET: Statut de l'environnement naturel, 4. (Unpublished report prepared for the PNUD/UNOPS HAI/92/001 project).

Sergile, F. E. and Woods, C. A. (2001) Status of conservation in Haiti: a 10-year retrospective. Pp.547–560 in C. A. Woods and F. E. Sergile, eds. *Biogeography of the West Indies: patterns and perspectives*. Boca Raton, Florida: CRC Press.

Sergile, F. E., Woods, C. A. and Paryski, P. (1992) Final report of the Macaya Biosphere Reserve Project. Gainesville, Florida: Florida Museum of Natural History. (Unpublished report).

Wingate, D. (1964) Discovery of breeding Black-capped Petrels in Hispaniola. *Auk* 81: 147–159.

Woods, C. A. (1983) Biological survey of Haiti: status of the endangered birds and mammals. *National Geographic Society Research Reports* 15: 759–768.

Woods, C. A. (1986) The West Indian Flamingo in Haiti: aerial surveys and review status. *Flamingo Research Specialist Group Newsletter* 3: 19–23.

Woods, C. A. and Harris, L. (1986) Stewardship Plan for the national parks of Haiti. Port-au-Prince, Haiti. (Unpublished report for USAID-Haiti).

Woods, C. A and Ottenwalder, J. A. (1986) The birds of the National Parks of Haiti. Port-au-Prince, Haiti. (Unpublished report for USAID-Haiti).

Woods, C. A and Ottenwalder, J. A. (1992) The natural history of the Southern Peninsula. Gainesville, Florida: Florida State Museum. (Unpublished report).

Woods, C. A., Sergile, F. E. and Ottenwalder, J. A. (1992) *Stewardship plan for the national parks and protected areas of Haiti*. Gainesville, Florida: Florida Museum of Natural History.

ACKNOWLEDGEMENTS

The author would like to thank Charles Woods (Bear Mountain Natural History Center), Philippe Bayard and Jean Vilmond Hilaire (SAH), Judex Edouarzin (Ministère de l'Environnement), Jamie Rhodes, Chris Rimmer (Vermont Center for Ecostudies), Eladio Fernandez (SOH), Jim Goetz (Cornell Laboratory of Ornithology) and Fondation Seguin.

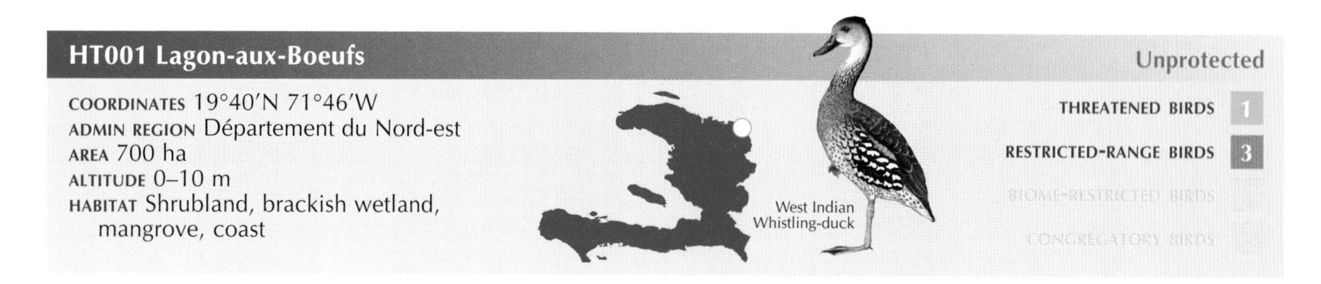

■ Site description

Lagon-aux-Boeufs IBA is a brackish (20 parts per thousand) estuarine lake on the coastal plain of Dauphin, north-easternmost Haiti. It lies east of Fort-Liberté and is bounded to the north by the Baie de Mancenille, east by the Massacre River and the town of Melliac, to the west by Dérac and south by the tertiary road to Melliac. The lake is fringed with mangroves and xerophytic shrubland set within the otherwise agricultural coastal plain. Around the lake, sisal, annual staple crops and grazing are commonplace, but rice is the major crop, irrigated by the nearby Massacre River and Maribaroux irrigation system. The lake is used by local fishermen, hunters (shooting waterbirds), and local population for watering cattle and washing clothes.

■ Birds

This IBA supports an important population of the Vulnerable West Indian Whistling-duck *Dendrocygna arborea*. Three Hispaniola EBA restricted-range species occur, namely Antillean Mango *Anthracothorax dominicus*, Broad-billed Tody *Todus subulatus* and Black-crowned Palm-tanager *Phaenicophilus palmarum*, although this total will no doubt rise with seasonal surveys. It is suspected that the Near Threatened Caribbean Coot *Fulica caribaea* and Piping Plover *Charadrius melodus* occur, but this requires confirmation. Large numbers of migratory ducks and shorebirds use the site which supports a population of Caribbean Flamingo *Phoenicopterus ruber*.

■ Other biodiversity

The Vulnerable American crocodile *Crocodylus acutus* and West Indian manatee *Trichechus manatus*, and globally threatened sea-turtles are all thought to occur.

■ Conservation

Lagon-aux-Boeufs IBA is state owned but unprotected. The area was included within a biosphere reserve nomination, and the Ministry of Environment identified the area as in need of protection (and as a recreational area) within a departmental environmental action plan, but no conservation action has been implemented. Approximately 20,000 people live around this wetland, living on fisheries, agriculture, charcoal production and occasionally boat rides. The resulting threats include water pollution, introduced predators (cats, dogs, mongooses and rats), surface water diversion (for irrigation and industrial needs), over-fishing, fuelwood extraction, hunting (especially coots *Fulica* spp., locally called "poule d'eau"), and conversion of habitat to agriculture.

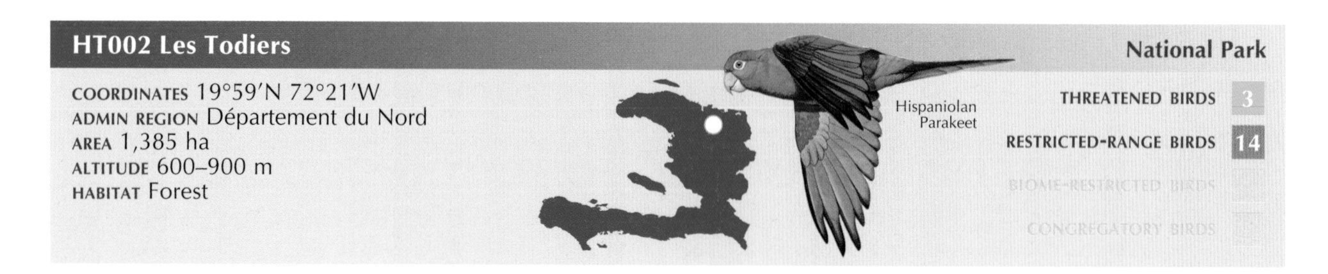

■ Site description

Les Todiers IBA is in the Massif du Nord on the Bonnet-à-l'Évêque mountain chain. It is situated within the Parc Historique la Citadelle, Sans-Souci, les Ramiers overlooking from its steep slopes the towns of Milot (a tourist and crafts town), Grande Rivière du Nord and Dondon (both farming communities) on the Plaine du Nord. The area comprises karst limestone outcroppings with low montane wet forest (with abundant tree ferns, epiphytes and melastomes) above 800 m, and subtropical moist forest (with a mix of broadleaf hardwoods and pine) lower down. The Plaine du Nord is one of the most important agricultural areas in Haiti, well known for its citrus, coffee, cocoa and bananas due to high and regular rainfall, and the remaining canopy trees.

■ Birds

Although poorly studied, this IBA supports a diverse avifauna including populations of 14 (of the 30) Hispaniola EBA restricted-range birds. Among the restricted-range birds are the Vulnerable Hispaniolan Parakeet *Aratinga chloroptera* and Hispaniolan Amazon *Amazona ventralis*. Although usually a lowland species in this part of Haiti, the parakeet occurs sympatrically with the Amazon parrots in this IBA. The Near Threatened Hispaniolan Palm Crow *Corvus palmarum* occurs.

■ Other biodiversity

The herpetofauna in this IBA is diverse, with reptiles such as *Anolis christophei*, *A. eugenegrahami* and *Sphaerodactylus lazelli* first discovered in the La Citadelle region. The Critically Endangered frogs *Eleutherodactylus poolei* and *E. schmidti limbensis* occur. Many endemic plants are found on the limestone.

■ Conservation

Les Todiers IBA is state owned and within the 2,200-ha Parc Historique la Citadelle, Sans-Souci, les Ramiers. This national park has been the focus of some ornithological surveys (although none since 1994) and historical monument restoration. Management plans have been proposed (including protection of historic monuments, promotion of tourism and biodiversity protection), and in spite of interest among local guides there has been little conservation action within this area. Insufficient legislation, commitment, funding, institutional capacity or awareness have left this park exposed to immense pressure from a poor and growing local population. The integrity of this park is deteriorating.

HT003 Coquillage–Pointe Est

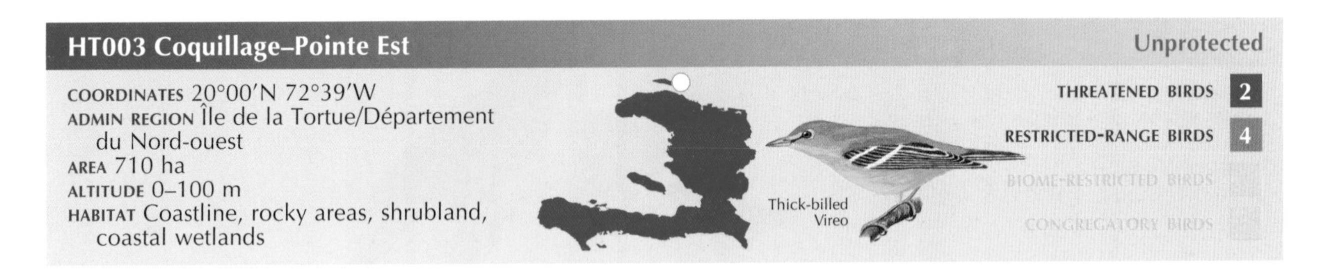

■ Site description

Coquillage–Point Est IBA is in south-easternmost Île de la Tortue, a 180-km² island off the north-west coast of Haiti. It lies between the slopes of Morne Ti Clos (154 m) to the north, the Canal de la Tortue to the south, Grand Sable to the west and Pointe Est. Coquillage is a densely populated locality. The IBA covers an area of coastal limestone cliffs, lagoons, marshes, mangrove woodland, coral reefs and white sand beaches. Inland is an area of rugged limestone with xerophytic woodland.

■ Birds

This IBA supports populations of four (of the 30) Hispaniola EBA restricted-range birds, namely Hispaniolan Lizard-cuckoo *Saurothera longirostris*, Antillean Mango *Anthracothorax dominicus*, Vervain Hummingbird *Mellisuga minima* and (the only area in Hispaniola for) Thick-billed Vireo *Vireo crassirostris*. Three Île de la Tortue endemic subspecies are present in this IBA: *V. crassirostris tortugae*, Bananaquit *Coereba flaveola nectarea* and Greater Antillean Bullfinch *Loxigilla violacea maurella*. The Near Threatened White-crowned Pigeon *Patagioenas leucocephala* and Plain Pigeon *P. inornata* occur. White-tailed Tropicbird *Phaethon lepturus* is present as a breeding species in unknown numbers.

■ Other biodiversity

Nothing recorded although the marine fauna almost certainly includes a number of globally threatened species such as sea-turtles.

■ Conservation

Coquillage–Point Est IBA has been proposed as part of a protected area (based on the biodiversity, historical and cultural interests, and its ecotourism potential), but there is currently no protection afforded this site. The IBA is used by (marine) fishermen, boat builders and farmers. The woodland is cut for lumber, fuelwood and charcoal (and then cultivated) and birds are hunted. With loss of vegetation, erosion is a problem which is leading to sedimentation of the wetland and reef system. The human population in this area is growing. Tourism and ecotourism is in its infancy on the island.

HT004 Les Grottes

■ Site description

Les Grottes (the caves) embraces a small section of Île de la Tortue, a 180-km² island on the north-west coast of Haiti. Located on the north-west (Atlantic) side of the island, the IBA includes La Grotte aux Bassins and the Trou d'Enfer, an area of limestone cliffs and terraces on the north-east slopes of Morne Monde and Morne Pois Congo. The limestone terraces are cultivated (plantains, bananas, coffee and beans) and grazed (goats and some cattle), but on the steep slopes between terraces there is xerophytic shrubland and woodland with many epiphytes.

■ Birds

This IBA supports populations of five (of the 30) Hispaniola EBA restricted-range birds, namely Hispaniolan Lizard-cuckoo *Saurothera longirostris*, Antillean Mango *Anthracothorax dominicus*, Vervain Hummingbird *Mellisuga minima*, Stolid Flycatcher *Myiarchus stolidus* and (the only area in Hispaniola for) Thick-billed Vireo *Vireo crassirostris*. Three Île de la Tortue endemic subspecies are present in this IBA: *V. crassirostris tortugae*, Bananaquit *Coereba flaveola nectarea* and Greater Antillean Bullfinch *Loxigilla violacea maurella*. The Near Threatened White-crowned Pigeon *Patagioenas leucocephala* and Plain Pigeon *P. inornata* occur.

■ Other biodiversity

Nothing recorded although restricted-range species are almost certainly present.

■ Conservation

Grottes aux Bassins et Trou d'Enfer IBA has been proposed as part of a protected area (based on the biodiversity, historical and cultural interests of the area, and its ecotourism potential), but there is currently no protection allocated this site. The woodland and shrubland is being destroyed as a result of fuelwood extraction and conversion to agriculture. Cave system alteration is increasing with human uses during planting seasons non-guided ecotourism tours.

■ Site description

Trou Caïman IBA (or Dlo Gaye) is a shallow, freshwater lake c.20 km north-east of Port-au-Prince, in the Plaine du Cul-de-Sac. Trou Caïman, together with Lac Azuéi (IBA HT006) and the Enriquillo wetlands in the Dominican Republic, form a fresh to salty water ecosystem of outstanding biological value. Reeds, sedges and cattails predominate in the north and east of the lake, grass fringes the western edge and there is dry shrubland on the southern edge. Mangroves growing on the north-eastern side are a remnant of the coastal vegetation of a shallow sea that separated Hispaniola into two Paleo-islands during the Pleistocene. The local population of c.22,000 people lives on cultivating surrounding land for sugarcane, sweet potatoes, beans etc. Over 150 fishermen work the lake, and artisans use the reeds and sedges to weave straw products.

■ Birds

This IBA is primarily a waterbird site. The Near Threatened Caribbean Coot *Fulica caribaea* occurs, and the Vulnerable West Indian Whistling-duck *Dendrocygna arborea* occurred historically, although there are no recent records. Flocks of Caribbean Flamingo *Phoenicopterus ruber* are present along with good (but unknown) numbers of waterbirds and shorebirds. The IBA also supports populations of 10 (of the 30) Hispaniola EBA restricted-range birds, four of which are globally threatened birds, namely the Vulnerable Hispaniolan Parakeet *Aratinga chloroptera*, Hispaniolan Amazon *Amazona ventralis* and White-necked Crow *Corvus leucognaphalus*, and the Near Threatened Hispaniolan Palm Crow *Corvus palmarum*.

■ Other biodiversity

Nothing recorded.

■ Conservation

Trou Caïman IBA is unprotected and no conservation actions have been undertaken, although the site was identified for protection over 20 years ago. Due to the development of sugar industry during colonial times, this area was heavily exploited for its hardwood to provide lumber for the railroad system and logs for energy. Current threats come from industrial development (using both land and water), hunting (of ducks and other waterbirds), pollution and the general unregulated direct and indirect impact of 20,000 people using the lake's resources. The ducks were monitored by Ducks Unlimited (2002–2005) and birdwatchers regularly visit the lake. Its proximity to the capital provides an excellent opportunity to develop the IBA as a "Watchable Wildlife Pond".

■ Site description

Lac Azuéi IBA (also called Etang Saumâtre) is Haiti's largest lake which lies south-east of Trou Caïman IBA (HT005), close to the Haiti–Dominican Republic border in the Neiba–Cul-de-Sac depression. Its waters are brackish. Reeds, sedges and cattails predominate in the north-west, adjacent to the wetland leading to the Trou Caïman lake. Grass fringes the western edge and there is dry shrubland on the karstic southern edge. Mangroves growing on the north-western side are a remnant of the coastal vegetation of a shallow sea that separated Hispaniola into two Paleo-islands during the Pleistocene. A number of springs provide drinking water to the populations of Malpasse and Fond Parisien. Surrounding settlements house 60,000 people supported by (irrigated) agriculture, hunting and fishing.

■ Birds

This important wetland supports a large population (up to 300) of the Near Threatened Caribbean Coot *Fulica caribaea*. Flocks of (up to 100) Caribbean Flamingo *Phoenicopterus ruber* are present along with good (but unknown) numbers of waterbirds and shorebirds. The Vulnerable West Indian Whistling-duck *Dendrocygna arborea* has been reported, but its continued presence needs to be confirmed.

■ Other biodiversity

The Vulnerable American crocodile *Crocodylus acutus* occurs, and the lake supports five endemic species of fish and an endemic turtle.

■ Conservation

Lac Azuéi IBA is state owned, and although listed by the government to become a protected area, it is not protected yet. The lake's waterbirds were surveyed by Ducks Unlimited (2002–2005) and birdwatchers visit the lake frequently. The crocodiles were surveyed in 1985. Due to the development of the indigo and sugar industry during colonial times, this area was heavily exploited for its hardwood to provide lumber for the railroad system and logs for energy. Rice, sugarcane and other staples are grown in the surrounding land, with vegetables under irrigation during the cool season. Over 300 fishermen work the lake, along with hunters (shooting ducks and flamingos). The slopes on the lower Massif de la Selle are being exploited for limestone building materials. Being very close to the transnational road and international border between Haiti and the Dominican Republic, the lake serves as a garbage "landfill" and as a toilet for passengers waiting to pass through immigration.

HT007 Aux Diablotins

National Park

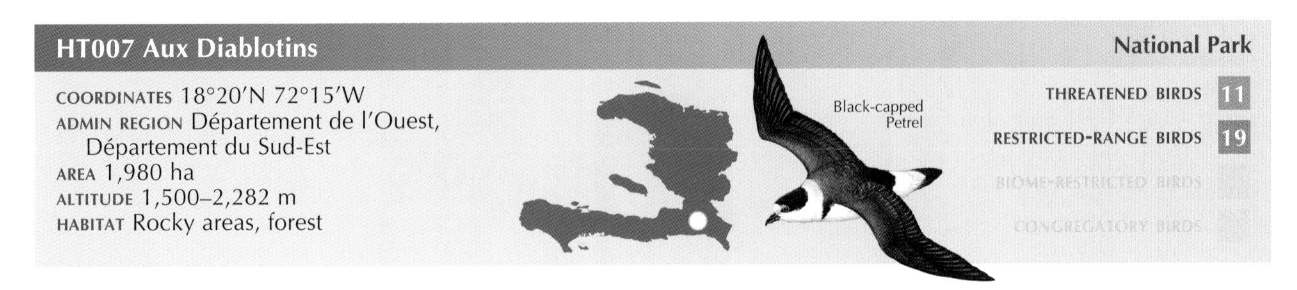

■ Site description
Aux Diablotins IBA extends along the escarpment that forms the northern boundary of La Visite National Park in the Massif de la Selle, south-eastern Haiti. The IBA embraces a narrow band c.20 km long from Morne d'Enfer (in the east) to Morne Kaderneau (in the west), including Morne La Visite, Morne Tête Opaque and Morne Cabaio (all of which are c.2,000–2,200 m high). It comprises the 1,500-ha of scarp face and cliffs (which drop down to 1,500 m in the north), and areas down to 1,600 m on the southern slopes. The scarp face supports montane broadleaf forest, with small remnant patches also on the southern slopes. There is pine (*Pinus occidentalis*) forest along the southern border. Almost all the forest (except on the scarp face) has been cleared for vegetable and corn cultivation. Aux Cornichons IBA (HT008) abuts this IBA to the south.

■ Birds
This IBA is home to one of the few known breeding sites of the Endangered Black-capped Petrel *Pterodroma hasitata*. Small numbers nest in colonies along the cliffs (Morne La Visite escarpment) north-east of Seguin. The IBA supports populations of 19 (of the 30) Hispaniola EBA restricted-range birds, eight of which are globally threatened, including the Endangered La Selle Thrush *Turdus swalesi* and Hispaniolan Crossbill *Loxia megaplaga*, the Vulnerable Hispaniolan Parakeet *Aratinga chloroptera*, Hispaniolan Amazon *Amazona ventralis*, Golden Swallow *Tachycineta euchrysea* and (Western) Chat Tanager *Calyptophilus frugivorus tertius*. The Vulnerable Bicknell's Thrush *Catharus bicknelli* winters in this IBA in significant numbers.

■ Other biodiversity
The Vulnerable Hispaniolan hutia *Plagiodontia aedium* occurs along with many plants that are endemic to the area.

■ Conservation
Aux Diablotins IBA is state-owned land along the northern side of Parc National La Visite. It has been identified as a core zone of the park due to the endemicity of the biodiversity, scenic view and water catchment importance. Various studies have been undertaken to document the biodiversity (especially birds in recent years) and develop management plans, but formal park management is non-existent. Local NGOs (e.g. Fondation Seguin) are working with communities in the park to try and reduce threats, although the socio-economic needs of the people are great. As a result there has been significant habitat loss and very little montane broadleaf forest remains.

HT008 Aux Cornichons

National Park

■ Site description
Aux Cornichons IBA is situated in the centre of La Visite National Park in the Massif de la Selle, south-eastern Haiti. The IBA lies north-west of Seguin and is bordered to the north by Ti Place and the Aux Diablotins IBA (HT007), on the east by Bois Pin Jean Noel and to the west by Roche Plate. La Scierie, which was the centre of the area's logging industry in the 1950s (when the pine forest was harvested), is on the south-eastern edge of the IBA. This sloping karst limestone area was formerly covered in wet montane broadleaf forest (with tree ferns, orchids and bromeliads), but only remnants remain, the main one (20 ha) being along the ravine at Berac.

■ Birds
This IBA supports populations of 17 (of the 30) Hispaniola EBA restricted-range birds, including eight globally threatened birds, including the Endangered La Selle Thrush *Turdus swalesi*, the Vulnerable Hispaniolan Parakeet *Aratinga chloroptera*, Golden Swallow *Tachycineta euchrysea* and (Western) Chat Tanager *Calyptophilus frugivorus tertius*, and the Near Threatened Hispaniolan Trogon *Priotelus roseigaster* and Hispaniolan Palm Crow *Corvus palmarum*. The Vulnerable Bicknell's Thrush *Catharus bicknelli* winters in this IBA in significant numbers and the Near Threatened Plain Pigeon *Patagioenas inornata* occurs. It is thought that Hispaniolan Amazon *Amazona ventralis* and White-winged Warbler *Xenoligea montana* have already been extirpated from this IBA.

■ Other biodiversity
The Vulnerable Hispaniolan hutia *Plagiodontia aedium* occurs along with many endemic plants including *Hypericum millefolium*, *Miconia rigidissima*, *Gesneria hypoclada*, *Siphocamylus caudatus*, *Ilex blancheana*, and numerous orchids and epiphytes.

■ Conservation
Aux Cornichons IBA is state-owned and protected within the 3,000-ha Parc National La Visite. It has been identified as a core zone of the park due to the endemicity of the biodiversity and water catchment importance. Aux Cornichons IBA is probably the most threatened remnant forest in Haiti. In spite of the legal protection, there is no formal park management and the forest is cleared by poor farmers to produce vegetables and herbs (some of which are proving to be invasive species), and is cut for fuelwood, both of which lead inevitably to erosion. Animals in the IBA are hunted, and plants are extracted to be sold as ornamentals in Pétion-Ville and Port-au-Prince.

HT009 Aux Bec-Croisés

National Park/Biosphere Reserve

COORDINATES 18°20′N 73°58′W
ADMIN REGION Département du Sud
AREA 2,455 ha
ALTITUDE 1,850 m
HABITAT Forest

THREATENED BIRDS 5
RESTRICTED-RANGE BIRDS 11
BIOME-RESTRICTED BIRDS
CONGREGATORY BIRDS

Hispaniolan Crossbill

■ Site description

Aux Bec-Croisés IBA is in the Massif de la Hotte, in the Southern Peninsula. Included in the core zone of the Macaya Biosphere Reserve, it lies north-east of Bois Musicien IBA (HT010). The IBA is situated in Plaine Boeuf on the Chaine Formond, overlooking the Grande Ravine du Sud. Settlements in the area include Ti Chien, Kay Tilus and Kay Ogile. This is an area of mixed wet broadleaf and pine (*Pinus occidentalis*) forest growing on a mosaic of volcanic and limestone soils. The broadleaf forest is known locally as "rak bwa".

■ Birds

This IBA supports populations of 11 (of the 30) Hispaniola EBA restricted-range birds, including the Endangered Hispaniolan Crossbill *Loxia megaplaga*, the Vulnerable Golden Swallow *Tachycineta euchrysea* and (Western) Chat Tanager *Calyptophilus frugivorus tertius*, and the Near Threatened Grey-crowned Palm-tanager *Phaenicophilus poliocephalus*. The Vulnerable Bicknell's Thrush *Catharus bicknelli* winters in this IBA in significant numbers.

■ Other biodiversity

Chaine Formond is the centre of biodiversity for the plant genus *Mecranium* and supports a number of plants endemic to the IBA such as *Ekmaniocharis* spp. Many Melastomataceae, ferns, bromeliads, *Peperomia* spp. and orchids are endemic.

■ Conservation

Aux Bec-Croisés IBA is state-owned land, and it shares its boundary with the main part of the Macaya Biosphere Reserve. Management of the area by University of Florida technicians in the early 1990s has not been continued, but the birds were surveyed in 2004 by Vermont Institute of Natural Science (in collaboration with the Société Audubon Haiti, Université Notre-Dame, and the Ministry of Environment). The area is subject to frequent fires (on the lower slopes), and forest is being lost to cultivation, lumber and fuelwood extraction, cutting of bamboo (*Arthrostylidum haisiense*), over-grazing and erosion. With a growing local population, the pressures on the forest are increasing.

HT010 Bois Musicien

National Park/Biosphere Reserve

COORDINATES 18°19′N 74°01′W
ADMIN REGION Département du Sud
AREA 1,060 ha
ALTITUDE 950–1,200 m
HABITAT Forest

THREATENED BIRDS 6
RESTRICTED-RANGE BIRDS 14
BIOME-RESTRICTED BIRDS
CONGREGATORY BIRDS

White-winged Warbler

■ Site description

Bois Musicien IBA is in the Massif de la Hotte, in the Southern Peninsula. Located on the Morne Cavalier mountains, it forms the south-western spur of the Macaya Biosphere Reserve. To the east is the plain of Durand and to the south is the harsh karstic zone of Soulette. Nearby localities of Durand, Portal and Formond connect to Cavalier by paths used intensively on market days, and during planting and harvest seasons. The IBA comprises diverse wet broadleaf forest ("rak bwa" with numerous ferns, orchids and bromeliads) on limestone karst, and cloud cover is common at higher altitudes. The forest is in fact a mosaic of habitats resulting from cultivation (of very small fields), wood gathering and livestock grazing.

■ Birds

This IBA is significant for a diversity of globally threatened and restricted-range birds. The area supports populations of 14 (of the 30) Hispaniola EBA restricted-range birds, including the Vulnerable Golden Swallow *Tachycineta euchrysea*, White-winged Warbler *Xenoligea montana* and (Western) Chat Tanager *Calyptophilus frugivorus tertius*. Three Near Threatened birds also occur. The bird diversity in this IBA is higher than elsewhere in the Macaya Biosphere Reserve. The area is important for wintering Neotropical migrants.

■ Other biodiversity

There are many endemic species in this area, most notably among the snails, *Anolis* lizards and *Eleutherodactylus* frogs, including the Critically Endangered *E. amadeus*, *E. corona* and *E. dolomedes*. The Endangered Hispaniolan solenodon *Solenodon paradoxus* and Vulnerable Hispaniolan hutia *Plagiodontia aedium* occur, as do many bats such as the Near Threatened moustached bat *Pteronotus quadridens* and Brazilian free-tailed bat *Tadarida brasiliensis constanzae*.

■ Conservation

Bois Musicien is within Macaya Biosphere Reserve. Rich families "control" (but do not own) the land in Plain Durand and Plain Formond, and very poor families (living in "ajoupas" or thatch huts) farm this land, or raise free-roaming livestock. Forest is being lost to cultivation, lumber and fuelwood extraction, over-grazing and erosion. The birds are additionally threatened by hunting. This IBA is the most accessible part of the Macaya Biosphere Reserve and thus the most impacted, and also the most biologically studied. Conservation management (1989–2001) enabled some regeneration of forest, but there is currently no management within the IBA.

JAMAICA

LAND AREA **10,829 km²** ALTITUDE **0–2,258 m**
HUMAN POPULATION **2,780,132** CAPITAL **Kingston**
IMPORTANT BIRD AREAS **15, totalling 3,113 km²**
IMPORTANT BIRD AREA PROTECTION **44%**
BIRD SPECIES **300**
THREATENED BIRDS **16** RESTRICTED-RANGE BIRDS **36**

CATHERINE LEVY AND SUSAN KOENIG
(WINDSOR RESEARCH CENTRE)

Karst limestone "cockpits" in Cockpit Country IBA.
(PHOTO: GEOEYE/JAMAICA FORESTRY DEPARTMENT, FROM IKONOS IMAGES)

INTRODUCTION

Jamaica is the third largest island in the Greater Antilles lying 145 km south of eastern Cuba and 161 km west of Haiti. It is 235 km long (east to west) and 35–82 km wide (north to south). Administratively, Jamaica is divided into 14 parishes, and the territory includes the Morant Cays (off the eastern end of the island) and Pedro Cays (off the south-west coast). The island is rugged with mountains and plateaus: much of the land is above 300 m. The highest point is Blue Mountain Peak in the Blue Mountains, a dramatically uplifted ridge-block of Cretaceous metamorphic rock which rises sharply from the coast. The eastern end of this block is capped in limestone, which forms the steep and extremely rugged John Crow Mountains. The Rio Grande, Jamaica's largest river (by surface-water runoff), separates the Blue and John Crow Mountains. The centre and centre-west of the island is composed of a massive limestone block with well-defined features of a karst landscape, including doline, polygonal (cockpit), and tower karst, large alluvial poljes (valleys), and many sinkholes and caves. The best-developed polygonal karst is found in Cockpit Country (the "type area" for cockpit karst). A portion of the Cockpit Country aquifer drains to the south-west, forming the Black River and Great Morass, the largest swampland ecosystem on Jamaica. In extreme western Jamaica, alluvial plains and rolling karst limestone are punctuated by a Cretaceous igneous outcrop capped in limestone, and known as Dolphin Head. Due to the island's geologic history of volcanic extrusion, subsidence, and tectonic uplift, the Blue and John Crow Mountains, Cockpit Country, and Dolphin Head are recognised as three "hotspots" of adaptive radiation and endemism. The island's forested mountains and hilly interior are incised by steep valleys, particularly in the east where erosion is now prevalent due to the removal of forest cover. In the centre and west, the limestone formations provide little surface water, and removal of vegetation exposes a thin red soil.

Jamaica has a tropical maritime climate. In the lowlands, the mean annual temperature is 26°C, but just 13°C at Blue Mountain Peak. Rainfall varies across the island with average annual precipitation greater than 500 cm in John Crow Mountains, 250 cm in the highest parts of Cockpit Country, and less than 75 cm in the Hellshire Hills—the driest part of the country. Although it rains in every month, the heaviest rains are from September to November, and again (but less heavily) in May and June. The driest months are January through March. Natural vegetation corresponds to geology, elevation and precipitation, and ranges from very wet and wet tropical forest, particularly on the north side of the Blue

Mountains (where a remnant of elfin forest remains), and on limestone (especially the John Crow Mountains and Cockpit Country), to dry scrub forest, and dry woodland along in coastal areas. The Forestry Department's land-use figures (up to 1998) broadly classify Jamaica as: c.30% forest (only 8% of which is minimally disturbed or "closed broadleaf" forest); 30% mixed land-use (including plantations and fields); and 39% non-forest land-use (buildings/infrastructure, wetlands, and bauxite and limestone-aggregates mining). Between 1989 and 1998, the greatest loss of forest land was in "disturbed broadleaf", and the greatest gain (44%) in "mixed land-use/cover" (including "partly forested and partly bauxite lands"). The expansion of bauxite mining accounts for much of this latter change in land-use.

■ Conservation

National laws for biodiversity conservation in Jamaica have lagged behind other legislation, but are now under scrutiny as part of the process to update the Protected Areas Systems Plan. The Wildlife Protection Act, originally passed in 1945 to regulate sports-hunting and fishing, has been enhanced by many regulations that attempt to address gaps, particularly in relation to protection of animals. However, this act does not address habitat protection or the conservation of flora. Habitat protection comes under the Natural Resources Conservation Authority Act (1991) which provides the legislative framework for a system of protected areas and paved the way for the establishment of marine parks and the Blue and John Crow Mountains National Park. The island also has over 150 forest reserves designated under the Forest Act (1996, and subsequent regulations) which provides for the preservation of forests, watershed protection, and ecotourism. Private lands declared as forest reserves can be entitled to property tax exemptions. Jamaica's protected areas portfolio is biased towards the forested mountains of the interior, leaving lowland and coastal ecosystems under-represented. Most of the remaining forested coastal areas are privately owned.

Recognition of Jamaica's unique biodiversity has come about slowly over the past 30 years despite an encouraging start when, after the first United Nations Conference on the Environment (Stockholm 1972), it was decided to amalgamate the various national environmental commissions into one agency—the Natural Resources Conservation Department (NRCD). This agency was felt to have only advisory capabilities, so it was expanded in 1991 to become the Natural Resources Conservation Authority (NRCA). NRCA is responsible for declaring and managing national parks, and enforces the requirement for project-related environmental impact assessments. The Forestry Department manages the island's forest estate. In 1988, Hurricane Gilbert's devastating effects revealed problems in natural forest management; this prompted the preparation of a National Forestry Action Plan (1990), followed by a new Forest Act (1996). The latter explicitly includes "conservation and sustainable management of forest", thus covering activities such as the protection of forest resources for ecosystem services and biodiversity. Conservation remains a challenge due to limited financial resources. Partly because of insufficient capacity within Forestry Department and NRCA and partly because of policy changes in the governance of natural resources, management of Jamaica's first national park (declared in 1990) was delegated to the NGO Jamaica Conservation and Development Trust (JCDT). Other NGOs to be mandated with protected area management are: the Caribbean Coastal Area Management (CCAM) Foundation with responsibility for the Portland Bight Protected Area; and the Montego Bay Marine Park Trust which has been given the mandate to manage the Montego Bay Marine Park. Elsewhere, the Jamaican iguana *Cyclura collei* project, lead by Dr Byron Wilson (University of the West Indies), provides a focus for research and conservation activities within the dry forest habitat (including a small forest reserve) of Hellshire. In Cockpit Country, the Forestry Department and Windsor Research Centre (Trelawny) are working together to facilitate the work of three Local Forest Management Committees—encouraging local community engagement in sustainable forest resource use and management.

Even though an attempt was made to set up one agency to "provide for the management, conservation, and protection of the natural resources of Jamaica", there are at least 34 pieces of legislation that refer to the environment, e.g. Land Acquisition Act (1947), Urban Development Act (1968), Maritime Areas Act (1996)—not all of which are administered by NRCA, but by other government agencies as well. Problems associated with conservation in Jamaica include poor coordination between the plethora of government institutions responsible for the various laws and regulations insufficient recognition of the value of biological diversity, insufficient funding, poor enforcement, incomplete or improper environmental impact assessments, and incomplete island-wide evaluation of landscape and biodiversity values. While these

Bauxite mining is driving habitat destruction across the centre of the island including Mount Diablo IBA. (PHOTO: SUSAN KOENIG/WINDSOR RESEARCH CENTRE)

Black River lower morass. (PHOTO: VAUGHAN TURLAND)

issues are inhibiting effective conservation action, there are a number of significant threats that are directly impacting Jamaica's unique biodiversity. Habitat loss and fragmentation are the greatest threats. With primary forest reduced to just 8% of the land area, multiple factors (e.g. increased predation, increased competition from invasive species, reduction of genetic variability etc.) impinge on the long-term survival prospects of the species populations that remain. Driving this habitat loss, degradation and fragmentation is the expansion of bauxite mining and limestone quarrying; residential, hotel and resort developments (particularly along the coasts); highways and roads; and, to a lesser extent, agriculture. Annual dry season (or drought period) fires (started intentionally) have a significant impact on woodlands and forests. Climate change models are predicting significantly drier summers in the Caribbean suggesting that fire risk will be of increasing concern. Another consequence of increasing habitat loss, degradation and fragmentation is the reduced resilience of the remaining forests to stochastic events such as hurricanes (or indeed the forecasted effects of global climate change). The last serious hurricane to hit almost the entire island was Hurricane Gilbert in 1988. More recently the trajectories of hurricanes Allan and Charley (2004), Emily (2005) and Dean (2007) carried them near or over Jamaica's south coast. It would be prudent for future protected area planning to consider coastal vulnerability to hurricanes.

Alien invasive species impacting Jamaica's native biodiversity include small Indian mongoose *Herpestes auropunctatus*, black and brown rat (*Rattus* spp.), dogs, cats, and feral pigs. The impact of these species has not been quantified although the mongoose has been identified as a causal factor in the possible extinction of the (Critically Endangered) ground-nesting Jamaican Petrel *Pterodroma caribbaea* and Jamaican Pauraque *Siphonorhis americana*. It was also thought to have contributed to the extinction of the endemic Jamaican iguana *Cyclura collei* until a small population was rediscovered in Hellshire Hills in 1990. Between 1996 and 2008, the Jamaican Iguana Recovery Group removed c.1,000 mongoose from the core iguana conservation zone and operates a trapping grid every day. Snares are used to trap and remove pigs from the core iguana area and from the adjacent coastal fringe. Recent research has indicated that pigs may be responsible for the loss of nearly all sea-turtle nests in a given season. Non-native psittacines imported for the pet trade pose a high threat to the endemic *Amazona* parrots and the native *Aratinga* parakeet through the introduction of disease, the potential for hybridisation, and competitive exclusion of nesting cavities. A temporary ban on their importation, grounded in concerns for the introduction of highly pathogenic strains of avian influenza, remains in effect as of 2008, and efforts are being directed to support the Veterinary Services Division in making the ban permanent. Invasive plants are also a threat. Where natural vegetation has been cleared, exotic species frequently out-compete native species, and forest fragmentation facilitates their colonisation of new areas. Aggressively invasive species which create biologically sterile monocultures in Jamaica include *Bambusa vulgaris*, Asian ferns *Nephrolepis* spp., *Pittosporum undulatum*, and *Alpinia allughas* (contributing to loss of native species in the swamp forest of the Black River Great Morass).

Birds

Of Jamaica's c.300 recorded bird species, 124 breed (including 12 that are introduced) and over 170 species occur as wintering Neotropical migrants, transients or vagrants. The Jamaican avifauna exhibits exceptional levels of endemism, with 36 restricted-range species defining the Jamaica Endemic Bird

Orangequit, a species and genus endemic to Jamaica. (PHOTO: HUGH VAUGHAN)

The Endangered Jamaican Blackbird or "Wildpine Sergeant", one of Jamaica's four endemic genera. (PHOTO: HUGH VAUGHAN)

Area, and 30 breeding species confined to the island (and primarily to natural forest and woodlands). A number of the restricted-range species are shared with neighbouring islands (e.g. Vervain Hummingbird *Mellisuga minima*, Stolid Flycatcher *Myiarchus stolidus*, Greater Antillean Elaenia *Elaenia fallax* and Golden Swallow *Tachycineta euchrysea* are all shared with Hispaniola). Five of the species endemic to Jamaica represent four endemic genera: two *Trochilus* spp. (streamertails), *Euneornis campestris* (Orangequit), *Loxipasser anoxanthus* (Yellow-shouldered Grassquit) and *Nesopsar nigerrimus* (Jamaican Blackbird). In fact 48 species are endemic to the island at the genus, species or subspecies level. Black-billed Streamertail *Trochilus scitulus* has the narrowest range of all the island endemics, being confined (but abundant) in the John Crow Mountains IBA (JM014).

The threat category and national population sizes of the threatened birds are listed in Table 1. Although 16 threatened species occur on Jamaica, two of these, Golden-winged Warbler *Vermivora chrysoptera* and Cerulean Warbler *Dendroica cerulea*, are only known as vagrants and have not been considered in the IBA analysis. The Jamaica Petrel *Pterodroma caribbaea* and Jamaican Pauraque *Siphonorhis americana* are classified as Critically Endangered, and neither has been seen with certainty for 130 years although there are possibilities that the pauraque persists in Hellshire Hills IBA (JM011) and the petrel could survive in John Crow Mountains IBA (JM014). The Jamaican Blackbird *N. nigerrimus* is considered Endangered as it occurs in small numbers and only inhabits moist forest with numerous bromeliads such as is found in the Blue and John Crow Mountains, Mount Diablo, and the larger IBAs of the Cockpit Country Conservation Area.

Table 1. Key bird species at Important Bird Areas in Jamaica.

Key bird species	Criteria		National population
West Indian Whistling-duck *Dendrocygna arborea*	VU ■	■	250–999
Jamaica Petrel *Pterodroma caribbaea*	CR ■	■	<50
Pied-billed Grebe *Podilymbus podiceps*		■	2,500
Magnificent Frigatebird *Fregata magnificens*		■	4,500
Brown Pelican *Pelecanus occidentalis*		■	250
Masked Booby *Sula dactylatra*		■	2,400–3,000
Brown Booby *Sula leucogaster*		■	6,000
Caribbean Coot *Fulica caribaea*	NT ■		50–100
Laughing Gull *Larus atricilla*		■	800
Royal Tern *Sterna maxima*		■	350
Sandwich Tern *Sterna sandvicensis*		■	350
Least Tern *Sterna antillarum*		■	250–600
Bridled Tern *Sterna anaethetus*		■	2,500–3,000
Sooty Tern *Sterna fuscata*		■	75,000–95,000
Brown Noddy *Anous stolidus*		■	10,000
White-crowned Pigeon *Patagioenas leucocephala*	NT ■		
Ring-tailed Pigeon *Patagioenas caribaea*	VU ■	■	2,500–9,999
Plain Pigeon *Patagioenas inornata*	NT ■		
Crested Quail-dove *Geotrygon versicolor*	NT ■	■	
Yellow-billed Amazon *Amazona collaria*	VU ■	■	10,000–19,999
Black-billed Amazon *Amazona agilis*	VU ■	■	15,000–19,999
Jamaican Lizard-cuckoo *Coccyzus vetula*		■	
Chestnut-bellied Cuckoo *Coccyzus pluvialis*		■	
Jamaican Owl *Pseudoscops grammicus*		■	
Jamaican Pauraque *Siphonorhis americana*	CR ■	■	<50
Jamaican Mango *Anthracothorax mango*		■	
Red-billed Streamertail *Trochilus polytmus*		■	60,000–120,000
Black-billed Streamertail *Trochilus scitulus*		■	60,000–120,000
Vervain Hummingbird *Mellisuga minima*		■	
Jamaican Tody *Todus todus*		■	
Jamaican Woodpecker *Melanerpes radiolatus*		■	
Jamaican Becard *Pachyramphus niger*		■	
Jamaican Elaenia *Myiopagis cotta*		■	
Greater Antillean Elaenia *Elaenia fallax*		■	
Jamaican Pewee *Contopus pallidus*		■	
Sad Flycatcher *Myiarchus barbirostris*		■	
Rufous-tailed Flycatcher *Myiarchus validus*		■	
Stolid Flycatcher *Myiarchus stolidus*		■	
Jamaican Vireo *Vireo modestus*		■	
Blue Mountain Vireo *Vireo osburni*	NT ■	■	
Jamaican Crow *Corvus jamaicensis*		■	
Golden Swallow *Tachycineta euchrysea*	VU ■	■	
Bahama Mockingbird *Mimus gundlachii*		■	
Rufous-throated Solitaire *Myadestes genibarbis*		■	
Bicknell's Thrush *Catharus bicknelli*	VU ■		90
White-chinned Thrush *Turdus aurantius*		■	
White-eyed Thrush *Turdus jamaicensis*		■	
Arrowhead Warbler *Dendroica pharetra*		■	
Jamaican Oriole *Icterus leucopteryx*		■	
Jamaican Blackbird *Nesopsar nigerrimus*	EN ■	■	2,500–9,999
Yellow-shouldered Grassquit *Loxipasser anoxanthus*		■	
Orangequit *Euneornis campestris*		■	
Jamaican Spindalis *Spindalis nigricephala*		■	
Jamaican Euphonia *Euphonia jamaica*		■	

All population figures = numbers of individuals.
Threatened birds: Critically Endangered ■; Endangered ■; Vulnerable ■; Near Threatened ■.
Restricted-range birds ■. **Congregatory birds** ■.

				Jamaica IBAs									
2	JM003	JM004	JM005	JM006	JM007	JM008	JM009	JM010	JM011	JM012	JM013	JM014	JM015
■	■	■	■	■	■	■		■	■	■	■	■	
■	■	■	■	■	■	■		■	■	■	■	■	
					■		■	■		■			■
					100–300			50–249	✓				
												✓	
					1,000–2,499								
					100		4,000	300					
					250								
							2,400–3,000						
							6,000						
					50								
					500		223						
					250		25						30–60
					250								
					250					50			
							1,500–2,000						1,000
							4,000–5,000						70,000–90,000
							4,500	600					4,500
✓	✓	✓	✓	✓	500	✓		✓	✓	✓		✓	
✓		✓	✓	50–100						✓	300–700	100–300	
		✓	✓	✓				✓					
		✓	✓	✓		✓				✓	✓	✓	
	✓	✓	10,000–16,000	✓		✓					✓	✓	
		✓	15,000–19,000	✓		✓						✓	
✓		✓	✓	✓		✓		✓	✓	✓	✓	✓	
		✓	✓	✓		✓				✓	✓	✓	
✓	✓	✓	✓		✓	✓		✓	✓	✓			
									✓				
✓	✓	✓	✓	✓	✓	✓		✓	✓	✓	✓	✓	
✓	✓	✓	✓	✓	✓	✓		✓	✓	✓	✓		
												60,000–120,000	
✓	✓	✓	✓	✓	✓	✓		✓	✓	✓	✓	✓	
✓	✓	✓	✓	✓		✓		✓	✓	✓	✓	✓	
✓	✓	✓	✓	✓	✓	✓		✓	✓	✓	✓	✓	
✓	✓	✓	✓	✓		✓				✓	✓	✓	
✓	✓	✓	✓	✓		✓		✓	✓	✓	✓	✓	
✓	✓	✓	✓			✓				✓	✓		
✓	✓	✓	✓	✓	✓	✓				✓	✓	✓	
✓	✓	✓	✓	✓	✓	✓		✓	✓	✓	✓	✓	
✓	✓	✓	✓	✓	✓	✓		✓	✓	✓	✓	✓	
✓	✓	✓	✓	✓	✓	✓		✓	✓	✓			
✓	✓	✓	✓	✓	✓	✓		✓	✓	✓	✓	✓	
		✓	✓	✓		✓					✓	✓	
	✓	✓	✓	✓		✓						✓	
			✓								✓		
								3,000–5,000	5,000				
✓	✓	✓	✓	✓		✓				✓	✓	✓	
											90		
✓	✓	✓	✓	✓	✓	✓				✓	✓	✓	
✓	✓	✓	✓	✓		✓				✓	✓	✓	
✓	✓	✓	✓	✓		✓				✓	✓	✓	
✓	✓	✓	✓	✓	✓	✓		✓	✓	✓	✓	✓	
			✓	✓		✓					✓	✓	
✓	✓	✓	✓	✓	✓	✓		✓	✓	✓	✓	✓	
✓	✓	✓	✓	✓	✓	✓				✓	✓	✓	
✓	✓	✓	✓	✓		✓		✓		✓	✓	✓	
✓	✓	✓	✓	✓	✓	✓		✓	✓	✓	✓	✓	

A significant percentage of the Caribbean's Brown Boobies nest in Jamaica.
(PHOTO: BRANDON HAY)

Thirteen seabird species nest on Jamaica and its offshore cays, and the island is regionally important for four of these, namely Masked Booby *Sula dactylatra* (over 50% of the Caribbean's nesting birds), Sooty Tern *Sterna fuscata* (c.30%) and Brown Noddy *Anous stolidus* (c.30%) and Brown Booby *Sula leucogaster* (c.20%). Given the serious decline in Jamaican seabird numbers, their nesting sites (which includes coastal areas and the offshore cays such as Morant and Pedro Cays) urgently need active conservation management, research and especially monitoring.

IMPORTANT BIRD AREAS

Jamaica's 15 IBAs—the island's international site priorities for bird conservation—cover 3,113 km², about 25% of Jamaica's land area. Many of the terrestrial IBAs overlap with forest reserves or crown lands to some extent, thus some form of protection is in place. However, only 44% of the area covered by the IBAs is under formal protection, and active management is minimal in many areas.

The IBAs have been identified on the basis of 53 key bird species (listed in Table 1) that variously meet the IBA criteria. These 53 species include 13 (of the 16) globally threatened birds, all 36 restricted-range species, and 14 congregatory waterbirds/seabirds. It was not possible to identify IBAs for significant (qualifying) populations of two of Jamaica's globally threatened birds, namely Piping Plover *Charadrius melodus* and Black Rail *Laterallus jamaicensis*. However, *C. melodus* is known to occur (although not in significant numbers) in Black River Great Morass IBA (JM007), Portland Ridge and Bight IBA (JM010) and Yallahs IBA (JM012), and *L. jamaicensis* has been recorded in Black River Great Morass IBA.

Significant populations of the majority of Jamaica's key bird species are found in two or more IBAs. However, for many of the congregatory species, significant (i.e. >1% of the global or Caribbean population of the species) populations are only found in one IBA (see Table 1). The Black River Great Morass IBA (JM007) and Pedro Cays and Bank IBA (JM009) support most of these populations, emphasising how critically important they are for the maintenance of Jamaica's waterbird and seabird populations.

At least 47 terrestrial areas have been identified as "potential Important Bird Areas" and the boundaries for many of these have been used in the preparation of the Protected Areas Master Plan. However, at present there is

Middle Cay, Pedro Cays and Bank IBA.
(PHOTO: BRANDON HAY)

Figure 1. Location of Important Bird Areas in Jamaica.

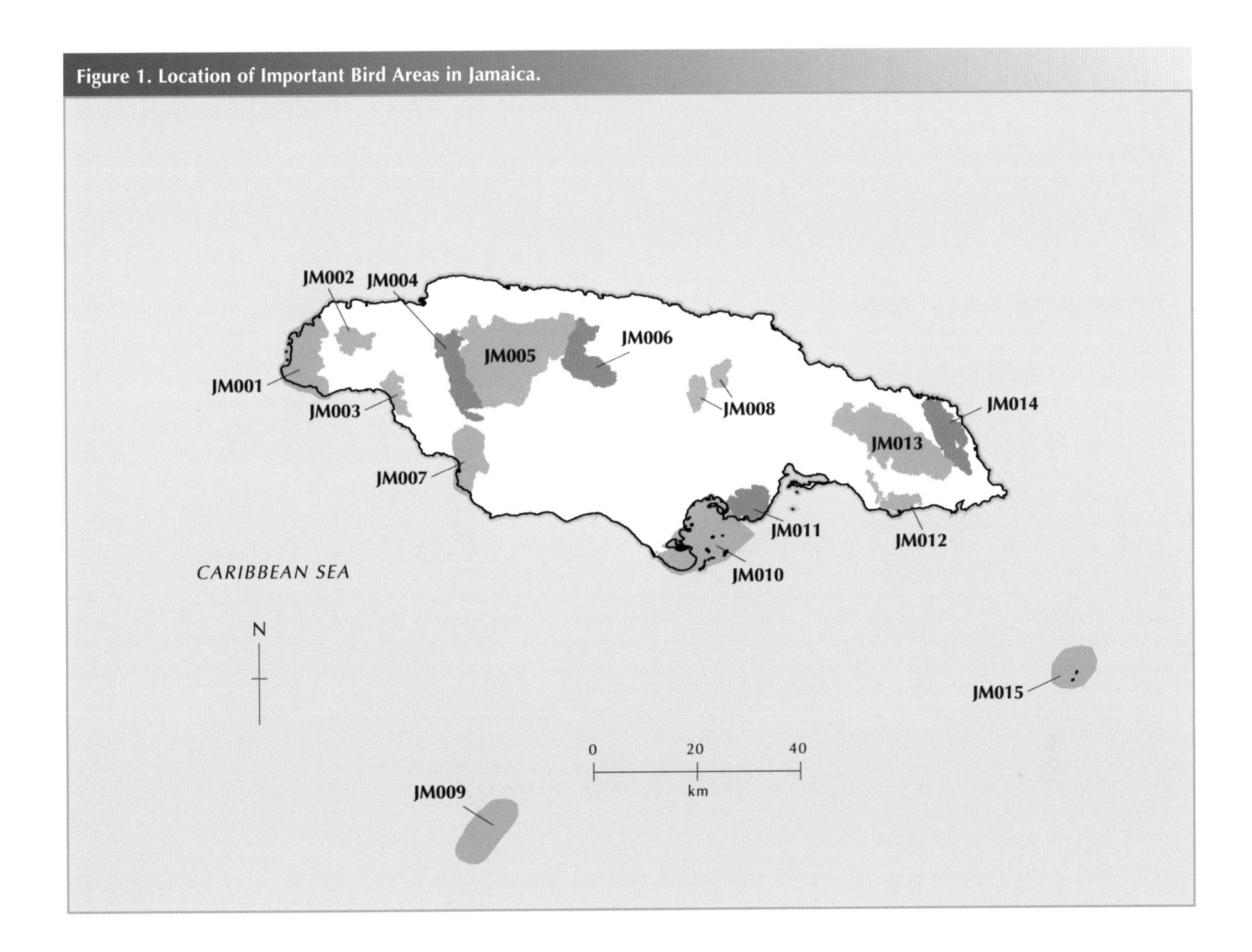

insufficient information concerning the occurrence of populations of key bird species at these sites for them to qualify as IBAs. This clearly presents field research objectives for the academic and conservation communities within Jamaica, namely to clarify (based on quantitative data) the international importance of Jamaica's 47 "potential" IBAs.

State, pressure and response scores have been collated for some of Jamaica's IBAs, but should be monitored annually at all IBAs to provide an objective status assessment and highlight management interventions that might be required to maintain these internationally important biodiversity sites.

KEY REFERENCES

Azan, S. and Webber, D. (2007) The characterization and classification of the Black River Upper Morass, Jamaica, using the three-parameter test of vegetation, soils and hydrology. *Aquatic Conservation: Marine and Freshwater Ecosystems* 17: 5–23.

Beard, J. S. (1955) The classification of tropical American vegetation types. *Ecology* 36: 89–100.

Bourne, W. R. P. (1965) The missing petrels. *Bull. Brit. Orn. Club* 85: 6.

Douglas, L. and Levy, C. (2002) An estimate of the number of Masked and Brown Boobies (*Sula dactylatra* and *S. leucogaster*) breeding on Southwest Cay, Pedro Cays, Jamaica. (Unpublished report).

Douglas, L. and Zonfrillo, B. (1997) First record of Audubon's Shearwater and Black-capped Petrel from Jamaica. *Gosse Bird Club Broadsheet* 69: 4–6.

Downer, A. and Sutton, R. (1990) *Birds of Jamaica: a photographic field guide*. Cambridge, U.K.: Cambridge University Press.

Dunkley, C. S. and Barrett, S. (2001) Case study of the Blue and John Crow Mountain National Park. Trinidad: Caribbean Natural Resources Institute. (CANARI Technical Report 282).

Fincham, A. G. (1997) *Jamaica underground: the caves, sinkholes and underground rivers of the island*. Kingston: University of the West Indies Press.

Gonsiska, P. and Koenig, S. E. (2007) Epiphyte surveys in Barrett Hut: Litchfield—Matteson's Run Forest Reserve. (Unpublished report).

Haynes-Sutton, A, and Hay, D. B. (2002) Survey of migratory ducks in Jamaican wetlands. Phase one: January–April 2001. Mandeville, Jamaica. (Unpublished report for Natural Resources Conservation Authority, National Environment and Planning Agency and Ducks Unlimited).

Gosse, P. H. (1848) *The birds of Jamaica*. London: John van Voorst.

Gosse, P. H. (1851) *A naturalist's sojourn in Jamaica*. London: Longmans.

Hedges, S. B. (1999) Distribution patterns of amphibians in the West Indies. Pp 211–254 in W. E. Duellman ed. *Regional patterns of amphibian distribution: a global perspective*. Baltimore: Johns Hopkins University Press.

Jamaica Conservation and Development Trust (2005) Blue and John Crow Mountains National Park: management plan (2005–2010). Kingston: Jamaica Conservation and Development Trust. (Unpublished report).

Koenig, S. E. (2008) Black-billed Parrot (*Amazona agilis*) population viability assessment (PVA): a science-based prediction for policy makers. *Orn. Neotrop.* 19 (suppl.): 135–149.

Koenig, S. E. (2008) Status and threat-risks to Jamaica's two endemic Amazon parrots. (Unpublished manuscript).

Koenig, S. E., Wunderle, J. M. and Enkerlin-Hoefflich, E. (2007) Vines and canopy contact: a route for snake predation on parrot nests. *Bird Conserv. Internatn.* 17: 79–91.

LACK, D. (1976) *Island* biology: illustrated by the land birds of Jamaica. Oxford: Blackwell Scientific Publications (Studies in Ecology 3).
LAWRENCE, V. M. (2005) Urban Development Corporation annual report 2004–2005. Kingston: Urban Development Corporation. (Unpublished report).
LEHNERT, M. S. (2008) The population biology and ecology of the homerus swallowtail *Papilio (Pterourus) homerus*, in the Cockpit Country, Jamaica. *J. Insect Conserv.* 12: 179–188.
MCCALLA, W. (2004) Protected Area Systems Plan: legal framework. Kingston. (Unpublished final report).
MCFARLANE, D. A., LUNDBERG, J. AND FINCHAM, A. G. (2002) A late Quaternary paleoecological record from caves of southern Jamaica, West Indies. *J. Cave and Karst Studies* 64: 117–125.
MINISTRY OF AGRICULTURE (2001) National forest management and conservation plan. Kingston: Forestry Department. (Unpublished report).
MORRISEY, M. (1989) *Our island, Jamaica*. London: Collins.
QUAMMEN, D. (1996) *The song of the Dodo*. New York: Scribner.
RAFFAELE, H. WILEY J., GARRIDO, O., KEITH, A. AND RAFFAELE, J. (1998) *A guide to the birds of the West Indies*. Princeton, New Jersey: Princeton University Press.
ROSENBERG, G. AND MURATOV, I. V. (2005) Status report on the terrestrial Mollusca of Jamaica. *Proc. Acad. Nat. Sci. Philadelphia* 155: 117–161.
SCHREIBER, E. A. AND LEE, D. S. EDS. (2000) *Status and conservation of West Indian seabirds*. Ruston, USA: Society of Caribbean Ornithology (Spec. Publ. 1).
STATISTICAL INSTITUTE OF JAMAICA (2007) Environmental statistics downloaded from: www.statinja.com/env_stats.html.
STRONG, A. M. AND JOHNSON, M. D. (2001) Exploitation of a seasonal resource by non-breeding Plain and White-crowned Pigeons: implications for conservation of tropical dry forests. *Wilson Bull.* 113: 73–77.
SVENSSON, S. (1983) Ornithological survey of the Negril and Black River Morasses, Jamaica. Appendix VI to Environmental feasibility study of peat mining in Jamaica. Kingston, Jamaica. (Unpublished report).
TUBERVILLE, T. D. AND BUHLMANN, K. A. (2005) Ecology of the Jamaican slider turtle (*Trachemys terrapen*), with implications for conservation and management. *Chelonian Conserv. and Biol.* 4: 908–915.
WILLIAMS, S. A. (2007) Strategic management plan for the Royal Palm reserve and the Negril Great Morass. Negril, Jamaica: Negril Area Environmental Protection Trust. (Unpublished report for BirdLife International and UNEP-GEF).
WILSON, B. (2008) Battling invasive predators to save the Jamaican Iguana. *Aliens of Xamayca* 1(2).
WILSON, B. S. AND VOGEL, P. (2000) A survey of the herpetofauna of the Hellshire Hills, Jamaica, including the rediscovery of the Blue-tailed Galliwasp (*Celestus duquesneyi* Grant). *Carib. J. Sci.* 36: 244–249.

ACKNOWLEDGEMENTS

The authors would like to thank Garfield "Jimmy" Basant, Marlon Beale (BirdLife Jamaica), Herlitz Davis (BirdLife Jamaica), Chandra Degia, Owen Evelyn (Forestry Department), John Fletcher (BirdLife Jamaica), Gary Graves (National Museum of Natural History, Smithsonian Institute), Brandon Hay, Ricardo Miller (NEPA), Michael Schwartz (Windsor Research Centre), Ann Sutton, Charles Swaby, Vaughan Turland (BirdLife Jamaica), and Byron Wilson (Jamaica Iguana Project, University of the West Indies) for their help in preparing and commenting on this chapter.

JM001 Negril

Environmental Protection Area/Nature Reserve

COORDINATES 18°19′N 78°19′W
ADMIN REGION Hanover, Westmoreland
AREA 27,740 ha
ALTITUDE 0–280 m
HABITAT Forest, inland wetland, coastline, mangrove, caves

West Indian Whistling-duck

THREATENED BIRDS 2
RESTRICTED-RANGE BIOMES 19
BIOME-RESTRICTED BIRDS
CONGREGATORY BIRDS ✓

■ Site description

Negril IBA is situated at the westernmost end of Jamaica. It follows the boundary of the Environmental Protection Area and embraces the entire Negril watershed including ecosystems in the Negril Great Morass (Jamaica's second largest wetland), the Royal Palm Reserve, and the (limestone) Fish River and Negril hills. The morass is bounded by the Fish River Hills to the east, the Negril Hills to the south, and Long Bay beach to the west. The Royal Palm Reserve is in the southern part of the morass with its southern boundary being the South Negril River which enters the Caribbean sea at Negril, Jamaica's third largest tourist resort. Private residential and commercial developments (including tourist developments) are found throughout the area.

■ Birds

This IBA is significant for 19 (of the 36) Jamaica EBA restricted-range birds. More than 90 Vulnerable West Indian Whistling-ducks *Dendrocygna arborea* now occur in the morass (especially the Royal Palm Reserve), and there is a notable population of Near Threatened White-crowned Pigeon *Patagieonas leucocephala*. Yellow-breasted Crake *Porzana flaviventer* is present and at least 17 species of Neotropical migratory birds use the IBA in winter.

■ Other biodiversity

The Vulnerable Macleay's mustached bat *Pteronotus macleayi*, Jamaican boa *Epicrates subflavus*, Jamaican slider *Trachemys terrapen* (an endemic freshwater turtle) and Jamaican kite swallowtail *Protographium marcellinus* occur. The Near Threatened (and endemic) morass royal palm *Roystonea princeps* and Caribbean endemic anchovy pear *Grias cauliflora* dominate the Royal Palm Reserve.

■ Conservation

Negril IBA is primarily state-owned land designated as an Environmental Protection Area under the jurisdiction of the Urban Development Corporation, the Petroleum Corporation of Jamaica and Ministry of Agriculture. The Negril Royal Palm Reserve covers c.121 ha, and is managed by Negril Area Environmental Protection Trust (NEPT). A strategic management plan has been developed by NEPT as part of regional BirdLife/UNEP-GEF project, and they coordinate awareness efforts in the Negril area. The IBA faces multiple, inter-related threats including: massive population growth linked to an expanding tourism industry; unregulated (and encroaching) development; wetlands drying out (as a result of river canalisation); fires; invasive plants and animals; cattle grazing; garbage dumping; flash flooding; and inappropriate agricultural practices.

Site description

Dolphin Head IBA is an isolated mountainous area in westernmost Jamaica. It includes the forested Dolphin Head, Raglan and Bath mountains. These limestone mountains were one of three "emerging islands" separated by seawater 10–15 million years ago, but now fully exposed to create the present-day island. The IBA supports well developed wet limestone forest (evergreen seasonal and closed broadleaf forest). At the heart of the IBA is natural, closed and disturbed forest and forestry plantations, and these are surrounded by a mosaic of mixed- and non-forest land-use, including bamboo, sugarcane, pasture, small family farms and rural communities.

Birds

This IBA is significant for 25 (of the 36) Jamaica EBA restricted-range birds, including the Vulnerable Ring-tailed Pigeon *Patagioenas caribaea*. At least 11 Neotropical migratory birds occur in the IBA, and seasonal altitudinal migration is pronounced among some of the resident species such as the Rufous-throated Solitaire *Myadestes genibarbis*.

Other biodiversity

Dolphin Head supports the highest density of endemic plant species in Jamaica. At least four animals (a freshwater crab, two fireflies and a snail) are endemic to the IBA. Globally threatened species include: earspot eleuth *Eleutherodactylus fuscus* (Critically Endangered), pallid eleuth *E. grabhami* (Endangered), Jamaican masked eleuth *E. luteolus* (Endangered), Jamaican bromeliad eleuth *E. jamaicensis* (Endangered), green bromeliad frog *Osteopilus wilderi* (Endangered) and the tree-roosting Jamaican fig-eating bat *Ariteus flavescens* (Vulnerable).

Conservation

This small, isolated IBA is a mix of private and state lands. A core forest area comprises three Forestry Department-managed reserves—Raglan Mountain (101 ha), Bath Mountain (121 ha) and Burnt Savanna (c.80 ha). Dolphin Head has been proposed for national park status. Efforts towards conservation and public education within the area are being undertaken by the local NGO Dolphin Head Trust. The forests have been depleted for over 300 years by the harvesting of fuelwood associated with sugarcane and slaked lime production. Illegal timber harvesting and clearance of hilltops for marijuana *Cannabis sativa* cultivation occurs within both the forest reserves and on private lands. Hilltop forest clearance has a profound negative impact on the avifauna. The alien invasive (and predatory) mongoose *Herpestes auropunctatus* and cane toad *Bufo marinus* occur throughout this IBA.

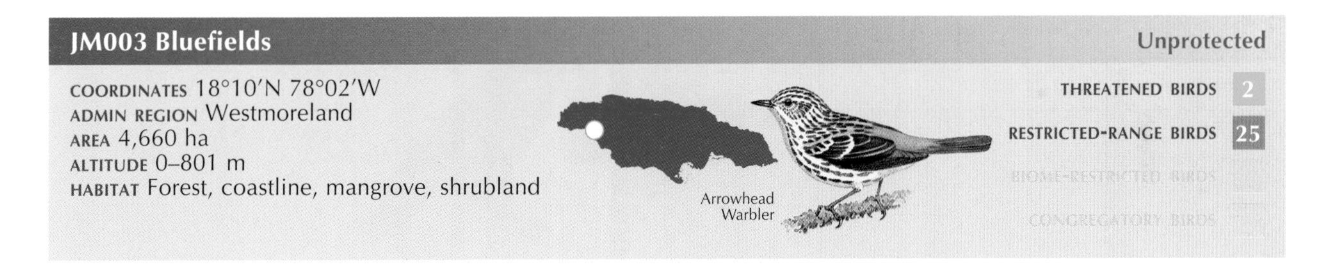

Site description

Bluefields IBA is a spectacularly scenic, rural area on the south-west coast of Jamaica. It comprises the large natural harbour of Bluefields Bay (visited by pirate and Governor of Jamaica Henry Morgan in 1670, and Captain Bligh in 1793) and Bluefields beach (a popular bathing area) behind which is a small wetland through which streams and the Bluefields River percolate. A limestone mountain-range rises steeply from the narrow coastal plain. There are remnants of pimento (Jamaican allspice) plantations throughout the hilly areas. Along the coast are narrow stands of mangrove. Wet forest is confined to the deep, humid gullies on the mountainsides, while the rest of the area supports dry forest and shrubland.

Birds

This IBA is significant for supporting 25 (of the 36) Jamaica EBA restricted-range birds (particularly in the forested gullies in the mountains), with densities of Jamaican Tody *Todus todus*, Arrowhead Warbler *Dendroica pharetra* and Jamaican Becard *Pachyramphus niger* being particularly high. The Vulnerable Yellow-billed Amazon *Amazona collaria* has been recorded, but the population is unknown. Brown Pelicans *Pelecanus occidentalis* roost in the area at night but appear to breed on small mangrove islands to the west of the IBA.

Other biodiversity

A small population of the Vulnerable Jamaican kite swallowtail *Protographium marcellinus* survives on the coastal plain and a number of other less common endemic butterflies occur, including Jamaican admiral *Adelpha abyla*, Shoumatoff's hairstreak *Nesiostrymon shoumatoffi*, Thersites swallowtail *Papilio thersites*, Hewiston's silver-spotted skipper *Epargyreus antaeus* and Butler's skipper *Astraptes jaira*. The Vulnerable Jamaican boa *Epicrates subflavus* is thought to be in the limestone areas.

Conservation

Bluefields IBA is an unprotected mix of private and state-owned lands. The local farming and fishing communities have been sensitised of the need for conservation through the Bluefields Peoples' Community Association, and fishermen have started to impose controls on illegal and poor fishing practices. On land, the main threat is illegal timber felling, slash-and-burn agriculture and uncontrolled housing development (mostly confined to the coastal plain and the main road that leads from Cave along the ridge of the mountains). The naturalist Philip Henry Gosse spent from 1844–1846 at Bluefields Great House studying the flora and fauna of the area, resulting in his two books "A Naturalist's Sojourn in Jamaica" and "The Birds of Jamaica".

JM004 Catapuda

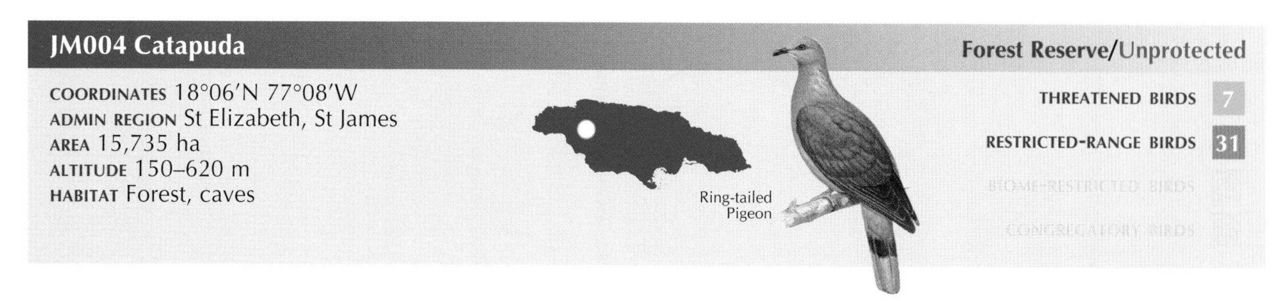

Ring-tailed Pigeon

■ Site description
Catadupa IBA is located in west-central Jamaica and comprises highly karstified (white limestone formation) mountains supporting good secondary forest, albeit with disturbed areas of cultivation and bamboo. The IBA includes (on its south side) the Lacovia Mountains. It forms part of the Cockpit Country Conservation Area and is just to the west of Cockpit Country IBA (JM005), from which it is separated by rural communities, agriculture, and Class B road networks. Several rivers (including the Great River) flow through the area giving rise to small pockets of alluvium. The mountain range is steep in places and supports disturbed broadleaf forest. Agriculture and small rural settlements occur in the less steep areas and alongside the rivers.

■ Birds
This IBA is significant for populations of 31 (of the 36) Jamaica EBA restricted-range birds, and seven threatened (Vulnerable and Near Threatened) species. It appears to be particularly notable for the Vulnerable Ring-tailed Pigeon *Patagioenas caribaea* and Yellow-billed Amazon *Amazona collaria*. Due to the relatively undisturbed nature of the forest in some sections of the IBA, the site is probably important for many wintering Neotropical migratory birds.

■ Other biodiversity
A small population of the Endangered Jamaican giant swallowtail *Papilio homerus* occurs in Elderslie (in the south-east part of the IBA): it is unknown whether gene flow is maintained between this population and the larger population found in Cockpit Country IBA. Snail diversity is high, and a recently recognised species, *Pleurodonte catadupae*, is endemic to the IBA. Ipswich is the type locality for the endemic plant *Gesnaria jamaicensis*.

■ Conservation
Catadupa IBA is a mix of private and state ownership. Included within it are a number of forest reserves, namely: Fyffe and Rankine, Mocho (a number of blocks) and Garlands (two blocks), as well as the Croydon Plantation—a private property—which has been accorded reserve status by the Forestry Department. The area is relatively poorly known biologically and targeted fieldwork is a priority. Threats include illegal cutting of trees and saplings, clearance for agriculture, and hunting. Importantly, Catadupa embraces three watersheds (Great River, Martha Brae and Black River) and provides connectivity with the main Cockpit Country IBA to the east. The railway station at the town of Catadupa has been declared a National Heritage Site.

JM005 Cockpit Country

Black-billed Amazon

■ Site description
Cockpit Country IBA is in west-central Jamaica within the Cockpit Country Conservation Area which includes Litchfield Mountain–Matheson's Run IBA (JM006) to the east, and Catadupa IBA (JM004) to the west, each separated by agricultural communities and roads. Cockpit Country comprises "cockpit karst limestone" and supports the largest contiguous block of wet limestone forest on Jamaica. Surface water is restricted to low-lying areas because of the limestone geology. However, the IBA includes the upper reaches of five major watersheds. The origins of Jamaica's two longest rivers—the Black River (which runs into Black River Great Morass IBA, JM007) and Great River—are in Cockpit Country.

■ Birds
This IBA supports populations of 33 (of the 36) Jamaica EBA restricted-range birds. It is the stronghold for Black-billed Amazon *Amazona agilis* (Vulnerable), with 90–95% of the global population in the IBA, and is particularly important for Ring-tailed Pigeon *Patagioenas caribaea* (Vulnerable). The Endangered Jamaican Blackbird *Nesopsar nigerrimus* occurs in isolated pockets of humid forest. Golden Swallow *Tachycineta euchrysea* (Vulnerable) was last reported in this IBA in 1982. This IBA supports populations of Jamaica's 67 resident breeding landbirds and 34 species of wintering Neotropical migrants.

■ Other biodiversity
Over 66 plant species are endemic to Cockpit Country. Globally threatened species include: Jamaican giant swallowtail *Papilio homerus* (Endangered), Jamaican kite swallowtail *Protographium marcellinus* (Vulnerable); Critically Endangered Cockpit eleuth *Eleutherodactylus griphus* and leaf mimic eleuth *E. sisyphodemus*, Vulnerable Jamaican boa *Epicrates subflavus*, Macleay's mustached bat *Pteronotus macleayi* and Jamaican slider *Trachemys terrapen*, and Jamaican flower bat *Phyllonycteris aphylla* (Endangered).

■ Conservation
Cockpit Country is a mix of private and state-owned lands. Seven forest reserves encompass c.29,000 ha within the IBA, the largest being Cockpit Country Forest Reserve (22,327 ha). Under guidance from the Forestry Department, three Local Forest Management Committees have been established to facilitate co-management for biodiversity conservation and watershed management. Cockpit Country is threatened by bauxite mining (prospecting licenses—now suspended—cover 75% of the area) and limestone quarrying. Conservationists are lobbying the government to declare the IBA "closed to mining". Secondary threats include clearing for agriculture and the encroachment of non-native plants. Illegal hunting occurs along access roads and trails.

Yellow-billed Amazon

■ Site description

Litchfield Mountain–Matheson's Run IBA is in west-central Jamaica. It forms the eastern flank of the Cockpit Country Conservation Area, with Cockpit Country IBA (JM005) situated to the west separated by rural communities, agriculture, and roads. The IBA is the source of two major rivers—the Lowe River and Cave River. It is an area of limestone karst supporting moist forest and a rich community of terrestrial and arboreal epiphytes. The forest is moderately disturbed, resulting from a long history of selective logging, plantations and continued extraction of saplings for yam support stakes. Communities (involved in large scale yam cultivation) have established around the periphery of the well-developed cockpit karst.

■ Birds

This IBA supports populations of 30 (of the 36) Jamaica EBA restricted-range birds. Of particular importance is the presence of the Endangered Jamaican Blackbird *Nesopsar nigerrimus*. Large numbers (c.50–100) of the Vulnerable Ring-tailed Pigeon *Patagioenas caribaea* congregate in yam fields adjacent to the closed-canopy forest in order to feed on the immature yam leaves.

■ Other biodiversity

Up to six globally threatened *Eleutherodactylus* and *Osteopilus* frogs and the Endangered Jamaican giant swallowtail *Papilio homerus* are presumed to occur in the IBA although surveys have not been conducted to confirm this. The south-eastern corner of this IBA supports some of the highest densities (>100 species/ha) of endemic snails anywhere in the world. It is likely that the limestone cliffs support site-endemic plant species.

■ Conservation

This IBA is a mix of private and state ownership. At the core of the IBA are two forest reserves—Litchfield–Matheson's Run (4,485 ha) and Hyde Hall Mountain (662 ha), both managed by the Forestry Department—and Brislington Crown Land (232 ha). Partly because of its proximity to the larger, well-known Cockpit Country IBA, the area has not been well surveyed for biodiversity. Bauxite mining is encroaching from the east and is the single most important threat. A prospecting license covering the whole IBA was suspended in 2007 following strong public and community opposition. Other threats include alien invasive plants which prevent natural forest regeneration; the large-scale harvesting of saplings for yam sticks; and invasion of Shiny Cowbirds *Molothrus bonariensis* (a probable brood parasite of *N. nigerrimus*) along corridor gaps.

West Indian Whistling-duck

■ Site description

Black River Great Morass IBA is the island's largest freshwater wetland and lies on the coastal flood plain of the Black River in south-west Jamaica. It consists of low marshland with limestone islands, and supports human habitation, grazing of livestock and cultivation. The lower morass comprises the 5,700-ha Ramsar site and is bounded on the west and north by roads linking the towns of Black River (St Elizabeth's capital), Middle Quarters and Lacovia, on the east by Santa Cruz Mountains, and on the south by the coast. The integrally-linked upper morass wetland of streams, ponds and dykes (from rice cultivation abandoned in the 1970s) is bordered by roads linking Lacovia, Santa Cruz, Braes River, Elim and Newton.

■ Birds

This IBA is significant as a stronghold for the Vulnerable West Indian Whistling-duck *Dendrocygna arborea* in Jamaica, and for important numbers of the Near Threatened Caribbean Coot *Fulica caribaea* and White-crowned Pigeon *Patagioenas leucocephala*. It also supports populations of 15 (of the 36) Jamaica EBA restricted-range birds. Large numbers of Pied-billed Grebe *Podilymbus podiceps* have been recorded, and regionally important populations of gulls and terns are found on the coast. There are records from this IBA of the rarely seen Spotted Rail *Pardirallus maculates*, and the Near Threatened Black Rail *Laterallus jamaicensis* and Piping Plover *Charadrius melodus*.

■ Other biodiversity

The morass supports important populations of the Endangered frog *Eleutherodactylus luteolus*, the Vulnerable American crocodile *Crocodylus acutus*, the endemic ticki ticki fish *Gambusia melapleura* and the endemic freshwater turtle *Pseudemys terrapen*. Of 92 species of flowering plants in the morass, 8% are endemic to Jamaica.

■ Conservation

The Great Morass is a mix of private and state ownership. The area is a game reserve, with the lower morass designated a Ramsar site (for which a management plan was prepared but never implemented). The IBA and its species face many threats including: illegal hunting; invasive mammalian predators; introduced tilapia, catfish and lobsters; large scale illegal cultivation of *Cannabis sativa* (with associated use of pesticides); industrial and agricultural pollution; invasive plants, e.g. *Alpinia allughas*; removal of trees for timber and fuel; fires in the reed-beds; over-harvesting of palm fronds and reeds; and infill for development at Parottee. Guided boat tours in the southern section of the wetland have exceeded carrying capacity.

Site description

Mount Diablo IBA is located in the centre of the island, near the community of Moneague and at the eastern end of a central limestone ridge that traverses east-central to western Jamaica. The doline and cockpit karst landscape once supported a "spinal forest" that blanketed over 60% of the island. The original native forest of Mount Diablo was dominated by the Jamaican endemic *Podocarpus purdieannus* (a large gymnospermous tree). During the early twentieth century, the area was logged intensively and large areas converted to blue mahoe *Hibiscus elatus* plantations. The *Podocarpus* is now very rare.

Birds

This IBA supports populations of 31 (of the 36) Jamaica EBA restricted-range birds. Importantly, the Endangered Jamaican Blackbird *Nesopsar nigerrimus* still occurs in the area (in spite of the decline in epiphytes associated with the loss of large trees), as do the Vulnerable Black-billed Amazon *Amazona agilis* (at low densities) and Yellow-billed Amazon *A. collaria*.

Other biodiversity

Four vascular plant species are endemic to Mount Diablo: *Dipazium montediabloense*, *Polystichum ambiguum*, *Lepanthes tubuliflora* and *Psychotria coeloneura*, none of which are on the 2004 IUCN Red List. However, based on the extreme habitat destruction occurring in this IBA, the population status for each should be evaluated immediately. The Jamaican giant swallowtail *Papilio homerus* (Endangered) has been extirpated from Mount Diablo within the past 80 years.

Conservation

Approximately 2,250 ha of the 7,150-ha Mount Diablo IBA is a state-owned forest reserve. The rest is held by private companies (e.g. bauxite companies) and individuals (<1,000 ha). Conversion of the forest for agriculture, forestry plantations, rural settlement and, within the past 50 years, open-pit bauxite mining, has left the forest severely fragmented and secondary in nature. The populations of forest-dependent species are presumed declining because of mining-associated habitat loss, but the forest reserves do serve as vital refugia. The severity and irreversibility of the bauxite mining requires immediate conservation attention to protect the remnant forest areas. Pits where mining was completed >10–15 years ago are typically vegetated with herbaceous plants or non-native ferns, but no regeneration of native woody tree species. Other threats include small-scale farming, cattle grazing, illegal timber extraction and illegal poaching of *Amazona collaria*, all of which are facilitated by the extensive network of mining roads.

JM009 Pedro Cays and Bank

Wildlife Sanctuary/Bird Sanctuary

COORDINATES 17°08′N 78°33′W
ADMIN REGION Kingston
AREA 23,345 ha
ALTITUDE 0–5 m
HABITAT Coral islands, reef, coast, shrubland

Masked Booby

CONGREGATORY BIRDS ✓

Site description

Pedro Cays and Bank IBA lies c.97 km south-west of Portland Point on the south coast of Jamaica (and 161 km from Kingston). It comprises a group of small isolated coralline islands emerging from the south-eastern edge of the Pedro Banks. There are four cays—North-East Cay, Middle Cay, South-West Cay and South Cay (now just an over-washed sandy beach)—and associated shallow reefs, rocks and shoals. South-West Cay is the largest. It is flat with a coast of calcareous sand, gravel or hurricane boulder beach. The vegetation comprises low bushes, shrubs and grass. Pedro Bank is Jamaica's main commercial and artisanal fishing ground.

Birds

This IBA is significant for globally and regionally important populations of seabirds. At least 25,000 birds breed on the cays, with the colonies of Magnificent Frigatebird *Fregata magnificens*, Masked Booby *Sula dactylatra*, Brown Booby *S. leucogaster*, Brown Noddy *Anous stolidus* and Bridled Tern *Sterna anaethetus* being particularly notable. Roseate Tern *S. dougallii* also nests on the cays, but not in significant numbers. The cays are used by Neotropical migrants as a stop-over site.

Other biodiversity

This IBA represents one of Jamaica's last remaining healthy marine ecosystems, supporting coral reefs, deep reefs, sea grass beds, and coral cays. Both the Critically Endangered hawksbill *Eretmochelys imbricata* and the Endangered loggerhead *Caretta caretta* turtles nest on the cays in this IBA. The area is the primary harvesting area for the largest export of queen conch *Strombus gigas* from the Caribbean region.

Conservation

Pedro Cays are state owned. The IBA is designated the Great Pedro Banks Wildlife Sanctuary, with South-West Cay a designated bird sanctuary. In 2004, the Pedro Bank was declared an underwater cultural heritage sit. The Morant and Pedro Cays Act makes provision for licensing of all fishing and the taking of turtles, turtle eggs, birds and bird eggs for the cays. However, intensive fishing and high human densities are endangering the survival of the bank as a viable and functioning ecosystem. Fishermen occupy North-East and Middle Cays, and the cays are regularly visited by fishermen from neighbouring countries. The Jamaica Defence Force operates a security post on Middle Cay, and The Nature Conservancy's Pedro Bank Management Project aims to manage some of the negative impacts. Mice *Mus musculus* are present and could be impacting the seabird populations.

JM010 Portland Ridge and Bight

Protected Area/Forest Reserve

COORDINATES 17°44′N 77°10′W
ADMIN REGION Clarendon
AREA 4,200 ha
ALTITUDE 0–150 m
HABITAT Forest, shrubland, coast, mangrove, caves

■ Site description
Portland Ridge and Bight IBA is mid-way along the south coast of Jamaica and forms the most southerly point on the island. Along with Hellshire Hills IBA (JM011) and Brazilletto Mountain to the east, the area is contained within the Portland Bight Protected Area. Portland Ridge is an area of relatively intact (but secondary) dry limestone forest on a peninsula that projects into the Caribbean Sea and protects the waters of Portland Bight. Portland Bight is a shallow marine and wetland area with well developed mangrove woodlands, salt flats, sandy beaches and offshore cays.

■ Birds
This IBA supports populations of 17 (of the 36) Jamaica EBA restricted-range birds, including a sizeable population the endemic subspecies of Bahama Mockingbird *Mimus gundlachi hillii*. Significant populations of the Vulnerable West Indian Whistling-duck *Dendrocygna arborea* and Near Threatened Plain Pigeon *Patagioenas inornata* and White-crowned Pigeon *P. leucocephala* occur although precise numbers are unknown. Regionally important numbers of Magnificent Frigatebird *Fregata magnificens* and Brown Noddy *Anous stolidus* nest on the Portland Bight cays. Shorebirds are reported as being "numerous" in this IBA.

■ Other biodiversity
The Critically Endangered frog *Eleutherodactylus cavernicola* is known only from two caves in Portland Ridge. The Vulnerable Jamaican boa *Epicrates subflavus* and Jamaican fruit-eating bat *Ariteus flavescens* occur, as do two thunder snakes *Trophidophis stullae* and *T. jamaicencis*—both Portland Ridge endemics. The "data deficient" blue-tailed galliwasp *Celestus duquesneyi* may still survive.

■ Conservation
This IBA is a mix of state and private land ownership. It is part of the 87,615-ha Portland Bight Protected Area. Conservation management is minimal, with hunting clubs providing some unofficial conservation. However, they are also responsible for replanting dry forest with "bird feeding trees", creating a semi-monoculture in some areas. Pigeons are hunted in August and September. Unplanned urban sprawl is occurring within the protected area (involving 30 towns or settlements) and natural resources (such as timber/charcoal, and marine products) are significantly exploited by the residents, including over 4,000 fishermen. Port Esquivel and Rocky Point port are within Portland Bight and handle alumina, oil and other bulk cargos. Recent fires and hurricanes have significantly impacted the dry forest.

JM011 Hellshire Hills

Protected Area/Forest Reserve

COORDINATES 17°53′N 76°57′W
ADMIN REGION St Catherine
AREA 9,400 ha
ALTITUDE 0–200 m
HABITAT Dry forest, shrubland, mangrove, caves

Jamaican Pauraque

THREATENED BIRDS 3
RESTRICTED-RANGE BIRDS 17
BIOME-RESTRICTED BIRDS
CONGREGATORY BIRDS

■ Site description
Hellshire Hills IBA is on the south coast of eastern Jamaica. The hills project into the Caribbean Sea, forming the north-east side of the Portland Bight bay, and supporting dry limestone forest and scrub. Portland Ridge and Bight IBA (JM010) is south-west of Hellshire, and together with Brazilletto Mountain these areas form the Portland Bight Protected Area (Jamaica's largest protected area). The IBA includes Great Goat Island, an uninhabited 1-km^2 limestone cay c.1 km offshore from the Hellshire Hills. Little Goat Island (which is flat, sandy and heavily impacted by man and animals) is "joined" to it by an impenetrable morass of mangrove swamp.

■ Birds
This IBA supports populations of 17 (of the 36) Jamaica EBA restricted-range birds, including the endemic subspecies of Bahama Mockingbird *Mimus gundlachi hillii*. The Near Threatened Plain Pigeon *Patagioenas inornata* occurs although numbers are unknown. The Critically Endangered Jamaican Pauraque *Siphonorhis americana*, last seen in 1860, is rumoured to persist in the Hellshire Hills. The mangroves provide nesting, roosting and feeding areas for sea and shorebirds.

■ Other biodiversity
The Vulnerable Jamaican hutia *Geocapromys brownii* and Jamaican fig-eating bat *Ariteus flavescens* occur. Hellshire is exceptionally important for reptiles: the Critically Endangered Jamaican iguana *Cyclura collei* was rediscovered in Hellshire in 1990, as was the "data deficient" blue-tailed galliwasp *Celestus duquesneyi* in 1997. The Vulnerable Jamaican boa *Epicrates subflavus* occurs, and a potentially new species of *Tropidophis* snake was recently found.

■ Conservation
This IBA is part of the much larger Portland Bight Protected Area. Management of the protected area was first delegated to a local NGO, the Caribbean Coastal Areas Management Foundation, and then (in 2006) to the Urban Development Corporation (UDC). UDC has undertaken significant planning, but continued implementation of work on housing developments/solutions in Caymanas and Hellshire is seemingly in conflict with the protected area designation. Conservation management is in its infancy in the area with almost no enforcement of environmental laws. The area is impacted by pig-hunters and people extracting logs, timber and poles. The dry forests are now mostly secondary in nature. The Iguana Recovery Project is working to conserve *Cyclura collei*, and inventory the area's herpetofauna.

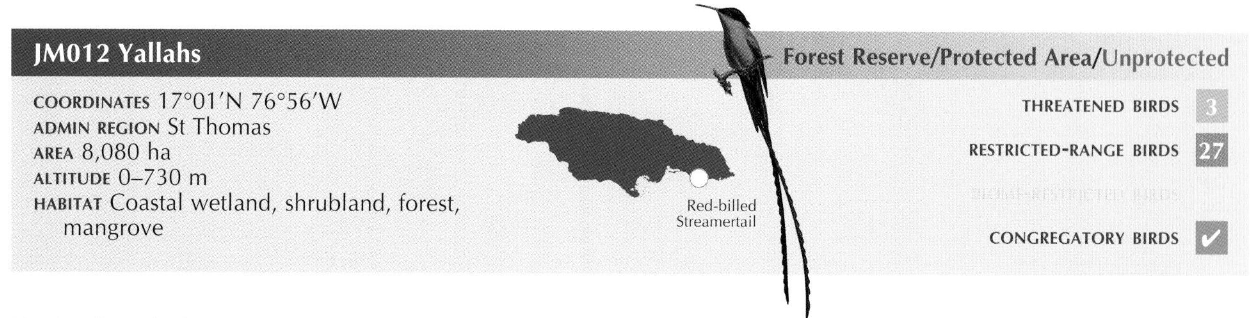

■ Site description

Yallahs IBA embraces an area along the south-east coast of Jamaica in the rain-shadow of the Blue Mountains IBA (JM013). It forms part of the watershed basins for the Yallahs and Morant rivers, both of which have wide, rocky channels in their lower reaches with deep deposits of alluvium. These rivers may become intermittent in dry months, but then have torrential flows after moderate rains. The IBA includes two natural salt ponds and surrounding mangroves. The larger pond (up to 1.5 m deep) covers 80 ha and is 10 times saltier than the ocean. The smaller pond is less saline. Vegetation is characterised by (degraded) xeric scrub and small patches of moister forest near to rivers or in the higher elevations on Yallahs Hill.

■ Birds

This IBA supports populations of 27 (of the 36) Jamaica EBA restricted-range birds, including the Vulnerable Ring-tailed Pigeon *Patagioenas caribaea* and Near Threatened Crested Quail-dove *Geotrygon versicolor*. The salt ponds are home to a regionally important breeding colony of Least Tern *Sterna antillarum*. The Near Threatened Plain Pigeon *Patagioenas inornata* and Piping Plover *Charadrius melodus* have been seen in the IBA, but not in significant numbers. The salt ponds support a wide diversity of migratory shorebirds and waterbirds.

■ Other biodiversity

Globally threatened species found within this IBA include the Vulnerable Jamaican kite swallowtail *Protographium marcellinus*, Jamaican fig-eating bat *Ariteus flavescens* and Jamaican boa *Epicrates subflavus*.

■ Conservation

Yallahs IBA is state owned, and includes the c.60-ha Lloyds Forest Reserve (thus also a Game Reserve). The ponds are designated as a protected area in the parish of St Thomas. However, illegal hunting occurs in and around the ponds regularly, and also along access roads and trails of the forest reserve. The major threat to the Yallahs is residential development and limestone and sand/gravel quarrying. There are also deposits of high-grade gypsum and marble in the area. Other threats include clearance for agriculture and the encroachment of non-native plant species. *Artemia* (brine shrimp) farming is being considered for the ponds, and water extraction (for Kingston) reduces the Yallahs and Negro rivers to intermittent streams in the dry season. Some conservation (e.g. mangrove and tree planting) and public awareness efforts are currently underway through NGOs (e.g. the Yallahs Development Area Committee) which may be the genesis of a Site Support Group for the IBA.

■ Site description

Blue Mountains IBA is a 16-km long mountain ridge of sharp peaks across the eastern part of Jamaica. Much of the "Grand Ridge" is over 1,800 m, the highest section being Blue Mountain Peak, comprising Middle Peak (2,256 m—Jamaica's highest point) and East Peak (2,246 m). Lesser peaks and ridges radiate from these. To the east of the Blue Mountains, separated by the Rio Grande, is the John Crow Mountains IBA (JM014), and to the west are the lower Port Royal Mountains. Tall, wet forest persists on the north slope (below 1,000 m). The rest of the IBA comprises upper montane forest which, however, has been much altered and is now used for forestry, coffee production or subsistence farming. These forests protect the watershed for Kingston.

■ Birds

This IBA supports significant populations of 29 (of the 36) Jamaica EBA restricted-range birds. It is the stronghold for the Endangered Jamaican Blackbird *Nesopsar nigerrimus*, and is important for the Vulnerable Ring-tailed Pigeon *Patagioenas caribaea* and Yellow-billed Amazon *Amazona collaria*. The Vulnerable Bicknell's Thrush *Catharus bicknelli* occurs in small numbers and Golden Swallow *Tachycineta euchrysea* was last recorded in Jamaica in this IBA in 1989.

■ Other biodiversity

Approximately 20% of all flowering plants are endemic to the IBA and 10 species are globally threatened. Five globally threatened frogs occur: *Eleutherodactylus alticola* and *E. orcutti* (both Critically Endangered), *E. andrewsi* and *E. nubicola* (both Endangered) and *E. glaucoreius* (Near Threatened). The Blue Mountain anole *Anolis reconditus* is endemic to the IBA, and the Vulnerable Jamaican fig-eating bat *Ariteus flavescens* occurs.

■ Conservation

This area is under private and state ownership, and most is within the Blue Mountains Forest Reserve, itself part of the larger Blue Mountain/John Crow Mountain National Park. The national park was the first protected area to be managed by a local NGO (Jamaica Conservation and Development Trust). Funding for management (e.g. ongoing reforestation efforts and bird monitoring) has been and continues to be a limiting factor. Commercial forestry started during the 1970s has been scaled down since Hurricane Gilbert destroyed many of the plantations in 1988. Clearance for agriculture permitted the expansion of invasive plants which are now the focus of control projects. The Forestry Department manages a dynamic conservation program on the north side of the IBA in the Buff Bay–Pencar area.

JM014 John Crow Mountains

Black-billed Streamertail

■ Site description

John Crow Mountains IBA forms the easternmost end of Jamaica. This mountain range comprises white limestone overlain by marine sandstones and shale resulting in an unusual landscape of sinkholes and outcrops. It rises gently from the east to a tilted plateau, and then drops abruptly along a steep escarpment to the west. The IBA is separated from the Blue Mountains IBA (JM013) to the west by the Rio Grande. The ranges join at Corn Puss Gap, the boundary of the parishes of Portland and St Thomas, for which this IBA is a major watershed. Below 900 m the vegetation is wet limestone forest (with tree-ferns and bromeliads), while the plateau supports montane limestone thicket. Cultivation is for cash crops including coffee, sugar cane and bananas.

■ Birds

This IBA supports populations of 29 (of the 36) Jamaica EBA restricted-range birds, including the Black-billed Streamertail *Trochilus scitulus* that occurs only in this IBA. Of the threatened species that occur, the populations of Jamaican Blackbird *Nesopsar nigerrimus* (Endangered), Ring-tailed Pigeon *Patagioenas caribaea* (Vulnerable) and Crested Quail-dove *Geotrygon versicolor* (Near Threatened) are particularly significant. In the 1960s it was said that the Critically Endangered Jamaica Petrel *Pterodroma caribeae* "still calls at night" in this IBA, and it may yet persist.

■ Other biodiversity

Globally threatened amphibians include: *Eleutherodactylus orcutti* (Critically Endangered), *E. andrewsi* (Endangered), *E. jamaicensis* and *E. pentasyringus* (both Vulnerable) and green bromeliad frog *Osteopilus wilderi* (Endangered). The Jamaican giant swallowtail *Papilio homerus* (Endangered) occurs albeit in numbers reduced by collecting and habitat disturbance. About 20% of the IBA's flowering plants are endemic to the area, 10 of which are Vulnerable.

■ Conservation

Most of the area is a forest reserve, and part of the Blue Mountain/John Crow Mountains National Park. The national park was the first protected area to be managed by a local NGO (Jamaica Conservation and Development Trust). Funding for management has been, and continues to be, a limiting factor. Threats include forest clearance for subsistence and commercial agriculture, invasion of exotic plants, and collecting of epiphytes for the local market. After Hurricane Gilbert (1988), white-tailed deer *Odocoileus virginianus* escaped from a collection and spread throughout the north side of the IBA.

JM015 Morant Cays and Bank

Sooty Tern

■ Site description

Morant Cays and Bank IBA lies 51 km south-south-east of Morant Point—the easternmost point of Jamaica. This offshore island group consists of three small islets grouped closely together (c.1.5–2 km apart) on the summit of an extensive, 7-km long crescent-shaped bank of coral, which rises from the seabed at 1,000 m. The cays are low-lying, sparsely vegetated (shrubs and grasses), and fronted by highly exposed reefs over which the sea constantly breaks. North-East Cay is sometimes divided into three parts as a result of the sea washing over connecting sand spits. It supports a fishermen's camp (with huts and water tank) and a lighthouse at Breezy Point, the easternmost point of the cay and of Jamaica.

■ Birds

This IBA is significant for breeding seabirds. The Sooty Tern *Sterna fuscata* colony is one of the largest in the Caribbean (up to 90,000 individuals). Numbers of Brown Noddy *Anous stolidus* and Royal Tern *S. maxima* are regionally important. Other breeding seabirds include Bridled Tern *S. anaethetus*, Magnificent Frigatebird *Fregata magnificens*, Laughing Gull *Larus atricilla*, and Brown Pelican *Pelecanus occidentalis*. The cays are also used as a stopover site by Neotropical migratory birds. Small numbers of Audubon's Shearwater's *Puffinus lherminieri* were discovered on South-east Cay in 1998 and their population should be assessed. The Endangered Black-capped Petrel *Pterodroma hasitata* was seen close by the IBA in December 1997—a first record for Jamaica.

■ Other biodiversity

The cays are important as a nesting site for globally threatened sea turtles, although the species involved are not recorded.

■ Conservation

The Morant Cays are state owned. Middle Morant Cay is a designated nature reserve and scientific reserve. The Morant and Pedro Cays Act makes provision for licensing of all fishing and the taking of turtles, turtle eggs, birds and bird eggs for the cays, and a fishermen's camp was established by the Department of Agriculture on the south side of North-East Cay. However, there has been a serious decline in recruitment within the seabird populations and Middle Cay has been selected for monitoring of egg removal. A base camp is established there for a month starting at the end of each April.

MARTINIQUE

LAND AREA **1,100 km²** ALTITUDE **0–1,397 m**
HUMAN POPULATION **391,000** CAPITAL **Fort-de-France**
IMPORTANT BIRD AREAS **10, totalling 545 km²**
IMPORTANT BIRD AREA PROTECTION **33%**
BIRD SPECIES **180**
THREATENED BIRDS **6** RESTRICTED-RANGE BIRDS **19**

VINCENT LEMOINE AND LIONEL DUBIEF
(SOCIÉTÉ D'ETUDE, DE PROTECTION ET D'AMÉNAGEMENT DE LA NATURE DE MARTINIQUE, SEPANMAR[1])

Forêts du Nord and de la Montagne Pelée IBA embraces some of the most significant populations of forest-dependent birds on the island. (PHOTO: SEPANMAR)

INTRODUCTION

Martinique is a mountainous, volcanic island surrounded by numerous small islets. It is one of smallest *département d'outre-mer* (DOM, overseas department) of France (and an outermost region of the European Union) and has been divided into four administrative regions, which in turn have been subdivided into 45 cantons and 34 communes. Martinique is one of the Windward Islands in the southern Lesser Antilles, c.40 km south of Dominica and 28 km north of St Lucia. The north of the island is very mountainous, dominated by the Pitons du Carbet (1,207 m) and Montagne Pelée (1,397 m), an active (and closely monitored) volcano. Across the rest of the island, the relief primarily comprises hills (called 'mornes'), the tallest of which reach 505 m (Montagne du Vauclin) and 478 m (Morne Larcher) in the Diamant commune. There is just one flat area—the plain of Le Lamentin in west-central Martinique—amongst the hills.

Martinique's climate is tropical (26°C yearly average) and humid (80% humidity in March/April and 87% in October/November). The mountainous massifs of the north have a cooler climate and more rain than the coasts. They form a barrier to the "Alizé" (the trade winds), causing clouds to build up resulting in increased rainfall on the eastern slopes. Average annual rainfall in these mountains is 5,000 mm, with a maximum of 10,000 mm on Montagne Pelée. The south of Martinique, with its low hills, receives 1,200 mm of rainfall a year. With these differences in rainfall and topography, a wide diversity of habitats cover the island, ranging from xerophytic shrublands and dry forest, to wet tropical forest, high altitude grasslands and savannas, and mangroves.

Conservation

A considerable portion of Martinique is protected and many sites benefit from several categories of protection. Presqu'île de la Caravelle (IBA MQ004) and Îlets de Sainte Anne (IBA MQ010) are protected under the designation of Réserve Naturelle (national nature reserve), covering 527 ha. Thirteen sites (totalling 105 ha) are designated under the category of Arrêté de Protection de Biotope (APB, habitat protection bylaw) which forbids activities including climbing, camping, fires, introduction of animals, undertaking excavations, and flying over the area below a 300 m ceiling. Three of these APB sites are considered IBAs, namely Pointe du Pain de Sucre (MQ003), Rocher du Diamant (MQ008) and L'Îlet Boiseau

1 Société d'Etude, de Protection et d'Aménagement de la Nature de Martinique is referred to throughout this chapter by the acronym SEPANMAR.

Le Rocher du Diamant IBA: protected under various designations, but the seabirds are still at risk of disturbance from fishermen and invasive alien predators. (PHOTO: SEPANMAR)

(part of IBA MQ005). Nine areas (covering 4,248 ha) are protected as Réserves de Chasse (hunting reserves, where hunting is forbidden). Five of these hunting reserves are totally or partly within Martinique's IBAs: Pointe Rouge (131 ha) and Presqu'île de la Caravelle (517 ha) cover part of IBA (MQ004); Forêt Domaniale des Pitons du Carbet (1,880 ha) covers 15% of Pitons du Carbet IBA (MQ002); Réserve de Chasse Macabou (144 ha) corresponds to Grand Macabou IBA (MQ009); and Génipa (250 ha) covers 7% of Mangrove de Fort de France IBA (MQ006).

Sites across the island have also been designated as Sites Classés et Sites Inscrits (under Classified and Registered Site Orders), classifications widely used for the protection of the wider countryside through the limitation of urbanisation and other development. Three Sites Classés cover 6,629 ha (each partly overlapping with IBAs), and 11 Sites Inscrits cover c.4,000 ha. Presqu'île de la Caravelle IBA (MQ004) has 3,100 ha "Classified" and 311 ha "Registered". On Montagne Pelée, 2,100 ha is "Classified" and corresponds to the Forêts du Nord et de la Montagne Pelée IBA (MQ001). Massif forestier entre Le Diamant et les Trois-Îlets (IBA MQ007) includes a number of "Classified" and "Registered" sites such as the 1,429-ha Classified sites of Mornes de la Pointe du Diamant and Mornes du Rocher du Diamant, and the "Registered" sites of Morne Champagne–Village des Anses d'Arlet (320 ha) and Petite Anse–Anse Cafard (126 ha). Rocher du Diamant is included in Rocher du Diamant IBA (MQ008). The Baie des Anglais (644 ha) and Etang des Salines–Savane des Pétrifications (577 ha), located in Sainte Anne commune, are "Registered" sites that make up a part of Îlets et falaises de Sainte Anne (IBA MQ010).

Nine of the IBAs are included wholly or partly within the protection of Littoral Conservation Areas belonging to the Conservatoire du Littoral (CEL, Coastal Protection Agency[2]) that forbids any development within 15 m from the coast (the littoral zone). Four of these Littoral Conservation Areas are critically important due to the presence of the Endangered White-breasted Thrasher *Ramphocinclus brachyurus*. All four form part of the Presqu'île de la Caravelle IBA (MQ004). CEL has also acquired land for conservation on the Habitation Blin (two areas, 16 and 26 ha), on Pointe Rouge (55 ha), and on the Presqu'île de la Caravelle (257 ha).

The Direction Régionale de l'Environnement (DIREN, Regional Department of Environment), is the representative in Martinique of the Ministère de l'écologie, de l'énérgie, du développement durable et de l'aménagement du territoire (Ministry of Ecology, Energy, Sustainable Development and Planning) and is based in Fort de France. Most of the forests are under the jurisdiction of and managed by the Office National des Forêt (ONF, French National Forest Office[3]) and Conservatoire du Littoral (CEL). The 700-km^2 Parc Naturel Régional de la Martinique (PNRM) also plays a role in the protection of habitats across Martinique. Environmental guards of the DIREN, ONF, CEL and PNRM, as well as the Office National de la Chasse et de la Faune Sauvage (ONCFS, the National Hunting and Wildlife Agency) enforce the protection of areas designated under the categories mentioned above.

2 The Conservatoire du Littoral (Coastal Protection Agency) is referred to throughout this chapter by the acronym CEL.
3 The Office National des Forêt (French National Forest Office) is referred to throughout this chapter by the acronym ONF.

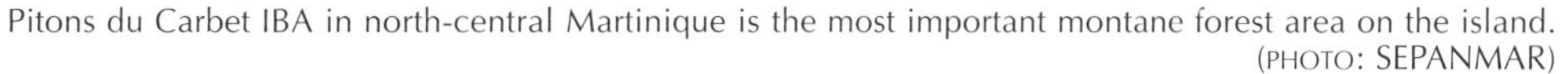

Pitons du Carbet IBA in north-central Martinique is the most important montane forest area on the island. (PHOTO: SEPANMAR)

There are no government or university-based avian research programmes active on Martinique. However, NGO naturalist associations such as SEPANMAR, Association Ornithologique de Martinique (AOMA) and Association Le Carouge are carrying out species monitoring for some birds.

Other than volcanic eruptions (the eruption of 1902 was responsible for the extinction of an endemic rodent *Megalomys demarestii*) and hurricanes, the main threats to biodiversity on Martinique are directly linked to human activities including urban encroachment, agriculture (agricultural land covers 30% of the island), pesticide use, harvesting/gathering, and alien invasive species. Rats (both *Rattus norvegicus* and *R. rattus*), mongoose *Herpestes auropunctatus* and domestic cats *Felix catus* affect the island's seabird colonies. In addition, the Shiny Cowbird *Molothrus bonariensis,* which has arrived through natural range expansion, is a nest-parasite of the Vulnerable Martinique Oriole *Icterus bonana.* Poaching is an increasing threat on the island, and numerous protected species (snipe, herons/egrets, pelicans, terns etc.) are regularly shot. Regulated hunting of "game" birds (of which there are 32 species) is permitted during a seven-month season. However, adherence to the season or to the listed birds is generally poor, and there is no monitoring of the numbers of birds (or species) shot, or of the population status of the listed game birds.

■ Birds

Of Martinique's 180 recorded bird species, about 74 breed (there are c.10 species whose breeding status remains to be confirmed), 21 are exotics introduced by man (one of which—Orange-winged Parrot *Amazona amazonica*—was a deliberate introduction), or have naturally colonised the island (Cattle Egret *Bubulcus ibis*, Eared Dove *Zenaida auriculata*, Bare-eyed Thrush *Turdus nudigenis,* and Shiny Cowbird *Molothrus bonariensis*), and 13 are summer migrants (of which 10 are seabirds). There are 19 restricted-range birds (see Table 1) of the 38 occurring in the Lesser Antilles Endemic Bird Area (EBA), including the Endangered White-breasted Thrasher *Ramphocinclus brachyurus*, and the only extant island endemic, the Vulnerable Martinique Oriole *Icterus bonana.*

The Blue-headed Hummingbird *Cyanophaia bicolor* is restricted to Martinique and neighbouring Dominica, while the Grey Trembler *Cinclocerthia gutturalis* and *R. brachyurus* are shared with St Lucia. The other restricted-range species on Martinique are found across a range of habitats, and occur in many parts of the island. A subspecies of Ringed Kingfisher *Megaceryle torquata stictipennis* is endemic to Martinique, Guadeloupe and Dominica, but is highly threatened and sightings on Martinique are becoming rare. Two Martinique endemic birds are now extinct, namely Martinique Amazon *Amazona martinicana* and Lesser Antillean Macaw *Ara guadeloupensis,* while three more widespread birds (Caribbean Flamingo *Phoenicopterus ruber*, Burrowing Owl *Athene cunicularia*, House Wren *Troglodytes aedon*) have disappeared from Martinique.

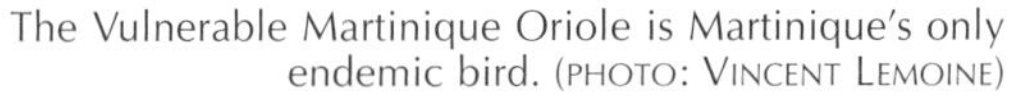
The Vulnerable Martinique Oriole is Martinique's only endemic bird. (PHOTO: VINCENT LEMOINE)

Sooty Tern is one of many seabird species that nest in Martinique. (PHOTO: VINCENT LEMOINE)

Only two globally threatened species have been considered in the IBA analysis (see Table 1). However, four other species have been recorded (albeit rarely and not in significant numbers) on the island: White-crowned Pigeon *Patagioenas leucocephala*, Caribbean Coot *Fulica caribaea*, Buff-breasted Sandpiper *Tryngites subruficollis* (all Near Threatened) and the Endangered Black-capped Petrel *Pterodroma hasitata*. The Endangered *R. brachyurus* is found in Martinique only in Presqu'île de la Caravelle IBA (MQ004) where its population is estimated at 200–400 individuals. The endemic and Vulnerable *I. bonana* is found in various habitats but it has been suggested that dry forests and mangroves are most important for the bird. The total population is estimated at 10,000–19,000 individuals. Deforestation on the island has facilitated the spread of the recently established Shiny Cowbird *Molothrus bonariensis*, which parasitises 75% of oriole nests annually. However, a recent decline in cowbird numbers has allowed for a slight oriole population recovery.

Eight species of waterbirds (five herons/egrets, one duck, one rail, and one grebe) are nesting residents, and c.40 (10 ducks and 30 shorebirds) are Neotropical migrants. However, there are no areas with large concentrations of waterbirds. Migratory flocks are modest in size, with birds grouping together in small numbers across the island—along the coast, in the marshes, in natural ponds, in the lagoons behind the mangroves, and in the converted hunting areas.

IMPORTANT BIRD AREAS

Martinique's 10 IBAs—the island's international site priorities for bird conservation—cover 54,514 ha (including marine areas), and c.39% of the country's land area. Most of Martinique IBAs have some protection (see "Conservation" above).

The IBAs have been identified on the basis of 26 key bird species (listed in Table 1) that variously meet the IBA criteria. These 26 species include two (of the 6) globally threatened birds (see "Birds" above), all 19 restricted-range species, and seven congregatory waterbirds/seabirds. Presqu'île de la

Caravelle IBA (MQ004) is critical for supporting the only population of the Endangered White-breasted Thrasher *Ramphocinclus brachyurus*, as well is populations of other restricted-range species including the Vulnerable Martinique Oriole *Icterus bonana*. Five other IBAs have been identified on the basis of supporting a significant portion (between 52 and 95%) of the Lesser Antilles EBA restricted-range species that occur on the island (including, in each case, the Vulnerable *I. bonana*). The remaining four IBAs support significant populations of seabirds, highlighting the importance of this species-group in terms of Martinique's international conservation responsibilities.

It is essential that hunting regulations are developed and applied for protected (e.g. Near Threatened White-crowned Pigeon *Patagioenas leucocephala* and Ruff-breasted Sandpiper *Tringytes subruficollis*) and restricted-range birds. It will be also important to educate hunters on the regulations they are legally obliged to adhere to, and on bird identification. To provide further protection for IBAs, further planning regulations will be needed, additional lands acquired, more visiting areas (e.g. Habitat Protection bylaw areas) designated, and new posts for environmental guardians created. Continued monitoring of the seabird colony IBAs for the presence of rats *Rattus* spp., with eradication actions taken where appropriate, will be essential in maintaining (and potentially increasing) the populations of Martinique's seabirds. Further public education actions are needed to raise awareness of the richness and fragility of the island's biodiversity.

The monitoring of globally threatened, restricted-range, hunted birds, seabirds and waterbirds should be continued

Table 1. Key bird species at Important Bird Areas in Martinique.

			Martinique IBAs									
Key bird species	Criteria	National population	MQ001	MQ002	MQ003	MQ004	MQ005	MQ006	MQ007	MQ008	MQ009	MQ010
Criteria			■ ■	■ ■	■	■ ■	■	■ ■	■ ■	■	■ ■	■
Audubon's Shearwater *Puffinus lherminieri*	■	150–240										146
Magnificent Frigatebird *Fregata magnificens*	■	50–100			100							
Brown Booby *Sula leucogaster*	■	<50								160		
Roseate Tern *Sterna dougallii*	■	1,170–1,320			750		300–600					
Bridled Tern *Sterna anaethetus*	■	480–1,230								300–360		130–240
Sooty Tern *Sterna fuscata*	■	3,300–36,000										7,000–24,000
Brown Noddy *Anous stolidus*	■	990–3,060								600–750		110–1,152
Bridled Quail-dove *Geotrygon mystacea*	■		✓	✓		✓						
Lesser Antillean Swift *Chaetura martinica*	■		✓	✓		✓						
Purple-throated Carib *Eulampis jugularis*	■		✓	✓		✓		✓	✓		✓	
Green-throated Carib *Eulampis holosericeus*	■		✓	✓		✓		✓	✓		✓	
Antillean Crested Hummingbird *Orthorhyncus cristatus*	■		✓	✓		✓		✓	✓		✓	
Blue-headed Hummingbird *Cyanophaia bicolor*	■		✓	✓								
Caribbean Elaenia *Elaenia martinica*	■		✓	✓		✓		✓	✓		✓	
Lesser Antillean Pewee *Contopus latirostris*	■		✓	✓		✓		✓	✓		✓	
Lesser Antillean Flycatcher *Myiarchus oberi*	■		✓	✓		✓		✓	✓		✓	
White-breasted Thrasher *Ramphocinclus brachyurus*	EN ■ ■	200–400				200–400						
Scaly-breasted Thrasher *Margarops fuscus*	■		✓	✓		✓		✓	✓		✓	
Pearly-eyed Thrasher *Margarops fuscatus*	■		✓	✓					✓			
Brown Trembler *Cinclocerthia ruficauda*	■		✓	✓								
Grey Trembler *Cinclocerthia gutturalis*	■		✓	✓								
Rufous-throated Solitaire *Myadestes genibarbis*	■		✓	✓								
Martinique Oriole *Icterus bonana*	VU ■ ■	10,000–19,999	✓	✓		✓		✓	✓		✓	
Lesser Antillean Bullfinch *Loxigilla noctis*	■		✓	✓		✓		✓	✓		✓	
Antillean Euphonia *Euphonia musica*	■		✓	✓					✓			
Lesser Antillean Saltator *Saltator albicollis*	■		✓	✓				✓	✓		✓	

All population figures = numbers of individuals.
Threatened birds: Endangered ■; Vulnerable ■. **Restricted-range birds** ■. **Congregatory birds** ■.

Pointe du Pain de Sucre IBA, one of Martinique's key seabird breeding sites. (PHOTO: SEPANMAR)

The Endangered White-breasted Thrasher, found in Martinique only in Presqu'île de la Caravelle IBA. (PHOTO: VINCENT LEMOINE)

Figure 1. Location of Important Bird Areas in Martinique.

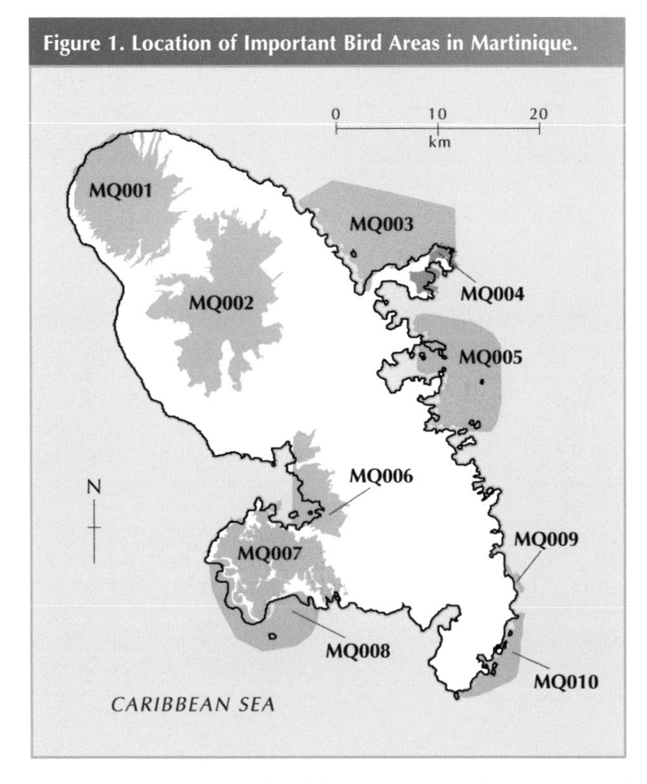

(or expanded) and used to inform the annual assessment of state, pressure and response scores at each IBA which will provide an objective status assessment for these internationally important biodiversity sites and highlight management interventions required to maintain their integrity.

KEY REFERENCES

BARRÉ, N. AND BENITO-ESPINAL, E. (1985) Oiseaux granivores exotiques implantés en Guadeloupe, à Marie-Galante et en Martinique (Antilles Françaises). *Oiseau Rev. Fr. Ornithol.* 55: 235–241.

BENITO-ESPINAL, E. (1990) *La grande encyclopédie de la Caraïbe: Fauna 1, les oiseaux*. Italy: Edition Sanoli.

BENITO-ESPINAL, E. AND HAUCASTEL, P. (2003) *Les oiseaux des Antilles et leur nid, Petites et Grandes Antilles*. Le Gosier, Guadeloupe: PLB Editions.

BON SAINT-CÔME, M. AND LE DRU, A. (1994) Liste des oiseaux de Martinique. (Unpublished report).

EVANS, P. G. H. (1990) *Birds of the eastern Caribbean*. London: Macmillan Education Ltd.

JAMARILLO, A. AND BURKE, P. (1999) *New World blackbirds: the icterids*. Princeton, New Jersey: Princeton University Press.

LEMOINE, V., DUBIEF, L. AND GENESSEAUX, V. (in press) French Antilles: Martinique. In P. E. Bradley and R. L. Norton eds. *Breeding seabirds of the Caribbean*. Gainesville, Florida : Univ. Florida Press.

LÉVESQUE, A. AND LARTIGES, A. (2000) Colombidés antillais: biologie, écologie, méthodes d'études—analyse bibliographique. Fort de France, Martinique: Office National de la Chasse et de la Faune Sauvage (ONCFS). (Unpublished report).

LÉVESQUE, A., VILLARD, P., BARRÉ, N., PAVIS, C. AND FELDMANN, P. (2005) Liste des oiseaux des Antilles Françaises. Petit Bourg, Guadeloupe: Association pour l'Etude et la protection des Vertébrés des petites Antilles (Unpublished AEVA report, 29).

MAILLARD, J. F. (2004) Orientations régionales de gestion de la faune sauvage et d'amélioration de la qualité des habitats (ORGHF), région de Martinique /Etat des lieux. Fort de France, Martinique: Office National de la Chasse et de la Faune Sauvage (ONCFS), Direction Régionale de l'Environnement (DIREN). (Unpublished report).

PASCAL, M., N. BARRÉ, P. FELDMANN, O. LORVELEC AND PAVIS, C. (1996) Faisabilité écologique d'un programme de piégeage de la mangouste dans la Caravelle (Martinique). Petit Bourg, Guadeloupe: Association pour l'Etude et la protection des Vertébrés des petites Antilles (Unpublished AEVA report, 12).

PINCHON, R. AND BON SAINT-CÔME, M. (1951) Notes et observations sur les oiseaux des Antilles Françaises. *Oiseau Rev. Fr. Ornithol.* 21: 229–277.

PINCHON, R. AND BON SAINT-CÔME, M. (1952) Note complémentaire sur l'avifaune des Antilles Françaises. *Oiseau Rev. Fr. Ornithol.* 22: 113–119.

RAFFAELE, H., WILEY J., GARRIDO, O., KEITH, A. AND RAFFAELE, J. (1998) *A guide to the birds of the West Indies*. Princeton, New Jersey: Princeton University Press.

SOUBEYRAN, Y. (2003) Evolution des populations de tourterelles à queue carrée, *Zenaida aurita*, en Martinique, Antilles Françaises, de 1985 à 2000. Fort de France, Martinique: Office National de la Chasse et de la Faune Sauvage (ONCFS). (Unpublished report).

VILLARD, P. AND FELDMANN, P. (1999) Restauration de la biodiversité dans les petites Antilles françaises: intérêt et faisabilité de l'installation d'une population naturelle viable d'un perroquet endémique des Petites Antilles en Martinique. Petit Bourg, Guadeloupe: Association pour l'Etude et la protection des Vertébrés des petites Antilles (Unpublished AEVA report, 20).

ACKNOWLEDGEMENTS

The authors would like to thank Mr Bernard Deceuninck (Ligue pour la Protection des Oiseaux, BirdLife in France) for his comments and valuable advice. We also give our sincere thanks to those who have provided much of the information: Mr Philippe Feldmann (CIRAD Montpellier), Mr Vincent Arenales del Campo and Mr Stéphane Defos (DIREN Martinique), Mr Philippe Richard and Mr Michel Tanasi (ONF Martinique), Mr Jean-François Maillard (ONCFS Martinique), Mrs Marie-Michèle Moreau and Ms Floriane Minguy (CEL Martinique), Ms Valérie Genesseaux (PNRM), M. Tayalay (Association Ornithologique de la Martinique, AOMA), and Mr Jean-Raphaël Gros-Desormeaux (Université des Antilles et de la Guyane en Martinique).

MQ001 Forêts du Nord and de la Montagne Pelée

Littoral Conservation Area/Classified Site/Unprotected

COORDINATES 14°49'N 61°10'W
ADMIN REGION Ajoupa-Bouillon, Basse-Pointe, Grand Rivière, Le Prêcheur, Macouba, Morne Rouge, Saint-Pierre
AREA 9,262 ha
ALTITUDE 0–1,395 m
HABITAT Forest, grassland

Martinique Oriole

THREATENED BIRDS 1
RESTRICTED-RANGE BIRDS 18
BIOME-RESTRICTED BIRDS
CONGREGATORY BIRDS

■ Site description

Forêts du Nord and de la Montagne Pelée IBA is situated at the northernmost end of the island. This forested massif embraces the Montagne Pelée (1,395 m) and several other peaks such as Piton Marcel (897 m) and Piton Mont Conil (1,026 m), and supports contiguous forest from the coast to the volcanic summits. Receiving exceptional levels of rainfall, the landscape is broken by numerous streams and river valleys. It is mostly forested, but also has areas of grassland and pasture, wetland and coast (including an offshore islet), and Creole gardens and plantations.

■ Birds

This IBA supports 18 (of the 19) Lesser Antilles EBA restricted-range birds including the Vulnerable Martinique Oriole *Icterus bonana*, the Blue-headed Hummingbird *Cyanophaia bicolor* and Grey Trembler *Cinclocerthia gutturalis*. It embraces some of the most significant populations of forest-dependent birds on the island. The Near Threatened White-crowned Pigeon *Patagioenas leucocephala* is present according to the hunters who shoot this increasingly rare species. Small numbers of seabirds breed on L'îlet La Perle. The Ringed Kingfisher *Megaceryle torquata stictipennis* sometimes occurs.

■ Other biodiversity

The Vulnerable Lesser Antillean iguana *Iguana delicatissima* occurs in the coastal forests, and globally threatened sea-turtles are present offshore. Taxa with ranges restricted to the Lesser Antilles include the pinktoe tarantula *Avicularia versicolor*, the coleopteran *Dynastes hercules braudrii* (endemic to Martinique), three bats and 69 tree species.

■ Conservation

Some of this IBA is privately owned. The coastal zone is protected within the Anse Couleuvre, Fond Moulin and Montagne Pelée Littoral Conservation Areas, and 2,100 ha on Montagne Pelée is designated a Site Classé (classified site). Strict protection and management have been advocated through the Schéma d'Aménagement Régional and the Schéma de Services Collectifs des Espaces Naturels et Ruraux; botanists have advocated for a nature reserve; and a marine reserve is being investigated. Forest on Montagne Pelée is managed by ONF, and CEL purchased 789 ha of forest for conservation purposes. A survey of the area's biodiversity is needed. The inaccessibility of the massif means human activity and impact is limited. However, the coastal zone is subject to heavy erosion, uncontrolled hunting, tourism development, invasive alien predators, and pollution are concerns.

MQ002 Pitons du Carbet

Hunting Reserve/Forêt Domaniale/Unprotected

COORDINATES 14°43'N 61°06'W
ADMIN REGION Gros Morne, Fonds St Denis, Fort-de-France, Lorrain, Marigot, Morne Rouge, Morne Vert, Schœlcher, St Joseph
AREA 12,423 ha
ALTITUDE 44–1,197 m
HABITAT Forest, grassland

Martinique Oriole

THREATENED BIRDS 1
RESTRICTED-RANGE BIRDS 18
BIOME-RESTRICTED BIRDS
CONGREGATORY BIRDS

■ Site description

Pitons du Carbet IBA is in north-central Martinique, and comprises the forested massif of the five Pitons du Carbet (the peaks of which are extinct volcanoes), and more than 10 other smaller hills. It is the most important montane forest area on the island. High rainfall has resulted in a dense network of streams and rivers throughout the mountains. There are several houses at mid-elevations, as well as numerous well-used walking trails. The IBA is used for recreational activities such as walking and canyoning, and hunting and "wild-harvesting" are permitted in certain areas.

■ Birds

This IBA is home to 18 (of the 19) Lesser Antilles EBA restricted-range birds including the Vulnerable Martinique Oriole *Icterus bonana*, the Blue-headed Hummingbird *Cyanophaia bicolour* and Grey Trembler *Cinclocerthia gutturalis.* Extrapolated (1996) abundance indices defined for one part of the massif suggest this IBA supports large populations of a number of the restricted-range forest species. The Ringed Kingfisher *Megaceryle torquata stictipennis* subspecies has been recorded within the IBA which is also a winter home to many Neotropical migratory birds

■ Other biodiversity

Amongst the endemic plants are the Endangered *Pouteria pallida*, and Vulnerable *Freziera cordata*, *Schefflera urbaniana* and *Inga martinicensis*.

■ Conservation

Pitons du Carbet IBA is state-owned. The 7,000-ha Forêt Domaniale of Pitons du Carbet is strictly protected under the Schéma d'Aménagement Régional. ONF has started a Réserve Biologique Intégrale initiative that will cover c.4,000 ha of the Forêt Domaniale. A Réserve de Chasse (hunting reserve) covers a further 1,880 ha within the Forêt Domaniale. Although used for recreation, the limited accessibility of the area, and fear to the snake *Bothrops lanceolatusi* mean that human disturbance and impact is minimal. ONF manages the forests, but very few trees are harvested from this area. ONF also has plantations of *Swietenia macrophylla* which have profoundly modified the natural forest structure in some areas. Shiny Cowbird *Molothrus bonariensis* could turn out to be a threat to *I. bonana* but seems to be decreasing on the island. There is no permanent research programme within the IBA, although SEPANMAR has maintained constant-effort bird ringing (banding) sites since 2003.

MQ003 Pointe du Pain de Sucre

Habitat Protection Bylaw

COORDINATES 14°47′N 60°56′W
ADMIN REGION Sainte-Marie, La Trinité
AREA 8,700 ha
ALTITUDE 0–78 m
HABITAT Rocky Areas

Roseate Tern

THREATENED BIRDS
RESTRICTED-RANGE BIRDS
BIOME-RESTRICTED BIRDS
CONGREGATORY BIRDS ✔

■ Site description

Pointe du Pain de Sucre IBA is located along the coast of north-east of Martinique, north of (but including the coast of) the Caravelle Peninsula (Presqu'île de la Caravelle). The IBA is focused on a volcanic headland that extends into the sea for 300 m. It is surrounded by cliffs, covers 21 ha, and ends at the "sugarloaf" of Le Pain de Sucre at the end of the headland. Le Pain de Sucre and the nearby shoreline are not inhabited or developed, and are only visited by a few fishermen and walkers. l'Ilet Sainte-Marie, l'Ilet Saint-Aubain and Rocher de la Caravelle are relatively small, low-lying islets included within the extensive marine area of this IBA. The vegetation comprises coastal, drought-resistant plants. The marine extension to this IBA incorporates the island's northern-most coral reefs.

■ Birds

Pointe du Pain de Sucre is a key seabird breeding site. The IBA supports globally significant populations of Roseate Tern *Sterna dougallii* (estimated to be c.250 pairs in 2006) and Common Tern *S. hirundo* (c.20 pairs on Petits Saints in 2001, but not seen there in 2007). Regionally important numbers of Magnificent Frigatebird *Fregata magnificens* roost on Le Rocher de la Caravelle, and the Brown Booby *Sula leucogaster* population on Gros Îlets and Petits Saints is also regionally important.

■ Other biodiversity

The lizard *Anolis roquet roquet* subspecies is endemic to Martinique, and is present on the islets.

■ Conservation

La Pointe Pain de Sucre IBA is state-owned. It is protected under an Arrêté de Protection de Biotope (habitat protection bylaw) which forbids access between 1 March and 31 August, and is managed as a Forêt Départementale du Littorale by ONF. The three islets are formally protected as a Littoral Conservation Area (under the terms of which no development is allowed within 15 m of the coast) belonging to CEL. A "Fish Preserve" forbids fishing within a marine zone between the towns of Sainte-Marie and Trinité (a 4-km stretch of coast) and extending 2 km out to sea from the coast of îlet Saint-Aubin. The main threats are human disturbance and invasive alien predators. A count of the *Sterna dougallii* colony was carried out in 2006 by SEPANMAR. For several years, SEPANMAR worked alongside ONF in proposing management interventions and monitoring Pointe Pain de Sucre.

MQ004 Presqu'île de la Caravelle

Nature Reserve/Littoral Conservation Area/Classified and Registered Sites

COORDINATES 14°45′N 60°53′W
ADMIN REGION La Trinité
AREA 960 ha
ALTITUDE 0–189 m
HABITAT Forest, grassland

White-breasted Thrasher

THREATENED BIRDS 2
RESTRICTED-RANGE BIRDS 12
BIOME-RESTRICTED BIRDS
CONGREGATORY BIRDS

■ Site description

Presqu'île de la Caravelle (Caravelle Peninsula) IBA is situated on a large peninsula (10 km long, 1 km wide) on the east coast of central Martinique. It consists of two main parts with identical habitat and less than 2 km between them, namely: the 900-ha Presqu'île de la Caravelle where Réserve Naturelle de la Caravelle lies; and the 60-ha Pointe Rouge to the west. Its unique location, pronounced cliffs and important forest cover harbour the only extant population of White-breasted Thrasher *Ramphocinclus brachyurus* present on the island. The ruins of Château Dubuc are within the reserve, and remnants of mangroves (a rare occurrence on the north Atlantic coast) persist at Pointe Rouge. This area is one of the driest in Martinique.

■ Birds

The dry forests of this IBA support Martinique's only population of the Endangered White-breasted Thrasher *Ramphocinclus brachyurus* which numbers c.200–400 individuals distributed in three discrete parts of the IBA. Twelve (of the 19) Lesser Antilles EBA restricted-range birds occur, including the Vulnerable Martinique Oriole *Icterus bonana*. Several seabird species nest on the cliffs (although not in significant numbers). About 80 bird species have been recorded in the IBA, c.65% of which are migrants (landbirds, shorebirds and seabirds).

■ Other biodiversity

Taxa endemic to Martinique include the snake *Bothrops lanceolatus*, a bat *Myotis martiniquensis*, a lizard *Anolis roquet roquet* and coleopteran *Dynastes hercules baudrii*.

■ Conservation

Presqu'île de la Caravelle IBA is a mix of private and state (e.g. CEL) owned land. Pointe Rouge is formally protected as a Littoral Conservation Area, and the Réserve Naturelle de la Caravelle (a national nature reserve within the broader Parc Naturel Régional de la Martinique, PNRM) covers much of the rest. PNRM manages this reserve. The whole of the Presqu'île de la Caravelle is a Site Classé (i.e. covered by Classified Site Order) which controls and limits urbanisation. Areas already urbanised are within a Site Inscrit (i.e. covered by a Registered Site Order), which is less constraining. Research on *R. brachyurus* by M. Tayalay (Association Ornithologique de Martinique, AOMA) located important populations of the species at the edges of the Réserve Naturelle de la Caravelle, leading to the extension of the reserve into Morne Pavillon (on private lands). Invasive alien (mammalian) predators are a concern, as is the Shiny Cowbird *Molothrus bonariensis*, a brood-parasite of *I. bonana*.

MQ005 Ilets Boiseau and Petit Piton

Littoral Conservation Area/Habitat Protection Bylaw/Unprotected

COORDINATES 14°40′N 60°52′W
ADMIN REGION Le François, Le Robert
AREA 7,300 ha
ALTITUDE 0–25 m
HABITAT Rocky areas, shrubland

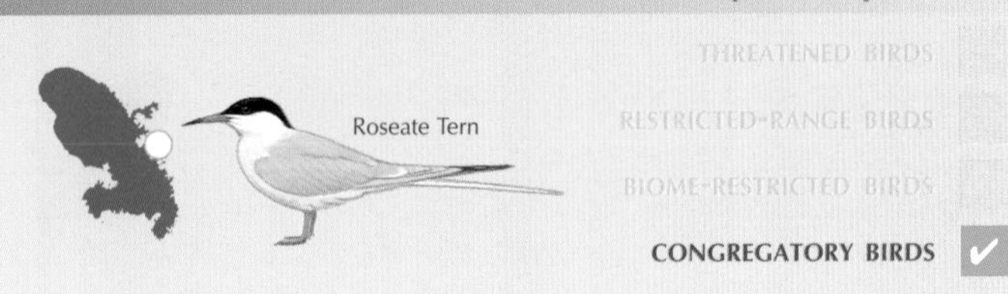

THREATENED BIRDS
RESTRICTED-RANGE BIRDS
BIOME-RESTRICTED BIRDS
CONGREGATORY BIRDS ✓

■ Site description
Ilets Boiseau and Petit Piton IBA comprises two small islets off the east coast of central Martinique, south of Presqu'île de la Caravelle (IBA MQ004). L'îlet Boisseau (0.5 ha) and L'îlet Petit-Piton (0.6 ha) are 1.5 km apart, and are the islets furthest from shore within the Baie du Robert. They are volcanic, range between 15 and 25 m high and are surrounded by cliffs. Neither islet is inhabited or managed, but they lie within an area used by many fishermen, yachtsmen and tourists. They support typical xerophytic coastal vegetation.

■ Birds
This IBA supports a globally significant breeding population of Roseate Tern *Sterna dougalli.* The colony is thought to move between the two islets, although L'îlet Boisseau is the most frequently occupied. In 2006, the colony was on L'îlet Petit Piton. Counts of *S. dougalli* in 1997 (carried out by George Tayala) suggested the presence of c.600 birds. A follow-up survey (by Lionel Dubief in 2006) found c.150–200 pairs. Bridled Tern *S. anaethetus* nests in small numbers (c.10 pairs on each islet).

■ Other biodiversity
The Martinique endemic lizard *Anolis roquet roquet* is found on l'îlet Boisseau along with a rare plant species, *Maclura tinctoria.*

■ Conservation
Ilets Boiseau and Petit Piton IBA is state-owned and managed by the ONF. L'îlet Boisseau is protected under an Arrêté de Protection de Biotope (habitat protection bylaw). L'îlet Petit Piton is unprotected. Human disturbance and invasive alien predators are the main threats. In 2005, the colony of *S. dougallii* was mysteriously abandoned on l'îlet Boisseau, although the bodies of adult terns were found. SEPANMAR conducted a count on L'îlet Petit Piton in 2006 in order to proposing measure for management and monitoring of the site. Also, SEPANMAR and wardens from the commune of Le Robert carried out some urgent rat *Rattus* spp. eradications as a forerunner to an extended future project. The wardens of Le Robert ensure the upkeep and management of certain islets, and help enforce the laws relevant to the islands. CEL delegated the ecological management responsibility for L'îlet Boisseau to the commune of Le Robert.

MQ006 Mangrove de Fort de France

Hunting Reserve/Unprotected

COORDINATES 14°33′N 61°00′W
ADMIN REGION Lamentin, Ducos, Rivière Salée, Trois-îlets
AREA 3,361 ha
ALTITUDE 0–17 m
HABITAT Wetlands, grassland

Martinique Oriole

THREATENED BIRDS 1
RESTRICTED-RANGE BIRDS 10
BIOME-RESTRICTED BIRDS
CONGREGATORY BIRDS

■ Site description
Mangrove de Fort-de-France IBA covers the mangroves of the Baie de Fort-de-France, the largest area of mangrove in Martinique (1,200 ha of the 1,850 ha that remain on the island). It also includes the largest wetland on Martinique comprising numerous natural and man-made habitats. The IBA is situated on the south-eastern part of the large Baie de Fort-de-France (on the west coast of Martinique), next to the built-up area of Fort-de-France and Le Lamentin. To the east lie the agricultural plains of Le Lamentin and Rivière Salée.

■ Birds
This IBA supports 10 (of the 19) of the Lesser Antilles EBA restricted-range birds including the Vulnerable Martinique Oriole *Icterus bonana.* Many resident waterbirds also occur, as do more than 90 migratory species that visit the mangroves and associated habitats. This is the most significant migratory waterbird stop-over site on the island.

■ Other biodiversity
The endemic lizard *Anolis roquet roquet* subspecies occurs.

■ Conservation
Mangrove de Fort-de-France IBA is mostly a state-owned Domaine Public Maritime managed by the Direction Départementale de l'Equipement (which is responsible for the land) and ONF (which manages 240 ha of coastal forest). CEL protects 20 ha through a land purchase. The brackish grasslands and artificial water bodies are mostly private. There is a 250-ha hunting reserve, although the ban on hunting is poorly observed. Main threats are development (e.g. commercial and industrial parks, extension of the airport and marinas), pollution, sedimentation, siltation of marine habitats, agricultural expansion, uncontrolled hunting and introduced mammals predators. Université des Antilles et de la Guyane in Martinique has studied the flora and problems of pollution and siltation, but no inventory of the birds has been undertaken. Since the 1990s, all conservation projects in this IBA have failed or been abandoned. In 2007, Parc Naturel Régional de Martinique ordered a feasibility study for the establishment of a regional nature reserve. A "Bay Agreement" project, looking to limit the pollution and siltation problems, is also being carried out by the Communauté d'Agglomération du Centre et de l'Est de la Martinique (CACEM). Finally, an avian nesting ecology study (and wintering Neotropical migrant inventory) is being developed for ONF by the study group BIOS.

MQ007 Massif forestier entre Le Diamant et les Trois-Îlets

Littoral Conservation Area/Classified Site

COORDINATES 14°30'N 61°02'W
ADMIN REGION Anses d'Arlet, Diamant, Trois-îlets
AREA 6,619 ha
ALTITUDE 0–478 m
HABITAT Forest, grassland

Martinique Oriole

THREATENED BIRDS 1
RESTRICTED-RANGE BIRDS 12
BIOME-RESTRICTED BIRDS
CONGREGATORY BIRDS

■ Site description
Massif forestier entre Le Diamant et les Trois-Îlets IBA is in south-west of Martinique, on the Caribbean coast south of the Baie de Fort-de-France. This forested massif (the highest point of which is Morne Larcher) has a rugged and varied relief including mornes, slopes, cliffs, rocks, ridges, ravines and ponds. The climate is one of the driest on Martinique: the village of Anses d'Arlet et du Diamant records annual rainfall of less than 1,200 mm. The IBA supports forest (of varying types depending on humidity), and is floristically diverse. There are also small areas of mangrove on the coastal fringe in the south-west.

■ Birds
This IBA supports 12 (of the 19) of the Lesser Antilles EBA restricted-range birds, including the Vulnerable Martinique Oriole *Icterus bonana*. The coast is visited by numerous seabirds as it is adjacent to the seabird breeding colonies of Rocher du Diamant IBA (MQ008). Ponds and rivers are visited regularly by resident waterbirds, and Neotropical migratory birds. The Endangered White-breasted Thrasher *Ramphocinclus brachyurus* occurred in the forests of Trois-Îlets until the end of the nineteenth century, but it has not been seen since then.

■ Other biodiversity
Endemic reptiles include the snake *Bothrops lanceolatus*, lizard *Anolis roquet roquet*, and gecko *Sphaerodactylus vicenti*.

■ Conservation
Massif forestier entre Le Diamant et les Trois-Îlets IBA is state-owned. Le Cap Salomon and Morne Larcher are Littoral Conservation Area (which prohibits development within 15 m of the coast. CEL owns two areas within the IBA: Morne Larcher (64 ha) and Cap Salomon (137 ha). The area from Morne Champagne to the village of Anses d'Arlet (320 ha) and Petite Anse to Anse Cafard (126 ha) are designated as Sites Classés (under Classified Site Orders). The whole area is designated as a Zone Naturelle d'Intérêt Ecologique Floristique et Faunistique (ZNIEFF) although this carries no legal weight. There is little human activity within the IBA, apart from the urban expansion at the boundaries (which needs to be monitored). However, some plants are selectively harvested, and this needs to be controlled or stopped. Invasive alien predators are abundant. Mr Gros-Désormeaux, a PhD student, recently completed studies of birds in this massif but the results are not yet available.

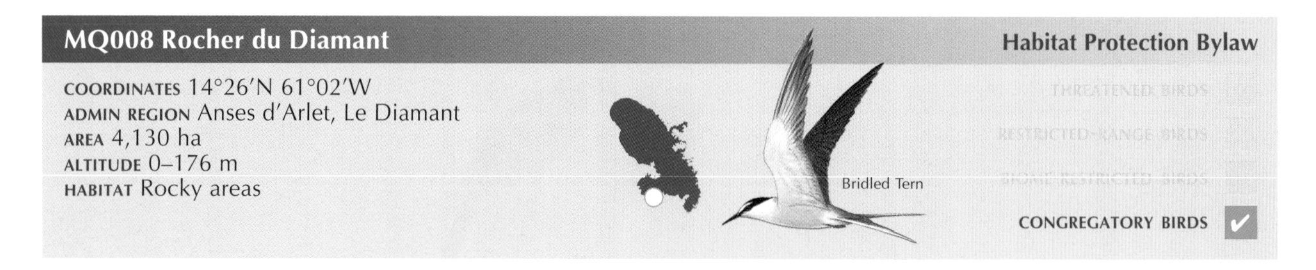

■ Site description
Le Rocher du Diamant IBA is on the Caribbean coast of south-western Martinique. It comprises a 6-ha conical-shaped volcanic islet lying c.2 km offshore from Pointe du Diamant and the Morne Larcher massif. The islet is surrounded by cliffs rising to 176 m, with caves and wave-cut platforms. Vegetation is sparse, and confined to fissures in the rocks, and the very small areas of flat ground. The islet is uninhabited, although it was occupied towards the end of the eighteenth century when the British installed artillery pieces. Le Rocher du Diamant is within traditional fishing grounds often visited by fishermen. The IBA includes marine areas around the islet, and extending along the coasts on the southern side of Le Diamant et les Trois-Îlets IBA (MQ007).

■ Birds
This IBA is a significant seabird breeding site that supports a globally important population of Bridled Tern *Sterna anaethetus* (100–120 pairs), and regionally significant populations of non-breeding Brown Booby *Sula leucogaster* (160 birds) and breeding Brown Noddy *Anous stolidus* (200–250 pairs). The site is one of the most important roosting areas for Magnificent Frigatebird *Fregata magnificens* (50 individuals) in Martinique. Other seabirds that nest include Red-billed Tropicbird *Phaethon aethereus* (15–20 pairs) and White-tailed Tropicbird *P. lepturus* (1 pair). Audubon's Shearwater *Puffinus lherminieri* nested regularly before the mid-twentieth century, but has not been recorded since.

■ Other biodiversity
The Critically Endangered snake *Liophis cursor* (endemic to Martinique) is found only on this islet.

■ Conservation
Le Rocher du Diamant IBA is state-owned and protected under an Arrêté de Protection de Biotope (habitat protection order). This bylaw is currently being modified to prohibit all access to the site. At the moment, fishermen frequently visit the island to store their equipment and fish with rods. Access restrictions take effect between 1 January and 31 August. The main threats are human disturbance and introduced alien species. The only recent studies of the islet are an ecological evaluation published in 1999 by the Centre de Recherche Géographie Développement Environnement (GEODE) de la Caraïbe (of the Université des Antilles et de la Guyane in Martinique), and an avifauna inventory published in 2004 by AMAZONA. Monitoring of animal populations on the islet is the responsibility of a working group of trained scientists.

MQ009 Grand Macabou

Littoral Conservation Area/Hunting Reserve

COORDINATES 14°29'N 60°49'W
ADMIN REGION Le Marin
AREA 157 ha
ALTITUDE 0–89 m
HABITAT Forest, grassland

Martinique Oriole

THREATENED BIRDS 1
RESTRICTED-RANGE BIRDS 10
BIOME-RESTRICTED BIRDS
CONGREGATORY BIRDS

■ Site description

Grand Macabou IBA is on the Atlantic coast of south-eastern Martinique, to the north of Îlets and falaises de Sainte Anne IBA (MQ010). It is bounded to the east by the sea and to the west by wire fencing keeping out wandering livestock. The IBA is formed from a plateau that slopes gently down towards the sea, and two hills. The area is traversed by two weakly flowing water courses, and there are some man-made water bodies: a 12-ha lake formed during quarrying; and two ponds created for livestock. Quarrying for sand has stopped. The beach and immediately adjacent areas are well visited. The IBA supports a mix of dry forest, shrubland, mangrove, and pasture.

■ Birds

This IBA is supports 10 (of the 19) of the Lesser Antilles EBA restricted-range birds including the Vulnerable Martinique Oriole *Icterus bonana*. The site is frequented by waterbirds and Neotropical migratory birds.

■ Other biodiversity

The Critically Endangered hawksbill *Eretmochelys imbricata* and leatherback *Dermochelys coriacea* turtles nest on the beach at Grand Macabou. The endemic subspecies of lizard *Anolis roquet roquet* and subspecies of gecko *Sphaerodactylus vicent* occur.

■ Conservation

Grand Macabou IBA is state-owned and comprises two main parts: 113 ha acquired by CEL, and the coastal forest (30 ha) along the edge of the CEL lands. The area is (in part) protected as a Littoral Conservation Area, and is also a Réserve de Chasse (hunting reserve, in which hunting is forbidden). Harvesting/gathering, camping and fires are also forbidden in this IBA. Human activity and disturbance is significant, most notably on the beach and in the forest behind it. Even though it is a hunting reserve, poaching still persists. Invasive alien mammal species occur. No research on birds is being undertaken within the IBA. The Office National de la Chasse et de la Faune Sauvage (ONCFS) will soon start monitoring the breeding sea-turtles.

MQ010 Îlets and falaises de Sainte Anne

Nature Reserve/Registered Site

COORDINATES 14°24'N 60°49'W
ADMIN REGION Sainte-Anne, Le Marin
AREA 1,600 ha
ALTITUDE 0–21 m
HABITAT Rocky areas

Audubon's Shearwater

THREATENED BIRDS
RESTRICTED-RANGE BIRDS
BIOME-RESTRICTED BIRDS
CONGREGATORY BIRDS ✓

■ Site description

Îlets and falaises de Sainte Anne (islets and cliffs of St Anne) IBA is made up of islets, cays and cliffs on the Atlantic coast of south-easternmost Martinique. Two flat, rocky islands of 0.25 ha comprise the Ilets Aux Chiens to the north, and to the south the rocky islet Table du Diable covers 0.4 ha. In the middle, four islets "Des îlets de Saint-Anne" form the Réserve Naturelle (national nature reserve) namely L'îlet Hardy (2.5 ha), l'Ilet Poirier (2 ha), L'îlet Burgaux (0.5 ha) and L'îlet Percé (0.5 ha). The coastal cliffs at Savane des Pétrifications extend for 1 km and are included in the IBA. The vegetation is poorly developed and consists of xerophytic coastal plants on limestone and volcanic substrate. The islets are all uninhabited but regularly visited by tourists.

■ Birds

This IBA supports more than 20,000 nesting seabirds, representing the most important site for seabirds in Martinique. Species with regionally significant populations are Sooty Tern *Sterna fuscata* (7,000–24,000 birds), Brown Noddy *Anous stolidus* (110–1,152 birds), Bridled Tern *Sterna anaethetus* (130–240 birds) and Audubon's Shearwater *Puffinus lherminieri* (146 birds). Neotropical migratory shorebirds use this IBA.

■ Other biodiversity

This nature reserve supports Martinique's only population of the crab *Gecarcinus ruricola*.

■ Conservation

Îlets and falaises de Sainte Anne IBA is state-owned. Access to the Réserve Naturelle is forbidden (other than for scientists) and it is illegal to approach within 50 m of the islets (a limit that will soon be extended to 100 m), or to fly over the area at less than 300 m (to avoid disturbing the seabirds). The îlets aux Chiens and the cliffs of Savane des Pétrifications are included in two "Sites Inscrits" (under Registered Site Orders). These ministerial orders allow the control of all forms of urbanisation or building, but do not allow for control of people visiting the bird colonies, and nor does it put in place specific (or proactive) conservation measures. The Table du Diable is not protected. Poaching and the anchoring of boats persist even though access is forbidden. However, the main threat is the presence of black rats *Rattus rattus*. After 1999 when all seabird breeding attempts failed, rat eradication has been undertaken every year. The Réserve Naturelle is administered by two organisations, the Parc Naturel Régional de la Martinique (PNRM) and ONF, which oversee scientific research, monitoring (e.g. seabirds) and management within the reserve. The islets outside of the reserve enjoy no such attention.

MONTSERRAT

LAND AREA **102 km²** ALTITUDE **0–914 m**
HUMAN POPULATION **4,819** CAPITAL **Plymouth** (defunct due to volcanic eruption)
IMPORTANT BIRD AREAS **3, totalling 16.5 km²**
IMPORTANT BIRD AREA PROTECTION **48%**
BIRD SPECIES **101**
THREATENED BIRDS **2** RESTRICTED-RANGE BIRDS **12**

GEOFF HILTON (ROYAL SOCIETY FOR THE PROTECTION OF BIRDS), LLOYD MARTIN AND JAMES 'SCRIBER' DALY (DEPARTMENT OF ENVIRONMENT, MONTSERRAT) AND RICHARD ALLCORN (FAUNA AND FLORA INTERNATIONAL)

The endemic Montserrat Oriole lost 60% of its forest habitat during the eruptions of the Soufriere Hills volcano. (PHOTO: JAMES MORGAN/DWCT)

INTRODUCTION

Montserrat is a UK Overseas Territory in the Leeward Islands towards the northern end of the Lesser Antilles, just 40 km south-west of Antigua and between the islands of Nevis and Guadeloupe. The island is about 16 km long and 11 km wide, and its volcanic origins are reflected in an extremely rugged topography. There are three major volcanic hill ranges—the Soufriere and South Soufriere Hills, the Centre Hills, and the Silver Hills. Prior to the eruption of the Soufriere Hills volcano (1995 to present: see below), Chances Peak was the highest point on the island, rising to 914 m, while the highest point in the Centre Hills—Katy Hill—reaches 741 m. There are also two smaller hills: Garibaldi Hill and St Georges Hill. There are a few, very small offshore islets. The coastline is mostly rocky and rather steep, with low cliffs in a few places in the north. A number of relatively small, sandy beaches are scattered around the island.

Montserrat has a tropical climate with average annual rainfall varying between c.1,100 and 2,100 mm as a result of the mountainous topography. The wet season extends from June to December and coincides with the Atlantic hurricane season. The natural vegetation over the great majority of the island is tropical forest. This ranges from dry deciduous forest in the lowlands, through semi-deciduous and evergreen wet forest in the hills, to montane elfin forest on the highest peaks. There are small areas of littoral woodland, and in the driest areas of the lowlands, the vegetation is xerophytic scrub, with numerous cacti. All but a few small forest patches were apparently cleared during the plantation era, and the bulk of the remaining forest is therefore secondary. In the Centre Hills, the largest remaining forest block, native trees are mixed with numerous large, non-native fruit trees—remnants of earlier agricultural endeavours. Substantial areas in the lowlands are now cleared for agriculture and settlement. In the Silver Hills, forest clearance and over-grazing has resulted in degraded scrub vegetation. The island has very few wetlands. Prior to the eruption, Foxes Bay Bird Sanctuary contained areas of saline lagoon and mangroves: this area was destroyed by heavy silt deposits of eroded volcanic debris. A very small, partly degraded saline lagoon and mangrove area remain at Carr's Bay.

The recent ecological and human history of the island is dominated by the eruption of the Soufriere Hills volcano, 1995–2008 (and ongoing). Explosive eruptions, ash-falls and pyroclastic flows have been frequent and devastating. The southern two thirds of the island have been evacuated, including the capital, Plymouth. The majority of

The ruins of Plymouth, in the shadow of the Soufriere Hills volcano that has dominated the island's recent history. (PHOTO: RICH YOUNG/DWCT)

the human population emigrated, with the population declining from c.12,701 (July 1994) to 2,726 (1998). Economic, administrative and civic life was massively disrupted by the mass emigration and the loss of the capital. The forests of the Soufriere/South Soufriere hills ranges were almost entirely destroyed by pyroclastic flows, leaving only a small remnant in the Roche's area. Most other areas of Montserrat, including the largest surviving forest block in the Centre Hills, were subject to repeated heavy ash-falls and acid rain. The remaining human population is now clustered in the north, around the fringes of the Centre Hills. Prior to the volcanic eruption, tourism (though not mass tourism) and agriculture were the mainstays of the economy. Subsequently, both sectors have been depressed, and reconstruction work has provided the main economic activity for the island's greatly reduced human population.

■ Conservation

The Department of Environment (DOE) of the Ministry of Agriculture, Lands, Housing and Environment (MALHE) has responsibility for biodiversity conservation. Enabling legislation for conservation is provided by the Forestry, Wildlife, National Parks and Protected Areas Ordinance which makes provision for the designation of protected areas and the protection of wildlife. However, as yet, there are no national parks on Montserrat. However, this legislation does not reflect recent research findings or the obligations of regional and international environmental agreements. With this in mind, a legislative review and revision has recently been conducted to produce modern, relevant, and enforceable environmental legislation. An advance draft of this legislation, which has benefited from broad based stakeholder consultation, is (July 2008) before the Legal Department for final drafting before submission to cabinet for approval.

The main conservation NGO is the Montserrat National Trust. It is mandated to preserve and protect the natural, historical and cultural heritage of Montserrat. To date, most funding for conservation work is received from international donor agencies and UK government funds such as the Overseas Territories Environment Project (OTEP) and The Darwin Initiative. Several UK-based NGOs (e.g. Royal Society for the Protection of Birds, RSPB, Durrell Wildlife Conservation Trust, DWCT, and Royal Botanic Gardens Kew) as well as the U.S. International Institute of Tropical Forestry have a long history of involvement in Montserrat's conservation, working in partnership with the Forestry Division of DOE and the National Trust. Academic researchers have also been active in recent years, perhaps most notably entomologists from Montana State University, bat

The Centre Hills support the island's largest remaining forest block. (PHOTO: GEOFF HILTON/RSPB)

experts from South Dakota State University and marine turtle experts from University of Exeter and ecologists from University of East Anglia.

Conservation actions have focused on the Centre Hills and the globally threatened species therein. In June 2005, the Darwin Centre Hills Project was launched with a primary goal to enable the people of Montserrat to effectively manage the Centre Hills and associated resources. Under the umbrella of this project, NGO-assisted efforts (working in collaboration with MALHE) have included detailed socio-economic assessments, in-depth biodiversity assessments, and area management planning (for a proposed national park). The Centre Hills Project concluded in March 2008 with the integration of pending and planned activities into the workplans of each of the project partners (which included Ministry of Agriculture, Lands, Housing, and Environment, Montserrat National Trust, Montserrat Tourist Board, RSPB, Royal Botanic Gardens, Kew and DWCT). The release of the Centre Hills Management Plan has amplified the need for more funding to be directed to Biodiversity conservation and use of environmental goods and services to ensure effective management. Additionally species action plans have been developed for the Montserrat Oriole *Icterus oberi* and mountain chicken *Leptodactylus fallax*, and other such plans are being developed for additional key species in the Centre Hills.

Montserrat's ecology has been radically altered by human activity since the arrival of Europeans. Massive forest clearance during the plantation era left only a tiny remnant of primary forest. Subsequently, much of the area of the main hill ranges reverted to secondary forest. Parts of the Silver Hills are heavily degraded by soil exhaustion and erosion. Much of the lowland and coastal areas of Montserrat have been converted to agriculture and settlement. Lowland forest is now relatively rare, and occurs primarily as narrow riparian strips. The devastating Hurricane Hugo hit the island in 1989 and caused massive tree fall and almost complete defoliation. This was followed six years later by the start of the volcanic eruptions, which have had a massive impact on the native wildlife. The impacts of these natural catastrophes have been exacerbated by the human habitat degradation that preceded them. Ash-fall is known to have had pronounced effects on the arthropod fauna, with knock-on effects on the food chain, although the direct impacts on birds are not well known. As a result of the volcanic eruptions, the human population is now entirely in the north of the island. There is much pressure to provide new housing and infrastructure to accommodate the relocation, and the return of emigrants. However, there is clearly a need to ensure that this development is environmentally sustainable, despite its urgency. Non-native species of mammal are widespread on Montserrat. Recent research on *Icterus oberi* and *Leptodactylus fallax* indicates that rats *Rattus* spp. are major predators of the native biota, and probably also affect vegetation dynamics. Feral cats, goats and pigs are also significant conservation problems. Invasive alien plants have not been well studied, but may also have significant impacts.

■ Birds

Of Montserrat's 101 recorded bird species, 47 are resident breeding land birds and 54 are Neotropical migrants (either passage migrants or winter visitors). However, the migrant landbirds are very scarce relative to the resident birds. Twelve of the resident land birds (see Table 1) are Lesser Antilles EBA restricted-range birds (of the 38 that define the EBA). The Lesser Antillean Flycatcher *Myiarchus oberi* is also a Lesser Antilles EBA restricted-range bird, but it is very rare on Montserrat and thought to be a non-breeding vagrant so has not been considered in the IBA analysis. Of the restricted-range birds, Lesser Antillean Bullfinch *Loxigilla noctis* is most abundant in the dry lowland forest areas (and is rather rare in the wetter forests of the Centre Hills), while the Bridled

Critically Endangered Montserrat Oriole—one of the focal species for conservation research and action in the Centre Hills. (PHOTO: CHRIS BOWDEN/RSPB)

The Vulnerable Forest Thrush is an elusive species, but is relatively common in the Centre Hills. (PHOTO: ALLAN SANDER)

Table 1. Key bird species at Important Bird Areas in Montserrat.

Key bird species	Criteria	National population	Montserrat IBAs MS001	MS002	MS003
		Criteria	■	■	■
			■	■	■
Bridled Quail-dove *Geotrygon mystacea*	■		✓	100–1,000	✓
Purple-throated Carib *Eulampis jugularis*	■		✓	12,000–42,000	
Green-throated Carib *Eulampis holosericeus*	■		✓	100–1,000	✓
Antillean Crested Hummingbird *Orthorhyncus cristatus*	■		✓	32,000–127,000	✓
Caribbean Elaenia *Elaenia martinica*	■		✓	✓	✓
Scaly-breasted Thrasher *Margarops fuscus*	■		✓	3,800–12,800	✓
Pearly-eyed Thrasher *Margarops fuscatus*	■		✓	20,000–36,000	✓
Brown Trembler *Cinclocerthia ruficauda*	■		✓	500–2,000	✓
Forest Thrush *Cichlherminia lherminieri*	VU ■ ■	3,100	✓	1,800–5,200	✓
Montserrat Oriole *Icterus oberi*	CR ■ ■	5,000		930–3,000	150–300
Lesser Antillean Bullfinch *Loxigilla noctis*	■		✓	✓	✓
Antillean Euphonia *Euphonia musica*	■		✓		

All population figures = numbers of individuals.
Threatened birds: Critically Endangered ■; Vulnerable ■. **Restricted-range birds** ■.

Quail-dove *Geotrygon mystacea*, Purple-throated Carib *Eulampis jugularis* and Brown Trembler *Cinclocerthia ruficauda* all occur on occasion in lowland dry forest but are much more abundant in the mesic and wet forest of the Centre Hills/South Soufriere Hills. The Montserrat Oriole *Icterus oberi* only occurs in mesic and wet forest. Antillean Euphonia *Euphonia musica* is very scarce and is primarily found in the lower fringes of the Centre Hills during the winter months and is thought to breed only in very small numbers. The Forest Thrush *Cichlherminia lherminieri* is commonest in dry forest (see below).

Two globally threatened birds are present in Montserrat: the Critically Endangered *Icterus oberi* and Vulnerable *Cichlherminia lherminieri* (see Table 1). The oriole—Montserrat's endemic national bird—was formerly found throughout the island's hill forests (at altitudes greater than c.150 m). However, c.60% of the forest it occupied (primarily the southern hills) was destroyed in the eruption of the Soufriere Hills volcano (1995–present). The main remaining population, in the Centre Hills (and in Roaches forest in the South Soufriere Hills), suffered a further decline of c.50% between 1997 and 2002. The population has been estimated at c.5,000 individuals, equating to a breeding population of c.1,000 pairs (based on breeding success and estimates of the number of immatures and floaters in the population). However, territory mapping work done in 2005 suggests a population of several hundred pairs, but certainly well under 1,000. Durrell Wildlife Conservation Trust has a captive breeding program (in Jersey) to safeguard the species from the risk of extinction in the wild and to provide birds for reintroduction should that become necessary in the future. *Cichlherminia lherminieri* is an elusive species that frequents dry hill forest (favouring the least disturbed forest which occurs on the higher slopes). Unlike the populations on Dominica, Guadeloupe and St Lucia the species is relatively common in the Centre Hills with numbers estimated to be in the thousands (it apparently recovered from the most severe period of volcanic ash-fall in 1996–1997). The Centre Hills probably represent the species' global centre of abundance. The Near Threatened Caribbean Coot *Fulica caribaea* survives in a pond at Brimm's Ghaut, but appears to have been extirpated by the volcanic eruption from the Foxes Bay wetland and other small ponds in the north of the island.

There are some seabird nesting colonies around the island, with White-tailed Tropicbird *Phaethon lepturus*, Brown Pelican *Pelecanus occidentalis* and Brown Booby *Sula leucogaster* thought to nest. Establishing the size of the populations at these colonies would be a valuable exercise and should perhaps be built into the broader biodiversity monitoring program for the island. Forest birds have been assessed and monitored in some detail (by members of the Forestry Division, the National Trust and RSPB) since the eruption in 1995. The Forest Bird Monitoring Programme started in 1997 (building on data gathered by Wayne Arendt in 1984 and 1990) and was supplemented by information gathered in the context of the Montserrat Oriole Emergency Conservation Programme (2001–2004). Density estimates for the more abundant forest birds have been made using point-data from a full census of the Centre Hills in 2004, and monitoring is on-going so that the effect of management actions (or further volcanic activity) can be accurately assessed.

Figure 1. Location of Important Bird Areas in Montserrat.

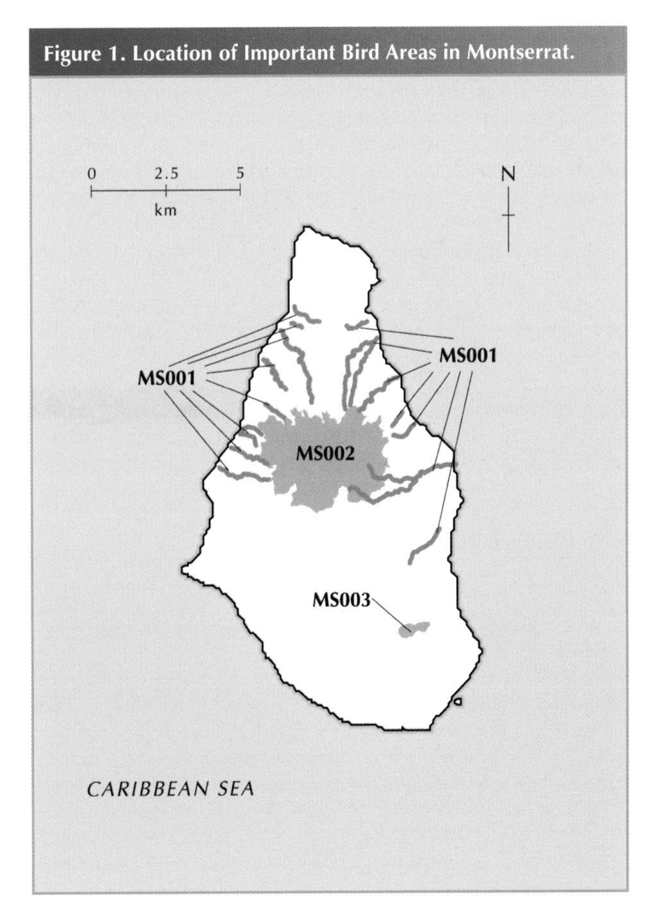

IMPORTANT BIRD AREAS

Montserrat's three IBAs—the island's international site priorities for bird conservation—cover 16.5 km², about 16% of Montserrat's land area. Only the Centre Hills IBA (MS002) is currently protected (as a forest reserve/protected forest, albeit mostly privately owned), and there are plans to designate this area as a national park. The other two IBAs are privately owned and unprotected although the Forestry, Wildlife, National Parks and Protected Arrears Act make provisions for their protection. The South Soufriere Hills IBA (MS003) is deep inside the volcanic exclusion zone and thus is no longer directly impacted by any human activity (although it is still heavily affected by invasive species introduced by humans).

The IBAs have been identified on the basis of 12 key bird species (listed in Table 1) that variously trigger the IBA criteria. These 12 species are all restricted-range birds, but include two that are globally threatened. For 11 of these species there are substantial populations in the Centre Hills IBA (the exception being Antillean Euphonia *Euphonia musica*: see above), and for Bridled Quail-dove *Geotrygon mystacea*, Purple-throated Carib *Eulampis jugularis*, Forest Thrush *Cichlherminia lherminieri*, Brown Trembler *Cinclocerthia ruficauda* and Montserrat Oriole *Icterus oberi* the majority of the Montserratian population occurs in the Centre Hills. The remaining species have substantial populations elsewhere in Montserrat (see above). The two remaining IBAs on Montserrat are the Northern Forested Ghauts (MS001) and the South Soufriere Hills (MS003). The former comprises discrete, small patches of lowland riparian and dry forest, mostly fringing the Centre Hills. The latter is a patch of wet and mesic forest in the exclusion zone of the south of the island, which is surrounded by pyroclastic flows. Both sites support most of the key bird species that the Centre Hills holds. However, the Northern Forested Ghauts IBA does not have *I. oberi*, and other key species such as *G. mystacea*, *Cichlherminia lherminieri* and *Cinclocerthia ruficauda* are extremely scarce there. Both sites are much smaller than the Centre Hills (only c.200 ha compared to 1,400 ha for the Centre Hills) emphasizing how critical the Centre Hills IBA for the maintenance of viable populations of Montserrat's key bird species. The Centre Hills and South Soufriere Hills IBAs between them embrace almost all of the remaining hill forest and along with it the entire population of the Critically Endangered oriole and the vast majority of the population of the Vulnerable *Cichlherminia lherminieri*. Based on current knowledge, no seabird or wetland sites on Montserrat qualify as IBAs.

The existing monitoring programs for the land birds are critical to maintain (and could be expanded to include seabirds) to determine the impact of management activities or future threats. The monitoring results should be used to inform the annual assessment of state, pressure and response scores at each of the territory's IBAs to provide an objective status assessment and highlight additional management interventions that might be required to maintain these internationally important biodiversity sites.

KEY REFERENCES

Adams, R., and Pedersen, S. (1999) The effects of natural disasters on bat populations on Montserrat, BWI: a 20 year history. *Amer. Zool.* 38(5): 52A.

Arendt, W. J. (1990) Impact of Hurricane Hugo on the Montserrat Oriole, other forest birds, and their habitat. Río Piedras, Puerto Rico: International Institute of Tropical Forestry, U.S. Dept. Agriculture, Forest Dept. (Unpublished report).

Arendt, W. J. and Arendt, A. I. (1984) Distribution, population size, status and reproductive ecology of the Montserrat oriole (*Icterus oberi*). Río Piedras, Puerto Rico: International Institute of Tropical Forestry, U.S. Dept. Agriculture, Forest Dept. (Unpublished report).

Arendt, W. J., Gibbons, D. W. and Gray, G. A. L. (1999) Status of the volcanically threatened Montserrat Oriole *Icterus oberi* and other forest birds in Montserrat, West Indies. *Bird Conserv. Internat.* 9: 351–372.

Blankenship, J. (1990) *The wildlife of Montserrat*. Plymouth, Montserrat: Montserrat National Trust.

Dalsgaard, B., Hilton, G. M., Gray, G. A. L., Aymer, L., Boatswain, J., Daley, J., Fenton, C., Martin, J., Martin, L., Murrain, P., Arendt, W. J., Gibbons, D. W. and Olesen, J. M. (2007). Impacts of a volcanic eruption on the forest bird community of Montserrat, Lesser Antilles. *Ibis* 149: 298–312.

DOE (2008) *Montserrat Centre Hills management plan, 2008–2010: enabling the effective conservation and management of natural resources within Montserrat's Centre Hills*. Brades, Montserrat and Sandy, U.K.: Dept. of the Environment, Ministry of Agriculture, Lands, Housing, and the Environment, and the Royal Society for the Protection of Birds.

Gibbons, D. W., Smith, K. W., Atkinson, P., Pain, D., Arendt, W. J., Gray, G., Hartley, J., Owen, A. and Clubbe, C. (1998) After the volcano: a future for the Montserrat Oriole? *RSPB Conservation Review* 12: 97–101.

Hilton, G. M. and Atkinson, P. W. (2001) The Montserrat Oriole: in trouble again. *Dodo* 37: 100.

Hilton, G. M., Atkinson, P. W., Gray, G. A. L., Arendt, W. J. and Gibbons, D. W. (2003) Rapid decline of the volcanically threatened Montserrat Oriole. *Biol. Conserv.* 111: 79–89.

Hilton, G. M., Gray, G. A. L., Fergus, E., Sanders, S. M., Gibbons, D. W., Bloxam, Q., Clubbe, C. and Ivie, M. (2005) *Species Action Plan for the Montserrat Oriole* Icterus oberi, *2005–2009*. Montserrat and Sandy, UK: Department of Agriculture and Royal Society for the Protection of Birds.

Hilton, G., Martin, L. and Daley, J. (2006) Montserrat. Pp.171–184 in S. M. Sanders, ed. *Important Bird Areas in the United Kingdom Overseas Territories*. Sandy, U.K.: Royal Society for the Protection of Birds.

Marske, K. A., Ivie, M. A. and Hilton, G. M. (in press). Effects of volcanic ash on the forest canopy insects of Montserrat, West Indies. *Environmental Entomology* 36: 817–825.

Raffaele, H. Wiley J., Garrido, O., Keith, A. and Raffaele, J. (1998) *A guide to the birds of the West Indies*. Princeton, New Jersey: Princeton University Press.

Sanders, S. M. (ed.) (2006) *Important Bird Areas in the United Kingdom Overseas Territories*. Sandy, U.K.: Royal Society for the Protection of Birds.

Siegel, A. (1983) *Birds of Montserrat*. Plymouth, Montserrat: Montserrat National Trust.

Young, R. P. (ed.) (2008). *A biodiversity assessment of the Centre Hills, Montserrat*. Jersey, Channel Islands: Durrell Wildlife Conservation Trust (Durrell Cons. Monogr. 1).

ACKNOWLEDGEMENTS

The authors would like to thank Gerard Gray and Stephen Mendes (Department of Environment, Ministry of Agriculture, Land, Housing and the Environment) for their assistance and input into this chapter. A large number of Forest Rangers have gathered the bird monitoring data used here including Philemon Murrain, James Boatswain, Calvin Fenton, John Martin, Lloyd Aymer and Jervain Greenaway. RSPB staff members Chris Bowden, Liz Mackley, Joah Madden and Mark Hulme also played an important role in data gathering. Alan Mills (Alan Fisher Consulting) and Ian Fisher (RSPB) developed Montserrat's biodiversity database; Richard Young and Matthew Morton (DWCT), and Colin Clubbe and Martin Hamilton (Royal Botanic Gardens Kew) lead the Centre Hills biodiversity assessments and other aspects of Montserrat conservation; Mike Ivie and Katie Marske made significant contributions to our understanding of the effect of ashfall on the ecology of the Centre Hills. Sarah Sanders (RSPB) co-ordinated the Centre Hills project, which was led within Montserrat by Stephen Mendes (DOE) and Carole McCauley. Calvin Fenton and Jervain Greenaway (DOE) led the research under the Centre Hills project.

MS001 Northern Forested Ghauts

■ Site description

The Northern Forested Ghauts IBA comprises a discontinuous series of steep, forested streams (known locally as ghauts) that originate in the Centre Hills IBA (MS002). The ghauts sustain a more or less continuous riparian fringe of native forest (c.50–150 m across) as they run through the open agricultural and residential lowlands of northern Montserrat. Several ghauts are still contiguous with the forest of the Centre Hills IBA. The watercourses of each ghaut are very small and there is no associated wetland habitat.

■ Birds

This IBA is important for the Vulnerable Forest Thrush *Cichlherminia lherminieri* and 11 (of the 12) Lesser Antilles EBA restricted-range birds. *Cichlherminia lherminieri* reaches densities comparable with those of the Centre Hills IBA in some of the wetter ghauts, and those with forest that is contiguous with the Centre Hills. Green-throated Carib *Eulampis holosericeus* occurs at densities twice as high as those in the Centre Hills, reflecting its preference for lower-altitude and forest-edge sites. The Critically Endangered Montserrat Oriole *Icterus oberi* is absent.

■ Other biodiversity

The endemic Montserrat anole *Anolis lividus* is thought to be common, and a number of other reptiles exist as endemic subspecies. The bat fauna is thought to be similar to the Centre Hills IBA (MS002), but the bats, like the reptiles and also the insects are poorly recorded.

■ Conservation

The ghauts are privately owned and unprotected. Rapid expansion of built areas in northern Montserrat, as a result of the abandonment of the south, has affected some ghauts, and is impacting the forest, both through direct habitat destruction and the increased presence of dogs, cats,goats, pigs and rodents. The small forest patches in the ghauts are frequently considered to be "wasteland", and there is some dumping of rubbish. Invasive alien species present in the IBA include the widespread feral goats (impacting plant communities), abundant rats (*Rattus* spp., likely having a significant ecological impact) and, being close to human habitation, pet dogs and cats, while feral cats are also fairly common (and may be important predators of some species). Domestic fowl are a rapidly increasing presence, and may be having strong ecological effects. Invasive alien plants may also be a threat, but have not been studied.

MS002 Centre Hills

■ Site description

Centre Hills IBA represents the largest forest area on Montserrat and the main water catchment for the inhabited north. It is an almost continuous block of steep, pathless hill forest in the centre of the island. A series of small, steep streams radiate from central ridges, some of which form the Northern Forested Ghauts IBA (MS001). Tropical deciduous forest in the drier lowlands develops with altitude into semi-deciduous, then tropical evergreen and eventually elfin forest on the summit of Katy Hill. The forest is mostly secondary and in a range of successional stages as a result of clearance for agriculture and the frequent passage of hurricanes.

■ Birds

This IBA is important as its mesic and wet forest supports the majority of the Critically Endangered *Icterus oberi* population. The Vulnerable Forest Thrush *Cichlherminia lherminieri* also occurs at relatively high densities, and with an estimated 900–2,600 pairs this IBA may well be the world stronghold for the species. All 12 Lesser Antilles EBA restricted-range species occur. Pearly-eyed Thrasher *Margarops fuscatus* densities are among the highest in its range.

■ Other biodiversity

The Critically Endangered (CR) Montserrat galliwasp *Diploglossus montisserrati* is known only from this IBA. Mountain chicken *Leptodactylus fallax* (also CR)—the world's second largest frog—occurs in this IBA and on Dominica. About 10% of all insect species are endemic to the IBA. The Endangered lignum vitae *Guaiacum officinale* and mahogany *Swietenia mahagoni*, and Vulnerable *S. macrophylla* and red cedar *Cedrela odorata* occur.

■ Conservation

The Centre Hills IBA, although largely privately owned, incorporates a forest reserve. The IBA extends beyond the forest reserve boundary to include the core range of *I. oberi*. A proposal to declare the IBA a national park is being prepared. Rats *Rattus* spp. and *M. fuscatus* predate nests of *I. oberi* and *C. lherminieri*, and feral pigs destroy stands of the oriole's preferred nest plant *Heliconia caribea*. Alien invasive animals are impacting the general ecology of the IBA. Ash falls from the volcano may continue to affect the ecology, and physically destroy *I. oberi* nests. Capping springs to supply water to the north of the island has reduced some stream flows and may have resulted in some loss of wet valley-bottom habitat. Small-scale encroachment around the fringes, both for housing and agricultural development, appears to be increasing.

MS003 South Soufriere Hills

Unprotected

COORDINATES 16°42′N 62°10′W
ADMIN REGION —
AREA 35 ha
ALTITUDE 200–750 m
HABITAT Evergreen and semi-deciduous forest

THREATENED BIRDS 2
RESTRICTED-RANGE BIRDS 10
BIOME-RESTRICTED BIRDS
CONGREGATORY BIRDS

■ Site description

The South Soufriere Hills IBA is in the south of the island. It comprises one small, isolated patch—Roche's Estate—of the original forests that once cloaked the slopes of the Soufriere and South Soufriere Hills. Despite being no more than 1.5 km from the Chances Peak volcano, this area of forest (on the eastern slope of the South Soufriere Hills) remained intact in spite of the pyroclastic flows from the volcano. The IBA is in the volcanic exclusion zone and has been little explored since 1997, but brief field visits in 2001 and 2002 confirmed the apparent good condition of the (albeit secondary) forest.

■ Birds

This IBA is important for the Critically Endangered Montserrat Oriole *Icterus oberi*, with densities apparently similar to those in the best parts of the Centre Hills IBA (MS002). Visits in 2001 and 2002 resulted in estimates of 50–100 pairs and records of fledglings. The Vulnerable Forest Thrush *Cichlherminia lherminieri* is also present in this IBA along with a total of 10 (of the 12) Lesser Antilles EBA restricted-range birds. Lesser Antillean Bullfinches *Loxigilla noctis* were recorded in exceptional numbers during 2001 and 2002.

■ Other biodiversity

Data on taxa other than birds are almost completely lacking from the post-eruption period, although many of the Centre Hills IBA forest species probably occur. An undescribed endemic long-horned grasshopper, an undescribed, endemic soldier beetle (*Cantharidae*) and two undescribed, endemic darkling ground beetle species (*Tenebrionidae*) have been found, suggesting that endemism is high.

■ Conservation

The South Soufriere Hills IBA is privately owned and deep inside the volcanic exclusion zone. As a result, human activity in the area has been minimal since 1997, and consequently little is known about the ecological or conservation status of this isolated forest. Rats *Rattus* spp. are present, and were abundant in 2002, and it is likely that feral livestock (possibly pigs and goats) also occur and may be at high and/or increasing densities. The area has presumably been impacted by ash falls since 1997. There are many non-native fruit trees and small, abandoned agricultural plots which favours the rats and the predatory Pearly-eyed Thrasher *Margarops fuscatus*. It is separated from other forest areas by pyroclastic flows which, if volcanic activity remains low, will likely be recolonised by vegetation rapidly enough to avert the threat of ecological and genetic isolation.

NAVASSA

LAND AREA **5 km²** MARINE AREA **1,476 km²** ALTITUDE **0–76 m**
HUMAN POPULATION **Uninhabited** CAPITAL **None; administered by USFWS Caribbean Islands National Wildlife Complex, Boquerón, Puerto Rico**
IMPORTANT BIRD AREAS **1, totalling 1,481 km²**
IMPORTANT BIRD AREA PROTECTION **100%**
BIRD SPECIES **58** THREATENED BIRDS **2** RESTRICTED-RANGE BIRDS **1**

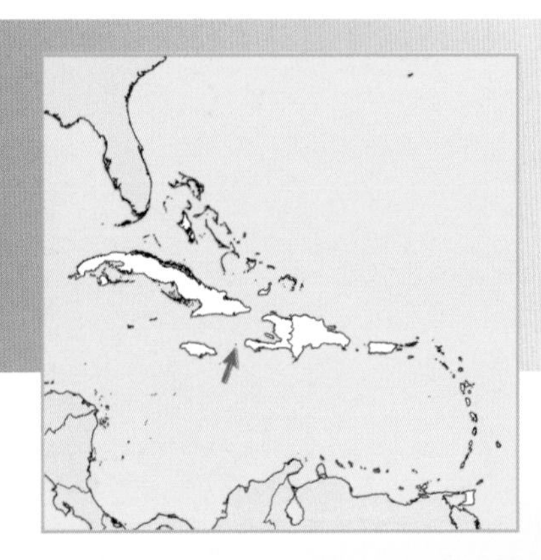

JOSEPH SCHWAGERL (U.S. FISH AND WILDLIFE SERVICE) AND
VERÓNICA ANADÓN-IRIZARRY (BIRDLIFE INTERNATIONAL)

Navassa Island Light (lighthouse) built in 1917.
(PHOTO: JEAN WIENER)

INTRODUCTION

Navassa Island is an unorganized unincorporated territory of the USA. It is grouped politically with other islands claimed under the Guano Islands Act of 1856 as one of the United States Minor Outlying Islands. Navassa is the only island of this group located in the Caribbean Sea; the others are all in the Pacific Ocean. The 500-ha island is situated 53 km south of Haiti, 136 km east of Jamaica and 152 km south of Cuba, and it rises abruptly from deep water with cliffs reaching heights of 20 m or more. There are no beaches, ports or harbours on Navassa and access (from offshore anchorages) is extremely hazardous. Dunning Hill, the highest point on the island at 76 m, is c.100 m south of the Navassa Island Light (lighthouse), which in turn is 400 m from the south-western coast or 600 m east of Lulu Bay.

The climate is tropical, as a result of which the island's primary vegetation cover is evergreen woodland/forest comprising four main tree species, namely short-leaf fig *Ficus citrifolia*, pigeon plum *Coccoloba diversifolia*, mastic *Sideroxylon foetidissimum* and poisonwood *Metopium brownei*. The island's upper plateau around Dunning Hill supports small, scattered areas of grassland which seem to be maintained by frequent fires. A second major habitat is the fan palm *Thrinax morrisii* forest that occurs in pure stands in the lower north-western part of the island although the species is also scattered throughout the upper plateau and ridges. There is also a palm *Pseudophoenix sargentii* var. *navassana* represented by only one (relatively healthy) individual which is located on the ridge to the east of the lighthouse. Snow cactus *Mammaillaria nivosa* plants are scattered across the limestone surface of the island.

Historically, Navassa has supported a herpetofauna thought to consist of eight endemic reptiles, of which *Cyclura onchiopsis, Leiocephalus eremitus, Tropidophilus bucculentus* and *Typhlops sulcatus* are presumably extinct as a result of habitat alteration during the guano-mining, human exploitation or depredation by introduced mammalian predators. The four extant species are abundant (see "Other biodiversity" below).

Navassa is critically important for its marine environment, including pristine coral reef ecosystems, which sustains foraging habitat for one of the largest colonies of Red-footed Booby *Sula sula* in the Caribbean. More than 300 marine species have been identified by biologists, including three new fish species.

Haitian fishermen and researchers camp, albeit infrequently, on the island which is otherwise uninhabited and closed to the public.

Navassa's coast has no beaches or harbours making access extremely hazardous. (PHOTO: JEAN WIENER)

■ Conservation

Navassa National Wildlife Refuge was established in 1999 by Department of the Interior Secretarial Order No. 3210. It is administered as part of the Caribbean Islands National Wildlife Refuge Complex, the headquarters for which are located in Boquerón, Puerto Rico. The refuge includes the island of Navassa and marine habitats up to 12 nautical miles (c.22 km) from the island. The U.S. Office of Insular Affairs retains authority for the island's political affairs and judicial authority is exercised directly by the nearest U.S. Circuit Court in Miami, Florida. Access to Navassa is hazardous and visitors need permission from the USFWS Office in Boquerón in order to enter refuge waters or to land on the island.

The island was discovered in 1498 by Christopher Columbus. In 1504, two Spaniards and several Indians who arrived on the island drank water contaminated with sea water and most in the group died. In 1857, Peter Duncan claimed Navassa under the Guano Islands Act as a possession of the USA for its guano deposits. Guano-mining operations were active from 1865 to 1898, and removed over 1 million tons. Navassa Phosphate Company of Baltimore built large mining facilities and railway tracks on the island, ruins of which can still be seen in Lulu Town. In 1898 the Spanish–American War forced the Phosphate Company to evacuate the island and file for bankruptcy, and thus the island was abandoned in 1901. The opening of the Panama Canal led to increased shipping in the area, and due to hazardous navigation past the island, the U.S. Lighthouse Service built the Navassa Island Light (lighthouse) in 1917. A lighthouse keeper and two assistants were assigned to live there until the U.S. Lighthouse Service installed an automatic beacon in 1929. The U.S. Coast Guard serviced the lighthouse twice each year until it was shut down in 1996 and the administration of the island was transferred to the Department of the Interior. A scientific expedition in 1998 described Navassa as a unique preserve of Caribbean biodiversity: the following year it became a national wildlife refuge to preserve and protect the biodiversity, health, heritage, and social and economic value of U.S. coral reef ecosystems and the marine environment. Scientific expeditions have continued.

The Republic of Haiti laid claim to sovereignty over Navassa in 1804, and they disputed the U.S. annexation of the island in 1857. The island has remained in the Haitian constitution since 1856. A socio-cultural assessment of the Navassa fisheries was carried out recently by Fondation pour la Protection de la Biodiversité Marine, a Haitian NGO. Haitians feel that they have been the only ones

Haitian fishermen at anchor in Lulu Bay. (PHOTO: JEAN WIENER)

Clean carcasses of Red-footed Booby at Lulu Bay. (PHOTO: JEAN WIENER)

harvesting Navassa's marine resources over the generations, and therefore feel a strong sense of ownership over the island. The fisheries at Navassa are critical to the livelihoods of fishermen in south-west Haiti, making them an important stakeholder of the island's resources. Traditionally, the fishermen used sailing and rowing boats but motorised vessels have become more readily available, shortening the travelling time to the island and increasing the size of catch they can return to Haiti with. In the recent assessment, none of the fishermen encountered were aware that Haitian or U.S. fisheries laws existed or were in effect at the island. Harsh conditions including heat, lack of food and water, difficulty landing on the island (no beach or other landing site) and remoteness have prevented its habitation by fishermen.

Local human impacts at Navassa have varied greatly over time. The former guano-mining and lighthouse maintenance operations impacted both the vegetation and marine environment. Other specific threats include fishing for shellfish, reef fish, spiny lobster, queen conch *Strombus gigas*, and the federally listed (Critically Endangered) hawksbill turtle *Eretmochelys imbricata*. Harvest of nesting and roosting seabirds, fires from transient campsites, fires for large-scale land clearance for attempted settlement, and introduced invasive alien species (goats, rats *Rattus* spp.) are also concerns. Also, fishing operations out of south-west Haiti and incursion by international trawlers purportedly from the Dominican Republic are suspected of targeting pelagic (economically valuable) fish species.

Establishing refuge management has been very difficult. In 2004, a conservation plan was developed participatively with many stakeholders to strategise ways to address the issue of fishing and difficulties of dealing with the remote refuge. This plan is being implemented with management activities including close monitoring of the artesanal fishing pressure by Haitian fishermen and other nationals entering refuge waters with fishing trawlers. The establishment of monitoring stations for research (of corals, water temperature and photography, vegetation etc.), eradication of invasive species and continued socio-cultural assessments are underway. The opening of a dialogue with the Haitian conservation community has been a necessary first step and needs to be developed to ensure the conservation of the tremendous fish and wildlife resources of Navassa Island National Wildlife Refuge.

Table 1. Key bird species at the Important Bird Area in Navassa.

Key bird species	Criteria	National population	Navassa IBA UM001
Criteria			■ ▪
Magnificent Frigatebird *Fregata magnificens*	▪	175	175
Red-footed Booby *Sula sula*	▪	5,000–7,000	5,000–7,000
White-crowned Pigeon *Patagioenas leucocephala*	■	150–300	150–300

All population figures = numbers of individuals.
Threatened birds: Near Threatened ■. **Congregatory birds** ▪.

Figure 1. Location of the Important Bird Area in Navassa.

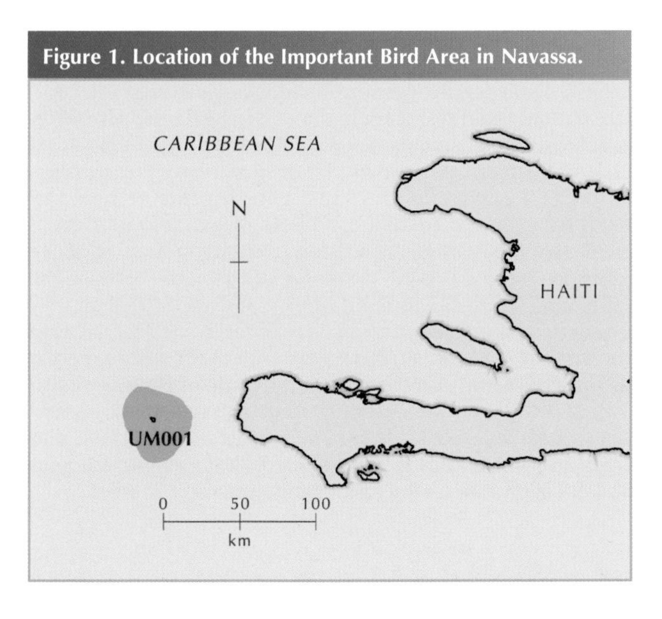

Birds

Fifty-eight (58) bird species have been recorded from Navassa including many Neotropical migratory landbirds. No restricted-range birds are known to breed on Navassa and thus, it is not included in any of the nearby Endemic Bird Areas (EBAs)—namely Cuba, Jamaica or Hispaniola. However, White-necked Crow *Corvus leucognaphalus*, a restricted-range bird from the Hispaniola EBA, and also a Vulnerable species, was seen during expeditions in 1998 and 2003, and a subspecies of Common Ground-dove *Columbina passerina navassae* is endemic to the island. Hundreds of Near Threatened White-crowned Pigeon *Patagioenas leucocephala* breed on the island.

Navassa is notable for its breeding seabirds, especially Red-footed Booby *Sula sula* and Magnificent Frigatebird *Fregata magnificens* (see Table 1). Studies of *S. sula* nestling mortality suggest a number of causal factors including adult defence of the nest, chicks being left unattended or human disturbance. Carcasses of *S. sula*, eaten clean by humans have been found under several trees and at a campsite at Lulu Bay. Birds are dislodged from their tree nests using 7-m bamboo poles. The meat is eaten during the fishermen's stays or used for commercial purposes. More exhaustive research to better estimate the *S. sula* and *F. magnificens* colonies and establish a baseline against which to monitor the populations has been scheduled for the next biological expedition to the island.

IMPORTANT BIRD AREAS

Navassa IBA (UM001) covers 148,100 ha of critical terrestrial and marine habitats that are legally protected as a national wildlife refuge. However, remoteness from USFWS administration in Puerto Rico and disputed sovereignty by Haiti have made enforcement of regulations and conservation management impractical. As a result, foreign nationals enter the refuge and harvest protected natural resources. Illegal commercial and subsistence fishing and hunting activities (including the breeding seabirds) have been documented during recent expeditions. In spite of these infringements and other threats, the significant populations of Red-footed Booby *Sula*

sula and Magnificent Frigatebird *Fregata magnificens* are thought to be have remained stable since the island became a refuge in 1999. Monitoring these seabird populations will help determine the true impact of the various threats and conservation management actions.

KEY REFERENCES

LOMBARD, C. D. (2006) Navassa Island National Wildlife Refuge Bird Observations. Boquerón, Puerto Rico: United States Fish and Wildlife Service. (Unpublished report).

MILLER, M. W. (2003). Status of reef resources of Navassa Island. Unpublished U.S. Department of Commerce – National Oceanic and Atmospheric Administration Technical Memorandum (NMFS-SEFSC-501).

MILLER, M. W., MCCLELLAN, D. B., WIENER, J. W. AND STOFFLE, B. (2007) Apparent rapid fisheries escalation at a remote Caribbean island. *Env. Conserv.* 34:92–94.

POWELL, R. (1999) Herpetology of Navassa Island, West Indies. *Carib. J. Sci.* 35: 1–13.

SILANDER, S., SCHWAGERL, J., CERVENY, K. AND GUDE, A. (2006) Navassa Island National Wildlife Refuge coordinated management strategy. Unpublished White Paper submitted to the Department of the Interior, Washington, D.C.

WIENER, J. W. (2005) Oral history and contemporary assessment of Navassa island fishermen. Port-au-Prince, Haiti: Fondation pour la Protection de la Biodiversité Marine, (Unpublished report).

YEARMAN, K. (1998) Navassa Island – a look at hell. *Bull. Illinois Geogr. Soc.* 40.

ACKNOWLEDGEMENTS

The authors would like to thank Susan Silander, Claudia Lombard (USFWS), Jean W. Wiener (Fondation pour la Protection de la Biodiversité Marine) and John Curnutt (USDA Forest Service Eastern Region) for contributing to this chapter.

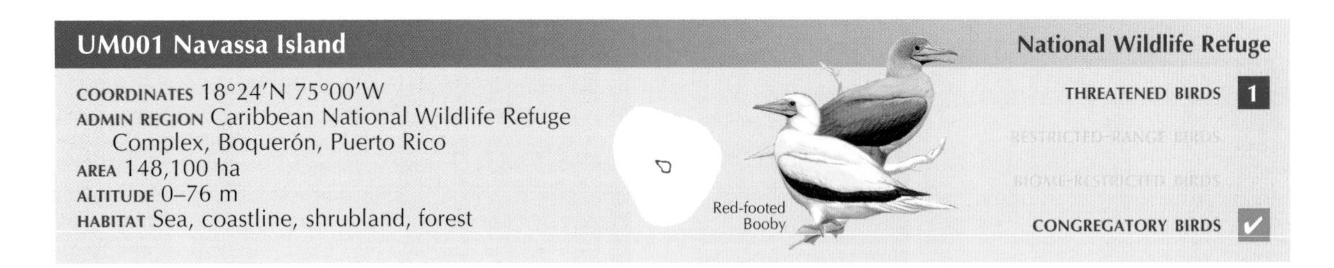

■ Site description

Navassa Island IBA is situated 53 km west of Haiti, 136 km east of Jamaica and 152 km south of Cuba. The IBA includes the 500-ha island and marine areas up to 22 km from it, thus covering 147,600 ha of open ocean. Navassa is a small, pear-shaped island plateau that rises abruptly from deep water. It is surrounded by a series of submarine coralline terraces. The karst dolomite terrain slopes from the lower north-western edge towards the south-eastern side and supports upland evergreen woodland and sparse shrubby vegetation. The island is surrounded by a submerged coral reef ecosystem and open sea.

■ Birds

This IBA is globally significant for its breeding colony of Red-footed Booby *Sula sula*, a large majority of which are immature, which is concentrated along the sheltered (leeward) north-western to southern perimeter of the island. The breeding population, conservatively estimated at 175 individuals, of Magnificent Frigatebird *Fregata magnificens* is regionally important, and small numbers of Brown Booby *Sula leucogaster* also breed on the island. Hundreds of Near Threatened White-crowned Pigeon *Patagioenas leucocephala* occur, and the Vulnerable White-necked Crow *Corvus leucognaphalus* has been recorded on the island, but is probably a transient visitor.

■ Other biodiversity

The Critically Endangered hawksbill turtle *Eretmochelys imbricata* occurs. Four endemic reptiles are abundant on the island, namely Navassa anole *Anolis longiceps*, Navassa gecko *Aristelliger cochranae*, Navassa dwarf gecko *Sphaerodactylus becki* and the Navassa galliwasp *Celestus badius*.

■ Conservation

This IBA is federally owned, and is the only oceanic and coral reef national wildlife refuge in the western Atlantic. However, it lacks on-the-ground management due to its remoteness and lack of resident personnel. There is a history of foreign nationals, primarily Haitians, entering the refuge and harvesting protected species. These illegal activities are a challenge: the island's disputed sovereignty means any move to prosecute could result in a diplomatic incident between the USA and Haiti. Socio-cultural assessments carried out by Fondation pour la Protection de la Biodiversité Marine have shown that Navassa's fisheries are critical for sustaining the fishing community on the south-western tip of Haiti. A multi-agency task force has been created to target management issues and reduce the threats to species and habitats, and USFWS will continue to monitor the island's seabird colonies and document its use by resident and migratory birds.

PUERTO RICO

LAND AREA **8,870 km²** ALTITUDE **0–1,338 m**
HUMAN POPULATION **3,944,259** CAPITAL **San Juan**
IMPORTANT BIRD AREAS **20, totalling 1,971 km²**
IMPORTANT BIRD AREA PROTECTION **10%**
BIRD SPECIES **354**
THREATENED BIRDS **6** RESTRICTED-RANGE BIRDS **23**

VERÓNICA MÉNDEZ-GALLARDO AND JOSÉ A. SALGUERO-FARÍA
(SOCIEDAD ORNITOLÓGICA PUERTORRIQUEÑA, INC.[1])

Mona Island IBA, off the coast of western Puerto Rico—important for its seabirds and Endangered Yellow-shouldered Blackbird population. (PHOTO: ENRIQUE A. SILVA RODRIGUEZ)

INTRODUCTION

Puerto Rico is the smallest and most easterly of the Greater Antilles lying 114 km east of the Dominican Republic and just 60 km west of the Virgin Islands (to USA). It is a Commonwealth and territory of the USA. Puerto Rico is a small archipelago of islands and cays such as Vieques, Culebra, Mona, Monito, Desecheo, Caja de Muertos amongst others. Its geographical location and geological history has had a profound influence on the rich diversity of its flora and fauna. It can be divided into three geomorphologic regions of the central mountainous interior, the karst, and coastal plains. Nearly 85% of the country lies below 500 m and less than 1% is above 1,000 m, the highest point being Cerro Punta, at 1,338 m above sea level. The mountainous region acts as a barrier to the moisture-rich trade winds which unload most of their humidity in the form of rain on the windward side of the mountains. Thus, the north and east is relatively wet (averaging 1,550 mm per year), whereas the south is fairly dry (910 mm annually). The west receives the remainder of the moisture from the trade winds and the humidity-soaked winds from the south. The windward side of the Luquillo Mountains may receive more than 5,000 mm of rain annually. Rainfall is also associated with tropical low-pressure systems which form or pass through the region between June and November (which is the period of highest rainfall).

Forests in Puerto Rico are considered to be subtropical and can be classified into 10 types depending on a combination of temperature, elevation and substrate: dry coastal limestone forests; dry coastal volcanic forests; dry and moist alluvial forests; dry forest; moist and wet forests on igneous rock; moist coastal forests on sandy bedrock; moist and wet forest on limestone; lowland moist forest on volcanic bedrock; montane moist forests on volcanic and granitic bedrock; montane wet forests on volcanic bedrock; montane wet and rain forests on granitic bedrock; and rainforests on volcanic bedrock. Puerto Rico's flora includes c.3,130 species with nearly 9% (240 spp.) endemism. Naturalised exotic species such as *Albizia procera*, *Spathodea campanulata* and *Bambusa vulgaris*, make up a high percentage of forest species, and represent an important component of the "new" forested landscape of Puerto Rico. In addition to forested areas, Puerto Rico also has diverse wetlands, including forested wetlands, such as mangroves, *Pterocarpus officinalis* and *Annona glabra* freshwater swamps,

1 The Sociedad Ornitológica Puertorriqueña, Inc. (Puerto Rican Ornithological Society, BirdLife in Puerto Rico) is a national NGO and is referred to throughout this chapter under the acronym SOPI.

brackish and freshwater herbaceous wetlands and hypersaline saltflats. Major wetland areas include Caño Tiburones, Laguna Tortuguero, Laguna Cartagena, San Juan and Jobos bays estuaries and Cabo Rojo saltflats.

With a c.3.9 million people (429 people/km^2) Puerto Rico has the highest population density in the Caribbean. Between the sixteenth to eighteenth centuries the population remained relatively small and stable, but after this is grew exponentially. This increase peaked in the second half of the twentieth century when the growth rate reach 72%. This growth rate and resultant high-density population has had a profound and detrimental affect on the island's natural resources.

■ Conservation

Approximately 60,800 ha (about 6.8% of the country's land area), have been designated by the government as conservation areas in Puerto Rico, but only 31,055 ha (less than 3.5%) have been acquired. The remainder are still in private or public ownership and thus threatened by development. Only land acquisition specifically for protection guarantees these areas' final conservation. Various state and federal agencies, as well as private institutions, manage these protected areas: the Department of Natural and Environmental Resources (DNER[2]) is responsible for more than 90% of the total; and the United States Fish and Wildlife Service (USFWS[3]), United States Forest Service (USFS) and Conservation Trust of Puerto Rico (CTPR) manages the rest. In addition, other public agencies such as the United States Geological Survey (USGS) and the Natural Resources Conservation Service (NRCS) are involved in protecting water and soil resources through several different programs. The National Oceanic and Atmospheric Administration (NOAA) are also involved in management and protection of marine resources.

El Yunque IBA in the Sierra de Luquillo, home to many restricted-range species including the Critically Endangered Puerto Rican Amazon and Vulnerable Elfin Woods Warbler. (PHOTO: SOPI)

These agencies also fund and coordinate research projects, many of which are geared toward avian studies. Locally, the DNER monitors a number of bird species, implements recovery plans, and several initiatives such as the Puerto Rico Critical Wildlife Areas (PRCWA), the Atlantic Coast Joint Venture (ACJV) through the Puerto Rico Waterfowl Focus Areas (PRWFA), the Puerto Rico Gap Analysis Project (PRGAP), and the Fisheries and Wildlife Management Plan (FWMP). In addition, all research projects in the Commonwealth must have a DNER permit. The PRGAP is a joint initiative of the DNER, the U.S. Forest Service International Institute of Tropical Forestry (IITF), the North Carolina Cooperative Fish and Wildlife Research Unit (NCSU), and the U.S. Geological Survey (USGS) Biological Resources Division. At present the DNER is working on the first "Programmatic Safe Harbor Agreement" with the USFWS for the conservation of Plain Pigeon *Patagioenas inornata*. The USFWS conducts surveys of national wildlife refuges throughout the Commonwealth where management usually centres on avian resources. The Caribbean Field Office of the USFWS coordinates funding for research, establishment of management initiatives for endangered bird species, and initiatives with landowners as partners of the Fish and Wildlife Program. The U.S. Forest Service (USFS) contributes to the conservation and protection of native birds through two divisions, the El Yunque National Forest (El Yunque) and the IITF.

At El Yunque, the USFS is involved in managing and studying the resident Puerto Rican Amazon *Amazona vittata* population (until recently, the only wild extant population) in coordination with the USFWS and the DNER. The IITF conducts studies on forest birds and the implication of forestry management practices on their populations. The USGS Biological Resources Division, which collaborates with the PRGAP, has also managed the Puerto Rico Breeding Bird Survey since 1997.

2 The Department of Natural and Environmental Resources is referred to throughout this chapter by the acronym DNER.
3 The United States Fish and Wildlife Service is referred to throughout this chapter by the acronym USFWS.

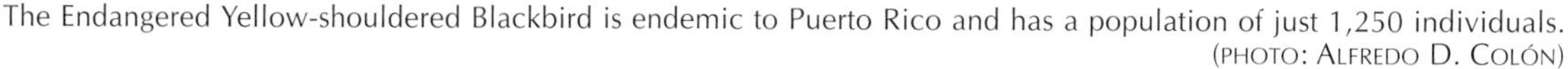
The Endangered Yellow-shouldered Blackbird is endemic to Puerto Rico and has a population of just 1,250 individuals. (PHOTO: ALFREDO D. COLÓN)

Puerto Rican Nightjar, Critically Endangered and the focus of research and conservation action by SOPI and other institutions. (PHOTO: MICHAEL J. MOREL)

For many years, universities have been an important source of avian scientific research in Puerto Rico. The University of Puerto Rico's campuses at Rio Piedras, Mayagüez, Humacao and Cayey have contributed to our knowledge of local avifauna for over 30 years. The Humacao campus was home to a captive breeding program for *Patagioenas inornata* until 2000, and important research on threatened taxa such as Yellow-shouldered Blackbird *Agelaius xanthomus*, Broad-winged Hawk *Buteo platypterus brunnescens*, Sharp-shinned Hawk *Accipiter striatus venator* and Elfin-woods Warbler *Dendroica angelae* have been implemented by the UPR-Mayagüez. The North Carolina State University's Cooperative Fish and Wildlife Research Unit (NCSU) and Mississippi State University's Cooperative Fish and Wildlife Research Unit (MSU) have also studied native avifauna.

The Sociedad Ornitológica Puertorriqueña, Inc. (SOPI, BirdLife in Puerto Rico) has been coordinating bird-focused conservation efforts, including education, research, land stewardship, and the description of the island's IBAs. Other SOPI projects include the Puerto Rico Shorebird Monitoring Network (Programa Red Limícola) which has volunteers surveying important shorebird sites; the Puerto Rico Breeding Bird Atlas (coordinated by the Puerto Rico Gap Analysis Project), the first atlas project in Latin America; and eBird, in collaboration with the Cornell Lab of Ornithology and the Conservation Trust of Puerto Rico (CTPR). SOPI is the BirdLife Species Guardian for the Critically Endangered Puerto Rican Nightjar *Caprimulgus noctitherus* and, as a result of funding from the British Birdwatching Fair, will implement research to provide additional information about the species' distribution and new breeding populations, and facilitating conservation actions. SOPI has also contributed with bird data, educational activities throughout the island and supported conservation efforts implemented by local community groups. It has also collaborated in bird conservation projects with DNER, USFWS, USDA and the Society for the Conservation and Study of Caribbean Birds (SCSCB). Among the initiatives with SCSCB, are the West Indian Whistling-Duck and Wetlands Conservation Project and the Caribbean Endemic Bird Festival. Volunteers coordinate and participate in four Christmas Bird Counts sponsored by National Audubon Society (BirdLife in the US) at Fajardo, San Juan, Arecibo and Cabo Rojo. Counts have been carried out for more than 30 years at the latter site.

The island has numerous laws and regulations to protect the island's precious natural resources. Among the most important are the New Wildlife Law (1999) and its associated regulations: Regulation for Management of Vulnerable and Endangered Species (2004) and Regulation for the Management of Wildlife, Exotic Species and Hunting in the Commonwealth of Puerto Rico (2004). Also, the DNER has been responsible for protecting the coastal littoral zone (areas under tidal influence) since its creation in 1972. As a US territory, Puerto Rico has to abide by federal laws such as the Endangered Species Act (1972), Migratory Bird Treaty Act (1918) and the Clean Water Act (1972). However, although these laws and regulations can be very strict they are rarely enforced properly due to lack of resources and government bureaucracy. Local and federal laws provide for private entities to set aside part of their land as conservation easements. The Conservation Trust of Puerto Rico (CTPR) has been working with this provision, encouraging and providing guidance to landowners willing to take advantage of the tax benefits associated with the conservation easements.

Conservation concerns on the island include habitat loss, lack of environmental education, the introduction of exotic species, illegal hunting and natural events. Habitat loss is the main threat for birds and other wildlife. Since the late twentieth century, there has been a major increase in urban expansion, with housing projects being built at a fast pace. Agricultural areas are being converted into housing projects and forested land is cleared for agriculture as well as for development. Major roads are being constructed through the island leading to more development and increased strain on natural resources, especially water. Forest fragmentation is a major concern as it promotes negative impacts from exotic species, diseases, illegal hunting, intentional fires, filling of wetlands and pollution. Ocean resources have also been impacted, reducing fish stocks and therefore affecting resident marine species. Puerto Ricans are, for the most part, unaware of their local flora and fauna. The lack of proper environmental education limits the capacity of environmentalists and land managers to conserve the natural resources of the island. The introduction of exotic species is a major concern as laws and regulations are in place but these are not enforced properly and illegal trafficking continues. In addition, some native bird species are trapped and exported, a fact that very few people are aware of. Hunting is strictly regulated and permitted only for certain species during the year, but enforcement is lacking in most areas. The number of non-licensed hunters exceeds that of the legal hunters, demonstrating the scale of the illegal hunting problem on the island. Puerto Rico lies on the path of many tropical storms and hurricanes. Since 1989, the island has endured two major hurricanes and a strong tropical storm. In 1989, Hurricane Hugo demonstrated the destructive force that these tropical systems may have on local endangered bird species, wiping out nearly half the wild population of Puerto Rican Amazon *Amazona vittata*.

Conservation needs for Puerto Rico are: land protection, education, biological information, enforcement and planning. Even though several laws exist to protect natural resources, land acquisition is the most effective way to protect important areas for birds and other wildlife. Money is a severe constraint but several NGOs and the DNER have stepped in and have been very active in finding funds for land acquisition. A major conservation need is the development of a strong educational program for land managers and their staff, environmental law enforcement personnel, lawmakers and the general public. Recent initiatives by SOPI, the SCSCB and the USFWS are raising public awareness on the importance of bird conservation. Most available information related to the ecological requirements of species is incomplete and often outdated (being based on data gathered 20–30 years ago). Efforts are being made to increase our knowledge, but information from most initiatives is not publicly available. Formerly, information was gathered mainly by the DNER and the USFWS but these agencies have since reorganised or changed their priorities. Current population trends in several species are very different from what had been published decades ago and many common species are now showing apparent declines. Also, previously unknown breeding species do not receive any conservation consideration even when immediate protection is warranted. A recent generation of biologists and an increasing group of volunteers have begun to fill the gaps through various projects including the Puerto Rico Gap Analysis Project and the Breeding Bird Atlas. The local

Table 1. Key bird species at Important Bird Areas in Puerto Rico.

Key bird species	Criteria	National population		PR001	PR002	PR003	PR004	PR005	PR006
			Criteria	■ ■	 ■ ■	■ ■	■ ■ ■	■ ■	■ ■
West Indian Whistling-duck *Dendrocygna arborea*	VU ■	60–90					30		
Masked Duck *Nomonyx dominicus*	■	90–150					50–249		
White-tailed Tropicbird *Phaethon lepturus*	■	450–750			237				
Magnificent Frigatebird *Fregata magnificens*	■	1,500–2,200		600–900					
Brown Pelican *Pelecanus occidentalis*	■	180–450							
Masked Booby *Sula dactylatra*	■	225–750		315					
Red-footed Booby *Sula sula*	■	3,000–9,000		3,600–7,200					
Brown Booby *Sula leucogaster*	■	2,400–4,500		1,800					
Black-necked Stilt *Himantopus mexicanus*	■								
Black-bellied Plover *Pluvialis squatarola*	■								
Semipalmated Plover *Charadrius semipalmatus*	■								
Wilson's Plover *Charadrius wilsonia*	■	300–600							
Snowy Plover *Charadrius alexandrinus*	■	30–60							
Short-billed Dowitcher *Limnodromus griseus*	■								
Greater Yellowlegs *Tringa melanoleuca*	■								
Lesser Yellowlegs *Tringa flavipes*	■								
Ruddy Turnstone *Arenaria interpres*	■								
Semipalmated Sandpiper *Calidris pusilla*	■								
Western Sandpiper *Calidris mauri*	■	708							
Least Sandpiper *Calidris minutilla*	■	723							
White-rumped Sandpiper *Calidris fuscicollis*	■	147							
Pectoral Sandpiper *Calidris melanotos*	■	168							
Stilt Sandpiper *Calidris himantopus*	■								
Royal Tern *Sterna maxima*	■	90–375							
Sandwich Tern *Sterna sandvicensis*	■	2,250–3,300							
Roseate Tern *Sterna dougallii*	■	1,800–2,400					250		
Least Tern *Sterna antillarum*	■	225–360							
Sooty Tern *Sterna fuscata*	■	150,000–180,000							
Brown Noddy *Anous stolidus*	■	3,000–6,000							
Plain Pigeon *Patagioenas inornata*	NT ■								
Bridled Quail-dove *Geotrygon mystacea*	■								
Puerto Rican Amazon *Amazona vittata*	CR ■ ■	51				25–27			
Puerto Rican Lizard-cuckoo *Coccyzus vieilloti*	■				✓	✓	✓	✓	
Puerto Rican Screech-owl *Megascops nudipes*	■				✓	✓	✓	✓	✓
Puerto Rican Nightjar *Caprimulgus noctitherus*	CR ■ ■	1,500–2,000						177	
Antillean Mango *Anthracothorax dominicus*	■				✓	✓	✓	✓	✓
Green Mango *Anthracothorax viridis*	■				✓	✓		✓	✓
Green-throated Carib *Eulampis holosericeus*	■					✓	✓		
Antillean Crested Hummingbird *Orthorhyncus cristatus*	■								
Puerto Rican Emerald *Chlorostilbon maugaeus*	■				✓	✓	✓	✓	✓
Puerto Rican Tody *Todus mexicanus*	■				✓	✓	✓	✓	✓
Puerto Rican Woodpecker *Melanerpes portoricensis*	■				✓	✓	✓	✓	✓
Lesser Antillean Pewee *Contopus latirostris*	■				✓	✓		✓	✓
Caribbean Elaenia *Elaenia martinica*	■								✓
Puerto Rican Flycatcher *Myiarchus antillarum*	■				✓	✓	✓	✓	✓
Puerto Rican Vireo *Vireo latimeri*	■				✓	✓	✓	✓	✓
Pearly-eyed Thrasher *Margarops fuscatus*	■				✓	✓	✓	✓	✓
Adelaide's Warbler *Dendroica adelaidae*	■				✓	✓	✓	✓	✓
Elfin-woods Warbler *Dendroica angelae*	VU ■ ■	1,830						250–999	
Yellow-shouldered Blackbird *Agelaius xanthomus*	EN ■ ■	1,250		260					<50
Puerto Rican Bullfinch *Loxigilla portoricensis*	■				✓	✓	✓	✓	✓
Puerto Rican Tanager *Nesospingus speculiferus*	■					✓		✓	
Puerto Rican Spindalis *Spindalis portoricensis*	■				✓	✓	✓	✓	✓
Antillean Euphonia *Euphonia musica*	■				✓	✓	✓	✓	✓

All population figures = numbers of individuals.
Threatened birds: Critically Endangered ■; Endangered ■; Vulnerable ■; Near Threatened ■. **Restricted-range birds** ■. **Congregatory birds** ■.

Puerto Rico IBAs												
008	PR009	PR010	PR011	PR012	PR013	PR014	PR015	PR016	PR017	PR018	PR019	PR020
■	■			■	■	■	■	■	■			
■	■	■	■	■	■	■	■	■	■	■		■
■	■		■		■						■	■
				35–40					35–70			
120												300
											225–600	
891					500							
213					382							
381					516							
294					93							
266												
					547							
					110							
000					1,059							
315					173							
372					5,506							
708					5,506							
723					3,666							
147												
168												
895					4,385							
											150–465	
395												
650	350											
			81–89									
											75–105,000	
											1,440–1,860	
						200	50–249					
												✓
								26–30				
✓	✓	✓				✓	✓	✓	✓	✓		
✓	✓	✓		✓	✓	✓	✓	✓	✓	✓		✓
✓	347											
✓	✓	✓	✓	✓	✓	✓	✓	✓	✓	✓		
✓	✓	✓	✓			✓	✓	100				
	✓		✓	✓	✓	✓		✓	✓	✓		✓
	✓				✓	✓		✓	✓	✓		✓
✓	✓	✓	✓	✓	✓	✓	✓	104		✓		
✓	✓	✓	✓		✓	✓	✓	50–130	✓	✓		
✓	✓	✓	✓	✓	✓	✓	✓	✓	✓	✓		✓
✓	✓	✓	✓			✓	✓			✓		
✓	✓				✓					✓		✓
✓	✓	✓	✓	✓	✓	✓	✓	8	✓	✓		✓
✓	88	✓	✓	✓	✓	✓	✓	✓				
✓	✓	✓	✓	✓	✓	✓	✓	28	✓	✓		✓
✓	✓	✓	✓		✓				✓	✓		✓
								414				
994					59							
✓	26	✓	✓		✓	✓	✓	<16	✓	✓		
		✓				✓	✓	40				
✓	✓	✓	✓	✓	✓	✓	✓	✓	✓	✓		
✓	✓	✓	✓		✓	✓	✓	✓	✓	✓		

Punta Ventanas in the Karso del Sur IBA: effective conservation of this important Puerto Rican Nightjar area is threatened by a wind energy project. (PHOTOS: F. GONZÁLEZ)

legislature must be pressured to assign more funds and resources instead of creating more regulations. This could be achieved by a strong educational campaign and the help of a "Conservation Champion", a charismatic person who could summon public interest and force politicians to respond accordingly. A significant factor hampering Puerto Rico's conservation efforts is lack of proper land-use planning which has been, until recently, implemented at very small scales, usually for individual projects or municipalities. There is a clear need for an island-wide land-use framework. The Commonwealth government recently presented such a land-use plan (for Puerto Rico) but developers, conservationists and government officials have not been able to work out their different opinions and finalisation of the plan seems to have lost momentum.

■ Birds

Of Puerto Rico's 354 recorded bird species, c.133 are known to breed and over 200 species occur as wintering Neotropical migrants, transients or vagrants. More than 45 exotic bird species have been reported in Puerto Rico and more than 35 are either well-established or have small breeding populations. The geographical position of Puerto Rico integrates a Greater Antillean and Lesser Antillean avifauna like no other island in the region. A number of the restricted-range species are shared with neighbouring islands such as Antillean Mango *Anthracothorax dominicus*, Lesser Antillean Pewee *Contopus latirostris* and Antillean Euphonia *Euphonia musica*. Of the 27 species within the Puerto Rico and Virgin Islands Endemic Bird Area (EBA), 23 are covered by the Puerto Rican IBA network. Although 27 species are known from Puerto Rico, two are now considered extinct: the White-necked Crow *Corvus leucognaphalus* and the Hispaniolan Parakeet *Aratinga chloroptera*. However, *A. chloroptera* has recently been considered as an introduced species (albeit a different subspecies), as individuals have been seen in various parts of the island.

A total of 16 breeding species are confined to the island, primarily to natural forest and woodlands, one of which—*Nesospingus*—represents a monotypic endemic genus. Waterbirds are an important component of the local avifauna representing more than 35% of the species recorded for the Commonwealth: 45 species are known to breed on the island.

Over the last 120 years a number of taxa have become extinct in Puerto Rico, namely: Limpkin *Aramus guarauna*, *Corvus leucognaphalus*, Culebra Island Parrot *Amazona vittata gracipiles* and *Aratinga chloroptera maugei*. Puerto Rican endemics that are heading in the same direction include the Critically Endangered Puerto Rican Amazon *Amazona vittata* and Puerto Rican Nightjar *Caprimulgus noctitherus*. The wild population of *A. vittata* is estimated at c.26–30 individuals in El Yunque National Forest (IBA PR016) and 25–27 in the Rio Abajo State Forest (IBA PR003). The *Caprimulgus noctitherus* population is estimated to be 1,500–2,000 individuals, confined to the drier forests of south-western Puerto Rico. Other globally threatened species (the threat categories and national populations for which are presented in Table 1) include the endemic Yellow-shouldered Blackbird *Agelaius xanthomus* (with a population of 1,250) and the Elfin-woods Warbler *Dendroica angelae* (with 1,830 individuals). At the island-level, various species are considered by the DNER and USFWS as threatened, namely: Snowy Plover *Charadrius alexandrinus* (10–20 breeding pairs); Masked Duck

Puerto Rican Woodpecker, endemic to the island and Puerto Rico's national bird. (PHOTO: ALFREDO D. COLÓN)

Punta Soldado in the Culebra IBA—one of Puerto Rico's major seabird sites. (PHOTOS: F. GONZÁLEZ)

Nomonyx dominicus (30–50 breeding pairs); Brown Pelican *Pelecanus occidentalis* (60–150 breeding pairs); Ruddy Duck *Oxyura jamaicensis* (200–600 breeding pairs); White-cheeked Pintail *Anas bahamensis* (200–600 breeding pairs); Caribbean Coot *Fulica caribaea* (200–400 breeding pairs); Roseate Tern *Sterna dougallii* (200–800 breeding pairs); and the migratory Piping Plover *Charadrius melodus*. Other species such as Least Tern *Sterna antillarum*, Least Grebe *Tachybaptus dominicus*, Wilson's Plover *Charadrius wilsonia*, American Oystercatcher *Haematopus palliatus*, and Willet *Catoptrophorus semipalmatus* should be considered locally as species of concern and may warrant higher conservation status. Two subspecies of birds are locally classified as critically endangered: Puerto Rican Sharp-shinned Hawk *Accipiter striatus venator* and Puerto Rican Broad-winged Hawk *Buteo platypterus brunnescens*.

IMPORTANT BIRD AREAS

Puerto Rico's 20 IBAs—the island's international site priorities for bird conservation—cover 1,969 km² including marine areas, about 22% of Puerto Rico's land area. Many of the terrestrial IBAs overlap to some extent with protected areas such as state forests, natural reserves and wildlife refuges, thus some form of protection is in place. However, only 10% of the area covered by the IBAs is under formal protection. The IBAs have been identified on the basis of 52 key bird species (listed in Table 1) that variously meet the IBA criteria. These 52 species include six globally threatened birds, all 23 (extant) restricted-range species, and 28 congregatory waterbirds/seabirds. The restricted-range species are well represented within the IBA network which also covers all the main populations of the globally threatened birds.

The IBA program in Puerto Rico is an initiative of the Sociedad Ornitológica Puertorriqueña, Inc. (SOPI). The program started initially in 2002 with visits (led by SOPI volunteers) to "potential" IBAs to document the presence of key bird species. A committee was established to evaluate the field survey and site information, and in June 2006 a national IBA workshop, in which state and federal personnel and NGO members participated, was held. Participants had the opportunity to review the information, discuss proposals and revise the criteria of the selected areas. The initial list of 26 proposed IBAs was consolidated to 18 selected sites. An additional two IBAs were proposed and have been included in the final inventory of 20 IBAs which embrace protected areas and privately owned lands. These 20 IBAs represent a critical network covering the full diversity of habitats for Puerto Rico's avifauna.

SOPI is continuing to implement conservation actions at IBAs through the provision of assistance to IBA Site Support Groups; monitoring IBAs through the Shorebird Monitoring Network, Christmas Bird Counts, and eBird; and collaboration with state and federal agencies. The results from the shorebird, threatened bird and seabird monitoring undertaken by SOPI, DNER, USFWS and others should be used to inform the annual assessment of state, pressure and response scores at each IBA in order to provide an objective status assessment and highlight management interventions

Cabo Rojo Salt Flats (Suroeste IBA) with the Sierra Bermeja and Laguna Cartagena IBA in the distance: two areas where SOPI is implementing collaborative conservation projects. (PHOTOS: F. GONZÁLEZ)

Figure 1. Location of Important Bird Areas in Puerto Rico.

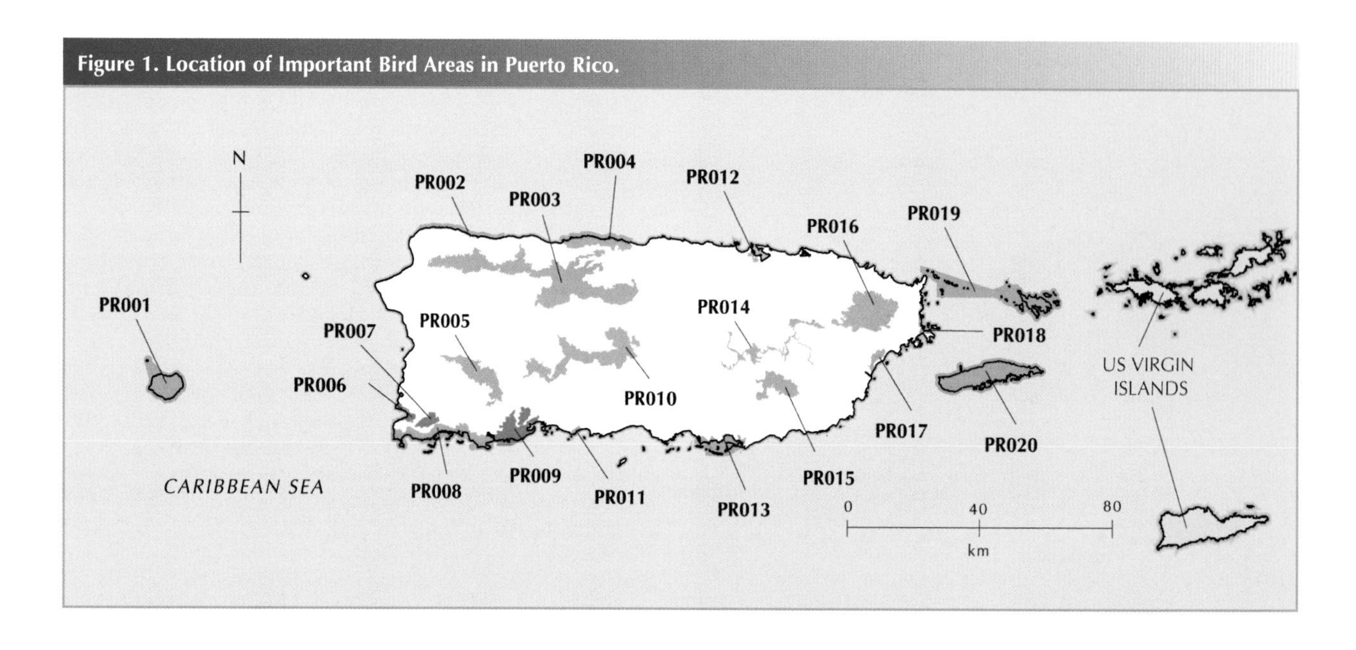

that might be required to maintain these internationally important biodiversity sites.

KEY REFERENCES

AUKEMA, J. E., CARLO, T. A., TOSSAS, A. G. AND ANADÓN-IRIZARRY, V. (2006) *A call to protect Sierra Bermeja for future generations*. San Juan, Puerto Rico: Sociedad Ornitológica Puertorriqueña, Inc.

BANCHS, E. (2006) Censos en las Salinas de Punta Cucharas. Ponce, Puerto Rico. (Unpublished data).

BONILLA, G., VÁZQUEZ, M. AND PÉREZ, E. (1992) Estatus, estimado poblacional y distribución de cuatro aves acuáticas nativas en Puerto Rico. Pp.135–148 in E. L. Cardona, ed. *XVIII Simposio de los Recursos Naturales*. Puerta de Tierra, Puerto Rico: Departamento de Recursos Naturales y Ambientales.

COLLAZO, J., HARRINGTON, B., GEAR, J. AND COLÓN, J. (1995) Abundance and distribution of shorebirds at the Cabo Rojo Salt Flats, Puerto Rico. *J. Field Orn.* 66: 424–438.

COLÓN, S. (2004) Aves de la zona del Caño Tiburones y Bosque de Cambalache. San Juan: Sociedad Ornitológica Puertorriqueña, Inc. (Unpublished data).

DÍAZ, R. AND PÉREZ, R. (1989) Assessment of the West Indian Whistling Duck *Dendrocygna arborea* in eastern Puerto Rico. San Juan: Department of Natural and Environmental Resources. (Unpublished report).

GARCÍA, M., CRUZ, J., VENTOSA, E. AND LÓPEZ, R. (2005) *Puerto Rico comprehensive wildlife conservation strategy*. San Juan: Department of Natural and Environmental Resources.

JOGLAR, R. L. EDS. (2005) *Biodiversidad de Puerto Rico: vertebrados terrestres y ecosistemas*. San Juan: Editorial del Instituto de Cultura Puertorriqueña.

LÓPEZ, T. AND VILLANUEVA, N. (2006) *Atlas ambiental de Puerto Rico*. San Juan: La Editorial Universidad de Puerto Rico.

MIRANDA, L., PUENTE, A. AND VEGA, S. (2000) First list of the vertebrates of Los Tres Picachos State Forest, Puerto Rico, with data on relative abundance and altitudinal distribution. *Carib. J. Sci.* 36: 117–126.

MORALES, A. (2002) Lista de aves observadas en los Acantilados de Guajacata, Quebradillas, Puerto Rico. San Juan: Sociedad Ornitológica Puertorriqueña, Inc. (Unpublished report).

RAFFAELE, H. A. (1989) *A guide to the birds of Puerto Rico and the Virgin Islands* (Revised edition). Princeton, New Jersey: Princeton University Press.

RIVERA, M. (1997) *Puerto Rican Broad-winged Hawk and Puerto Rican Sharp-shinned Hawk* Buteo platypterus brunnescens *and* Accipiter striatus venator *Recovery Plan*. Cabo Rojo, Puerto Rico: US Fish and Wildlife Service (Boquerón Field Office).

SALGUERO, J. (2006) Censos en Bahía de Jobos, Salinas. San Juan: Sociedad Ornitológica Puertorriqueña, Inc. (Unpublished Programa Red Limícola data).

SCHREIBER, E. A. AND LEE, D. S. EDS. (2000) *Status and conservation of West Indian Seabirds*. Ruston, USA: Society of Caribbean Ornithology (Spec. Publ. 1).

SOPI (2008) Puerto Rico breeding bird atlas. San Juan: Sociedad Ornitológica Puertorriqueña, Inc. (Unpublished report).

SORRIÉ, B. A. (1975) Observations on the birds of Vieques Island, Puerto Rico. *Carib. J. Science* 15: 89–103.

US FISH AND WILDLIFE SERVICE (2006) Environmental Assessment: reintroduction of the Puerto Rican Parrot to Rio Abajo Commonwealth Forest, Puerto Rico. Río Grande, Puerto Rico: US Fish and Wildlife Service (Río Grande Field Office).

US FISH AND WILDLIFE SERVICE (2007) Refugio Nacional de Vida Silvestre de Vieques: plan abarcador de conservación y declaración de impacto ambiental. Vieques, Puerto Rico: US Fish and Wildlife Service.

VENTOSA, E., CAMACHO, M., CHABERT, J., SUSTACHE, J. AND DÁVILA, D. (2005) *Puerto Rico critical wildlife areas*. San Juan: Department of Natural and Environmental Resources.

VENTOSA, E., CAMACHO, M., CHABERT, J., SUSTACHE, J. AND DÁVILA, D. (2005) *Puerto Rico waterfowl focus areas*. San Juan: Department of Natural and Environmental Resources.

VILELLA, F. J. AND ZWANK, P. J. (1993) Geographic distribution and abundance of the Puerto Rican Nightjar. *J. Field Orn.* 64: 233–238.

WAIDE, R. B. (1995) Status and conservation of the Elfin-woods Warbler *Dendroica angelae* in the Luquillo Experimental Forest. (Unpublished final report submitted to the US Fish and Wildlife Service).

WUNDERLE, J. M., WAIDE, R. B. AND FERNÁNDEZ, J. (1988) Seasonal abundance of shorebirds in the Jobos Bay estuary in southern Puerto Rico. *J. Field Orn.* 60: 329–339.

ACKNOWLEDGEMENTS

The authors would like to thank Sergio Colón for contributions including revision of bird lists and edition of IBA profiles; Javier Mercado and Jorge Saliva for their contributions and revision of bird lists; Joel Mercado for the preparation of the maps; José Colón for editing IBA profiles; DNER, USFWS, USFS, USDA and CTPR personnel for their collaboration and assistance; SOPI's past IBA program coordinators—Adrianne Tossas and Verónica Anadón-Irizarry—for establishing and building the program; and SOPI's past and present Board, members and volunteers for all their help, enthusiasm and dedication to the conservation of Puerto Rican birds.

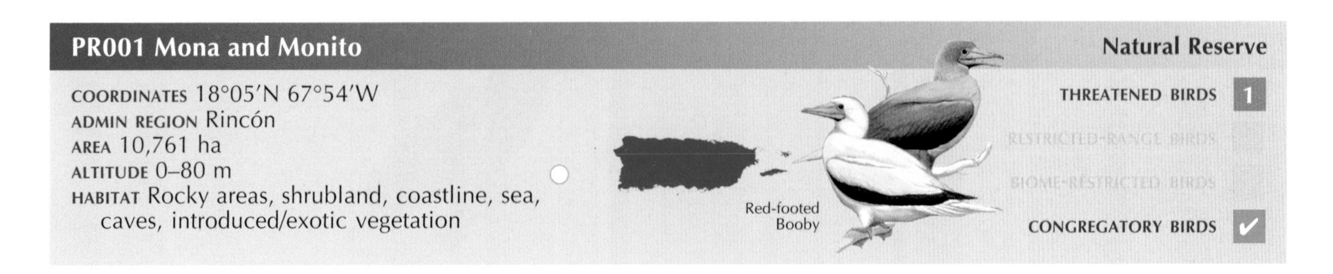

■ Site description

Mona and Monito IBA is situated c.66 km south-west of Mayagüez, Puerto Rico and 61 km south-east of Punta Espada, Dominican Republic, in the centre of the Mona Passage. Mona is a 5,700-ha island composed of emergent coral rock plateaus surrounded by sea cliffs and beaches, with caves found throughout. Monito is a 14-ha islet lying 5 km north-west of Mona, and is very difficult to access. The islands are extremely dry, and support four types of subtropical dry forest: cacti scrub, highland forest, coastal forest, and central lowland forest. The islands are uninhabited (except for guards and visitors) and their isolated location has limited human interventions.

■ Birds

This IBA is home to large numbers of seabirds including regionally significant breeding populations of Magnificent Frigatebird *Fregata magnificens*, Red-footed Booby *Sula sula*, Masked Booby *S. dactylatra* and Brown Booby *S. leucogaster*. Other seabirds observed in the IBA include Audubon's Shearwater *Puffinus lherminieri* and Sooty Tern *Sterna fuscata*. A subspecies of the Endangered Yellow-shouldered Blackbird *Agelaius xanthomus monensis* is endemic to the IBA, with an estimated population of 260 individuals.

■ Other biodiversity

Mona Island supports a number of endemic taxa including the Endangered Mona blind snake *Typhlops monensis*, and the Vulnerable Mona island rock iguana *Cyclura cornuta stejnegeri* and Mona coqui *Eleutherodactylus monensis*. The Mona dwarf gecko *Sphaerodactylus monensis* and a number of reptile subspecies are also endemic. Four species of globally threatened (Critically Endangered and Endangered) sea-turtles nest on Mona Island's beaches. The Endangered Monito gecko *Sphaerodactylus micropithecus* is endemic to Monito Island.

■ Conservation

Mona and Monito islands IBA is state owned and legally protected as a natural reserve. It is mainly used for conservation, although hunting, fishing, scuba-diving, and nature tourism activities are allowed to a limited degree. Feral pigs and goats are present on the islands and, in an attempt to control these populations, access is permitted to hunters during a prescribed season. The DNER has undertaken eradication programs for black rat *Rattus rattus* on Monito Island. The NGO Amigos de Amoná, Inc. promotes the sustainable-use of these islands' resources. Threats include water pollution, invasive species, overfishing, and human disturbance.

■ Site description

The Acantilados del Noroeste (north-west cliffs) IBA covers the coastline from the municipality of Aguadilla to Camuy in north-west Puerto Rico, and includes the area where the Guajataca river flows into the sea at Quebradillas. It comprises the karstic fringe of the island's northern karst zone and has sea cliffs, canyons, caves and coastal sand-dunes. The cliffs provide protection for nesting seabirds. The area is subject to strong winds and waves. The IBA includes marine areas up to 1 km from the shoreline.

■ Birds

The coastal cliffs of this IBA are home to a regionally significant breeding population (237 birds) of the White-tailed Tropicbird *Phaethon lepturus*. The IBA also supports populations of 15 (of the 23) Puerto Rico and the Virgin Islands EBA restricted-range species (11 of which are island endemics) which inhabit the coastal forests adjacent to the rocky areas and dunes. The Endangered Yellow-shouldered Blackbird *Agelaius xanthomus* has been seen in this IBA, but it does not breed.

■ Other biodiversity

The endemic butterfly *Atlantea tulita* inhabits this IBA. Some shrubs present are the pricklybush *Oplonia spinosa*, rosewood *Drypetes ilicifolia* and the endemic *Manilkara pleeana*. The endemic shrub *Ottoschulzia rhodoxylon*, classified as Endangered by USFWS and Critically Endangered by DNER, is also found at this IBA.

■ Conservation

Acantilados del Noroeste IBA comprises privately-owned lands, and is unprotected. The area is used primarily for tourism, recreation and fisheries. Habitat loss and degradation is the main threat and results from housing and industrial development, invasive species, water pollution and human disturbance.

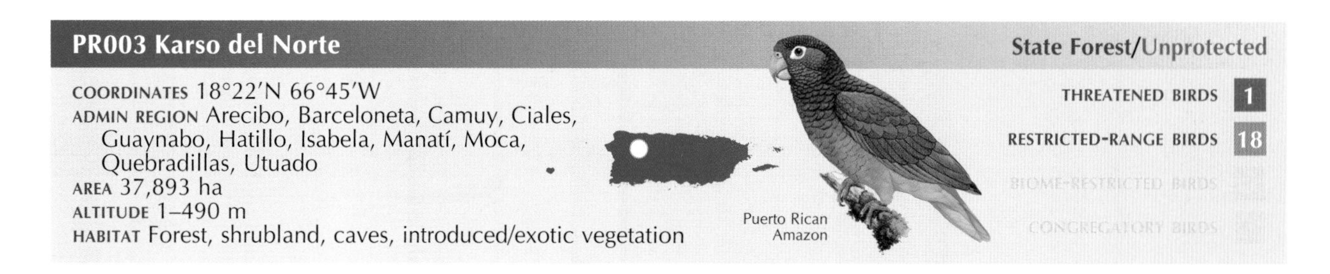

■ Site description

The Karso del Norte (northern karst) IBA stretches across north-central Puerto Rico, from Camuy in the west to Guaynabo in the east. It lies to the south of Acantilados del Noroeste (PR002) and Caño Tiburones (PR004). The IBA embraces the karstic fringe of the northern limestone formations, and contains the largest freshwater aquifer and the most extensive underground cave system on the island. It also supports the island's most extensive mature moist forest areas.

■ Birds

This IBA supports populations of 18 (of the 23) Puerto Rico and the Virgin Islands EBA restricted-range species (12 of which are island endemics). The Critically Endangered Puerto Rican Amazon *Amazona vittata* was reintroduced into this IBA (starting in 2007). The 25–27 parrots now present in the area represent a second wild population of the bird (the other being in El Yunque IBA, PR016). The endemic subspecies of Broad-winged Hawk *Buteo platypterus brunnescens* (classified as critically endangered by DNER) occurs in this IBA.

■ Other biodiversity

The Vulnerable Desmarest's fig-eating bat *Stenoderma rufum*, Critically Endangered bronze coqui *Eleutherodactylus richmondi* and Endangered melodius coqui *E. wightmanae* all occur. Trees (of which there are more than 200 species) and shrubs include the Critically Endangered *Pleodendron macranthum*, *Banara vanderbiltii* and *Auerodendron pauciflorum*, and the Endangered *Goetzea elegans*. The endemic Puerto Rican boa *Epicrates inornatus* and crustacean *Alloweckelia gurneii* have also been recorded.

■ Conservation

Karso del Norte IBA includes the state-owned and legally protected Rio Abajo and Cambalache state forests. However, the majority of land in the IBA is privately owned. Land-use within the IBA includes conservation, research, tourism, recreation and agriculture. The José Vivaldy Puerto Rican Parrot Aviary, an initiative of the DNER and USFWS, was developed within the IBA to facilitate the reintroduction and recovery of the parrot. The NGO Ciudadanos del Karso promotes the conservation of Puerto Rico's karst forest, and North Carolina State University undertakes research on the birds throughout this karst region. The main threats within this IBA are habitat loss and degradation, especially as a result of housing and industrial infrastructure development, invasive species and human disturbance.

■ Site description

Caño Tiburones IBA is located on the north coast between the Manatí and Arecibo rivers, and is the largest estuarine wetland in Puerto Rico. It is located over an underground aquifer fed by springs from Karso del Norte (IBA PR003) resulting in a wide diversity of freshwater wetland habitats. The IBA includes marine areas up to 1 km offshore, and thus embraces the seabird nesting cays of Peñón de Mera (and others) to the north of the wetland.

■ Birds

This IBA is a significant waterbird and seabird site supporting c.30 Vulnerable West Indian Whistling-duck *Dendrocygna arborea*, and regionally important populations of Masked Duck *Nomonyx dominicus* and Roseate Tern *Sterna dougallii*. The terns breed on the cays to the north of Caño Tiburones and at Hacienda La Esperanza. The wetlands are home to large numbers of duck, herons, egrets and shorebirds. White Ibis *Eudocimus albus* have also recently been found breeding in the IBA (representing the first breeding record for Puerto Rico), and a number of vagrant waterbirds have been recorded. The shrubland supports 14 (of the 23) Puerto Rico and the Virgin Islands EBA restricted-range species (10 of which are island endemics), and the Endangered Yellow-shouldered Blackbird *Agelaius xanthomus* has been recorded visiting the IBA.

■ Other biodiversity

The herpetofauna includes the endemic common coqui *Eleutherodactylus coqui* and whistling coqui *E. cochranae*, the Puerto Rican slider *Trachemys stejnegeri* and Nichol's dwarf gecko *Sphaerodactylus nicholsi townsendi*.

■ Conservation

The majority of lands within the Caño Tiburones IBA are state and NGO owned (and legally protected within the Caño Tiburones Natural Reserve and Hacienda La Esperanza Natural Reserve), although some are held by private landowners. Land uses include conservation, research, water resources management, hunting, fisheries, tourism and recreation. It is monitored as part of SOPI's Shorebird Monitoring Network. Some NGOs promoting the conservation of this IBA are Ciudadanos en Defensa del Ambiente (CEDDA), Sierra Club, Comité de Amigos y Vecinos del Barrio Islote and the Sociedad Protectora de Tortugas. Threats include habitat loss and degradation, housing and industrial development, invasive species, illegal hunting, soil and water pollution caused by the presence of a nearby landfill site, and human disturbance.

PR005 Maricao and Susúa

State Forest/Unprotected

COORDINATES 18°07′N 66°57′W
ADMIN REGION Maricao, San Germán, Yauco, Sabana Grande
AREA 8,555 ha
ALTITUDE 60–900 m
HABITAT Forest, shrubland, caves, introduced/exotic vegetation

Elfin-woods Warbler

THREATENED BIRDS 2
RESTRICTED-RANGE BIRDS 18
BIOME-RESTRICTED BIRDS
CONGREGATORY BIRDS

■ Site description

Maricao and Susúa IBA is in the Cordillera Central of central south-western Puerto Rico, in the municipalities of Maricao, Sabana Grande, San Germán and Yauco. The forests of this IBA grow on serpentine soils, setting them apart from other forests on the island, and resulting in high floristic and faunal diversity.

■ Birds

This IBA is home to 18 (of the 23) Puerto Rico and the Virgin Islands EBA restricted-range species (14 of which are island endemics), including significant populations of the Critically Endangered Puerto Rican Nightjar *Caprimulgus noctitherus* (177 individuals) and Vulnerable Elfin-woods Warblers *Dendroica angelae*. The IBA is one of only two areas known to support populations of this rare warbler (the other being El Yunque IBA, PR016). It also has the largest population of the endemic Sharp-shinned Hawk *Accipiter striatus venator* subspecies (classified as critically endangered by DNER).

■ Other biodiversity

Trees present in Susúa include the Endangered *Stahlia monosperma*, and the endemic *Polygala cowelli*, *Ottoschulzia rhodoxylon*, and *Calophyllum cabala*. The Puerto Rican helmet orchid *Cranichis ricartii* (classified as endangered by the DNER), is also present. The endemic blue-tailed ground lizard *Ameiva wetmorei* occurs.

■ Conservation

Maricao and Susúa IBA incorporates the state-owned and legally protected Maricao State Forest and Susúa State Forest, but the IBA also includes privately-owned (and therefore unprotected) areas including state forest buffer zones and corridors. Lands are used for conservation, research, tourism and recreation. The main threats at this IBA are habitat loss and degradation, primarily resulting from infrastructure and housing development, invasive species, water pollution and human disturbance.

PR006 Guaniquilla and Boquerón

Nature Reserve/Wildlife Refuge/Unprotected

COORDINATES 18°00′N 67°10′W
ADMIN REGION Cabo Rojo
AREA 1,152 ha
ALTITUDE 0–50 m
HABITAT Coastlne, forest, shrubland, wetland, introduced/exotic vegetation

Yellow-shouldered Blackbird

THREATENED BIRDS 1
RESTRICTED-RANGE BIRDS 16
BIOME-RESTRICTED BIRDS
CONGREGATORY BIRDS

■ Site description

Guaniquilla and Boquerón IBA is in south-westernmost Puerto Rico, in the municipality of Cabo Rojo. It includes part of Boquerón village to the east and Punta Melones to the south. It is separated from the Suroeste IBA (PR008, which lies to the south) by Combate village. This IBA is characterised by dry coastal forest, beaches, saltflats and mangroves swamps. At Guaniquilla there are karst formations inside the lagoons, and also caves (including one named after the pirate Cofresí).

■ Birds

This IBA supports populations of 16 (of the 23) Puerto Rico and the Virgin Islands EBA restricted-range species (10 of which are island endemics), including the Endangered Yellow-shouldered Blackbird *Agelaius xanthomus*. The blackbird has a population of less than 50 individuals in this IBA, diminished by a combination of habitat loss and Shiny Cowbird *Molothrus bonariensis* brood-parasitism. It is possible that the Critically Endangered Puerto Rican Nightjar *Caprimulgus noctitherus* occurs in these forests and this should be the focus of future research. Migratory shorebirds such as Black-bellied Plover *Pluvialis squatarola* and Greater Yellowlegs *Tringa melanoleuca* frequent the lagoons.

■ Other biodiversity

The endemic Cook's anole *Anolis cooki* (classified as endangered by DNER, and confined to south-west Puerto Rico), inhabits the dry forests in this IBA.

■ Conservation

Guaniquilla and Boquerón IBA lands are mainly state and NGO owned, and are legally protected within the Punta Guaniquilla Nature Reserve and Boquerón Wildlife Refuge. However, the IBA does contain some privately-owned land. Lands are used for conservation, research, hunting, tourism, recreation and pasture. The Yellow-shouldered Blackbird Recovery Project, run by DNER, monitors this species on a regular basis. SOPI's Shorebird Monitoring Network also operates at this site. Hunting is permitted during the designated season in the wildlife refuge. Major threats include: habitat lost and degradation, housing and industrial development, invasive species, water pollution and human disturbance.

PR007 Sierra Bermeja and Laguna Cartagena

National Wildlife Refuge/Unprotected

COORDINATES 17°59′N 67°06′W
ADMIN REGION Cabo Rojo, Lajas
AREA 1,979 ha
ALTITUDE 11–301 m
HABITAT Shrubland, grassland, forest, wetlands, introduced/exotic vegetation

Puerto Rican Nightjar

THREATENED BIRDS 3
RESTRICTED-RANGE BIRDS 18
BIOME-RESTRICTED BIRDS
CONGREGATORY BIRDS

■ Site description
Sierra Bermeja and Laguna Cartagena IBA is located in south-western Puerto Rico. It is surrounded by Guaniquilla and Boquerón IBA (PR006) to the west and Suroeste IBA (PR008) to the south and east. Part of the Maguayo community is within the IBA's eastern boundary. Sierra Bermeja (1,537 ha) is a series of hills (the highest being Cerro Mariquita) over the oldest rocks on the Caribbean Plate, south of Laguna Cartagena (428 ha). It is an area of high biodiversity, geological importance and low housing development. Laguna Cartagena is fed by rainfall and runoff from irrigation channels of the Lajas Valley.

■ Birds
This IBA supports populations of 18 (of the 23) Puerto Rico and the Virgin Islands EBA restricted-range species (12 of which are island endemics), including a population of 12–16 Critically Endangered Puerto Rican Nightjar *Caprimulgus noctitherus*, and the Endangered Yellow-shouldered Blackbird *Agelaius xanthomus*. Laguna Cartagena is home to a population of 30 Vulnerable West Indian Whistling-duck *Dendrocygna arborea*. It is also an important nesting site for waterbirds, and number of vagrants been recorded in this IBA.

■ Other biodiversity
Flora at this IBA includes shrubs such as the Critically Endangered *Eugenia woodburyana*, the Endangered *Stahlia monosperma*, and *Vernonia proctorii*, *Aristida portoricensis* and *Ottoschulzia rhodoxylon* (all of which are considered endangered by USFWS). The mistletoe *Dendrophthora bermejae* is restricted to just a few areas in the south-west of the IBA.

■ Conservation
Sierra Bermeja and Laguna Cartagena IBA includes the federally-owned and legally protected Laguna Cartagena National Wildlife Refuge, but other parts of the IBA are under private ownership and remain unprotected. Lands are used for agriculture, hunting, conservation, research, pasture, tourism and recreation. A conservation plan (entitled "A call to protect Sierra Bermeja for future generations") has been developed by SOPI. The NGO Comité Caborrojeños Pro Salud y Ambiente is leading a reforestation project and promotes the conservation of the Sierra Bermeja. Main threats are habitat loss and degradation, housing and industrial development, invasive species such as patas monkey *Erythrocebus patas* and rhesus macaque *Macaca mulatta*, soil and water pollution, diversion of runoff, and human disturbance.

PR008 Suroeste

National Wildlife Refuge/Natural Reserve/State Forest/Unprotected

COORDINATES 17°57′N 67°03′W
ADMIN REGION Cabo Rojo, Lajas
AREA 13,600 ha
ALTITUDE 5–155 m
HABITAT Forest, shrubland, grassland, wetlands, coastline, rocky areas, sea, introduced/exotic vegetation

Snowy Plover

THREATENED BIRDS 2
RESTRICTED-RANGE BIRDS 18
BIOME-RESTRICTED BIRDS
CONGREGATORY BIRDS ✓

■ Site description
Suroeste IBA covers an area along the south coast of south-western Puerto Rico, extending from Cabo Rojo, east through La Parguera to Guánica. It includes marine areas up to 1 km from the coast and includes a number of offshore cays. The IBA supports various ecosystems including dry coastal forest (with cacti, mesquite etc.), saltflats, saline lagoons and mangrove swamps. However, the forests are mostly secondary, having been affected by agricultural practices.

■ Birds
This IBA supports populations of 18 (of the 23) Puerto Rico and the Virgin Islands EBA restricted-range species (12 of which are island endemics), including unknown numbers of the Critically Endangered Puerto Rican Nightjar *Caprimulgus noctitherus*, and almost 80% of the known Endangered Yellow-shouldered Blackbird *Agelaius xanthomus* population. This IBA is also significant for its waterbirds. Over 20,000 shorebirds (including Neotropical migratory species) congregate on the Cabo Rojo Salt Flats, and large numbers also present on the Papayo Salt Flats in Lajas. Regionally significant populations of Brown Pelican *Pelecanus occidentalis*, Sandwich Tern *Sterna sandvicensis*, Roseate Tern *S. dougallii*, Wilson's Plover *Charadrius wilsonia*, Snowy Plover *C. alexandrinus* and Stilt Sandpiper *Calidris himantopus* also occur. The Near Threatened Piping Plover *Charadrius melodus* is an annual visitor to the IBA (in small numbers).

■ Other biodiversity
The coastal waters are inhabited by Critically Endangered leatherback *Dermochelys coriacea* and hawksbill *Eretmochelys imbricata*, and Endangered green *Chelonia mydas* turtles, and also the Vunerable West Indian manatee *Trichechus manatus*.

■ Conservation
Suroeste IBA embraces the Cabo Rojo National Wildlife Refuge, Papayo Salt Flats Natural Reserve, La Parguera Natural Reserve and the Boquerón State Forest—all variously state-, federal- and NGO-owned legally protected areas. However, parts of the IBA are privately-owned and unprotected. Land is used for agriculture, fisheries, conservation, tourism, recreation and pasture. DNER coordinates the Yellow-shouldered Blackbird Recovery Project in this IBA and SOPI's Shorebird Monitoring Network operates at a number of wetland sites within the IBA. The local NGO Comité Caborrojeños Pro Salud y Ambiente promotes conservation and education projects. Threats include development of industrial and housing infrastructure, invasive species such as *Erythrocebus patas* and *Macaca mulatta*, water pollution and human disturbance.

PR009 Karso del Sur

State Forest/Natural Reserve/Biosphere Reserve/Unprotected

COORDINATES 17°58′N 66°50′W
ADMIN REGION Guayanilla, Guánica, Yauco, Guayanilla
AREA 8,162 ha
ALTITUDE 1–240 m
HABITAT Forest, shrubland, coastline, rocky areas, wetlands, introduced/exotic vegetation

■ Site description

Karso del Sur IBA covers the dry forest zone from Guánica to Ponce along the southern coast of Puerto Rico. It includes the Guánica Dry Forest (biosphere reserve) containing a well preserved and exceptionally diverse dry subtropical forest with xerophytic vegetation on karst soils. The IBA includes marine areas (and a number of cays) up to 1 km from the coast.

■ Birds

The forests and shrubland in this IBA are home to 19 (of the 23) Puerto Rico and the Virgin Islands EBA restricted-range species (11 of which are island endemics), including the largest known population (c.20% of the total) of the Critically Endangered Puerto Rican Nightjar *Caprimulgus noctitherus*. The IBA also supports a regionally significant breeding population of Roseate Tern *Sterna dougallii*. Small numbers of Brown Pelican *Pelecanus occidentalis* nest on Don Luis Cay—one of the few nesting locations for the species in Puerto Rico.

■ Other biodiversity

The herpetofauna includes the Critically Endangered Puerto Rican crested toad *Bufo lemur*, and the endemic blue-tailed ground lizard *Ameiva wetmorei*. The IBA has more than 700 plant species including the Critically Endangered Woodbury's stopper *Eugenia woodburyana*, Vahl's boxwood *Buxus vahlii* and Puerto Rico manjack *Cordia rupicola*, and the Endangered lignumvitae *Guaiacum officinale* and *Stahlia monosperma*.

■ Conservation

Karso del Sur IBA includes the state-owned and legally protected Guanica State Forest (Puerto Rico's second UNESCO-designated biosphere reserve) and Bahia Ballena Natural Reserve, but also privately owned lands. The land in this IBA is variously used for conservation, research, pasture, tourism and recreation. Since 1973, annual bird migration studies have been carried out in the Guánica State Forest. The NGO Coalición Pro Bosque Seco Ventanas Verraco, is promoting the integration of Punta Verraco, Ventanas and Cerro Toro (currently in private hands) into the Guánica protected area. A wind energy project threatens the effective conservation of this IBA. Habitat loss and degradation is occurring as a result of housing and industrial development, invasive species, water pollution and human disturbance.

PR010 Cordillera Central

State Forest/Protected Natural Area/Unprotected

COORDINATES 18°10′N 66°37′W
ADMIN REGION Orocovis, Ciales, Jayuya, Juana Díaz, Ponce, Utuado, Adjuntas, Peñuelas, Guayanilla, Yauco
AREA 18,250 ha
ALTITUDE 220–1,338 m
HABITAT Forest, introduced/exotic vegetation

Puerto Rican Tanager

THREATENED BIRDS
RESTRICTED-RANGE BIRDS 16
BIOME-RESTRICTED BIRDS
CONGREGATORY BIRDS

■ Site description

The Cordillera Central IBA embraces the forested mountains of central Puerto Rico, creating an ecological corridor between five protected areas. These forests, classified as subtropical humid forest are characterised by high annual rainfall and relatively low temperatures. The IBA includes the mountains of Cerro Punta (1,338 m) and Monte Guilarte (1,204 m), and several north-flowing rivers, such as the Arecibo River, traverse the IBA.

■ Birds

This IBA supports populations of 16 (of the 23) Puerto Rico and the Virgin Islands EBA restricted-range species (12 of which are island endemics), including the Puerto Rican Tanager *Nesospingus speculiferus* (an endemic, monotypic genus) which occurs in relatively few of the IBAs. The forests are important for a number of wintering Neotropical migratory warbler species (including Magnolia Warbler *Dendroica magnolia*, Cape May Warbler *D. tigrina* and Black-throated Green Warbler *D. virens*), and the endemic Sharp-shinned Hawk *Accipiter striatus venator* subspecies (classified as critically endangered by DNER) occurs in the Toro Negro and Guilarte state forests within the IBA.

■ Other biodiversity

The herpetofauna at this IBA include the Critically Endangered Eneida's coqui *Eleutherodactylus eneidae*, the Endangered melodius coqui *E. wightmanae*, and the common coqui *E. coqui*, barred anole *Anolis stratulus*, emerald anole *A. evermanni* and Puerto Rican boa *Epicrates inornatus*.

■ Conservation

Cordillera Central IBA includes state owned and legally protected lands in the form of Toro Negro State Forest, Guilarte State Forest, Bosque del Pueblo, Río Encantado Protected Natural Area, and Tres Picachos State Forest, but also areas in under private ownership. Land-uses include conservation, research, agriculture and recreation, with a number of urban areas sited within the IBA. The NGO Casa Pueblo leads activities focused on protection, education and natural resources management at Bosque del Pueblo. The major threats are habitat loss and degradation from housing and industrial development, invasive species, water pollution and human disturbances.

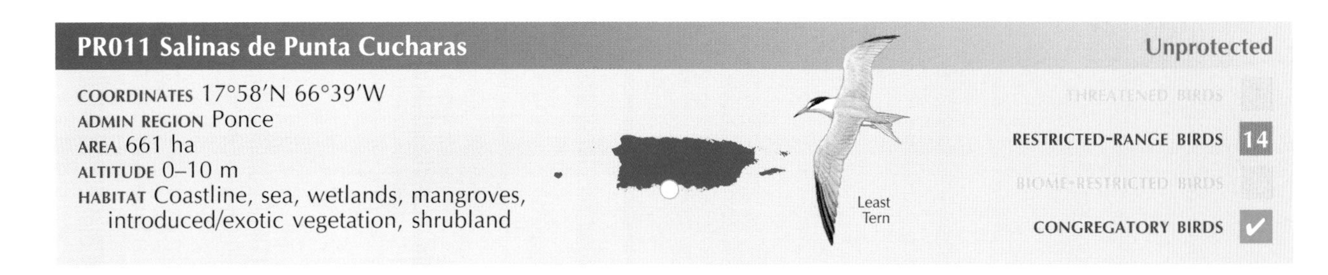

■ Site description

Punta Cucharas IBA is on the coast of central-south Puerto Rico, on the south-western limits of Ponce municipality. It is an area of dry coastal forest, saltflats, open water and mangroves, and includes marine areas up to 1 km from the coast and thus embraces Arena cay. The salinas are bounded to the north by the P.R. State Road #2, and to the west and east by tourist and urban developments.

■ Birds

This IBA supports a regionally significant breeding population of Least Tern *Sterna antillarum*. Other waterbirds occur (including shorebirds, Royal Tern *S. maxima* and White-cheeked Pintail *Anas bahamensis*) but not in significant numbers. The shrublands and mangroves are home to 14 (of the 23) Puerto Rico and the Virgin Islands EBA restricted-range species (10 of which are island endemics). The Endangered Yellow-shouldered Blackbird *Agelaius xanthomus* has occurred as a visitor to this IBA. Antillean Nighthawks *Chordeiles gundlachii* nest at the edge of the saltflats.

■ Other biodiversity

The herpetofauna recorded in this IBA includes the common coqui *Eleutherodactylus coqui*, dryland grass anole *Anolis poncensis*, crested anole *A. cristatellus*, common grass anole *A. pulchellus* and Puerto Rican ground lizard *Ameiva exsul*.

■ Conservation

Salinas de Punta Cucharas IBA includes lands owned privately, and none of it is protected. The area is used for tourism, fisheries and recreation. The NGO Amigos de la Laguna leads conservation efforts and educational activities at the site. The IBA is also monitored as part of SOPI's Shorebird Monitoring Network. Main threats include habitat loss and degradation at the hands of housing and industrial infrastructure development, invasive species, water pollution and human disturbances. Housing, infrastructure and highway development on the northern side of the IBA could limit the flow of freshwater into the wetland.

■ Site description

Ciénaga Las Cucharillas IBA lies just to the west of San Juan on the north coast of Puerto Rico. It is an estuarine system on the north-western side of the Bahía de San Juan, the largest bay on the island, and contains a system of bays, channels, and lagoons connected to the Atlantic Ocean. The IBA includes: Secret Lagoon, Bahía de San Juan, Caño Martín Peña, San José Lagoon, Los Corozos Lagoon, Suarez Channel, La Torrecilla Lagoon, Piñones Lagoon, Condado Lagoon, and the San Antonio Channel. Vegetation includes large areas of mangrove, dragonsblood tree *Pterocarpus officinalis*, and cattail *Typha domingensis* marsh.

■ Birds

This IBA is home to a significant breeding population (35–40 individuals) of the Vulnerable West Indian Whistling-duck *Dendrocygna arborea*. Other waterbirds occur, including nesting populations of Ruddy Duck *Oxyura jamaicensis* and the Near Threatened Caribbean Coot *Fulica caribaea* (both classified as vulnerable by the DNER), but not in significant numbers. The IBA also supports nine (of the 23) Puerto Rico and the Virgin Islands EBA restricted-range species (six of which are island endemics). The Endangered Yellow-shouldered Blackbird *Agelaius xanthomus* has recorded visiting the IBA.

■ Other biodiversity

The endemic common coqui *Eleutherodactylus coqui* and barred anole *Anolis stratullus* occur.

■ Conservation

Ciénaga Las Cucharillas IBA includes lands owned privately, and none of it is protected. However, it is under the management of the San Juan Bay Estuary Consortium (itself managed by the local government) which has conservation, management and education as its major goals. The consortium has identified Ciénaga Las Cucharillas and the Secret Lagoon as having the highest diversity of waterbirds within the San Juan Bay estuary. The IBA has been proposed for acquisition in order to ensure its eventual conservation and restoration. The NGOs Ciudadanos Unidos en contra de la Contaminación (CUCCO) and Corredor del Yaguazo have been leading conservation initiatives at the site. Threats include habitat loss and degradation from housing and industrial infrastructure development, invasive species, air, soil and water pollution, and human disturbance.

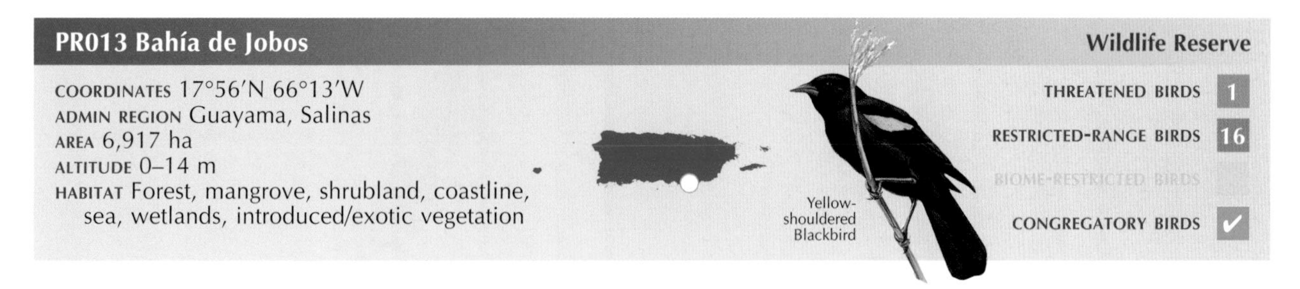

Site description

Bahía de Jobos IBA is on the south coast of eastern Puerto Rico in the municipalities of Guayama and Salinas. It is the second largest bay in the island, and the IBA includes small mangrove islands to the south that are surrounded by sea-grass beds, coral reefs and channels between the cays. It is mainly an estuarine system, fed by groundwater and freshwater springs from the easternmost section of the Cordillera Central. The bay includes hypersaline lagoons, saltponds and other wetland habitats.

Birds

This wetland IBA hosts over 20,000 shorebirds on an annual basis. Migratory birds such as Semipalmated Sandpiper *Calidris pusilla*, Western sandpiper *C. mauri*, Least Sandpiper *C. minutilla* and Stilt Sandpiper *C. himantopus* are the most numerous species, but many others occur and shorebirds like Wilson's Plover *Charadrius wilsonia* and Willet *Catoptrophorus semipalmatus* breed. The IBA supports populations of 16 (of the 23) Puerto Rico and the Virgin Islands EBA restricted-range species (nine of which are island endemics), including an important population of the Endangered Yellow-shouldered Blackbird *Agelaius xanthomus* which nests on the cays. The Near Threatened Plain Pigeon *Patagioenas inornata* has been observed in the Aguirre Forest, although not in significant numbers.

Other biodiversity

The Vulnerable West Indian manatee *Trichechus manatus*, Critically Endangered hawksbill *Eretmochelys imbricata* and Endangered green *Chelonia mydas* turtles frequent marine areas within the IBA.

Conservation

Bahía de Jobos IBA is state owned and legally protected as a wildlife reserve. The area is used for conservation, research, tourism, conservation, fisheries, agriculture and pasture. The reserve's environmental monitoring program has carried out research on abiotic parameters, biodiversity and land-use patterns as well as community education and participation projects. The IBA is included within SOPI's Shorebird Monitoring Network and thus the shorebirds are monitored on a regular basis. Although protected, the IBA is threatened by habitat loss and degradation from housing and industrial development, invasive species, water pollution and human disturbances.

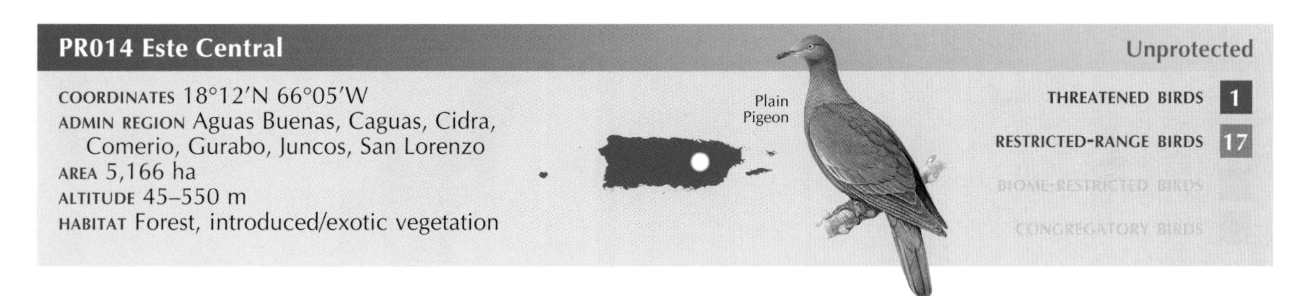

Site description

Este Central IBA is in the central eastern part of the island, to the north of Carite IBA (PR015) and west of El Yunque (PR016). The IBA boundary follows the watershed of the Cidra River and reservoir. It is an area of freshwater systems, humid forest and high rainfall. The forests provide important habitat for the Plain Pigeon *Patagioenas inornata*.

Birds

This IBA supports populations of 17 (of the 23) Puerto Rico and the Virgin Islands EBA restricted-range species (12 of which are island endemics) including Puerto Rican Lizard-cuckoo *Coccyzus vieilloti*, Puerto Rican Screech-owl *Megascops nudipes*, Green Mango *Anthracothorax viridis*, Puerto Rican Emerald *Chlorostilbon maugaeus*, Puerto Rican Tody *Todus mexicanus*, Puerto Rican Woodpecker *Melanerpes portoricensis*, Puerto Rican Flycatcher *Myiarchus antillarum*, Puerto Rican Vireo *Vireo latimeri*, Puerto Rican Bullfinch *Loxigilla portoricensis* and Puerto Rican Spindalis *Spindalis portoricensis*. A significant population (200 birds) of the Near Threatened *Patagioenas inornata* occurs within the IBA.

Other biodiversity

The endemic common grass anole *Anolis pulchelus* and crested anole *A. cristatellus* occur.

Conservation

Este Central IBA is privately owned and unprotected, leaving this important catchment area exposed to expanding housing and infrastructure developments which are being built at the expense of the forest. Invasive species, storms and human disturbance are also threatening the habitat and birds within this IBA.

■ Site description

Carite IBA is in central-eastern Puerto Rico, and covers an area of subtropical humid forest in the watersheds of the Loiza, Patillas and Plata rivers. The latter is the longest river (97 km) in Puerto Rico. Este Central IBA (PR014) lies to the north. The Carite forests support more than 200 species of tree including candle tree *Dacryodes excelsa*, leatherwood *Cyrilla racemiflora*, mahogany *Swietenia macrophylla*, Caribbean pine *Pinus caribaea*, assai palm *Euterpe globosa*, Puerto Rican royal palm *Roystonea borinquena* and the endemic shrub *Micropholis garciniifolia*.

■ Birds

This IBA supports populations of 15 (of the 23) Puerto Rico and the Virgin Islands EBA restricted-range species (11 of which are island endemics). A significant population (249 birds) of the Near Threatened Plain Pigeon *Patagioenas inornata* occurs within the IBA. The Vulnerable Elfin-woods Warbler *Dendroica angelae* was suspected to be present in this IBA, but recent research has been unable to confirm its presence. The endemic subspecies of Broad-winged Hawk *Buteo platypterus brunnescens* and the Sharp-shinned Hawk *Accipiter striatus venator* (both classified as critically endangered by the DRNA, and endangered by USFWS) which are confined to humid montane forest, both occur in this IBA.

■ Other biodiversity

The Endangered Hedrick's coqui *Eleutherodactylus hedricki* and melodius coqui *E. wightmanae* both occur. The IBA used to support populations of the Critically Endangered golden coqui *E. jasperi*, Eneida's coqui *E. eneidae* and web-footed coqui *E. karlschmidti*, but these species are currently presumed extinct in this forest.

■ Conservation

Much of Carite IBA is state owned and legally protected as a state forest, but some areas under private ownership are included within the IBA. Land use is mainly for conservation, research and tourism. Private lands in the IBA (such as in the state forest buffer zones) are threatened by habitat loss from housing and infrastructure development, and human disturbance. Invasive species are a problem for the whole IBA.

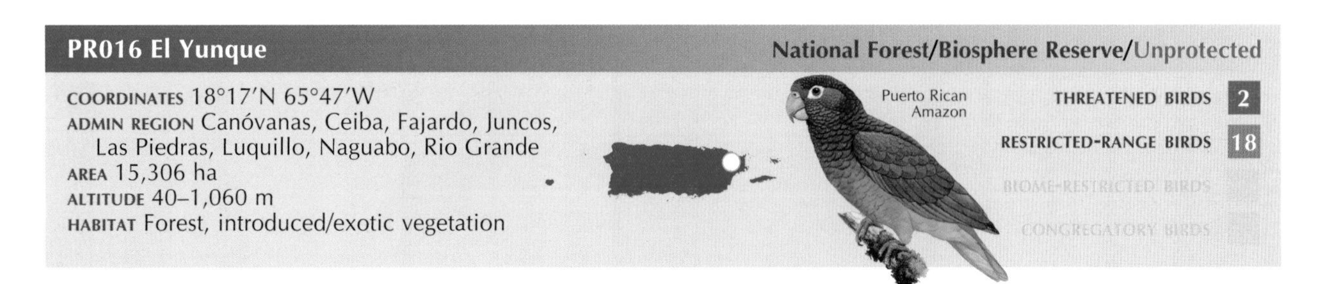

■ Site description

El Yunque IBA is in the Sierra de Luquillo, at the eastern end of the island. It lies to the west of Ceiba and Naguabo IBA (PR018) and north of Carite IBA (PR015). The IBA includes areas down to 40 m elevation, but is primarily a montane area with high rainfall (ranging between 2,000 to 5,000 mm per year). El Yunque IBA is the watershed for four main rivers namely Rio Mameyes (Puerto Rico's last pristine river), Rio Blanco, Rio Fajardo and Rio Espíritu Santo. The main vegetation types include tabonuco, palo colorado, sierra palm, and cloud forests.

■ Birds

This IBA supports one of only two wild populations of the Critically Endangered Puerto Rican Amazon *Amazona vittata* (the other being of recently reintroduced birds at Karso del Norte IBA, PR003). The 18 individuals at El Yunque is the remnant of a population reduced by habitat destruction, removal of chicks, and hurricanes. The IBA supports 18 (of the 23) Puerto Rico and the Virgin Islands EBA restricted-range species (12 of which are island endemics), including the parrot and the Vulnerable Elfin-woods Warbler *Dendroica angelae* (which occurs only in this IBA and the Maricao and Susúa IBA, PR007). The largest populations of the endemic Broad-winged Hawk *Buteo platypterus brunnescens* and the Sharp-shinned Hawk *Accipiter striatus venator* subspecies (both classified as critically endangered by DRNA, and endangered by USFWS) are in this IBA.

■ Other biodiversity

The Critically Endangered webbed-footed coqui *Eleutherodactylus karlschmidti* and the endemic Puerto Rican boa *Epicrates inornatus* occur. There are 23 species of tree endemic to this forest, and shrubs include the Critically Endangered *Ternstroemia subsessilis* and *Styrax portoricensis*.

■ Conservation

Most of El Yunque IBA is state owned and legally protected (since 1876) as a national forest, although some privately owned lands are included within the IBA. The IBA is a UNESCO-designated biosphere reserve. Land is used for conservation, research, tourism and recreation. The Puerto Rican Parrot Recovery Program, led by USFWS, USDA and DRNA, and with the Iguaca Aviary, has coordinated activities aimed at the parrot's long-term conservation. Organisations leading conservation efforts in the national forest's buffer zone are Coalición Pro Corredor Ecológico del Noreste and Liga de Conciencia Ambiental del Este. Threats come from housing and infrastructure development at the edge of the forest, habitat degradation, invasive species and hurricanes.

PR017 Humacao

■ Site description

Humacao IBA is on the coast of eastern Puerto Rico, bordered by the village of Punta Santiago to the east and Humacao town to the south-west. It extends to Naguabo municipality in the north. The IBA consists of brackish lagoons, coastal mangroves and beaches.

■ Birds

This IBA is home to a significant breeding population (35–70 individuals) of the Vulnerable West Indian Whistling-duck *Dendrocygna arborea*. Many other waterbirds occur, including White-cheeked Pintail *Anas bahamensis* (classified as vulnerable by DNER) and Purple Gallinule *Porphyrio martinica* but not in significant numbers. The IBA also supports 13 (of the 23) Puerto Rico and the Virgin Islands EBA restricted-range species (eight of which are island endemics), namely: Puerto Rican Lizard-cuckoo *Coccyzus vieilloti*, Puerto Rican Screech-owl *Megascops nudipes*, Green Mango *Anthracothorax viridis*, Puerto Rican Tody *Todus mexicanus*, Puerto Rican Woodpecker *Melanerpes portoricensis*, Puerto Rican Flycatcher *Myiarchus antillarum*, Adelaide's Warbler *Dendroica adelaidae*, Puerto Rican Bullfinch *Loxigilla portoricensis* and Puerto Rican Spindalis *Spindalis portoricensis*.

■ Other biodiversity

The beaches of this IBA are used for nesting by the Critically Endangered leatherback *Dermochelys coriacea* and hawksbill *Eretmochelys imbricata*, and Endangered loggerhead *Caretta caretta* turtles. Other reptiles include the endemic Puerto Rican slider *Trachemys stejnegeri stejnegeri*, crested anole *Anolis cristatellus* and common grass anole *A. pulchellus*. Puerto Rico's largest stand of bloodwood *Pterocarpus officinalis* grows in the freshwater swamps of this IBA.

■ Conservation

Humacao IBA is state owned and legally protected as the Humacao Wildlife Refuge and Humacao Pterocarpus Forest Natural Reserve. The area is used for conservation, research, tourism, recreation and hunting. The Programa de Educación Comunal de Entrega y Servicio (PECES) has an agreement with the DNER and Tourism Company to promote ecotourism micro-enterprise development at the reserve. Principal threats come from the increasing salinity of the lagoons and residual pesticides from agriculture. Other threats include encroaching infrastructure and housing development, invasive species, water pollution, storms and human disturbance. Hunting is only permitted at certain times of the year in designated areas and is regulated by the DNER.

PR018 Ceiba and Naguabo

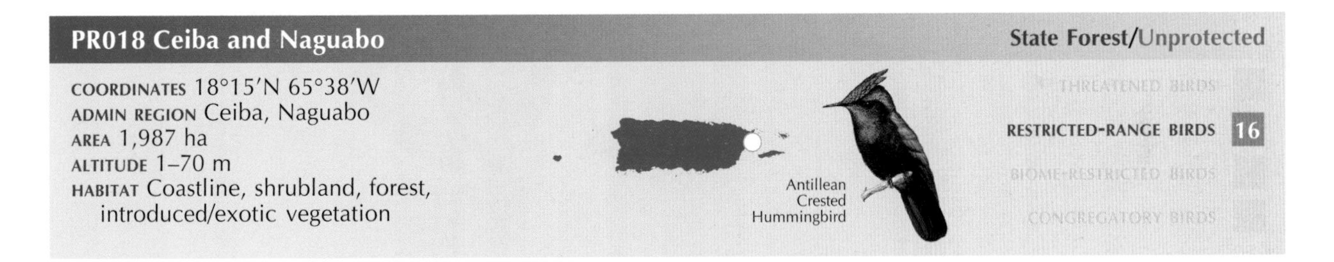

■ Site description

Ceiba and Naguabo IBA is on the easternmost coast of mainland Puerto Rico. It lies east of El Yunque (IBA PR016) and stretches from the coast of Fajardo in the north, south along the coast to Naguabo municipality. The IBA comprises dry coastal forest, mangroves, sandy beaches and coral reefs.

■ Birds

This IBA supports populations of 16 (of the 23) Puerto Rico and the Virgin Islands EBA restricted-range species (10 of which are island endemics) including the Antillean Crested Hummingbird *Orthorhyncus cristatus* which in Puerto Rico is mainly confined to the south-east. The population of the Endangered Yellow-shouldered Blackbird *Agelaius xanthomus* has declined within this IBA in recent years, but the area is still considered as important habitat for the bird.

■ Other biodiversity

The Vulnerable West Indian manatee *Trichechus manatus* is present in the coastal waters of this IBA.

■ Conservation

Ceiba and Naguabo IBA is owned by the state and is legally protected within the Ceiba State Forest. Land is used for conservation, research, tourism, recreation and, in the past, by the military. The Yellow-shouldered Blackbird Recovery Project monitors this species within the IBA. The Alianza Pro Desarrollo de Ceiba (APRODEC) is a consortium of several organisations which are working on projects related to sustainable tourism, alternatives for cultural and social development, collective transport systems and the development of a scientific research centre. Threats include encroaching housing and infrastructure development, invasive species, water pollution, tropical storms, hurricanes and human disturbance.

PR019 Culebra

National Wildlife Refuge/Natural Reserve/Unprotected

COORDINATES 18°20′N 65°20′W
ADMIN REGION Culebra
AREA 23,704 ha
ALTITUDE 1–195 m
HABITAT Forest, shrubland, rocky areas, sea, wetlands, shrubland, introduced/exotic vegetation

Sooty Tern

THREATENED BIRDS
RESTRICTED-RANGE BIRDS
BIOME-RESTRICTED BIRDS
CONGREGATORY BIRDS ✓

■ Site description

Culebra IBA embraces the cays of La Cordillera, from Las Cabezas de San Juan to the east of Fajardo on mainland Puerto Rico, and extending 27 km east to the 2,800-ha island-municipality of Culebra and its satellite cays, rocky islets and coral reefs. The IBA includes Los Farallones, Icacos, Ratones, Diablo, la Blanquilla, Cuchara, Lobo, Yerba, Luis Peña, Matojo, Del Agua, Ratón, Lobito, Alcarraza, Noreste, Molinos, Geñiqui, Culebrita, Pelá, Tiburón and Pelaíta cays; Hermanos and Barriles reefs; Palominito and Palomino islands; and Punta Soldado. The IBA boundary extends 1 km out to sea from the cays and the shoreline.

■ Birds

The IBA is major seabird site, supporting a globally important population of is Sooty Tern *Sterna fuscata* (75,000–105,000 birds), and regionally significant numbers of Royal Tern *S. maxima* (150–465 birds) and Brown Noddy *Anous stolidus* (1,440–1,860 birds). Other seabirds nesting on Culebra include Brown Booby *Sula leucogaster*, Black Noddy *A. minutus* and Sandwich Tern *Sterna sandvicensis*. The restricted-range Puerto Rican Flycatcher *Myiarchus antillarum*, Adelaide's Warbler *Dendroica adelaidae* and two hummingbirds, Green-throated Carib *Eulampis holosericeus* and Antillean Crested Hummingbird *Orthorhyncus cristatus* all occur on the island.

■ Other biodiversity

The Mona boa *Epicrates monensis granti* (classified as endangered by USFWS) occurs. *Peperomia wheeleri* is a shrub endemic to Culebra, and Britton sebucan *Leptocereus grantianus* (classified as endangered by DNER) is also present.

■ Conservation

Culebra IBA includes state and federally owned, legally protected areas (namely Culebra National Wildlife Refuge and La Cordillera Natural Reserve), and privately owned, unprotected lands. Land use is for conservation, research, tourism, recreation and in the past included military use. Educational projects, environmental interpretation, law enforcement and wildlife monitoring takes place in areas managed by USFWS. The organisation CORALations specialises in coral reef conservation, educational activities and the conservation of Culebra. It coordinates coral restoration projects and sea-turtle monitoring. Threats to the IBA include habitat loss and degradation, housing and infrastructure development, invasive species, soil contamination, tropical storms, and human disturbances.

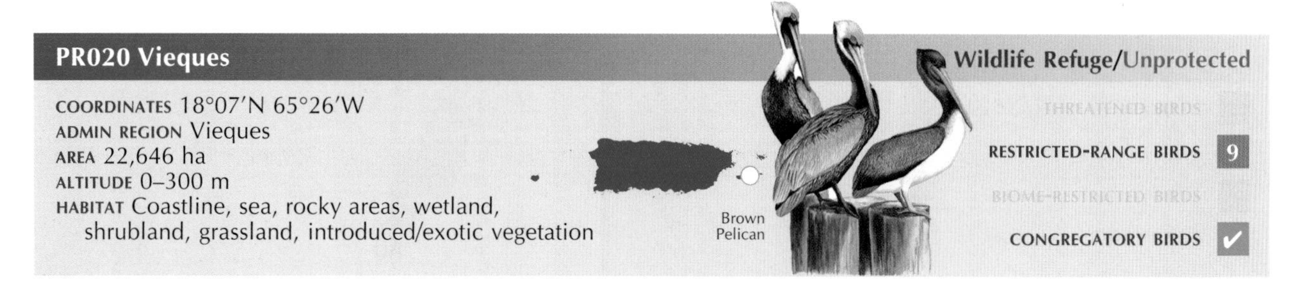

PR020 Vieques

Wildlife Refuge/Unprotected

COORDINATES 18°07′N 65°26′W
ADMIN REGION Vieques
AREA 22,646 ha
ALTITUDE 0–300 m
HABITAT Coastline, sea, rocky areas, wetland, shrubland, grassland, introduced/exotic vegetation

THREATENED BIRDS
RESTRICTED-RANGE BIRDS 9
BIOME-RESTRICTED BIRDS
CONGREGATORY BIRDS ✓

■ Site description

Vieques IBA is a 13,500-ha island-municipality located 13 km east of the Puerto Rico mainland, south of Culebra. It is an island of rolling hills, with an east–west oriented centre ridge. The highest point is Monte Pirata at 300 m. The IBA includes cays, islets and open sea, up to 1 km from the shore of the main island. The human population of Vieques reaches 10,000 individuals at its peak.

■ Birds

This IBA supports a regionally significant population of Brown Pelican *Pelecanus occidentalis* (listed as endangered by DNER). The nesting colony of 300 birds at Conejo Cay represents one of only three nesting sites in Puerto Rico. The IBA is home to nine (of the 23) Puerto Rico and the Virgin Islands EBA restricted-range species (three of which are island endemics), including the Antillean Crested Hummingbird *Orthorhyncus cristatus* (which in Puerto Rico is mainly confined to the south-east), Puerto Rican Woodpecker *Melanerpes portoricensis*, Puerto Rican Screech-owl *Megascops nudipes* and Puerto Rican Flycatcher *Myiarchus antillarum*. Three quail-doves coexist in Vieques: Key West Quail-dove *Geotrygon chrysia*, Bridled Quail-dove *G. mystacea* and Ruddy Quail-dove *G. montana*. It is the only known site for *G. mystacea* in Puerto Rico.

■ Other biodiversity

Three species of sea-turtle nest in Vieques: the Critically Endangered leatherback *Dermochelys coriacea* and hawksbill *Eretmochelys imbricata*, and Endangered loggerhead *Caretta caretta*. The Vulnerable West Indian Manatee *Trichechus manatus* occurs. The herpetofauna includes the endemic Mona boa *Epicrates monensis*, Puerto Rican giant anole *Anolis cuvieri* (classified as endangered by USFWS), and Gunther's white-lipped frog *Leptodactylus albilabris*. Plants include the endemic Endangered Thomas' lidflower *Calyptranthes thomasiana* and beautiful goetzea *Goetzea elegans*.

■ Conservation

Vieques IBA includes state and federally owned, legally protected lands (namely the Cerro El Buey Wildlife Refuge and Vieques National Wildlife Refuge), but it also includes privately-owned, unprotected areas. Land is used for conservation, research, tourism, recreation and, in the past, for military purposes. The Comité Pro Desarrollo y Rescate de Vieques promotes initiatives to clean up soils contaminated by military exercises as well as healthcare and sustainable development projects. Threats to this IBA are housing and infrastructure development, inappropriate recreational uses, invasive species, soil contamination (from military use), tropical storms, fires and human disturbances.

SABA

LAND AREA **13 km²** ALTITUDE **0–887 m**
HUMAN POPULATION **1,420** CAPITAL **The Bottom**
IMPORTANT BIRD AREAS **1, totalling 20 km²**
IMPORTANT BIRD AREA PROTECTION **65%**
BIRD SPECIES **87**
THREATENED BIRDS **0** RESTRICTED-RANGE BIRDS **8**

NATALIA COLLIER AND ADAM BROWN
(ENVIRONMENTAL PROTECTION IN THE CARIBBEAN)

Saba coastline at Spring Bay. (PHOTO: BERT DENNEMAN)

INTRODUCTION

Saba is small, round island in the northern Lesser Antilles. It is situated c.45 km south-west of St Maarten, and c.25 km north-west of St Eustatius, and together these three islands form the Windward Islands of the Netherlands Antilles[1]. Saba is an extinct volcanic peak, rising steeply to 887 m at the top of Mount Scenery, and with a coastline dominated by agglomerate cliffs. Coral reefs surround most of the island. Saba's climate is generally dry, with an average of 1,000 mm of rain falling predominantly between August and November. Vegetation in the interior of the island comprises scrub and grassland which transitions to secondary rainforest and tree-fern brakes at mid elevations, and ultimately elfin woodland at the top of the mountain. There are no terrestrial wetlands on the island. Ecotourism is a significant part of Saba's economy: scuba-diving on the island's reefs and hiking up Mount Scenery are among the primary reasons for tourists to visit the island.

1 At some point in the near future the "Netherlands Antilles" will be dissolved. St Maarten and Curaçao will become separate countries within the Kingdom of the Netherlands (similar to the status currently enjoyed by Aruba). The islands of Bonaire, Saba and St Eustatius will be linked directly to the Netherlands as overseas territories.

Conservation

The Netherlands Antilles have a draft Island Nature Protection Ordinance which must be approved by each island's government in order to facilitate the creation of island-specific conservation legislation. This process is ongoing within the Saba government, but at the present time there is no legislation in place for the designation of terrestrial protected areas or for the conservation of species. However, legislation does exist for marine areas and the 1,300-ha Saba National Marine Park was legally established in 1987. The park encompasses areas (including the seabed and overlying waters) from the high-tide mark to a depth of 60 m around the entire island. The park is administered by the NGO Saba Conservation Foundation, and is one of the few self-sustaining marine parks anywhere in the world, with revenue raised through visitor fees, souvenir sales, and donations.

No terrestrial areas on Saba are legally protected. However, there is *de facto* (and government-recognised) conservation management at a number of sites. The Saba Conservation Foundation, by way of a donation from the Thissell family, owns 35 ha of land on the north side of the island at the former sulphur mine, and manages this as a park. The island government provides an in-kind subsidy to support the management of this area, as well as for the Saba

Seabird signage along one of the trails managed by the Saba Conservation Foundation. (PHOTO: BERT DENNEMAN)

Red-billed Tropicbird. (PHOTO: BRENDA AND DUNCAN KIRKBY)

Conservation Foundation's other responsibility, namely the maintenance and repair of the public hiking trail system. A management plan has been prepared for "Thissell Park" and legislation to formally designate the area as a National Land Park will be submitted to the island government as soon as the Island Nature Protection Ordinance has been prepared. The "Elfin Forest reserve" is an 8.6 ha plot of montane cloud-forest at the top of Mount Scenery that the island government intends to claim title to by prescriptive rights, and in due course designate as a National Land Park (also to be managed by the Saba Conservation Foundation).

Other than the management of areas outlined above, conservation actions on Saba have included a feral cat sterilisation program (that has sterilised at least 200 cats), and a government sponsored rodent control program (active in localised areas). However, it is not clear if either of these efforts is having a positive conservation impact. A multi-year Red-billed Tropicbird *Phaethon aethereus* nest productivity and site/mate fidelity study at a colony near the Fort Bay landfill was undertaken by Martha Walsh McGehee of the NGO Island Conservation Effort (which is no longer active on Saba). Environmental Protection in the Caribbean (EPIC) has continued checks of the tropicbird study area when possible, and has also conducted searches for Audubon's Shearwater *Puffinus lherminieri*. Ethan Temeles (Amherst College, USA) has recently studied the Purple-throated Carib *Eulampis jugularis* and its *Heliconia*-based diet.

The primary threats to Saba's birds, particularly the burrow-nesting seabirds, are alien invasive species. Evidence of rats has been found in areas reported to be shearwater nesting sites, and feral cats are also a serious concern. Goats roam freely, trampling nests and possibly consuming unattended eggs, as well as impacting the vegetation through uncontrolled grazing. Current eradication efforts are too limited to reduce the populations of these invasive animals. However, the Exotic Species Ordinance and Ordinance on the Identification and Registration of Livestock and Domestic Animals represent positive legislative efforts to recognise and control these threats.

■ Birds

Of the 87 species of bird recorded from Saba, just 26 breed, and 36 are regular Neotropical migratory birds (although Saba is too small to hold significant populations of these migrants). Eight (of the 38) Lesser Antilles EBA restricted-range birds occur on the island, although none of these is endemic to Saba. A ninth restricted-range species, the Antillean Euphonia *Euphonia musica* has not been recorded on the island since 1952 and is probably extirpated. The Bridled Quail-dove *Geotrygon mystacea* may also be heading for extinction on the island, having declined dramatically over the last 10 years as a result of hurricane impacts and predation.

It is for the breeding seabirds—Red-billed Tropicbird *Phaethon aethereus* and Audubon's Shearwater *Puffinus lherminieri*—that Saba is most noted (see Table 1 for national population estimates). *Puffinus lherminieri* is the national bird of Saba and is familiar to residents across the island, although predation from rats and cats could be significantly impacting the population (as it could be with the population of *Phaethon aethereus*). Assessing the population of the shearwater on the island is difficult due to the extent of breeding habitat, the lack of an obvious peak breeding season (birds are known to be present between at least December and May) and the nature of the terrain (e.g. steep dirt "cliffs" that are unsuitable for rope work). The use of monitoring technology, such as autonomous audio recorders, may provide more consistent and unbiased data than nest searches or the call/playback method.

Table 1. Key bird species at the Important Bird Area in Saba.

Key bird species	Criteria	National population	Saba IBA AN006
		Criteria	■ ■
Audubon's Shearwater *Puffinus lherminieri*	■	1,000	1,000
Red-billed Tropicbird *Phaethon aethereus*	■	2,250–3,000	2,250–3,000
Bridled Quail-dove *Geotrygon mystacea*	■		✓
Purple-throated Carib *Eulampis jugularis*	■		✓
Green-throated Carib *Eulampis holosericeus*	■		✓
Antillean Crested Hummingbird *Orthorhyncus cristatus*	■		✓
Caribbean Elaenia *Elaenia martinica*	■		✓
Scaly-breasted Thrasher *Margarops fuscus*	■		✓
Lesser Antillean Bullfinch *Loxigilla noctis*	■		✓

All population figures = numbers of individuals.
Restricted-range birds ■. Congregatory birds ■.

IMPORTANT BIRD AREAS

The Saba coastline IBA (AN006)—the island's site priority for bird conservation—has been identified on the basis of nine key bird species (listed in Table 1) that variously trigger the IBA criteria. The IBA covers 2,000 ha of critical terrestrial and marine habitats that support the entire island's population of breeding seabirds, and also the full complement of the restricted-range birds that still occur on the island. The lack of any legal protection for terrestrial areas is a concern that must be addressed to facilitate pro-active conservation of Saba's terrestrial biodiversity (including breeding seabirds).

At present, it seems that the globally significant populations of Audubon's Shearwater *P. lherminieri*, Red-billed Tropicbird *P. aethereus* and the restricted-range Bridled Quail-dove *G. mystacea* in the IBA are declining (or are at least limited) as a result of predation from cats and rats (exacerbated by trampling and grazing from goats). Saba is small enough that complete eradication of some invasive species may be feasible, given sufficient funding, time and local support. In anticipation of both formal protection of terrestrial habitats within the IBA, and a possible eradication program, there is an urgent need to determine the population of *P. lherminieri* and *G. mystacea*, and to continue monitoring the population of *P. aethereus*. Monitoring these populations within the IBA should be used to inform the assessment of state, pressure and response scores at each IBA in order to provide objective status assessments and inform management decisions (such as the necessity for invasive species control) that might be required to maintain this internationally important biodiversity site.

Rainforest Ravine at "Thissell Park". (PHOTO: BERT DENNEMAN)

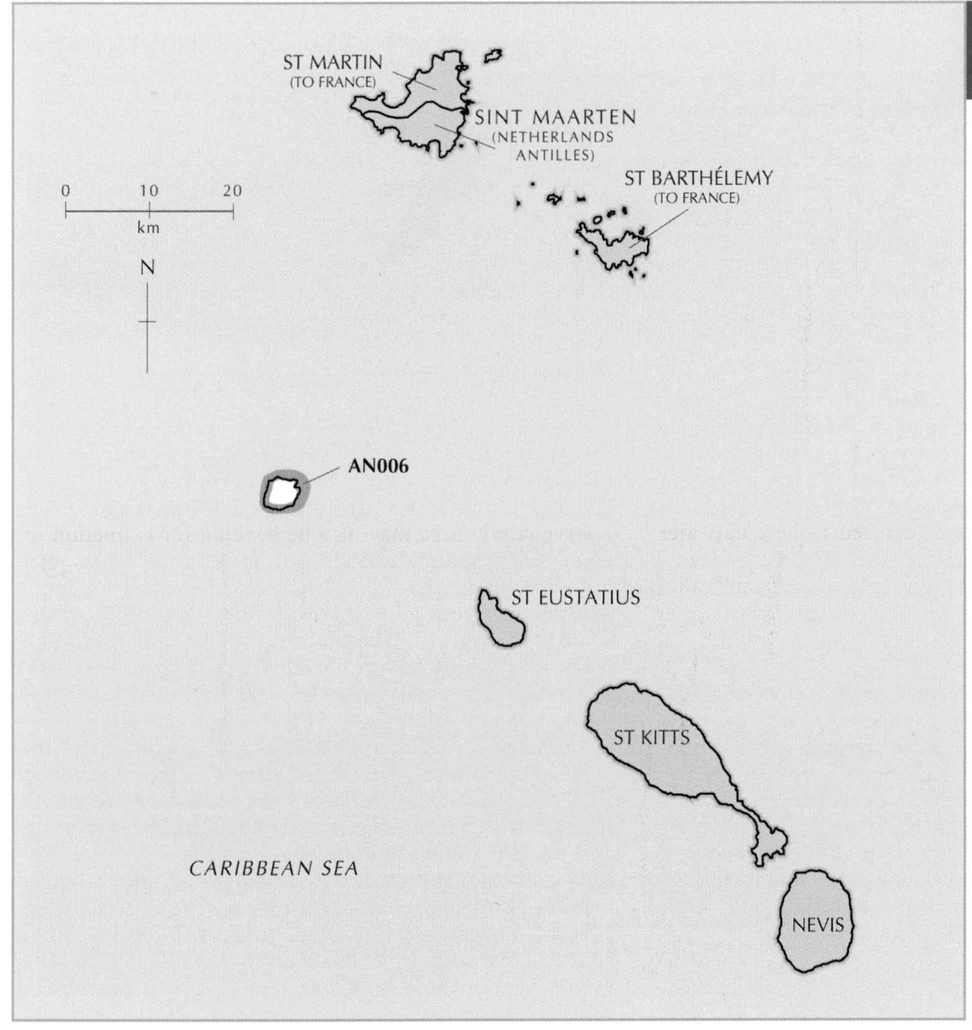

Figure 1. Location of the Important Bird Area in Saba.

KEY REFERENCES

Augustinus, P. G. E. F., Mees, R. P. R. and Prins, M. (1985) Biotic and abiotic components of the landscapes of Saba (Netherlands Antilles). *Uitgave Natuurwetenschappelijke Studiekring voor Suriname en de Nederlandse Antillen* 115.

Brown, A. C. and Collier, N. (in press) The Netherlands Antilles I: Saint Maarten, Saba and Saint Eustatius. In Bradley P. E. and Norton, R. L. eds. *Breeding seabirds of the Caribbean.* Gainesville, Florida: Univ. Florida Press.

Collier, N. C., Brown, A. C. and Hester, M. (2002) Searches for seabird breeding colonies in the Lesser Antilles. *El Pitirre* 15(3): 110–116.

van Halewyn, R. and Norton, R. L. (1984) The status and conservation of seabirds in the Caribbean. Pp 169–222 in J. P. Croxall, P. G. H. Evans and R. W. Schreiber, eds. *Status and conservation of the world's seabirds.* Cambridge, U.K.: International Council for Bird Preservation (Techn. Publ. 2).

Hoogerwerf, A. (1977) Notes on the birds of St. Martin, Saba, and St. Eustatius. *Studies on the Fauna of Curaçao and other Caribbean islands* 54(176): 60–123.

Lee, D. S. (2000) Status and conservation priorities for Audubon's Shearwaters in the West Indies. Pp 25–30 in E. A. Schreiber and D. S. Lee, eds. *Status and conservation of West Indian Seabirds.* Ruston, USA: Society of Caribbean Ornithology (Spec. Publ. 1).

Rojer, A. (1997) Biological inventory of Saba. Willemstad, Curaçao: CARMABI Foundation (Unpublished report).

Stuffers, A. L. (1956) The vegetation of the Netherlands Antilles. *Studies on the Flora of Curaçao and other Caribbean Islands* 1(15).

Voous, K. H. (1955) *The birds of the Netherlands Antilles.* Curaçao: Uitg. Natuurwet. Werkgroep Ned. Ant.

Voous, K. H. (1983) *Birds of the Netherlands Antilles.* Zutphen, The Netherlands: De Walburg Pers.

Walsh-McGehee, M. (2000) Status and conservation priorities for White-tailed and Red-billed Tropicbirds in the West Indies. Pp 31–38 in E. A. Schreiber and D. S. Lee, eds. *Status and conservation of West Indian Seabirds.* Ruston, USA: Society of Caribbean Ornithology (Spec. Publ. 1).

Westermann, J. H. and Kiel, H. (1961) *The geology of Saba and St. Eustatius, with notes on the geology of St. Kitts, Nevis and Montserrat (Lesser Antilles).* Utrecht, The Netherlands: Uitg. Natuur Wetenschappelijke Studiekring Voor Suriname en de Nederlandse Antillen 24.

ACKNOWLEDGEMENTS

The authors would like to thank Bert Denneman (Vogelbescherming Nederlands), Nicole Esteban (STENAPA), Tom van't Hof, and Kalli de Meyer (Dutch Caribbean Nature Alliance) for their help in reviewing this chapter.

■ Site description

Saba coastline IBA includes all land areas from the coast to 400 m inland around the perimeter of this small island, and all sea areas up to 1 km from the coast. It also includes the rainforest ravine at "Thissell Park" (site of a former sulphur mine) and the Elfin Forest reserve at the top of Mount Scenery. The coastline comprises rocky cliffs, 100-m high and over. The only human settlements along the coast are at Fort Bay, where a dock, several buildings, a rock quarry, and landfill exist.

■ Birds

This IBA is significant for all seven of the Lesser Antilles EBA restricted-range birds that occur in the vegetated ghauts, the rainforest ravine and elfin forest. However, it is the seabirds that best characterise this IBA. A population of 750–1,000 pairs of Red-billed Tropicbird *Phaethon aethereus* nest around the island in coastal cliffs and xeric, rocky hills. The nesting population of Audubon's Shearwater *Puffinus lherminieri* in the IBA is thought to be c.1,000 individuals although this is very hard to estimate with accuracy. "Thissell Park" is the only confirmed, current breeding site.

■ Other biodiversity

The Critically Endangered hawksbill *Eretmochelys imbricata* and Endangered green *Chelonia mydas* turtles occur, as does the Endangered red-bellied racer *Alsophis rufiventris* (endemic to Saba and St Eustatius). The island endemic lizard *Anolis sabanus* occurs, as does a restricted-range gecko *Sphaerodactylus sabanus*. The bat *Natalus stramineus stramineus* is endemic to Saba.

■ Conservation

This IBA is a mix of state and private ownership. A proposed National Land Park would encompass the 35-ha Saba Conservation Foundation-owned land at "Thissell Park". It would also include the Elfin Forest reserve, but legislation is not yet in place to allow for legal protection of these terrestrial areas. Saba National Marine Park covers 1,300 ha of sea around the entire coast. Saba Conservation Foundation oversees the management of these areas. Free roaming goats cause erosion and trample seabird nesting burrows. Introduced predators (rats and cats) consume nest contents and attack fledged seabirds. A multi-year tropicbird study (by the NGO Island Conservation Effort) at a colony near the Fort Bay landfill recently concluded. EPIC has continued to monitor the study area when possible, and has also searched for nesting shearwaters.

ST BARTHÉLEMY

LAND AREA 25 km² ALTITUDE 0–104 m
HUMAN POPULATION 8,450 CAPITAL Gustavia
IMPORTANT BIRD AREAS 3, totalling 10.5 km²
IMPORTANT BIRD AREA PROTECTION 75%
BIRD SPECIES c.80
THREATENED BIRDS 0 RESTRICTED-RANGE BIRDS 3

ANTHONY LEVESQUE, ALAIN MATHURIN (AMAZONA) AND
FRANCIANE LE QUELLEC (RÉSERVE NATURELLE DE ST BARTHÉLEMY)

Îlet Tortue IBA, just off the coast of north-east St Barthélemy. (PHOTO: ASSOCIATION GRENAT)

INTRODUCTION

St Barthélemy (also known as St Barts) is an archipelago of 20 islands in which the main island covers 24 km². It has recently become a *collectivité d'outre-mer* (COM, overseas collectivity) of France (and not an integral part of the European Union), and thus is now administratively separated from Guadeloupe (to France). St Barthélemy was first claimed by France in 1648, but was subsequently sold to Sweden in 1784. Sweden sold it back to France in 1878, but this period of occupancy left its mark with many street and town names (and national coat of arms) having Swedish origins. The island and its satellites are exposed parts of the Anguilla Bank (along with neighbouring islands of St Martin and Anguilla). St Barthélemy lies 20 km east-south-east of the island of St Martin. The island and its satellites are formed from uneven, rocky limestone, but the relief is low and the trade winds pass over without much precipitation falling. The climate is tropical maritime with a dry season (called "Lent") from December to May, and a more humid "winter" season from June to November. The resulting vegetation is xerophytic, much of which has been degraded from overgrazing by goats. The main island has a number of brackish and saline ponds (at Grand Cul-de-Sac and Petit Cul-de-Sac, and Saint-Jean) that support small numbers of resident and migratory waterbirds, and the island's last stands of mangrove.

The satellite islands of St Barthélemy range in size from L'île Fourchue, the largest at 24 ha, to those that are less than a hectare. Most of them have sea cliffs that rise to 20 m and above (up to 104 m). Historically goats have been left to run wild on these islands and consequently, as is the case on the mainland, large areas have been overgrazed.

■ Conservation

Conservation efforts in St Barthélemy are still administered through the Regional Directorate of the Environment (DIREN, part of the French Ministry of Ecology and Sustainable Development) based in Guadeloupe. It provides environmental expertise for the development of local planning documents, land management, balancing the environment and economic development, raising environmental awareness and protecting resources and natural spaces.

Several forms of protected areas exist on St Barthélemy. The St Barthélemy National Nature Reserve (Réserve Naturelle) is a marine area that covers 1,200 ha, including 275 ha of coral reef. It comprises five public marine zones located mainly in the north and north-west of the main island, and is managed by the Grenat Association. The uninhabited islets

Île Bonhomme, one of St Barthélemy's many offshore islands that have been severely overgrazed by goats. (PHOTO: RÉSERVE NATURELLE ST BARTH)

have been identified as Littoral Conservation Areas, and are integrated into the Schéma Régional d'Aménagement (SAR). Biotope Protection Orders cover 5.5 ha around Saint-Jean and 16 ha of the Grand Cul-de-Sac and Petit Cul-de-Sac ponds, and many of the migratory birds that occur on these ponds form part of a Ministerial Protection Order. Zone Naturelles d'Intérêt Ecologique Faunistique et Floristique (ZNIEFF) designations, which have no legal status, cover 37 ha that embraces the island's five ponds, and 42 ha of unique xerophytic vegetation at Pointe à Toiny (on the east of the island).

A number of threats impinge on St Barthélemy's birds and biodiversity. Seabirds are affected by the poaching of adults and young birds, and egg collection. Urban encroachment, agricultural expansion, pesticide use and uncontrolled grazing by goats and cows is resulting in habitat loss and degradation, and soil erosion. Alien invasive predators including black rat *Rattus rattus*, small Indian mongoose *Herpestes auropunctatus* and domestic cat *Felis catus* are all impacting bird populations. An integrated protection scheme covering the seabird colonies should be put in place during the breeding season. The protection would consist of a ban on approaching closer than 100 m to the coasts of the islets, and landing would be prohibited. On the recent establishment of St Barthélemy as an overseas collectivity, a promise was made to regenerate the environment within the islands. A seabird conservation and restoration plan would provide an excellent focus for the implementation of such a promise. Restoration efforts will need to focus on goat eradications on the islets (to allow for the assisted recolonisation of the vegetation), and predator eradication (where necessary).

■ Birds

Approximately 80 species are thought to have been recorded on St Barthélemy. The island's small land area and xerophytic vegetation that is often degraded from overgrazing by goats means that breeding landbirds are poorly represented. About 15 landbird species breed, including four species restricted to the Lesser Antilles Endemic Bird Area (EBA), namely Antillean Crested Hummingbird *Orthorhyncus cristatus*, Green-throated Carib *Eulampis holosericeus*, Caribbean Elaenia *Elaenia martinica* and Lesser Antillean Bullfinch *Loxigilla noctis*. Small numbers of waterbirds (of six species) are present on the island's ponds, and the island's avifauna increases with the arrival of Neotropical migrants from North America each winter. However, the main bird interest on St Barthélemy is from the 13 regularly breeding species of seabird that form colonies on many of the satellite islands. The species include those listed (with national population estimates, and estimates of numbers in the IBAs) in Table 1. Also breeding are Red-billed Tropicbird *Phaethon aethereus* (140–280 pairs), White-tailed Tropicbird *P. lepturus* (20–40 pairs), Sooty Tern *Sterna fuscata* (150–300 pairs), Bridled Tern *S. anaethetus* (75–100 pairs), Brown Noddy *Anous stolidus* (98 to 120 pairs), Least Tern *S. antillarum* (15–20 pairs), and Brown Pelican *Pelecanus occidentalis* (18 pairs). Both Audubon's Shearwater *Puffinus lherminieri* and Roseate Tern *S. dougallii* are rare or occasional breeding birds. All of the satellite islands have the strong potential to support larger numbers of seabirds. However, on some of the islands extensive goat grazing has led to erosion which is hindering the recolonisation of vegetation needed by the nesting seabirds. Other pressures (tourist disturbance, pollution etc.) are also suppressing the populations of St Barthélemy's seabirds.

IMPORTANT BIRD AREAS

St Barthélemy's three IBA—the country's international site priorities for bird conservation – cover 1,050 ha (including marine areas), and about 0.4% of the country's land area. The IBAs have been identified on the basis of four key seabird species (listed in Table 1) that variously meet the IBA criteria,

Figure 1. Location of Important Bird Areas in St Barthélemy.

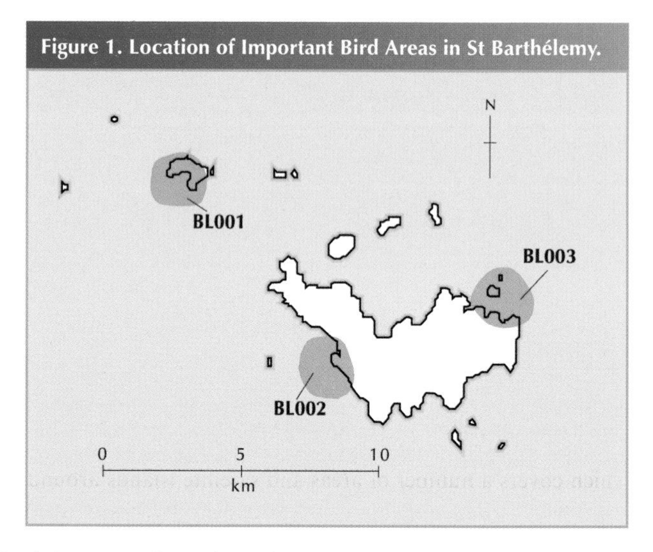

More than 1% of the Caribbean populations of both Brown Booby and Royal Tern nest within St Barthélemy's IBAs. (PHOTOS: VINCENT LEMOINE)

Les Petite Saints, part of IBA BL002, just south of the entrance to Gustavia harbour. (PHOTO: ANTHONY LEVESQUE)

Table 1. Key bird species at IBAs in St Barthélemy.

Key bird species	Criteria	National population	Criteria	St Barthélemy IBAs BL001	BL002	BL003
				■	■	■
Brown Booby *Sula leucogaster*	■	450–525		180	130	
Laughing Gull *Larus atricilla*	■	1,320				450
Royal Tern *Sterna maxima*	■	90–150				150
Common Tern *Sterna hirundo*	■	90–105			60	

All population figures = numbers of individuals.
Congregatory birds ■.

and thus the IBAs all represent globally significant seabird colonies, and all are on St Barthélemy's satellite islands. Mainland St Barthélemy supports populations of three Lesser Antilles EBA restricted-range species. However, these species are well represented within the wider IBA network throughout the Lesser Antilles and thus no IBA has been identified specifically for them in St Barthélemy.

The cliffs and land up to 15 m from the coasts of these IBAs are within Littoral Conservation Areas, giving them some legal protection from development. Two of the IBAs are also within the marine Réserve Naturelle de St Barthélemy which covers a number of areas and satellite islands around the coast of the "mainland", and also offers some protection from disturbance. However, the IBAs (and the seabirds within them) are still threatened by disturbance and undocumented levels of egg collecting. It is unknown whether rats *Rattus* spp. are present within the IBAs, and this should be a priority to assess, both within the IBAs and on the other satellite islands. Eradication efforts where rats are present would be feasible and would boost the recovery of seabird populations while they are also benefitting from the relatively recent removal of goats from some breeding islands. An assessment of the presence of goats on the satellite islands would also be worthwhile as a precursor to the development of broad seabird conservation strategy for St Barthélemy.

The monitoring of the seabirds within St Barthélemy's IBAs should be continued and used to inform the assessment of state, pressure and response scores at each IBA in order to provide objective status assessments and inform management decisions that might be required to maintain (or improve) these internationally important biodiversity sites. Monitoring the seabird populations on the other satellite islands will also be important to determine the extent to which species might, between years, use a larger network of breeding locations to sustain the local population. Such monitoring will also be essential to assess the impact of any future management activities (such as rat or goat eradications) on these islands.

KEY REFERENCES

LEBLOND, G. (2003) Les oiseaux marins nicheurs de Guadeloupe, de Saint-Martin et de Saint-Barthélemy. (Unpublished report: BIOS/DIREN, 56 p.).

LUREL, F. (1994) Typologie et cartographie de la végétation naturelle de l'île de Saint Barthélemy. Université des Antilles et de la Guyane. (Unpublished report available from the Office National des Forêts).

RAFFAELE, H., WILEY J., GARRIDO, O., KEITH, A. AND RAFFAELE, J. (1998) *A guide to the birds of the West Indies*. Princeton, New Jersey: Princeton University Press.

ACKNOWLEDGEMENTS

The authors would like to thank Alison Duncan (Ligue pour la Protection des Oiseaux, BirdLife in France), Gilles Leblond (BIOS) and Félix Lurel for their input to and comments on this chapter, and Hervé Vitry and Gaël Thébault who accompanied us during our field survey.

BL001 Petite Islette

Littoral Conservation Area/National Nature Reserve

COORDINATES 17°57′N 62°54′W
ADMIN REGION —
AREA 325 ha
ALTITUDE 0–33 m
HABITAT Rocky areas, shrubland, coast

Brown Booby

THREATENED BIRDS
RESTRICTED-RANGE BIRDS
BIOME-RESTRICTED BIRDS
CONGREGATORY BIRDS ✓

■ Site description

Petite Islette IBA lies just to the south-west of l'île Fourchue, itself c.5 km north-west of the main island of St Barthélemy. The uninhabited islet is just 1.3 ha, and it is surrounded by near-vertical sea cliffs rising to 33 m. The IBA includes marine areas up to 1 km from Petite Islette. The island itself is extremely dry, with the parched stony ground colonised by drought-resistant xeric vegetation, and grassland dominated by various species of cactus, bushes and a few small trees.

■ Birds

This IBA is regionally significant for its Brown Booby *Sula leucogaster* colony, with more than 60 pairs found nesting in 2002. A survey in 2007 confirmed the species' continued presence with at least 80 adults and some young birds counted on the islet. Other species present were Laughing Gull *Larus atricilla*, an American Oystercatcher *Haematopus palliatus*, Magnificent Frigatebird *Fregata magnificens* and Zenaida Dove *Zenaida aurita*.

■ Other biodiversity

The Vulnerable Lesser Antillean iguana *Iguana delicatissima* occurs along with the Anguilla Bank endemic lizards, namely the ground lizard *Ameiva pleei* and *Anolis gingivinus*. Clumps of Royen's tree cactus *Pilosocereus royenii*, prickly pear *Opuntia tricantha*, and Turk's cap cactus *Melocactus intortus* are present.

■ Conservation

Petite Islette is privately owned by Serge and Carole Beal. The island's cliffs are formally protected as a Littoral Conservation Area belonging to the Conservatoire du littoral that forbids any development within 15 m of the coast. The island is within the Domaine Public Maritime (DPM) of the St Barthélemy National Nature Reserve (Réserve Naturelle) that surrounds much of l'île Fourchue, and this prohibits the anchoring of boats. Consequently, landings on the island are limited. The principle threat to the island was the presence of goats, but these were removed some years ago. It is unknown if rats *Rattus* spp. are present on the island. The collection of seabird eggs might still be an occasional activity, and this should be monitored. There is also potential disturbance from plant collectors taking cacti from the island. A census of breeding seabirds was carried out in 2001–2002 by Gilles Leblond and a complementary visit made in 2007 by Anthony Levesque (AMAZONA). An inventory of the flora was undertaken by Félix Lurel in 2006.

BL002 Les Petit Saints et Gros Islets

Littoral Conservation Area/National Nature Reserve/Unprotected

COORDINATES 17°54′N 62°51′W
ADMIN REGION —
AREA 360 ha
ALTITUDE 0–20 m
HABITAT Rocky areas, shrubland

Common Tern

THREATENED BIRDS
RESTRICTED-RANGE BIRDS
BIOME-RESTRICTED BIRDS
CONGREGATORY BIRDS ✓

■ Site description

Les Petits Saints et Gros Islets IBA comprises two sets of rocky islets that lie 500 m offshore from the entrance to Gustavia harbour on the western side of St Barthélemy. Les Gros Islets consist of two volcanic islands of 0.6 and 0.2 ha, that rise to c.20 m. They lie just north (500 m) of Les Petits Saints which are a small group of rocky mounds to the south of the Gustavia harbour entrance. Xerophytic vegetation clings to the cliffs and rock crevices on these islands, including the rare herbaceous plant *Pappophorum papiferum* as well as a species of *Cyperacee* and *Cleome*. The IBA includes marine areas up to 1 km from the islets.

■ Birds

This IBA supports a globally significant breeding population of Common Tern *Sterna hirundo*. About 20 pairs were found on Les Petits Saints in 2001, but this population has not been re-assessed since then. A regionally important population of Brown Booby *Sula leucogaster* also breeds, with c.100 birds (various age classes) on Les Gros Îlets, and c.20 found on Les Petits Saints in 2007. Other seabirds breeding on these islands include Bridled Tern *Sterna anaethetus*, Royal Tern *S. maximus*, Brown Noddy *Anous stolidus*, Red-billed Tropicbird *Phaeton aetherus* (several pairs), Magnificent Frigatebird *Fregata magnificens*, and Laughing Gull *Larus atricilla*.

■ Other biodiversity

No globally threatened or endemic taxa have been recorded.

■ Conservation

Les Petits Saints et Gros Islets IBA is state-owned. Les Petits Saints cliffs are formally protected as a Littoral Conservation Area belonging to the Conservatoire du littoral that forbids any development within 15 m of the coast. Gros Islets are within the Domaine Public Maritime (DPM) of the St Barthélemy National Nature Reserve (Réserve Naturelle) that surrounds the Pain de Sucre island, and this prohibits the anchoring of boats. Consequently, landings on the island are limited. It is unknown if rats *Rattus* spp. are present on the island. The collection of seabird eggs might still be an occasional activity, and this should be monitored. A census of breeding seabirds was carried out in 2001–2002 by Gilles Leblond and a complementary visit made in 2007 by Anthony Levesque (AMAZONA). An inventory of the flora was undertaken by Félix Lurel in 2006.

BL003 Îlet Tortue

Littoral Conservation Area/National Nature Reserve

COORDINATES 17°55′N 62°48′W
ADMIN REGION —
AREA 370 ha
ALTITUDE 0–37 m
HABITAT Rocky areas, shrubland

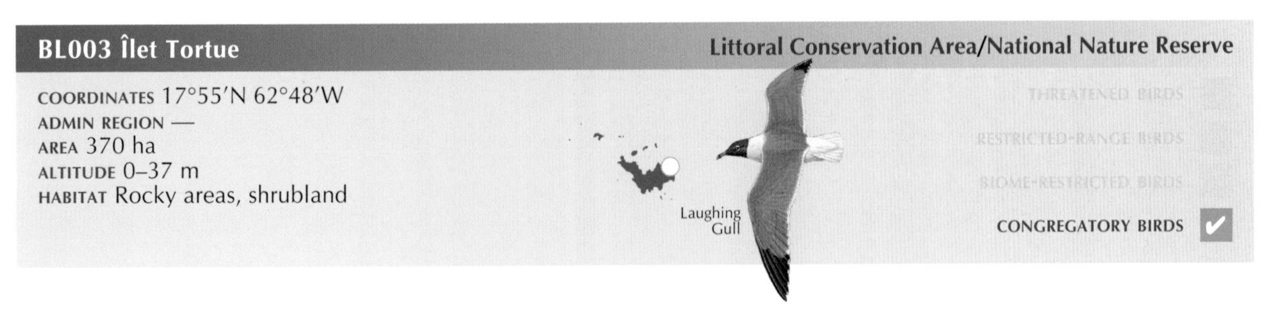

■ Site description

Îlet Tortue IBA lies just offshore from l'Anse du Grand Cul-de-Sac at the north-eastern end of St Barthélemy. This uninhabited island covers 7 ha, and is 35 m at its highest point. It is vegetated in a mosaic of grasses, bushes and small trees that rarely get above 3 m (due to the constant strong winds), with agaves and cacti like the Turk's cap cactus *Melocactus intortus* and Royen's tree cactus *Pilosocereus royenii*. Goats were present on the island until recently, as a consequence of which the vegetation is still recovering and erosion remains a problem. The IBA includes marine areas up to 1 km from the island.

■ Birds

This IBA supports a significant seabird colony. A globally important breeding population of Laughing Gull *Larus atricilla* and regionally significant Royal Tern *Sterna maximus* population have remained stable between surveys in 2002 and 2007. Red-billed Tropicbird *Phaeton aethereus* also breed, with less than 10 pairs observed in 2007.

■ Other biodiversity

Îlet Tortue is one of the largest and best vegetated islands in St Barthélémy and represents a refuge for a number of rare species.

■ Conservation

L'îlet Tortue IBA is a privately owned, belonging to Henri Gréaux. The island's cliffs are formally protected as a Littoral Conservation Area belonging to the Conservatoire du littoral that forbids any development within 15 m of the coast. The island is within the legally protected zone of the St Barthélemy National Nature Reserve (Réserve Naturelle) that covers marine areas off the north-eastern tip of the main island. This prohibits the anchoring of boats, shooting and fishing (as disincentives to land) but there is occasional landing on the island and egg collecting remains unregulated. Terrestrial areas are not covered by the reserve. It is unknown if rats *Rattus* spp. are present on the island, but the goat population was recently removed. However, restoration of goat affected areas would reduce erosion and sedimentation of the nearby marine environment. A census of breeding seabirds was carried out in 2001–2002 by Gilles Leblond and a complementary visit made in 2007 by Anthony Levesque (AMAZONA). An inventory of the flora was undertaken by Félix Lurel in 2006.

ST EUSTATIUS

LAND AREA **21 km^2** ALTITUDE **0–802 m**
HUMAN POPULATION **3,000** CAPITAL **Oranjestad**
IMPORTANT BIRD AREAS **2, totalling 14.86 km^2**
IMPORTANT BIRD AREA PROTECTION **100%**
BIRD SPECIES **54**
THREATENED BIRDS **0** RESTRICTED-RANGE BIRDS **8**

NATALIA COLLIER AND ADAM BROWN
(ENVIRONMENTAL PROTECTION IN THE CARIBBEAN)

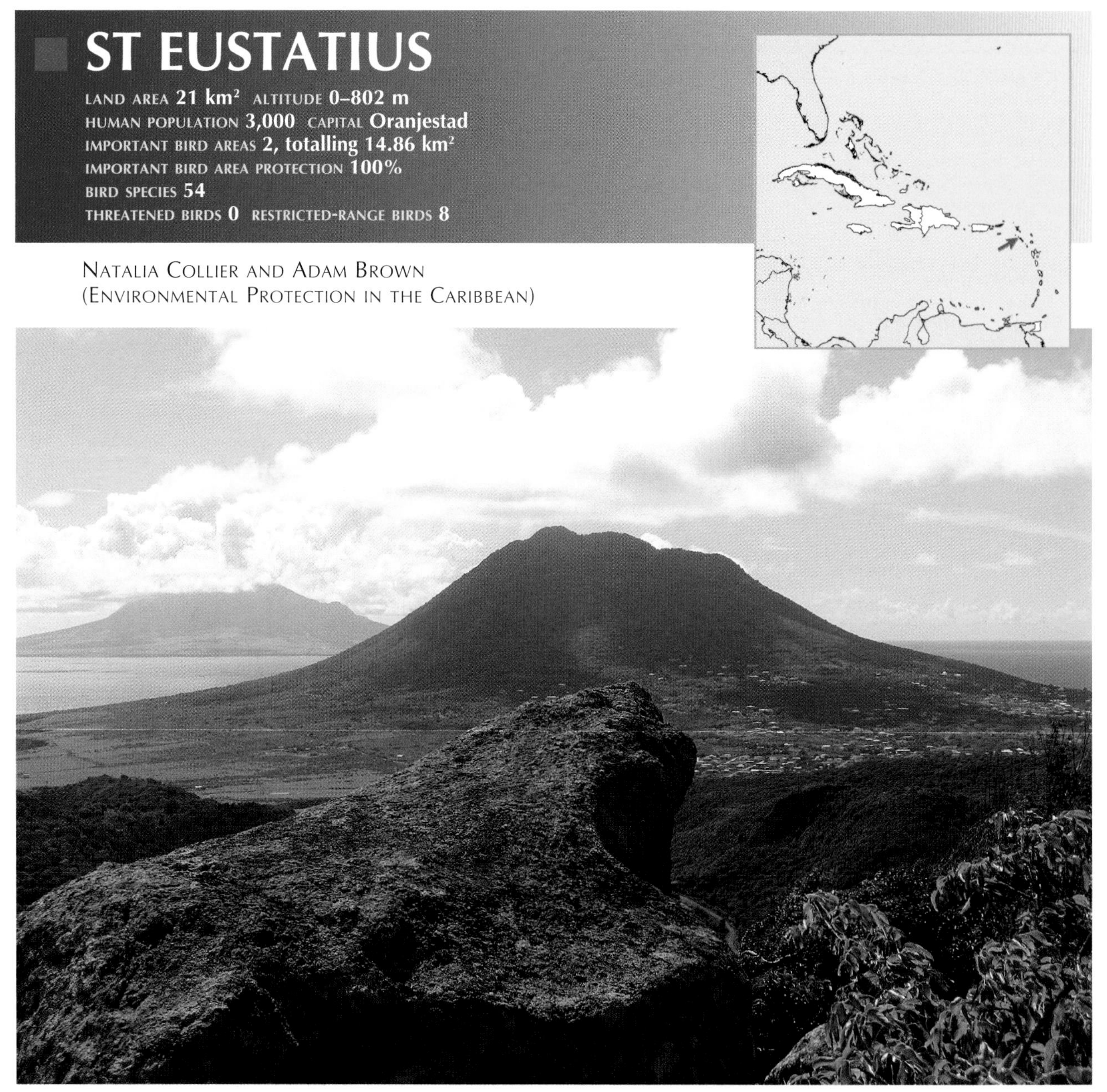

The Quill from Boven hill, with St Kitts in the distance. (PHOTO: NICOLE ESTEBAN)

INTRODUCTION

St Eustatius (or Statia) is a small, volcanic island in the northern Lesser Antilles. It is situated c.13 km north-west of St Kitts and Nevis and c.25 km south-east of Saba. Together with Saba and St Maarten (which is c.45 km to the north) these three islands form the Windward Islands of the Netherlands Antilles[1]. Statia is saddle-shaped: the 602-m high Mount Mazinga (locally called "the Quill") is a young volcano at the south-east end of the island, and a denuded, dormant volcano—the "Northern Hills" comprising Signal Hill, Little Mountain and Boven Mountain—is at the north-west end. Between these two volcanic formations is a low sloping plain (where the majority of the island's population lives) with a sand beach on the north-east and a rocky beach on the south-east coasts. The remainder of the coastline is steep cliffs or xeric vegetated slopes. Coral reef surrounds much of the island. The interior vegetation of Statia is composed primarily of thorn woodland and grassland, but secondary evergreen and elfin forest are found within the volcanic caldera of the Quill. There are no ponds or other terrestrial wetlands on Statia. The climate is generally dry with an average of 986 mm of rain falling predominately between August and November.

■ Conservation

The Statia government developed the National Nature Conservation Ordinance (based on the draft Netherlands Antilles Island Nature Protection Ordinance) during the mid-1990s which provided the legislative framework to designate protected areas and develop a national parks system. Three parks were designated in 1997–1998, the management authority for which was delegated to a local NGO, St Eustatius National Parks Foundation (STENAPA). The St Eustatius Marine Park encompasses 27.5 km^2 (including the seabed and overlying waters) from the high-tide mark to a depth of 30 m around the entire island, within which are no-take and no-anchor zones. Scuba-diving and mooring fees support the administration and management of the park. The major terrestrial park (the Netherlands Antilles first) is the Quill–Boven National Park which consists of two sub-sectors that equate to the terrestrial parts of the two Statia IBAs and cover 41% (865 ha) of Statia's land area. The Quill is actively managed by STENAPA, but

1 At some point in the near future the "Netherlands Antilles" will be dissolved. St Maarten and Curaçao will become separate countries within the Kingdom of the Netherlands (similar to the status currently enjoyed by Aruba). The islands of Bonaire, Saba and St Eustatius will be linked directly to the Netherlands as overseas territories.

Antillean Crested Hummingbird and Green-throated Carib—two of Statia's restricted-range birds. (PHOTOS: BRENDA AND DUNCAN KIRKBY)

the Boven sub-sector, while legally protected is currently the subject of a land dispute and is unmanaged.

There is no direct bird conservation work underway in Statia although STENAPA has facilitated a number of repeat bird surveys since 2003. STENAPA is also sponsoring school educational programs (with a focus on marine ecosystems) that are increasing environmental awareness among the island's youth.

Although historically a major point of trade prior to the 1800s, Statia is not currently under threat of major tourist developments, possibly due to its lack of calm, sandy beaches. However, other threats are having an impact on the island's ecosystems. Severe overgrazing has resulted in large areas denuded of vegetation, and proper management and containment of livestock will be needed to reduce the resultant desertification. Enforcement of an animal registry program has begun and (at least within the Quill) should decrease grazing pressure. The island's refuse landfill is within a coastal ravine, meaning that during storm surges and high winds this trash is dispersed across the coastal zone, possibly resulting in entanglement or consumption of debris by seabirds and other marine life. A sanitary refuse disposal site is urgently needed. Oiling of wildlife is a concern during spills from the oil transfer station which needs to improve its oil spill monitoring and response efforts. Statia's biodiversity is also at risk from introduced predators including dogs, cats, rats and mice.

■ Birds

Statia's 54 recorded bird species comprise 26 breeding species and 28 Neotropical migrants (although Statia is so small it does not hold significant populations of these migrants). Eight (of the 38) Lesser Antilles EBA restricted-range birds occur on the island, although none of these is endemic to Statia. A ninth restricted-range species, the Scaly-breasted Thrasher *Margarops fuscus*, was seen at the Quill in 2003 having gone unrecorded for 76 years. Its status as a breeding species remains to be confirmed. Eurasian Collared-dove *Streptopelia decaocto* was recorded for the first time on the island in 2003.

A population of 100–200 Red-billed Tropicbirds *Phaethon aethereus* was estimated from part of the Boven IBA (AN007) in 2003, suggesting that the island breeding population may be higher. Small numbers of other seabirds including White-tailed Tropicbird *P. lepturus* (<20) and Brown Booby *Sula leucogaster* (c.10) have been recorded in the vicinity of "White Wall" within the Quill IBA (AN008). Audubon's Shearwater *Puffinus lherminieri* was historically recorded on the island,

Red-billed Tropicbird.
(PHOTO: BRENDA AND DUNCAN KIRKBY)

Table 1. Key bird species at Imortant Bird Areas in St Eustatius.

Key bird species	Criteria	National population	Criteria	St Eustatius IBAs AN007	AN008
			Restricted-range ■	■	■
			Congregatory ■	■	
Red-billed Tropicbird *Phaethon aethereus*	■ (congregatory)	100–200		150	
Bridled Quail-dove *Geotrygon mystacea*	■				✓
Purple-throated Carib *Eulampis jugularis*	■				✓
Green-throated Carib *Eulampis holosericeus*	■			✓	✓
Antillean Crested Hummingbird *Orthorhyncus cristatus*	■				✓
Caribbean Elaenia *Elaenia martinica*	■			✓	✓
Pearly-eyed Thrasher *Margarops fuscatus*	■			✓	✓
Brown Trembler *Cinclocerthia ruficauda*	■				✓
Lesser Antillean Bullfinch *Loxigilla noctis*	■			✓	✓

All population figures = numbers of individuals.
Restricted-range birds ■. Congregatory birds ■.

and one resident has reported still hearing them call at night, but no breeding has been confirmed.

IMPORTANT BIRD AREAS

Statia's two IBAs—the island's international priority sites for bird conservation—cover 41% of the island's land area. Both IBAs are formally designated as protected areas—the terrestrial components are covered within the Quill–Boven National Park, and the marine component within the St Eustatius Marine Park. The IBAs have been identified on the basis of nine species that variously trigger the IBA criteria for restricted-range birds and congregatory birds. The Quill IBA (AN008) embraces the island's forest-dependent species (the restricted-range species), while Boven IBA (AN007) supports nesting habitat for *P. aethereus*.

Resolving the land dispute within Boven IBA appears to be critical to enabling effective management of grazing and thus the successful recovery of vegetation within the park. It would also facilitate the potential control of goats, cats and rats that almost certainly represent limiting factors for the breeding population of *P. aethereus*. Enforcement of an animal registry program within the Quill IBA should decrease the incidence of grazing in the park, but both goats and chickens are often present around and within the volcano and presumably impact the native fauna. Surveys to assess the population of each of the IBA trigger species should be a priority. Such surveys should be combined with annual monitoring of state, pressure and response scores at each IBA to provide an objective status assessment and highlight management interventions that might be required to maintain these internationally important biodiversity sites.

Boven IBA.
(PHOTO: NICOLE ESTEBAN)

KEY REFERENCES

COLLIER N. C. AND BROWN, A. C. (2003) Surveys of sea and terrestrial birds in Sint Eustatius. Florida, USA: Environmental Protection in the Caribbean (Unpublished report to St Eustatius National Parks)

VAN HALEWYN, R. AND NORTON, R. L. (1984) The status and conservation of seabirds in the Caribbean. Pp 169–222 in J. P. Croxall, P. G. H. Evans and R. W. Schreiber, eds. *Status and conservation of the world's seabirds*. Cambridge, U.K.: International Council for Bird Preservation (Techn. Publ. 2).

HOOGERWERF, A. (1977) Notes on the birds of St. Martin, Saba, and St. Eustatius. *Studies on the Fauna of Curaçao and other Caribbean islands* 54(176): 60–123.

RAFFAELE, H. WILEY J., GARRIDO, O., KEITH, A. AND RAFFAELE, J. (1998) *A guide to the birds of the West Indies*. Princeton, New Jersey: Princeton University Press.

ROJER, A. (1997) Biological inventory of Sint Eustatius. Curaçao, Netherlands Antilles: Carmabi Foundation (Unpublished report)

STUFFERS, A. L. (1956) The vegetation of the Netherlands Antilles. *Studies on the Flora of Curaçao and other Caribbean Islands* 1(15).

VOOUS, K. H. (1955)*The birds of the Netherlands Antilles*. Curaçao: Uitg. Natuurwet. Werkgroep Ned. Ant.

VOOUS, K. H. (1983) *Birds of the Netherlands Antilles*. Zutphen, The Netherlands: De Walburg Pers.

WALSH-MCGEHEE, M. (2000) Status and conservation priorities for White-tailed and Red-billed Tropicbirds in the West Indies. Pp 31–38 in E. A. Schreiber and D. S. Lee, eds. *Status and conservation of West Indian Seabirds*. Ruston, USA: Society of Caribbean Ornithology (Spec. Publ. 1).

ACKNOWLEDGEMENTS

The authors would like to thank Bert Denneman (Vogelbescherming Nederlands), Nicole Esteban (STENAPA) and Kalli de Meyer (Dutch Caribbean Nature Alliance) for their help in reviewing this chapter.

Figure 1. Location of Important Bird Areas in St Eustatius.

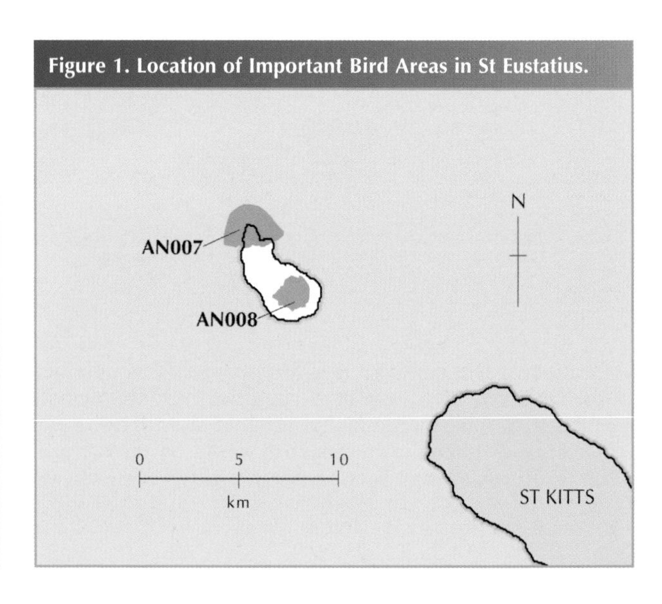

■ Site description

Boven IBA is an area of xeric, uninhabited rocky hills in the north-west peninsula of St Eustatius that represents about 25% of the island's land area. Boven, Venus, Gilboa Hill, Signal Hill and Bergje comprise the "Northern Hills". These receive much less rainfall than the higher Quill IBA (AN008) resulting in a predominance of *Acacia* thorn scrublands and grassland. Rocky outcroppings are scattered through the hills. The coastline of Boven IBA is cliffs and rocky shore, from which this IBA extends 1 km out to sea. Zeelandia beach (and adjacent sandy cliffs) is at the south-east end of the area, and an oil transfer station is situated on the western border.

■ Birds

This IBA supports a globally significant population of Red-billed Tropicbird *Phaethon aethereus* and is important for four (of the 8) Lesser Antilles EBA restricted-range birds. *Phaethon aethereus* nest at Zeelandia beach and in the hills above the airport. Surveys in 2003 by Environmental Protection in the Caribbean (which did not cover all of the IBA due to access issues and limits on time) resulted in an estimate of 100–200 breeding individuals.

■ Other biodiversity

Restricted-range reptiles in the IBA include Lesser Antillean iguana *Iguana delicatissima* (Vulnerable), lizards *Ameiva erythrocephala*, *Anolis bimaculatus* and *A. wattsi*, and geckos *Sphaerodactylus sputator* and *S. sabanus*. The Near Threatened tree bat *Ardops nichollsi montserratensis* has a restricted range, and the endemic Statia morning glory *Ipomoea sphenophylla* can be found growing in the IBA.

■ Conservation

This mixed-ownership IBA is one sub-sector (525 ha) of the Quill–Boven National Park (the Netherlands Antilles' first park). The NGO park service, STENAPA, is attempting to manage Boven and limit grazing, although these efforts have met with resistance due to ancestral (disputed) claims to the land. Grazing continues within the protected area although vegetation has recovered from previous periods of clearance for agriculture and cattle rearing. Recreational use of the area has been hampered due to the land disputes, and visitors are rare. *Phaethon aethereus* are threatened by introduced alien predators (rats and cats), trampling of nest burrows by goats, erosion due to overgrazing and potential oiling due to spills from oil transfer station. The St Eustatius Marine Park surrounds the island and includes the 491-ha extension of this IBA.

■ Site description

The Quill IBA is a dormant volcano at the south-eastern end of St Eustatius. Rising to over 600 m—the highest point on the island—it dominates the landscape. The IBA follows the national park boundaries which include the volcanic cone above 250 m (vegetated with thorn scrub transitioning to semi-evergreen seasonal forest on the north-west slope), the crater (which supports evergreen seasonal forest), the rim of the volcano (a small portion of which supports elfin forest), and the "White Wall" (a limestone formation on the southern slope of the volcano that drops down to sea-level). The IBA extends 1 km out to sea. There are no human settlements within the IBA.

■ Birds

This IBA is significant for supporting populations of all eight Lesser Antilles EBA restricted-range birds found on the island. A Scaly-breasted Thrasher *Margarops fuscus* was observed at the Quill in 2003 for the first time since 1927 although its status as a breeding bird (and thus a potential ninth restricted-range species) is unknown. Seabirds breed on the coast at White Wall, but the numbers are not thought to be significant internationally.

■ Other biodiversity

The restricted-range Antillean iguana *Iguana delicatissima* (Vulnerable) and red-bellied racer *Alsophis rufiventris* (Endangered), and the island endemic Statia morning glory *Ipomoea sphenophylla* are present in the IBA.

■ Conservation

This state-owned IBA is one sub-sector (340 ha) of the Quill–Boven National Park (the Netherlands Antilles' first park). STENAPA—the island's NGO park service—oversees conservation management for the Quill. Hiking trails are well maintained by volunteers and staff, and botanical interests, such as the endemic Statia morning glory, are protected. Grazing by untended goats has been problematic in the past. However, enforcement of an animal registry program has begun and should decrease the incidence of grazing within the park. The only known recent avian survey of the Quill was conducted by Environmental Protection in the Caribbean in March 2003. However, access to the crater forest was restricted due to trail repairs. The St Eustatius Marine Park surrounds the island and includes the 130-ha marine extension of this IBA.

ST KITTS & NEVIS

LAND AREA **261 km²** ALTITUDE **0–1,156 m**
HUMAN POPULATION **42,700** CAPITAL **Basseterre**
IMPORTANT BIRD AREAS **3, totalling 65.7 km²**
IMPORTANT BIRD AREA PROTECTION **90%**
BIRD SPECIES **196**
THREATENED BIRDS **0** RESTRICTED-RANGE BIRDS **10**

NATALIA COLLIER AND ADAM BROWN
(ENVIRONMENTAL PROTECTION IN THE CARIBBEAN)

St Kitts Central Forest Reserve. (PHOTO: KATE ORCHARD)

INTRODUCTION

The Federation of St Kitts and Nevis comprises two islands in the Leeward Islands at the northern end of the Lesser Antilles. The islands lie c.13 km south-east of St Eustatius, 67 km west of Antigua and 25 km north-west of Redonda (to Antigua and Barbuda). St Kitts is the larger island (168 km²) supporting c.75% of the population, and is separated from Nevis (93 km²) by a shallow 3-km channel called "the Narrows". Nevis is south-east of St Kitts. The islands constitute the smallest nation in the Americas in terms of both area and population, and were also among the first islands in the Caribbean to be settled by Europeans.

Both islands are volcanic in origin. The northern end of St Kitts is dominated by the dormant volcano Mount Liamunga (1,156 m) which is cloaked in tropical moist forest at higher elevations while the lower slopes have been cleared for agriculture (sugarcane production ended in 2007). Streams flow down the numerous ghauts on the sides of the volcano. The northern coastline of St Kitts is primarily cliffs of up to 15 m high. The south-east of the island comprises low hills supporting dry thorn-forest along a tapering peninsula used for grazing livestock and for tourism development. The southern coastline is characterised by sandy or rocky shores with saline coastal ponds. Nevis is a conical-shaped island formed by an extinct volcano that rises to 985 m at Nevis Peak, and is similarly vegetated to northern St Kitts. The island is fringed on three sides by long sandy beaches, and protected by (intermittent) fringing coral reefs. St Kitts and Nevis have a tropical maritime climate. Average annual rainfall is 1,300 mm although low-lying areas can receive as little as 200 mm and it can reach over 3,800 mm on the volcanoes (up to 6,000 mm has been recorded). Most of the precipitation arrives during hurricane season in late summer and fall. Modern land-use (on St Kitts) includes agricultural plantations in the lowlands, local market agriculture, tourism development (especially from Basseterre to Frigate Bay and the South-east Peninsula), fishing, and light manufacturing.

Conservation

On St Kitts the Central Forest Reserve was created as a result of the Forestry Ordinance (1904) which legislates for the protection of areas above 300 m. This ordinance was enacted to prevent further deforestation (at the hands of the sugarcane industry) and thereby also protect soil and water sources. The Wild Birds Protection Ordinance (1913) prohibits the hunting of 18 species of bird, and established a regulated hunting season for another nine species, reflecting the number of bird species known on the island at this time. The National Conservation and Environment Protection Act (1987) legislates for further protection of c.90 species. However, the extent to which these laws are enforced is unknown. There are no known officially protected areas in Nevis, but there is also no hunting of any species on the island. The Brimstone Hill Fortress National Park (on the west slope of Mount Liamuiga, but outside the Central Forest Reserve) is a World Heritage Site.

There is little systematic bird research or bird-specific conservation action currently being undertaken on the islands. Environmental Protection in the Caribbean (EPIC) conducted a survey of the South-east Peninsula Ponds and Booby Island in May 2004, and Steadman *et al.* (1997) provides data from their research on St Kitts in the 1970s and 1980s. Birds have been documented informally for many years on Nevis (e.g. through the maintenance and provision of birding lists for visitors). The NGO Nevis Historical and Conservation Society are active on the island, especially with education and awareness efforts and community conservation action.

Much of St Kitts was converted (by the British) to agricultural land dominated by sugarcane. St Kitts became Britain's most productive Caribbean supplier of sugar. Even though lands were abandoned as the sugar market fluctuated in the 1830s, much of the island remained cleared for sugarcane cultivation through the nineteenth century. Pesticides (such as DDT) were used on the sugarcane fields during the 1960s and impacted the bird populations of St Kitts. Nevis had abandoned the mass monoculture by this time. In spite of the clearances for agriculture, upper elevation forests have been spared and are largely protected. However, throughout all areas (including these montane forests) alien invasive predators are a problem. Of particular concern are the African green monkey *Cercopithecus aethiops* (brought by the French as a pet and established as a wild population since the 1700s) and the small Indian mongoose *Herpestes auropunctatus*, brought by the British from Jamaica in 1884 to reduce rat damage to sugarcane. These two species have undoubtedly reduced populations of native amphibians, reptiles and birds. Other invasives such as cows, goats and white-tailed deer are widespread, but their presence along pond shorelines (e.g. in the South-east Peninsula) means that they impact nesting shorebirds and terns through trampling. These ponds should be fenced to prevent trampling from these livestock. Development for the tourism industry (which takes little account of global biodiversity priorities) poses a huge threat to the future ecological integrity of the South-east Peninsula, and presumably other areas. The government has recently undertaken a campaign to remove unfenced and feral livestock from the peninsula in preparation for tourism development of the South-east Peninsula.

■ Birds

Over 190 species of birds have been documented (as certainly or hypothetically) occurring on St Kitts and Nevis, 37 are breeding residents (23 of which are landbirds). The majority are Neotropical migrants (or vagrants). Ten (of the 38) Lesser Antilles EBA restricted-range birds occur on both of the islands, none of which is endemic to the country. A number of these species (especially Bridled Quail-dove *Geotrygon mystacea* and Antillean Euphonia *Euphonia musica*) are restricted to the moist forested slopes and ghauts of the volcanoes. St Kitts did support an endemic subspecies of the Puerto Rican Bullfinch *Loxigilla portoricensis* (one of the Puerto Rico and the Virgin Islands EBA restricted-range birds). The "St Kitts" Bullfinch *L. p. grandis* was last recorded in 1929. Its apparent extinction was

Bridled Quail-dove.
(PHOTO: JIM JOHNSON)

Table 1. Key bird species at Important Bird Areas in St Kitts and Nevis.

Key bird species	Criteria	National population	St Kitts and Nevis IBAs KN001	KN002	KN003
Criteria			■	■	■
Brown Pelican *Pelecanus occidentalis*	■	168		168	
Laughing Gull *Larus atricilla*	■	375			375
Least Tern *Sterna antillarum*	■	200–250		195	15–255
Bridled Tern *Sterna anaethetus*	■	180			180
Bridled Quail-dove *Geotrygon mystacea*	■		✓		
Purple-throated Carib *Eulampis jugularis*	■		✓		
Green-throated Carib *Eulampis holosericeus*	■		✓		
Antillean Crested Hummingbird *Orthorhyncus cristatus*	■		✓		
Lesser Antillean Flycatcher *Myiarchus oberi*	■		✓		
Scaly-breasted Thrasher *Margarops fuscus*	■		✓		
Pearly-eyed Thrasher *Margarops fuscatus*	■		✓		
Brown Trembler *Cinclocerthia ruficauda*	■		✓		
Lesser Antillean Bullfinch *Loxigilla noctis*	■		✓		
Antillean Euphonia *Euphonia musica*	■		✓		

All population figures = numbers of individuals.
Restricted-range birds ■. **Congregatory birds** ■.

probably due to a combination of habitat loss (especially for sugarcane) exacerbated by forest-damaging hurricanes, and predation by non-native mammals, including monkeys, mongoose, cats, and rats. However, further searches for this taxon are probably warranted.

No globally threatened birds are present in significant numbers on the islands. The Vulnerable West Indian Whistling-duck *Dendrocygna arborea* may have nested on St Kitts in the past and was been seen "occasionally" up until the 1980s. The Near Threatened Piping Plover *Charadrius melodus* was recorded on St Kitts for the first time in 1988, and the species does occur on Nevis, albeit involving very small numbers. Waterbirds, seabird and shorebirds in the Federation are concentrated in the salt ponds of the South-east Peninsula IBA (KN002), on Booby Island IBA (KN003), and in Nevis' coastal lagoons and ponds. However, there seem to be considerable fluctuations and declines in the numbers of birds breeding. For example, 21 pairs of Snowy Plover *C. alexandrinus nivosus* were counted at Little and Great Salt Pond in 1985, but none were seen in 2004, then small numbers have been recorded since. Similarly, Roseate Tern *Sterna dougallii* breeds on the South-east Peninsula with 100–200 pairs estimated in the mid-1990s, but just 12 individuals found there in 1998 and 2004. There is a clear need for a regular census in order to better estimate species populations, breeding sites, and local movements. Magnificent Frigatebirds *Fregata magnificens* and Laughing Gulls *Larus atricilla* are reported to nest in the Hurricane Hill–Newcastle area, but in unknown numbers. Least Tern *Sterna antillarum* are reported to breed at the north end of White Bay, but again, numbers are unknown. A survey of nesting seabirds on Nevis is long overdue.

Figure 1. Location of Important Bird Areas in St Kitts and Nevis.

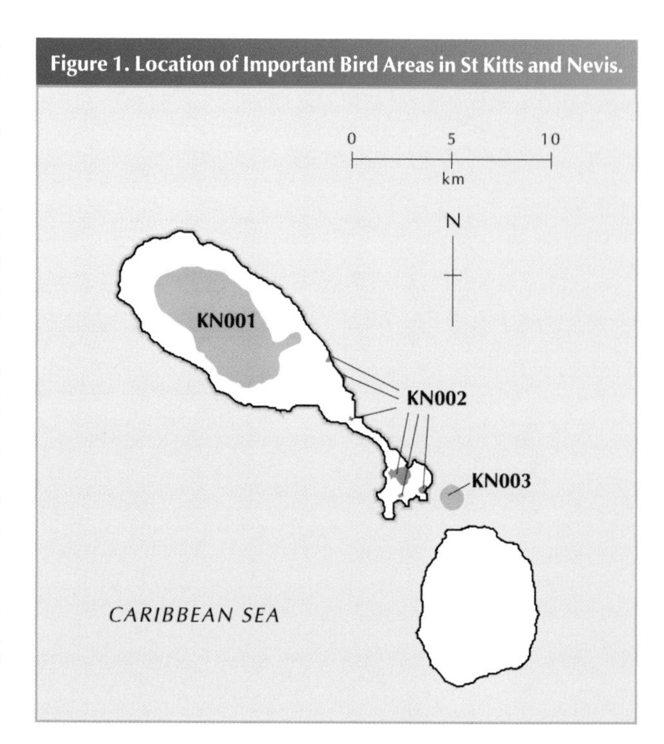

IMPORTANT BIRD AREAS

St Kitts and Nevis' three IBAs—the nation's international priority sites for bird conservation—cover 6,575 ha (including marine areas), and c.24% of the country's land area. While the St Kitts Central Forest Reserve IBA (KN001) is protected, neither of the congregatory bird IBAs receives any form of protection or conservation management at the current time.

The three IBAs have been identified on the basis of 15 key bird species that variously trigger the IBA criteria. The restricted-range landbirds are represented only in the Central Forest Reserve IBA and almost nothing is known about their status or abundance within the reserve. For example, the Bridled Quail-dove *Geotrygon mystacea* was hunted (legally) on St Kitts, but little is known of what impact this had, or whether hunting still occurs. Many of the restricted-range species occur also on Nevis. With further details of their distribution and status it might be appropriate to define a terrestrial IBA for the island (and a second internationally-recognised area for these species within the Federation).

In order to maintain the Federation's biodiversity, there is an urgent need to instigate protective measures for the South-east Peninsula Ponds IBA, and for Booby Island IBA. The annual take by hunters of *G. caniceps* on St Kitts and its current status need to be determined. Similarly there is a need for some baseline assessments of congregatory bird populations on Nevis, and for regular monitoring of these same species groups on St Kitts. Such monitoring should be used to inform the assessment of state, pressure and response scores at each of the Federation's IBAs in order to provide objective status assessments and highlight management interventions that might be required to maintain these internationally important biodiversity sites.

South-east Peninsula with Nevis on the horizon. (PHOTO: KATE ORCHARD)

Great Salt Pond, South-east Peninsula. (PHOTO: KATE ORCHARD)

KEY REFERENCES

ARENDT, W. J. (1985) Wildlife assessment of the Southeastern Peninsula, St Kitts, West Indies. Puerto Rico: USDA Forest Service, Institute of Tropical Forestry. (Unpublished report).

BURDON, K. J. (1920) *A handbook of St Kitts-Nevis*. London: The West India Committee (Government of St. Kitts-Nevis).

CHILDRESS, R. B. AND HUGHES, B. (2001) The status of the West Indian Whistling-duck (*Dendrocygna arborea*) in St Kitts-Nevis, January–February 2000. *El Pitirre* 14: 107–112.

DANFORTH, S. T. (1936) The birds of St Kitts and Nevis. *Trop. Agric.* 13: 213–217.

FAABORG, J. (1985) Ecological constraints on West Indian bird distributions. Pp 621–653 in P. A. Buckley, M. S. Foster, E. S. Morton, R. S. Ridgely and F. G. Buckley, eds. *Neotropical ornithology*. Washington, D.C.: American Ornithologists' Union (Ornithol. Mon. 36).

VAN HALEWYN, R. AND NORTON, R. L. (1984) The status and conservation of seabirds in the Caribbean. Pp 169–222 in J. P. Croxall, P. G. H. Evans and R. W. Schreiber, eds. *Status and conservation of the world's seabirds*. Cambridge, U.K.: International Council for Bird Preservation (Techn. Publ. 2).

HILDER, P. (1989) *The birds of Nevis*. Charlestown, Nevis: Nevis Historical and Conservation Society.

RAFFAELE, H. WILEY J., GARRIDO, O., KEITH, A. AND RAFFAELE, J. (1998) *A guide to the birds of the West Indies*. Princeton, New Jersey: Princeton University Press.

SALIVA, J. E. (2000) Conservation priorities for Roseate Terns in the West Indies. Pp 87–95 in E. A. Schreiber and D. S. Lee, eds. *Status and conservation of West Indian Seabirds*. Ruston, USA: Society of Caribbean Ornithology (Spec. Publ. 1).

STEADMAN, D. W., NORTON, R. L., BROWNING, M. R. AND ARENDT, W. J. (1997) The birds of St Kitts, Lesser Antilles. *Carib. J. Sci.* 33(1–2): 1–20.

STIMMELMAYR, R. (2007) At last a Least: Least Tern (*Sterna antillarium antillarium*) nesting on the Atlantic coast of St Kitts, Lesser Antilles. (Unpublished report for St Christopher Heritage Society).

STREET, H. M. (1994) Birds seen at St. Kitts, March 12–19, 1994. Ancaster, Canada. (Unpublished report).

TOWLE E., ARCHER, A., BUTLER, P. J., COULIANOS, K. E., GOODWIN, M. H., GOODWIN, S. T., JACKSON, I., NICHOLSON, D., RAINEY, W. E., TOWLE, J. A., WERNICKE, W. AND LIVERPOOL, N. J. O. (1986) Environmental assessment report on the proposed southeast peninsula access road, St Kitts, W.I. St Thomas, USVI: Island Resources Foundation. (Unpublished report).

VITTERY, A. (2006) The ornithological and ecological importance of wetland sites and the eco-tourism potential of birds in St Kitts. (Unpublished report for St Christopher Heritage Society).

ACKNOWLEDGEMENTS

The authors would like to thank Brooks Childress, Randolph Edmead (Dept. Physical Planning and Environment), Jim Johnson (Walk Nevis), Kate Orchard (St Christopher Heritage Society), Michael Ryan and Raphaela Stimmelmayr for their contributions to this chapter.

■ Site description

St Kitts Central Forest Reserve IBA comprises the north–south mountain range in the north of St Kitts. It embraces all areas over 300 m elevation from Mount Liamuiga—the highest point on the island—south to Mount Olivee. Canyons, or ghauts, radiate from the peak, and tropical forest (including areas of palm brake and elfin forest) covers the majority of the IBA. The Brimstone Hill Fortress National Park (and World Heritage Site) is located on the western slope of Mount Liamuiga but is not included within the IBA boundary.

■ Birds

This IBA is significant for its populations of restricted-range species, with all 10 of the Lesser Antilles EBA birds occurring, a number of which are confined to these forests on the island. Specific locations and populations for these species are unknown although it has been reported that all except the Green-throated Carib *Eulampis holosericeus* are common in undisturbed moist forest on St Kitts. Six Neotropical migrant species are known from the tropical moist forest on the island.

■ Other biodiversity

No endemic or threatened plant species are reported for St Kitts. Restricted-range bats and herpetofauna are likely to exist within the IBA.

■ Conservation

This IBA is state owned and protected as the Central Forest Reserve. This reserve is part of the Organization of Eastern Caribbean States Protected Areas and Associated Livelihoods Project which aims to develop sustainable resource extraction which benefits local residents. There is no ongoing avian research taking place in the IBA. Human settlement is limited to the lower elevations—where sugarcane was farmed—while villages are confined to coastal areas. Although various laws protect many bird species from hunting, the degree to which these laws are enforced is unknown. The Bridled Quail-dove *Geotrygon mystacea* is considered a game bird, but it is not known if hunting is posing a threat to the species' long-term survival on the island.

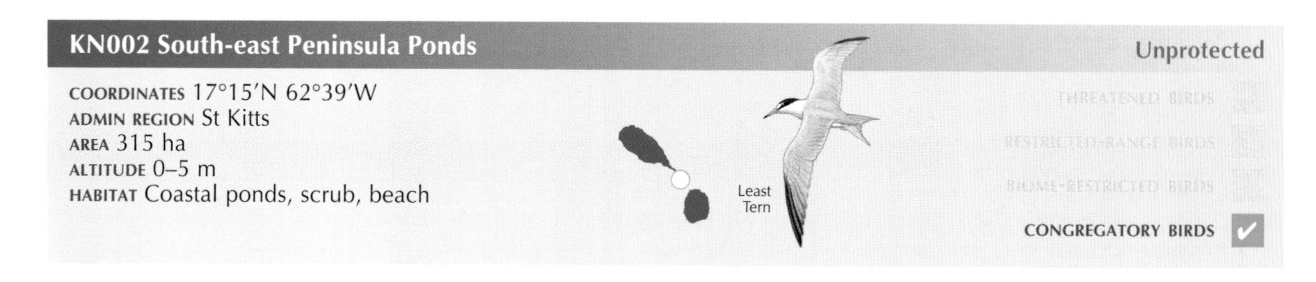

■ Site description

South-east Peninsula Ponds IBA encompasses an arid area of low hills, eight salt ponds, coastal cliffs, and beaches on the South-east Peninsula of St Kitts. The peninsula is c.15 km long, widening to c.4 km wide at the south-eastern tip although the narrowest point is less than 1 km wide. A road runs the length of the peninsula which supports thorn forest dominated by *Acacia* spp. and grassland. Ponds of importance to birds on the peninsula include Greatheeds Pond and beach, Half Moon, Friar's Bay, Great Salt, Major's Bay, Mosquito Bay, Little Salt, and Frigate Bay Ponds. The IBA is delimited by boundaries 30 m from the high water line of each pond.

■ Birds

South-east Peninsula Ponds IBA is important for breeding seabirds and waterbirds. The populations of Least Tern *Sterna antillarum* and Brown Pelican *Pelecanus occidentalis* are regionally significant. The *S. antillarum* colonies have used a number of sites within the IBA suggesting that all eight ponds are important for this population. The current status of Roseate Tern *S. dougallii* on the peninsula (juveniles were seen in the mid-1990s) needs to be determined. Up to 2,000 shorebirds and hundreds of ducks have been recorded.

■ Other biodiversity

The Critically Endangered leatherback *Dermochelys coriacea* and hawksbill *Eretmochelys imbricata*, and Endangered green *Chelonia mydas* turtles are known to nest in this IBA.

■ Conservation

The South-east Peninsula Ponds IBA is not protected, although building is prohibited on steep slopes. There is no ongoing research or conservation work in the IBA, although EPIC has recently undertaken periodic bird surveys. The St Kitts Historical Society has conducted sea turtle conservation activities. Tourism is currently concentrated in the northern section of the peninsula, which is dominated by resorts, a golf course, and restaurants. Existing developments on the peninsula are limited to snack bars and restaurants on the relatively isolated beaches. It is unknown if hunting is a threat to the breeding seabirds, but grazing is a serious threat to the nesting *S. antillarum*. Colonies are marked with a high density of cattle tracks and it is likely nests are frequently trampled. A mega resort (Christophe Harbour), which will impact the entire peninsula, is already in the planning stages. It plans to include a golf course, multiple hotels and villas, and will have to dredge an unidentified pond for a mega-yacht marina.

■ Site description

Booby Island IBA is a small (1-ha), circular islet located in "the Narrows", approximately half-way between the islands of St. Kitts and Nevis. A shoreline comprising large rocks encircles a steep hillside that supports a mix of dense, brushy vegetation and rocky outcroppings. The islet is uninhabited. The IBA includes all marine areas up to 1 km from the island.

■ Birds

This IBA is significant for its breeding seabirds—c.425 pairs of various species. The colony of Laughing Gulls *Larus atricilla* is globally important, while the Bridled Terns *Sterna anaethetus* are of regional significance. Other species that breed on Booby Island include Red-billed Tropicbird *Phaethon aethereus*, Roseate Tern *S. dougallii*, Sooty Tern *S. fuscata* and Brown Noddy *Anous stolidus*. It is the only recorded breeding site within the Federation of St Kitts and Nevis for a number of these species.

■ Other biodiversity

Nothing recorded.

■ Conservation

Booby Island is unprotected and appears to be the property of the Federation (i.e. it is state owned). There is no easy boat access to the island and visitors must swim onto the rocks. Fishermen are reported to collect seabird eggs at this site, especially those of *L. atricilla*. The impact of this is unknown. No mammals were recorded during the one known survey of the island in 2004.

ST LUCIA

LAND AREA **616 km²** ALTITUDE **0–950 m**
HUMAN POPULATION **170,650** CAPITAL **Castries**
IMPORTANT BIRD AREAS **5, totalling 155 km²**
IMPORTANT BIRD AREA PROTECTION **70%**
BIRD SPECIES **162**
THREATENED BIRDS **6** RESTRICTED-RANGE BIRDS **23**

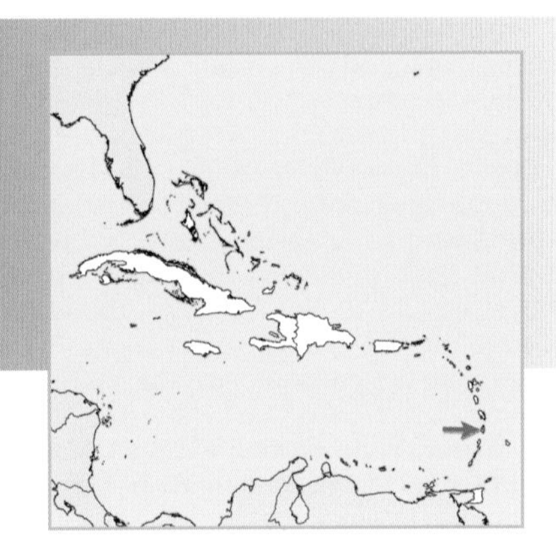

DONALD ANTHONY AND ALWIN DORNELLY
(FORESTRY DEPARTMENT, MINISTRY OF AGRICULTURE)

The Pitons. (PHOTO: DONALD ANTHONY)

INTRODUCTION

St Lucia is in the Windward Islands of the Lesser Antilles and lies between Martinique (to France), 28 km to the north, and St Vincent, 31 km to the south. It is 45 km long (north to south) and 21 km at its widest. St Lucia is a mountainous, volcanic island with a main axial ridge that stretches from La Sorciere in the north to Saltibus in the south. Mount Gimie is the island's tallest mountain. Two spectacular pitons (volcanic plugs) rise from the sea in the south-west of the island—Gros Piton (798 m) and Petit Piton (743 m). The Pitons are a Caribbean landmark and are designated a World Heritage Site due to their unique beauty and geology. The island's forested mountainous and hilly interior is incised by steep valleys resulting from the numerous streams that emanate in this rugged terrain although there are also some broad, fertile valleys. St Lucia's tropical climate has two seasons: a dry season from December to May, and a wet season from June to November, although there appears to be increasing variance from this norm. Rainfall is highest in the wet, mountainous interior and lowest in the dry coastal zone resulting in wet tropical forest (primary and secondary) cloaking the main ridge with scrub forest and dry woodland (mostly now degraded) along the coast. A small area of elfin woodland is found at the top of Mount Gimie. Land use can be classified as forest (c.35%), agriculture (c.29%, and primarily permanent crops such as banana), residential, commercial and industrial.

Conservation

Government legislation for biodiversity conservation in St Lucia includes the 1980 Wildlife Protection Act (which provides for the protection of wildlife and the establishment of wildlife reserves), the 1984 Fisheries Act (which provides for the creation of marine reserves), and the 1983 Forest, Soil and Water Conservation Act (which contains provisions governing the declaration of forest reserves and protected forests on private land).

Almost 35% of St Lucia is still under some form of forest cover. Over a third of this (c.7,690 ha) is protected within the Government Forest Reserve IBA (LC002). The remaining forest is mostly privately owned. There is no protection afforded the dry coastal forest. The St Lucia National Trust (SLNT)—a quasi-governmental organisation under the 1975 National Trust Act—has management authority for some offshore islets, and a few parks and protected areas amounting to c.255 ha. The St Lucia iguana *Iguana iguana*, St Lucia whiptail *Cnemidophorus vanzoi* and White-breasted Thrasher *Ramphocinclus brachyurus* projects—lead by Durrell Wildlife

Conservation Trust in collaboration with the Forestry department—are providing a focus for research activities within the unprotected dry forest habitat (especially on the east coast). These initiatives started in 2002 with funding from Durrell. Forestry Department and Durrell are implementing a monitoring program for the Endangered *R. brachyurus* throughout its St Lucian range, but with a particular focus on the habitat changes resulting from the hotel and resort development at Praslin Bay—the centre of the species' abundance. A pilot project using distance-sampling to determine the population of the Vulnerable St Lucia Parrot *Amazona versicolor* was field-tested by staff of the Forestry Department and Durrell during 2007. Both the parrot and thrasher surveys include components to estimate populations of other selected forest bird species.

Agricultural expansion (especially banana cultivation), residential, hotel and resort developments, and roads are the main causes of deforestation and habitat degradation of St Lucia's dry forest, wet tropical forest, mangroves, littoral vegetation and wetlands (and thus are the main threats to the IBAs). Most livestock are tethered but sheep, goats and cattle do have a localised impact, especially on the dry woodlands which are also sometimes affected by dry-season fires lit by farmers and others. Of potentially greater impact to the island's biodiversity are the alien invasive predators including mongoose *Herpestes auropunctatus*, rats *Rattus* spp., pigs *Sus scrofa*, and Giant African snail *Achalina* sp. The actual impact of these alien invasives has not been quantified although the mongoose has been identified as a cause for the probable extinction of the (Critically Endangered) ground-dwelling Semper's Warbler *Leucopeza semperi*. Hurricanes are an ever-present threat: the last one to hit the island was Hurricane Allen in 1980 which damaged or destroyed over 80% of the island's forest. The increasing destruction, degradation and fragmentation of habitats is significantly reducing their resilience to the impacts of stochastic events such as hurricanes, and also the forecasted effects of global climate change.

Golf course development, Praslin Bay.
(PHOTO: HESTER WHITEHEAD)

■ Birds

Over 160 species of bird have been recorded from St Lucia, 97 of which breed. The island is internationally important for its resident populations of six globally threatened bird species and also its restricted-range birds. The threat category and estimated national population sizes of the globally threatened birds are listed in Table 1. Semper's Warbler *Leucopeza semperi* is considered Critically Endangered but is only

Table 1. Key bird species at Important Bird Areas in St Lucia.

			St Lucia IBAs				
			LC001	LC002	LC003	LC004	LC005
Key bird species	Criteria	National population	■ ■ ■	■ ■	■ ■ ■	■ ■ ■	■ ■
Masked Duck *Nomonyx dominicus*	■	<50	<50				
Red-billed Tropicbird *Phaethon aethereus*	■	50–249	<50				<50
Royal Tern *Sterna maxima*	■	50–249	<50		<50	<50	<50
Roseate Tern *Sterna dougallii*	■	50–249					50–100
Bridled Tern *Sterna anaethetus*	■	250–999					250–500
Sooty Tern *Sterna fuscata*	■	20,000–49,999					20,000–49,999
Bridled Quail-dove *Geotrygon mystacea*	■			✓		✓	
St Lucia Amazon *Amazona versicolor*	VU ■ ■	800–1,200		800–1,000		✓	
Lesser Antillean Swift *Chaetura martinica*	■			✓			
Purple-throated Carib *Eulampis jugularis*	■		✓	✓	✓	✓	
Green-throated Carib *Eulampis holosericeus*	■			✓		✓	
Antillean Crested Hummingbird *Orthorhyncus cristatus*	■		✓	✓	✓	✓	
Caribbean Elaenia *Elaenia martinica*	■		250–999	1,000–2,499	<50	<50	
Lesser Antillean Pewee *Contopus latirostris*	■		250–999	2,500–9,999	<50	<50	
Lesser Antillean Flycatcher *Myiarchus oberi*	■		✓	✓	✓	✓	
White-breasted Thrasher *Ramphocinclus brachyurus*	EN ■ ■	1,200–1,600	100			1,010–1,130	
Scaly-breasted Thrasher *Margarops fuscus*	■		✓	✓	✓	✓	
Pearly-eyed Thrasher *Margarops fuscatus*	■			✓		✓	
Brown Trembler *Cinclocerthia ruficauda*	■		✓	✓	✓		
Grey Trembler *Cinclocerthia gutturalis*	■		✓	✓	✓	✓	
Rufous-throated Solitaire *Myadestes genibarbis*	■			✓			
Forest Thrush *Cichlherminia lherminieri*	VU ■ ■	<50		✓			
St Lucia Warbler *Dendroica delicata*	■		✓	✓	✓		
Semper's Warbler *Leucopeza semperi*	CR ■ ■	<50		?			
St Lucia Oriole *Icterus laudabilis*	NT ■ ■	2,500–9,999	100–300	2,500–9,999	100	100–200	
Lesser Antillean Bullfinch *Loxigilla noctis*	■		✓	✓	✓		
St Lucia Black Finch *Melanospiza richardsoni*	EN ■ ■	2,500–9,999	500	2,500–9,999	200	✓	✓
Antillean Euphonia *Euphonia musica*	■		✓	✓	✓	✓	
Lesser Antillean Saltator *Saltator albicollis*	■		✓	✓	✓		

All population figures = numbers of individuals.
Threatened birds: Critically Endangered ■; Endangered ■; Vulnerable ■; Near Threatened ■. **Restricted-range birds** ■. **Congregatory birds** ■.

St Lucia's Endangered White-breasted Thrasher.
(PHOTO: GREGORY GUIDA)

The endemic St Lucia Warbler.
(PHOTO: GREGORY GUIDA)

possibly still extant. The Endangered White-breasted Thrasher *Ramphocinclus brachyurus* is better known, being confined to the dry forests on the east coast, although its habitat is currently unprotected and is a major conservation priority for the island. The Near Threatened, migratory Buff-breasted Sandpiper *Tryngites subruficollis* has also been recorded from St Lucia on three occasions but has not been used to identify IBAs. Twenty-three (23) (of the 38) Lesser Antilles EBA restricted-range birds occur in this country. Five of these are endemic to the island, namely St Lucia Amazon *A. versicolor*, St Lucia Warbler *Dendroica delicata*, *L. semperi*, St Lucia Black Finch *Melanospiza richardsoni* and St Lucia Oriole *Icterus laudabilis*. A number of endemic subspecies of birds are present on the island, the most threatened being the Rufous Nightjar *Caprimulgus rufus otiosus* which is local and uncommon in the unprotected dry forests of the east coast. The Lesser Antillean Flycatcher *Myiarchus oberi santaeluciae* is also uncommon, but is relatively secure within the well protected wet forests.

IMPORTANT BIRD AREAS

St Lucia's five IBAs—the island's international site priorities for bird conservation—cover 15,505 ha or 25% of the island's land area. The critical habitat components of three of these IBAs are formally designated as protected areas. The remaining two IBAs, North-east coast (LC001) and Mandele Dry Forest (LC004) are unprotected in terms of their critical forest habitat, but both enjoy some protection of their marine and wetland areas. Approximately 70% of the land area of St Lucia's IBAs is formally protected, and the Government

Figure 1. Location of Important Bird Areas in St Lucia.

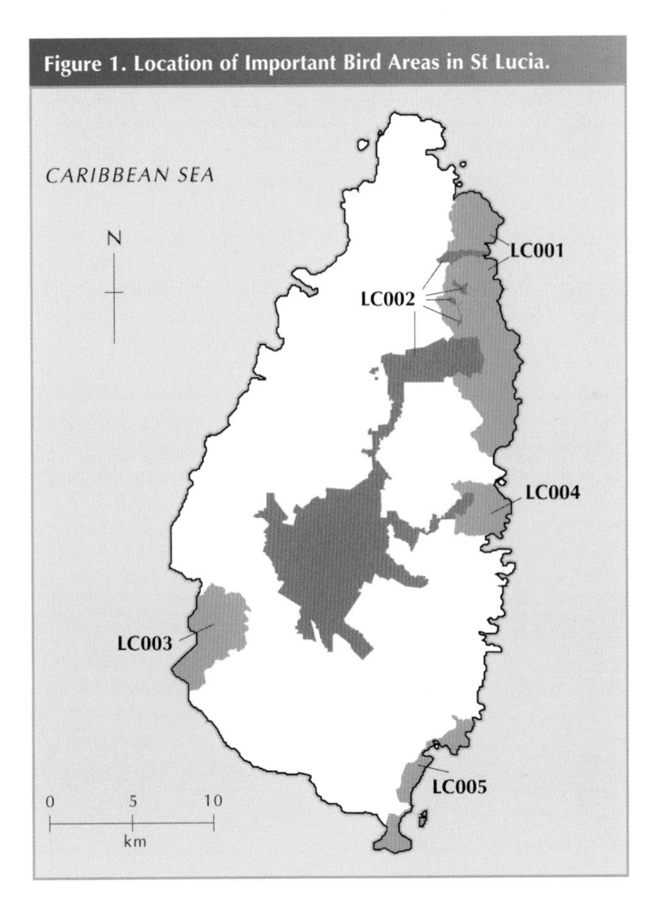

Dry forest at Louvet River, Mandele Dry Forest (IBA LC004). (PHOTO: MATTHEW MORTON, DWCT)

Forest Reserve IBA (LC002) protects 30% of St Lucia's remaining forest.

The IBAs have been identified on the basis of 29 key bird species that variously trigger the IBA criteria. The majority of these birds occur in two or more of the IBAs. However, some of the moist forest dependent species are found only in the Government Forest Reserve IBA. This protected IBA supports the entire breeding population of St Lucia Amazon *Amazona versicolor*, and a large percentage of the St Lucia Black Finch *Melanospiza richardsoni* population. It also protects all but one of the restricted-range birds that occur on St Lucia. Of greatest concern is the White-breasted Thrasher *Ramphocinclus brachyurus*. It occurs in two IBAs (North-east coast and Mandele Dry Forest, which encompass c.97.5% of the species' St Lucia population), but the bird's dry forest habitat in these IBAs is totally unprotected.

There is an urgent need for the formal protection of critical dry forest habitat (probably through purchasing habitat presently in private hands) within the two IBAs that encompass the range of *R. brachyurus*. Such protection would need to be an integral part of a sensitively designed strategic development plan for the east coast dry forests—a plan that takes into account the needs of this region's unique biodiversity. Little is know of the populations of St Lucia's globally threatened or restricted-range birds (apart from *A. versicolor* and *R. brachyurus*). Establishing the population status and subsequent monitoring of these priority species is a critical need, and could perhaps be done as an extension to the ongoing St Lucia Amazon and White-breasted Thrasher projects. An assessment of the impacts of invasive alien species (including Shiny Cowbird *Molothrus bonairensis*, a brood-parasite of the St Lucia Oriole *Icterus laudabilis*) would also provide important inputs to site management strategies. State, pressure and response scores at each IBA should be monitored annually to provide an objective status assessment and highlight management interventions that might be required to maintain these internationally important biodiversity sites.

KEY REFERENCES

ANON. (2000) *St Lucia biodiversity country report*. Castries, St Lucia: Ministry of Agriculture, Forestry and Fisheries.

ANTHONY, D. (1977) Inventory of flora and fauna of Gros Piton, St Lucia. St Lucia: World Wildlife Fund /USAID /ENCORE (Unpublished report).

ANTHONY, D. (2005) Inventory of the flora and fauna of the Chassin/La Sorciere Area, St Lucia. St Lucia: St. Lucia Rain Forest Tram Inc. (Unpublished report)

COX, C. (1999) A rapid inventory of the flora and fauna of Petit Piton and the ridge between the Pitons. St Lucia: St Lucia World Heritage Committee (Unpublished report).

DE BEAUVILLE, S. (2004) Lessons learnt during the development and implementation of the Coastal Zone Management Project (2001–2002), St Lucia. Castries, St Lucia: Department of Fisheries. (Unpublished report)

ESPEUT, P. (2006) Opportunities for sustainable livelihoods in one protected area in each of the six independent OECS territories. Unpublished report for the OECS Protected Areas and Sustainable Livelihoods (OPAAL) Project. OECS Contract Number OECS/121/05.

GEOGHAGEN, T. AND SMITH, A. H. (1998) *Conservation and sustainable livelihoods: collaborative management of the Mankòtè mangrove, St Lucia*. Vieux Fort, St Lucia: Caribbean Natural Resources Institute. (Community Participation in Forest Management: Case Study Series, 1).

GONZALES, O. J. AND DONALD, Z. R. (1996) Tropical dry forest of St Lucia West Indies: vegetation and soil properties. *Biotropica* 28(4b): 618–626

HOWARD, R. A. (1979) *Flora of the Lesser Antilles—Windward and Leeward islands, 1–4*. Jamaica plain, Massachusetts: Arnold Arboretum, Harvard University.

HUDSON, L., RENARD, Y. AND ROMULUS, G. (1992) A system of protected areas for St Lucia. Castries, St Lucia: St Lucia National Trust (Unpublished report).

ISAAC, C. L., PIERRE, L. L. AND ANTHONY, D. (1990) A report on plant inventory, ethnobotanical study and community resource use survey from Dennery Knob to Grande Anse in St Lucia. Castries, St Lucia: St Lucia National Trust (Unpublished report).

JENNINGS-CLARK, S. AND D'AUVERGNE, C. (1990) An assessment of the coastal marine habitats of Savannes Bay, St Lucia: mangroves, seagrass and reef. St Lucia: Department of Fisheries, Ministry of Agriculture, Forestry, Fisheries and the Environment (Unpublished report).

KEITH, A. R. (1997) *The birds of St. Lucia, West Indies: an annotated check-list*. Tring, UK: British Ornithologists' Union (BOU Check-list No 15).

PIERRE, L. L. (1988) Report on the status of the flora and fauna of the area of the proposed Piton National Park. Castries, St Lucia: St Lucia National Trust (Unpublished report)

RAFFAELE, H. WILEY J., GARRIDO, O., KEITH, A. AND RAFFAELE, J. (1998) *A guide to the birds of the West Indies*. Princeton, New Jersey: Princeton University Press.

TOWLE, J. A. AND TOWLE, E. L., EDS. (1991) *St Lucia country environmental profile*. St Michael, Barbados: Caribbean Conservation Association.

ACKNOWLEDGEMENTS

The authors would like to thank Lyndon John, Michael Bobb and Michael Andrew (Forestry Department, Ministry of Agriculture), and Matthew Morton (Durrell Wildlife Conservation Trust) for their help in drafting this chapter.

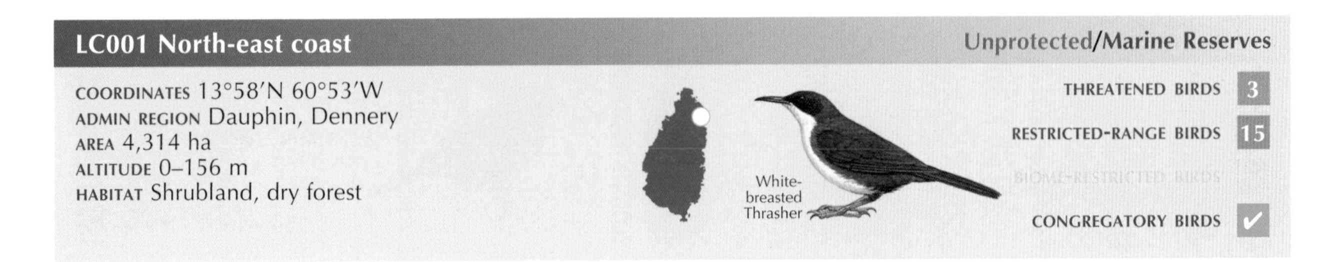

■ Site description

The North-east coast IBA stretches from the Dennery Knob westward to the Marquis Estate just outside the periphery of the Government Forest Reserve IBA (LC002). It then runs north-east, adjacent to the Northern Range extension of the Government Forest Reserve IBA to Grand Anse. This IBA covers a considerable portion of St Lucia's tropical dry forest life zone and is characterised by low canopy, scrub forest. The IBA comprises a mix of agriculture, pastureland and undeveloped secondary forest cover.

■ Birds

This IBA is important for the Endangered White-breasted Thrasher *Ramphocinclus brachyurus* (holding c.7.5% of the St Lucia population), the Endangered St Lucia Black Finch *Melanospiza richardsoni*, and 15 (of the 23) Lesser Antilles EBA restricted-range birds. It is the last stronghold for the endemic subspecies of Rufous Nightjar *Caprimulgus rufus otiosus*, and supports populations of the Lesser Antillean Flycatcher *Myiarchus oberi santaeluciae* and House Wren *Troglodytes aedon martinicensis*. A pond at Grande Anse is the only known location where Masked Duck *Nomonyx dominicus* breeds on the island. Red-billed Tropicbird *Phaethon aethereus* may nest on the sea cliffs.

■ Other biodiversity

The endemic St Lucia boa constrictor *Constrictor orophias*, St Lucia viper *Bothrops caribbaeus*, and St Lucia race of the green iguana *Iguana iguana* occur. The Grand Anse beach is currently the most important nesting ground for three globally threatened turtle species. The adjacent Louvet beach is important for nesting iguanas. The St Lucia muskrat *Megalomys luciae*—thought to be extinct—may actually be present in this area.

■ Conservation

Five different marine reserves cover the coastal zone of this IBA but the terrestrial areas, which are mostly privately owned (e.g. Louvet and Grand Anse estates), are unprotected. The large estates are potential targets for development (following the trend set for east-coast tourist development by the Praslin Bay resort in the Mandele Dry Forest IBA). Deforestation for agriculture threatens much of this IBA, and leads to fragmentation which in turn increases the impact of the numerous invasive alien predators. Latanyé palm *Coccothrinax barbadensis* and a number of other plant species are over-exploited in the IBA. Forest Department and Durrell Wildlife Conservation Trust are working on *R. brachyurus* and the iguana in this IBA.

■ Site description

The Government Forest Reserve IBA straddles the central, volcanic mountain massif running from the north to the south of the island. The reserve, which supports high canopy forest throughout, is divided into management units: Northern Range (the Morne La Sorciere forest in the north, adjacent to the communities of Desbarras, Des Chassin, Forestiere and Marc), Dennery Range (east/central and adjacent to the communities Aux-Leon, Denniere Riviere, Grand Riviere and Dennery), Millet Range (west/central and bounded by the communities Millet, Ravine Poisson, Anse-la-Raye and Cannaries), Soufriere Range (south/western and adjacent to Fond St Jacques and Saltibus) and Quilesse Range (adjacent to the communities Desruisseaux and Tirocher in the south-east). This IBA abuts the North-east coast (LC001) and Mandele (LC004) IBAs.

■ Birds

This IBA is important for four globally threatened birds and 22 (of the 23) Lesser Antilles EBA restricted-range birds. It embraces the breeding range of the Vulnerable St Lucia Amazon *Amazona versicolor* (with c.800–1,000 birds in the IBA), and a large population of the Endangered St Lucia Black Finch *Melanospiza richardsoni*. The Vulnerable Forest Thrush *Cichlherminia lherminieri* is very rarely reported in St Lucia, but one at Morne La Sorciere in August 2007 (with subsequent records) confirms its continued presence in this IBA. The last confirmed sighting of the Critically Endangered Semper's Warbler *Leucopeza semperi* was in 1972—if it survives it will be within this IBA.

■ Other biodiversity

The endemic St Lucia boa *Constrictor orophias*, St Lucia viper *Bothrops caribbaeus*, St Lucia anole *Anolis luceae* and St Lucia pygmy gecko *Sphaerodactylus microlepis* occur in this IBA, as do at least nine endemic plants.

■ Conservation

The Government Forest Reserve IBA is a state-owned protected area that covers over a third of the island's remaining forest. There are no human settlements within the IBA although the reserve is used (to a limited degree) for tourism and recreational activities. Invasive alien (mammalian) predators are present in the reserve and presumably impact the native fauna. The forest was significantly damaged by Hurricane Allen in 1980. Research and conservation actions by Forestry Department and Durrell Wildlife Conservation Trust have focused on *A. versicolor* since the 1970s.

LC003 Pitons

■ Site description

The Pitons IBA comprises two steep, cone-shaped mountains rising abruptly side by side from the sea on the south-west coast of St Lucia near the coastal town of Soufriere. The Pitons are a spectacular area of iconic coastal scenery—familiar throughout the region. Gros Piton (3 km wide at its base, and 777 m high) and Petit Piton (1 km wide and 730 m high) are cores of two lava-dome volcanoes, and are joined by the Piton Mitan ridge. The Pitons support a productive forest ecosystem and an offshore fringing reef.

■ Birds

This IBA is important for its population of the Endangered St Lucia Black Finch *Melanospiza richardsoni*, and for the 14 (of the 23) Lesser Antilles EBA restricted-range birds that occur (including the Near Threatened St Lucia Oriole *Icterus laudabilis*). The endemic subspecies of Lesser Antillean Flycatcher *Myiarchus oberi santaeluciae* and House Wren *Troglodytes aedon martinicensis* also occur. A small but regionally important population of Royal Tern *Sterna maxima* breeds.

■ Other biodiversity

The endemic St Lucia anole *Anolis luceae*, St Lucia pygmy gecko *Sphaerodactylus microlepis*, St Lucia boa *Constrictor orophias*, and St Lucia viper (or fer-de-lance) *Bothrops caribbaeus* occur. The Pitons are floristically diverse with many endemic plants represented.

■ Conservation

This IBA is protected as the Pitons Management Area (PMA) which has also been designated a World Heritage Site. The Soufriere Marine Management Area protects part of the marine system. The lands are crown (state) owned. The communities of Fond Jens Libre and Chateau Belair are within the PMA, as are two resorts—Ladera and the Jalousie Hilton. The area is threatened by development pressures from tourism and dry-season fires (originating from farmers or campers). Invasive alien (mammalian) predators are present and presumably impacting the native fauna. The threat of hurricanes is ever present and that of volcanic eruption—though remote—cannot be ruled out.

LC004 Mandele Dry Forest

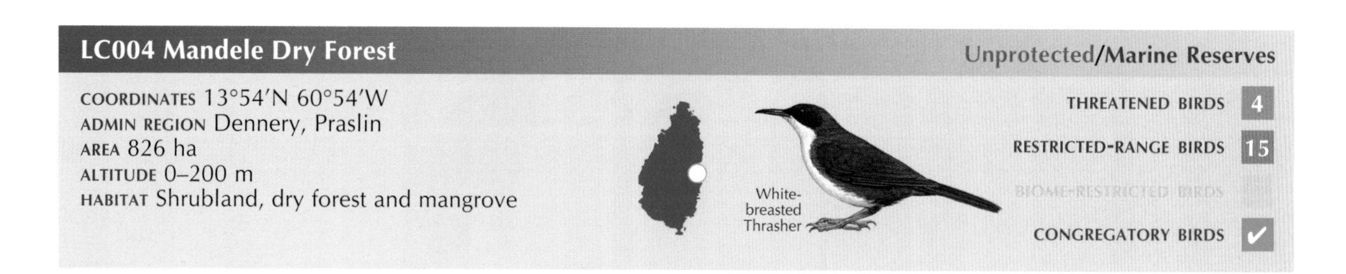

■ Site description

The Mandele Dry Forest IBA is located on and inland from the east coast of St Lucia, bounded by Ravine Pascal to the north, Ravine Bourge to the south and on the north-west by the Dennery Range extension (the Dennery Water Works Forest Reserve) of the Government Forest Reserve IBA (LC002). The IBA consists of flat coastal areas and foothills rising towards the Dennery Range, and includes the large Praslin Bay estate (currently being developed), Praslin Island and Frigate Islands.

■ Birds

This IBA is important for three globally threatened bird species. It is the centre of abundance for the Endangered White-breasted Thrasher *Ramphocinclus brachyurus*, holding over 90% of the St Lucia population (c.1,200 individuals). The Endangered St Lucia Black Finch *Melanospiza richardsoni* is also present, as are seasonal foraging flocks of the Vulnerable St Lucia Amazon *Amazona versicolor*, and a total of 15 (of the 23) Lesser Antilles EBA restricted-range birds. The endemic subspecies of Lesser Antillean Flycatcher *Myiarchus oberi santaeluciae* is present. A small population of Royal Tern *Sterna maxima* is regionally important.

■ Other biodiversity

The Vulnerable St Lucia whiptail *Cnemidophorus vanzoi* was successfully translocated (in 1997) to Praslin Island by the Forestry Department and Durrell Wildlife Conservation Trust. The endemic St Lucia boa *Constrictor orophias* and high densities of St Lucia viper *Bothrops caribbaeus* occur.

■ Conservation

The Mandale Dry Forest IBA is a mix of private and crown (state) ownership, but there is no protection afforded the important dry forest and shrubland habitats. Frigate Islands Nature Reserve and Praslin Mangroves Marine Reserve cover some coastal and marine areas. Provision for a nature reserve (for *R. brachyurus*) is part of the development plan for the Praslin Bay hotel and resort complex which will however result in significant loss of critical thrasher habitat. Crown lands around Bordelais may have the potential for future conservation management. Land along the Praslin River is under cultivation and indiscriminate harvesting of Latanyé palm *Coccothrinax barbadensis* for broom handles is slowly depleting the dry forest vegetation within the IBA. Invasive alien (mammalian) predators are present and presumably impacting the native fauna. Forestry Department and Durrell Wildlife Conservation Trust are undertaking research and conservation actions for *R. brachyurus* in this IBA.

■ Site description

Pointe Sable National Park IBA is located along the south and south-east coasts of St Lucia. Terrestrially it comprises a narrow coastal strip and the Moule-à-Chique peninsula. The terrain is low to undulating, the highest point being at Moule-à-Chique—the southernmost tip of the island. The coastal zone is proportionately larger, consisting of long sandy beaches, the Savannes Bay and Mankòtè Mangroves, Scorpion Island, the Maria Islands, and several coral reefs and near-shore islands. St Lucia's largest wetlands are within this IBA. The area also includes a number of historical buildings including old fortresses, a lighthouse and a World War II radar tracking station.

■ Birds

This IBA is important for its breeding seabirds. Over 20,000 Sooty Terns *Sterna fuscata* breed, as do 250–500 Bridled Terns *S. anaethetus*. Roseate Tern *S. dougallii*, Royal Tern *S. maxima* and Red-billed Tropicbird *Phaethon aethereus* breed in regionally important numbers. The Endangered St Lucia Black Finch *Melanospiza richardsoni* occurs in the forested areas although the population is unknown. The mangroves and wetlands (especially the swamp at the northern end of the IBA) are important for waterbirds and Neotropical migrants.

■ Other biodiversity

This IBA supports an important herpetofauna including five endemics, two of which are globally threatened. The Endangered St Lucia racer *Liophis ornatus* and Vulnerable St Lucia whiptail *Cnemidophorus vanzoi* are endemic to the Maria Islands (Maria Major is rat- and mongoose-free).

■ Conservation

Pointe Sable National Park is a mix of private and crown (state) lands that include a number of reserves (e.g. Mankòtè Mangrove and Savannes Bay marine reserves/ Ramsar sites). The Maria Islands nature reserve is managed by the St Lucia National Trust. The marine reserves are crown lands—managed by the government in collaboration with non-governmental groups. The IBA is being impacted by rapid commercial development; unauthorised harvesting of mangrove for charcoal; upland deforestation (causing siltation of the near-shore environment); mining and quarrying discharges; inappropriate disposal of household waste; and alien invasive predators. Pressure from coastal tourist developments in an area of high unemployment and poverty presents a huge challenge in managing the area for sustainable conservation. Forestry Department and Durrell Wildlife Conservation Trust are working on the St Lucia whiptail in this IBA.

ST MAARTEN

LAND AREA 33 km² ALTITUDE 0–425 m
HUMAN POPULATION 39,000 CAPITAL Philipsburg
IMPORTANT BIRD AREAS 5, totalling 8.15 km²
IMPORTANT BIRD AREA PROTECTION 55%
BIRD SPECIES 164
THREATENED BIRDS 1 RESTRICTED-RANGE BIRDS 5

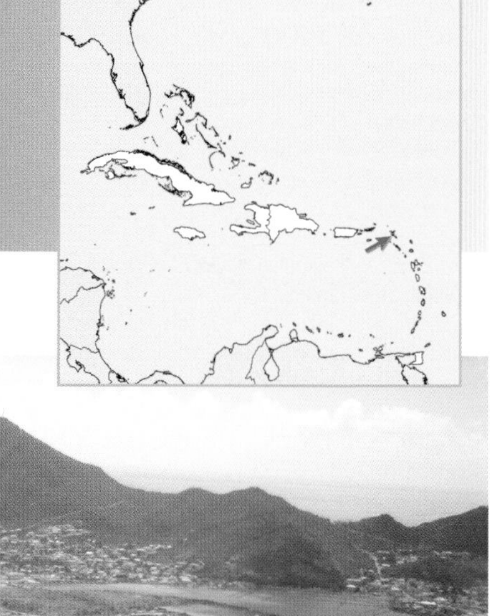

NATALIA COLLIER AND ADAM BROWN
(ENVIRONMENTAL PROTECTION IN THE CARIBBEAN)

Great Salt Pond and Fresh Pond in Philipsburg. (PHOTO: JAY HAVISER)

INTRODUCTION

The island of St Martin is situated just 8 km south of Anguilla (to UK) and 20 km west-north-west of St Barthelemy (to France). The northern, French half is called St Martin and is an overseas department of the French Republic. The southern, Dutch half is called St Maarten. St Maarten, Saba and St Eustatius together form the Windward Islands of the Netherlands Antilles[1]. The Dutch and the French have shared the island—the smallest land mass in the world to be divided between two governments—for almost 350 years.

The centre of the island (across which the political boundary runs) is composed of a mountainous spine rising to 425 m. The coastal areas are a mixture of flat lands and low hills punctuated by numerous ponds, primarily of high salinity. Coastal shorelines are characterised by sand or rock beaches with cliffs in between. Simpson Bay Lagoon, one of the largest lagoons in the Lesser Antilles, is a dominant feature of the island and a major yachting centre. The St Maarten side of the lagoon is extremely polluted and the shoreline almost completely developed. The terrestrial vegetation is thorny woodland, dominated by scrub in the lowlands and low forest in the mountains (with small patches of the original semi-evergreen forest on the highest ridges). Average annual rainfall is 1,770 mm, much of which arrives during the hurricane season in late summer and fall. St Maarten is densely populated, and the population increases greatly during the influx of seasonal visitors and tourists which form the base of the economy.

Conservation

Conservation in St Maarten sits within the framework of a number of "ordinances" that provide legislation to: prevent the destruction of valuable flora and fauna (although there is no associated list of species considered valuable); prohibit construction above 200 m; and establish marine protected areas. The government has drafted, but not yet approved, new nature protection legislation (the Marine Park Ordinance). The St Maarten Marine Park (managed by the Nature Foundation of St Maarten, but still not formally designated by government) surrounds the Dutch side of the island from the coastal waters and beaches to the 60-m depth contour. The park includes Pelikan Rock IBA (AN005) and three smaller islets.

1 At some point in the near future the "Netherlands Antilles" will be dissolved. St Maarten and Curaçao will become separate countries within the Kingdom of the Netherlands (similar to the status currently enjoyed by Aruba). The islands of Bonaire, Saba and St Eustatius will be linked directly to the Netherlands as overseas territories.

Pelikan Rock IBA.
(PHOTO: NATALIA COLLIER)

Environmental Protection in the Caribbean has conducted bird research and monitoring on the island (with a particular focus on wetlands and seabirds) for the past seven years. Research has included pond water quality testing. The Nature Foundation of St Maarten has implemented a program of planting mangroves at Little Bay Pond IBA (AN001) and Fresh Pond IBA (AN002). Awareness of environmental issues is on the increase due to the work of local organisations and the conspicuous loss of natural areas and wildlife. However, the conservation concerns of the general public have not yet translated into government action.

The primary threat within St Maarten is the development of land and ponds. St Maarten is a duty-free island and the "island area" is rapidly developing free of sufficient environmental legislation. Ponds and lagoons are routinely filled in and altered (e.g. Great Salt Pond IBA—AN003—is being used for landfill and filled to create parking) while building takes place on steep hillsides. Development permits are issued without thorough environmental impact assessments, despite public outcry, and permit restrictions are difficult to impose. Without enforceable legislation, these destructive practices are likely to continue. Other threats include disturbance (which is significant due to the high concentration of residents and tourists) due to watercraft and introduced alien species. There is almost no control of exotic species introductions. Introduced predators include dogs, cats, rats, mice, raccoon *Procyon lotor*, mongoose *Herpestes auropunctatus* and green monkey *Chlorocebus sabaeus*. Groups of monkeys have been observed in the higher mountainous areas and could expand to lower elevations (as has happened on St Kitts).

■ Birds

Of the 164 bird species recorded from St Maarten, 39 species are resident (and breed) on the island, although the majority of species are Neotropical migrants (or vagrants). Five (of the 38) Lesser Antilles EBA restricted-range birds occur in St Maarten, none of which is endemic to the country. Three other restricted-range species—Bridled Quail-dove *Geotrygon mystacea*, Purple-throated Carib *Eulampis jugularis* and Scaly-breasted Thrasher *Margarops fuscus*—are found in the montane forests on the St Martin side of the border and probably occur. The country is important for waterbirds (in spite of the severe alteration and destruction of wetland habitats), with 50 species recorded (18 of which breed). This is partly due to the presence of two low-salinity ponds which are unique within the region.

Little Bay Pond IBA. (PHOTO: BEVERLY MAE NISBETH, NATURE FOUNDATION ST MAARTEN)

Table 1. Key bird species at Important Bird Areas in St Maarten.

			St Maarten IBAs				
			AN001	AN002	AN003	AN004	AN005
Key bird species	**Criteria**	**National population**	**Criteria:** ■	■ ■	■	■ ■	■
Brown Pelican *Pelecanus occidentalis*	■	285				180	105
Caribbean Coot *Fulica caribaea*	NT ■	75		53			
Laughing Gull *Larus atricilla*	■	6,100			5,800		300
Royal Tern *Sterna maxima*	■	141					90
Green-throated Carib *Eulampis holosericeus*	■		✓	✓		✓	
Antillean Crested Hummingbird *Orthorhyncus cristatus*	■		✓	✓		✓	
Caribbean Elaenia *Elaenia martinica*	■		✓	✓		✓	
Pearly-eyed Thrasher *Margarops fuscatus*	■		✓	✓		✓	
Lesser Antillean Bullfinch *Loxigilla noctis*	■		✓	✓		✓	

All population figures = numbers of individuals.
Threatened birds: Near Threatened ■. **Restricted-range birds** ■. **Congregatory birds** ■.

Fort Amsterdam IBA.
(PHOTO: BEVERLY MAE NISBETH, NATURE FOUNDATION ST MAARTEN)

Significant numbers of the Near Threatened Caribbean Coot *Fulica caribaea* breed at Fresh Pond IBA (AN002) and Little Bay Pond IBA (AN001) and sometimes hybridise with American Coot *F. americana*. Studies to determine nesting success, especially of *F. caribaea*, are needed to better understand the primary threats facing wetland species. Similarly, the Brown Pelican *Pelecanus occidentalis* colony at Fort Amsterdam IBA (AN004) should be studied further to determine the factors influencing fluctuations in the breeding population there. National population estimates for these waterbirds are given in Table 1.

IMPORTANT BIRD AREAS

St Maarten's IBAs—the country's international site priorities for bird conservation—cover 815 ha (including marine areas), and about c.36% of the country's land area. Of the five IBAs, Pelikan Rock (AN005) is protected within the St Maarten Marine Park, and Fort Amsterdam (AN004) is protected as a historical site. The IBAs have been identified on the basis of eight key bird species (listed in Table 1) that variously trigger the IBA criteria. They are centred on wetland and marine sites being primarily significant for their populations four congregatory waterbird and seabird species (including the Near Threatened Caribbean Coot *Fulica caribeae*. However, shrublands in three IBAs support populations of the four restricted-range species known to occur in the country. With further targeted field research, three additional restricted-range birds that occur in the montane forest across the border in St Martin would be expected to be found in the semi-evergreen forest remnants on the St Maarten side. Such a discovery could warrant the identification of an additional IBA for these forest dependent species.

The wetland IBAs of Little Bay Pond, Fresh Pond and Great Salt Pond all face similar, multiple threats such as land reclamation for development, inappropriate development, use for landfill, pollution and contamination from runoff and sewage, inappropriate water management (e.g. maintaining water levels by pumping in sea water), alien invasive predators and disturbance. The Nature Foundation of St Maarten and Environmental Protection in the Caribbean (EPIC) have variously planted mangroves, constructed bird observation towers and installed educational signage at Little Bay Pond and Fresh Pond IBAs, but it is clear that enforced legislation is critical if the biological integrity of these wetlands is to be maintained in the long term.

The protection afforded Fort Amsterdam (Historical Site) and Pelikan Rock (Marine Park) IBAs appears to be preventing site-based threats although factors outside of these areas are having negative impacts such as disturbance to the mainland pelican nesting colony from jet skis, dive boats, and parasail boats. Over-fishing, oil spills, and plastics entanglement are constant threats to the marine-based seabirds and waterbirds. The regular monitoring of the waterbirds at St Maarten's IBAs (e.g. as has been undertaken by EPIC) and the monitoring of the other key bird species should be used to inform the assessment of state, pressure and response scores at each IBA in order to provide objective status assessments and inform management decisions (should the necessary legislation be enacted) that might be required to maintain these internationally important biodiversity sites.

Figure 1. Location of Important Bird Areas in St Maarten.

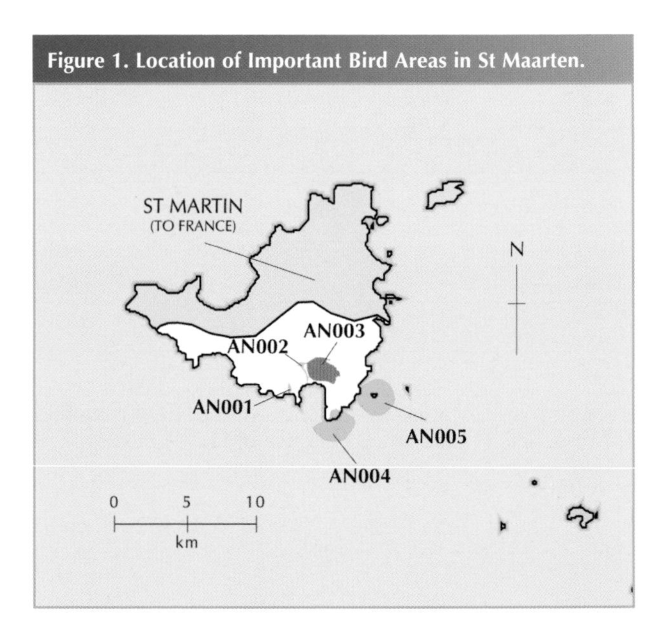

KEY REFERENCES

Brown, A. C. and Collier, N. (2001) Pond surveys of Sint Maarten/ St Martin. Riviera Beach, USA: Environmental Protection in the Caribbean (Unpublished report).

Brown, A. C. and Collier, N. (2002) Pond surveys of St Martin. Riviera Beach, USA: Environmental Protection in the Caribbean (Unpublished report: EPIC contribution 4).

Brown, A. C. and Collier, N. (2003) Pond surveys of St Martin. Riviera Beach, USA: Environmental Protection in the Caribbean (Unpublished report: EPIC contribution 10).

Brown, A. C. and Collier, N. (2004) Pond surveys of St Martin. Riviera Beach, USA: Environmental Protection in the Caribbean (Unpublished report: EPIC contribution 23)..

Brown, A. C. and Collier, N. (2004) A new breeding population of *Oxyura jamaicensis* (Ruddy Duck) on St Martin, Lesser Antilles. *Caribbean Journal of Science* 40(2): 259–263.

Brown, A. C. and Collier, N. (2005) Pond surveys of St Martin. Riviera Beach, USA: Environmental Protection in the Caribbean (Unpublished report).

Brown, A. C. and Collier, N. (in press) The Netherlands Antilles I: Saint Maarten, Saba and Saint Eustatius. In Bradley P. E. and Norton, R. L. eds. *Breeding seabirds of the Caribbean*. Gainesville, Florida: Univ. Florida Press.

Collier, N. (2006) Raising awareness of St Maarten's wetland wildlife. *Birds Caribbean* 4:18.

Collier, N. C., Brown, A. C. and Hester, M. (2002) Searches for seabird breeding colonies in the Lesser Antilles. *El Pitirre* 15(3): 110–116.

Collazo, J. A., Saliva, J. E. and Pierce, J. (2000) Conservation of the Brown Pelican in the West Indies. Pp 39–45 in E. A. Schreiber and D. S. Lee, eds. *Status and conservation of West Indian Seabirds*. Ruston, USA: Society of Caribbean Ornithology (Spec. Publ. 1).

Danforth, S. T. (1930) Notes on the birds of St Martin and St Eustatius. *Auk* 47: 44–47.

Norton, R. L. and White, A. (2001) West Indies. *North American Birds* 55(4): 493–494.

van Halewyn, R. and Norton, R. L. (1984) The status and conservation of seabirds in the Caribbean. Pp 169–222 in J. P. Croxall, P. G. H. Evans and R. W. Schreiber, eds. *Status and conservation of the world's seabirds*. Cambridge, U.K.: International Council for Bird Preservation (Techn. Publ. 2).

Hoogerwerf, A. (1977) Notes on the birds of St Martin, Saba, and St Eustatius. *Studies on the Fauna of Curaçao and other Caribbean islands* 54(176): 60–123.

Pinchon, R. (1976) *Faune des Antilles Françaises: les oiseaux*. Fort-de-France, Martinique: M. Ozanne and Cie.

Raffaele, H. Wiley J., Garrido, O., Keith, A. and Raffaele, J. (1998) *A guide to the birds of the West Indies*. Princeton, New Jersey: Princeton University Press.

Rojer, A. (1997) Biological inventory of St Maarten. Curaçao, Netherlands Antilles: Carmabi Foundation (Unpublished report)

Voous, K. H. (1955) The birds of St Martin, Saba and St. Eustatius. *Studies on the Fauna of Curaçao and other Caribbean island* 6:1–82.

Voous, K. H. (1955) *The birds of the Netherlands Antilles*. Curaçao: Uitg. Natuurwet. Werkgroep Ned. Ant.

Voous, K. H. (1983) *Birds of the Netherlands Antilles*. Zutphen, The Netherlands: De Walburg Pers.

Voous, K. H. and Koelors, H. J. (1967) Checklist of birds of St Martin, Saba, and St. Eustatius. *Ardea* 55:115–137.

ACKNOWLEDGEMENTS

The authors would like to thank Bert Denneman (Vogelbescherming Nederlands) Beverly Mae Nisbeth (Nature Foundation of St Maarten) and Kalli de Meyer (Dutch Caribbean Nature Alliance) for their help in reviewing this chapter.

AN001 Little Bay Pond

Unprotected

COORDINATES 18°01′N 63°04′W
ADMIN REGION St Maarten
AREA 8 ha
ALTITUDE 0 m
HABITAT Inland wetland, mangrove, shrubland

Caribbean Coot

THREATENED BIRDS 1
RESTRICTED-RANGE BIRDS 5
BIOME-RESTRICTED BIRDS
CONGREGATORY BIRDS

■ Site description

Little Bay Pond IBA is near the capital city of Philipsburg, in the middle of the south coast of St Maarten. It is c.2.5 km in diameter and has low (4–8 parts per thousand) salinity. It is bordered by aquatic grasses and red, black, and white mangrove trees, with surrounding areas supporting shrubland. A busy road runs along one side of the pond, above which is a residential development. A new development is being built in the corner nearest Little Bay beach. The pond is encircled by a hiking path which connects to the sea at the rocky shore of Little Bay. A small outlet runs from the pond into the ocean.

■ Birds

This IBA is significant for its population of the Near Threatened Caribbean Coot *Fulica caribaea*. Up to 22 birds have been recorded and some pairs breed. A number of other waterbird species breed at the site. All five Lesser Antilles EBA restricted-range birds occur around Little Bay Pond.

■ Other biodiversity

Green iguanas *Iguana iguana* occur, but no threatened or endemic species have been recorded.

■ Conservation

Little Bay Pond IBA has no legal protection. It is owned by a foreign development corporation and is up for sale. Local opposition, inspired by the aesthetic and ecological value of the pond, has so far halted plans to turn the pond into a marina, but a new hotel and condominium development is being built next to Little Bay beach. The legality of owning this pond has been questioned as all ponds are supposed to be public land. Regular waterbird population counts have been conducted each winter (and during spring/summer 2004) since 2001 by Environmental Protection in the Caribbean (EPIC). The Nature Foundation of St Maarten planted mangrove trees at the site which have thrived. EPIC and the Nature Foundation, with funding from Royal Caribbean, constructed a bird observation tower and educational signage along the hiking path. Monthly educational mangrove/bird walks are held in winter. The water is high in nutrients from sewage outflow from surrounding areas, sometimes resulting in eutrophication and fish die-offs. Pollution runoff from adjacent roads is problematic. Predators such as cats, dogs, rats, and mongoose frequent the area.

■ Site description

Fresh Pond IBA is within the capital city of Philipsburg and comprises a large pond (c.2.5 km by 1 km) just to the west of Great Salt Pond IBA (AN003). It is a low-salinity pond (2–3 parts per thousand) that is bordered by aquatic grasses and red mangrove trees. Artificial islands at each end of the pond are vegetated with mangroves and coconut trees and provide popular nesting sites for waterbirds. A busy bridge bisects the pond and heavy traffic and development encircle the area.

■ Birds

This IBA is significant for its population of Near Threatened Caribbean Coot *Fulica caribaea*. Over 50 coots have been recorded, and nesting does occur. All five Lesser Antilles EBA restricted-range birds occur. Fresh Pond IBA also supports populations of many waterbirds including 180+ Snowy Egret *Egretta thula*, Pied-billed Grebe *Podilymbus podiceps*, Great Egret *Casmerodius albus*, White-cheeked Pintail *Anas bahamensis* and Ruddy Duck *Oxyura jamaicensis*.

■ Other biodiversity

Green iguanas *Iguana iguana* occur, but no threatened or endemic species have been recorded.

■ Conservation

The open water of Fresh Pond is state-owned and "designated" public space. The surrounding land is privately owned and totally developed. Regular waterbird population counts have been conducted each winter (and during spring/summer 2004) since 2001 by Environmental Protection in the Caribbean (EPIC). The Nature Foundation of St Maarten planted mangrove trees at the site, which have thrived in the low-salinity, high-nutrient waters. EPIC and the Nature Foundation, with funding from Royal Caribbean, constructed a bird observation tower and educational signage. A sewage treatment plant on the shore of Fresh Pond causes concern regarding contamination. The water is high in nutrients from sewage outflow from surrounding areas, sometimes resulting in eutrophication and fish die-offs. Trash is prevalent among the shoreline vegetation. Pollution runoff from adjacent roads is problematic. Predators such as cats, dogs, rats, and mongoose frequent the area, and human disturbance is an issue at this urban location.

■ Site description

Great Salt Pond IBA is in south-central St Maarten, on the outskirts of the capital Philipsburg. It is the largest pond on the island and is bordered on all sides by busy roads. Fresh Pond IBA (AN002) lies just to the west. Great Salt Pond is highly saline (27–38 parts per thousand) as a result of which there is little visible vegetation. The borders of the pond comprise roadside grass and urban development. The pond was previously used for salt extraction and remnant rock walls, which are now important roost and nesting areas, still remain. The pond's primary use now is as landfill and land reclamation.

■ Birds

This IBA is significant for its population of Laughing Gull *Larus atricilla*. Up to 5,800 gulls congregate at the IBA prior to the breeding season. It is unclear if this congregation occurred historically or if the gulls now assemble to feed from the landfill. About 50 pairs of Black-necked Stilt *Himantopus mexicanus* breed—the only species confirmed to do so within the IBA.

■ Other biodiversity

Nothing recorded.

■ Conservation

The open water of Great Salt Pond is state owned and "designated" public space. However, the IBA faces multiple threats. Land "reclamation" is used to create parking areas, carnival grounds, and other facilities within the pond. Trash from the landfill located within the pond and from the town is blown into the pond. Fires at the landfill occur several times a year. Pollution leaching from the landfill has caused ecological collapse, resulting in massive midge infestations which have plagued Philipsburg each year since 2006. The government undertook an intensive six-week pesticide application program to alleviate the infestation. Water levels are artificially controlled using seawater, which has resulted in flooded nests and altered salinity levels. Regular waterbird population counts have been conducted each winter (and during spring/summer 2004) since 2001 by Environmental Protection in the Caribbean.

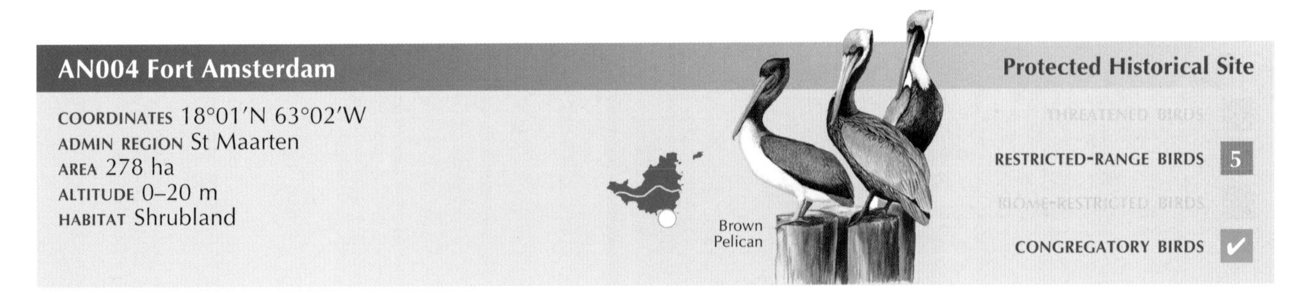

■ Site description

Fort Amsterdam IBA is a 2.5-km long peninsula of land in south-east St Maarten. On the top of the peninsula are ruins of a fort (a registered historical site), the slopes below which support 2 m-high thorny *Acacia macracantha* and *A. tortuosa* shrubland. The bay on the east side of the peninsula is a major cruise ship port, and a tourist resort is situated just 500 m away to the north.

■ Birds

This IBA is significant for Brown Pelican *Pelecanus occidentalis*. The breeding population varies greatly between years, but up to 50 pairs breed on the western side of the point, with 10 pairs on the eastern side making this a regionally important colony. Birds nest as close as 10 m from the fort ruins. All five Lesser Antilles EBA restricted-range birds occur at this IBA.

■ Other biodiversity

The island-endemic lizard *Anolis pogus* may be present in the IBA.

■ Conservation

Fort Amsterdam is privately owned and a zoned historical site, which affords it legal protection from development. Public access to the IBA is controlled by the resort's security checkpoint, and the thorny vegetation restricts visitors from accessing the pelican nesting area. The waters surrounding the peninsula are used heavily (and increasingly) by watercraft (including jet skis, dive boats, and parasail boats) which causes disturbance to the pelican nesting colony. Weekly/monthly population counts of the pelicans have been conducted every winter since 2001 by Environmental Protection in the Caribbean. A nesting success survey was also conducted in 2001.

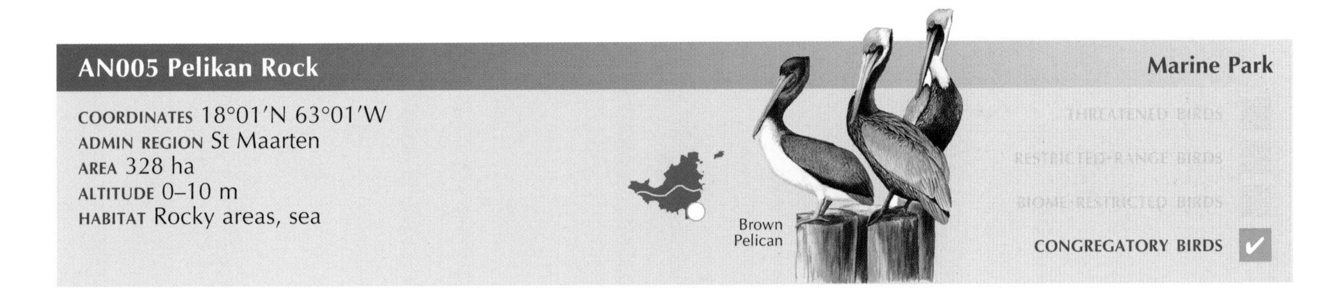

■ Site description

Pelikan Rock is a small, rocky islet about 1.5 km off the south-east coast of St Maarten. It is about 1.2 ha in size. The shoreline is rocky and difficult to access by sea, with rocky cliffs rising 6–10 m up to a grass- and low shrub-covered plateau. The IBA includes a 1-km seaward extension around the island.

■ Birds

This IBA is notable for its breeding waterbirds. The population of Laughing Gull *Larus atricilla* (100 pairs) is globally important, while those of Brown Pelican *Pelecanus occidentalis* (35 pairs) and Royal Tern *Sterna maxima* (47 pairs) are regionally so. Bridled Tern *S. anaethetus* also nests (c.15 pairs), and Brown Noddy *Anous stolidus* and Audubon's Shearwater *Puffinus lherminieri* have both been reported breeding, although surveys by Environmental Protection in the Caribbean in 2004 failed to find any.

■ Other biodiversity

Nothing recorded.

■ Conservation

Pelikan Rock is state owned and managed by the Nature Foundation of St Maarten as part of the St Maarten Marine Park. Fishing, anchoring, the use of jet skis and ship traffic is prohibited within the marine park, although moorings have been installed and the area is popular for diving. Access to the islet itself is prohibited and human visitation is rare. There is no evidence of rodents or other predators on the islet. Threats to birds using the IBA are found primarily outside the IBA and include issues such as over-fishing, oil spills, and plastics entanglement. EPIC has conducted weekly or monthly population counts of *P. occidentalis* from the mainland every winter since 2001. A ground-truthing survey was conducted on the islet in June 2004.

ST MARTIN

LAND AREA **56 km²** ALTITUDE **0–425 m**
HUMAN POPULATION **36,000** CAPITAL **Marigot**
IMPORTANT BIRD AREAS **3, totalling 8.9 km²**
IMPORTANT BIRD AREA PROTECTION **67%**
BIRD SPECIES **164**
THREATENED BIRDS **0** RESTRICTED-RANGE BIRDS **8**

NATALIA COLLIER AND ADAM BROWN
(ENVIRONMENTAL PROTECTION IN THE CARIBBEAN)

Pic Paradis and the lowlands of St Martin. (PHOTO: NATALIA COLLIER/EPIC)

INTRODUCTION

The island of St Martin is situated just 8 km south of Anguilla (to UK) and 20 km west-north-west of St Barthélemy (to France). The northern, French half is called St Martin and is in the process of becoming a *collectivité d'outre-mer* (COM, overseas collectivity) of France (and not an integral part of the European Union)[1]. The southern, Dutch half is called St Maarten. The French and the Dutch have shared the island—the smallest land mass in the world to be divided between two governments—for almost 350 years.

The centre of the island (across which the political boundary runs) is composed of a mountainous spine rising to 425 m (at Pic Paradis). The coastal areas are characterised by low hills or flat areas punctuated by numerous ponds (most of which are highly saline). Shorelines are either sand or rocky beaches with interspersed cliffs. Simpson Bay Lagoon, one of the largest lagoons in the Lesser Antilles, is a dominant feature of the island and a major yachting centre. The French side of the lagoon is less polluted than the St Maarten side and the yachting industry is less developed. St Martin supports two significant areas of mangrove. One is located in the Simpson Bay Lagoon, the other at Etang aux Poissons. The terrestrial vegetation is thorny woodland, dominated by scrub in the lowlands and low forest cover in the mountains (with small patches of the original semi-evergreen forest on the highest ridges). Average annual rainfall is 1,770 mm, much of which arrives during the hurricane season in late summer and fall. St Martin is less densely populated than the Dutch half of the island, but the population increases greatly during the influx of seasonal visitors and tourists which form the base of the economy.

■ Conservation

There are two main protected areas in the French section of the island. The national Nature Reserve of St Martin (Réserve Naturelle, created in 1998) covers 154 ha of terrestrial habitat representing c.3% of the land area and including beaches, mangrove forests, and saline ponds. It also embraces 2,796 ha of marine habitats, including those around Tintamarre Island (IBA MF003). The second area is owned

1 St Martin has been in the process of changing its status since July 2007 from a *département d'outre-mer* (DOM, overseas department) of France (and an outermost region of the European Union, EU) to a *collectivité d'outre-mer* (COM, overseas collectivity) of France (and not an integral part of the EU). Currently its status with respect to the EU is unclear. Long-term it may become an overseas country and territory of France.

Searching for Red-billed Tropicbird nests on Tintamarre IBA. (PHOTO: NATALIA COLLIER/EPIC

by the "Conservatoire du Littoral" and embraces 14 pools and ponds (totalling 200 ha and including Grand Etang IBA, MF001) protected by a local decree (Arrêté Préfectoral de Protection de biotope). The office of the Réserve Naturelle manages these protected areas. The Loterie Farm ecotourism site (see Pic Paradis IBA, MF002) contains secondary dry forest in a canyon with a seasonal creek. The lower reaches have been converted to an aerial rope course for tourists but the upper area remains a relatively pristine (forested) hiking area. Although forest at Loterie Farm has no legal protection, it is expected to remain undeveloped.

Environmental Protection in the Caribbean (EPIC) has conducted bird research and monitoring on the island for the past seven years. Research has included pond water quality testing. Nesting success of Red-billed Tropicbird *Phaethon aethereus*, Least Tern *Sterna antillarum* and Wilson's Plover *Charadrius wilsonia* has been monitored for one season. EPIC coordinates this work with funding from private donors and local companies, and many volunteer hours. The French government and the Réserve Naturelle also sponsor periodic wetland and marine bird surveys (often conducted by Gilles LeBlond from Guadeloupe).

The primary threat to biodiversity on St Martin is development, most of which is associated with the tourism industry. Unauthorised filling in or encroachment of wetlands (pond and lagoons) occurs in some areas, and building takes place on steep hillsides (leading to erosion and siltation of wetlands). Land based pollution via sewage and refuse is a visible problem, and disturbance from watercraft, 4x4 vehicles, and other recreational activities is significant due to the high concentration of residents and tourists. Nest predation by introduced species appears to be a limiting factor to reproductive success on St Martin, and populations of forest birds, particularly game birds such as Bridled Quail-dove *Geotrygon mystacea*, are reduced through hunting.

The main obstacles to protection are inadequate legislation and insufficient enforcement of existing regulations. Funding is not available for essential projects such as sewage treatment and predator eradication. The eradication of invasive predators from offshore islands and wetlands would reduce a significant threat to breeding species, particularly *S. antillarum* and Audubon's Shearwater *Puffinus lherminieri*. Also, given the pressures already placed on avian populations by the limited remaining habitat, hunting should be prohibited or further restricted. Sewage systems need to be upgraded or repaired in order to reduce contaminated effluent entering local waters.

■ Birds

Of the 164 bird species recorded from St Martin, c.50 are Neotropical migrants. The most important habitats for the birds in St Martin (including the Neotropical migrants) are the saline ponds, mangroves, and secondary dry forest of the mountains. Eight (of the 38) Lesser Antilles EBA restricted-range birds occur in St Martin, none of which is endemic to the island. Scaly-breasted Thrasher *Margarops fuscus* and Bridled Quail-dove *Geotrygon mystacea* are only recently confirmed as present on the island, with both occurring in the montane forests on Pic Paradis (IBA MF002) near the border with St Maarten. The island is important for

Loterie Farm ecotourism site and the Pic Paradis IBA. (PHOTO: NATALIA COLLIER/EPIC)

Table 1. Key bird species at Important Bird Areas in St Martin.

Key bird species	Criteria	National population	MF001	MF002	MF003
			St Martin IBAs		
Criteria			■ ■	■	■ ■
Red-billed Tropicbird *Phaethon aethereus*	■	200			180
Least Tern *Sterna antillarum*	■	250	246		
Brown Noddy *Anous stolidus*	■	450			420
Bridled Quail-dove *Geotrygon mystacea*	■			✓	
Purple-throated Carib *Eulampis jugularis*	■			✓	
Green-throated Carib *Eulampis holosericeus*	■		✓	✓	✓
Antillean Crested Hummingbird *Orthorhyncus cristatus*	■		✓	✓	✓
Caribbean Elaenia *Elaenia martinica*	■		✓	✓	✓
Scaly-breasted Thrasher *Margarops fuscus*	■			✓	
Pearly-eyed Thrasher *Margarops fuscatus*	■		✓	✓	✓
Lesser Antillean Bullfinch *Loxigilla noctis*	■		✓	✓	✓

All population figures = numbers of individuals.
Restricted-range birds ■. Congregatory birds ■.

waterbirds (in spite of the severe alteration and destruction of wetland habitats), with 50 species recorded (13 of which breed). The pond Grand Etang (IBA MF001) is an important breeding area for Least Tern *Sterna antillarum* and shorebirds while the offshore island of Tintamarre (IBA MF003) supports important seabird populations including Red-billed Tropicbird *Phaethon aethereus*, terns, noddys and Audubon's Shearwater *Puffinus lherminieri*. National population estimates for the key seabird species are given in Table 1.

IMPORTANT BIRD AREAS

St Martin's three IBAs—the country's international site priorities for bird conservation—cover 890 ha (including marine areas), and about c.6% of the country's land area. Only portions of Tintamarre IBA (MF003) are protected within the national Réserve Naturelle St Martin, and the Grand Etang IBA (MF001) is protected by local decree. The IBAs have been identified on the basis of 11 key bird species (listed in Table 1) that variously meet the IBA criteria. Both Grand Etang IBA (MF001) and Tintamarre IBA (MF003) are primarily congregatory bird sites—identified on the basis of their breeding seabird populations, although IBAs also support many of the country's restricted-range birds. However, Pic Paradis IBA (MF001) is the main terrestrial IBA in the country and its forests are home to the more specialised (forest-dependent) species such as Scaly-breasted Thrasher *Margarops fuscus,* Bridled Quail-dove *Geotrygon mystacea* and Purple-throated Carib *Eulampis jugularis*.

The IBAs are threatened by multiple factors which are described in the IBA profiles below. However, they can all be addressed given the necessary commitment and some funding. Making the Grand Etang tern colony inaccessible to dogs would reduce predation; re-routing the power cables would eliminate unnecessary mortality; and providing sewage treatment facilities for the surrounding communities would eliminate the contamination of this wetland. The nesting seabird populations on Tintamarre Island would benefit from a highly feasible rat eradication program, and also the removal of goats. The more vulnerable game birds (e.g. *G. mystacea*) in the country's montane forests would benefit from better regulated hunting practices (or indeed a cessation of hunting particular species).

The regular monitoring of the birds at St Martin's IBAs (e.g. as has been undertaken by EPIC) and the monitoring of the other key bird species should be used to inform the assessment of state, pressure and response scores at each IBA in order to provide objective status assessments and inform management decisions that might be required to maintain these internationally important biodiversity sites.

Figure 1. Location of Important Bird Areas in St Martin.

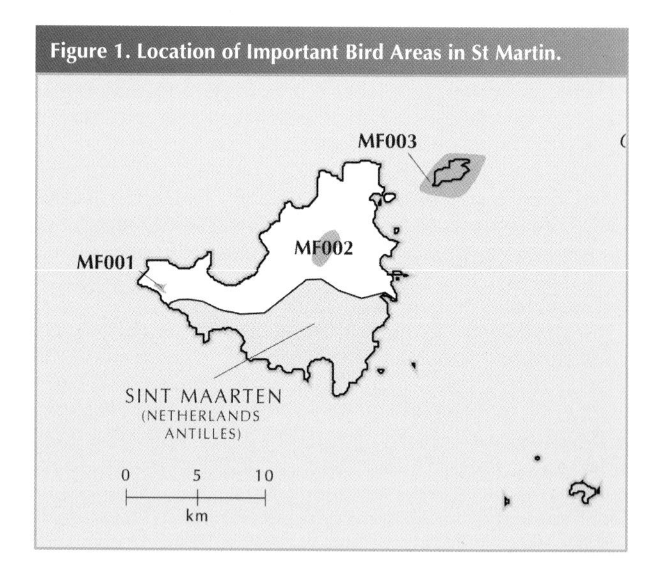

Purple-throated Carib, dependent on the Pic Paradis forest.
(PHOTO: GREGORY GUIDA)

KEY REFERENCES

Breuil, M. (2002) Histoire naturelle des amphibiens et reptiles terrestres de l'archipel Guadeloupéen: Guadeloupe, Saint-Martin, Saint-Barthélemy. *Patrimoines Naturels* 54: 1–339.

Brown, A. C. and Collier, N. (2003) Occurrence of an over-wintering Chestnut-sided Warbler (*Dendroica pennsylvanica*) on St Martin, Lesser Antilles. *Carib. J. Orn.* 16: 66–67.

Brown, A. C. and Collier, N. (2003) Recent colonization of St Martin by the Scaly-breasted Thrasher (*Margarops fuscus*). *Carib. J. Orn.* 16: 24–25.

Brown, A. C. and Collier, N. (2003) Terrestrial bird studies on St Martin: winter of 2003. Riviera Beach, USA: Environmental Protection in the Caribbean (Unpublished report: EPIC contribution 14).

Brown, A. C. and Collier, N. (2004) Terrestrial bird studies on St Martin: winter of 2004. Riviera Beach, USA: Environmental Protection in the Caribbean (Unpublished report: EPIC contribution 24).

Brown, A. C. and Collier, N. (2004) New and rare bird records from St Martin, West Indies. *Cotinga* 25: 52–58.

Brown, A. C. and Collier, N. (2005) Terrestrial bird studies on St Martin: winter of 2005. Riviera Beach, USA: Environmental Protection in the Caribbean (Unpublished report: EPIC contribution 25).

Danforth, S. T. (1930) Notes on the birds of St Martin and St Eustatius. *Auk* 47: 44–47.

van Halewyn, R. and Norton, R. L. (1984) The status and conservation of seabirds in the Caribbean. Pp 169–222 in J. P. Croxall, P. G. H. Evans and R. W. Schreiber, eds. *Status and conservation of the world's seabirds*. Cambridge, U.K.: International Council for Bird Preservation (Techn. Publ. 2).

Hoogerwerf, A. (1977) Notes on the birds of St Martin, Saba, and St Eustatius. *Studies on the Fauna of Curaçao and other Caribbean islands* 54(176): 60–123.

Leblond, G. (2005) Evaluation scientifique des vertébrés terrestres (amphibiens, reptiles, oiseaux et mammifères) des Etangs de Saint Martin. (Unpublished report: BIOS, 56 p.).

Pinchon, R. (1976) *Faune des Antilles Françaises: les oiseaux*. Fort-de-France, Martinique: M. Ozanne and Cie.

Raffaele, H. Wiley J., Garrido, O., Keith, A. and Raffaele, J. (1998) *A guide to the birds of the West Indies*. Princeton, New Jersey: Princeton University Press.

Rojer, A. (1997) Biological inventory of St Maarten. Curaçao, Netherlands Antilles: Carmabi Foundation (Unpublished report).

Voous, K. H. (1954) Het ornithologisch onderzoek van de Nederlandse Antillen tot 1951. *Ardea* 41:23–34.

Voous, K. H. (1955) The birds of St Martin, Saba and St. Eustatius. *Studies on the Fauna of Curaçao and other Caribbean island* 6:1–82.

Voous, K. H. (1955) *The birds of the Netherlands Antilles*. Curaçao: Uitg. Natuurwet. Werkgroep Ned. Ant.

Voous, K. H. (1983) *Birds of the Netherlands Antilles*. Zutphen, The Netherlands: De Walburg Pers.

Voous, K. H. and Koelors, H. J. (1967) Checklist of birds of St Martin, Saba, and St. Eustatius. *Ardea* 55:115–137.

ACKNOWLEDGEMENTS

The authors would like to thank Alison Duncan and Bernard Deceuninck (Ligue pour la Protection des Oiseaux, BirdLife in France), Louis Redaud (DIREN), Andy Caballero, Beverly Mae Nisbeth (Nature Foundation St Maarten) and Nicolas Maslach (Réserve Naturelle de Saint-Martin) for their review comments and input to this chapter, and Nicolas Maslach, Gilles LeBlond, Loterie Farm, and the many volunteers who have made our fieldwork possible.

MF001 Grand Etang

Arrêté Préfectoral de Protection de Biotope

COORDINATES 18°03′N 63°09′W
ADMIN REGION —
AREA 18 ha
ALTITUDE 0–5 m
HABITAT Inland wetland, shrubland

Least Tern

THREATENED BIRDS
RESTRICTED-RANGE BIRDS 5
BIOME-RESTRICTED BIRDS
CONGREGATORY BIRDS ✓

■ Site description

Grand Etang IBA is a large saline pond located in a residential area of the gated Lowlands subdivision of St Martin (at the western end of the island). The pond is highly saline (>100 parts per thousand) and supports little aquatic vegetation. Vegetation begins c.5 m from the waterline and comprises *Acacia*-dominated thorn scrub. The pond is surrounded by scrub-covered low hills interspersed with roads and large residential estates. A dirt road runs next to the shore on one side of the pond, and an exclusive resort borders this road. In some areas, the gardens of upmarket houses back onto the pond. At the western end of Grand Etang, a small pond is separated from the larger pond by a narrow strip of land.

■ Birds

This IBA is significant for its regionally important population of Least Tern *Sterna antillarum*. Over 80 pairs have been recorded breeding on a sandy spit located in the smaller pond of Grand Etang. Wilson's Plover *Charadrius wilsonia* also nests in this area. The shrubland surrounding the pond supported populations of five (of the eight) Lesser Antilles EBA restricted-range birds.

■ Other biodiversity

Nothing recorded.

■ Conservation

Grand Etang IBA is afforded some legal protection under a local decree "Arrêté Préfectoral de Protection de Biotope". EPIC has monitored this IBA (and the nesting terns) regularly since 2001, and sponsor free educational bird walks at the site each year. Heavy rains increase the water level of Grand Etang and have flooded c.10% of *S. antillarum* nests in the past. Introduced predators are also a major concern for nesting birds: at least 5% of all tern nests are predated by dogs and rats (*Rattus* spp.). Birds collide with (and are periodically found dead beneath) the power lines that cross between the ponds. Garbage is dumped in the pond and often burnt along the shore, and the resort which borders the pond periodically allows its septic system to empty across the road into the pond.

MF002 Pic Paradis

■ Site description

Pic Paradis IBA is in central St Martin on the forested western side of the island's highest mountain. The IBA extends from 300 m to the summit, with the ridgeline forming the southern edge of the IBA. To the north, the IBA ends at the Pic Paradis road. The Loterie Farm ecotourism site borders the IBA below 300 m. The IBA comprises secondary dry forest (dominated by mango *Mangifera* and *Ficus* spp.) characterised by almost complete canopy cover and little understorey or ground cover. A seasonal boulder creek flows for much of the year, depending on rainfall levels. The lower parts of the IBA support a restaurant and adventure rope course (part of the Loterie farm enterprise), while the upper elevations are undeveloped hiking areas.

■ Birds

This IBA supports all eight of the Lesser Antilles EBA restricted-range birds. Recent mist-netting studies confirmed Scaly-breasted Thrasher *Margarops fuscus* as a breeding species in St Martin (in 2002) and also provided the first record of Bridled Quail-dove *Geotrygon mystacea* for the island (in 2006).

■ Other biodiversity

The island-endemic lizard *Anolis pogus* is present in the IBA.

■ Conservation

Pic Paradis IBA is privately owned, much of it by the Loterie Farm ecotourism site. There is no legal protection for the IBA although Loterie Farm intends to preserve the forest portions of its property. Land use plans for the other landowners are not known. Winter bird ringing and monitoring undertaken by EPIC since 2001 has resulted in many new island and regional records of species (e.g. *G. mystacea*). School groups and the general public are taught about birds during visits to the study site. Hunting is permitted in the IBA but does not appear to be regulated and could be affecting populations of species such as the quail-dove. Invasive mammalian predators (such as mongoose *Herpestes auropunctatus* and African green monkey *Cercopithecus aethiops*) occur throughout the IBA and presumably impact some bird species. Hiking trails are used by paying individuals and guided tours for groups.

MF003 Tintamarre

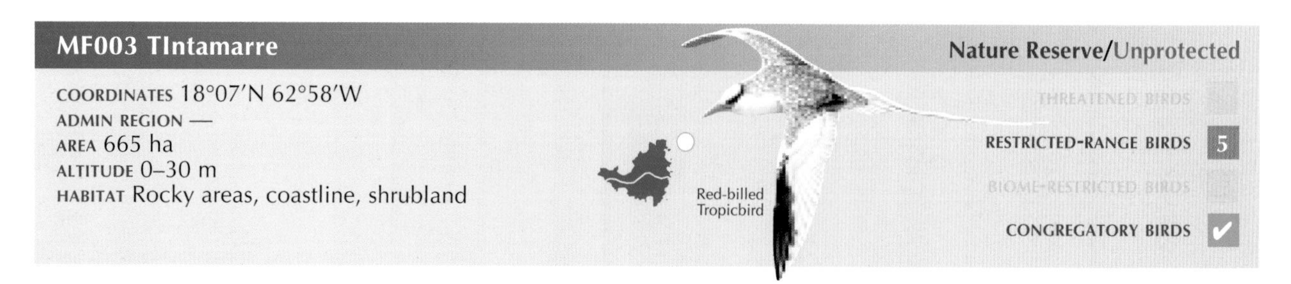

■ Site description

Tintamarre IBA is a 100-ha island (also known as Flat Island) situated 3 km from north-easternmost point of St Martin. The IBA includes marine areas up to 1 km from the island. Tintamarre is essentially flat, but with coastal cliffs (on the west coast) rising to 30 m. Vegetation comprises scrub (up to 3 m high). There is an abandoned air strip in the centre of the island and one unused residence. Human activity is limited to the grazing of goats and recreation. The recreational activities are heavily concentrated on the southern beach, where tourists are brought on day trips to snorkel and sunbathe. Private yachts also anchor at this beach.

■ Birds

This IBA is significant for its breeding seabirds. The Red-billed Tropicbird *Phaethon aethereus* is globally significant, while that of Brown Noddy *Anous stolidus* is regionally so. An estimated 60 pairs of *Phaethon aethereus* nest on the western cliffs where Audubon Shearwater *Puffinus lherminieri* breeds in unknown numbers. Small numbers of Roseate Tern *Sterna dougallii* and Bridled Tern *S. anaethetus* nest on the eastern side of the island with *A. stolidus*. American Oystercatcher *Haematopus palliatus* may also breed. Five (of the eight) Lesser Antilles EBA restricted-range birds occur on the island.

■ Other biodiversity

The island's beaches are sea-turtle nesting grounds. The Critically Endangered hawksbill *Eretmochelys imbricata* and leatherback *Dermochelys coriacea* turtles and Endangered loggerhead *Caretta caretta*, green *Chelonia mydas* and Olive Ridley *Lepidochelys olivacea* turtles all occur although it is unclear which nest. The Vulnerable Lesser Antillean iguana *Iguana delicatissima* occurs on Tintamarre.

■ Conservation

The coastline of Tintamarre Island is part of the national Réserve Naturelle St Martin which extends 82 m inland and prohibits wildlife disturbance, plant collection, mining, fishing, jet skiing, and littering. The island's interior is privately owned, but building is prohibited. EPIC has conducted summer and winter seabird surveys of the island. Staff of the nature reserve monitor sea-turtles and whales, and the government periodically hires avian researchers to survey the island. Predation of seabirds by rats (*Rattus* spp.) is the primary threat on the island. A proposal for rat eradication has been developed but has yet to be funded. Trampling of nests by goats is also a concern, and grazing by goats has almost certainly changed the island's vegetation. Coastal zone disturbance is a potential threat to nesting seabirds.

ST VINCENT & THE GRENADINES

LAND AREA **389 km²** ALTITUDE **0–1,234 m**
HUMAN POPULATION **102,250** CAPITAL **Kingstown**
IMPORTANT BIRD AREAS **15, totalling 179 km²**
IMPORTANT BIRD AREA PROTECTION **31%**
BIRD SPECIES **152**
THREATENED BIRDS **6** RESTRICTED-RANGE BIRDS **14**

LYSTRA CULZAC-WILSON (AVIANEYES)

Ashton Lagoon IBA on Union Island in the southern Grenadines.
(PHOTO: GREGG MOORE)

INTRODUCTION

St Vincent and the Grenadines is a multi-island nation in the Windward Islands of the Lesser Antillean chain. St Vincent is the main island (c.29 km long and 18 km wide, making up c.88% of the nation's land area) and lies furthest north, c.35 km south-south-west of St Lucia. The chain of Grenadine islands (comprising numerous islands, islets, rocks and reefs) extends south for 75 km towards the island of Grenada, with Union Island being the most southerly. Other major islands of the (St Vincent) Grenadines are Bequia (which is the largest), Mustique, Canouan, Mayreau, Palm (Prune) Island and Petit St Vincent. The country is divided into six parishes, five of which (Charlotte, Saint Andrew, Saint David, Saint George and Saint Patrick)) cover the main island of St Vincent, the sixth being the Grenadines. The capital, Kingstown (in St George parish on the south-east coast) supports c.25% of the country's population, while the Grenadines are home to about 8%.

St Vincent and the Grenadines were formed volcanically. The island of St Vincent is divided by a central mountain range which starts in the north with La Soufriere (1,234 m)—an active volcano and the island's highest point. The Morne Garu mountain range (with Richmond peak, 1,077 m and Mount Brisbane, 932 m) lies to the south of La Soufriere, and then Grand Bonhomme (970 m), Petit Bonhomme (756 m) and Mount St Andrew (736 m) are south of this. A large number of very steep lateral ridges emanate from the central massif culminating in high, rugged and almost vertical cliffs on the (eastern) leeward coast, while the windward coast is more gently sloping, with wider, flatter valleys. In contrast to St Vincent, the Grenadines have a much gentler relief, with the mountain peaks on these islands rising to150–300 m. There are no perennial streams in the Grenadines (although there is a spring on Bequia), and unlike much of the mainland, these islands are surrounded by fringing reefs and white sand beaches.

St Vincent's tropical climate has two distinct seasons: a dry season from December to May; and a rainy season from May through October. The average annual rainfall is 3,800 mm inland, and 2,000 mm on the coast. However, the forested interior of St Vincent can receive as much as 5,100 mm, while the Grenadines may receive as little as 460 mm. Natural vegetation corresponds to elevation, geology and rainfall, and includes rainforest (mostly between 300 and 500 m), elfin woodland and montane forest (above 500 m), palm break (between the rainforest and montane forest, and in disturbed areas), and mangrove (of which there is just c.50 ha in the country, most of which is on Union Island with some on Mustique). The country is about 29% forested, with natural forest comprising 70% of this, and planted forest and agro-forest representing c.25% and 5% respectively. Although these forests are some of the most extensive unaltered tropical forests in the Lesser Antilles, they are being lost at a rate of 3–5% annually, due primarily to encroachment of banana cultivation and illegal farming. Tourism and agriculture are the major contributors to the country's economy. However, agriculture relies almost exclusively on banana plantations/industry and thus this sector is highly vulnerable to global economic fluctuations and natural disasters. St Vincent and the Grenadines have suffered considerably from natural disasters. In 1902, La Soufriere volcano erupted and killed c.2,000 people. It erupted again in 1979, this time without loss of life, but on both occasions extensive damage was caused to agricultural lands and thus the economy. Hurricanes hit the island hard in 1980 and 1987, destroying (amongst other things) banana and coconut plantations.

■ Conservation

St Vincent and the Grenadines' National Parks Act (2002) is the country's most comprehensive piece of protected area legislation under which a System of Protected Areas and Heritage Sites (SPAHS) has been developed to protect and manage existing and proposed protected areas. As a program, SPAHS has a comprehensive set of management aims including: scientific research, wilderness protection and landscape maintenance, preservation of species and genetic diversity, maintenance of environmental services, protection of specific natural features, promotion of recreation and tourism, education, sustainable use of natural ecosystems and maintenance of cultural and traditional attributes. However, the program is awaiting funding before full implementation can take place and, in the interim, the conservation of the country's biodiversity is being undertaken in a piecemeal fashion by several government agencies and statutory bodies (e.g. National Parks, Rivers and Beaches Authority. The main agencies and government departments involved biodiversity conservation include the National Parks Unit (NPU, a statutory body affiliated with the Ministry of Tourism, and the agency responsible for implementing SPAHS); Ministry of Health and the Environment (through its Environmental Services Unit, ESU, which is responsible for environmental monitoring, regulation and education but is currently not fully established or staffed); Ministry of Agriculture, Forestry and Fisheries (within which Forestry Department coordinates the protection and management of the country's forests and wildlife [including birds], conducts environmental education, the biennial census of St Vincent Amazon *Amazona guildingii* and a captive-breeding programme for the species); Central Water and Sewerage Authority; and the Central Planning Unit (which, through its Physical Planning Unit, prepares development plans and administers planning regulations).

The legislation that speaks directly to the protection of birds is the Wildlife Protection Act (1987) which provides authority for the establishment of bird sanctuaries and wildlife reserves. The Act provides full protection for over 75 species of birds, but allows shorebirds and gamebirds to be hunted during an October–February open season. There are many other pieces of legislation that offer indirect protection to birds through the protection of habitats and biodiversity as a whole. These include the Marine Parks Act (which makes provision for the declaration of marine parks); the National Parks Act (allowing the establishment of national parks); the Mustique Company Ltd Act (which declares Mustique to be a conservation area); and others such as the Beach Protection Act, Fisheries Act, and the Forest Resource Conservation Act. Under these various Acts, 36 protected areas have been established (three forest reserves, 23 wildlife reserves, one marine park, one marine reserve and seven marine conservation areas). However, SPAHS proposes a system of 47 protected areas (one national park, eight forest reserves, 16 wildlife reserves, three natural landmarks, seven cultural landmarks, one protected landscape/seascape, five marine parks, three marine reserves and three marine conservation areas) and will result in the reclassification or re-designation of a number of the existing protected areas to remove duplication or to change management objectives.

Trinity Falls within the Richmond Forest Reserve IBA, a proposed forest reserve under SPAHS. (PHOTO: LYSTRA CULZAC-WILSON)

Major bird conservation actions in St Vincent and the Grenadines are generally implemented by the Forestry Department. Through its environmental education unit, pupils and community personnel are provided with information on the country's birds (mainly endemic) and their importance. Forestry Department also conducts guided tours to the Vermont Nature Trail (within the St Vincent Parrot Reserve) and other bird habitats, and manages the *A. guildingii* captive breeding programme at the Nicholl's Wildlife Complex in the Botanic Gardens. This programme is supported by the international St Vincent Parrot Conservation Consortium. The only national bird conservation NGO in the country is AvianEyes which aims to support nature conservation through birding, and conducts research, environmental education (e.g. as part of the Society for the Conservation and Study of Caribbean Birds', SCSCB's, West Indian Whistling-duck and Wetlands Conservation Program), and leads birding tours. In the Grenadines, the Mustique Company has stipulated protection for all its birds, and has developed a self-guided trail (with viewing hide) around the Lagoon wetland. On Union Island, a trail managed by the local NGO Union Island Ecotourism Movement is part of a conservation initiative for part of the Ashton Wetland (although the NGO does not have legal ownership of the area). This same wetland is the focus of a restoration project that will include the establishment of the site as a "Watchable Wildlife Pond" (under SCSCB's West Indian Whistling-duck and Wetlands Conservation Program). In spite of these efforts and initiatives, there is a clear need for: greater education and awareness (at all levels of society); the implementation of SPAHS; strengthened legislation and rigorous enforcement; and strengthened capacity for bird conservation.

Habitat loss and fragmentation due to squatting for housing, agriculture, illegal marijuana *Cannabis sativa* farming and development are major factors threatening biodiversity in St Vincent and the Grenadines. Deforestation has been identified as a main factor impacting the country's national bird—the Vulnerable *A. guildingii*. A proposed "cross-country road" that would bisect the centre of the parrot's range (and primary rainforest habitats) would result in a new axis for deforestation across the centre of St Vincent. It would also provide increased access to the parrots for poachers and hunters. Poaching has been identified as one of the main threats to *A. guildingii*, with birds removed to (illegally) supply the international pet trade. Hunting parrots as a source of food is an ongoing (although declining) threat. Wetland habitats (including beaches, mangroves, and marshland) are suffering as a result of developments such as hotels and marinas, but also due to illegal removal of beach (and dune) sand for the construction industry, and cutting of mangroves for charcoal production. The fragmentation of habitats and degradation of coastal ecosystems is making the country increasingly vulnerable to the impacts of natural disasters such as hurricanes, tropical storms, storm surges and heavy rains. Specific threats to birds include the removal of unfledged Scaly-naped Pigeon *Patagioenas squamosa* from the nest for meat, collection of seabird eggs (and taking adult seabirds for food), incidental poisoning of birds with agrochemicals (especially pesticides associated with the banana industry), legal but unregulated or monitored hunting of waterbirds, and predation from alien invasive mammals (mongoose *Herpestes auropunctus*, rats *Rattus rattus* and *R. norvegicus*, mouse *Mus muscalus* and opossum *Didelphis marsupialis*).

■ Birds

Over 150 species of bird have been recorded from St Vincent and the Grenadines, 95 of which breed on the islands. Lesser Antilles EBA restricted-range birds (of which there are 38) are represented by 14 species (see Table 1), two of which—St Vincent Parrot *Amazona guildingii* and Whistling Warbler *Catharopeza bishopi*—are endemic to the main island of St Vincent. The Grenada Flycatcher *Myiarchus nugator* and Lesser Antillean Tanager *Tangara cucullata* are restricted to St Vincent and Grenada. A subspecies of Rufous-throated Solitaire *Myadestes genibarbis sibilans* is endemic to St Vincent, as is a subspecies of House Wren *Troglodytes aedon musicus*. Just two globally threatened species have been considered in the IBA analysis (see Table 1). However, six species have been recorded from the islands. The Critically Endangered Eskimo Curlew *Numenius borealis* was noted as a "rare migrant" prior to 1943; the Near Threatened Buff-breasted Sandpiper *Tryngites subruficollis* is a very rare migrant; the Near Threatened Caribbean Coot *Fulica caribaea* appears to be a recent colonist (post-1970s) and although it breeds now in Mustique (and possibly elsewhere), the population is unknown; and the Near Threatened Piping Plover *Charadrius melodus* is recorded only as a vagrant. The two species that feature prominently in the IBA analysis are the two St Vincent island endemics—the Endangered *C. bishopi*, and the Vulnerable *A. guildingii*.

Catharopeza bishopi is endemic to mainland St Vincent where it is found primarily within the Colonarie and Perseverance valleys and at Richmond peak. It is most abundant in primary, elfin and palm brake forests (mostly between 300 and 600 m) of which there are c.80 km^2 that (in June–August 1988) supported an estimated 1,500–2,500 territorial males. There have been no estimates of the population in the last 20 years. Forest loss from illegal human activities and particularly eruptions of La Soufriere volcano is the main threat. Eruptions in 1902 and 1979 had a devastating effect on the warbler's habitats on and around La Soufriere—after 1902 the species was seemingly extinct in the northern mountains. Potential confusion between the call of *C. bishopi* and the Brown Trembler *Cinclocerthia ruficauda* should be assessed, and with this in mind, a new population density estimate for the warbler made.

Amazona guildingii is St Vincent and the Grenadines' national bird. It is confined to mature rainforest between 125 and 1,000 m, mostly in the upper reaches of the Buccament, Cumberland, Colonarie, Congo–Jennings–Perseverance and Richmond valleys, though birds do stray into nearby farmland and plantations to forage. The Forestry Department conducts a biennial census of the parrot, and in 2004 the population was estimated at 734 individuals. In 1987 the 4,400-ha St Vincent Parrot Reserve was established to protect the species, and in 2005 a 5-year Species Conservation Plan was developed for the bird. There is an ongoing captive breeding programme for the parrot at the Nicholl's Wildlife Complex in the Botanic Gardens.

St Vincent and the Grenadines supports populations of 76 species of waterbirds (including seabirds). Three species of seabird breed on St Vincent (White-tailed Tropicbird *Phaethon lepturus*, Roseate Tern *Sterna dougallii* and Brown Noddy *Anous stolidus*), and an additional nine species nest on

The Endangered Whistling Warbler is endemic to mainland St Vincent. (PHOTO: ALLAN SANDER)

Grenada Flycatcher is endemic to St Vincent and the Grenadines, and Grenada. (PHOTO: GREGORY GUIDA)

Table 1. Key bird species at Important Bird Areas in St Vincent and the Grenadines.

Key bird species	Criteria	National population		VC001	VC002	VC003	VC004	VC005	VC006
			Criteria	■ ■	■ ■	■ ■	■ ■	■ ■	■ ■
Magnificent Frigatebird *Fregata magnificens*	■	350–1,000							
Brown Pelican *Pelecanus occidentalis*	■	<50							
Red-footed Booby *Sula sula*	■	9,000							
Brown Booby *Sula leucogaster*	■	500–1,000							
Laughing Gull *Larus atricilla*	■	500–1,000							
Royal Tern *Sterna maxima*	■	500–1,000							
Sandwich Tern *Sterna sandvicensis*	■	100–250							
Roseate Tern *Sterna dougallii*	■	200–500							
Sooty Tern *Sterna fuscata*	■	10,000							
St Vincent Amazon *Amazona guildingii*	VU ■ ■	800		✓	43	✓	175	142	164
Lesser Antillean Swift *Chaetura martinica*	■			✓	✓	✓	✓	✓	✓
Purple-throated Carib *Eulampis jugularis*	■			✓	✓	✓	✓	✓	✓
Green-throated Carib *Eulampis holosericeus*	■			✓	✓	✓	✓	✓	✓
Antillean Crested Hummingbird *Orthorhyncus cristatus*	■			✓	✓	✓	✓	✓	✓
Caribbean Elaenia *Elaenia martinica*	■				✓	✓	✓	✓	✓
Grenada Flycatcher *Myiarchus nugator*	■			✓	✓	✓	✓	✓	✓
Scaly-breasted Thrasher *Margarops fuscus*	■			✓	✓	✓	✓	✓	✓
Brown Trembler *Cinclocerthia ruficauda*	■			✓	✓	✓	✓	✓	✓
Rufous-throated Solitaire *Myadestes genibarbis*	■			✓	✓	✓	✓	✓	✓
Whistling Warbler *Catharopeza bishopi*	EN ■ ■	3,000–5,000		✓	✓	✓	✓	✓	✓
Lesser Antillean Bullfinch *Loxigilla noctis*	■			✓	✓	✓	✓	✓	✓
Lesser Antillean Tanager *Tangara cucullata*	■			✓	✓	✓	✓	✓	✓
Antillean Euphonia *Euphonia musica*	■			✓	✓	✓	✓	✓	✓

All population figures = numbers of individuals.
Threatened birds: Endangered ■; Vulnerable ■. **Restricted-range birds** ■. **Congregatory birds** ■.

Dalaway Forest Reserve IBA, a protected stronghold for the St Vincent Amazon.
(PHOTO: LYSTRA CULZAC-WILSON)

…s IBAs

VC009	VC010	VC011	VC012	VC013	VC014	VC015
■	■			■		■
		■	■	■	■	
		119	250–500		250–500	
				<50		
		9,000				
		600	100–250		250–500	
		400	100–250		250–500	
			250–500	50–100	250–500	
				50–100		
			100–250		100–250	
					10,000	
✓	✓			✓		✓
✓						
✓	✓			✓		✓
✓	✓			✓		✓
✓	✓					
✓	✓			✓		✓
✓	✓			✓		✓
✓	✓			✓		✓

uninhabited or undisturbed islets in the Grenadines (namely Red-billed Tropicbird *Phaethon aethereus*, Magnificent Frigatebird *Fregata magnificens*, Masked Booby *Sula dactylatra*, Red-footed Booby *S. sula*, Brown Booby *S. leucogaster*, Laughing Gull *Larus atricilla*, Royal Tern *Sterna maxima*, Bridled Tern *S. anaethetus* and Sooty Tern *S. fuscata*). The current breeding status of Audubon's Shearwater *Puffinus lherminieri* in the country is unknown (although it certainly used to breed). In fact the current status and population of most of the country's seabirds is poorly known, although poaching of seabird eggs by fishermen is a common (but neither regulated nor policed) tradition practiced on the smaller islets, and could be significantly impacting on a number of species. Similarly, the populations of waterbirds (ducks, shorebirds) are poorly known, but many are listed as game birds that can be hunted between 1 October and 28 February. This hunting is not policed or regulated—numbers of individuals of each species shot and therefore the impact on species populations is unknown.

IMPORTANT BIRD AREAS

St Vincent and the Grenadines' 15 IBAs—the country's international site priorities for bird conservation—cover 179 km² (including marine areas) and about 35% of the land area. They have been identified on the basis of 23 key bird species (listed in Table 1) that variously trigger the IBA criteria. These 23 species include two (of the six: see above) globally threatened birds, all 14 restricted-range species and nine congregatory waterbirds/seabirds. Ten of the IBAs are on St Vincent and five are scattered throughout the Grenadines islands. Of the St Vincent island IBAs, seven are contiguous with each other in the forested interior where they form the proposed (under SPAHS) Central Forest Reserve. These seven

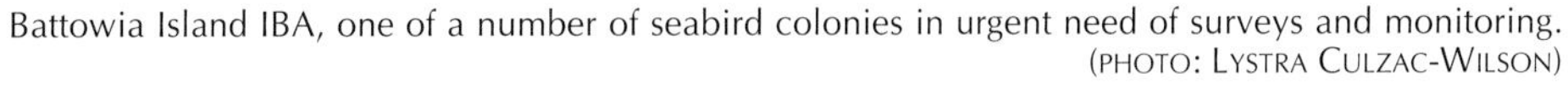

Battowia Island IBA, one of a number of seabird colonies in urgent need of surveys and monitoring. (PHOTO: LYSTRA CULZAC-WILSON)

Figure 1. Location of Important Bird Areas in St Vincent and the Grenadines

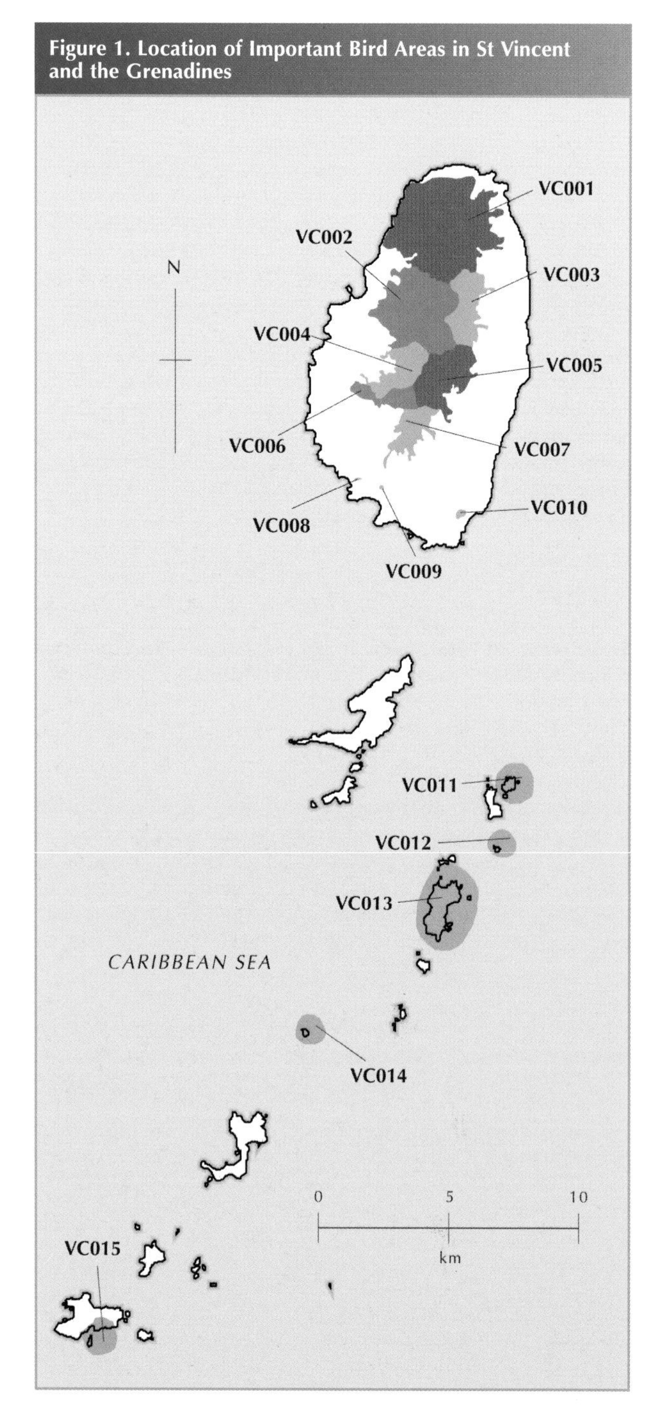

IBAs (which comprise the existing Cumberland Forest Reserve, five individual proposed forest reserves and a proposed national park) embrace the majority of the country's remnant primary rainforest, secondary forest, elfin woodland and palm brake, and thus significant portions of the ranges of endemic species such as the St Vincent Parrot *Amazona guildingii*, Whistling Warbler *Catharopeza bishopi* and St Vincent blacksnake *Chironius vincenti*. They also cover watersheds that produce over 90% of the country's potable water and a significant proportion of its hydroelectric power. However, only Cumberland Forest Reserve IBA (VC004) and Dalaway Forest Reserve IBA (VC006) are formally protected at the present time. Protection for the remaining IBAs and for the whole Central Forest Reserve (which encompasses prime gaps in the range of *A. guildingii* that existed after the creation of the St Vincent Parrot Reserve in 1987) requires the implementation of SPAHS. Thus, only 31% of the area covered by the St Vincent and Grenadines IBAs is currently under formal protection and active management/regulation is minimal.

Five of the country's IBA have been documented for the Grenadines. They include four entire islands that are significant for their congregatory waterbirds and seabirds and which are formally designated as protected areas. However, they all variously suffer from poaching of seabird eggs, illegal burning of vegetation or disturbance, and law enforcement is essentially non-existent. There is very little information related to the seabird populations on these (and indeed other Grenadines islands). Estimates have derived from fishermen and other boat operators (some involved in the annual poaching of eggs) and thus require verification before monitoring can start. Mustique Island IBA (VC013) with its Lagoon wetland, and Ashton Wetland IBA (VC015)—both in the Grenadines—are the country's largest wetlands and together represent 80% of the nation's wetland habitat. As with the seabirds, very little has been recorded concerning species presence and abundance at these or other wetlands in the country. There is a clear and urgent need for surveys of wetlands and seabird colonies to establish a baseline against which to monitor and from which additional IBAs could possibly be described.

The existing parrot monitoring program (implemented by Forestry Department) could be usefully expanded to include field assessments (surveys and subsequent monitoring) for *Catharopeza bishopi* and potentially the seabird populations. All monitoring results should be used to inform the annual assessment of state, pressure and response scores at each of the country's IBAs to provide an objective status assessment and highlight management interventions that might be required to maintain these internationally important biodiversity sites.

KEY REFERENCES

Andrle, R. F. and Andrle, P. R. (1976) The Whistling Warbler of St Vincent, West Indies. *Condor* 78: 236–243.

AvianEyes Birding Group (1999) Birds of Mustique: a report of a bird study conducted on Mustique, 22–24 January, 1999. (Unpublished report).

Butler, P. J. (1988) The St Vincent Parrot *Amazona guildingii*: road to recovery. Philadelphia, Penn.: RARE Center for tropical Conservation (Unpublished report).

Carr, M., Foster, J., Gittings, T. and Morris, R. (1988) Distribution and abundance of St Vincent's Whistling Warbler and other endemics. Norwich, UK: University of East Anglia (Unpublished expedition report).

Caribbean Conservation Association (1991) *St Vincent and the Grenadines: country environmental profile*. St Michael, Barbados: Caribbean Conservation Association.

Christian, C. S. (1993) The challenge of parrot conservation in St Vincent and the Grenadines. *J. Biogeogr.* 20: 463–469.

Collar, N. J., Gonzaga, L. P., Krabbe, N., Madroño Nieto, A., Naranjo, L. G., Parker, T. A. and Wege, D. C. (1992) *Threatened birds of the Americas: the ICBP/IUCN Red Data Book*. Cambridge, UK: International Council for Bird Preservation.

Culzac-Wilson, L. (2005) *Species conservation plan for the St Vincent Parrot* Amazona guildingii. Puerto de la Cruz, Tenerife: Loro Parque Fundación.

Daudin, J. (2003) *A natural history monograph of Union Island.* Martinique: Desormeaux.

Department of Tourism (2006) *The ins and outs of St Vincent and the Grenadines*. St Thomas, Barbados: Miller Publishing Company, Ltd.

Evans, P. G. H. (1990) *Birds of the eastern Caribbean*. London: Macmillan Education Ltd.

Faaborg, J. R. and Arendt, W. J. (1985) *Wildlife assessments in the Caribbean*. San Juan, Puerto Rico: Institute of Tropical Forestry.

Forestry Department (2004) St Vincent Parrot census, 2004. Kingstown, St Vincent and the Grenadines: Forestry Department, Ministry of Agriculture, Forestry and Fisheries. (Unpublished report).

FROST, M. D., HAYES, F. E. AND SUTTON, A. H. (in press) Saint Vincent, the Grenadines and Grenada. In P. E. Bradley and R. L. Norton eds. *Breeding seabirds of the Caribbean*. Gainesville, Florida: Univ. Florida Press.

GOREAU, T. J. AND SAMMONS, N. (2003) Water quality in Ashton Harbour, Union Island, St Vincent and the Grenadines: environmental impacts of marina and recommendations for ecosystem and fisheries restoration. Downloaded from www.globalcoral.org/Water%20Quality%20in%20Ashton%20Harbour.htm (23 January 2006).

VAN HALEWYN, R. AND NORTON, R. L. (1984) The status and conservation of seabirds in the Caribbean. Pp 169–222 in J. P. Croxall, P. G. H. Evans and R. W. Schreiber, eds. *Status and conservation of the world's seabirds*. Cambridge, U.K.: International Council for Bird Preservation (Techn. Publ. 2).

IVOR JACKSON AND ASSOCIATES (2004) Master Plan: System of Protected Areas and Heritage Sites, St Vincent and the Grenadines. St John's, Antigua: Ivor Jackson and Associates (Unpublished report).

JEGGO, D. (1990) Preliminary international studbook: St Vincent Parrot *Amazona guildingii*. Jersey: Jersey Wildlife Preservation Trust. (Unpublished report).

JUNIPER, T. AND PARR, M. (1998) *A guide to the parrots of the world*. Robertsbridge, UK: Pica press.

LACK, D., LACK, E., LACK, P. AND LACK, A. (1973) Birds of St Vincent. *Ibis* 115: 46–52.

OVERING, J. AND CAMBERS, G. (1995) Mustique environmental inventory, 1. (Unpublished report for the Mustique Company).

OVERING, J. AND CAMBERS, G. (2004) 2004 update on environmental management in Mustique. (Unpublished report for the Mustique Company).

RAFFAELE, H. WILEY J., GARRIDO, O., KEITH, A. AND RAFFAELE, J. (1998) *A guide to the birds of the West Indies*. Princeton, New Jersey: Princeton University Press.

REID, COLLINS AND ASSOCIATES (1994) Watershed management plan for the Colonarie river basin. Vancouver: Reid, Collins and Associates. (Unpublished report to the Ministry of Agriculture, Industry and Labour—Forestry Division).

SIMMONS AND ASSOCIATES, INC. (2000) National Biodiversity Strategy and Action Plan for St Vincent and the Grenadines. St Michael, Barbados: Simmons and Associates, Inc. (Unpublished report).

ACKNOWLEDGEMENTS

The author would like to thank Adrian Bailey (Forestry Department), Amos Glasgow (Forestry Department), Andrew Wilson (National Parks Unit), AvianEyes Birding Group, Carlton Thomas (Forestry Department), Cynthia Robertson (Surveys Department), Cornelius Richards (Forestry Department), Darnley Hazell (The Mustique Company), Edward Bess (Bequia), Father Mark Da Silva (Mayreau), Fitzroy Springer (Forestry Department), Kyron Bynoe (Bequia), Lennox Quammie (The Mustique Company), Matthew Harvey (Union Island Environmental Movement), Orton King (Old Hegg Turtle Sanctuary, Bequia), Peter Ernst (The Mustique Company), and Rudolph Ragguette (retired fisherman/current tour operator, Mustique), for their contributions to this chapter.

■ Site description

La Soufrière National Park IBA encompasses St Vincent's youngest volcano, and much of the northern quarter of the island (including part of the north-west coast). La Soufrière is an active volcano (and one of the country's main tourist attractions), historically first erupting in 1718, and most recently in 1979. As a result of the volcanic eruptions the surrounding area supports a unique successional ecosystem with a mixture of secondary rainforest and volcanic pioneer vegetation. Various tributaries and waterfalls emanate from the steep slopes, while the foothills support some agriculture (particularly banana cultivation) and, at lower elevations, several forest plantations established by the Forestry Department. High rainfall, along with loose volcanic deposit makes the area vulnerable to erosion and landslides.

■ Birds

This IBA supports populations of 13 (of the 14) Lesser Antilles EBA restricted-range birds, including the Endangered Whistling Warbler *Catharopeza bishopi* (considered "occasional") and the Vulnerable St Vincent Amazon *Amazona guildingii* (considered "rare"). Following the 1979 eruption, the entire population of parrots disappeared, returning around 1999. The IBA is important for Rufous-throated Solitaire *Myadestes genibarbis*, particularly in the higher elevation, well-forested parts of the park.

■ Other biodiversity

The Critically Endangered St Vincent blacksnake *Chironius vincenti* and the Endangered (endemic) tree frog *Eleutherodactylus shrevei* occur, as do a number of other endemic reptiles, namely the lizards *Anolis griseus* and *A. vincentiana*, and the regionally endemic congo snake *Mastigodryas bruesi*. A number of island-endemic plants are present in the IBA.

■ Conservation

La Soufrière National Park is state-owned. It is a proposed national park under the System of Protected Areas and Heritage Sites, and also part of the proposed Central Forest Reserve. Eruptions of La Soufrière have previously destroyed large tracts of forest with the impacts (including on the avifauna) lasting many decades (e.g. the disappearance of *A. guildingii* in 1979). Volcanic activity is monitored by the Seismic Unit (Ministry of Agriculture) in association with the University of the West Indies in Trinidad. Rats (*Rattus* spp.), opossum *Didelphis marsupialis* and mongoose *Herpestes auropunctatus* are all present in the area and presumably impacting bird populations. Marijuana *Cannabis sativa* is illegally farmed (at the expense of natural forest) within this proposed reserve. Illegal squatting and hunting also occur.

VC002 Richmond Forest Reserve

■ Site description

Richmond Forest Reserve IBA is in west-central St Vincent, immediately south of and abutting La Soufrière National Park (IBA VC001) and west of Mount Pleasant Forest Reserve (IBA VC003). It is an area of rugged terrain and spectacular mountain scenery (including St Vincent's second highest peak—Richmond) overlooking the Caribbean Sea that supports several of the country's rivers and waterfalls of recreational importance. The IBA is within a major hydropower catchment area. Forest is predominantly rainforest with montane forest and elfin forest on the upper slopes, and secondary and dry scrub forests at lower elevations. Loose volcanic material is washed towards the coast by the Richmond River, then collected and sold on the beach by locals (for use in the construction industry).

■ Birds

This IBA supports populations of all 14 Lesser Antilles EBA restricted-range birds, including the Endangered Whistling Warbler *Catharopeza bishopi* and Vulnerable St Vincent Amazon *Amazona guildingii*. Deforestation (particularly for marijuana *Cannabis sativa* cultivation) has reduced the population of the parrot in this reserve: an estimated 43 individuals are thought to be present. The Brown Trembler *Cinclocerthia ruficauda*, Scaly-breasted Thrasher *Margarops fuscus*, Rufous-throated Solitaire *Myadestes genibarbis sibilans* and Antillean Euphonia *Euphonia musica* are confined to the higher elevations, primarily above 305 m.

■ Other biodiversity

The Critically Endangered St Vincent blacksnake *Chironius vincenti* and the Endangered (endemic) tree frog *Eleutherodactylus shrevei* occur. Endemic flora in the IBA includes *Begonia rotundifolia*, the epiphytic *Peperomia cuneata* and *P. vincentiana*, forest orchid *Epidendrum vincentinum* and the giant fern *Cyathea tenera*.

■ Conservation

Richmond Forest Reserve IBA is a state-owned, proposed forest reserve under the System of Protected Areas and Heritage Sites. It is also part of the proposed Central Forest Reserve. Illegal marijuana *Cannabis sativa* cultivation has been responsible for a significant loss of forest habitat. This habitat loss, combined with heavy rains and steep terrain, has resulted in increased rates of soil erosion and the area's susceptibility to landslides. Illegal hunting of opossum *Didelphis marsupialis*, agouti *Dasyprocta agouti* and armadillo *Dasypus novemcinctus* occurs.

VC003 Mount Pleasant Forest Reserve

■ Site description

Mount Pleasant Forest Reserve IBA is in north-central St Vincent, south-east of and abutting La Soufrière National Park (IBA VC001), and east of Richmond Forest Reserve (IBA VC002). It is bordered to the south by Colonarie Forest Reserve (VC005). The IBA is characterised by an undulating landscape supporting a rich diversity of flora and fauna and many scenic mountain vistas. Rainforest (some of which is secondary) and palm brake are the dominant vegetation formations. The reserve encompasses the Perseverance water catchment that produces c.5% of the country's annual water output. Agricultural activities are concentrated along the IBA's eastern boundary.

■ Birds

This IBA supports populations of all 14 Lesser Antilles EBA restricted-range birds, including the Endangered Whistling Warbler *Catharopeza bishopi* and Vulnerable St Vincent Amazon *Amazona guildingii*. Other species of note include Common Black-hawk *Buteogallus anthracinus*, Scaly-naped Pigeon *Patagioenas squamosa*, Short-tailed Swift *Chaetura brachyura* and the threatened endemic race of House Wren *Troglodytes aedon musicus*.

■ Other biodiversity

The Critically Endangered St Vincent blacksnake *Chironius vincenti* occurs, as do a number of other endemic reptiles, namely the lizards *Anolis griseus* and *A. trinitatus*, and the regionally endemic congo snake *Mastigodryas bruesi*. Several endemic plants are found including *Begonia rotundifolia*, the epiphytic *Peperomia cuneata* and *P. vincentiana*, forest orchid *Epidendrum vincentinum* and giant fern *Cyathea tenera*.

■ Conservation

Mount Pleasant Forest Reserve IBA is a state-owned, proposed forest reserve under the System of Protected Areas and Heritage Sites. It is also part of the proposed Central Forest Reserve. Illegal marijuana *Cannabis sativa* cultivation has been responsible for a significant loss of forest habitat. This habitat loss, combined with heavy rains and steep terrain, has resulted in increased rates of soil erosion and the area's susceptibility to landslides. Illegal (agriculture-related) squatting occurs. A major concern is the continued poaching of *Amazona guildingii*.

Site description

Cumberland Forest Reserve IBA is in central St Vincent, bordered to the north by Richmond Forest Reserve (IBA VC002), to the east by Colonarie Forest Reserve (IBA VC005), and to the south by Dalaway Forest Reserve (IBA VC006). The IBA supports portions of the last remaining stand of primary rainforest and montane forest (including elfin forest) in the country. The reserve also has secondary and coastal scrub forest at lower elevations, and some plantation forests (e.g. *Hibiscus elatus* and *Pinus caribaea*). Cumberland Forest Reserve's numerous rivers and streams (an important source of hydroelectric power) add great diversity to these species-rich, relatively intact forests. Farming is practiced in the valley areas but does not adversely affect the forest.

Birds

This IBA supports populations of all 14 Lesser Antilles EBA restricted-range birds, including the Endangered Whistling Warbler *Catharopeza bishopi* and Vulnerable St Vincent Amazon *Amazona guildingii*. It is a stronghold for *A. guildingii*, which numbers around 175 individuals (about a third of the total population) and can be found foraging within valleys and near agricultural lands. The parrots also move between the mountain peaks surrounding the valleys. Other species of note include Scaly-naped Pigeon *Patagioenas squamosa* and Short-tailed Swift *Chaetura brachyura*.

Other biodiversity

The Critically Endangered St Vincent blacksnake *Chironius vincenti* and the Endangered (endemic) tree frog *Eleutherodactylus shrevei* occur. Other endemic reptiles include the lizards *Anolis griseus* and *A. trinitatus*, and the regionally endemic congo snake *Mastigodryas bruesi*. Several endemic plants are found.

Conservation

Cumberland IBA is a state-owned, currently established forest reserve, and is also part of the Central Forest Reserve as proposed under the System of Protected Areas and Heritage Sites. Also proposed is a scenic nature trail that will traverse a variety of biodiverse habitats, and will be supported by overnight camping and lodging facilities. Illegal hunting of opossum *Didelphis marsupialis*, agouti *Dasyprocta agouti* and armadillo *Dasypus novemcinctus* occurs within the reserve and is a major concern along with the poaching of *Amazona guildingii* nests. Also, burrows of *D. novemcinctus* (e.g. under trees) often result in soil exposure, tree falls and increased erosion, and this is quite evident within the reserve.

Site description

Colonarie Forest Reserve IBA is in central St Vincent encompassing the mid and upper reaches of the Colonarie watershed. The Colonarie River is the longest watercourse (with the second largest catchment area) on St Vincent and is a main source of potable water and hydro-electricity. The IBA is bordered to the north by Richmond (IBA VC002) and Mount Pleasant (IBA VC003) forest reserves and to the west by Cumberland Forest Reserve (IBA VC004). In the upper reaches of the reserve, above 300 m, watercourses have cut deeply into ash agglomerates and basaltic bedrock resulting in a landscape of irregular, complex and steeply-sloping landforms. However, much of the IBA is covered by primary rainforest. The steepness of the terrain (combined with high rainfall) is responsible for the high rates of erosion and landslide hazards.

Birds

This IBA supports populations of all 14 Lesser Antilles EBA restricted-range birds, including the Endangered Whistling Warbler *Catharopeza bishopi* and Vulnerable St Vincent Amazon *Amazona guildingii*. It is a traditional stronghold for *A. guildingii*, which numbered 142 individuals in 2004 (about 25% of the total population). Other species of note include Common Black-hawk *Buteogallus anthracinus*, Scaly-naped Pigeon *Patagioenas squamosa*, Short-tailed Swift *Chaetura brachyura* and the threatened endemic race of House Wren *Troglodytes aedon*.

Other biodiversity

The Critically Endangered St Vincent blacksnake *Chironius vincenti* occurs. Other endemic reptiles include the lizards *Anolis griseus* and *A. trinitatus*, and the regionally endemic congo snake *Mastigodryas bruesi*. Several endemic plants are found.

Conservation

Colonarie Forest Reserve IBA is a state-owned, proposed forest reserve under the System of Protected Areas and Heritage Sites. It is also part of the proposed Central Forest Reserve. Encroachment, land settlement and (agriculture-related) squatting are serious threats to the remaining primary forest. The lower slopes of the IBA have, in the past, seen intensive exploitation for agriculture. Deforestation for agriculture (cultivation and livestock) and charcoal production is a major concern as it leads to erosion, slope instability and sedimentation in the streams. The lack of riparian buffers exacerbates these problems. River poisoning to harvest crayfish is a common Easter-time practice in nearby communities causing the death of much of the aquatic fauna.

Site description

Dalaway Forest Reserve IBA is in central south-west St Vincent, south of and abutting Cumberland Forest Reserve (IBA VC004) and west of Colonarie Forest Reserve (IBA VC005). The IBA encompasses forested watersheds that supply c.45% of the country's potable water. The forest comprises both primary and secondary rainforest and montane forest. Several forest plantations were established by the Forestry Department as a soil conservation initiative on abandoned farmlands. Thus agricultural crops can be found interspersed between secondary vegetation.

Birds

This IBA supports populations of all 14 Lesser Antilles EBA restricted-range birds, including the Endangered Whistling Warbler *Catharopeza bishopi* and Vulnerable St Vincent Amazon *Amazona guildingii*. It supports prime habitat for *A. guildingii* and is another stronghold for the species (with 164 individuals estimated as present in 2004). Other species of note include Scaly-naped Pigeon *Patagioenas squamosa*, Short-tailed Swift *Chaetura brachyura* and the threatened endemic race of House Wren *Troglodytes aedon*.

Other biodiversity

The Critically Endangered St Vincent blacksnake *Chironius vincenti* occurs. Other endemic reptiles include the lizard *Anolis griseus* and the regionally endemic congo snake *Mastigodryas bruesi*. Several endemic plants are found including *Begonia rotundifolia*, the epiphytic *Peperomia cuneata* and *P. vincentiana*, forest orchid *Epidendrum vincentinum* and giant fern *Cyathea tenera*.

Conservation

Dalaway Forest Reserve IBA is a state-owned, proposed forest reserve under the System of Protected Areas and Heritage Sites. It is also part of the proposed Central Forest Reserve. Over 90% of the IBA falls within the designated 4,401-ha St Vincent Parrot Reserve. The 3.5-km Vermont Nature Trail (in the south of the IBA) is a major tourist attraction, allowing visitors to observe the parrot in its natural habitat. Educational tours, and patrols by the Forestry Department staff are regular and ongoing, and have limited the impact of threats such as poaching of parrots or hunting of other animals. Boundary encroachment and squatting are periodic problems, and armadillo *Dasypus novemcinctus*, cats and dogs are all having an impact on the area.

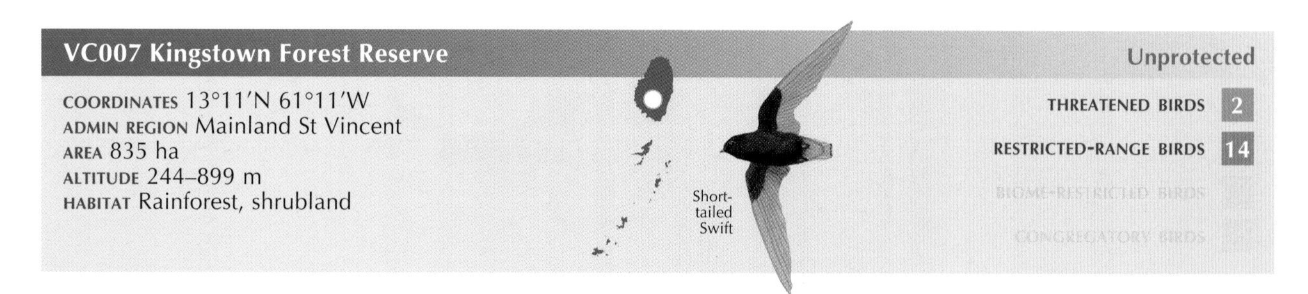

Site description

Kingstown Forest Reserve IBA is in south-central St Vincent, adjoining and immediately south of Colonarie Forest Reserve (IBA VC005) and Dalaway Forest Reserve (IBA VC006). The IBA encompasses four watersheds (together producing over 25% of the country's potable water) and includes the island's highest southerly peak—Mount St Andrew—which overlooks the capital. This IBA, which maintains primary and secondary rainforest and dry scrub woodland, contains portions of St Vincent Amazon *Amazona guildingii* habitat and range that were omitted during the establishment of the St Vincent Parrot Reserve in 1987.

Birds

This IBA supports populations of all 14 Lesser Antilles EBA restricted-range birds, including the Endangered Whistling Warbler *Catharopeza bishopi* and Vulnerable *A. guildingii*. Other species of note include Common Black-hawk *Buteogallus anthracinus*, Scaly-naped Pigeon *Patagioenas squamosa*, Short-tailed Swift *Chaetura brachyura* and the threatened endemic race of House Wren *Troglodytes aedon*.

Other biodiversity

The Critically Endangered St Vincent blacksnake *Chironius vincenti* occurs. Other endemic reptiles include the lizards *Anolis griseus* and *A. trinitatus*, and the regionally endemic congo snake *Mastigodryas bruesi*. Several endemic plants are found including *Begonia rotundifolia*, the epiphytic *Peperomia cuneata* and *P. vincentiana*, forest orchid *Epidendrum vincentinum* and giant fern *Cyathea tenera*.

Conservation

Kingstown Forest Reserve IBA is a state-owned, proposed forest reserve under the System of Protected Areas and Heritage Sites. It is the most southerly part of the proposed Central Forest Reserve. Due to its close proximity to the capital (and thus c.25% of the human population) human activities (including illegal squatting, hunting, farming and marijuana *Cannabis sativa* cultivation) have encroached across the forest reserve's boundaries.

■ Site description

Campden Park Forest Reserve IBA is in south-west St Vincent, near the coast adjacent to the Campden Park River. It shares its western borders with the compound of the Forestry Department's headquarters. The majority of the remaining boundary lies adjacent to human settlement and boundary encroachment has been a serious challenge to the Forestry Department. The area comprises dry scrub woodland with dominant tree species including *Bursera simaruba*, *Swietenia mahogani* and *Tabebuia pallida*.

■ Birds

This IBA supports seven (of the 14) Lesser Antilles EBA restricted-range birds, namely those species that associate with the relatively dry tropical woodland found in the reserve such as Caribbean Elaenia *Elaenia martinica*, Grenada Flycatcher *Myiarchus nugator* and Lesser Antillean Tanager *Tangara cucullata*.

■ Other biodiversity

The endemic lizards *Anolis griseus* and *A. trinitatus*, and the regionally endemic congo snake *Mastigodryas bruesi* occur.

■ Conservation

Campden Park Forest Reserve IBA is a state-owned, proposed forest reserve under the System of Protected Areas and Heritage Sites. Being adjacent to the Forestry Department headquarters this site is often used for staff field training and public education. However, due its proximity to human settlements, squatting, bushfires and use of the site by domestic animals (cats, dogs and chickens) is common. The iguana *Iguana iguana* is sometimes hunted within the IBA.

■ Site description

Botanic Gardens Natural Landmark IBA is in southern St Vincent, at the foot of Mount St Andrew, the island's highest southern peak. It includes the Botanic Gardens (8 ha), Government House grounds (the 10-ha residence of the Governor General) and the Nicholls Wildlife Complex (a breeding aviary for St Vincent Amazon *Amazona guildingii* established in 1988). The Botanic Gardens were established in 1765 as an outstation of the Royal Botanic Gardens Kew (London) and are the oldest in the western hemisphere. The vegetation is a mixture of dry scrub woodland, plantation forest, horticultural and agricultural crops, and also the exotic plants of the Botanic Gardens themselves.

■ Birds

This IBA supports populations of eight (of the 14) Lesser Antilles EBA restricted-range birds, primarily those affiliated with dry forest habitats, but also (due to the IBAs location at the foot of Mount St Andrew) several species more normally found in rainforest (e.g. Purple-throated Carib *Eulampis jugularis*). Other species of note include Common Black-hawk *Buteogallus anthracinus*, Scaly-naped Pigeon *Patagioenas squamosa* the threatened endemic race of House Wren *Troglodytes aedon*.

■ Other biodiversity

The endemic lizard *Anolis griseus* and the regionally endemic congo snake *Mastigodryas bruesi* occur.

■ Conservation

The Botanic Gardens Natural Landmark IBA is state-owned, and both an existing bird sanctuary and wildlife reserve. It is a proposed natural landmark under the System of Protected Areas and Heritage Sites. The aviary currently houses 36 *A. guildingii* and annually produces an average of two chicks. The Botanic Gardens are one of the most visited sites by tourists on the island. However, inadequate awareness and operating procedures means that nesting birds are disturbed and snakes (none of which are poisonous) often killed.

VC010 King's Hill Forest Reserve

Site description

King's Hill Forest Reserve IBA is in southernmost St Vincent, just north of the highway at Stubbs Bay. It is an area of light sandy soils supporting dry woodland. The IBA is the oldest forest reserve in St Vincent and the second oldest in the Western Hemisphere, established in 1791 to "attract the clouds and rain" and thus provide rain for surrounding areas that were otherwise dry and windswept. Due to its age and historical significance (it contains a vegetation sample plot established by Beard in 1949), the site is sometimes used by the Forestry Department for research, school visits and dendrology training. However, use of the site is restricted and legally requires the permission from the Director of the Forestry Department.

Birds

This IBA supports populations of seven (of the 14) Lesser Antilles EBA restricted-range birds, primarily those often affiliated with dry forest habitats, including Caribbean Elaenia *Elaenia martinica*, Grenada Flycatcher *Myiarchus nugator* and Lesser Antillean Tanager *Tangara cucullata*. Other species of note include Common Black-hawk *Buteogallus anthracinus* and Scaly-naped Pigeon *Patagioenas squamosa*.

Other biodiversity

The endemic lizards *Anolis griseus* and *A. vincentiana*, and the regionally endemic congo snake *Mastigodryas bruesi* occur. It is possible that endemic plant species are also present.

Conservation

King's Hill Forest Reserve IBA is state-owned and designated as a forest reserve, wildlife reserve and bird sanctuary. Being located close to the highway the forest is easily accessible to local residents and tourists, and human pressure is a significant concern. All species are protected year-round within this site but illegal hunting is a major issue and wild yams and other plant material are harvested, threatening the integrity of the ecosystem. It is thought that domesticated animals (cats, dogs, chickens and small ruminants) use the IBA. Dry season fires are always a threat and forest management should take this into account.

VC011 Battowia Island

Site description

Battowia Island IBA (which includes the islet Battowia Bullet) is the most easterly of the Grenadine islands and is immediately adjacent to the island of Baliceaux. It lies c.10 km south-east of Bequia (the largest Grenadine Island), c.8 km north-east of Mustique (IBA VC013) and c.5 km north-east of All Awash island IBA (VC012). Because of its location, Battowia is constantly exposed to the Atlantic Ocean and therefore the island is sparsely covered along the coast (north, west and east), particularly during the dry season. Vegetation comprises coastal grassland interspersed with patchy coastal dry scrub woodland, the healthiest stands of which are found within the few sheltered coves on the west coast and central and southern slopes. The 60-ha island is steep-sided making access difficult in most areas except the south. The IBA includes marine areas up to 1 km from the island.

Birds

This IBA is regionally significant for its breeding population of c.3,000 pairs of Red-footed Booby *Sula sula*. Battowia is a key roosting and nesting site for a number of seabird species including Magnificent Frigatebird *Fregatta magnificens*, gulls and boobies. However, as with many of the other Grenadines islands, data on their populations are non-existent. The island could prove to be of global importance for its seabirds. It appears to support the second largest seabird colony in the country.

Other biodiversity

It is believed that the regionally endemic congo snake *Mastigodryas bruesi* occurs on the island. However, there is little information available concerning the biodiversity.

Conservation

Battowia Island IBA is privately owned but designated a wildlife reserve. A number of management objectives have been identified and include habitat and wildlife protection, biodiversity conservation, research, recreation, education and heritage tourism. However, the extent to which any of these are being implemented is unknown. Some hunting occurs on the island: young seabirds are taken for their meat and eggs are taken. The impact of this "harvest" has not been quantified. A population of goats is present on the island and will be impacting both the vegetation and the seabirds. It is unknown if rats *Rattus* spp. are present, but this should be determined. Birdwatchers and photographers visit the island on a private tour operated from Mustique—it is unknown if this causes any disturbance to the seabirds.

Site description

All Awash Island IBA is a small (3 ha), uninhabited island in the northern Grenadines, c.3 km south of the island of Baliceaux, and 5–6 km north-east of Mustique (IBA VC013). The vegetation comprises degraded and wind-swept dry scrub woodland although it is particularly sparse along the water's edge in the north and east of the island which, as a result, is relatively accessible. A saddle separates the thicker westerly vegetation from the more sparsely-vegetated eastern sector. Marine areas up to 1 km from the island are included within the IBA.

Birds

This IBA is significant for its seabirds, and probably represents the country's third largest seabird colony. Globally important numbers of Roseate Tern *Sterna dougallii* nest along with regionally significant populations of Magnificent Frigatebird *Fregata magnificens*, Brown Booby *Sula leucogaster*, Laughing Gull *Larus atricilla* and Royal Tern *Sterna maxima*. Outside of the breeding season, seabirds forage in the surrounding waters, and use the island for roosting.

Other biodiversity

Nothing recorded.

Conservation

All Awash Island IBA is a state-owned wildlife reserve. However, the nesting seabirds attract annual raids by fishermen, who begin to frequent the site from around April to poach seabird eggs. This is believed to have a considerable impact on the breeding success of the species involved. The presence of humans on the island causes disturbance to all the breeding seabirds resulting in possible nest abandonment. The waters surrounding the island are a scuba-diving attraction which may also cause disturbance.

Site description

Mustique Island IBA is in the northern Grenadines, c.28 km west-south-west of Barbados. The island (565 ha) is 4 km long (north to south) and up to 2.4 km wide with hills rising to 150 m (the Southern Hills and Central Cambell Hills). It supports three small (natural) wetlands: Lagoon (the second largest wetland in the country), Bird Sanctuary and Macaroni; several man-made ponds; and coastal dry scrub vegetation. Surrounding the Lagoon wetland (which has been well protected by the Mustique Company and remained relatively intact since the 1950s) is a mixed mangrove woodland, and a large stretch of sea-grass and coral reefs.

Birds

This IBA supports populations of six (of the 14) Lesser Antilles EBA restricted-range birds (including Grenada Flycatcher *Myiarchus nugator* and Lesser Antillean Bullfinch *Loxigilla noctis*) and regionally significant numbers of breeding Brown Pelican *Pelecanus occidentalis*, Royal Tern *Sterna maxima* and Sandwich Tern *S. sandvicensis*. Other seabird species breed (e.g. *Sula* spp.) but the populations are unknown. Similarly, the Near Threatened Caribbean Coot *Fulica caribaea* is a common breeder at the Bird Sanctuary, but the numbers involved are unknown.

Other biodiversity

The Critically Endangered hawksbill *Eretmochelys imbricata* and leatherback *Dermochelys coriacea* turtles nest on several of Mustique's beaches, and the Endangered green turtle *Chelonia mydas* forages within its waters. A population of the regionally endemic congo snake *Mastigodryas bruesi* occurs (although its status is unknown).

Conservation

Mustique Island IBA is a privately-owned conservation area. Custodianship of the island was granted to the Mustique Company under the Mustique Company Limited Act (1989). Regulations of the Act afford protection to the island and marine areas to at least 1 km offshore. Pet cats roam freely on the island and there is a growing feral population estimated to be in the hundreds which will be impacting the island's breeding waterbirds and seabirds. Cat numbers appear to be increasing in spite of an ongoing neutering programme. Opossum *Didelphis marsupialis* and iguana *Iguana iguana* are illegally hunted on the island, and introduced exotic (invasive) plants are impacting the native flora. Soil erosion caused by house and garden construction is a problem. A rise in sea level will seriously impact coastal areas which are also hurricane-prone. Hurricanes such as Lenny (1999) and Ivan (2004) damaged east coast coral reefs and beaches.

Sooty Tern

■ Site description

Petit Canouan IBA is a small (15 ha), uninhabited island located c.8 km north-north-east of Canouan in the middle of the Grenadines island chain. Originally supporting dry scrub woodland, the island has in recent times been converted mostly to coastal grassland, with few shrubs intermixed. Parts of the island have been left bare and eroded. The conversion to grassland is a result of the annual burning regime implemented by fishermen to facilitate poaching of seabird eggs (Petit Canouan is known among poachers as the island for the "egg birds"). The island is identified as a scuba-diving site under the System of Protected Areas and Heritage Sites. The IBA includes marine areas up to 1 km from the island.

■ Birds

This IBA is significant for its breeding seabirds with up to 20,000 individuals breeding on the island in the late 1990s. Abundance information is poor, but the numbers of Laughing Gull *Larus atricilla* and Roseate Tern *Sterna dougallii* are thought to be globally important, while those of Magnificent Frigatebird *Fregata magnificens*, Brown Booby *Sula leucogaster*, Royal Tern *S. maxima* and Sooty Tern *S. fuscata* (thought to be c.10,000) are regionally so. Seabird numbers appear to have declined dramatically over the last 5–10 years.

■ Other biodiversity

Nothing recorded.

■ Conservation

Petit Canouan IBA is a state-owned wildlife reserve. However, enforcement of this protective designation is very limited. Law enforcement and biodiversity awareness is generally poor on Grenadine islands. Without regulation, the practice of collecting huge quantities of seabird eggs continues unimpeded and provides a rewarding return for the fishermen involved. Vegetation is burnt just prior to the first eggs being laid (March/April) and has significantly degraded the landscape and exposed the already poor soils to increased erosion. It is not known whether rats *Rattus spp.* are present on the island, but this should be determined as a matter of priority.

Grenada Flycatcher

■ Site description

Ashton Wetland IBA is part of Ashton Harbour on the south coast of Union Island, at the southern end of the Grenadines. The wetland supports a diverse (25-ha) stand of mangrove, tidal mud flats, salt ponds, sea and some dry scrub forest. On the seaward side of the mangroves were (originally) diverse sea-grass beds and coral reefs (fringing, patch, and barrier reef). Much of this coral lagoon system (the largest in the Grenadines) has been devastated by an unfinished but now abandoned marina development. The marina developed connects the mainland to an offshore island (Frigate Island) that was previously important for birds.

■ Birds

This IBA supports populations of six (of the 14) Lesser Antilles EBA restricted-range birds, including Grenada Flycatcher *Myiarchus nugator* and Lesser Antillean Tanager *Tangara cucullata*. The wetland itself supports good numbers of migratory shorebirds and also species such as Laughing Gull *Larus atricilla*, Magnificent Frigatebird *Fregata magnificens* and Brown Pelican *Pelecanus occidentalis*. However, the numbers involved are not known but should be assessed at the next opportunity.

■ Other biodiversity

The recently discovered (2005) endemic gecko *Gonatodes daudini* appears to be endemic to Union Island and may occur in the IBA. The regionally endemic congo snake *Mastigodryas bruesi* occurs on Union Island although its distribution and abundance are unknown.

■ Conservation

Ashton Wetland IBA is state owned. The harbour is a marine conservation area designated under the Fisheries Act of 1986. Despite official designation (and a damning environmental impact assessment) a 300-boat marina project began at the lagoon in 1994, compromising the entire ecosystem. The development company declared bankruptcy in 1995 and the half completed marina was abandoned. Ashton Wetland is threatened by a range of factors including: grazing by cows and goats (released during the "let-go season" to allow them to feed freely during periods of drought and food scarcity); development pressures; cutting of trees for fuelwood; pollution from run-off; and indiscriminate dumping of garbage. Ashton lagoon is the focus of a restoration project being proposed by the Society for the Conservation and Study of Caribbean Birds and that also aims to develop the wetland as a Watchable Wildlife Pond.

TRINIDAD & TOBAGO

LAND AREA **5,128 km²** ALTITUDE **0–940 m**
HUMAN POPULATION **1,300,000** CAPITAL **Port-of-Spain**
IMPORTANT BIRD AREAS **7, totalling 1,062 km²**
IMPORTANT BIRD AREA PROTECTION **80%**
BIRD SPECIES **468**
THREATENED BIRDS **6** RESTRICTED-RANGE BIRDS **2** BIOME-RESTRICTED BIRDS **5**

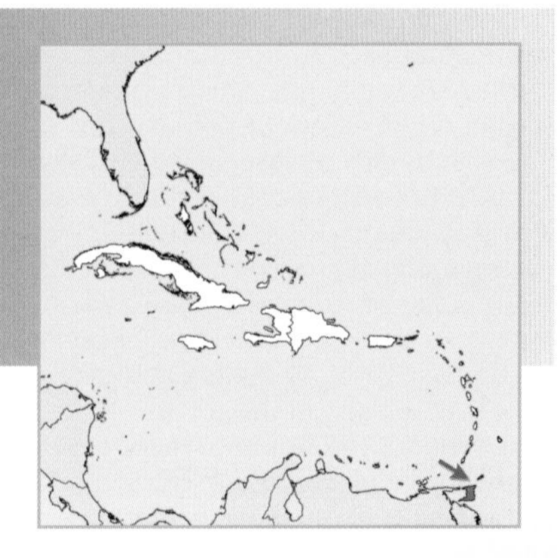

GRAHAM WHITE
(TRINIDAD AND TOBAGO FIELD NATURALISTS CLUB)

Victoria-Mayaro Forest Reserve IBA. (PHOTO: GRAHAM WHITE)

INTRODUCTION

The Republic of Trinidad and Tobago is a twin-island state located where the Lesser Antilles island chain meets mainland South America. Trinidad is the larger of the two islands and has an area of 4,828 km². It is highly industrialised and hosts 95% of the population. Economic activities are mainly energy-based due to national reserves of oil and natural gas. The major population centres are on the west coast at Port-of-Spain and San Fernando. Tobago, 19 km to the north-east of Trinidad has an area of just over 300 km². It has a degree of local governance through the Tobago House of Assembly and is sparsely populated, with tourism and fishing representing the major economic activities. Trinidad has three mountain ranges: the Northern and Southern Ranges run along the north and south coasts, and the Central Range runs diagonally from north-east to south-west. The Northern Range rises to 940 m at El Cerro Del Aripo, while the Central and Southern ranges are more diffuse, rising to 307 m and 303 m at Tamana Hill and Trinity Hills respectively. The basins between the ranges are generally low-lying, with swamps where the larger rivers meet the sea at Caroni (a mangrove-dominated swamp) on the west coast and Nariva (dominated by herbaceous swamp and swamp-forest) on the east. Bordering the Caroni Swamp and extending down the west coast of Trinidad are extensive coastal mudflats. The south and east coast are characterised by long sandy beaches and occasional headlands. The north and north-east coasts are steep and rocky with occasional sandy beaches. Due to the outflow from the Orinoco the waters are generally turbid. Tobago is characterised by the Central Ridge which forms the "backbone" of the north-eastern half of the island. The highest point is at 549 m but there are no well defined peaks. The western end of Tobago is low-lying with a coastal plain draining into two coastal wetlands. The coast of Tobago comprises sandy beaches alternating with rocky headlands. While still influenced by the Orinoco, the waters off Tobago are much clearer with an extensive off-shore coral reef and fringing reefs in many of the bays.

While politically associated with the West Indies, biogeographically Trinidad and Tobago is distinctly South American, although Tobago does have some Antillean influences. The mountains in northern Trinidad represent an eastward extension of the coastal cordillera of Venezuela, and at the closest point Trinidad is only 11 km from Venezuela's Paria Peninsula. Indirect evidence suggests that the two were connected as recently as 2,200 years ago. Due to this close affinity with South America, Trinidad and Tobago have a continental flora and fauna characterised by high species richness and low levels of endemism. Both Islands lie on the continental shelf and are influenced by the outflow of the Orinoco River and the South Equatorial Current, resulting in comparatively nutrient-rich and low salinity coastal waters. The islands have a seasonal tropical climate with a wet season from May to December and a dry season from January to April. Annual rainfall is heavily influenced by topography. Average annual precipitation in Trinidad is 2,200 mm (ranging from 3,500 mm towards the eastern end of the Northern Range, to 1,300 mm at the westernmost points of the island). On Tobago, average annual precipitation is 1,427 mm in the west to 2,363 mm in the north-east. Trinidad and Tobago lie at the edge of the Atlantic Hurricane Belt with the last major hurricane being Hurricane Flora in 1963.

Originally, Trinidad and Tobago would have been almost entirely forested, and both islands still maintain comparatively large areas of forest cover, recently estimated at 44–48%, although arguably just 15% is in a natural state. Forest has been cleared to make way for development and agriculture, with traditional crops of cocoa, coffee and citrus once occupying large tracts of land. Many of these areas are now semi-abandoned and support mature secondary-growth forest. In Trinidad, evergreen seasonal forest predominates in the lowlands where rainfall is high. Moving towards the south and west of the island, the dry season is more pronounced and the forest grades through semi-evergreen seasonal forest to deciduous forest (this latter being best developed along the north-west peninsula and on the offshore-islands). In the Northern Range (above 240 m) there is generally no seasonal drought and lower montane rainforest occurs. This grades into montane rainforest above c.540 m, and finally, above 870 m are small patches of elfin woodland. In Tobago, the natural vegetation of the lowlands is seasonal forest although this has almost all been removed for urban development and agriculture. The Main Ridge on Tobago supports rainforest which, however, is still recovering (structurally) from the devastating Hurricane Flora in 1963.

■ Conservation

The legal framework for protection of wildlife and natural areas in Trinidad and Tobago has recently been reformulated. The former system was characterised by a wide range of regulations managed by a number of different government agencies leading to gaps and overlaps. The new system, progressively being implemented, provides for a single government agency—the Environmental Management Authority (EMA)—to coordinate all environmental management activities. The EMA, together with the Forestry Division (which is responsible for the management of protected areas in Trinidad), both fall within the Ministry of Public Utilities and the Environment[1]. Protected areas in Tobago are the responsibility of the Tobago House of Assembly (THA) and governed by the Tobago House of Assembly Act. The country has an extensive system of formally designated protected areas, including wildlife or game sanctuaries, nature conservation reserves, scientific reserves, forest reserves, Ramsar sites as well as historic sites, natural landmarks and recreation parks. The protected areas are however small, disparate and difficult to enforce based on existing regulations. The Forests Act makes provision for declaring areas as "Prohibited Areas" and this is the legal mechanism through which critical elements of biodiversity have been protected. The Prohibited Areas-provision has been used to designate especially sensitive areas (e.g. sea-turtle nesting beaches) in order to regulate human activities until effective management can be initiated. The new Environmental Management Act provides authority for designating Environmentally Sensitive Areas and Environmentally Sensitive Species, and requires consideration be given to the environmental impacts of large developmental projects. State-owned forest amounts to c.37% of the land area, and "protected forest" (including wildlife sanctuaries, nature reserves, wind-belt reserves and forest above 90 m) comprises c.11%. Unfortunately, despite the network of designated sites, there is a general lack of respect for the legislation, and the enforcement agencies are hampered by inadequate human and financial resources to properly carry out their mandate. For example, in Trinidad they are not equipped with boats which are capable of patrolling marine areas such as Soldado Rock and Saut d'Eau.

Major conservation actions in Trinidad and Tobago generally involve government agencies, especially the Wildlife Section of the Forestry Division, the EMA and, in Tobago, the Department of Natural Resources and the Environment of the Tobago House of Assembly. Research efforts are mainly through the University of the West Indies (UWI), with input from visiting researchers and the Wildlife Section, while the Institute of Marine Affairs (IMA) conducts research on the coastal and marine environment. However, a wide range of NGOs and individuals make contributions mainly through research, public education or advising policy development. NGOs with a bird focus include the Trinidad and Tobago Field Naturalists' Club (TTFNC), the Asa Wright Nature Centre (AWNC) and the Pointe-a-Pierre Wildfowl Trust (P-a-PWT). TTFNC provides a forum for education and appreciation of the natural environment. It publishes a journal (*Living World*) which focuses on the natural history of Trinidad and Tobago and the wider Caribbean, and sponsors the Trinidad and Tobago Rare Bird (Records) Committee. AWNC is a private nature reserve with visitor accommodation in the Arima Valley. The centre actively manages an area of 526 ha which includes one colony of Oilbirds *Steatornis caripensis* and support local conservation initiatives through education and publications. P-a-PWT has the objective of captive-breeding and release of threatened waterfowl in Trinidad. It makes a major contribution to public awareness and education on conservation issues, especially where wetlands are involved. A particularly encouraging development is the contribution being made by community-based groups, including Nature Seekers Inc. and Grande

The Critically Endangered Trinidad Piping-guan. (PHOTO: ALFREDO COLÓN ARCHILLA)

1 As of November 2007, the Forestry Division was returned to the Ministry of Agriculture, Lands and Marine Resources and the Division of the Environment placed in the Ministry of Planning, Housing and the Environment

Riviere Environmental Awareness Trust, to the conservation of sea-turtles and the Trinidad Piping-guan *Pipile pipile* (with support from the Wildlife Section).

Pipile pipile has been the focus of 25 years of basic research, public awareness and some community-based efforts by Nature Seekers Inc., Grande Riviere Environmental Awareness Trust, the Wildlife Section, RARE Center, Caribbean Union College, UWI and Glasgow University. The species has been designated an Environmentally Sensitive Species which, under the Environmental Management Act, requires that a national action plan for its protection be implemented. A national plan is being prepared by the EMA through the UWI-coordinated "Pawi Project". The White-tailed Sabrewing *Campylopterus ensipennis*, which went unrecorded for 10 years after the devastating Hurricane Flora in 1963, has also received attention in terms of basic research concerning its distribution, reproduction and behaviour. Research conducted by Caribbean Union College and UWI has received support from EMA and the Tobago House of Assembly.

Habitats in Trinidad and Tobago are being impacted from many quarters. The current economic wealth of the country has placed a severe pressure on land (including wetland) for housing, commercial and industrial development. As well as the loss of habitat, these developments (e.g. heavy industry along the west coast) adversely affect the quality of the surrounding environment (both terrestrial and marine) through pollution. There are currently plans to accelerate the economic development of the country. During the dry season use of fire to clear vegetation either for agriculture or "aesthetic" purposes is a common practice (despite legislation prohibiting fires) often with little or no attempt to manage the fires. This leads to a degradation of the natural forest to open scrub vegetation dominated by grasses, vines and other early successional species, and to erosion problems. About 8% of the country's forest was lost to fire during the period 1990–2000. Squatting is a major problem on state lands, even within protected areas.

In addition to habitat loss and degradation, hunting birds is popular on both islands. Seabird colonies are threatened by poachers who collect the adult birds for meat (and presumably also take the eggs). While both Little Tobago and St Giles Islands are legally protected, poaching on St Giles is a regular occurrence and the breeding seabird population has declined. Management of Little Tobago is far more successful and demonstrates what could be accomplished if St Giles received the same attention. In Trinidad the seabird colonies at Soldado Rock and Saut d'Eau are infrequently visited and the current status of seabird nesting is not certain (see "Important Bird Areas" below). There is evidence however that the colony at Soldado has been abandoned, possibly due to poaching. Apart from the seabirds, species at particular risk from hunting include the Critically Endangered Trinidad Piping-guan *Pipile pipile*, all ducks, night-herons, Limpkin *Aramus guarauna*, Scarlet Ibis *Eudocimus ruber* (in Trinidad) and Rufous-vented Chachalaca *Ortalis ruficauda* (in Tobago). *Pipile pipile* makes a notoriously easy target as it can appear oblivious to human presence even at close range. Hunting, coupled with destruction of freshwater marsh, is largely responsible for the country's small waterbird populations, and threatens the continued breeding of *E. ruber*. In Trinidad there is a strong tradition of keeping cage-birds (including species of parrot, finches, euphonias and an oriole). The preferred species of seedeaters and seed-finches have now been so severely depleted that no viable population is thought to exist in the wild and birds are imported from the mainland. A prized Lesser Seed-finch *Oryzoborus angolensis* can sell for up to US$2,000 and current legislation permits the keeping of these birds. It is unlikely that a wild population would survive while current attitudes prevail.

■ Birds

In Trinidad and Tobago 468 bird species have been recorded. In Trinidad this includes 227 year-round resident species and 168 regular seasonal migrants (17 of which breed). A further 50 species are listed as wanderers or vagrants. In Tobago, 240 species have been recorded of which 92 are year-round residents, 122 are regular seasonal migrants (17 of which breed), and 22 are vagrants. The species composition is reflective of the South American origins with hummingbirds, trogons, woodcreepers, ovenbirds, antbirds, manakins and tanagers well represented. The only endemic bird is the

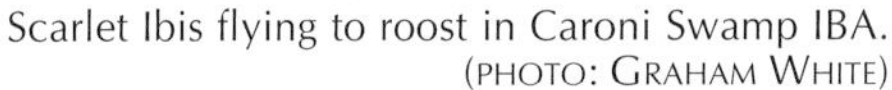

Scarlet Ibis flying to roost in Caroni Swamp IBA. (PHOTO: GRAHAM WHITE)

Table 1. Key bird species at Important Bird Areas in Trinidad and Tobago.

Key bird species	Criteria	National population	TT001	TT002	TT003	TT004	TT005	TT006	TT007
			Trinidad and Tobago IBAs						
Criteria			■	■	■ ■ ■	■ ■ ■	■ ■ ■	■	■ ■
Rufous-vented Chachalaca *Ortalis ruficauda*	■						✓		✓
Trinidad Piping-guan *Pipile pipile*	CR ■ ■ ■	70–200			70–200	✓			
Audubon's Shearwater *Puffinus lherminieri*	■	1,000–5,000						250–500	500–1,000
Scarlet Ibis *Eudocimus ruber*	■	10,000–25,000	50–100	15,000					
Black-crowned Night-heron *Nycticorax nycticorax*	■	500–1,000	50–100	50–100					
Yellow-crowned Night-heron *Nyctanassa violacea*	■	500–1,000	250–500	250–500					
Cattle Egret *Bubulcus ibis*	■	10,000–25,000	500–1,000	5,000–10,000					
Tricoloured Heron *Egretta tricolor*	■	500–1,000	100–250	400–600					
Little Blue Heron *Egretta caerulea*	■	1,000–5,000	100–250	1,000–1,500					
Snowy Egret *Egretta thula*	■	1,000–5,000	500–1,000	2,000–3,000					
Red-billed Tropicbird *Phaethon aethereus*	■	500–1,000						100–250	250–500
Magnificent Frigatebird *Fregata magnificens*	■	1,000–5,000						1,000–2,000	
Brown Pelican *Pelicanus occidentalis*	■	1,000–5,000	1,500–2,000						
Masked Booby *Sula dactylatra*	■	50–100						80	
Red-footed Booby *Sula sula*	■	1,000–5,000						1,000–1,500	250–500
Brown Booby *Sula leucogaster*	■	500–1,000						50–100	250–500
Laughing Gull *Larus atricilla*	■	5,000–10,000	4,000–5,000						150–624
Bridled Tern *Sterna anaethetus*	■	100–250							129
Yellow-billed Tern *Sternula superciliaris*	■	250–500	250–500						
Large-billed Tern *Phaetusa simplex*	■	500–1,000	400–600						
Royal Tern *Thalasseus maximus*	■	500–1,000	100–250						
Black Skimmer *Rynchops niger*	■	750–1,200	750–1,250						
Semipalmated Plover *Charadrius semipalmatus*	■	1,000–5,000	1,000–2,000						
Whimbrel *Numenius phaeopus*	■	1,000–5,000	500–1,000						
Lesser Yellowlegs *Tringa flavipes*	■	5,000–10,000	500–1,000						
Short-billed Dowitcher *Limnodromus griseus*	■	1,000–5,000	1,000–2,000						
Semipalmated Sandpiper *Calidris pusilla*	■	5,000–10,000	5,000–8,000						
Western Sandpiper *Calidris mauri*	■	5,000–10,000	5,000–8,000						
White-tailed Sabrewing *Campylopterus ensipennis*	NT ■ ■ ■	800–1,200					800–1,200		
Copper-rumped Hummingbird *Amazilia tobaci*	■				✓	✓	✓		✓
Venezuelan Flycatcher *Myiarchus venezuelensis*	■						✓		

All population figures = numbers of individuals.
Threatened birds: Critically Endangered ■; Near Threatened ■. **Restricted-range birds** ■. **Biome-restricted birds** ■. **Congregatory birds** ■.

Trinidad Piping-guan *Pipile pipile*. However, there are 36 endemic subspecies of birds currently recognised in the country.

The threat category and national population sizes of the globally threatened birds are listed in Table 1. Although six globally threatened species occur in Trinidad and Tobago, four of these have not been considered in the Important Bird Area analysis. The Near Threatened Caribbean Coot *Fulica caribaea* and Buff-breasted Sandpiper *Tryngites subruficollis* are both rare on the islands while the Endangered Red Siskin *Carduelis cucullata* has not been recorded since 1960 (and prior to that, 1926). The Near Threatened Olive-sided Flycatcher *Contopus cooperi* is a regular winter visitor from North America, frequenting the Northern Range forests, but in very small numbers (perhaps c.30 individuals). However, Trinidad does support the only population of the Critically Endangered *P. pipile* (thought to number just 70–200 birds), and the Near Threatened White-tailed Sabrewing *Campylopterus ensipennis* occurs in the Main Ridge forest on Tobago where the population is recovering from near-extirpation as a result of Hurricane Flora (in 1963). The species is also found in the coastal cordillera and Paria Peninsula in Venezuela.

A number of biome-restricted species (in addition to *P. pipile* and *C. ensipennis*) are resident in the country. On Tobago, the Rufous-vented Chachalaca *Ortalis ruficauda* is abundant (and recognised as the national bird), while the Venezuelan Flycatcher *Myiarchus venezuelensis* is less common. The Copper-rumped Hummingbird *Amazilia tobaci* is particularly abundant in residential areas of both Trinidad and Tobago. These three species are confined to the Northern South America biome. The Rufous-shafted Woodstar *Chaetocercus jourdanii*—a Northern Andes biome-restricted bird—is occasionally observed in Trinidad with one undocumented but credible record of nesting.

The small islands off both Trinidad and Tobago have supported important nesting concentrations of seabirds including Audubon's Shearwater *Puffinus lherminieri*, Red-billed Tropicbird *Phaethon aethereus*, Red-footed Booby *Sula sula*, Brown Booby *S. leucogaster*, Magnificent Frigatebird *Fregata magnificens*, Sooty Tern *Sterna fuscata* and Brown Noddy *Anous stolidus*. However, poaching, and disturbance have significantly reduced the numbers involved and some colonies have been abandoned although source populations do persist and could lead to re-colonisation if protection can be afforded the nesting areas. The Caroni Swamp and associated coastal mudflats support significant congregations of Scarlet Ibis *Eudocimus ruber* (up to 15,000 birds), shorebirds (10,000–20,000) and seabirds (mainly 5,000–6,000 Laughing Gulls *Larus atricilla*). However, waterbirds are otherwise generally rare in the country. Although 16 species of duck have been recorded the only ones observed on a regular basis in numbers are wintering Blue-Winged Teal *Anas discors*, and resident Black-bellied Whistling-duck *Dendrocygna autumnalis*, White-cheeked Pintail *A. bahamensis* and Masked

Black Skimmers and Western Sandpipers congregating on the West Coast Mudflats IBA. (PHOTO: GRAHAM WHITE)

Duck *Oxyura dominica.* Post-breeding dispersal (presumably from Venezuela) results in small numbers of Fulvous Whistling-duck *D. bicolor*, White-faced Whistling-duck *D. viduata* and the occasional Comb Duck *Sarkidiornis melanotos.*

IMPORTANT BIRD AREAS

Trinidad and Tobago's seven IBAs—the island's international site priorities for bird conservation—cover 1,062 km² (including marine areas), and about 18% of the islands' land area. Most of the IBAs (with the exception of the West Coast Mudflats, TT001) are under some form of formal protective designation, representing 80% of the area covered by the IBAs. However, management and enforcement (as outlined above) are insufficient to ensure the long-term survival of the key species currently found within their boundaries.

The IBAs have been identified on the basis of 31 key bird species (listed in Table 1) that variously trigger the IBA criteria. These 31 species includes two globally threatened birds (both of which are also restricted-range and biome-restricted species), the three additional regularly occurring biome-restricted birds, and 26 congregatory waterbirds/ seabirds (a number of which are listed because the combined total of the individuals present triggers the ">20,000 waterbirds" criterion). None of the key species occurs in more than two IBAs in the country, emphasising the critical importance of each individual IBA. Of particular note is Caroni Swamp IBA (TT002) in supporting the largest concentrations of waterbirds; the St Giles Islands (TT006) and Little Tobago Island (TT007) IBAs for supporting most of the seabirds; Main Ridge IBA (TT005) for harbouring the only population of *Campylopterus ensipennis* and the biome-restricted species; and the Northern Range IBA (TT003) for the only confirmed population of the Critically Endangered *Pipile pipile.*

The Northern Range IBA comprises six almost contiguous sub-units, defined by river catchments, together with all remaining lands above 500 m. These sub-units reflect different levels of state protection, private ownership and consequent levels of settlement and forest degradation. *Pipile pipile* is currently regularly observed within private lands at the northern edge of Matura National Park (embracing 9,000 ha from 0–500 m, and including the south-flowing Salybia and Rio Seco rivers, and the north-flowing Shark River and Grande Riviere) and along the western ridge of Morne Bleu. There are also records from the north-flowing Madamas River catchment (4,700 ha, from 0–600 m). The species probably survives in other eastern Northern Range watersheds within the IBA including the Quare River catchment (which includes El Cerro del Aripo and the Hollis Reservoir), and the Matelot, Oropuche and Matura river catchments. Surveys for *P. pipile* throughout this IBA are urgently needed (as is confirmation of its survival in the Victoria-Mayaro Forest Reserve IBA, TT004), and must be followed up with regular monitoring and redoubled efforts to eliminate hunting as a threat to this Critically Endangered endemic.

Figure 1. Location of Important Bird Areas in Trinidad and Tobago.

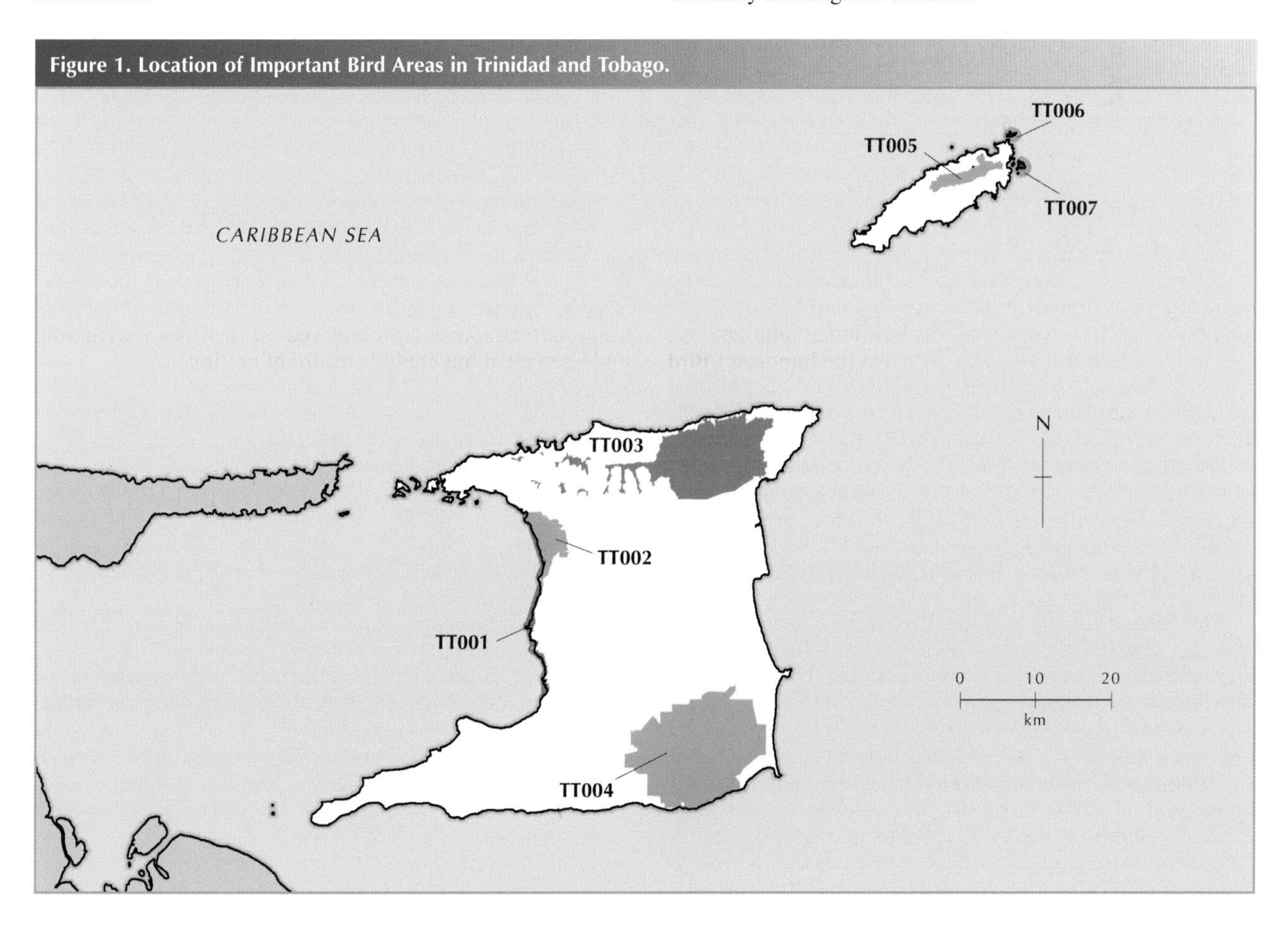

Little Tobago Island IBA. (PHOTO: GRAHAM WHITE)

Two small islands off Trinidad's coast hosted seabird colonies in the past, but their current status is in doubt and they have not been considered IBAs at this point. Soldado Rock is a small (c.0.6 ha and rising to 35 m) island, located offshore from Icacos Point, midway between Venezuela and Trinidad. The island supported a major breeding colony of Sooty Tern *Sterna fuscata* and Brown Noddy *Anous stolidus*, with c.2,500 pairs of each species from 1960 to 1982. Small numbers of Magnificent Frigatebird *Fregata magnificens* and Brown Pelican *Pelecanus occidentalis* currently roost on the island but the status of other seabirds is not known. It is likely that the colony has been abandoned either due to disturbance or due to general depletion of the fish stocks within the Gulf or Paria. Saut d'Eau island (c.10 ha) is located off Trinidad's north coast. In the past it hosted up to 100 breeding pairs of *P. occidentalis* but the current status of seabird breeding is not known.

Public education and awareness will continue to play an important role in engendering an appreciation of both the needs of conservation and a general respect for the environment. This will be essential if the impacts of hunting are to be reduced, both within the forests and in relation to the dwindling seabird colonies. Existing seabird colonies urgently need to be protected and efforts made to determine how local communities can be included in and benefit from the protection of these assets. Surveys are needed to determine which seabird species continue to breed in significant numbers (both in the IBAs, and in the other offshore islands as mentioned above). Field assessments (surveys and subsequent monitoring) for seabirds, for *P. pipile*, and for the other key species (especially *Campylopterus ensipennis* and *Eudocimus ruber*) should be used to inform the annual assessment of state, pressure and response scores at each of the country's IBAs to provide an objective status assessment and highlight management interventions that might be required to maintain these internationally important biodiversity sites.

Two sites that are essential for the maintenance of Trinidad's biodiversity but which do not meet the IBA criteria are Nariva Swamp and the Aripo Savannah. Nariva Swamp is Trinidad's largest freshwater marsh and critical for crakes, bitterns and rails. The site is potentially important for migrating waterfowl and has been designated as a Ramsar site. However, the combined waterbird populations probably do not exceed 10,000 birds. The swamp forest and stands of moriche palm within the Nariva Swamp and surrounding the Aripo Savannah are critical for five resident species: Rufescent Tiger-heron *Tigrisoma lineatum*, Red-bellied Macaw *Orthopsittaca manilata*, Moriche Oriole *Icterus chrysocephalus*, Sulphury Flycatcher *Tyrannopsis sulphurea* and, to a lesser extent, Fork-tailed Palm-swift *Tachornis squamata*. Attempts have been made to re-establish populations of the locally extirpated Blue-and-yellow Macaw *Ara ararauna* in this area.

KEY REFERENCES

AGARD, J. B. R. AND GOWRIE, M. (2003) Environmental Vulnerability Index (EVI): provisional indices and profiles for Trinidad and Tobago. Port-of-Spain: Environmental Management Authority (Unpublished State of the Environment Report 2001 and 2002).

ALKINS, M. E. (1979) *The mammals of Trinidad.* St Augustine, Trinidad: University of the West Indies (Dept. Zoology Occas. Paper 2).

ANON. (1999) Action plan for marsh restoration in the Caroni Swamp, Trinidad. St Augustine, Trinidad: University of the West Indies. (Unpublished report to Inter-American Development Bank and Wildlife Section, Forestry Division Government of Trinidad and Tobago).

ANON. (2005) National report on the implementation of the Ramsar Convention on wetlands: Trinidad and Tobago. Port-of-Spain: Government of Trinidad and Tobago. (Unpublished report to the Ramsar Convention).

BACON, P. R. (1970) *The ecology of Caroni Swamp, Trinidad.* Trinidad: Central Statistical Office Printing Unit (Government of Trinidad and Tobago).

BACON, P. B. AND FFRENCH, R. P. (1972) *The wildlife sanctuaries of Trinidad and Tobago.* Trinidad and Tobago: Wildlife Conservation Committee (Min. Agriculture, Lands and Fisheries).

BEARD, J. S. (1944) The natural vegetation of the island of Tobago. *Ecol. Monogr.* 14:135–163.

BEARD, J. S. (1946) *The natural vegetation of Trinidad.* Oxford: Clarendon Press (Oxford Forestry Memoirs).

BILDSTEIN, K. L. (1990) Status, conservation and management of the Scarlet Ibis *Eudocimus ruber* in the Caroni Swamp, Trinidad, West Indies. *Biol. Conserv.* 54:61–78

DINSMORE, J. J. (1972) Avifauna of Little Tobago Island. *Quart. J. Fla. Acad. Sci.* 35: 55–71.

DINSMORE, J. J. AND FFRENCH, R. P. (1969) Birds of St Giles Island, Tobago. *Wilson Bull.* 81: 460–463.

FFRENCH, R. P. (1977) Birds of the Caroni Swamp and marshes. *Living World, J. Trinidad & Tobago Field Nat. Club* 1977–78: 42–44.

ffrench, R. (1990) The birds and other vertebrates of Soldado Rock. *Living World, J. Trinidad & Tobago Field Nat. Club* 1989–1990: 16–20.

ffrench, R. (1991) *A guide to the birds of Trinidad and Tobago.* Ithaca, New York: Comstock Publishing Associates.

ffrench, R. P. and Haverschmidt, F. (1970) The Scarlet Ibis in Suriname and Trinidad. *Living Bird* 9: 147–165

EMA (2001) Biodiversity strategy and action plan for Trinidad and Tobago. Port-of-Spain: Institute of Marine Affairs, Government of Trinidad and Tobago. (Unpublished report).

Hardy, J. D (1982) Biogeography of Tobago, West Indies, with special reference to reptiles and amphibians: a review. *Bull. Maryland Herpetological Soc.* 18: 37–143

Hayes, F. E. (2006) Trinidad Piping-Guan (*Aburria pipile*). Pp. 33–35 in D. M. Brooks, eds. *Conserving cracids: the most threatened family of birds in the Americas.* Houston, Texas: Houston Museum of Science (Misc. Publ. 6).

Hayes, F. E. and Bodnar, S. (in press) Breeding seabirds of Trinidad and Tobago. In Bradley P. E. and Norton, R. L. eds. *Breeding seabirds of the Caribbean.* Gainesville, Florida: Univ. Florida Press.

Hayes, F.E., White, G., Kenefick, M. and Kilpatrick, H. (2004) Seasonal variation in gull populations along Trinidad's west coast, Trinidad and Tobago. *Living World, J. Trinidad & Tobago Field Nat. Club* 2004: 3–5.

ITTO (2003) Achieving the ITTO Objective 2000 and sustainable forest management in Trinidad and Tobago. Panama City: International Tropical Timber Council. (Unpublished report from the ITTO 34th session).

Kenny, J. S. (1995) *Views from the bridge: a memoir on the freshwater fishes of Trinidad.* Barataria, Trinidad: Trinprint Ltd.

Morrison, R. I. G. and Ross, R. K. (1989) *Atlas of Nearctic shorebirds on the coast of South America.* Ottawa: Canadian Wildlife Service (Special Publ.).

Murphy, J. C. (1997) *Amphibians and reptiles of Trinidad and Tobago.* Florida: Krieger Publishing Company.

Murphy, J. C. (2008) An update of the amphibians and reptiles of Trinidad and Tobago. Downloaded from http://blog.jcmnaturalhistory.com/?page_id=46 (4 June 2008).

Nelson, H. P., Devenish, E., Erdmann, E. and Lucas, F. B. (2004) Madamas Valley expedition Trinidad: a preliminary assessment of species richness of a proposed national park. (Unpublished report to the Rufford Small Grants Foundation).

Phillip, D. A. T. and Ramnarine, I. W. (2001) *An illustrated guide to the freshwater fishes of Trinidad.* St Augustine, Trinidad: Dept. Life Sciences, University of the West Indies

Snow D. W. (1985) Affinities and recent history of the avifauna of Trinidad and Tobago. Pp 238–46 in P. Buckley *et al.* (eds.) *Neotropical Ornithology.* American Ornithologists' Union (Orn. Monogr. 36).

Van den Eynden, V. (2006) Review of endemic plants of Trinidad and Tobago. Downloaded from http://www.sta.uwi.edu/fsa/maturanp/ (17 May 2008).

White, G. and Kenefick, M. (2004) The avifauna of the Brickfield mudflats. *Living World, J. Trinidad & Tobago Field Nat. Club* 2004: 6–11.

ACKNOWLEDGEMENTS

The author would like to thank Floyd Hayes (Pacific Union College, USA) and Martyn Kenefick (Trinidad and Tobago Rare Bird Committee) for their considerable assistance with this chapter; Daveka Boodram (UWI) for information on White-tailed Sabrewing; Alesha Naranjit and Kerrie Naranjit (Pawi Study Group) for information on Trinidad Piping-guan; Nadra Nathai-Gyan (Wildlife Section, Ministry of Agriculture, Lands and Marine Resources) for assistance with wildlife legislation and population counts of Scarlet Ibis; and Alfredo Colón for the use of his photograph.

TT001 West Coast Mudflats

Unprotected/National Park

COORDINATES 10°29′N 61°29′W
ADMIN REGION Trinidad
AREA 4,770 ha
ALTITUDE 0 m
HABITAT Coastline, mudflats, mangrove

Western Sandpiper

THREATENED BIRDS
RESTRICTED-RANGE BIRDS
BIOME-RESTRICTED BIRDS
CONGREGATORY BIRDS ✓

Site description

West Coast Mudflats IBA extends along the west coast of Trinidad from Port-of-Spain in the north to the Godineau River in the south, and forms the eastern fringe of the Gulf of Paria. Outflow from the Orinoco River makes the waters of the gulf brackish (with salinity falling to 10–25 parts per thousand in the wet season). At its northern end, the IBA abuts the coastal mangroves of the Caroni Swamp IBA (TT002). The coastal fringe supports some small mangrove swamps, small fishing (and shrimping) villages and residential areas. However, much of the coastline is occupied by heavy industry including oil and natural gas facilities, iron and steel, methanol and desalination plants and power stations.

Birds

This IBA is significant for its congregatory waterbirds. Large, regionally important wintering congregations of Laughing Gull *Larus atricilla*, Royal Tern *Sterna maxima* and Black Skimmer *Rynchops niger* gather on the mudflats and around the fishing boats and depots. The IBA also supports flocks of transient or over-wintering Neotropical migratory shorebirds numbering between 10,000 and 20,000 birds (predominantly Semipalmated *Calidris pusilla* and Western *C. mauri* sandpipers). Herons and ibises that roost in the mangroves often feed on the adjacent mudflats. Regionally important numbers of Brown Pelicans *Pelecanus occidentalis* roost in the mangroves.

Other biodiversity

Nothing recorded.

Conservation

The shoreline is a mix of state and private ownership. The section of mudflats bordering Caroni Swamp is contiguous with the Caroni Swamp IBA and receives protection, but in general the shoreline is not protected. Several areas adjacent to industrial installations are not accessible. There have been attempts at planting mangrove along the coast to reduce coastal erosion. A community-group in the village of Brickfield received assistance from the United Nations Development Programme to protect a short, but important stretch of coastline for the benefit of the birdlife, but to date little tangible benefit has occurred. The main threats are the continued industrialisation of the west coast, and the resultant (further) reduction in water quality within the gulf. Current plans include the construction of an offshore aluminium smelter on an artificial island south of San Fernando. Ibises and herons are hunted along the margins of the mangrove. Rising sea-level is likely to have a profound impact on this coastal ecosystem.

Scarlet Ibis

Site description

Caroni Swamp IBA is on the west coast of Trinidad. The Caroni River, the main watercourse entering the swamp, runs along the north of the swamp which is bordered to the north and north-east by highways, residential and commercial development. Towards the south and east of the swamp, sugarcane cultivation has given way in recent years to mixed agriculture and housing. On the western margin, and extending down the west coast of Trinidad, is the extensive West Coast Mudflats IBA (TT001). This estuarine swamp is dominated by mangrove forest with small areas of herbaceous marsh and reed beds along the landward margin (remnants of a freshwater marsh which used to occupy the eastern third of the swamp).

Birds

This IBA is significant for waterbirds. It is the major roosting and breeding site of Scarlet Ibis *Eudocimus ruber* in Trinidad. In 1963 the breeding population was 2,500 pairs. After 1970 the species only roosted in the IBA, but currently 500 pairs are breeding on the eastern edge of the swamp. Up to 15,000 birds roost in the swamp. The Caroni Swamp hosts 10,000 to 15,000 resident and migratory herons. A large roost of migratory Neotropic Cormorant *Phalacrocorax brasilianus* has recently established. The swamp was once a major waterbird site, but this is no longer the case.

Other biodiversity

Nothing recorded.

Conservation

Most of the state-owned Caroni Swamp is a prohibited area with entry by permit only. It is managed as a National Park and Wildlife Sanctuary by the Forestry Division, and is a Ramsar site. The area has a management plan. An initiative to restore areas of freshwater marsh to encourage more breeding *Eudocimus ruber* has made slow progress due to cost constraints. Tour boats take visitors on a daily basis to see the ibises as they return to roost. Drainage and irrigation schemes attempting to facilitate rice cultivation in the 1920s and 1940s failed: the embankments eroded, brackish water infiltrated the swamp, and today, freshwater habitat is limited to a few hectares around the swamp's margins. Freshwater swamp is critical for the breeding ibises. Due to abstraction for irrigation, little water enters the swamp in the dry season, and that which does is often highly polluted. In spite of the area's protection, *E. ruber* and night-herons are favoured targets for hunters in the swamp.

TT003 Northern Range

National Park/Forest Reserve/Wildlife Sanctuary/Unprotected

COORDINATES 10°45′N 61°13′W
ADMIN REGION Trinidad
AREA 36,570 ha
ALTITUDE 100–925 m
HABITAT Montane forest, elfin forest

Trinidad Piping-guan

THREATENED BIRDS 1
RESTRICTED-RANGE BIRDS 1
BIOME-RESTRICTED BIRDS 2
CONGREGATORY BIRDS

Site description

The Northern Range IBA comprises all lands above 500 m at the eastern end of the mountains that form the northern edge of Trinidad, including the catchments of the Oropuche, Matura, Matelot, Quare and Madamas rivers and the Matura National Park. The range is dissected by major river valleys (used for recreation), each with access roads and settlements. Valleys in the east (where rainfall is highest) are largely agricultural (mostly coffee, shade-grown cacao and citrus), although many are abandoned and have reverted to secondary forest. Patches of elfin forest are found on the highest peaks. A road runs along the north coast providing access to villages and beaches. The major residential and commercial area (on the southern slopes and western foothills) is not within the IBA.

Birds

This IBA supports the only confirmed population (but see Victoria-Mayaro IBA, TT005) of the Critically Endangered Trinidad Piping-guan *Pipile pipile*. Birds occur throughout the eastern part of the range with recent sightings at Grande Riviere, Madamas, and the Morne Bleu ridge. It probably survives in other watersheds, but there are no recent records. Trinidad's "Andean" avifauna is largely restricted to the higher parts of this IBA, along with unknown numbers (possibly up to 30) of wintering Near Threatened Olive-sided Flycatchers *Contopus cooperi*.

Other biodiversity

The Critically Endangered golden tree-frog *Phylodytes auratus* is endemic to the highest peaks of the IBA. Other frogs include the Endangered *Eleutherodactylus urichi* and Vulnerable *Mannophryne trinitatis*, both national endemics. The luminous lizard *Proctoporus schrevei* and the snake *Leptophis stimsoni* are endemic to the IBA.

Conservation

Forest reserves and the Matura National Park cover 57% of the IBA. The El Tucuche summit area is protected within the Northern Range Wildlife Sanctuary, and the Quare River receives protection to safeguard the Hollis Reservoir water-supply. Management authorities have insufficient resources or personnel to enforce the protection of these lands, some of which are privately owned. Basic research, public awareness and some community-based efforts have focused on conserving *P. pipile*. Threats include bush fires and clearance for both agriculture and housing. Hunting is common, and a particular threat to *P. pipile*. A proposed eastward extension of the coastal road will open up the Madamas watershed to development.

TT004 Victoria-Mayaro Forest Reserve

Forest Reserve/Wildlife Sanctuary

COORDINATES 10°11′N 61°10′W
ADMIN REGION Trinidad
AREA 52,395 ha
ALTITUDE 0–303 m
HABITAT Forest

Trinidad Piping-guan

THREATENED BIRDS 1
RESTRICTED-RANGE BIRDS 1
BIOME-RESTRICTED BIRDS 2
CONGREGATORY BIRDS

■ Site description

Victoria-Mayaro Forest Reserve IBA is a large area embracing the south-east corner of Trinidad. It represents one of the largest tracts of intact forest in the country, although there are a few farms and settlements within the reserve and a number of roads and pipelines (including a major natural gas pipeline) passing through it. The area includes the steep-sided Trinity Hills on the south-east coast (representing the eastern end of the Southern Range). Forest in the IBA is seasonal evergreen, some areas of which are dominated by the canopy-forming mora *Mora excelsa*. Non-forested areas include some agricultural fields and one area of pasture. Habitation surrounding the reserve includes a fishing village to the east, several oil installations, and service providers to the oil and natural gas industries. The IBA is favoured by hunters.

■ Birds

This IBA supports a wide diversity of resident forest birds, but is significant for the Critically Endangered Trinidad Piping-guan *Pipile pipile*. In 1972 the Trinity Hills were thought to hold the last surviving *P. pipile* population. At this time there were no recent records from the Northern Range IBA (TT004). The last credible record of *P. pipile* in the area was in 2000, but the reserve is infrequently visited by birdwatchers. The biome-restricted Copper-rumped Hummingbird *Amazilia tobaci* is common.

■ Other biodiversity

The Trinidad and Tobago endemic frog *Eleutherodactylus urichi* is found it the Trinity Hills and surrounding forest. The forest is of national importance for terrestrial mammals.

■ Conservation

This IBA is state-owned and designated a forest reserve which also embraces the 7,500-ha Trinity Hills Wildlife Sanctuary. Research has been conducted on tree diversity and the impact of different methods for harvesting timber species. Some studies have been conducted on the fauna, often in support of environmental impact assessments associated with the oil and gas industry. A large natural gas pipeline was recently constructed which, together with existing roads, fragments the formerly contiguous forest and facilitates gradual habitat destruction from illegal logging or expansion of agricultural plots. The survival of any population of *P. pipile* is likely to be severely threatened by hunters, and this urgently requires attention. Much of this IBA is leased to the state-owned petroleum company for exploration and extraction of oil, which in part provides some degree of protection against shifting agriculture and bush fires.

TT005 Main Ridge

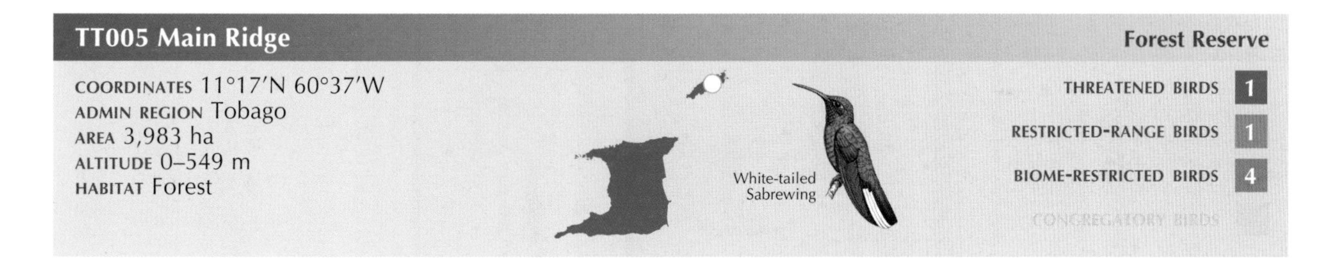

■ Site description

Main Ridge IBA forms the "backbone" of the north-eastern half of Tobago. There is no well defined peak along the ridge, but the highest points are near the mid-point, and at Pigeon Peak near the northern end. The IBA boundary coincides with that of the forest reserve which is surrounded by forested or (often abandoned) agricultural lands. One main road crosses the ridge but there are no settlements in the IBA. Soil moisture is adequate even in the dry season to support tropical forest, specifically: lowland rainforest (up to 240–360 m); lower-montane rainforest (which dominates); and xerophytic rainforest (on the drier soils in the south-west of the IBA).

■ Birds

This IBA is significant for biome-restricted birds. All four biome-restricted species in Trinidad and Tobago (Northern South America and Northern Andes biomes) occur in this IBA including the Venezuelan Flycatcher *Myiarchus venezuelensis* and the Near Threatened White-tailed Sabrewing *Campylopterus ensipennis* (also a restricted-range species in the Tobago secondary EBA). The Tobago population of *C. ensipennis* was thought to have been extirpated by Hurricane Flora in 1963, but has since recovered to a population estimated at 800–1,200 birds, most of which inhabit the Main Ridge IBA.

■ Other biodiversity

The Tobago-endemic frogs *Manophryne olmonae* and (the Vulnerable) *Pristimantis charlotvillensis* occur, as does the Endangered *Eleutherodactylus urichi* (endemic to Trinidad and Tobago) and the Vulnerable *Hyalinobatrachium orientale*. Tobago-endemic reptiles occurring in the IBA include the ocellated gecko *Gonatodes ocellatus* and the snakes *Mastigodryas boddaerti dunni* and *Erythrolamprus ocellatus*. Tobago's 16 endemic plant species probably occur in the IBA.

■ Conservation

This IBA has been formally protected since 1765 as the Main Ridge Forest Reserve—a state-owned area managed by the Tobago House of Assembly. Research on *C. ensipennis* has been conducted through the University of the West Indies since the early 1990s, and is continuing with funding from the EMA. The people of Tobago appear to show a general appreciation of their environment which, coupled with the importance of the local eco-tourist industry (much of which is bird focused), helps to curb the threat of hunting and widespread habitat destruction. Bush fires are common during the dry season and contribute to significant habitat loss. Hurricane Flora left little or none of the original forest canopy intact, but the forest has been recovering since.

■ Site description

St Giles Islands IBA comprises one main island (St Giles, c.29 ha of steeply-sloping land rising to just over 100 m) and several outlying rocks c.1 km off the north-eastern end of Tobago. The main island of St Giles is cloaked in cactus, low scrub and a few deciduous trees up to 9 m tall. This vegetation represents a remnant of the deciduous seasonal forest once found throughout the lower elevations of Tobago. On the island of St Giles it is highly influenced by wind exposure and the nutrients generated by nesting seabirds. There is no human settlement on the islands and the topography makes development unlikely. The surrounding seas are generally rough and landing is hazardous.

■ Birds

This IBA is significant for its breeding seabird colonies. Globally important numbers of Red-billed Tropicbird *Phaethon aethereus* are present, as are regionally important numbers of Audubon's Shearwater *Puffinus lherminieri*, Magnificent Frigatebird *Fregata magnificens*, Masked Booby *Sula dactylatra* and Red-footed Booby *S. sula*. Other seabirds such as Brown Booby *S. leucogaster* and Brown Noddy *Anous stolidus* also breed. However, recent population counts are below previous records.

■ Other biodiversity

The terrestrial fauna of St Giles has not been studied but it may be similar to that of Little Tobago IBA (TT008).

■ Conservation

The St Giles Islands are a state-owned game sanctuary, but enforcement is minimal. The inhospitable terrain and hazardous waters do offer some form of protection from human disturbance. These factors also make it very difficult to survey the seabird populations—low numbers recently recorded may in part be due to insufficient sampling effort. However, there is a tradition in Tobago of eating seabirds, especially at harvest festivals. *Sula leucogaster* and *F. magnificens* are the preferred species with poachers regularly visiting the islands and collecting birds. This has presumably played a part in the apparent population declines. Invasive species (e.g. rats) which could potentially prey on nesting seabirds are a potential threat. It is unknown if they are already present and impacting the seabird populations, but this should be determined as soon as possible. Conservation action focused on the issues of poaching could facilitate the recovery of this site's seabird populations.

■ Site description

Little Tobago Island IBA is a small island (113 ha) c.2 km off the coast at Speyside in north-east Tobago. It is almost completely forested, and represents the largest remnant of deciduous seasonal forest once found throughout the lower elevations of Tobago. The forest is secondary in nature as the island was once cultivated, although it had fully regenerated (to a climax state) by 1944. Palms dominate, and the aroid *Anthurium hookeri* is abundant as a ground flora and as an epiphyte. Along exposed coasts the canopy becomes progressively lower and windswept with cacti and succulents occurring. The coastline comprises steep slopes and cliffs. There is no permanent habitation on the island but the tropicbirds and marine life attract regular day visitors.

■ Birds

This IBA is significant for nesting seabirds and biome-restricted birds. Two (of the 4) Northern South America biome-restricted species occur, namely Rufous-vented Chachalaca *Ortalis ruficauda* and Copper-rumped Hummingbird *Amazilia tobaci*. However, the island is renowned for its seabirds. Globally important populations of Red-billed Tropicbird *Phaethon aethereus* and Laughing Gull *Larus atricilla* breed, and the numbers of Audubon's Shearwater *Puffinus lherminieri*, Brown Booby *Sula leucogaster*, Red-footed Booby *S. sula* and Bridled Tern *Sterna anaethetus* are regionally significant. Sooty Tern *S. fuscata* and Brown Noddy *Anous stolidus* are also present.

■ Other biodiversity

Three restricted-range reptiles occur: ocellated gecko *Gonatodes ocellatus* and the snake subspecies *Mastigodryas boddaerti dunni* are endemic to Tobago and Little Tobago. The ground-lizard subspecies *Bachia heteropa alleni* is endemic to Tobago, Grenada and the Grenadines.

■ Conservation

The state-owned Little Tobago Island has been protected as a game sanctuary since 1928. It is currently managed by the Department of Natural Resources and the Environment (Tobago House of Assembly). Access is restricted without an authorised tour guide, most of who come from the local community and who also provide protection from hunters trying to get to the island. In the mid-1960s extensive studies and population estimates were conducted on the birds of Little Tobago. Since then however, objective fieldwork has been sporadic. A recent survey of the island's vertebrate fauna was conducted by the University of the West Indies. No mammals were detected although the seabird populations appear to have declined since the 1960s. Major threats include the potential introduction of mammalian predators and hurricanes.

TURKS & CAICOS ISLANDS

LAND AREA **500 km²** ALTITUDE **0–49 m**
HUMAN POPULATION **21,750** CAPITAL **Cockburn Town, Grand Turk**
IMPORTANT BIRD AREAS **9, totalling 2,470 km²**
IMPORTANT BIRD AREA PROTECTION **69%**
BIRD SPECIES **204**
THREATENED BIRDS **3** RESTRICTED-RANGE BIRDS **4**

MIKE PIENKOWSKI (UK OVERSEAS TERRITORIES CONSERVATION FORUM, AND TURKS AND CAICOS NATIONAL TRUST)

Caribbean Flamingos on the old saltpans at Town Salina, in the capital, Grand Turk. (PHOTO: MIKE PIENKOWSKI)

INTRODUCTION

The Turks and Caicos Islands (TCI), a UK Overseas Territory, lie north of Hispaniola as a continuation of the Bahamas Islands chain. The Caicos Islands are just 50 km east of the southernmost Bahamian islands of Great Inagua and Mayaguana. The Turks and Caicos Islands are on two shallow (mostly less than 2 m deep) banks—the 5,334 km² Caicos Bank and the 254-km² Turks Bank—with deep ocean between them. There are further shallow banks, namely Mouchoir, Silver and Navidad that extend for 250 km to the south-east, but without islands. They are important for whales and probably for feeding seabirds. The Bahamas lie on separate banks to the north-east, but share some aspects of the geography. The largest islands lie along the northern edge of the Caicos Bank, and comprise, from west to east: West Caicos (c.10 km by 3 km); Providenciales (c.23 km by 10 km); the small cays of Mangrove, Donna, Little Water, Water, Pine, Fort George, Stubbs, Dellis, and Parrot; North Caicos (c.25 km by 15 km); Middle Caicos (c.25 km by 15 km); East Caicos (c.20 km by 15 km); and South Caicos (c.9 km by 5 km), together with some smaller cays. Several very small cays, important for breeding seabirds, lie on the southern edge of the Caicos Bank, c.30–50 km south of the larger islands. Providenciales, North, Middle and South Caicos are inhabited, and resorts are being developed on many of the small island. The smaller Turks Bank holds the inhabited islands of Grand Turk (10 km by 3 km) and Salt Cay (6 km by 2 km), as well as numerous smaller cays.

The Turks Bank islands plus South Caicos (the "salt islands") were used to supply salt from about 1500. They were inhabited by the 1660s when the islands were cleared of trees to facilitate salt production by evaporation. By about 1900, Grand Turk was world famous for its salt. The industry closed down during the twentieth century with the last production ceasing in 1975, leaving behind the salinas or saltpans. After depopulation in the late fifteenth century, the Caicos Islands remained largely uninhabited until the eighteenth century (although up until this period, their shallow, complex channels and creeks made them a safe haven for many famous pirates). From 1787, land in the Caicos Islands was given to "Empire Loyalists" who had lost their lands in the newly created USA. The resulting plantations (initially mainly for cotton) cleared much of the high forest which appears to have survived until that time in largely pristine state. The plantation period lasted just a few decades (although it left a substantial archaeological heritage) and the last record of cotton exportation is in 1812. The major change of the twentieth century was the land grant of much of Providenciales to US development companies, leading to the establishment of holiday resorts and private homes on the island. Providenciales is now the commercial centre (and centre of population) of the territory, with the government remaining largely at Grand Turk.

The climate is generally warm and dry, with occasional heavy rain. Average annual rainfall on Grand Turk is 500 mm while the Caicos group averages c.1,000 mm with the wettest month being May. Over half of the TCI land area is wetland such as intertidal creeks, lagoons, flats, salinas and marshes. Only a small area of land is under cultivation. The terrestrial habitats, where these have not been destroyed by bulldozing prior to building, are in various stages of recovery to high tropical dry forest. This process is slow, and thorny scrub forest is the most widespread at present. Scattered through the dry forest are many ponds of various sizes and of differing hydrology—the freshwater ones being particularly important for birds. In some areas there is rocky shore, either cliffs or low rocks, but generally the coast consists of sandy beaches backed by low scrub (and sand-dunes in some places). Where the dry ground grades into wetter ground towards the southern side of the Islands, particularly on North and Middle Caicos, there are some areas of "Caicos pine" (the national tree) *Pinus caribaea*. Pine woodland occurs in extensive stands intermingled with other seasonally or temporally flooded habitats (such as marsh, mangrove inlets, saltmarsh, mudflats etc.).

■ Conservation

The protection of natural areas in TCI is provided for through two main legislative means, the National Parks Ordinance (1992) and the National Trust Ordinance (1992). The former defines four categories of protected area: national parks (nearer the UK, than the IUCN model, with some emphasis on recreation); nature reserves (with emphasis on nature conservation, and with visiting restrictions if appropriate); sanctuaries (nature areas with a presumption against visiting); and areas of historic interest (for human heritage sites). The National Trust Ordinance established the Turks and Caicos National Trust (TCNT) as a statutory but independent non-governmental membership body. It is responsible for safeguarding the environmental, cultural and historical heritage of the islands for present and future generations. One of the Trust's statutory roles is holding environmentally important land in trust for the country, and the TCI government has started transferring some such lands to the Trust to hold and manage. TCNT plays a unique role in its partnership approach to sustainable conservation, the protection of the natural environment and promotion of environmental awareness and responsibility. It is supported by membership fees, private sponsorship and project grants and fulfils its mission by implementing a range of sustainable projects and initiatives, some of which are revenue generating and used to finance new programs. The Department of Environmental and Coastal Resources (DECR) within the Ministry of Natural Resources, is the TCI government department responsible for nature conservation, fisheries and related matters.

Although statutory protected areas have existed since early 1990s, resource limitations have delayed much progress on the management of these areas by official bodies. However, the conservation fund resourced by a 1% addition to the (now 10%) visitor tax aims to provide the resources necessary for protected area management. Management plans have been developed and are being implemented (by the government) for three of TCI's marine national parks in the seas adjacent to Providenciales and West Caicos. TCNT has focused on terrestrial and wetland areas including the effective management of Little Water Cay, as well as several historic sites. It has also (in a collaborative venture) developed, and is starting to implement, a management plan (available at www.ukotcf.org) for the North, Middle and East Caicos Ramsar Site and its surroundings, including IBAs TC001–005.

Several projects implemented since 1998 have been designed to cover as much of TCI as possible with bird (and biodiversity) surveys. Where feasible these have been done to cover different seasons and multiple years. For example, systematic surveys through the Middle Caicos woodlands and wetlands, as well as areas on North Caicos and Grand Turk, have been undertaken on several occasions each year. Systematic efforts have been made to fill gaps in coverage, leading to several specially organised visits to Salt Cay, the seabird cays, East Caicos, and other sites. These surveys have highlighted yet more gaps, conservation priorities and also conservation opportunities such as the superb education and awareness potential of the salinas right inside the town on Grand Turk.

■ Birds

Of over 204 recorded species on TCI, 58 are known to breed and 110 occur as regular passage or wintering migrants. Four of the seven Bahamas Endemic Bird Area restricted-range

North Caicos mangroves, flats, pine and dry forest within the North, Middle and East Caicos Ramsar Site. (PHOTO: MIKE PIENKOWSKI)

species occur, namely Bahama Woodstar *Calliphlox evelynae*, Bahama Mockingbird *Mimus gundlachii*, Pearly-eyed Thrasher *Margarops fuscatus* and Thick-billed Vireo *Vireo crassirostris*. An endemic subspecies of the Thick-billed Vireo *V. crassirostris stalagmium* is restricted to the Caicos Bank. A number of other species, characteristic of adjacent islands are also present (or have arrived as vagrants) on the islands. These include Cuban Crow *Corvus nasicus*, otherwise endemic to Cuba, which is a permanent resident on the Caicos Islands, as is an endemic subspecies of the Greater Antillean Bullfinch *Loxigilla violacea ofella* (on Middle and East Caicos, with occasional records from North Caicos). Cuban vagrants (or irregular visitors) have included Endangered Blue-headed Quail-dove *Starnoenas cyanocephala* and Giant Kingbird

Table 1. Key bird species at Important Bird Areas in the Turks and Caicos Islands.

Key bird species	Criteria	National population	TC001	TC002	TC003	TC004	TC005	TC006	TC007	TC008	TC009
		Criteria	■	■ ■ ■	■ ■ ■	■ ■	■ ■ ■	■	■	■	■ ■
West Indian Whistling-duck *Dendrocygna arborea*	VU ■			48	✓	✓	✓				
White-cheeked Pintail *Anas bahamensis*	■				1,000						
Caribbean Flamingo *Phoenicopterus ruber*	■				3,000						
Reddish Egret *Egretta rufescens*	■				400		300				
White-tailed Tropicbird *Phaethon lepturus*	■	900		150						120	
Brown Pelican *Pelecanus occidentalis*	■				150		150		60		30
Sandhill Crane *Grus canadensis*	■				3						
Black-bellied Plover *Pluvialis squatarola*	■				2,500						
Wilson's Plover *Charadrius wilsonia*	■				100		90		90		90
Short-billed Dowitcher *Limnodromus griseus*	■				3,200				4,000		
Greater Yellowlegs *Tringa melanoleuca*	■				1,000				1,000		
Lesser Yellowlegs *Tringa flavipes*	■				5,000				6,000		
Least Sandpiper *Calidris minutilla*	■				6,000						
Laughing Gull *Larus atricilla*	■			150	900		150		900	450	900
Gull-billed Tern *Sterna nilotica*	■	300		50	60		60				
Royal Tern *Sterna maxima*	■			120	150		60	30	40		
Sandwich Tern *Sterna sandvicensis*	■				150			200	60		60
Roseate Tern *Sterna dougallii*	■	600						600			
Common Tern *Sterna hirundo*	■						600				
Least Tern *Sterna antillarum*	■	3,000		50	100		90		2,520		840
Bridled Tern *Sterna anaethetus*	■	10,500						3,000		6,900	
Sooty Tern *Sterna fuscata*	■	144,000								132,000	
Brown Noddy *Anous stolidus*	■	81,000						33,000		22,200	
Bahama Woodstar *Calliphlox evelynae*	■		✓	✓	✓	400	✓				✓
Thick-billed Vireo *Vireo crassirostris*	■		✓		✓	6,700	✓				✓
Bahama Mockingbird *Mimus gundlachii*	■		✓	✓	✓	200	✓				✓
Pearly-eyed Thrasher *Margarops fuscatus*	■		✓		✓	✓					

All population figures = numbers of individuals.
Threatened birds: Vulnerable ■. **Restricted-range birds** ■. **Congregatory birds** ■.

The Turks and Caicos Islands support endemic subspecies of Thick-billed Vireo and Greater Antillean Bullfinch. (PHOTO: MIKE PIENKOWSKI)

Large numbers of shorebirds, such as these Lesser Yellowlegs and Stilt Sandpipers, pass through or overwinter in the islands. (PHOTO: MIKE PIENKOWSKI)

Tyrannus cubensis, while the Hispaniolan endemic Green-tailed Warbler *Microligea palustris* has been recorded on Bush Cay. All of these landbirds rely on the under-valued dry woodlands.

Globally threatened birds include the Vulnerable West Indian Whistling-duck *Dendrocygna arborea*, significant numbers of which occur in a number of areas although the territorial population is unknown. The Near Threatened Piping Plover *Charadrius melodus* occurs, but in very small numbers and normally only single individuals are seen. The Near Threatened Kirtland's Warbler *Dendroica kirtlandii* is also recorded regularly as a wintering migrant, but only in very small numbers.

Studies on seabirds in TCI have largely been limited to surveys and censuses with incomplete coverage of irregular but major breeding in e.g. inland marsh and pond areas of several of the Caicos Islands. However, the small cays of the Caicos and Turks Banks, as well as some cliffs and stacks of the main islands are important breeding sites for significant numbers of seabirds. Thirteen species of seabird breed, 11 of which are migrants with at least some of the population absent for part of the year. The populations of Bridled Tern *Sterna anaethetus*, Sooty Tern *S. fuscata*, Least Tern *S. antillarum*, Brown Noddy *Anous stolidus* and Laughing Gull *Larus atricilla* are particularly noteworthy. Individual wetlands in TCI vary on a seasonal and annual basis depending on weather conditions. However, with so much wetland habitat available, the network of wetlands is able to support large numbers of breeding, passage and wintering waterbirds (ducks, shorebirds, herons and egrets etc.). The territorial populations of these waterbirds are poorly documented but deserve further survey and monitoring attention.

IMPORTANT BIRD AREAS

TCI's nine IBAs—the territory's international priority sites for bird conservation—cover 2,470 km² (including large marine areas), and about 15% of the islands' land area. The IBAs have been identified on the basis of 27 key bird species (listed in Table 1) that variously trigger the IBA criteria. These 27 species comprise all four restricted-range species, 22 congregatory waterbirds/seabirds, and the Vulnerable West Indian Whistling-duck *Dendrocygna arborea*—the only globally threatened bird that occurs in sufficient numbers to trigger the IBA criteria. Insufficient numbers of Near Threatened Piping Plover *Charadrius melodus* and Kirtland's Warbler *Dendroica kirtlandii* occur to identify IBAs for their conservation, although both do occur in IBAs.

The nine identified IBAs include two areas of forest (in one case including important ponds) that support the restricted-range birds; two groups of seabird breeding cays; two areas of former saltpans and adjacent creeks; and three other wetland areas (which include some related habitats including cliffs and other coastal types). These IBAs are on or near North, Middle and East Caicos, Grand Turk, Salt Cay and several small cays. No IBAs have been identified on West Caicos, Providenciales or South Caicos. However, West Caicos needs further survey and assessment (especially in light of the major development in progress on this formerly uninhabited island) as some areas may qualify as IBAs. Similarly, several areas of Providenciales (e.g. Pigeon Pond and Frenchman's Creek Nature Reserve, Juba Creek and Flamingo Lake) may qualify as IBAs, but further survey data

Wade's Green–Teren Hill IBA on North Caicos, one of the two dry forest IBAs identified in the territory. (PHOTO: MIKE PIENKOWSKI)

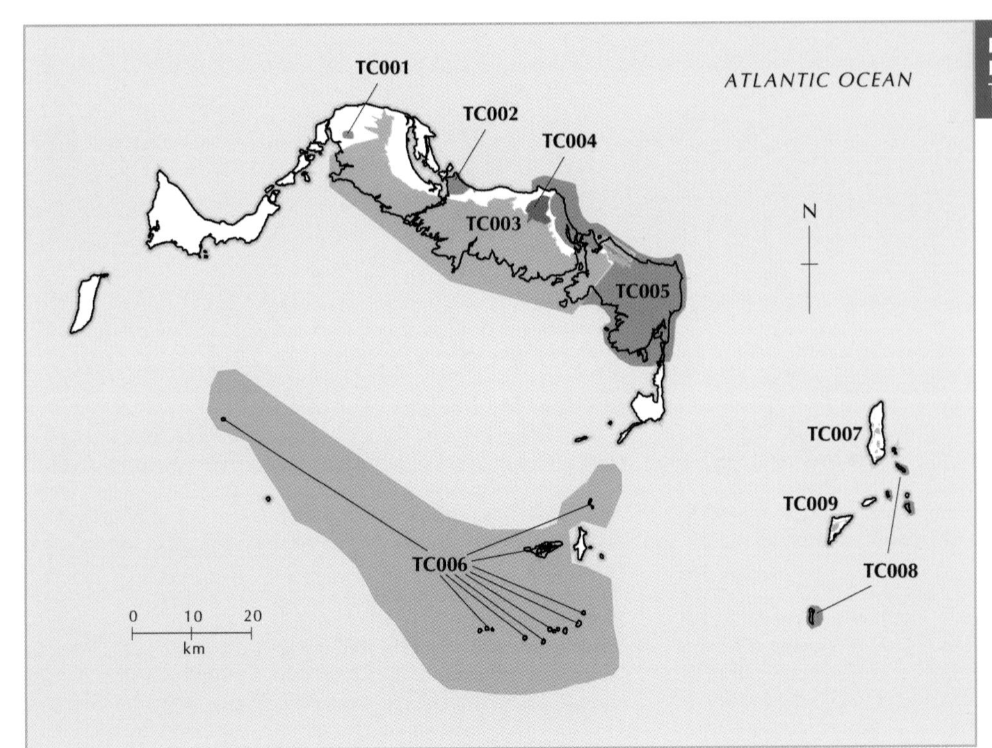

Figure 1. Location of Important Bird Areas in the Turks and Caicos Islands.

are needed. On South Caicos, reinstatement of the tidal flow around Boiling Hole would no doubt lead to the (re-) establishment of the extensive saltpan area as an internationally important site for waterbirds. Further seabird surveys may also point to IBA status for South Caicos' Admiral Cockburn Nature Reserve (Long Cay, Middleton Cay and Six Hills Cay).

There is a clear need for further survey work to fill information gaps, but also for the maintenance of the monitoring work that has started for the seabirds. Any such survey and monitoring results should be used to inform the annual assessment of state, pressure and response scores at each IBA in order to provide an objective status assessment and highlight management interventions that might be required to maintain these internationally important biodiversity sites.

KEY REFERENCES

Bond, J. (1985) *Birds of the West Indies.* London: William Collins Sons and Co Ltd.

Bradley, P. E. (1995) *The birds of the Turks and Caicos Islands: the official checklist.* Providenciales, Turks and Caicos Islands: Turks and Caicos National Trust.

Buden, D. W. (1987) *The birds of the Southern Bahamas: an annotated check-list.* London: British Ornithologists' Union (BOU Check-list 8).

Ground, R. (2001) *The birds of the Turks and Caicos.* Providenciales, Turks and Caicos Islands: Turks and Caicos National Trust.

Hallet, B. (2006) *Birds of the Bahamas and the Turks and Caicos Islands.* London: MacMillan Caribbean.

Haney, J. C., Lee, D. S. and Walsh-McGehee, M. (1998) A quantitative analysis of winter distribution and habitats of Kirtland's warblers in the Bahamas. *Condor* 100: 201–217.

Pavlidis, S. J. (1999) *The Turks and Caicos guide: a cruising guide to the Turks and Caicos Islands.* Port Washington, Wisconsin: Seaworthy Publications.

Pienkowski, M. ed. (2002) *Plan for Biodiversity Management and Sustainable Development around Turks and Caicos Ramsar Site.* www.ukotcf.org (Publications)

Pienkowski, M. W. (2006) Turks and Caicos Islands. Pp.247–272 in S. M. Sanders, ed. *Important Bird Areas in the United Kingdom Overseas Territories.* Sandy, U.K.: Royal Society for the Protection of Birds.

Pienkowski, M. W. (in press) The Turks and Caicos Islands. In Bradley P. E. and Norton, R. L. eds. *Breeding seabirds of the Caribbean.* Gainesville, Florida: Univ. Florida Press.

Raffaele, H. Wiley J., Garrido, O., Keith, A. and Raffaele, J. (1998) *A guide to the birds of the West Indies.* Princeton, New Jersey: Princeton University Press.

Scott, D. A. and Carbonell, M. (1986) *A directory of Neotropical wetlands.* IUCN, Cambridge.

Sealey, N. E. (1994) *Bahamian landscapes: an introduction to the geography of the Bahamas.* Media Publishing, Nassau: Media Publishing.

Walsh-McGehee, M., Lee, D. S. and Wunderle, J. M. (1998) A report of aquatic birds encountered in December from the Caicos Islands. *Bahamas J. Sci.* 6: 28–33.

White, A. W. (1998) *A birder's guide to the Bahama Islands (including Turks and Caicos).* Colorado Springs, Colorado: American Birding Association, Inc.

ACKNOWLEDGEMENTS

The author would like to thank the following for observations, advice, discussion or comment on many points: Bryan Naqqi Manco and Ethlyn Gibbs-Williams (Turks and Caicos National Trust); local residents, notably Cardinal Arthur, Alton Higgs and the late Telford Outton, particularly in the community meetings undertaken during the research work by the Turks and Caicos National Trust and the UK Overseas Territories Conservation Forum; Tony Murray, Dr Geoff Hilton and Tim Cleaves (then volunteers organised by RSPB); Patricia Bradley, Richard and Dace Ground, Dr Tony Hutson, Ann Pienkowski and Tony White. Some information has been gathered during or incidentally to work partly supported by the UK Government's Darwin Initiative and Overseas Territories Environment Programme, as well as the UK Overseas Territories Conservation Forum. The co-operation of the Department of Environmental Resources, in particular Michelle Fulford-Gardiner, Judith Garland-Campbell and Rob Wild, is appreciated in several regards, including the provision of permits to survey the statutory sanctuaries. For help in boat access to these difficult sites, thanks are due to Captains Alessio Girotti and Allen-Ray Smith, and able crews Phillip Garneau and Bryan Naqqi Manco.

■ Site description

Wade's Green–Teren Hill IBA is close to the settlement of Kew in north-western North Caicos Island. Wade's Green is an old Loyalist plantation with the manor house high up on a hill overlooking much of the island. Teren Hill, a second historic plantation ruin, is in close proximity. This IBA represents the most important remaining high "gallery" forest area in the territory. High forest was the dominant vegetation prior to c.1800, and this is one of the few areas where it has regenerated.

■ Birds

This IBA supports populations of all four Bahamas EBA restricted-range birds, namely Bahama Woodstar *Calliphlox evelynae*, Thick-billed Vireo *Vireo crassirostris*, Bahama Mockingbird *Mimus gundlachii* and Pearly-eyed Thrasher *Margarops fuscatus*. It is probably the most important area in TCI for the thrasher, and holds significant populations of Key-west Quail-dove *Geotrygon chrysia*, Western Spindalis *Spindalis zena*, and Cuban Crow *Corvus nasicus*. The globally threatened Blue-headed Quail-dove *Starnoenas cyanocephala* (a Cuban endemic) has been recorded (occasionally) in this IBA in 2001–2002, presumably as a vagrant.

■ Other biodiversity

It is an important site for a endemic reptile species, namely the Caicos barking gecko *Aristelliger hechti* (recently rediscovered), curly tail *Leiocephalus psammodromus*, Caicos Islands reef gecko *Sphaerodactylus caicosensis* and the snake, Caicos Islands trope boa *Tropidophis greenwayi*, as well as a number of endemic subspecies.

■ Conservation

Wade's Green–Teren Hill IBA is partly on crown lands. Due to its historic importance TCI Executive Council agreed a 99-year lease to TCNT for the central buildings area at Wades Green in 1999 and TCNT now manages the site. However, the rest of the area needs nature reserve status and TCNT management to secure the IBA's biodiversity interest. The area falls within the remit of the TCNT Biodiversity Management Plan.

■ Site description

Fish Ponds and Crossing Place Trail IBA is situated in the western part of the north coast of Middle Caicos. It includes Fish Ponds, Crossing Place Trail, Indian Cave, and Blowing and Juniper Holes. The IBA is characterised by limestone cliffs that slope inland to ponds, which are connected to the sea under the cliffs. There are several sea-caves, and a dry inland cave within the site, Indian Cave. Offshore are a number of small cays. Fish Ponds comprise some of the most important wetlands in the area that are not included within the Ramsar site (see TC003). Crossing Place Trail is the traditional route along the Caicos Islands. The Middle Caicos section is particularly important culturally and scenically.

■ Birds

This IBA is significant for supporting a population of the Vulnerable West Indian Whistling-duck *Dendrocygna arborea*, and two (of the four) Bahamas EBA restricted-range birds, namely Bahama Woodstar *Calliphlox evelynae* and Bahama Mockingbird *Mimus gundlachii*. Regionally important populations of White-tailed Tropicbird *Phaethon lepturus*, terns *Sterna* spp. and Laughing Gull *Larus atricilla* breed. Unknown numbers of Audubon's Shearwater *Puffinus lherminieri* breed on the offshore cays. Up to 500 Caribbean Flamingo *Phoenicopterus ruber* are present and small numbers of migrant Sandhill Crane *Grus canadensis* have been recorded.

■ Other biodiversity

The endemic Drury's hairstreak butterfly *Strymon acis leucostricha* occurs. The endemic reptile species curly tail *Leiocephalus psammodromus*, Caicos Islands reef gecko *Sphaerodactylus caicosensis* and the snake, Caicos Islands trope boa *Tropidophis greenwayi*, occur, as well as a number of endemic subspecies.

■ Conservation

Fish Ponds and Crossing Place Trail IBA is on crown lands. The Crossing Place Trail has interim protection against development under the Planning Regulations legislation, but more substantive protection (e.g. as a nature reserve) is required, both against built development and any impacts of the proposed causeway to link North and Middle Caicos. The track built to the existing North Caicos ferry cuts across some of the Fish Ponds and has disrupted water flow, resulting in deoxygenation and mass fish deaths, of great concern to local residents. This needs addressing, as noted in the TCNT Biodiversity Management Plan.

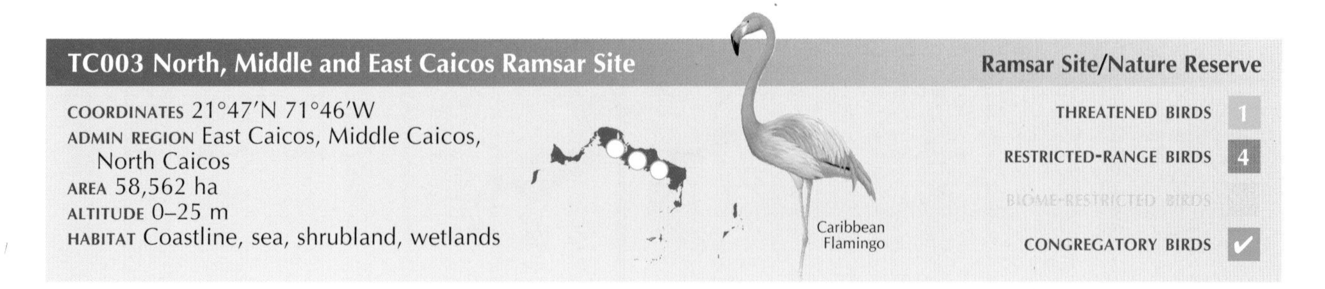

■ Site description

North, Middle and East Caicos Ramsar Site IBA stretches along the wetland (mainly south-west) sides of North Caicos, Middle Caicos and part of East Caicos. It is a wetland complex including a range of wetland and some linked dry-land ecosystems with important natural transitions between them. The submerged mangrove swamps, algal flats, sea-grass beds and several inlet cays are particularly noteworthy. The IBA boundary is the same as that of the Ramsar site.

■ Birds

This IBA is significant for supporting a population of the Vulnerable West Indian Whistling-duck *Dendrocygna arborea*, and all four Bahamas EBA restricted-range birds. However, this site is most notable for its waterbirds. Over 20,000 waterbirds occur, and the populations of a number of these (e.g. Caribbean Flamingo *Phoenicopterus ruber*, Reddish Egret *Egretta rufescens*, and a number of shorebirds) are globally significant. Small numbers of migrant Sandhill Crane *Grus canadensis* and the Near Threatened Kirtland's Warbler *Dendroica kirtlandii* have been recorded wintering in the IBA. The endemic subspecies of Greater Antillean Bullfinch *Loxigilla violacea ofella* occurs, as does Western Spindalis *Spindalis zena* and Cuban Crow *Corvus nasicus*.

■ Other biodiversity

The IBA is an important site for feeding of immatures or breeding by a number of globally threatened sea-turtles, including the Endangered green turtle *Chelonia mydas* and loggerhead turtle *Caretta caretta* and Critically Endangered hawksbill turtle *Eretmochelys imbricata*. On land the endemic reptile species curly tail *Leiocephalus psammodromus*, Caicos Islands reef gecko *Sphaerodactylus caicosensis* and the snake, Caicos Islands trope boa *Tropidophis greenwayi*, occur, as well as a number of endemic subspecies.

■ Conservation

The North, Middle and East Caicos Ramsar Site IBA is mainly on crown lands and is included in the TCNT Biodiversity Management Plan. The area is a statutory nature reserve (number 17) and includes the overlapping Vine Point (Man O' War Bush and Ocean Hole) Nature Reserve (number 22). Working with the local community, the TCNT and its partners have produced and started to implement a "Plan for biodiversity management and sustainable development around the Turks and Caicos Ramsar site". The strict protection afforded by the statutory nature reserve should be implemented fully and extended to include the ecologically linked dry-land and ponds in Middle Caicos, more fully to East Caicos, and reef areas on North and East Caicos.

TC004 Middle Caicos Forest

Unprotected

COORDINATES 21°48'N 71°42'W
ADMIN REGION Middle Caicos
AREA 1,374 ha
ALTITUDE 0–15 m
HABITAT Forest, shrubland, wetlands

Thick-billed Vireo

THREATENED BIRDS 1
RESTRICTED-RANGE BIRDS 4
BIOME-RESTRICTED BIRDS
CONGREGATORY BIRDS

■ Site description

Middle Caicos Forest IBA lies between the settlements of Lorimers and Bambarra on the north-east side of the island. It is an area of forest that was cleared during the plantation era and is now at various stages of recovery to higher forest and thus supports a good range of scrub and woodland types. There are also different types of permanent and temporary wetlands in the IBA. Some important plantation ruins remain.

■ Birds

This IBA is significant for supporting the territory's largest population of the Vulnerable West Indian Whistling-duck *Dendrocygna arborea* and populations of all four Bahamas EBA restricted-range birds, namely Bahama Woodstar *Calliphlox evelynae*, Thick-billed Vireo *Vireo crassirostris*, Bahama Mockingbird *Mimus gundlachii* and Pearly-eyed Thrasher *Margarops fuscatus* (this latter species being uncommon). Cuban Crow *Corvus nasicus* and Key-west Quail-dove *Geotrygon chrysia* occur, as does the endemic subspecies Greater Antillean Bullfinch *Loxigilla violacea ofella*. The majority of TCI records for wintering Near Threatened Kirtland's Warbler *Dendroica kirtlandii* are from this IBA

■ Other biodiversity

Endemic reptile species, namely the Caicos barking gecko *Aristelliger hechti* (recently rediscovered), curly tail *Leiocephalus psammodromus*, Caicos Islands reef gecko *Sphaerodactylus caicosensis* and the snake, Caicos Islands trope boa *Tropidophis greenwayi*, occur, as well as a number of endemic subspecies. The Near Threatened Cuban fruit-eating bat *Brachyphylla nana* occurs.

■ Conservation

Middle Caicos Forest IBA is a mixture of crown and private ownership. The area is unprotected, but its importance is recognised through inclusion in the TCNT Biodiversity Management Plan. Several field roads in the IBA have been re-opened by the TCNT for the development of interpretative trails. However, the area needs statutory nature reserve status and the government has approved TCNT's management of parts of the area, but related land transfers to TCNT are still awaited. Appropriate crown lands should be transferred to conservation ownership and management as soon as possible, and negotiations with private land owners should focus on raising awareness of the biodiversity value of the area, its conservation and appropriate access for visitors. The IBA is important to local people for plants still used for traditional purposes.

Site description

East Caicos and adjacent areas IBA embraces the totally uninhabited island of East Caicos and adjacent areas of eastern and north-eastern Middle Caicos. The East Caicos portion is a complex of inter-related scrub, woodland, ponds, caves, marshes, flats and other wetlands, adjoining an existing Ramsar site (IBA TC003) which covers only a small part of the island. On Middle Caicos the IBA includes Long Bay (on the north-east shore) and the creeks and flats at Lorimers and Increase, together with the coral reef off Middle and East Caicos. The area at the eastern end of Middle Caicos and around Joe Grant Cay is a complex of cays, creeks and marshes extending to Windward Going Through, and adjoining the existing Ramsar site.

Birds

This IBA is significant for the Vulnerable West Indian Whistling-duck *Dendrocygna arborea* and populations of three (of the four) Bahamas EBA restricted-range birds, namely Bahama Woodstar *Calliphlox evelynae*, Thick-billed Vireo *Vireo crassirostris* and Bahama Mockingbird *Mimus gundlachii*. The IBA supports important populations of a number of waterbirds including globally significant numbers of Reddish Egret *Egretta rufescens* and Common Tern *Sterna hirundo*. Cuban Crow *Corvus nasicus* and the endemic subspecies Greater Antillean Bullfinch *Loxigilla violacea ofella* occur. The Near Threatened Kirtland's Warbler *Dendroica kirtlandii* has been recorded wintering in this IBA, as have several Near Threatened Piping Plovers *Charadrius melodus*.

Other biodiversity

This IBA is important for the Critically Endangered hawksbill *Eretmochelys imbricata*, and Endangered green *Chelonia mydas* and leatherback *Caretta caretta* turtles. Most of TCI's nesting beaches are thought to be on East Caicos and Long Bay. A number of endemic reptiles occur, namely the curly tail *Leiocephalus psammodromus*, Caicos Islands reef gecko *Sphaerodactylus caicosensis* and the snake, Caicos Islands trope boa *Tropidophis greenwayi* as well as several endemic subspecies. The East Caicos cave system is probably important for bats and endemic invertebrates.

Conservation

East Caicos and adjacent areas IBA is unprotected but urgently needs statutory nature reserve status as recommended in the TCNT Biodiversity Management Plan. Quantitative information on the IBA's birds and other biodiversity is limited and additional survey information is needed. The currently uninhabited area (on East Caicos) has been previously threatened by major resort development, and some threats remain.

Site description

Caicos Bank Southern Cays IBA comprises several small cays that extend south from South Caicos on south-eastern edge of the Caicos Bank. From north to south these cays include Fish Cay, a small rocky cay between South Caicos and the Ambergris Cays; Little Ambergris Cay which consists of several small cays surrounding a central lagoon; Bush and Seal Cays, small rocky cays near the south-eastern extremity of Caicos Bank; and French Cay, a small sandy cay several kilometres to the west of Seal Cays along the southern edge of the Caicos Bank.

Birds

This IBA is a significant seabird breeding site with up to 40,000 birds present. The populations of Roseate Tern *Sterna dougallii* (primarily Fish Cay), Bridled Tern *S. anaethetus* (Bush Cay, with some on Fish Cay) and Brown Noddy *Anous stolidus* (11,000 pairs on French Cay) are globally important. Sandwich *S. sandvicensis* and Royal *S. maxima* terns occur in regionally important numbers. The Green-tailed Warbler *Microligea palustris* (a Hispaniola EBA restricted-range birds) has been recorded on Bush Cay as a vagrant.

Other biodiversity

The Critically Endangered Turks and Caicos rock iguana *Cyclura carinata carinata* occurs, as do a number of endemic reptiles, namely the curly tail *Leiocephalus psammodromus*, Caicos Islands reef gecko *Sphaerodactylus caicosensis* and the snake, Caicos Islands trope boa *Tropidophis greenwayi*. The waters, reef and beaches in this IBA are important to the Critically Endangered hawksbill *Eretmochelys imbricata*, and Endangered green *Chelonia mydas* and leatherback *Caretta caretta* turtles.

Conservation

Caicos Bank Southern Cays IBA is crown land. The southernmost French, Bush and Seal Cays are a statutory sanctuary (number 24) although this status is not widely realised or enforced, and there are many unauthorised landings from yachts, and boats carrying illegal immigrants. Fish and Little Ambergris cays are TCNT nature reserves, managed by the Trust on a 99-year lease from the TCI government. However, these cays deserve national-level statutory nature reserve status to better ensure their long-term conservation.

TC007 Grand Turk Salinas and Shores

Unprotected

COORDINATES 21°27′N 71°08′W
ADMIN REGION Grand Turk
AREA 268 ha
ALTITUDE 0–2 m
HABITAT Coastal wetland, salina, beach, sea

Lesser Yellowlegs

THREATENED BIRDS
RESTRICTED-RANGE BIRDS
BIOME-RESTRICTED BIRDS
CONGREGATORY BIRDS ✓

■ Site description

Grand Turk Salinas and Shores IBA embraces all the major wetland areas on Grand Turk Island. About 40% of Grand Turk is wetland—abandoned salinas (or saltpans) from the salt production industry, saltwater inlets (or creeks) and nearby shores. Some of these wetlands are viewable from the centre of TCI's capital (Cockburn Town, or "Grand Turk"). The salinas and wetlands include Town Salina, North Wells, South Wells, North Creek, South Creek, Great Salina, Hawkes Pond Salina and Hawkes Nest Salina. The creeks support tidal mudflats.

■ Birds

This IBA is significant for waterbirds—with important breeding and wintering populations of a range of species. Large numbers of herons, egrets, ducks, shorebirds and terns use these wetlands, but the breeding population of Least Tern *Sterna antillarum* is globally important, as are the large numbers of wintering Short-billed Dowitcher *Limnodromus griseus*, Greater Yellowlegs *Tringa melanoleuca* and Lesser Yellowlegs *T. flavipes*. The breeding population of Laughing Gull *Larus atricilla* is also globally significant. The Near Threatened Piping Plover *Charadrius melodus* has been recorded.

■ Other biodiversity

The Critically Endangered hawksbill *Eretmochelys imbricata*, and Endangered green *Chelonia mydas* and leatherback *Caretta caretta* turtles all nest on some of the beaches in this IBA.

■ Conservation

Grand Turk Salinas and Shores IBA is a mixture of crown and private land ownership. Saltpans are regarded as waste-land rather than the unique resource that they are, and are being in-filled. The TCI Government Development Manual outlines the requirement for an environmental impact assessment for any development in a salina, but the Department of Planning does not enforce this need. With the constant threat of development impinging on the integrity of these wetlands there is a strong case for designating them as a statutory nature reserve (for both their historic interest and biodiversity value). Town Salina would be an ideal focal area for development as an awareness-raising/ educational "watchable wildlife pond".

■ Site description

Turks Bank Seabird Cays IBA comprises the small, rocky cays south-east of Grand Turk, starting with Long Cay which is c.2.5 km south-east of Grand Turk, and ending in the south with Big Sand Cay—the south-easternmost point of the territory. The IBA includes Long Cay, East (formerly Pinzon) Cay, Penniston Cay and Big Sand Cay. Some of the cays have sandy beaches, especially at Big Sand Cay.

■ Birds

This IBA supports over 100,000 breeding seabirds. Globally important populations of Laughing Gull *Larus atricilla* (primarily East Cay), Bridled Tern *Sterna anaethetus* (primarily Penniston Cay), Sooty Tern *S. fuscata* (on Big Sand Cay) and Brown Noddy *Anous stolidus* (mostly Long Cay, but some also on Penniston) breed. Numbers of White-tailed Tropicbird *Phaethon lepturus* are regionally significant. East (or Pinzon) Cay supports unknown (but probably large) numbers of Audubon's Shearwater *Puffinus lherminieri* (which probably also breed on other cays in the IBA). Small numbers of Brown Booby *Sula leucogaster* and Magnificent Frigatebird *Fregata magnificens* breed on Penniston Cay.

■ Other biodiversity

The Critically Endangered Turks and Caicos rock iguana *Cyclura carinata carinata* is common on Long Cay and Big Sand Cay. The endemic curly tail lizard *Leiocephalus psammodromus* also occurs on some of the cays. The shores are likely to be nesting beaches for the Critically Endangered hawksbill *Eretmochelys imbricata*, and Endangered green *Chelonia mydas* and leatherback *Caretta caretta* turtles.

■ Conservation

The Turks Bank Seabird Cays IBA is crown land and the cays are variously designated as at different levels of protection. Long Cay is a statutory sanctuary (number 25). Other small cays near Grand Turk constitute the Grand Turk Cays Land and Sea National Park (statutory national park number 7). However, Penniston Cay and East Cay are wrongly classified in the protected areas system and need transferring from national park status to a statutory sanctuary. Gibbs Cay should remain a part of the national park. Big Sand Cay is a statutory sanctuary (number 23). The cays are subjected to disturbance and there appear to be problems enforcing the sanctuary status of the cays.

■ Site description

Salt Cay is the southernmost inhabited island in the Turks, about 10 km south-west of Grand Turk. Salt Cay Creek and Salinas IBA embraces the natural creek area on the south-east side of Salt Cay and the abandoned salinas (saltpans) throughout the island, many of which are viewable from the roads in and around the settlement of Balfour Town. The salinas support many historic relics of the abandoned salt production industry.

■ Birds

This IBA supports a globally significant breeding population of Least Tern *Sterna antillarum*, and regional important numbers of Brown Pelican *Pelecanus occidentalis*, Wilson's Plover *Charadrius wilsonia*, Laughing Gull *Larus atricilla* and Sandwich Tern *Sterna sandvicensis*. Large numbers of (wintering and migratory) shorebirds occur, with up to 2,500 wintering Stilt Sandpiper *Calidris himantopus* recorded. Three (of the four) Bahamas EBA restricted-range birds, namely Bahama Woodstar *Calliphlox evelynae*, Thick-billed Vireo *Vireo crassirostris* and Bahama Mockingbird *Mimus gundlachii*, occur.

■ Other biodiversity

The shores are probably important for nesting Critically Endangered hawksbill *Eretmochelys imbricata*, and Endangered green *Chelonia mydas* and leatherback *Caretta caretta* turtles. The Critically Endangered Turks and Caicos rock iguana *Cyclura carinata carinata* occurs but is rare.

■ Conservation

Salt Cay Creek and Salinas IBA is a mixture of crown and private land ownership. Parts of the saltpans are designated as a statutory area of historic interest (number 32). Salt Cay Creek is recognised locally as an informal sanctuary. However, this IBA needs nature reserve status with the boundary including both Salt Cay Creek and the salinas. With critical wetland areas right within the town, this would be an ideal focal area for development as an awareness-raising/ educational "watchable wildlife pond" (combining the historic interest with the wildlife feature).

US VIRGIN ISLANDS

LAND AREA **353 km²** ALTITUDE **0–477 m**
HUMAN POPULATION **106,000** CAPITAL **Charlotte Amalie**
IMPORTANT BIRD AREAS **9, totalling 62 km²**
IMPORTANT BIRD AREA PROTECTION **90%**
BIRD SPECIES **210**
THREATENED BIRDS **4** RESTRICTED-RANGE BIRDS **7**

JIM CORVEN (BRISTOL COMMUNITY COLLEGE)

Great Pond IBA on St Croix.
(PHOTO: LISA D. YNTEMA)

INTRODUCTION

The US Virgin Islands (USVI), an organised, unincorporated United States territory, are at the eastern end of the Greater Antillean chain of islands in the northern Caribbean Sea, and comprise three major islands and more than 50 offshore cays. As an archipelago, the Virgin Islands are politically divided between USVI (the south-western group of islands) and the British Virgin Islands (which stretch out to the north-east). However, St Croix (the largest of the three main islands at 217 km²) lies about 65 km to the south of the rest of the Virgin Islands and c.100 km south-east of mainland Puerto Rico. The two northern islands of St Thomas (74 km², and c.60 km east of mainland Puerto Rico) and St John (50 km²) are very hilly (high points of 477 m and 387 m respectively) with limited flat areas. St Croix is a much flatter and drier island, with fewer bays and offshore cays. Offshore cays collectively make up c.3% of the territory's area (12 km²).

The USVI climate is categorised as subtropical or dry tropical. Average annual rainfall is 750 mm in coastal areas and up to 1,400 mm at higher elevations. Annual rainfall patterns are erratic, but on average, peak rainfall periods are September–December with a brief wet "season" in May or June. The dry season is January–April. The resultant vegetation on the islands and cays is primarily dry forest, including closed-canopy forest, woodland and shrubland. Subtropical moist forest (with a continuous canopy and an abundant herbaceous understorey) is found in the uplands (where annual rainfall is above 1,200 mm), along drainage ghauts or streams and in some coastal basin areas (accounting for c.10–15% of the land area). USVI also supports grasslands, a variety of wetlands, and a full range of coastal zone habitats. However, all have been subject to human disturbance or development as well as natural effects of tropical storms. From the times of pre-ceramic aboriginal settlement to the present day, these ecosystems have been in a constant state of regeneration and succession. Virtually no primary forest exists on the islands. It has been replaced with secondary "new forest" that has regenerated following the cessation of plantation agriculture. Wetland habitats include salt ponds, salt flats, mangrove wetlands (the only legally protected habitat in USVI), mixed swamp, and freshwater ponds (all of which are man-made for livestock or landscaping). Despite their extremely high value to both humans and wildlife, many of the natural wetlands have been destroyed or lost to development.

Conservation

There are two primary pieces of legislation that apply to the protection of wildlife in USVI: the Virgin Islands Indigenous and Endangered Species Act (1990), and the federal US Endangered Species Act (1972) under which Brown Pelican *Pelecanus occidentalis*, Piping Plover *Charadrius melodus* and Roseate Tern *Sterna dougallii* are listed. A proposed modification of the territorial legislation (VI Endangered and Indigenous Species Act), with lists of USVI species considered under various categories of threat, is presented in the "Comprehensive Wildlife Conservation Strategy". The Division of Fish and Wildlife (DFW) is the key public agency for research and monitoring of birds and habitats in the USVI. They have directed or collaborated in extensive studies on all the islands that have documented the status of seabirds and, to a lesser extent other species. Working with US Dept. of Agriculture, DFW has undertaken eradications of rats and goats from important seabird breeding colonies on DFW-managed cays. They are also looking at exotic plant management with the National Park Service (NPS), and collaborate with US Coast Guard, US Fish and Wildlife Service (USFWS), US Geological Survey, The Nature Conservancy and others to implement their work. The NPS is responsible for all land and marine areas within the St John Park which it monitors regularly and within which it has completed numerous biological inventories. The University of the Virgin Islands (UVI) provides local logistics for a number of offshore projects, plus GIS and data management expertise, and technical advice. The author has also established landbird surveys based on numerous point counts. With support from Cornell Lab of Ornithology, a public website (www.ebird.org/usvi) has been made available for researchers and birders to submit observations and retrieve data regarding birds of the USVI. Other conservation and research initiatives on the islands include work by the Island Resources Foundation (IRF); field station facilities managed by the Clean Islands International at the Virgin Islands Environmental Resource Station; Christmas Bird Counts and Breeding Bird Surveys coordinated by the Audubon Society of St John and by unaffiliated individuals on St Croix; educational birding trips (for school groups and general public) sponsored by the Audubon Club of St Thomas and the St Croix Environmental Association (SEA); and numerous other conservation and awareness activities conducted by SEA, USFWS, the St Croix East End Marine Park and the Environmental Association of St Thomas.

Although the only wildlife habitat that is legally protected in the USVI is mangrove woodland, there are numerous protected areas provided by both territorial and federal governments, and private organisations including The Nature Conservancy, Island Resources Foundation and SEA. All federally owned cays are protected within the Virgin Islands National Park, Hassel Island National Monument, Buck Island National Monument, or as National Wildlife Refuges. The territorial government has designated Flat Cay, Little Flat Cay, Saba Island, and Turtledove Cay as wildlife reserves. More recently, legislation has been drafted that would result in all 33 cays owned by the territorial government being designated as wildlife sanctuaries. The 18 privately-owned cays are not protected and are therefore vulnerable to development and disturbance. However, USVI's small islands or cays are considered major seabird nesting habitat because of their natural isolation from land predators and (mostly) human disturbance. Projects have been implemented to eliminate rats (*Rattus* spp.), invasive plants, and goats from a number of these islands, benefiting not only the seabirds but also the sea-turtles that nest on the beaches of several cays, and other reptiles such as the globally threatened St Croix ground lizard *Ameiva polops* and the endemic Virgin Islands tree boa *Epicrates monensis granti*.

The islands' human population is growing and has caused the extensive loss and degradation of natural habitats, most severely on densely populated St Thomas, but with other islands suffering too. For example, Krause Lagoon—once the largest (260 ha) and most important mangrove wetland on the south side of St Croix—was destroyed in the 1960s to make way for industrial expansion. Industrial complexes and residential and commercial development are destroying and fragmenting native forest. Residential development, hotels, resorts, and marinas

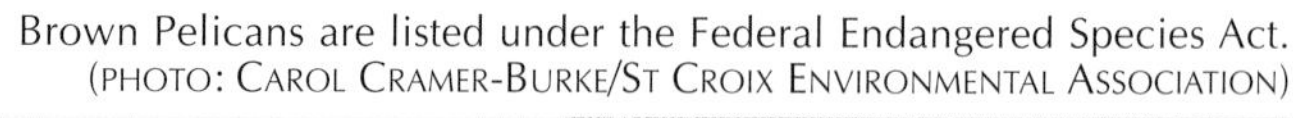
Brown Pelicans are listed under the Federal Endangered Species Act. (PHOTO: CAROL CRAMER-BURKE/ST CROIX ENVIRONMENTAL ASSOCIATION)

have been constructed on coastal wetlands, and marine recreational activities have damaged vital mangroves, coral reefs, and sea-grass beds. Human development has increased air and water pollution, and encouraged the introduction of exotic plants (for landscaping) and animal pests (especially domestic cats, dogs, and chickens that roam or become feral). Many domestic septic systems are not maintained or do not function and contaminate bays with sewage. This human-induced destruction and fragmentation has left ecosystems less able to absorb natural threats such as hurricanes (e.g. Hugo in 1989 and Marilyn 1995) and tropical storms. With rapid changes taking place to the natural environment there is a clear need for further habitat protection, increasing the level of environmental awareness (at government and public levels), monitoring the populations of key bird species, enhancing institutional resources and capacity, identifying critical conservation areas (not yet covered by the IBA network), and implementing conservation actions such as invasive species eradications and species reintroductions.

■ Birds

Of more than 210 species of birds recorded from USVI just 60 are resident breeding species and over 130 are migrants of which most are Nearctic breeders. Of the c.60 Nearctic–Neotropical landbird migrants recorded from the islands, 30 are vagrants. About 20 species of warbler over-winter on the islands, primarily in the mature continuous forest on St John (especially the upper elevations of north-facing dry forests), the Bordeaux Mountain moist forests on St Thomas, and the forested north-west quarter of St Croix. Seven of the 27 Puerto Rico and the Virgin Islands EBA restricted-range birds still occur in the islands (see Table 1), none of which is endemic to USVI. However, a number of restricted-range species have been extirpated from USVI including Puerto Rican Screech-owl *Megascops nudiceps* (last recorded in the 1930s) and Antillean Mango *Anthracothorax dominicus* (last recorded during the 1960s), probably reflecting the loss of primary forest throughout the islands.

Young White-crowned Pigeons—significant numbers breed at Great Pond IBA, but the territory's population is unknown. (PHOTO: CLAUDIA C. LOMBARD/USFWS)

Viable populations of two threatened species occur in USVI, namely the Near Threatened Caribbean Coot *Fulica caribaea* and White-crowned Pigeon *Patagioenas leucocephala* (see Table 1). The population of *P. leucocephala* is unknown on USVI (although along with the Bridled Quail-dove *Geotrygon mystacea* it appears to have partially recovered in recent years). *Fulica caribaea* occurs, albeit in small numbers, as a breeding species on St Croix with non-breeding birds found on the other main islands. They are outnumbered almost 2:1 by American Coot *F. americana*, with which they form mixed colonies. Populations of both species have almost certainly declined due to habitat destruction, degradation and, in some areas, hunting. Other threatened species include the Near Threatened Piping Plover *Charadrius melodus* which is accidental on the islands. The Vulnerable West Indian Whistling-duck *Dendrocygna arborea* formerly bred on St Croix, but unregulated hunting and poaching have extirpated

Table 1. Key bird species at Important Bird Areas in the US Virgin Islands.

			US Virgin Islands IBAs								
			VI001	VI002	VI003	VI004	VI005	VI006	VI007	VI008	VI009
		Criteria (threatened)								■	■
		Criteria (restricted-range)		■	■		■	■	■	■	■
Key bird species	**Criteria**	**National population** / Criteria (congregatory)	■			■			■	■	■
Red-billed Tropicbird *Phaethon aethereus*	■	675–900	300								
Brown Pelican *Pelecanus occidentalis*	■	900–1,050	300–600						900	25–150	
Masked Booby *Sula dactylatra*	■	75–210	300								
Brown Booby *Sula leucogaster*	■	>3,000	750–1,200								
Caribbean Coot *Fulica caribaea*	NT ■	35–70								34	
Laughing Gull *Larus atricilla*	■	6,000–9,000				2,400–3,000				250	
Royal Tern *Sterna maxima*	■	195–480				60–1,200				28	
Sandwich Tern *Sterna sandvicensis*	■	150–2,100				150–750					
Roseate Tern *Sterna dougallii*	■	2,320–6,775				60–1,800					
Least Tern *Sterna antillarum*	■	900–975								70–215	210–275
Bridled Tern *Sterna anaethetus*	■	1,200–3,000				150–600					
Sooty Tern *Sterna fuscata*	■	>90,000				60,000–120,000					
Brown Noddy *Anous stolidus*	■	1,800–2,400				300–1,200					
White-crowned Pigeon *Patagioenas leucocephala*	NT ■										100–175
Bridled Quail-dove *Geotrygon mystacea*	■								✓		
Green-throated Carib *Eulampis holosericeus*	■			✓	✓		✓	✓	✓	✓	✓
Antillean Crested Hummingbird *Orthorhyncus cristatus*	■			✓	✓		✓	✓	✓	✓	✓
Caribbean Elaenia *Elaenia martinica*	■						✓		✓	✓	✓
Puerto Rican Flycatcher *Myiarchus antillarum*	■								✓		
Pearly-eyed Thrasher *Margarops fuscatus*	■			✓	✓		✓	✓	✓	✓	
Lesser Antillean Bullfinch *Loxigilla noctis*	■							✓	✓		

All population figures = numbers of individuals.
Threatened birds: Near Threatened ■. **Restricted-range birds** ■.**Congregatory birds** ■.

The US Virgin Islands support a breeding population of around 900 Least Terns. (PHOTO: LISA D. YNTEMA)

it from USVI. Two individuals seen on the island in October 2002 (possibly vagrants from Puerto Rico) were the first documented in USVI since 1941. A single sighting was reported on St Thomas in April 2006. St Croix has sufficient wetlands to support reintroduction of this species. The 2005 Comprehensive Wildlife Conservation Strategy for USVI lists current and proposed endangered species, many of which are the key species listed in Table 1.

USVI is important for its breeding seabirds, with 15 species known to breed, mostly on the offshore cays. However, these seabird populations are threatened by predation from introduced rats *Rattus* spp., trampling by goats, human disturbance and illegal egging. Indirectly they are also being impacted by the depletion of fish stocks (from over fishing), bird entanglements in fishing lines, invasive plant species, and habitat loss. The Division of Fish and Wildlife (DFW) has eradicated rats from five cays (Saba Island, Dutchcap Cay, Congo Cay, Buck Island and Capella Island) and continues to monitor the presence of rats. DFW and USFWS have initiated a collaborative project to assess the status of Least Tern *Sterna antillarum* populations on St Croix and produce management recommendations. A *Sterna antillarum* nesting colony in the national park on St John is monitored and protected by the park staff.

IMPORTANT BIRD AREAS

USVI's nine IBAs—the territory's international priority sites for bird conservation—cover 62 km^2 (including marine areas). The majority (seven) of USVI's IBAs are protected under various designations. However, this protection does not seem to have afforded the sites immunity from a wide range of threats (described under each of the IBA profiles below) that still impinge on their long-term integrity. Increased enforcement of existing legislation appears to be necessary if the populations of birds at these IBAs (that make the sites internationally important) are to thrive.

The IBAs have been identified on the basis of 21 key bird species (listed in Table 1) that variously meet the IBA criteria. These 21 species include two globally threatened birds, all seven extant restricted-range species, and 12 congregatory seabirds and waterbirds. Only St John IBA (VI007) embraces populations of all the restricted-range species, Bridled Quail-dove *Geotrygon mystacea* and Puerto Rican Flycatcher *Myiarchus antillarum* not being found in any other IBA in the islands. The globally threatened species are only found in criteria-triggering populations in one IBA each—namely Southgate and Green Cay IBA (VI008) for the Caribbean Coot *Fulica caribaea*, and Great Pond IBA (VI009) for the White-crowned Pigeon *Patagioenas leucocephala*. With further information (e.g. concerning the numbers of breeding *P. leucocephala* on Ruth Island, St Croix) additional IBAs for these species may be identified. Five of the IBAs have been identified on the basis of their globally and regionally significant seabird and waterbird populations. However, other IBAs such as Magens Bay (VI005) support unknown (but seemingly large) numbers of seabirds. With further survey work these sites may prove to qualify as IBAs for their seabird and waterbird populations as well.

Southgate Pond IBA in St Croix, important for a range of congregatory waterbirds. (PHOTO: LISA D. YNTEMA)

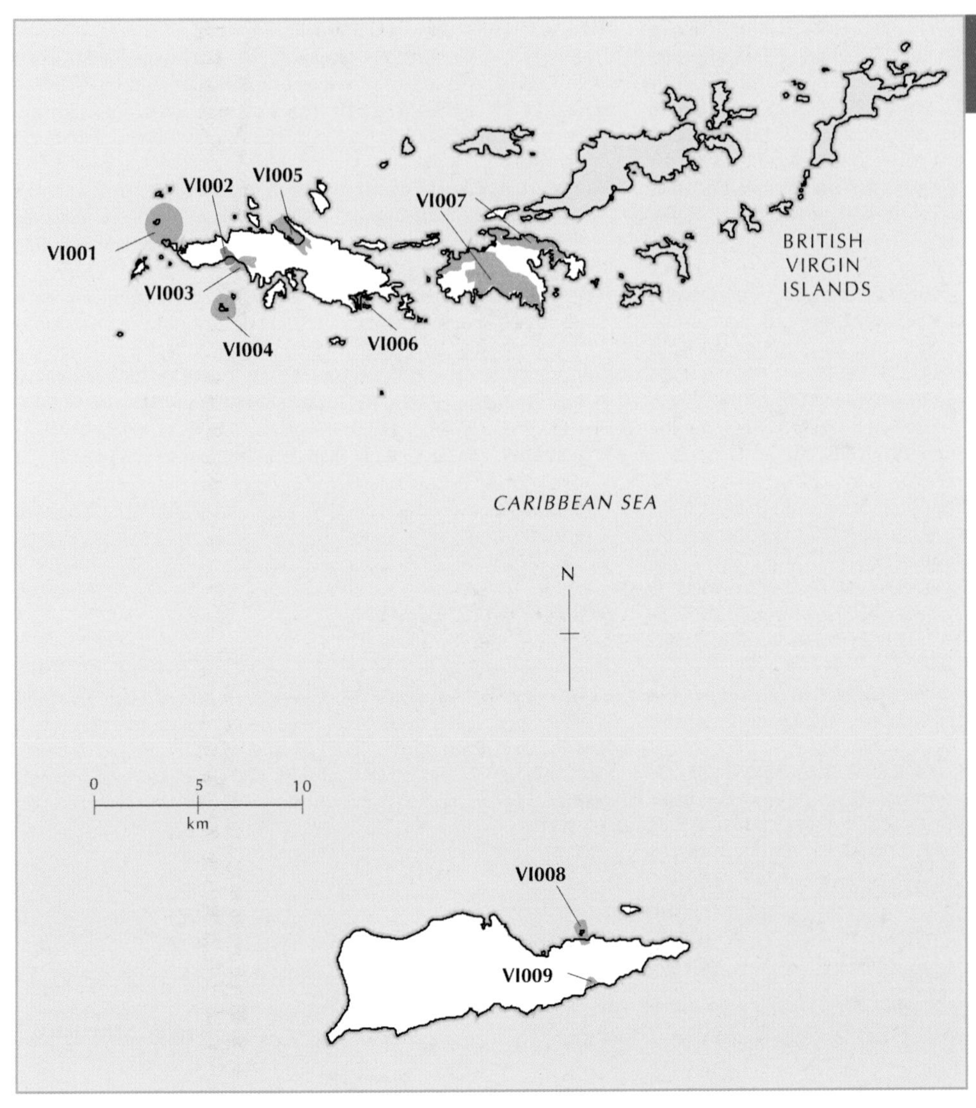

Figure 1. Location of Important Bird Areas in the US Virgin Islands.

The existing monitoring programs for seabirds and some other species could usefully be expanded to fill population status data gaps for key species (and any additional birds on the USVI Endangered species list) at IBAs, and then monitor their status. Monitoring results should be used to inform the annual assessment of state, pressure and response scores at each of the territory's IBAs to provide an objective status assessment and highlight management interventions that might be required to maintain these internationally important biodiversity sites.

KEY REFERENCES

Boulon, R. H. and Griffin, D. M. (1999) *Shoreline guide to the U.S. Virgin Islands.* St Thomas, USVI: Dept. Planning and Natural Resources, Div. Fish and Wildlife.

Dammann, A. E. and Nellis, D. W. (1992) *A natural history atlas to the cays of the U.S. Virgin Islands.* Sarasota, Florida: Pineapple Press, Inc.

Gaines, A. G. and Crawford, R. E. (2004) *Southgate Pond: geology and ecology of a tropical coastal pond.* Woods Hole, Mass.: The Coast and Harbor Institute (SCR Techn. Report 2).

Gibney, E., Thomas, T., O'Reilly, R. and Devine, B. (2000) *USVI vegetation community classification system.* St Thomas, USVI: University of the Virgin Islands Conservation Data Center [available at: http://cdc.uvi.edu/reaweb/vegbody.html].

Gladfelter, W. B. and Gladfelter, E. H. (2004) Birds *of Southgate Coastal Reserve: the effects of season and pond water levels as controlling factors in community structure.* Woods Hole, Mass.: Woods Hole Oceanographic Institute (SCR Techn. Report 2).

Island Resources Foundation (1993) Great Pond and Great Pond Bay, Area of Particular Concern and Area of Preservation and Restoration: a comprehensive analytical study. St Thomas, USVI: Dept. Planning and Natural Resources. (Unpublished report).

Island Resources Foundation (1993) Southgate Pond/Chenay Bay, Area of Particular Concern and Area of Preservation and Restoration: a comprehensive analytical study. St Thomas, USVI: Dept. Planning and Natural Resources. (Unpublished report).

Lombard, C. D. (2007) Nesting ecology and conservation of Least Terns in St Croix, U.S. Virgin Islands. Raleigh, North Carolina: North Carolina State University. Unpublished Master of Science thesis).

McLaughlin, M. ed. (1976) *Shoreline and marine associations of the Virgin Islands.* St Thomas, USVI: Island Resources Foundation (DCCA Environmental Fact Sheet 4).

McNair, D. B. (2006) Review of the status of American *(Fulica americana)* and Caribbean Coot *(Fulica caribaea)* in the United States Virgin Islands. *North American Birds* 59: 680–686.

McNair, D. B. (2006) Historical breeding distribution and abundance of the White-crowned Pigeon *(Patagioenas leucocephala)* on St. Croix, US Virgin Islands. *J. Carib. Orn.* 19: 1–7.

McNair, D. B. and Cramer-Burke, C. (2006) Breeding ecology of American and Caribbean Coots at Southgate Pond, St Croix: use of woody vegetation. *Wilson Bull.* 118: 208–217.

McNair, D. B. and Lombard, C. D. (2004) Population estimates, habitat associations, and management of *Ameiva polops* (Cope) at Green Cay, U.S. Virgin Islands. *Carib. J. Sci.* 40: 353–361.

McNair, D. B. and Lombard, C. D. (2006) Ground versus above-ground nesting of columbids on the satellite cays of St Croix, U.S. Virgin Islands. *J. Carib. Orn.* 19: 8–11.

McNair, D. B. and Sladen, F. W. (2007) Historical and current status of the Cattle Egret in the U.S. Virgin Islands, and management considerations. *J. Carib. Orn.* 20: 7–16.
McNair, D. B., Yntema, L. D. and Cramer-Burke, C. (2006) Use of waterbird abundance for saline wetland site prioritization on St Croix, U.S. Virgin Islands. *Carib. J. Sci.* 42: 220–230.
McNair, D. B., Yntema, L. D., Cramer-Burke, C. and Fromer, S. L. (2006) Recent breeding records and status review of the Ruddy Duck *(Oxyura jamaicensis)* on St Croix, U.S. Virgin Islands. *J. Carib. Orn.* 19: 91–96.
McNair, D. B., Yntema, L. D., Lombard, C. D., Cramer-Burke, C. and Sladen, F. W. (2005) Records of rare and uncommon birds from recent surveys on St Croix, United States Virgin Islands. *North American Birds* 59: 539–551.
Pierce, J. J. (1996) Survey of cay nesting avifauna in the U. S. Virgin Islands. St Thomas, USVI: Div. Fish and Wildlife, Dept. Planning and Natural Resources (Final report for the Pittman-Robertson Wildlife Restoration Aid Grant W5-11, Study 2).
Pierce, J. J. (in press) United States Virgin Islands. In Bradley P. E. and Norton, R. L. eds. *Breeding seabirds of the Caribbean*. Gainesville, Florida: Univ. Florida Press.
Platenberg, R. J., Hayes, F. E., McNair, D. B. and Pierce, J. J. (2005) *Comprehensive wildlife conservation strategy for the U.S. Virgin Islands*. St Thomas, USVI: Div. Fish and Wildlife, Dept. Planning and Natural Resources.
Raffaele, H. A. (1989) *A guide to the birds of Puerto Rico and the Virgin Islands* (Revised edition). Princeton, New Jersey: Princeton University Press.
Raffaele, H. Wiley J., Garrido, O., Keith, A. and Raffaele, J. (1998) *A guide to the birds of the West Indies*. Princeton, New Jersey: Princeton University Press.
Rennis, D. (2005) Jurisdictional delineation of the Southgate Coastal Reserve, St Croix, U.S. Virgin Islands. Christiansted, St Croix: St Croix Environmental Association (Unpublished report).
Sladen, F. W. (1992) Abundance and distribution of waterbirds in two types of wetland on St. Croix, U.S. Virgin Islands. *Ornitología Caribeña* 3: 35–42.
The Nature Conservancy (2005) A survey of the plants, birds, reptiles and amphibians at the Magen's Bay Preserve, St Thomas, U.S. Virgin Islands. St Thomas, USVI: Div. Fish and Wildlife, Dept. Planning and Natural Resources.
Weiss, M. P. and Gladfelter, W. B. (1978) A pre-Columbian conch midden, St Croix, U.S. Virgin Islands. *J. Virgin Islands Archaeological Soc.* 6: 23–30.
Woodbury, R. O. and Vivaldi, J. L. (1982) *The vegetation of Green Cay*. Boquerón, Puerto Rico: Caribbean Islands National Wildlife Refuge (US Fish and Wildlife Service Report CI-0076-1).
Wiley, J. W. (2002) Ecology, behavior, and conservation of the St Croix Ground Lizard (*Ameiva Polops*)—an endangered species. Unpublished proposal to Div. Fish and Wildlife, St Thomas, USVI.

ACKNOWLEDGEMENTS

The author would like to thank Carol Cramer-Burke (St Croix Environmental Association), Claudia Lombard (US Fish and Wildlife Service), Douglas McNair (Sapphos Environmental, Inc.), Judy Pierce (DPNR, Division of Fish and Wildlife) and Lisa Yntema for compiling individual IBA profiles and for contributing to this chapter.

VI001 North-west Cays

Wildlife Sanctuary

COORDINATES 18°22′N 65°02′W
ADMIN REGION St Thomas
AREA 1,185 ha
ALTITUDE 0 m
HABITAT Rocky cliffs, coast, shrubland, dry woodland

Masked Booby

THREATENED BIRDS
RESTRICTED-RANGE BIRDS
BIOME-RESTRICTED BIRDS
CONGREGATORY BIRDS ✓

■ Site description

North-west Cays IBA comprises Cockroach, Sula and Dutchcap Cays (totalling 22 ha) off the north-west coast of St Thomas. Cockroach and Sula are located c.10 km off the north-west coast, while Dutchcap Cay is closer to the main island, c.4 km off Botany Point, the westernmost end of St Thomas. There are steep cliffs on all sides of the islands except the eastern end of Cockroach where a flat shield slopes into the sea, leaving smaller Sula Cay separated by a large crevice. Dutchcap is dome-shaped with steep cliffs on the north and east faces. There are no sandy beaches or coastal plains on any of the cays, and no permanent water sources. Offshore from the cays are modest-sized coral reefs.

■ Birds

This IBA is important for seabirds and waterbirds. More than 100 pairs of Red-billed Tropicbird *Phaethon aethereus* breed in this IBA, with the population on Cockroach Cay being the largest in the US Virgin Islands. Breeding populations of Masked Booby *Sula dactylatra*, Brown Booby *S. leucogaster*, and Brown Pelican *Pelecanus occidentalis* are regionally important. Other seabirds breed on the islands and Dutchcap is the only nesting site for Red-footed Booby *S. sula* in the territory.

■ Other biodiversity

Critically Endangered hawksbill *Eretmochelys imbricata* and Endangered green *Chelonia mydas* turtles are common in the waters around the cays. A number of non-threatened terrestrial reptiles occur.

■ Conservation

The North-west Cays are state owned and subject to management and protection by the Department of Planning and Natural Resources—Division of Fish and Wildlife (DPNR-DFW). Sanctuary signs on the important seabird cays are maintained to limit foot traffic into the seabird colonies and to inform the public of the conservation restrictions, but their effectiveness is questionable. Insufficient enforcement of environmental laws is impeding effective conservation management. Goats introduced to Dutchcap by fishermen were eradicated in 2003. Control of rats has also taken place on Dutchcap and post-control monitoring occurs annually. Nesting seabirds are monitored annually by DPNR-DFW. Routine patrols have been maintained at the important breeding sites when colonies are most vulnerable.

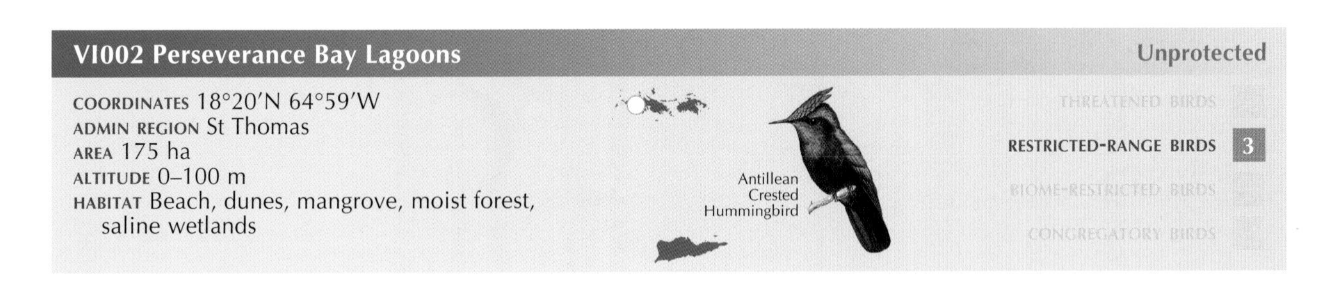

■ Site description

Perseverance Bay Lagoons IBA is on the south-west coast of St Thomas. The bay is along the coast to the west of John Brewer's Bay IBA (VI003). The site includes the sandy beach of Perseverance Bay, two conjoined and mangrove-fringed salt ponds/ lagoons (<2 ha each) about 50 m inland from the beach, and a flat moist-forest area extending c.100 m north (towards the main road) where it transitions into a steep slope of rocky terrain. Between the beach and ponds are sand dunes supporting littoral woodland. A narrow footpath from the road to the lagoons is maintained by local fishermen, birdwatchers, and beach visitors. A 2–3 m wide ghaut runs down the slope, coinciding with a footpath at several points. The derelict ruins of a sugar cane estate refinery lie within the IBA.

■ Birds

This IBA is important for range restricted species with three (of the seven) Puerto Rico and the Virgin Islands EBA restricted-range birds occurring, namely Green-throated Carib *Eulampis holosericeus*, Antillean Crested Hummingbird *Orthorhyncus cristatus* and Pearly-eyed Thrasher *Margarops fuscatus*. The forest areas provide good habitat for wintering Neotropical migrant warblers (and many resident species) and the salt ponds support a wide range of waterbirds.

■ Other biodiversity

No threatened or endemic species are known to occur.

■ Conservation

Perseverance Bay is privately owned (except the beachfront area) and completely undeveloped at present, although subject to potential development by the owner(s). Clearance of vegetation for any purpose in the upper reaches of the IBA would subject the slopes and lagoons to severe erosion and sedimentation. Current use of the IBA is very limited because of difficult access, but the isolated beach and near-shore coral reefs makes for an attractive site for beach-goers and snorkelling. Maintenance and improvement of the mangrove forest and littoral woodland is vital to conserving the salt ponds. Browsing by the introduced white-tailed deer *Odocoileus virginianus* may be a threat for understorey vegetation, especially the mangroves nearest the ponds. The introduction of the invasive wild pineapple has had a profound impact on the area's plant ecology and may have long-term consequences for the native vegetation.

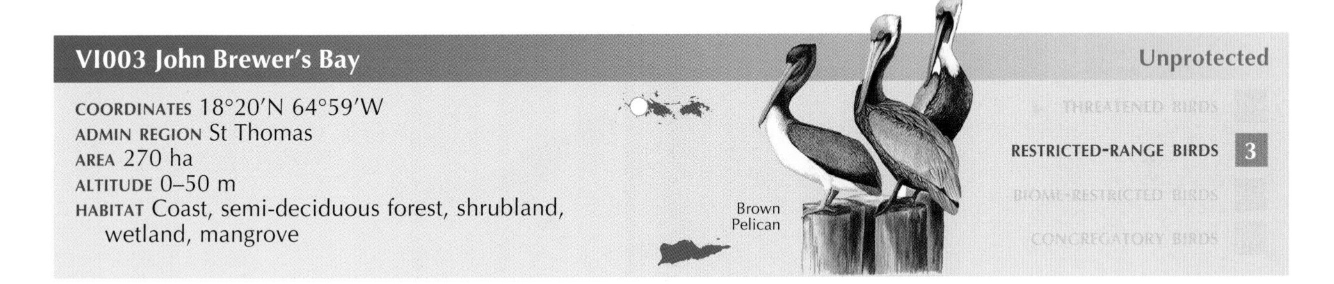

■ Site description

John Brewer's Bay IBA is on the south-central coast of St Thomas from Black Rock Point to the western end of the Cyril King airport runway, and includes the University of the Virgin Islands (UVI) campus. It includes the sandy beach of Brewer's Beach, Range Cay, Range Cay lagoon, the marsh along the north side of the airport, and the arid semi-deciduous forested catchment inland from the bay. The uplands extend inland and up the slopes above the UVI campus. The marsh by the airport drains into a shallow, tidal mangrove lagoon that is open to the bay. Perseverance Bay Lagoons IBA (VI002) is the next bay along the coast to the west.

■ Birds

This IBA is important for range restricted birds, with three (of the seven) Puerto Rico and the Virgin Islands EBA restricted-range species occurring. Unknown numbers of Brown Pelican *Pelecanus occidentalis*, Brown Booby *Sula leucogaster*, Magnificent Frigatebird *Fregata magnificens* and terns (*Sterna* spp.) nest on the offshore cays.

■ Other biodiversity

The Endangered green turtle *Chelonia mydas* is common in the bay which also supports coral reef and sea grass beds.

■ Conservation

John Brewer's Bay IBA is state owned by UVI, and subject to both development and protection by the university. The waters of Brewer's Bay are managed by the Department of Planning and Natural Resources with restricted boat moorings and anchorages. The beach is managed by UVI as a public beach. The narrow marsh above the lagoon has restricted access for reasons of airport security. Beach litter has been a chronic problem in the past. Extensive short-grass landscaping on the UVI campus results in soil erosion during heavy rains which then flows into Brewer's Bay turning it brown with sediment and covers the remnants of the coral reef near Range Cay. Any further clearance of forest or construction on steep slopes will threaten the biodiversity of the bay through sedimentation. A municipal sewage outfall extending southward from near the airport may be impacting the bay which is also impacted by oil and chemical run-off from the airport. Invasive alien predators are present. The effects of tropical storms and hurricanes could be mitigated by the maintenance and improvement of the mangrove and beach woodlands—vital to stabilising and protecting this area.

Sooty Tern

■ Site description

Saba Island and Cays IBA is located off the south-central coast of St Thomas, c.2 km south-south-west of the western end of the Cyril King airport runway extension. The IBA includes Saba Island plus the smaller Turtledove Cay and Flat Cay (totalling c. 14 ha). The smaller cays are flat and low, while Saba Island rises to about 80 m. Saba has two salt ponds, a coral-rubble shoreline on its north side (with a protected, sandy cove in the north-west) and rocky cliffs on the south. A shallow sandbar projects north to Turtledove Cay. Offshore of both Saba and Flat Cay are modest sized coral reefs and sea-grass beds.

■ Birds

This IBA is important for breeding seabirds and waterbirds. Between 20,000 and 40,000 pairs of Sooty Tern *Sterna fuscata* nest on these islands, as do globally significant numbers of Roseate Tern *S. dougallii*, Bridled Tern *S. anaethetus* and Laughing Gull *Larus atricilla*. A number of other tern species breed, as do small numbers of Audubon's Shearwater *Puffinus lherminieri*. The IBA supports a wide diversity of waterbirds.

■ Other biodiversity

Critically Endangered hawksbill *Eretmochelys imbricata* and Endangered green *Chelonia mydas* turtles are common in the waters around the cays. The locally threatened slipperyback skink *Mabuya sloanii* occurs, as do a number of reptiles endemic to the "Puerto Rican bank".

■ Conservation

Saba Island and Cays IBA is state owned, designated as a wildlife sanctuary, and managed by the Department of Planning and Natural Resources—Division of Fish and Wildlife (DPNR-DFW). A nature trail was developed on Saba some years ago behind the beach near the west pond, but this has now been abandoned. A bird observation blind is located near the east pond shore and overlooks the pond and its environs. Entry beyond the beach and bird blind is by special-use permit only. Current use of the cays is very limited to diving/snorkelling the coral areas and occasional visitors to the beach. The municipal sewage outfall that extends southward from the vicinity of the airport releases waste that can flow towards the island, cays, and coral reefs. The threat of invasive alien plants and animals establishing themselves is ever present. The house mouse *Mus musculus* is known to be present and presumably predates seabird eggs and chicks.

VI005 Magens Bay

Area of Particular Concern

COORDINATES 18°22'N 64°56'W
ADMIN REGION St Thomas
AREA 570 ha
ALTITUDE 0–75 m
HABITAT Beach, wetland, shrubland, mangrove, deciduous dry and moist forests

THREATENED BIRDS
RESTRICTED-RANGE BIRDS 4
BIOME-RESTRICTED BIRDS
CONGREGATORY BIRDS

Pearly-eyed Thrasher

■ Site description

Magens Bay IBA is on the north-central coast of St Thomas, opening to the Atlantic in a north-westerly direction. The site comprises a sandy beach, coastal wetland with mangroves, and woodlands further inland. The 129-ha Magens Bay Preserve is within the IBA and includes an arboretum, public beach and nature trail. The IBA is also home to some unique archaeological remains. Surrounding the area are various developments (including residential), a golf course and other recreational activities.

■ Birds

This IBA is important for range restricted birds, with four (of the seven) Puerto Rico and the Virgin Islands EBA restricted-range species occurring. Near Threatened White-crowned Pigeons *Patagioenas leucocephala* nest in the mangroves although numbers are unknown. Also, unknown numbers of Brown Pelican *Pelecanus occidentalis*, Brown Booby *Sula leucogaster*, Magnificent Frigatebird *Fregata magnificens* and terns (*Sterna* spp.) nest on the offshore cays.

■ Other biodiversity

Critically Endangered hawksbill *Eretmochelys imbricata* and Endangered green *Chelonia mydas* turtles have been seen in Magens Bay. The Endangered yellow mottled coqui (or mute frog) *Eleutherodactylus lentus* is found in the woodlands. Notable plants include Egger's cockspur *Erythrina eggersii* (Endangered), *Chrysophyllum pauciflorum* (Vulnerable), and the range restricted Bull's foot orchid *Psychilis macconelliae*.

■ Conservation

Magens Bay Preserve (also designated an Area of Particular Concern) is co-owned and managed by The Nature Conservancy, the Magens Bay Authority and the government (which owns areas outside the preserve). Efforts have been made to restore the arboretum and archaeological sites in the preserve. However, current use and development is oriented towards public recreation rather than conservation or education. Landscaping along the beach has not considered removal of invasive or exotic species. Loss of soil stabilising vegetation on the steep slopes above the preserve to development has already resulted in damaging erosion and runoff into the bay. Mangroves are critical to the stabilisation and filtering of slope runoff and are under growing threat from invasive plants where disturbance and openings have occurred. Magens Bay is exposed to tropical storms that could damage the beach, mangroves, or forests. Maintenance and restoration of the natural vegetation is vital to stabilising and protecting this area long-term.

VI006 Mangrove Lagoon

Wildlife Sanctuary and Marine Preserve

COORDINATES 18°19′N 64°52′W
ADMIN REGION St Thomas
AREA 225 ha
ALTITUDE 0–3 m
HABITAT Mangrove, shrubland, rocky coast, cays

White-crowned Pigeon

THREATENED BIRDS
RESTRICTED-RANGE BIRDS 4
BIOME-RESTRICTED BIRDS
CONGREGATORY BIRDS

■ Site description

Mangrove Lagoon IBA is on the south-eastern coast of St Thomas. It embraces several mangrove cays such as Bovoni Cay, Patricia Cay and Cas Cay and the lagoons (bays) around them. The cays are surrounded by 1–2 m deep lagoons that open to the sea through several channels. The St Thomas shoreline to this IBA starts in the south with volcanic cliffs and then extends into a flat mangrove-lined area which is broken by marinas with docks and moorings, and a major landfill (dump) at Bovoni along the north-west shoreline. Turpentine Run (an intermittent stream) flows into the lagoon next to a horse race track. The cays are rocky and sparsely vegetated with brush, sea grape, manchineel and organ pipe cactus.

■ Birds

This IBA is important for EBA species, with four (of the seven) Puerto Rico and the Virgin Islands EBA restricted-range birds occurring, namely Green-throated Carib *Eulampis holosericeus*, Antillean Crested Hummingbird *Orthorhyncus cristatus*, Pearly-eyed Thrasher *Margarops fuscatus* and Lesser Antillean Bullfinch *Loxigilla noctis*. Between 20 and 30 Near Threatened White-crowned Pigeons *Patagioenas leucocephala* nests throughout mangroves. A range of waterbird species occurs.

■ Other biodiversity

No endemic or threatened species are known to occur although the marine life in the surrounding waters is diverse.

■ Conservation

The Mangrove Lagoon and Cas Cay Wildlife Sanctuary and Marine Preserve protect this (primarily) state-owned IBA. Boat moorings are restricted, and although access to the lagoons is open, it is limited to visitors with boats and kayaks. Turpentine Run discharges it's contaminated (with variable levels of heavy metals, raw sewage, and pathological bacteria) waters into the lagoon. The Bovoni dump—the largest dump on St Thomas may also leak contaminants into the lagoon. However, an advanced sewage treatment plant at Bovoni appears to have significantly reduced the issue of human sewage discharge into coastal waters. The marina contains numerous medium to large yachts, but no provision for the sewage and bilge wastes that are pumped from the boats into the lagoon. Informal monitoring of the mangroves and birds has been conducted by the tour guides of the eco-tour company located between the mouth of Turpentine Run and the marinas.

VI007 St John

National Park/Nature Preserve

COORDINATES 18°20′N 64°45′W
ADMIN REGION St John
AREA 2,978 ha
ALTITUDE 0–387 m
HABITAT Wetland, coast, mangrove, woodland, shrubland, forest

Bridled Quail-dove

THREATENED BIRDS
RESTRICTED-RANGE BIRDS 7
BIOME-RESTRICTED BIRDS
CONGREGATORY BIRDS ✓

■ Site description

St John is the smallest of the US Virgin Islands, over half of which is designated as a national park. The island has many freshwater and salt ponds, mangrove woodlands, coral reefs, rocky and sandy beaches and upland areas that support (mostly second-growth) moist, semi-evergreen forests, dry deciduous forests, thorn woodlands, and coastal scrublands. There are also areas of small agricultural plots and pasture lands, and a number of small towns. Almost the entire island was clear-cut to make way for sugarcane production during the colonial era.

■ Birds

This IBA is important for range restricted species and waterbirds. All seven Puerto Rico and the Virgin Islands EBA restricted-range birds occur in this IBA. A regionally important population of up to 300 pairs of Brown Pelican *Pelecanus occidentalis* nest. A wide range of seabirds nest on the offshore cays although numbers are not known. Similarly, the Near Threatened White-crowned Pigeon *Patagioenas leucocephala* nest in St John's mangroves and forest but numbers are not known.

■ Other biodiversity

Critically Endangered hawksbill *Eretmochelys imbricata* and Endangered green turtles *Chelonia mydas* may nest on St John's beaches. The Virgin Islands tree boa *Epicrates monensis granti* is an endemic, threatened subspecies. Threatened plants include the Critically Endangered Woodbury's machaonia *Machaonia woodburyana*, and the Endangered Eggers' cockspur *Erythrina eggersii*, St Thomas lidflower *Calyptranthes thomasiana* and St Thomas prickly-ash *Zanthoxylum thomasianum*.

■ Conservation

St John IBA is a mix of private and state ownership. The Virgin Islands National Park covers 2,835 ha (with the 168-ha Maho Bay Estate to be added when funds are available) and is managed by the US National Park Service. The Nature Conservancy owns and manages the 9-ha Coral Bay Preserve, and the Island Resources Foundation has the 17-ha Nancy Spire Nature Preserve. Current land-use and development outside the parks is oriented towards tourism, public recreation, and resort development. Development has been intense and continues to threaten valuable habitats. Introduced invasive alien plants, insects, fungi etc. are proving significant problems for conservationists. Introduced mammals are likewise impacting the native flora and fauna. Erosion and sedimentation is a continuing problem that demonstrates the critical need for maintenance and increased restoration of the mangrove wetlands, coral reefs, and beach-area vegetation.

VI008 Southgate and Green Cay

■ Site description

Southgate and Green Cay IBA is in north-east St Croix. Southgate is on the coast, and Green Cay lies about 400 m off the coast from Southgate. Southgate Coastal Reserve (41 ha) encompasses a salt pond and associated wetlands, littoral deciduous woodland along a beach berm, and upland grassland. A man-made causeway contains Southgate Pond on the western side and a beach berm on its northern side separates it from the sea. Southgate is bordered to the west by Green Cay Marina (and an adjacent gated community and Tamarind Reef Hotel), to the east by Chenay Bay Beach Resort, and to the south by Route 83. Green Cay is an uninhabited 6-ha island, volcanic in origin, saddle shaped (rising to 21 m in the south) and mostly surrounded by steep cliffs.

■ Birds

This IBA is significant for supporting up to 34 Near Threatened Caribbean Coot *Fulica caribaea*, and regionally important breeding populations of Brown Pelican *Pelecanus occidentalis* and Least Tern *Sterna antillarum*. In 2003, 357 pairs of *S. antillarum* were breeding but the nests were decimated by feral dogs. Significant numbers of *Pelecanus occidentalis*, Laughing Gull *Larus atricilla*, *S. antillarum* and Royal Tern *S. maxima* congregate at Southgate Pond. Four (of the seven) Puerto Rico and the Virgin Islands EBA restricted-range birds occur at this IBA.

■ Other biodiversity

The largest known population of the Critically Endangered St Croix ground lizard *Ameiva polops*—extirpated from St Croix—persists on Green Cay. In 2006, probable nests of the Critically Endangered hawksbill *Eretmochelys imbricata* (42 nests), Critically Endangered leatherback turtle *Dermochelys coriacea* (10) and Endangered green turtles *Chelonia mydas* (61) were documented from Southgate Coastal Reserve. Hawksbill and green turtles also nest on Green Cay.

■ Conservation

This IBA is a mix of private and federal ownership. Southgate Coastal Reserve was established by the St Croix Environmental Association which is waiting for a permit to develop visitation facilities in the reserve. Green Cay National Wildlife Refuge was established to protect the St Croix ground lizard *Ameiva polops* and important bird nesting habitat– it is closed to the public. The IBA is also variously designated as part of the Caribbean Barrier Resource System, an Area of Particular Concern, and is within and/or adjacent to the St Croix East End Marine Park. Green Cay is administered under the Caribbean Islands National Wildlife Refuge Complex. Invasive alien predators (including feral dogs, mongooses, rats and cats) are all impacting the birds in this IBA. Other threats are from human disturbance, invasive alien plant species, and hurricanes.

VI009 Great Pond

■ Site description

Great Pond IBA is a mangrove-fringed, saline lagoon situated on the south-eastern shore of St Croix. The IBA includes the pond and the adjoining, vegetated baymouth bar on its southern edge which separates the pond from Great Pond Bay. A narrow channel in the pond's south-east corner connects it to Great Pond Bay. Pond levels (and area) fluctuate as a result of rainfall and tidal flow, which in turn results in a wide range of salinity levels. At low water levels, mudflats are exposed around much of the pond, particularly along the western border. Mudflats are surrounded on the west and north sides by gently sloping, fallow pastures of dry grassland with mixed thorny scrub.

■ Birds

This IBA is important for threatened birds, restricted-range birds and congregatory species. Fifty-five pairs of Near Threatened White-crowned Pigeons *Patagioenas leucocephala* have been recorded breeding in the mangroves, and three (of the seven) Puerto Rico and the Virgin Islands EBA restricted-range birds occur. A regionally important 134 pairs of Least Tern *Sterna antillarum* have bred at this IBA (although the numbers breeding and their success depends on seasonal and annual fluctuations in water levels), and more than 200 birds regularly congregate post-breeding (before migrating). As many as 50 individuals winter and up to 10 pairs of Wilson's Plover *Charadrius wilsonia* breed, making this the most important site for the species on St Croix.

■ Other biodiversity

Critically Endangered hawksbill *Eretmochelys imbricata* and Endangered green *Chelonia mydas* turtles have nested along the ocean side of the baymouth bar.

■ Conservation

Great Pond IBA is territorially owned, while the surrounding lands belong to Golden Resorts, LLLP. The pond is protected within an Area of Particular Concern, the boundaries of which extend to the east and west of the pond, and south out to sea. Great Pond is also included within the Caribbean Barrier Resource System and the St Croix East End Marine Park. Great Pond IBA is threatened by development pressures (e.g. the Golden Resort proposal for a mega-resort), compounded by the lack of an adequate and enforceable buffer zone. Human disturbance, illegal dumping and invasive alien predators also impact the birds and habitats. Nesting shorebirds and terns are susceptible to rainfall run-off after heavy rains (which can also cause erosion and siltation) and the mangroves have been impacted by hurricanes in the past.

APPENDICES

Appendix 1a. IUCN Red List categories and criteria for threat status

The IUCN Red List is widely recognised as the most authoritative, objective and comprehensive system for evaluating the global conservation status of species and categorising them according to their risk of extinction. The latest release included assessments for over 40,000 species, spanning every country of the world, and including vertebrates (including all 9,990 birds), invertebrates, plants and fungi (see http://www.iucnredlist.org/). The IUCN Red List uses quantitative criteria based on population size, rate of decline, and area of distribution to assign species to one of the categories of relative extinction risk, ranging from "Extinct" to "Least Concern" (see below). Species in the categories Critically Endangered, Endangered, Vulnerable and Near Threatened have been considered in the identification of IBAs in the Caribbean, and together have been generically referred to as "threatened" in this publication. This table shows the IUCN Red List criteria used to determine the degree of threat for the species that trigger the "species of conservation concern" (A1) category of the IBA criteria.

IUCN RED LIST CATEGORIES

EXTINCT (EX)
Example, Cuban Macaw *Ara tricolor* is considered Extinct as there is no reasonable doubt that the last individual has died.

CRITICALLY ENDANGERED (CR)
Example, Ridgway's Hawk *Buteo ridgwayi* is considered to be facing an extremely high risk of extinction in the wild.

ENDANGERED (EN)
Example, Black-capped Petrel *Pterodroma hasitata* is considered to be facing a very high risk of extinction in the wild.

VULNERABLE (VU)
Example, Martinique Oriole *Icterus bonana* is considered to be facing a high risk of extinction in the wild.

NEAR THREATENED (NT)
Example, Grey-crowned Palm-tanager *Phaenicophilus poliocephalus* has been evaluated against the IUCN Red List criteria but does not qualify for Critically Endangered, Endangered or Vulnerable now, although it is close to qualifying for or is likely to qualify for a globally threatened category in the near future.

LEAST CONCERN (LC)
Example, Jamaican Tody *Todus todus* has been evaluated against the criteria and does not qualify for Critically Endangered, Endangered, Vulnerable or Near Threatened as it is widespread and abundant.

... Appendix 1a continued overleaf

Appendix 1a ... continued.

IUCN Red List categories and criteria (2001) Produced by BirdLife International

Type of criteria	Main criteria	Sub-criteria	Qualifiers	Codes
A. REDUCTION IN POPULATION SIZE	**1.** Reduction =90% in 10 years or 3 generations (**CR**) Reduction =70% in 10 years or 3 generations (**EN**) Reduction =50% in 10 years or 3 generations (**VU**)	**1.** Reduction in the past (observed, estimated, inferred or suspected), where the causes are clearly reversible AND understood AND ceased, *based on a-e opposite*:	**a.** Direct observation	**A1a**
			b. Index of abundance	**A1b**
			c. Decline in area of occupancy, extent of occurrence, and/or quality of habitat	**A1c**
			d. Actual or potential levels of exploitation	**A1d**
			e. Effects of introduced taxa, hybridization, pathogens, pollutants, competitors or parasites	**A1e**
	2. Reduction =80% in 10 years or 3 generations (**CR**) Reduction =50% in 10 years or 3 generations (**EN**) Reduction =30% in 10 years or 3 generations (**VU**)	**2.** Reduction in the past (observed, estimated, inferred or suspected), where the reduction or its causes may not be reversible OR understood OR have ceased, *based on a-e opposite*:	**a.** As a above	**A2a**
			b. As b above	**A2b**
			c. As c above	**A2c**
			d. As d above	**A2d**
			e. As e above	**A2e**
	3. Reduction =80% in 10 years or 3 generations (**CR**) *to 100 years max* Reduction =50% in 10 years or 3 generations (**EN**) *to 100 years max* Reduction =30% in 10 years or 3 generations (**VU**) *to 100 years max*	**3.** Reduction in the future (projected or suspected), *based on b-e opposite*	**b.** As b above	**A3b**
			c. As c above	**A3c**
			d. As d above	**A3d**
			e. As e above	**A3e**
	4. Reduction =80% in 10 years or 3 generations (**CR**) *to 100 years max* Reduction =50% in 10 years or 3 generations (**EN**) *to 100 years max* Reduction =30% in 10 years or 3 generations (**VU**) *to 100 years max*	**4.** Reduction includes the past and the future (observed, estimated, inferred, projected or suspected) where the reduction or its causes may not be reversible OR understood OR have ceased, *based on a-e opposite*:	**a.** As a above	**A2a**
			b. As b above	**A2b**
			c. As c above	**A2c**
			d. As d above	**A2d**
			e. As e above	**A2e**
B. SMALL RANGE fragmented, declining or fluctuating	**1.** Extent of occurrence estimated <100 km² (**CR**) *with at least two of a,b or c:* Extent of occurrence estimated <5,000 km² (**EN**) *with at least two of a, b or c:* Extent of occurrence estimated <20,000 km² (**VU**) *with at least two of a, b or c:*	**a.** Severely fragmented or At 1 location (**CR**) At =5 locations (**EN**) At =10 locations (**VU**)	None	**B1a**
		b. Continuing decline (observed, inferred or projected) *in any of i-v opposite*:	**i.** Extent of occurrence	**B1bi**
			ii. Area of occupancy	**B1bii**
			iii. Area, extent and/or quality of habitat	**B1biii**
			iv. Number of locations or subpopulations	**B1biv**
			v. Number of mature individuals	**B1bv**
		c. Extreme fluctuations *in any of i-iv opposite:*	**i.** Extent of occurrence	**B1ci**
			ii. Area of occupancy	**B1cii**
			iii. Area, extent and/or quality of habitat	**B1ciii**
			iv. Number of locations or subpopulations	**B1civ**
	2. Area of occupancy estimated <10 km² (**CR**) *with at least two of a, b or c:* Area of occupancy estimated <500 km² (**EN**) *with at least two of a, b or c:* Area of occupancy estimated <2000 km² (**VU**) *with at least two of a, b or c:*	**a.** As a above	None	**B2a**
		b. As b above *in any of i-v opposite*	**i.** Extent of occurrence	**B2bi**
			ii. Area of occupancy	**B2bii**
			iii. Area, extent and/or quality of habitat	**B2biii**
			iv. Number of locations or subpopulations	**B2biv**
			v. Number of mature individuals	**B2bv**
		c. As c above *in any of I-iv opposite*	**i.** Extent of occurrence	**B2ci**
			ii. Area of occupancy	**B2cii**
			iii. Area, extent and/or quality of habitat	**B2ciii**
			iv. Number of locations or subpopulations	**B2civ**
C. SMALL POPULATION declining or fluctuating	Population <250 mature individuals (**CR**) *and either 1 or 2* Population <2,500 mature individuals (**EN**) *and either 1 or 2* Population <10,000 mature individuals (**VU**) *and either 1 or 2*	**1.** Continuing decline =25% in 3 years or 1 generation (**CR**) *to 100 years max* Continuing decline =20% in 5 years or 2 generations (**EN**) *to 100 years max* Continuing decline =10% in 10 years or 3 generations (**VU**) *to 100 years max*	None	**C1**
		2. Continuing decline (observed, projected or inferred) and *a and/or b opposite:*	**a.i.** all sub-pops=50 (**CR**) all sub-pops =250 (**EN**) all sub-pops =1,000 (**VU**)	**C2ai**
			a.ii =90% mature individuals in 1 sub-pop (**CR**) =95% mature individuals in 1 sub-pop (**EN**) all individuals in 1 sub-pop (**VU**)	**C2aii**
			b. Extreme fluctuations in number of mature individuals	**C2b**
D1. VERY SMALL POPULATION	Population <50 mature individuals (**CR**) Population <250 mature individuals (**EN**) Population <1,000 mature individuals (**VU**)	None	None	**D1**
D2. VERY SMALL RANGE	Area of occupancy typically <20 km² or typically <6 locations (**VU** only—capable of becoming CR or EX in v. short time)	None	None	**D2**
E. QUANTITATIVE ANALYSIS	Probability of extinction in wild >50% in 10 years or 3 gens (**CR**) *to 100 years max* Probability of extinction in wild >20% in 20 years or 5 gens (**EN**) *to 100 years max* Probability of extinction in wild is 10% in 100 years (**VU**)	None	None	**E**

Appendix 1b. Species of global conservation concern in the Caribbean

IUCN Red List category of threat
■ Critically Endangered (CR)

Common name	Scientific name
Trinidad Piping-guan	*Pipile pipile*
Jamaica Petrel	*Pterodroma caribbaea*
Cuban Kite	*Chondrohierax wilsonii*
Ridgway's Hawk	*Buteo ridgwayi*
Grenada Dove	*Leptotila wellsi*
Puerto Rican Amazon	*Amazona vittata*
Jamaican Pauraque	*Siphonorhis americana*
Puerto Rican Nightjar	*Caprimulgus noctitherus*
Ivory-billed Woodpecker	*Campephilus principalis*
Bachman's Warbler	*Vermivora bachmanii*
Semper's Warbler	*Leucopeza semperi*
Montserrat Oriole	*Icterus oberi*
CR Total	**12**

IUCN Red List category of threat
■ Endangered (EN)

Common name	Scientific name
Bermuda Petrel	*Pterodroma cahow*
Black-capped Petrel	*Pterodroma hasitata*
Gundlach's Hawk	*Accipiter gundlachi*
Zapata Rail	*Cyanolimnas cerverai*
Blue-headed Quail-dove	*Starnoenas cyanocephala*
Imperial Amazon	*Amazona imperialis*
Bay-breasted Cuckoo	*Coccyzus rufigularis*
Giant Kingbird	*Tyrannus cubensis*
Cuban Palm Crow	*Corvus minutus*
Zapata Wren	*Ferminia cerverai*
White-breasted Thrasher	*Ramphocinclus brachyurus*
La Selle Thrush	*Turdus swalesi*
Red Siskin	*Carduelis cucullata*
Hispaniolan Crossbill	*Loxia megaplaga*
Whistling Warbler	*Catharopeza bishopi*
Jamaican Blackbird	*Nesopsar nigerrimus*
Yellow-shouldered Blackbird	*Agelaius xanthomus*
Cuban Sparrow	*Torreornis inexpectata*
St Lucia Black Finch	*Melanospiza richardsoni*
EN total	**19**

IUCN Red List category of threat
■ Vulnerable (VU)

Common name	Scientific name
West Indian Whistling-duck	*Dendrocygna arborea*
Ring-tailed Pigeon	*Patagioenas caribaea*
Grey-headed Quail-dove	*Geotrygon caniceps*
Cuban Parakeet	*Aratinga euops*
Hispaniolan Parakeet	*Aratinga chloroptera*
Yellow-billed Amazon	*Amazona collaria*
Hispaniolan Amazon	*Amazona ventralis*
Black-billed Amazon	*Amazona agilis*
Yellow-shouldered Amazon	*Amazona barbadensis*
St Lucia Amazon	*Amazona versicolor*
Red-necked Amazon	*Amazona arausiaca*
St Vincent Amazon	*Amazona guildingii*
Fernandina's Flicker	*Colaptes fernandinae*
White-necked Crow	*Corvus leucognaphalus*
Bahama Swallow	*Tachycineta cyaneoviridis*
Golden Swallow	*Tachycineta euchrysea*
Bicknell's Thrush	*Catharus bicknelli*
Forest Thrush	*Cichlherminia lherminieri*
Cerulean Warbler	*Dendroica cerulea*
Elfin-woods Warbler	*Dendroica angelae*
White-winged Warbler	*Xenoligea montana*
Martinique Oriole	*Icterus bonana*
Chat Tanager	*Calyptophilus frugivorus*
VU total	**23**

IUCN Red List category of threat
■ Near Threatened (NT)

Common name	Scientific name
Northern Bobwhite	*Colinus virginianus*
Black Rail	*Laterallus jamaicensis*
Caribbean Coot	*Fulica caribaea*
Buff-breasted Sandpiper	*Tryngites subruficollis*
Piping Plover	*Charadrius melodus*
White-crowned Pigeon	*Patagioenas leucocephala*
Plain Pigeon	*Patagioenas inornata*
Crested Quail-dove	*Geotrygon versicolor*
Cuban Amazon	*Amazona leucocephala*
Least Pauraque	*Siphonorhis brewsteri*
White-tailed Sabrewing	*Campylopterus ensipennis*
Bee Hummingbird	*Mellisuga helenae*
Hispaniolan Trogon	*Priotelus roseigaster*
Guadeloupe Woodpecker	*Melanerpes herminieri*
Olive-sided Flycatcher	*Contopus cooperi*
Blue Mountain Vireo	*Vireo osburni*
Hispaniolan Palm Crow	*Corvus palmarum*
Cuban Solitaire	*Myadestes elisabeth*
Golden-winged Warbler	*Vermivora chrysoptera*
Barbuda Warbler	*Dendroica subita*
Kirtland's Warbler	*Dendroica kirtlandii*
Vitelline Warbler	*Dendroica vitellina*
St Lucia Oriole	*Icterus laudabilis*
Painted Bunting	*Passerina ciris*
Grey-crowned Palm-tanager	*Phaenicophilus poliocephalus*
NT total	**25**

Total number of species of global conservation concern in the Caribbean	**79**

Note: This list is based on the 2006 IUCN Red List and was used by national IBA coordinators and authors as the basis for identifying IBAs against the Species of global conservation concern criterion (A1). However, BirdLife International (on behalf of IUCN) updates the Red List for birds on an annual basis. These updates (there were several relevant to the Caribbean in 2008) and factsheets for all of these "threatened" species are available at the BirdLife Data Zone http:/ /www.birdlife.org/datazone/species.

Appendix 2. Restricted-range species in Caribbean Endemic Bird Areas

EBA 025 Cuba

Countries Cuba
Key habitats Dry forest, pine forest, arid scrub, wetlands
Area 110,000 km²
Altitude 0–1,500 m

Common name	Scientific name	Threat category
Zapata Rail	*Cyanolimnas cerverai*	EN
Cuban Macaw	*Ara tricolor*	EX
Thick-billed Vireo	*Vireo crassirostris*	LC
Cuban Palm Crow	*Corvus minutus*	EN
Zapata Wren	*Ferminia cerverai*	EN
Cuban Gnatcatcher	*Polioptila lembeyei*	LC
Bahama Mockingbird	*Mimus gundlachii*	LC
Olive-capped Warbler	*Dendroica pityophila*	LC
Yellow-headed Warbler	*Teretistris fernandinae*	LC
Oriente Warbler	*Teretistris fornsi*	LC
Red-shouldered Blackbird	*Agelaius assimilis*	LC
Cuban Sparrow	*Torreornis inexpectata*	EN

EBA 026 Bahamas

Countries Bahamas, Turks and Caicos Islands
Key habitats Arid scrub, pine forest, dry forest
Area 14,000 km²
Altitude 0–60 m

Common name	Scientific name	Threat category
Brace's Emerald	*Chlorostilbon bracei*	EX
Bahama Woodstar	*Calliphlox evelynae*	LC
Thick-billed Vireo	*Vireo crassirostris*	LC
Bahama Swallow	*Tachycineta cyaneoviridis*	VU
Bahama Mockingbird	*Mimus gundlachii*	LC
Pearly-eyed Thrasher	*Margarops fuscatus*	LC
Olive-capped Warbler	*Dendroica pityophila*	LC
Bahama Yellowthroat	*Geothlypis rostrata*	LC

EBA 027 Jamaica

Countries Jamaica
Key habitats Lowland and montane rain forest, limestone forest
Area 11,000 km²
Altitude 0–2,000 m

Common name	Scientific name	Threat category
Ring-tailed Pigeon	*Patagioenas caribaea*	VU
Crested Quail-dove	*Geotrygon versicolor*	NT
Yellow-billed Amazon	*Amazona collaria*	VU
Black-billed Amazon	*Amazona agilis*	VU
Chestnut-bellied Cuckoo	*Coccyzus pluvialis*	LC
Jamaican Lizard-cuckoo	*Coccyzus vetula*	LC
Jamaican Owl	*Pseudoscops grammicus*	LC
Jamaican Pauraque	*Siphonorhis americana*	CR
Jamaican Mango	*Anthracothorax mango*	LC
Red-billed Streamertail	*Trochilus polytmus*	LC
Black-billed Streamertail	*Trochilus scitulus*	LC
Vervain Hummingbird	*Mellisuga minima*	LC
Jamaican Tody	*Todus todus*	LC
Jamaican Woodpecker	*Melanerpes radiolatus*	LC
Jamaican Becard	*Pachyramphus niger*	LC
Jamaican Elaenia	*Myiopagis cotta*	LC
Greater Antillean Elaenia	*Elaenia fallax*	LC
Jamaican Pewee	*Contopus pallidus*	LC
Sad Flycatcher	*Myiarchus barbirostris*	LC
Rufous-tailed Flycatcher	*Myiarchus validus*	LC
Stolid Flycatcher	*Myiarchus stolidus*	LC
Jamaican Vireo	*Vireo modestus*	LC
Blue Mountain Vireo	*Vireo osburni*	NT
Jamaican Crow	*Corvus jamaicensis*	LC

EBA 027 Jamaica ... continued

Common name	Scientific name	Threat category
Golden Swallow	*Tachycineta euchrysea*	VU
Bahama Mockingbird	*Mimus gundlachii*	LC
Rufous-throated Solitaire	*Myadestes genibarbis*	LC
White-chinned Thrush	*Turdus aurantius*	LC
White-eyed Thrush	*Turdus jamaicensis*	LC
Arrowhead Warbler	*Dendroica pharetra*	LC
Jamaican Oriole	*Icterus leucopteryx*	LC
Jamaican Blackbird	*Nesopsar nigerrimus*	EN
Yellow-shouldered Grassquit	*Loxipasser anoxanthus*	LC
Orangequit	*Euneornis campestris*	LC
Jamaican Spindalis	*Spindalis nigricephala*	LC
Jamaican Euphonia	*Euphonia jamaica*	LC

EBA 028 Hispaniola

Countries Dominican Republic, Haiti
Key habitats Lowland and montane rain forest, dry forest, pine forest
Area 76,000 km²
Altitude 0–3,000 m

Common name	Scientific name	Threat category
Ridgway's Hawk	*Buteo ridgwayi*	CR
Hispaniolan Parakeet	*Aratinga chloroptera*	VU
Hispaniolan Amazon	*Amazona ventralis*	VU
Bay-breasted Cuckoo	*Coccyzus rufigularis*	EN
Hispaniolan Lizard-cuckoo	*Coccyzus longirostris*	LC
Ashy-faced Owl	*Tyto glaucops*	LC
Least Pauraque	*Siphonorhis brewsteri*	NT
Hispaniola Nightjar	*Caprimulgus ekmani*	LC
Antillean Mango	*Anthracothorax dominicus*	LC
Hispaniolan Emerald	*Chlorostilbon swainsonii*	LC
Vervain Hummingbird	*Mellisuga minima*	LC
Hispaniolan Trogon	*Priotelus roseigaster*	NT
Narrow-billed Tody	*Todus angustirostris*	LC
Broad-billed Tody	*Todus subulatus*	LC
Antillean Piculet	*Nesoctites micromegas*	LC
Hispaniolan Woodpecker	*Melanerpes striatus*	LC
Greater Antillean Elaenia	*Elaenia fallax*	LC
Hispaniolan Pewee	*Contopus hispaniolensis*	LC
Stolid Flycatcher	*Myiarchus stolidus*	LC
Thick-billed Vireo	*Vireo crassirostris*	LC
Flat-billed Vireo	*Vireo nanus*	LC
Hispaniolan Palm Crow	*Corvus palmarum*	NT
White-necked Crow	*Corvus leucognaphalus*	VU
Palmchat	*Dulus dominicus*	LC
Golden Swallow	*Tachycineta euchrysea*	VU
Pearly-eyed Thrasher	*Margarops fuscatus*	LC
Rufous-throated Solitaire	*Myadestes genibarbis*	LC
La Selle Thrush	*Turdus swalesi*	EN
Antillean Siskin	*Carduelis dominicensis*	LC
Hispaniolan Crossbill	*Loxia megaplaga*	EN
Green-tailed Warbler	*Microligea palustris*	LC
White-winged Warbler	*Xenoligea montana*	VU
Black-crowned Palm-tanager	*Phaenicophilus palmarum*	LC
Grey-crowned Palm-tanager	*Phaenicophilus poliocephalus*	NT
Chat Tanager	*Calyptophilus frugivorus*	VU
Antillean Euphonia	*Euphonia musica*	LC

Appendix 2 ... continued.

■ EBA 029 Puerto Rico and the Virgin Islands

Countries Puerto Rico, British Virgin Islands, US Virgin Islands
Key habitats Lowland and montane rain forest, dry forest, mangroves
Area 9,400 km²
Altitude 0–1,200 m

Common name	Scientific name	Threat category
Bridled Quail-dove	*Geotrygon mystacea*	LC
Hispaniolan Parakeet	*Aratinga chloroptera*	VU ■
Puerto Rican Amazon	*Amazona vittata*	CR ■
Puerto Rican Lizard-cuckoo	*Coccyzus vieilloti*	LC
Puerto Rican Screech-owl	*Megascops nudipes*	LC
Puerto Rican Nightjar	*Caprimulgus noctitherus*	CR ■
Antillean Mango	*Anthracothorax dominicus*	LC
Green Mango	*Anthracothorax viridis*	LC
Green-throated Carib	*Eulampis holosericeus*	LC
Antillean Crested Hummingbird	*Orthorhyncus cristatus*	LC
Puerto Rican Emerald	*Chlorostilbon maugaeus*	LC
Puerto Rican Tody	*Todus mexicanus*	LC
Puerto Rican Woodpecker	*Melanerpes portoricensis*	LC
Caribbean Elaenia	*Elaenia martinica*	LC
Lesser Antillean Pewee	*Contopus latirostris*	LC
Puerto Rican Flycatcher	*Myiarchus antillarum*	LC
Puerto Rican Vireo	*Vireo latimeri*	LC
White-necked Crow	*Corvus leucognaphalus*	VU ■
Pearly-eyed Thrasher	*Margarops fuscatus*	LC
Adelaide's Warbler	*Dendroica adelaidae*	LC
Elfin-woods Warbler	*Dendroica angelae*	VU ■
Yellow-shouldered Blackbird	*Agelaius xanthomus*	EN ■
Puerto Rican Bullfinch	*Loxigilla portoricensis*	LC
Lesser Antillean Bullfinch	*Loxigilla noctis*	LC
Puerto Rican Tanager	*Nesospingus speculiferus*	LC
Puerto Rican Spindalis	*Spindalis portoricensis*	LC
Antillean Euphonia	*Euphonia musica*	LC

■ EBA 030 Lesser Antilles

Countries Anguilla, Antigua and Barbuda, Barbados, Dominica, Grenada, Guadeloupe, Martinique, Montserrat, Netherlands Antilles, St Kitts and Nevis, St Lucia, St Vincent and the Grenadines
Key habitats Lowland and montane rain forest, dry forest, elfin forest
Area 6,300 km²
Altitude 0–1,500 m

Common name	Scientific name	Threat category
Grenada Dove	*Leptotila wellsi*	CR ■
Bridled Quail-dove	*Geotrygon mystacea*	LC
St Lucia Amazon	*Amazona versicolor*	VU ■
Red-necked Amazon	*Amazona arausiaca*	VU ■
St Vincent Amazon	*Amazona guildingii*	VU ■
Imperial Amazon	*Amazona imperialis*	EN ■
Lesser Antillean Swift	*Chaetura martinica*	LC
Purple-throated Carib	*Eulampis jugularis*	LC
Green-throated Carib	*Eulampis holosericeus*	LC
Antillean Crested Hummingbird	*Orthorhyncus cristatus*	LC
Blue-headed Hummingbird	*Cyanophaia bicolor*	LC
Guadeloupe Woodpecker	*Melanerpes herminieri*	NT ■
Caribbean Elaenia	*Elaenia martinica*	LC
Lesser Antillean Pewee	*Contopus latirostris*	LC
Grenada Flycatcher	*Myiarchus nugator*	LC
Lesser Antillean Flycatcher	*Myiarchus oberi*	LC
White-breasted Thrasher	*Ramphocinclus brachyurus*	EN ■
Scaly-breasted Thrasher	*Margarops fuscus*	LC
Pearly-eyed Thrasher	*Margarops fuscatus*	LC
Brown Trembler	*Cinclocerthia ruficauda*	LC
Grey Trembler	*Cinclocerthia gutturalis*	LC
Rufous-throated Solitaire	*Myadestes genibarbis*	LC
Forest Thrush	*Cichlherminia lherminieri*	VU ■

EBA 030 Lesser Antilles ... continued

Common name	Scientific name	Threat category
Barbuda Warbler	*Dendroica subita*	NT ■
St Lucia Warbler	*Dendroica delicata*	LC
Plumbeous Warbler	*Dendroica plumbea*	LC
Whistling Warbler	*Catharopeza bishopi*	EN ■
Semper's Warbler	*Leucopeza semperi*	CR ■
Montserrat Oriole	*Icterus oberi*	CR ■
Martinique Oriole	*Icterus bonana*	VU ■
St Lucia Oriole	*Icterus laudabilis*	NT ■
Puerto Rican Bullfinch	*Loxigilla portoricensis*	LC
Lesser Antillean Bullfinch	*Loxigilla noctis*	LC
Barbados Bullfinch	*Loxigilla barbadensis*	LC
St Lucia Black Finch	*Melanospiza richardsoni*	EN ■
Lesser Antillean Tanager	*Tangara cucullata*	LC
Antillean Euphonia	*Euphonia musica*	LC
Lesser Antillean Saltator	*Saltator albicollis*	LC

■ SA 014 Cayman Islands

Countries Cayman Islands
Key habitats Mangroves, coastal lagoons, dry forest, scrub
Area 262 km²
Altitude 0–43 m

Common name	Scientific name	Threat category
Grand Cayman Thrush	*Turdus ravidus*	EX
Caribbean Elaenia	*Elaenia martinica*	LC
Thick-billed Vireo	*Vireo crassirostris*	LC
Yucatan Vireo	*Vireo magister*	LC
Vitelline Warbler	*Dendroica vitellina*	NT ■
Jamaican Oriole	*Icterus leucopteryx*	LC

■ SA 015 Netherlands Antilles

Countries Aruba, Curaçao, Bonaire
Key habitats Xerophytic vegetation
Area 925 km²
Altitude 0–375 m

Common name	Scientific name	Threat category
Yellow-shouldered Amazon	*Amazona barbadensis*	VU ■
Pearly-eyed Thrasher	*Margarops fuscatus*	LC
Caribbean Elaenia	*Elaenia martinica*	LC

■ SA 016 Trinidad

Countries Trinidad and Tobago
Key habitats Lowland and montane rain forest
Area 4,828 km²
Altitude 0–940 m

Common name	Scientific name	Threat category
Trinidad Piping-guan	*Pipile pipile*	CR ■

■ SA 017 Tobago

Countries Trinidad and Tobago
Key habitats Lowland and montane rain forest
Area 300 km²
Altitude 0–576 m

Common name	Scientific name	Threat category
White-tailed Sabrewing	*Campylopterus ensipennis*	NT ■

Appendix 3. Biome-restricted species in Caribbean biomes

Greater Antilles (GAN)

Description Includes the four main islands of the Greater Antilles and other associated island groups.
Countries Cuba, Dominican Republic, Haiti, Jamaica, Puerto Rico, Bahamas, Cayman Islands, Virgin Islands, Turks and Caicos Islands, Navassa Island.

Common name	Scientific name	Threat category
West Indian Whistling-duck	*Dendrocygna arborea*	VU
Cuban Kite	*Chondrohierax wilsonii*	CR
Gundlach's Hawk	*Accipiter gundlachi*	EN
Ridgway's Hawk	*Buteo ridgwayi*	CR
Zapata Rail	*Cyanolimnas cerverai*	EN
Ring-tailed Pigeon	*Patagioenas caribaea*	VU
Plain Pigeon	*Patagioenas inornata*	NT
Grey-headed Quail-dove	*Geotrygon caniceps*	VU
Crested Quail-dove	*Geotrygon versicolor*	NT
Key West Quail-dove	*Geotrygon chrysia*	LC
Blue-headed Quail-dove	*Starnoenas cyanocephala*	EN
Cuban Parakeet	*Aratinga euops*	VU
Hispaniolan Parakeet	*Aratinga chloroptera*	VU
Cuban Amazon	*Amazona leucocephala*	NT
Yellow-billed Amazon	*Amazona collaria*	VU
Hispaniolan Amazon	*Amazona ventralis*	VU
Black-billed Amazon	*Amazona agilis*	VU
Puerto Rican Amazon	*Amazona vittata*	CR
Chestnut-bellied Cuckoo	*Coccyzus pluvialis*	LC
Bay-breasted Cuckoo	*Coccyzus rufigularis*	EN
Jamaican Lizard-cuckoo	*Coccyzus vetula*	LC
Puerto Rican Lizard-cuckoo	*Coccyzus vieilloti*	LC
Great Lizard-cuckoo	*Coccyzus merlini*	LC
Hispaniolan Lizard-cuckoo	*Coccyzus longirostris*	LC
Ashy-faced Owl	*Tyto glaucops*	LC
Puerto Rican Screech-owl	*Megascops nudipes*	LC
Bare-legged Owl	*Gymnoglaux lawrencii*	LC
Cuban Pygmy-owl	*Glaucidium siju*	LC
Jamaican Owl	*Pseudoscops grammicus*	LC
Antillean Nighthawk	*Chordeiles gundlachii*	LC
Jamaican Pauraque	*Siphonorhis americana*	CR
Least Pauraque	*Siphonorhis brewsteri*	NT
Cuban Nightjar	*Caprimulgus cubanensis*	LC
Hispaniola Nightjar	*Caprimulgus ekmani*	LC
Puerto Rican Nightjar	*Caprimulgus noctitherus*	CR
Antillean Palm-swift	*Tachornis phoenicobia*	LC
Jamaican Mango	*Anthracothorax mango*	LC
Antillean Mango	*Anthracothorax dominicus*	LC
Green Mango	*Anthracothorax viridis*	LC
Red-billed Streamertail	*Trochilus polytmus*	LC
Black-billed Streamertail	*Trochilus scitulus*	LC
Cuban Emerald	*Chlorostilbon ricordii*	LC
Hispaniolan Emerald	*Chlorostilbon swainsonii*	LC
Puerto Rican Emerald	*Chlorostilbon maugaeus*	LC
Bahama Woodstar	*Calliphlox evelynae*	LC
Bee Hummingbird	*Mellisuga helenae*	NT
Vervain Hummingbird	*Mellisuga minima*	LC
Cuban Trogon	*Priotelus temnurus*	LC
Hispaniolan Trogon	*Priotelus roseigaster*	NT
Cuban Tody	*Todus multicolor*	LC
Narrow-billed Tody	*Todus angustirostris*	LC
Puerto Rican Tody	*Todus mexicanus*	LC
Jamaican Tody	*Todus todus*	LC
Broad-billed Tody	*Todus subulatus*	LC
Antillean Piculet	*Nesoctites micromegas*	LC
Puerto Rican Woodpecker	*Melanerpes portoricensis*	LC
Hispaniolan Woodpecker	*Melanerpes striatus*	LC
Jamaican Woodpecker	*Melanerpes radiolatus*	LC

Greater Antilles (GAN) ... continued

Common name	Scientific name	Threat category
West Indian Woodpecker	*Melanerpes superciliaris*	LC
Cuban Green Woodpecker	*Xiphidiopicus percussus*	LC
Fernandina's Flicker	*Colaptes fernandinae*	VU
Jamaican Becard	*Pachyramphus niger*	LC
Jamaican Elaenia	*Myiopagis cotta*	LC
Greater Antillean Elaenia	*Elaenia fallax*	LC
Greater Antillean Pewee	*Contopus caribaeus*	LC
Jamaican Pewee	*Contopus pallidus*	LC
Hispaniolan Pewee	*Contopus hispaniolensis*	LC
Loggerhead Kingbird	*Tyrannus caudifasciatus*	LC
Giant Kingbird	*Tyrannus cubensis*	EN
Sad Flycatcher	*Myiarchus barbirostris*	LC
Rufous-tailed Flycatcher	*Myiarchus validus*	LC
La Sagra's Flycatcher	*Myiarchus sagrae*	LC
Stolid Flycatcher	*Myiarchus stolidus*	LC
Puerto Rican Flycatcher	*Myiarchus antillarum*	LC
San Andres Vireo	*Vireo caribaeus*	VU
Cuban Vireo	*Vireo gundlachii*	LC
Thick-billed Vireo	*Vireo crassirostris*	LC
Jamaican Vireo	*Vireo modestus*	LC
Flat-billed Vireo	*Vireo nanus*	LC
Puerto Rican Vireo	*Vireo latimeri*	LC
Blue Mountain Vireo	*Vireo osburni*	NT
Hispaniolan Palm Crow	*Corvus palmarum*	NT
Jamaican Crow	*Corvus jamaicensis*	LC
Cuban Crow	*Corvus nasicus*	LC
White-necked Crow	*Corvus leucognaphalus*	VU
Cuban Palm Crow	*Corvus minutus*	EN
Palmchat	*Dulus dominicus*	LC
Bahama Swallow	*Tachycineta cyaneoviridis*	VU
Golden Swallow	*Tachycineta euchrysea*	VU
Cuban Martin	*Progne cryptoleuca*	LC
Zapata Wren	*Ferminia cerverai*	EN
Cuban Gnatcatcher	*Polioptila lembeyei*	LC
Bahama Mockingbird	*Mimus gundlachii*	LC
Cuban Solitaire	*Myadestes elisabeth*	NT
White-chinned Thrush	*Turdus aurantius*	LC
White-eyed Thrush	*Turdus jamaicensis*	LC
La Selle Thrush	*Turdus swalesi*	EN
Antillean Siskin	*Carduelis dominicensis*	LC
Olive-capped Warbler	*Dendroica pityophila*	LC
Vitelline Warbler	*Dendroica vitellina*	NT
Arrowhead Warbler	*Dendroica pharetra*	LC
Elfin-woods Warbler	*Dendroica angelae*	VU
Bahama Yellowthroat	*Geothlypis rostrata*	LC
Green-tailed Warbler	*Microligea palustris*	LC
Yellow-headed Warbler	*Teretistris fernandinae*	LC
Oriente Warbler	*Teretistris fornsi*	LC
White-winged Warbler	*Xenoligea montana*	VU
Jamaican Oriole	*Icterus leucopteryx*	LC
Greater Antillean Oriole	*Icterus dominicensis*	LC
Jamaican Blackbird	*Nesopsar nigerrimus*	EN
Cuban Blackbird	*Dives atroviolaceus*	LC
Red-shouldered Blackbird	*Agelaius assimilis*	LC
Tawny-shouldered Blackbird	*Agelaius humeralis*	LC
Yellow-shouldered Blackbird	*Agelaius xanthomus*	EN
Cuban Sparrow	*Torreornis inexpectata*	EN
Greater Antillean Grackle	*Quiscalus niger*	LC
Cuban Bullfinch	*Melopyrrha nigra*	LC
Cuban Grassquit	*Tiaris canorus*	LC

Appendix 3 ... continued.

Greater Antilles (GAN) ... continued

Common name	Scientific name	Threat category
Yellow-shouldered Grassquit	*Loxipasser anoxanthus*	LC
Puerto Rican Bullfinch	*Loxigilla portoricensis*	LC
Greater Antillean Bullfinch	*Loxigilla violacea*	LC
Orangequit	*Euneornis campestris*	LC
Puerto Rican Tanager	*Nesospingus speculiferus*	LC
Black-crowned Palm-tanager	*Phaenicophilus palmarum*	LC
Grey-crowned Palm-tanager	*Phaenicophilus poliocephalus*	NT ■
Chat Tanager	*Calyptophilus frugivorus*	VU ■
Hispaniolan Crossbill	*Loxia megaplaga*	EN ■
Hispaniolan Spindalis	*Spindalis dominicensis*	LC
Puerto Rican Spindalis	*Spindalis portoricensis*	LC
Jamaican Spindalis	*Spindalis nigricephala*	LC
Jamaican Euphonia	*Euphonia jamaica*	LC

■ Lesser Antilles (LAN)

Description Includes the island chain from Anguilla to Grenada and Barbados.
Countries Anguilla, Antigua and Barbuda, Barbados, Dominica, Grenada, Guadeloupe, Martinique, Montserrat, Saba, St Eustatius, St Maarten, St Martin, St Kitts and Nevis, St Lucia, St Vincent and the Grenadines.

Common name	Scientific name	Threat category
Grenada Dove	*Leptotila wellsi*	CR ■
St Lucia Amazon	*Amazona versicolor*	VU ■
Red-necked Amazon	*Amazona arausiaca*	VU ■
St Vincent Amazon	*Amazona guildingii*	VU ■
Imperial Amazon	*Amazona imperialis*	EN ■
Lesser Antillean Swift	*Chaetura martinica*	LC
Purple-throated Carib	*Eulampis jugularis*	LC
Blue-headed Hummingbird	*Cyanophaia bicolor*	LC
Guadeloupe Woodpecker	*Melanerpes herminieri*	NT ■
Grenada Flycatcher	*Myiarchus nugator*	LC
Lesser Antillean Flycatcher	*Myiarchus oberi*	LC
White-breasted Thrasher	*Ramphocinclus brachyurus*	EN ■
Scaly-breasted Thrasher	*Margarops fuscus*	LC
Brown Trembler	*Cinclocerthia ruficauda*	LC

Lesser Antilles (LAN) ... continued

Common name	Scientific name	Threat category
Grey Trembler	*Cinclocerthia gutturalis*	LC
Forest Thrush	*Cichlherminia lherminieri*	VU ■
Adelaide's Warbler	*Dendroica adelaidae*	LC
Barbuda Warbler	*Dendroica subita*	NT ■
St Lucia Warbler	*Dendroica delicata*	LC
Plumbeous Warbler	*Dendroica plumbea*	LC
Whistling Warbler	*Catharopeza bishopi*	EN ■
Semper's Warbler	*Leucopeza semperi*	CR ■
Montserrat Oriole	*Icterus oberi*	CR ■
Martinique Oriole	*Icterus bonana*	VU ■
St Lucia Oriole	*Icterus laudabilis*	NT ■
St Lucia Black Finch	*Melanospiza richardsoni*	EN ■
Lesser Antillean Tanager	*Tangara cucullata*	LC
Lesser Antillean Saltator	*Saltator albicollis*	LC

■ Northern South America (NSA)

Description Includes lowland areas north of rivers Orinoco and Meta, west to the Gulf of Urabá and east to include the Orinoco delta as well as offshore islands in southern Caribbean, Netherlands Antilles, Trinidad and Tobago and several small islands belonging to Venezuela.
Countries (Caribbean) Aruba, Curaçao and Bonaire, Trinidad and Tobago.

Common name	Scientific name	Threat category
Rufous-vented Chachalaca	*Ortalis ruficauda*	LC
Bare-eyed Pigeon	*Patagioenas corensis*	LC
Yellow-shouldered Amazon	*Amazona barbadensis*	VU ■
Copper-rumped Hummingbird	*Amazilia tobaci*	LC
Venezuelan Flycatcher	*Myiarchus venezuelensis*	LC

■ Northern Andes (NAN)

Description All montane regions from coastal cordillera in Venezuela (including Northern Range in Trinidad), south to Porculla pass and river Marañón in Peru.
Countries (Caribbean) Trinidad and Tobago.

Common name	Scientific name	Threat category
Trinidad Piping-guan	*Pipile pipile*	CR ■
White-tailed Sabrewing	*Campylopterus ensipennis*	NT ■

Appendix 4. Congregatory species populations and thresholds

Common name	Scientific name	Total population estimate*	Total 1% threshold*	Caribbean population estimate	Caribbean 1% threshold	A4i	A4ii	Threat category
Fulvous Whistling-duck	*Dendrocygna bicolor*	800,000	8,000			✓		LC
White-faced Whistling-duck	*Dendrocygna viduata*	1,000,000	10,000			✓		LC
West Indian Whistling-duck	*Dendrocygna arborea*	10,000	100	10,000	100	✓		VU
Black-bellied Whistling-duck	*Dendrocygna autumnalis*	1,350,000	13,500			✓		LC
Comb Duck	*Sarkidiornis melanotos*	62,500	630			✓		LC
Wood Duck	*Aix sponsa*	28,000	280			✓		LC
Gadwall	*Anas strepera*	654,000	6,500			✓		LC
American Wigeon	*Anas americana*	1,112,500	11,100			✓		LC
American Black Duck	*Anas rubripes*	2,040	20			✓		LC
Mallard	*Anas platyrhynchos*	11,100	110			✓		LC
Blue-winged Teal	*Anas discors*							LC
Cinnamon Teal	*Anas cyanoptera*	263,250	2,600			✓		LC
Northern Shoveler	*Anas clypeata*	1,436,000	14,400			✓		LC
White-cheeked Pintail	*Anas bahamensis*	578,500	5,800	75,000	750	✓		LC
Northern Pintail	*Anas acuta*	1,024,000	10,200			✓		LC
Common Teal	*Anas crecca*	648,000	6,500			✓		LC
Canvasback	*Aythya valisineria*	104,200	1,000			✓		LC
Redhead	*Aythya americana*	118,400	1,200			✓		LC
Ring-necked Duck	*Aythya collaris*	294,000	2,900			✓		LC
Lesser Scaup	*Aythya affinis*	1,180,000	11,800			✓		LC
Red-breasted Merganser	*Mergus serrator*	4,500	45			✓		LC
Masked Duck	*Nomonyx dominicus*			750-1,000	10			LC
Ruddy Duck	*Oxyura jamaicensis*	325,000 (155,250)	3,300 (1,600)			✓		LC
Common Loon	*Gavia immer*	580,000	5,800			✓		LC
Northern Fulmar	*Fulmarus glacialis*	20,000,000	200,000				✓	LC
Bermuda Petrel	*Pterodroma cahow*	142	1.42	180	2		✓	EN
Black-capped Petrel	*Pterodroma hasitata*	5,000	50	5,000	50		✓	EN
Jamaica Petrel	*Pterodroma caribbaea*	50	0.5	50	1		✓	CR
Corys Shearwater	*Calonectris diomedea*	600,000	6,000				✓	LC
Great Shearwater	*Puffinus gravis*	15,000,000	150,000				✓	LC
Sooty Shearwater	*Puffinus griseus*	20,000,000	200,000				✓	NT
Manx Shearwater	*Puffinus puffinus*	1,000,000	10,000				✓	LC
Audubons Shearwater	*Puffinus lherminieri*	500,000	5,000	8,000	80		✓	LC
Wilsons Storm-petrel	*Oceanites oceanicus*	6,000,000	60,000		60,000		✓	LC
Leachs Storm-petrel	*Oceanodroma leucorhoa*	8,000,000	80,000		100,000		✓	LC
Least Grebe	*Tachybaptus dominicus*	69,000	690	10,000	100	✓		LC
Pied-billed Grebe	*Podilymbus podiceps*	25,000	250			✓		LC
Caribbean Flamingo	*Phoenicopterus ruber*	294,990	3,000			✓		LC
Caribbean Flamingo	*Phoenicopterus ruber*	294,990	3,000			✓		LC
Wood Stork	*Mycteria americana*	44,500	450			✓		LC
White Ibis	*Eudocimus albus*	100,000	1,000			✓		LC
Scarlet Ibis	*Eudocimus ruber*	125,000	1,300			✓		LC
Glossy Ibis	*Plegadis falcinellus*	62,500	630			✓		LC
Roseate Spoonbill	*Platalea ajaja*	144,350	1,400			✓		LC
Rufescent Tiger-heron	*Tigrisoma lineatum*	82,500	830			✓		LC
Boat-billed Heron	*Cochlearius cochlearius*	800,000	8,000			✓		LC
American Bittern	*Botaurus lentiginosus*	1,490,000	14,900			✓		LC
Stripe-backed Bittern	*Ixobrychus involucris*	500,000	5,000			✓		LC
Least Bittern	*Ixobrychus exilis*	64,450	650			✓		LC
Black-crowned Night-heron	*Nycticorax nycticorax*	625,750	6,300			✓		LC
Yellow-crowned Night-heron	*Nyctanassa violacea*	550,000	5,500			✓		LC
Green Heron	*Butorides virescens*	800,000	8,000			✓		LC
Cattle Egret	*Bubulcus ibis*	3,250,000	32,500			✓		LC
Great Blue Heron	*Ardea herodias*					✓		LC
Great Egret	*Casmerodius albus*	950,000	9,500			✓		LC
Reddish Egret	*Egretta rufescens*	23,500	240			✓		LC
Tricoloured Heron	*Egretta tricolor*	112,500	1,100			✓		LC
Little Blue Heron	*Egretta caerulea*	967,500	9,700			✓		LC
Snowy Egret	*Egretta thula*	1,825,000	18,300			✓		LC
Little Egret	*Egretta garzetta*	30-60	1			✓		LC
Red-billed Tropicbird	*Phaethon aethereus*	7,500	75	4,300	43		✓	LC

Appendix 4 ... continued.

Common name	Scientific name	Total population estimate*	Total 1% threshold*	Caribbean population estimate	Caribbean 1% threshold	A4i	A4ii	Threat category
White-tailed Tropicbird	*Phaethon lepturus*	50,000	500	6,000	60		✓	LC
Magnificent Frigatebird	*Fregata magnificens*	200,000	2,000	9,600	96		✓	LC
American White Pelican	*Pelecanus erythrorhynchos*		1,800			✓		LC
Brown Pelican	*Pelecanus occidentalis*	290,885	2,900	3,000	30	✓		LC
Northern Gannet	*Morus bassanus*	530,000	5,300				✓	LC
Masked Booby	*Sula dactylatra*	200,000	2,000	1,200	12		✓	LC
Red-footed Booby	*Sula sula*	600,000	6,000	18,200	182		✓	LC
Brown Booby	*Sula leucogaster*	200,000	2,000	13,300	133		✓	LC
Neotropic Cormorant	*Phalacrocorax brasilianus*	2,629,000	26,300			✓		LC
Double-crested Cormorant	*Phalacrocorax auritus*	34,225	340			✓		LC
Turkey Vulture	*Cathartes aura*	4,500,000	45,000				✓	LC
Merlin	*Falco columbarius*	550,000	5,500				✓	LC
Peregrine Falcon	*Falco peregrinus*	55,000	550				✓	LC
Osprey	*Pandion haliaetus*	460,000	4,600				✓	LC
Mississippi Kite	*Ictinia mississippiensis*	190,000	1,900				✓	LC
Northern Harrier	*Circus cyaneus*	1,300,000	13,000				✓	LC
Broad-winged Hawk	*Buteo platypterus*	1,800,000	18,000				✓	LC
Black Rail	*Laterallus jamaicensis*	32,500	330			✓		NT ■
Grey-necked Wood-rail	*Aramides cajanea*	500,000	5,000			✓		LC
Zapata Rail	*Cyanolimnas cerverai*	625	6			✓		EN ■
Spotted Rail	*Pardirallus maculatus*	5,000	50			✓		LC
Purple Gallinule	*Porphyrio martinica*	450,000	4,500			✓		LC
Common Moorhen	*Gallinula chloropus*	1,000,000	10,000			✓		LC
American Coot	*Fulica americana*	1,800,000	18,000			✓		LC
Caribbean Coot	*Fulica caribaea*	50,000	500	40,000	400	✓		NT ■
Sandhill Crane	*Grus canadensis*	650	7	300	3	✓		LC
Limpkin	*Aramus guarauna*	1,000,000	10,000			✓		LC
American Oystercatcher	*Haematopus palliatus*	62,800	630			✓		LC
Black-necked Stilt	*Himantopus mexicanus*	325,000	3,300			✓		LC
American Avocet	*Recurvirostra americana*	250,000	2,500			✓		LC
Southern Lapwing	*Vanellus chilensis*	2,000,000	20,000			✓		LC
American Golden Plover	*Pluvialis dominica*	200,000	2,000			✓		LC
Grey Plover	*Pluvialis squatarola*	140,000	1,400			✓		LC
Semipalmated Plover	*Charadrius semipalmatus*	120,000	1,200			✓		LC
Wilsons Plover	*Charadrius wilsonia*	82,300 (77,500)	820 (780)	6,000	60	✓		LC
Killdeer	*Charadrius vociferus*	500,000	5,000			✓		LC
Piping Plover	*Charadrius melodus*	1,283	15			✓		NT ■
Kentish Plover	*Charadrius alexandrinus*	18,448	180			✓		LC
Collared Plover	*Charadrius collaris*	15,000	150			✓		LC
Common Snipe	*Gallinago gallinago*	1,000,000	10,000			✓		LC
South American Snipe	*Gallinago paraguaiae*	121,000	1,200			✓		LC
Short-billed Dowitcher	*Limnodromus griseus*	122,400	1,200			✓		LC
Long-billed Dowitcher	*Limnodromus scolopaceus*	80,000	800			✓		LC
Hudsonian Godwit	*Limosa haemastica*	70,000	700			✓		LC
Marbled Godwit	*Limosa fedoa*	85,750	860			✓		LC
Whimbrel	*Numenius phaeopus*	52,800	530			✓		LC
Upland Sandpiper	*Bartramia longicauda*	350,000	3,500			✓		LC
Greater Yellowlegs	*Tringa melanoleuca*	80,000	800			✓		LC
Lesser Yellowlegs	*Tringa flavipes*	320,000	3,200			✓		LC
Solitary Sandpiper	*Tringa solitaria*	142,500	1,400			✓		LC
Spotted Sandpiper	*Actitis macularius*	120,000	1,200			✓		LC
Willet	*Catoptrophorus semipalmatus*	209,000 (18,000)	2,100 (180)		2,500	✓		LC
Ruddy Turnstone	*Arenaria interpres*	160,000	1,600		2,350	✓		LC
Red Knot	*Calidris canutus*	32,500	330		4,000	✓		LC
Sanderling	*Calidris alba*	240,000	2,400			✓		LC
Semipalmated Sandpiper	*Calidris pusilla*	2,000,000	20,000			✓		LC
Western Sandpiper	*Calidris mauri*	1,750,000	17,500			✓		LC
Least Sandpiper	*Calidris minutilla*	560,000	5,600			✓		LC
White-rumped Sandpiper	*Calidris fuscicollis*	1,120,000	11,200			✓		LC
Bairds Sandpiper	*Calidris bairdii*	300,000	3,000			✓		LC
Pectoral Sandpiper	*Calidris melanotos*	450,000	4,500			✓		LC

Appendix 4 ... continued.

Common name	Scientific name	Total population estimate*	Total 1% threshold*	Caribbean population estimate	Caribbean 1% threshold	A4i	A4ii	Threat category
Dunlin	*Calidris alpina*	77,500	780			✓		LC
Stilt Sandpiper	*Calidris himantopus*	779,000	7,800			✓		LC
Buff-breasted Sandpiper	*Tryngites subruficollis*	30,000	300			✓		NT ■
Wilsons Phalarope	*Steganopus tricolor*	1,500,000	15,000			✓		LC
Red-necked Phalarope	*Phalaropus lobatus*	2,500,000	25,000			✓		LC
Red Phalarope	*Phalaropus fulicarius*	750,000	7,500			✓		LC
Ring-billed Gull	*Larus delawarensis*	1,020,000	10,200			✓		LC
Great Black-backed Gull	*Larus marinus*	72,000	720			✓		LC
Herring Gull	*Larus argentatus*	148,000	1,500			✓		LC
Bonapartes Gull	*Larus philadelphia*	78,000	780			✓		LC
Laughing Gull	*Larus atricilla*	662,100 (22,500)	6,600 (230)	15,000	150	✓		LC
Gull-billed Tern	*Sterna nilotica*	50,000	500	600	6	✓		LC
Caspian Tern	*Sterna caspia*	89,298	890			✓		LC
Royal Tern	*Sterna maxima*	74,950	749.5	1,250	13	✓		LC
Sandwich Tern	*Sterna sandvicensis*	105,900	1,100	5,100	51	✓		LC
Cayenne Tern	*Sterna sandvicensis eurygnatha*	21,950	220	11,950	120			LC
Roseate Tern	*Sterna dougallii*	24,750 (12,450)	250 (130)	10,000	100	✓		LC
Common Tern	*Sterna hirundo*	749,220 (720)	7,500 (7)	150	2	✓		LC
Arctic Tern	*Sterna paradisaea*	300,000	20,000			✓		LC
Forsters Tern	*Sterna forsteri*	72,060	720			✓		LC
Least Tern	*Sterna antillarum*	66,725	670	4,500	45	✓		LC
Yellow-billed Tern	*Sterna superciliaris*	62,500	630			✓		LC
Bridled Tern	*Sterna anaethetus*	20,000	200	12,000	120	✓		LC
Sooty Tern	*Sterna fuscata*	2,855,500	28,600	500,000	5,000	✓		LC
Black Tern	*Chlidonias niger*	450,000	4,500			✓		LC
Large-billed Tern	*Phaetusa simplex*	62,500	630			✓		LC
Brown Noddy	*Anous stolidus*	500,000	5,000	28,000	280	✓		LC
Black Noddy	*Anous minutus*	15,000	150	200	2	✓		LC
Black Skimmer	*Rynchops niger*	141,600	1,400	91,600	916	✓		LC
Great Skua	*Catharacta skua*	15,000	150				✓	LC
Pomarine Jaeger	*Stercorarius pomarinus*	75,000	750				✓	LC
Parasitic Jaeger	*Stercorarius parasiticus*	750,000	7,500				✓	LC
Long-tailed Jaeger	*Stercorarius longicaudus*	300,000	3,000				✓	LC
Oilbird	*Steatornis caripensis*	300,000	3,000				✓	LC
Sand Martin	*Riparia riparia*	46,000,000	460,000				✓	LC
Barn Swallow	*Hirundo rustica*	190,000,000	1,900,000				✓	LC

Notes:
* Total population estimates and thresholds are Neotropical in the case of waterbirds (A4i) and global for non-waterbirds (A4ii).
Numbers in brackets refer to Neotropical breeding population estimates and thresholds.
All population figures = numbers of individuals. With original estimates in pairs, populations were derived on the basis of 1 pair = 3 indiviuals